T 357 .D56 1999
Dimensioning and tolerancing
 handbook

Y0-CKO-204

DATE DUE

ILL KPS WMS			
635309IL			
6/11/03			

DEMCO 38-297

**NEW ENGLAND INSTITUTE
OF TECHNOLOGY
LEARNING RESOURCES CENTER**

Dimensioning and Tolerancing Handbook

Dimensioning and Tolerancing Handbook

Paul J. Drake, Jr.

McGraw-Hill
New York   San Francisco   Washington, D.C.   Auckland   Bogatá
Caracas   Lisbon   London   Madrid   Mexico City   Milan
Montreal   New Delhi   San Juan   Singapore
Sydney   Tokyo   Toronto

McGraw-Hill

*A Division of The **McGraw-Hill** Companies*

Copyright © 1999 by Paul J. Drake, Jr. All rights reserved. Printed in the United States of America. Except as permitted under the United States Copyright Act of 1976, no part of this publication may be reproduced or distributed in any form or by any means, or stored in a data base or retrieval system, without the prior written permission of the publisher.

3 4 5 6 7 8 9 0 QM/QM 0 9 8 7 6 5 4 3 2 1 0

ISBN 0-07-018131-4

The sponsoring editor of this book was Linda Ludewig, and the production supervisor was Pamela Pelton.

Printed and bound by Quebecor/Martinsburg.

This book is printed on recycled, acid-free paper containing a minimum of 50% recycled, de-inked fiber.

McGraw-Hill books are available at special quality discounts to use as premiums and sales promotions, or for use in corporate training programs. For more information, please write to the Director of Special Sales, McGraw-Hill, Professional Publishing, Two Penn Plaza, New York, NY 10121-2298. Or contact your local bookstore.

> Information contained in this work has been obtained by The McGraw-Hill Companies, Inc. ("McGraw-Hill") from sources believed to be reliable. However, neither McGraw-Hill nor its authors guarantees the accuracy or completeness of any information published herein and neither McGraw-Hill nor its authors shall be responsible for any errors, omissions, or damages arising out of use of this information. This work is published with the understanding that McGraw-Hill and its authors are supplying information but are not attempting to render engineering or other professional services. If such services are required, the assistance of an appropriate professional should be sought.

Contents

Foreword xxi
About the Editor xxii
Contributors xxiii
Preface xxv
Acknowledgments xxix

Part 1 History/Lessons Learned

Chapter 1: Quality Thrust .... Ron Randall

1.1	Meaning of Quality	1-1
1.2	The Evolution of Quality	1-2
1.3	Some Quality Gurus and their Contributions	1-2
1.3.1	W. Edwards Deming	1-2
1.3.2	Joseph Juran	1-3
1.3.3	Philip B. Crosby	1-4
1.3.4	Genichi Taguchi	1-5
1.4	The Six Sigma Approach to Quality	1-6
1.4.1	The History of Six Sigma	1-6
1.4.2	Six Sigma Success Stories	1-7
1.4.3	Six Sigma Basics	1-7
1.5	The Malcolm Baldrige National Quality Award (MBNQA)	1-9
1.6	References	1-10

Chapter 2: Dimensional Management .... Robert H. Nickolaisen, P.E.

2.1	Traditional Approaches to Dimensioning and Tolerancing	2-1
2.1.1	Engineering Driven Design	2-2
2.1.2	Process Driven Design	2-2
2.1.3	Inspection Driven Design	2-2
2.2	A Need for Change	2-3
2.2.1	Dimensional Management	2-3
2.2.2	Dimensional Management Systems	2-3
2.2.2.1	Simultaneous Engineering Teams	2-4
2.2.2.2	Written Goals and Objectives	2-4
2.2.2.3	Design for Manufacturability (DFM) and Design for Assembly (DFA)	2-5
2.2.2.4	Geometric Dimensioning and Tolerancing (GD&T)	2-6
2.2.2.5	Key Characteristics	2-6
2.2.2.6	Statistical Process Control (SPC)	2-6
2.2.2.7	Variation Measurement and Reduction	2-7
2.2.2.8	Variation Simulation Tolerance Analysis	2-7
2.3	The Dimensional Management Process	2-8
2.4	References	2-10
2.5	Glossary	2-10

v

Chapter 3: Tolerancing Optimization Strategies .... Gregory A. Hetland, Ph.D.

3.1	Tolerancing Methodologies	3-1
3.2	Tolerancing Progression (Example # 1)	3-1
3.2.1	Strategy # 1 (Linear)	3-2
3.2.2	Strategy # 2 (Combination of Linear and Geometric)	3-5
3.2.3	Strategy # 3 (Fully Geometric)	3-6
3.3	Tolerancing Progression (Example # 2)	3-6
3.3.1	Strategy # 1 (Linear)	3-8
3.3.2	Strategy # 2 Geometric Tolerancing (⊕) Regardless of Feature Size	3-11
3.3.3	Strategy # 3 (Geometric Tolerancing Progression At Maximum Material Condition)	3-12
3.3.4	Strategy # 4 (Tolerancing Progression "Optimized")	3-13
3.4	Summary	3-15
3.5	References	3-15

Part 2 Standards

Chapter 4: Drawing Interpretation .... Patrick J. McCuistion, Ph.D

4.1	Introduction	4-1
4.2	Drawing History	4-2
4.3	Standards	4-2
4.3.1	ANSI	4-2
4.3.2	ISO	4-3
4.4	Drawing Types	4-3
4.4.1	Note	4-3
4.4.2	Detail	4-3
4.4.2.1	Cast or Forged Part	4-4
4.4.2.2	Machined Part	4-4
4.4.2.3	Sheet Stock Part	4-4
4.4.3	Assembly	4-4
4.5	Border	4-4
4.5.1	Zones and Center Marks	4-4
4.5.2	Size Conventions	4-13
4.6	Title Blocks	4-13
4.6.1	Company Name and Address	4-13
4.6.2	Drawing Title	4-13
4.6.3	Size	4-13
4.6.4	FSCM/CAGE	4-13
4.6.5	Drawing Number	4-14
4.6.6	Scale	4-14
4.6.7	Release Date	4-14
4.6.8	Sheet Number	4-14
4.6.9	Contract Number	4-14
4.6.10	Drawn and Date	4-14
4.6.11	Check, Design, and Dates	4-14
4.6.12	Design Activity and Date	4-15
4.6.13	Customer and Date	4-15
4.6.14	Tolerances	4-15
4.6.15	Treatment	4-15
4.6.16	Finish	4-15
4.6.17	Similar To	4-15
4.6.18	Act Wt and Calc Wt	4-15
4.6.19	Other Title Block Items	4-15
4.7	Revision Blocks	4-16
4.8	Parts Lists	4-16
4.9	View Projection	4-16

4.9.1	First-Angle Projection	4-16
4.9.2	Third-Angle Projection	4-16
4.9.3	Auxiliary Views	4-16
4.10	Section Views	4-16
4.10.1	Full Sections	4-19
4.10.2	Half Sections	4-19
4.10.3	Offset Sections	4-19
4.10.4	Broken-Out Section	4-19
4.10.5	Revolved and Removed Sections	4-22
4.10.6	Conventional Breaks	4-22
4.11	Partial Views	4-23
4.12	Conventional Practices	4-23
4.12.1	Feature Rotation	4-23
4.12.2	Line Precedence	4-23
4.13	Isometric Views	4-24
4.14	Dimensions	4-25
4.14.1	Feature Types	4-25
4.14.2	Taylor Principle / Envelope Principle	4-25
4.14.3	General Dimensions	4-26
4.14.4	Technique	4-27
4.14.5	Placement	4-27
4.14.6	Choice	4-28
4.14.7	Tolerance Representation	4-28
4.15	Surface Texture	4-28
4.15.1	Roughness	4-29
4.15.2	Waviness	4-29
4.15.3	Lay	4-29
4.15.4	Flaws	4-29
4.16	Notes	4-29
4.17	Drawing Status	4-30
4.17.1	Sketch	4-30
4.17.2	Configuration Layout	4-30
4.17.3	Experimental	4-30
4.17.4	Active	4-30
4.17.5	Obsolete	4-30
4.18	Conclusion	4-30
4.19	References	4-31

Chapter 5: Geometric Dimensioning and Tolerancing .... Walter M. Stites
.... Paul Drake, P.E.

5.1	Introducing Geometric Dimensioning and Tolerancing (GD&T)	5-1
5.1.1	What is GD&T?	5-2
5.1.2	Where Does GD&T Come From?—References	5-2
5.1.3	Why Do We Use GD&T?	5-4
5.1.4	When Do We Use GD&T?	5-8
5.1.5	How Does GD&T Work?—Overview	5-9
5.2	Part Features	5-9
5.2.1	Nonsize Features	5-10
5.2.2	Features of Size	5-10
5.2.2.1	Screw Threads	5-11
5.2.2.2	Gears and Splines	5-11
5.2.3	Bounded Features	5-11
5.3	Symbols	5-11
5.3.1	Form and Proportions of Symbols	5-12
5.3.2	Feature Control Frame	5-14
5.3.2.1	Feature Control Frame Placement	5-14
5.3.2.2	Reading a Feature Control Frame	5-16
5.3.3	Basic Dimensions	5-17

5.3.4	Reference Dimensions and Data	5-18
5.3.5	"Square" Symbol	5-18
5.3.6	Tabulated Tolerances	5-18
5.3.7	"Statistical Tolerance" Symbol	5-18
5.4	Fundamental Rules	5-18
5.5	Nonrigid Parts	5-19
5.5.1	Specifying Restraint	5-20
5.5.2	Singling Out a Free State Tolerance	5-20
5.6	Features of Size—The Four Fundamental Levels of Control	5-20
5.6.1	Level 1—Size Limit Boundaries	5-20
5.6.2	Material Condition	5-23
5.6.2.1	Modifier Symbols	5-24
5.6.3	Method for MMC or LMC	5-25
5.6.3.1	Level 2—Overall Feature Form	5-26
5.6.3.2	Level 3—Virtual Condition Boundary for Orientation	5-33
5.6.3.3	Level 4—Virtual Condition Boundary for Location	5-34
5.6.3.4	Level 3 or 4 Virtual Condition Equal to Size Limit (Zero Tolerance)	5-35
5.6.3.5	Resultant Condition Boundary	5-37
5.6.4	Method for RFS	5-38
5.6.4.1	Tolerance Zone Shape	5-38
5.6.4.2	Derived Elements	5-38
5.6.5	Alternative "Center Method" for MMC or LMC	5-43
5.6.5.1	Level 3 and 4 Adjustment—Actual Mating/Minimum Material Sizes	5-43
5.6.5.2	Level 2 Adjustment—Actual Local Sizes	5-45
5.6.5.3	Disadvantages of Alternative "Center Method"	5-46
5.6.6	Inner and Outer Boundaries	5-46
5.6.7	When do we use a Material Condition Modifier?	5-47
5.7	Size Limits (Level 1 Control)	5-48
5.7.1	Symbols for Limits and Fits	5-48
5.7.2	Limit Dimensioning	5-49
5.7.3	Plus and Minus Tolerancing	5-49
5.7.4	Inch Values	5-49
5.7.5	Millimeter Values	5-49
5.8	Form (Only) Tolerances (Level 2 Control)	5-50
5.8.1	Straightness Tolerance for Line Elements	5-51
5.8.2	Straightness Tolerance for a Cylindrical Feature	5-52
5.8.3	Flatness Tolerance for a Single Planar Feature	5-52
5.8.4	Flatness Tolerance for a Width-Type Feature	5-52
5.8.5	Circularity Tolerance	5-53
5.8.5.1	Circularity Tolerance Applied to a Spherical Feature	5-55
5.8.6	Cylindricity Tolerance	5-55
5.8.7	Circularity or Cylindricity Tolerance with Average Diameter	5-56
5.8.8	Application Over a Limited Length or Area	5-57
5.8.9	Application on a Unit Basis	5-57
5.8.10	Radius Tolerance	5-58
5.8.10.1	Controlled Radius Tolerance	5-59
5.8.11	Spherical Radius Tolerance	5-59
5.8.12	When Do We Use a Form Tolerance?	5-60
5.9	Datuming	5-61
5.9.1	What is a Datum?	5-61
5.9.2	Datum Feature	5-61
5.9.2.1	Datum Feature Selection	5-61
5.9.2.2	Functional Hierarchy	5-63
5.9.2.3	Surrogate and Temporary Datum Features	5-64
5.9.2.4	Identifying Datum Features	5-65
5.9.3	True Geometric Counterpart (TGC)—Introduction	5-67
5.9.4	Datum	5-69
5.9.5	Datum Reference Frame (DRF) and Three Mutually Perpendicular Planes	5-69

Section	Title	Page
5.9.6	Datum Precedence	5-69
5.9.7	Degrees of Freedom	5-72
5.9.8	TGC Types	5-74
5.9.8.1	Restrained versus Unrestrained TGC	5-75
5.9.8.2	Nonsize TGC	5-75
5.9.8.3	Adjustable-size TGC	5-75
5.9.8.4	Fixed-size TGC	5-77
5.9.9	Datum Reference Frame (DRF) Displacement	5-80
5.9.9.1	Relative to a Boundary of Perfect Form TGC	5-81
5.9.9.2	Relative to a Virtual Condition Boundary TGC	5-83
5.9.9.3	Benefits of DRF Displacement	5-83
5.9.9.4	Effects of All Datums of the DRF	5-83
5.9.9.5	Effects of Form, Location, and Orientation	5-83
5.9.9.6	Accommodating DRF Displacement	5-83
5.9.10	Simultaneous Requirements	5-86
5.9.11	Datum Simulation	5-89
5.9.12	Unstable Datums, Rocking Datums, Candidate Datums	5-89
5.9.13	Datum Targets	5-91
5.9.13.1	Datum Target Selection	5-91
5.9.13.2	Identifying Datum Targets	5-92
5.9.13.3	Datum Target Dimensions	5-94
5.9.13.4	Interdependency of Datum Target Locations	5-95
5.9.13.5	Applied to Features of Size	5-95
5.9.13.6	Applied to Any Type of Feature	5-97
5.9.13.7	Target Set with Switchable Precedence	5-99
5.9.14	Multiple Features Referenced as a Single Datum Feature	5-100
5.9.14.1	Feature Patterns	5-100
5.9.14.2	Coaxial and Coplanar Features	5-103
5.9.15	Multiple DRFs	5-103
5.10	Orientation Tolerance (Level 3 Control)	5-103
5.10.1	How to Apply It	5-103
5.10.2	Datums for Orientation Control	5-104
5.10.3	Applied to a Planar Feature (Including Tangent Plane Application)	5-104
5.10.4	Applied to a Cylindrical or Width-Type Feature	5-106
5.10.4.1	Zero Orientation Tolerance at MMC or LMC	5-107
5.10.5	Applied to Line Elements	5-107
5.10.6	The 24 Cases	5-109
5.10.7	Profile Tolerance for Orientation	5-109
5.10.8	When Do We Use an Orientation Tolerance?	5-109
5.11	Positional Tolerance (Level 4 Control)	5-113
5.11.1	How Does It Work?	5-113
5.11.2	How to Apply It	5-114
5.11.3	Datums for Positional Control	5-116
5.11.4	Angled Features	5-117
5.11.5	Projected Tolerance Zone	5-117
5.11.6	Special-Shaped Zones/Boundaries	5-121
5.11.6.1	Tapered Zone/Boundary	5-121
5.11.6.2	Bidirectional Tolerancing	5-122
5.11.6.3	Bounded Features	5-126
5.11.7	Patterns of Features	5-127
5.11.7.1	Single-Segment Feature Control Frame	5-127
5.11.7.2	Composite Feature Control Frame	5-129
5.11.7.3	Rules for Composite Control	5-131
5.11.7.4	Stacked Single-Segment Feature Control Frames	5-134
5.11.7.5	Rules for Stacked Single-Segment Feature Control Frames	5-136
5.11.7.6	Coaxial and Coplanar Features	5-136
5.11.8	Coaxiality and Coplanarity Control	5-137

5.12	Runout Tolerance	5-138
5.12.1	Why Do We Use It?	5-138
5.12.2	How Does It Work?	5-138
5.12.3	How to Apply It	5-139
5.12.4	Datums for Runout Control	5-140
5.12.5	Circular Runout Tolerance	5-141
5.12.6	Total Runout Tolerance	5-143
5.12.7	Application Over a Limited Length	5-143
5.12.8	When Do We Use a Runout Tolerance?	5-144
5.12.9	Worst Case Boundaries	5-145
5.13	Profile Tolerance	5-145
5.13.1	How Does It Work?	5-145
5.13.2	How to Apply It	5-145
5.13.3	The Basic Profile	5-147
5.13.4	The Profile Tolerance Zone	5-147
5.13.5	The Profile Feature Control Frame	5-149
5.13.6	Datums for Profile Control	5-149
5.13.7	Profile of a Surface Tolerance	5-149
5.13.8	Profile of a Line Tolerance	5-149
5.13.9	Controlling the Extent of a Profile Tolerance	5-150
5.13.10	Abutting Zones	5-153
5.13.11	Profile Tolerance for Combinations of Characteristics	5-153
5.13.11.1	With Positional Tolerancing for Bounded Features	5-153
5.13.12	Patterns of Profiled Features	5-154
5.13.12.1	Single-Segment Feature Control Frame	5-154
5.13.12.2	Composite Feature Control Frame	5-154
5.13.12.3	Stacked Single-Segment Feature Control Frames	5-155
5.13.12.4	Optional Level 2 Control	5-155
5.13.13	Composite Profile Tolerance for a Single Feature	5-156
5.14	Symmetry Tolerance	5-156
5.14.1	How Does It Work?	5-157
5.14.2	How to Apply It	5-159
5.14.3	Datums for Symmetry Control	5-159
5.14.4	Concentricity Tolerance	5-160
5.14.4.1	Concentricity Tolerance for Multifold Symmetry about a Datum Axis	5-160
5.14.4.2	Concentricity Tolerance about a Datum Point	5-161
5.14.5	Symmetry Tolerance about a Datum Plane	5-161
5.14.6	Symmetry Tolerancing of Yore (Past Practice)	5-161
5.14.7	When Do We Use a Symmetry Tolerance?	5-162
5.15	Combining Feature Control Frames	5-162
5.16	"Instant" GD&T	5-163
5.16.1	The "Dimension Origin" Symbol	5-163
5.16.2	General Note to Establish Basic Dimensions	5-163
5.16.3	General Note in Lieu of Feature Control Frames	5-164
5.17	The Future of GD&T	5-164
5.18	References	5-166

Chapter 6: Differences Between US Standards and Other Standards

Alex Krulikowski
Scott DeRaad

6.1	Dimensioning Standards	6-1
6.1.1	US Standards	6-2
6.1.2	International Standards	6-2
6.1.2.1	ISO Geometrical Product Specification Masterplan	6-4
6.2	Comparison of ASME and ISO Standards	6-5
6.2.1	Organization and Logistics	6-5
6.2.2	Number of Standards	6-5
6.2.3	Interpretation and Application	6-5

6.2.3.1	ASME	6-6
6.2.3.2	ISO	6-6
6.3	Other Standards	6-27
6.3.1	National Standards Based on ISO or ASME Standards	6-27
6.3.2	US Government Standards	6-28
6.3.3	Corporate Standards	6-28
6.3.4	Multiple Dimensioning Standards	6-29
6.4	Future of Dimensioning Standards	6-30
6.5	Effects of Technology	6-30
6.6	New Dimensioning Standards	6-30
6.7	References	6-30

Chapter 7: Mathematical Definition of Dimensioning and Tolerancing Principles .... Mark A. Nasson

7.1	Introduction	7-1
7.2	Why Mathematical Tolerance Definitions?	7-1
7.2.1	Metrology Crisis (The GIDEP Alert)	7-2
7.2.2	Specification Crisis	7-3
7.2.3	National Science Foundation Tolerancing Workshop	7-3
7.2.4	A New National Standard	7-4
7.3	What are Mathematical Tolerance Definitions?	7-4
7.3.1	Parallel, Equivalent, Unambiguous Expression	7-4
7.3.2	Metrology Independent	7-4
7.4	Detailed Descriptions of Mathematical Tolerance Definitions	7-4
7.4.1	Introduction	7-4
7.4.2	Vectors	7-5
7.4.2.1	Vector Addition and Subtraction	7-5
7.4.2.2	Vector Dot Products	7-6
7.4.2.3	Vector Cross Products	7-6
7.4.3	Actual Value / Measured Value	7-7
7.4.4	Datums	7-8
7.4.4.1	Candidate Datums / Datum Reference Frames	7-8
7.4.4.2	Degrees of Freedom	7-8
7.4.5	Form tolerances	7-9
7.4.5.1	Circularity	7-9
7.4.5.2	Cylindricity	7-12
7.4.5.3	Flatness	7-13
7.5	Where Do We Go from Here?	7-14
7.5.1	ASME Standards Committees	7-14
7.5.2	International Standards Efforts	7-14
7.5.3	CAE Software Developers	7-14
7.6	Acknowledgments	7-15
7.7	References	7-15

Chapter 8: Statistical Tolerancing .... Vijay Srinivasan, Ph.D

8.1	Introduction	8-1
8.2	Specification of Statistical Tolerancing	8-2
8.2.1	Using Process Capability Indices	8-2
8.2.2	Using RMS Deviation Index	8-4
8.2.3	Using Percent Containment	8-5
8.3	Statistical Tolerance Zones	8-5
8.3.1	Population Parameter Zones	8-6
8.3.2	Distribution Function Zones	8-7
8.4	Additional Illustrations	8-7
8.5	Summary and Concluding Remarks	8-9
8.6	References	8-10

Part 3 Design

Chapter 9: Traditional Approaches to Analyzing Mechanical Tolerance Stacks .... Paul Drake

9.1	Introduction	9-1
9.2	Analyzing Tolerance Stacks	9-1
9.2.1	Establishing Performance/Assembly Requirements	9-1
9.2.2	Loop Diagram	9-3
9.2.3	Converting Dimensions to Equal Bilateral Tolerances	9-5
9.2.4	Calculating the Mean Value (Gap) for the Requirement	9-7
9.2.5	Determine the Method of Analysis	9-8
9.2.6	Calculating the Variation for the Requirement	9-9
9.2.6.1	Worst Case Tolerancing Model	9-9
9.2.6.2	RSS Model	9-12
9.2.6.3	Modified Root Sum of the Squares Tolerancing Model	9-18
9.2.6.4	Comparison of Variation Models	9-22
9.2.6.5	Estimated Mean Shift Model	9-23
9.3	Analyzing Geometric Tolerances	9-24
9.3.1	Form Controls	9-25
9.3.2	Orientation Controls	9-26
9.3.3	Position	9-27
9.3.3.1	Position at RFS	9-27
9.3.3.2	Position at MMC or LMC	9-27
9.3.3.3	Virtual and Resultant Conditions	9-28
9.3.3.4	Equations	9-28
9.3.3.5	Composite Position	9-32
9.3.4	Runout	9-33
9.3.5	Concentricity/Symmetry	9-33
9.3.6	Profile	9-34
9.3.6.1	Profile Tolerancing with an Equal Bilateral Tolerance Zone	9-34
9.3.6.2	Profile Tolerancing with a Unilateral Tolerance Zone	9-35
9.3.6.3	Profile Tolerancing with an Unequal Bilateral Tolerance Zone	9-35
9.3.6.4	Composite Profile	9-36
9.3.7	Size Datums	9-36
9.4	Abbreviations	9-37
9.5	Terminology	9-39
9.6	References	9-39

Chapter 10: Statistical Background and Concepts .... Ron Randall

10.1	Introduction	10-1
10.2	Shape, Locations, and Spread	10-2
10.3	Some Important Distributions	10-2
10.3.1	The Normal Distribution	10-2
10.3.2	Lognormal Distribution	10-6
10.3.3	Poisson Distribution	10-8
10.4	Measures of Quality and Capability	10-10
10.4.1	Process Capability Index	10-10
10.4.2	Process Capability Index Relative to Process Centering (Cpk)	10-12
10.5	Summary	10-14
10.6	References	10-14
10.7	Appendix	10-15

Chapter 11: Predicting Assembly Quality (Six Sigma Methodologies to Optimize Tolerances) .... Dale Van Wyk

11.1	Introduction	11-1
11.2	What is Tolerance Allocation?	11-1
11.3	Process Standard Deviations	11-2
11.4	Worst Case Allocation	11-5
11.4.1	Assign Component Dimensions	11-6
11.4.2	Determine Assembly Performance, P	11-7
11.4.3	Assign the process with the largest si to each component	11-8
11.4.4	Calculate the Worst Case Assembly, t_{wc6}	11-8
11.4.5	Is $P \geq t_{wc6}$?	11-9
11.4.6	Estimating Defect Rates	11-10
11.4.7	Verification	11-12
11.4.8	Adjustments to Meet Quality Goals	11-13
11.4.9	Worst Case Allocation Summary	11-13
11.5	Statistical Allocation	11-13
11.5.1	Calculating Assembly Variation and Defect Rate	11-15
11.5.2	First Steps in Statistical Allocation	11-15
11.5.3	Calculate Expected Assembly Performance, P_6	11-15
11.5.4	Is $P \geq P_6$?	11-16
11.5.5	Allocating Tolerances	11-17
11.5.6	Statistical Allocation Summary	11-20
11.6	Dynamic RSS Allocation	11-20
11.7	Static RSS analysis	11-23
11.8	Comparison of the Techniques	11-24
11.9	Communication of Requirements	11-25
11.10	Summary	11-26
11.11	Abbreviations	11-26
11.12	References	11-27

Chapter 12: Multi-Dimensional Tolerance Analysis (Manual Method) .... Dale Van Wyk

12.1	Introduction	12-1
12.2	Determining Sensitivity	12-2
12.3	A Technique for Developing Gap Equations	12-4
12.4	Utilizing Sensitivity Information to Optimize Tolerances	12-12
12.5	Summary	12-13

Chapter 13: Multi-Dimensional Tolerance Analysis (Automated Method) .... Kenneth W. Chase, Ph.D.

13.1	Introduction	13-1
13.2	Three Sources of Variation in Assemblies	13-2
13.3	Example 2D Assembly – Stacked Blocks	13-3
13.4	Steps in Creating an Assembly Tolerance Model	13-4
13.5	Steps in Analyzing an Assembly Tolerance Model	13-12
13.5.5.1	Percent rejects	13-21
13.5.5.2	Percent Contribution Charts	13-22
13.5.5.3	Sensitivity Analysis	13-24
13.5.5.4	Modifying Geometry	13-24
13.6	Summary	13-26
13.7	References	13-27

Chapter 14: Minimum-Cost Tolerance Allocation .... Kenneth W. Chase, Ph.D.

14.1	Tolerance Allocation Using Least Cost Optimization	14-1
14.2	1-D Tolerance Allocation	14-1
14.3	1-D Example: Shaft and Housing Assembly	14-3
14.4	Advantages / Disadvantages of the Lagrange Multiplier Method	14-7
14.6	2-D and 3-D Tolerance Allocation	14-8
14.5	True Cost and Optimum Acceptance Fraction	14-8
14.7	2-D Example: One-way Clutch Assembly	14-9
14.7.1	Vector Loop Model and Assembly Function for the Clutch	14-10
14.8	Allocation by Scaling, Weight Factors	14-10
14.8.1	Proportional Scaling by Worst Case	14-11
14.8.2	Proportional Scaling by Root-Sum-Squares	14-11
14.8.3	Allocation by Weight Factors	14-11
14.9	Allocation by Cost Minimization	14-12
14.9.1	Minimum Cost Tolerances by Worst Case	14-13
14.9.2	Minimum Cost Tolerances by RSS	14-14
14.10	Tolerance Allocation with Process Selection	14-15
14.11	Summary	14-16
14.12	References	14-17
14.13	Appendix: Cost-Tolerance Functions for Metal Removal Processes	14-18

Chapter 15: Automating the Tolerancing Process .... Charles Glancy
James Stoddard
Marvin Law

15.1	Background Information	15-2
15.1.1	Benefits of Automation	15-2
15.1.2	Overview of the Tolerancing Process	15-2
15.2	Automating the Creation of the Tolerance Model	15-3
15.2.1	Characterizing Critical Design Measurements	15-3
15.2.2	Characterizing the Model Function	15-4
15.2.2.1	Model Definition	15-4
15.2.2.2	Model Form	15-5
15.2.2.3	Model Scope	15-5
15.2.3	Characterizing Input Variables	15-6
15.3	Automating Tolerance Analysis	15-6
15.3.1	Method of System Moments	15-6
15.3.3	Distribution Fitting	15-8
15.3.2	Monte Carlo Simulation	15-8
15.4	Automating Tolerance Optimization	15-9
15.5	Automating Communication Between Design and Manufacturing	15-9
15.5.1	Manufacturing Process Capabilities	15-10
15.5.1.1	Manufacturing Process Capability Database	15-10
15.5.1.2	Database Administration	15-11
15.5.2	Design Requirements and Assumptions	15-11
15.6	CAT Automation Tools	15-12
15.6.1	Tool Capability	15-12
15.6.2	Ease of Use	15-12
15.6.3	Training	15-13
15.6.4	Technical Support	15-13
15.6.5	Data Management and CAD Integration	15-13
15.6.6	Reports and Records	15-13
15.6.7	Tool Enhancement and Development	15-14
15.6.8	Deployment	15-14
15.7	Summary	15-14
15.8	References	15-14

Chapter 16: Working in an Electronic Environment .... Paul Matthews

16.1	Introduction	16-1
16.2	Paperless/Electronic Environment	16-2
16.2.1	Definition	16-2
16.3	Development Information Tools	16-3
16.3.1	Product Development Automation Strategy	16-3
16.3.2	Master Model Theory	16-4
16.3.3	Template Design	16-7
16.3.3.1	Template Part and Assembly Databases	16-7
16.3.3.2	Template Features	16-8
16.3.3.3	Templates for Analyses	16-9
16.3.3.4	Templates for Documentation	16-9
16.3.4	Component Libraries	16-9
16.3.5	Information Verification	16-10
16.4	Product Information Management	16-11
16.4.1	Configuration Management Techniques	16-11
16.4.2	Data Management Components	16-12
16.4.2.1	Workspace	16-12
16.4.2.2	Product Vault	16-12
16.4.2.3	Company Vault	16-12
16.4.3	Document Administrator	16-13
16.4.4	File Cabinet Control	16-13
16.4.5	Software Automation	16-13
16.5	Information Storage and Transfer	16-13
16.5.1	Internet	16-13
16.5.2	Electronic Mail	16-14
16.5.3	File Transfer Protocol	16-14
16.5.4	Media Transfer	16-15
16.6	Manufacturing Guidelines	16-15
16.6.1	Manufacturing Trust	16-15
16.6.2	Dimensionless Prints	16-15
16.6.2.1	Sheetmetal	16-16
16.6.2.2	Injection Molded Plastic	16-17
16.6.2.3	Hog Out Parts	16-17
16.6.2.4	Castings	16-18
16.6.2.5	Rapid Prototypes	16-18
16.7	Database Format Standards	16-19
16.7.1	Native Database	16-19
16.7.2	2-D Formats	16-19
16.7.2.1	Data eXchange Format (DXF)	16-19
16.7.2.2	Hewlett-Packard Graphics Language (HPGL)	16-20
16.8	3-D Formats	16-20
16.8.1	Initial Graphics Exchange Specification (IGES)	16-20
16.8.2	STandard for the Exchange of Product (STEP)	16-20
16.8.3	Virtual Reality Modeling Language (VRML)	16-20
16.8.4	STereoLithography (STL)	16-21
16.9	General Information Formats	16-21
16.9.1	Hypertext Markup Language (HTML)	16-21
16.9.2	Portable Document Format (PDF)	16-22
16.10	Graphics Formats	16-22
16.10.1	Encapsulated PostScript (EPS)	16-22
16.10.2	Joint Photographic Experts Group (JPEG)	16-22
16.10.3	Tagged Image File Format (TIFF)	16-22
16.11	Conclusion	16-23
16.12	Appendix A IGES Entities	16-23

Part 4 Manufacturing

Chapter 17: Collecting and Developing Manufacturing Process Capability Models .... Michael D. King

17.1	Why Collect and Develop Process Capability Models?	17-1
17.2	Developing Process Capability Models	17-2
17.3	Quality Prediction Models - Variable versus Attribute Information	17-3
17.3.1	Collecting and Modeling Variable Process Capability Models	17-3
17.3.2	Collecting and Modeling Attribute Process Capability Models	17-7
17.3.3	Feature Factoring Method	17-7
17.3.4	Defect Weighting Methodology	17-7
17.4	Cost and Cycle Time Prediction Modeling Variations	17-8
17.5	Validating and Checking the Results of Your Predictive Models	17-9
17.6	Summary	17-11
17.7	References	17-11

Part 5 Gaging

Chapter 18: Paper Gage Techniques .... Martin P. Wright

18.1	What is Paper Gaging?	18-1
18.2	Advantages and Disadvantages to Paper Gaging	18-2
18.3	Discrimination Provided By a Paper Gage	18-3
18.4	Paper Gage Accuracy	18-3
18.5	Plotting Paper Gage Data Points	18-4
18.6	Paper Gage Applications	18-4
18.6.1	Locational Verification	18-5
18.6.1.1	Simple Hole Pattern Verification	18-5
18.6.1.2	Three-Dimensional Hole Pattern Verification	18-8
18.6.1.3	Composite Positional Tolerance Verification	18-10
18.6.2	Capturing Tolerance From Datum Features Subject to Size Variation	18-12
18.6.2.1	Datum Feature Applied on an RFS Basis	18-12
18.6.2.2	Datum Feature Applied on an MMC Basis	18-12
18.6.2.3	Capturing Rotational Shift Tolerance from a Datum Feature Applied on an MMC Basis	18-16
18.6.2.4	Determining the Datum from a Pattern of Features	18-19
18.6.3	Paper Gage Used as a Process Analysis Tool	18-21
18.7	Summary	18-23
18.8	References	18-23

Chapter 19: Receiver Gages — Go Gages and Functional Gages .... James D. Meadows

19.1	Introduction	19-1
19.2	Gaging Fundamentals	19-2
19.3	Gage Tolerancing Policies	19-3
19.4	Examples of Gages	19-4
19.4.1	Position Using Partial and Planar Datum Features	19-4
19.4.2	Position Using Datum Features of Size at MMC	19-6
19.4.3	Position and Profile Using a Simultaneous Gaging Requirement	19-9
19.4.4	Position Using Centerplane Datums	19-12
19.4.5	Multiple Datum Structures	19-14
19.4.6	Secondary and Tertiary Datum Features of Size	19-17

19.5	Push Pin vs. Fixed Pin Gaging	19-20
19.6	Conclusion	19-20
19.7	References	19-20

Part 6 Precision Metrology

Chapter 20: Measurement Systems Analysis .... Gregory A. Hetland, Ph.D.

20.1	Introduction	20-1
20.2	Measurement Methods Analysis	20-2
20.2.1	Measurement System Definition (Phase 1)	20-2
20.2.1.1	Identification of Variables	20-2
20.2.1.2	Specifications of Conformance	20-3
20.2.1.3	Measurement System Capability Requirements	20-3
20.2.2	Identification of Sources of Uncertainty (Phase 2)	20-3
20.2.2.1	Machine Sources of Uncertainty	20-4
20.2.2.2	Software Sources of Uncertainty	20-4
20.2.2.3	Environmental Sources of Uncertainty	20-5
20.2.2.4	Part Sources of Uncertainty	20-5
20.2.2.5	Fixturing Sources of Uncertainty	20-5
20.2.2.6	Operator Sources of Uncertainty	20-6
20.2.3	Measurement System Qualification (Phase 3)	20-6
20.2.3.1	Plan the Capabilities Studies	20-6
20.2.3.2	Production Systems	20-7
20.2.3.3	Calibrate the System	20-7
20.2.3.4	Conduct Studies and Define Capabilities	20-8
20.2.4	Quantify the Error Budget (Phase 4)	20-8
20.2.4.1	Plan Testing (Isolate Error Sources)	20-8
20.2.4.2	Analyze Uncertainty	20-9
20.2.5	Optimize Measurement System (Phase 5)	20-9
20.2.5.1	Identify Opportunities	20-9
20.2.5.2	Attempt Improvements and Revisit Testing	20-9
20.2.5.3	Revisit Qualification	20-10
20.2.6	Implement and Control Measurement System (Phase 6)	20-10
20.2.6.1	Plan Performance Criteria	20-10
20.2.6.2	Plan Calibration and Maintenance Requirements	20-11
20.2.6.3	Implement System and Initiate Control	20-11
20.2.6.4	CMM Operator Competencies	20-11
20.2.6.5	Business Issue	20-12
20.3	CMM Performance Test Overview	20-17
20.3.1	Environmental Tests (Section 1)	20-17
20.3.1.1	Temperature Parameters	20-17
20.3.1.2	Other Environmental Parameters	20-20
20.3.2	Machine Tests (Section 2)	20-21
20.3.2.1	Probe Settling Time	20-21
20.3.2.2	Probe Deflection	20-24
20.3.2.3	Other Machine Parameters	20-27
20.3.2.4	Multiple Probes	20-27
20.3.3	Feature Based Measurement Tests (Section 3)	20-28
20.3.3.1	Number of Points Per Feature	20-30
20.3.3.2	Other Geometric Features	20-34
20.3.3.3	Contact Scanning	20-34
20.3.3.4	Surface Roughness	20-35
20.4	CMM Capability Matrix	20-35
20.5	References	20-38

Part 7 Applications

Chapter 21: Predicting Piecepart Quality .... Dan A. Watson, Ph.D.

21.1	Introduction	21-1
21.2	The Problem	21-2
21.3	Statistical Framework	21-3
21.3.1	Assumptions	21-3
21.3.2	Internal Feature at MMC	21-5
21.3.3	Internal Feature at LMC	21-7
21.3.4	External Features	21-8
21.3.5	Alternate Distribution Assumptions	21-8
21.4	Non-Size Feature Applications	21-9
21.5	Example	21-9
21.6	Summary	21-10
21.7	References	21-11

Chapter 22: Floating and Fixed Fasteners .... Paul Zimmermann

22.1	Introduction	22-1
22.2	Floating and Fixed Fasteners	22-1
22.2.1	What is a Floating Fastener?	22-4
22.2.2	What is a Fixed Fastener?	22-4
22.2.3	What is a Double-Fixed Fastener?	22-4
22.3	Geometric Dimensioning and Tolerancing (Cylindrical Tolerance Zone Versus +/- Tolerancing)	22-5
22.4	Calculations for Fixed, Floating and Double-fixed Fasteners	22-8
22.5	Geometric Dimensioning and Tolerancing Rules/Formulas for Floating Fastener	22-8
22.5.1	How to Calculate Clearance Hole Diameter for a Floating Fastener Application	22-8
22.5.2	How to Calculate Counterbore Diameter for a Floating Fastener Application	22-9
22.5.3	Why Floating Fasteners are Not Recommended	22-10
22.6	Geometric Dimensioning and Tolerancing Rules/Formulas for Fixed Fasteners	22-10
22.6.1	How to Calculate Fixed Fastener Applications	22-10
22.6.2	How to Calculate Counterbore Diameter for a Fixed Fastener Application	22-10
22.6.3	Why Fixed Fasteners are Recommended	22-11
22.7	Geometric Dimensioning and Tolerancing Rules/Formulas for Double-fixed Fastener	22-11
22.7.1	How to Calculate a Clearance Hole	22-11
22.7.2	How to Calculate the Countersink Diameter, Head Height Above and Head Height Below the Surface	22-11
22.7.3	What Are the Problems Associated with Double-fixed Fasteners?	22-13
22.8	Nut Plates: Floating and Nonfloating (see Fig. 22-14)	22-14
22.9	Projected Tolerance Zone	22-15
22.9.1	Comparison of Positional Tolerancing With and Without a Projected Tolerance Zone	22-16
22.9.2	Percent of Actual Orientation Versus Lost Functional Tolerance	22-18
22.10	Hardware Pages	22-18
22.10.1	Floating Fastener Hardware Pages	22-20
22.10.2	Fixed Fastener Hardware Pages	22-21
22.10.3	Double-fixed Fastener Hardware Pages	22-23
22.10.4	Counterbore Depths - Pan Head and Socket Head Cap Screws	22-25
22.10.5	Flat Head Screw Head Height - Above and Below the Surface	22-26
22.11	References	22-26

Contents   xix

Chapter 23: Fixed and Floating Fastener Variation .... Chris Cuba

23.1	Introduction	23-1
23.2	Hole Variation	23-2
23.3	Assembly Variation	23-4
23.4	Fixed and Floating Fasteners	23-4
23.4.1	Fixed Fastener Assembly Shift	23-5
23.4.2	Fixed Fastener Assembly Shift Using One Equation and Dimension Loop	23-6
23.4.3	Fixed Fastener Equation	23-7
23.4.4	Fixed Fastener Gap Analysis Steps	23-7
23.4.5	Floating Fastener Gap Analysis Steps	23-8
23.5	Summary	23-9
23.6	References	23-10

Chapter 24: Pinned Interfaces .... Stephen Harry Werst

24.1	List of Symbols (Definitions and Terminology)	24-1
24.2	Introduction	24-2
24.3	Performance Considerations	24-2
24.4	Variation Components of Pinned Interfaces	24-3
24.4.1	Type I Error	24-3
24.4.2	Type II Error	24-3
24.5	Types of Alignment Pins	24-4
24.6	Tolerance Allocation Methods - Worst Case vs. Statistical	24-6
24.7	Processes and Capabilities	24-6
24.8	Design Methodology	24-7
24.9	Proper Use of Material Modifiers	24-10
24.10	Temperature Considerations	24-11
24.11	Two Round Pins with Two Holes	24-11
24.11.1	Fit	24-12
24.11.2	Rotation Errors	24-12
24.11.3	Translation Errors	24-13
24.11.4	Performance Constants	24-13
24.11.5	Dimensioning Methodology	24-14
24.12	Round Pins with a Hole and a Slot	24-14
24.12.1	Fit	24-14
24.12.2	Rotation Errors	24-16
24.12.3	Translation Errors	24-17
24.12.4	Performance Constants	24-17
24.12.5	Dimensioning Methodology	24-17
24.13	Round Pins with One Hole and Edge Contact	24-18
24.13.1	Fit	24-19
24.13.2	Rotation Errors	24-20
24.13.3	Translation errors	24-20
24.13.4	Performance Constants	24-20
24.13.5	Dimensioning Methodology	24-20
24.14	One Diamond Pin and One Round Pin with Two Holes	24-23
24.14.1	Fit	24-23
24.14.2	Rotation and Translation Errors	24-24
24.14.3	Performance Constants	24-24
24.14.4	Dimensioning Methodology	24-24
24.15	One Parallel-Flats Pin and One Round Pin with Two Holes	24-26
24.15.1	Fit	24-26
24.15.2	Rotation and Translation Errors	24-27
24.15.3	Performance Constants	24-27
24.15.4	Dimensioning Methodology	24-28
24.16	References	24-29

Chapter 25: Gage Repeatability and Reproducibility (GR&R) Calculations .... Gregory A. Hetland, Ph.D.

25.1	Introduction 25-1
25.2	Standard GR&R Procedure 25-1
25.3	Summary 25-7
25.4	References 25-7

Part 8   The Future

Chapter 26: The Future .... Several contributors

Figures	.... F-1
Tables	.... T-1
Index	.... I-1

Foreword

Between the covers of this remarkable text one can experience, at near warp speed, a journey through the cosmos of subject matter dealing with dimensioning and tolerancing of mechanical products. The editor, as one of the contributing authors, has aptly summarized the content broadly as "about product variation." The contained chapters proceed then to wend their way through the various subjects to achieve that end. Under the individual pens of the authors, the wisdom, experience, writing style, and extensive research on each of the concerned topics presents the subject details with a unique richness. The authors, being widely renowned and respected in their fields of endeavor, combine to present a priceless body of knowledge available at the fingertips of the reader.

If not a first, this text surely is one of the best ever compiled as a consolidation of the contained related subjects. While possibly appearing a little overwhelming in its volume, the book succeeds in putting the reader at ease through the excellent subject matter arrangement, sequential flowing of chapters, listing of contents, and a complete index. The details of each chapter are self-explanatory and present "their story" in an enlightening, albeit challenging sometimes, individual style. Collectively, the authors and their respective chapters seem to reflect considerations and lessons learned from the past, inspiration and creativity for the state-of-the-art of the present, and insightful visions for the future. This text then equally represents a kind of status report of the various involved technologies, guidance and instruction for absorbing and implementing technical content, and some direction to the future path of progress.

Reflecting upon the significant contribution this text adds to the current state of progress on the contained subjects, a feeling of confidence prevails that there is no fear for the future— to the contrary, only a relish for the enlarging opportunities time will provide. Congratulations to the editor, Paul Drake, for his insight in conceiving this text and to all the authors and contributors. Your product represents a major achievement in its addition to the annals of product engineering literature. It is also a record of our times and a glimpse of the future. It is a distinct pleasure to endorse this text with added thanks for all the dedicated energy expended in behalf of this project and the professions involved. Your work will bring immediate returns and will also instill a pride of accomplishment on behalf of yourselves, our country, and the global community of industrial technology.

Lowell W. Foster
Lowell W. Foster Associates, Inc.
Minneapolis, Minnesota

About the Editor

Paul Drake is a Principal Engineer with Honors at the Raytheon Systems Company where he trains and consults in variation management, GD&T and Six Sigma mechanical tolerancing. He began the Mechanical Tolerancing and Performance Sigma Center for Excellence at Raytheon (formerly Texas Instruments, Inc.) in 1995. This center develops and deploys dimensioning and tolerancing best practices within Raytheon. As a member of the Raytheon Learning Institute, Paul has trained more than 3,500 people in GD&T and mechanical tolerancing in the past 12 years. He has also written numerous articles and design guides on optical and mechanical tolerancing,

Paul has ASME certification as a Senior Level GD&T Professional. He is a Subject Matter Expert (SME3) to ASME's Statistical Tolerancing Technical Subcommittee, a member of ASME's Geometric Dimensioning and Tolerancing Committee, a Six Sigma Blackbelt, and a licensed professional engineer in Texas. He holds two patents related to mechanical tolerancing.

Paul resides in Richardson, Texas, with his wife Jane and their three children.

Contributors

Timothy V. Bogard
Sigmetrix
Dallas, Texas
Chapter 26

Kenneth W. Chase, Ph.D.
Brigham Young University
Provo, Utah
Chapters 13, 14, and 26

Tom S. Cheek, Jr., Ph.D
Six Sigma Design Institute
Dallas, Texas
Reviewer

Chris Cuba
Raytheon Systems Company
McKinney, Texas
Chapter 23

Gordon Cumming
Raytheon Systems Company
McKinney, Texas
Reviewer

Don Day
Monroe Community College
Rochester, NY
Chapter 26

Scott DeRaad
General Motors Corporation
Ann Arbor, Michigan
Chapter 6

Paul Drake
Raytheon Systems Company
Plano, Texas
Chapters 5, 9, and 26

Charles Glancy
Raytheon Systems Company
Dallas, Texas
Chapter 15

Gregory A. Hetland, Ph.D.
Hutchinson Technology Inc.
Hutchinson, Minnesota
Chapters 3, 20, 25, and 26

Michael D. King
Raytheon Systems Company
Plano, Texas
Chapter 17

Alex Krulikowski
General Motors Corporation
Westland, Michigan
Chapter 6

Marvin Law
Raytheon Systems Company
Dallas, Texas
Chapter 15

Percy Mares
Boeing
Huntington Beach, California
Reviewer

Paul Matthews
Ultrak
Lewisville, TX
Chapter 16

Patrick J. McCuistion, Ph.D
Ohio University
Athens, Ohio
Chapter 4

James D. Meadows
Institute for Engineering & Design, Inc.
Hendersonville, Tennessee
Chapter 19

Jack Murphy
Raytheon Systems Company
Dallas, Texas
Reviewer

Mark A. Nasson
Draper Laboratory
Cambridge, Massachusetts
Chapter 7

Al Neumann
Technical Consultants, Inc.
Longboat Key, Florida
Chapter 26

Robert H. Nickolaisen, P.E.
Dimensional Engineering Services
Joplin, Missouri
Chapter 2

Ron Randall
Ron Randall & Associates, Inc.
Dallas, Texas
Chapters 1 and 10

Vijay Srinivasan, Ph.D
IBM Research and Columbia University
New York
Chapter 8

Walter M. Stites
AccraTronics Seals Corp.
Burbank, California
Chapter 5

James Stoddard
Raytheon Systems Company
Dallas, Texas
Chapter 15

Dale Van Wyk
Raytheon Systems Company
McKinney, Texas
Chapters 11 and 12

Stephen Harry Werst
Raytheon Systems Company
Dallas, Texas
Chapter 24

Robert Wiles
Datum Inspection Services
Phoenix, Arizona
Reviewer

Bruce A. Wilson
Aerospace Industry
St. Louis, Missouri
Chapter 26

Martin P. Wright
Behr Climate Systems, Inc.
Fort Worth, Texas
Chapter 18

Paul Zimmermann
Raytheon Systems Company
McKinney, Texas
Chapter 22

Dan A. Watson, Ph.D.
Texas Instruments Incorporated
Dallas, Texas
Chapter 21

Preface

This book is about transitioning from mechanical product design to manufacturing. The cover graphic illustrates two distinct phases of product development. The gear drawing (computer model) represents a concept that is perfect. The manufactured gear is imperfect. A major barrier in the journey from conceptual ideas to tangible products is *variation*. Variation can occur in the manufacturing of products, as well as in the processes that are used to develop the products.

This book is about mechanical product variation: how we understand it, how we deal with it, and how we control it. As the title suggests, this book focuses on *documenting* mechanical designs (dimensioning) and *understanding* the *variation* (tolerancing) within the product development process. If we accept all product variation into our design, our products may not function as intended. If we throw away parts with too much variation, our product costs will increase.

This book is about how we balance product variation with customer requirements. We generally deal with product variation in three ways.
- We *accept* product variation in our designs;
- We *control* product variation in our processes; or
- We *screen out* manufactured parts that have more variation than the design will allow.

Many experts refer to this balance between design requirements and manufacturing variation as *dimensional management*. I prefer to call it *variation management*. After all, variation is usually the primary contributor to product cost.

In order to manage variation we must understand how variation impacts the mechanical product development process.

This book is process driven. This book is not *just* a collection of related topics. At the heart of this book is the variation management process. Fig. P-1 shows a generic product development process, and captures the key activities we put in place to manage product variation. Your product development process may be similar in some areas and different in others, but I believe Fig. P-1 captures the essence of the design process.

Fig. P-1 does not try to document everything in the variation management process. This information is contained within the chapters. The purpose of Fig. P-1 is twofold; first, it gives a birds-eye view of the process to help the reader understand the "big-picture," and second, it is a starting point to show the reader where each chapter in the book fits into this process.

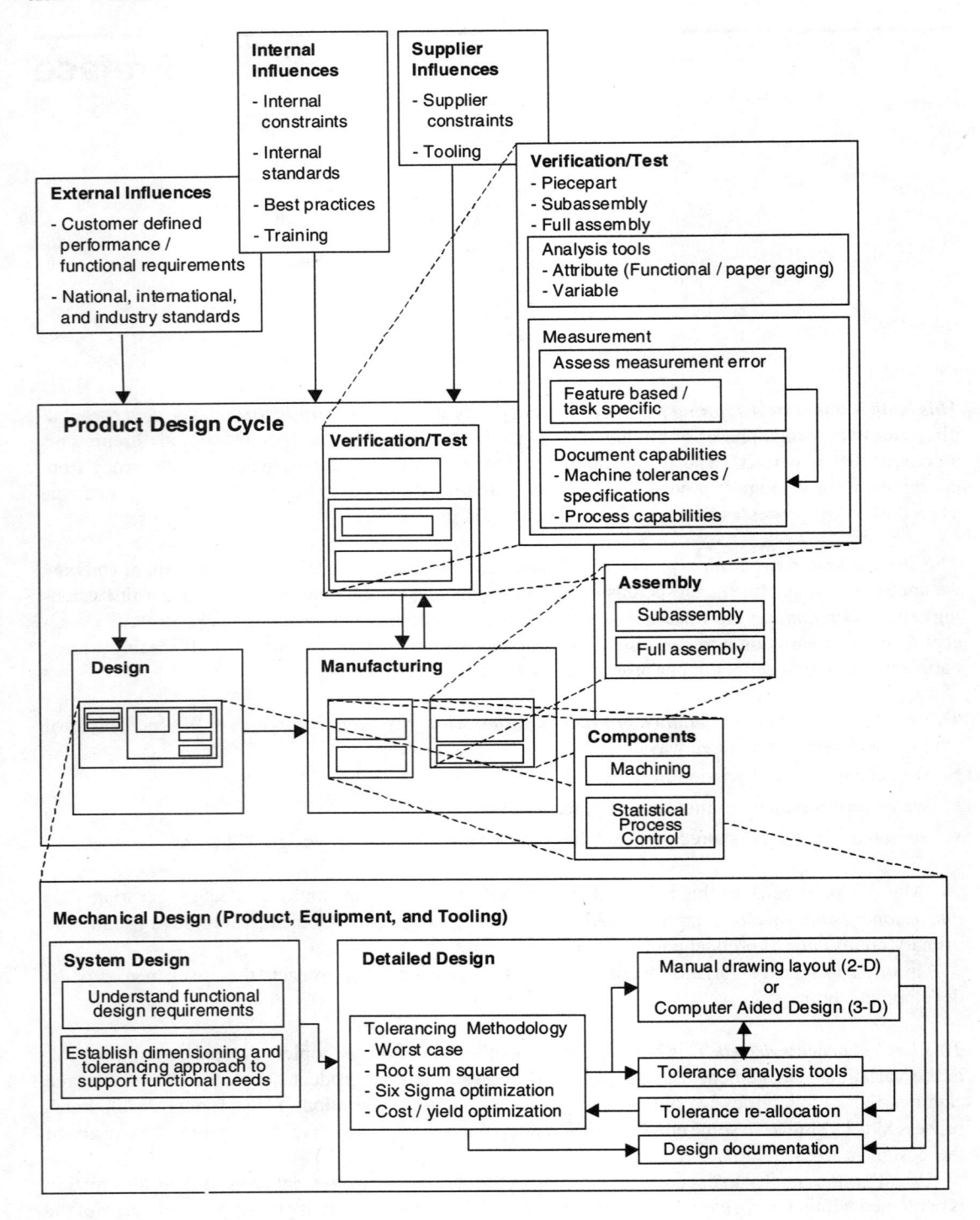

Figure P-1 Product development process

Each chapter of this book is linked to the product development process. The book is divided into seven parts that map to the process. Each chapter details the activities associated with the variation management process. By no means does this book capture everything. Although there is a wealth of information here, there is an endless amount of information that we could add. Likewise, new techniques, processes, and technologies will continue to evolve.

Although each chapter is a piece of the variation management puzzle, each chapter can stand alone. In practice, however, it is important to understand how each piece of the puzzle relates to others.

This book is about assessing design risk. If we understand the sources of product variation, and we understand the process(es) to manage them, we are well on our way to designing competitive products that meet customer requirements. If we capture the sources of variation and input these into the design process, we can assess the risk of meeting the manufacturing requirements as well as the performance of our designs.

Several experts contributed to this book. Each chapter reflects a wealth of experience from its author(s), many of whom are nationally and internationally recognized experts in their fields. This book could not contain the depth of information that it contains, without so many qualified contributors.

The audience for this book is very broad. Because it looks at the entire process of managing product variation, the audience for this book is large and very diverse. As a minimum, however, I suggest that everyone read the first chapter and the last chapter. Chapter 1 is a high-level historical perspective of where product quality has focused in the past. Chapter 26 is a compilation of where we think we will be in the future. Chapters 2 through 25 tell us how we are getting there today.

I appreciate any comments you have. Please send them to me at pdrake@mechsigma.com.

Paul Drake

Acknowledgments

I am grateful to the authors for their personal sacrifices and time they dedicated to this project. I am especially grateful to four people who have influenced my personal life, my career, and the writing of this book.

- Jane Drake, my wife, for her tireless editing and unwavering support
- Dr. Greg Hetland for his vision of the *big picture*
- Walt Stites for his meticulous detail and understanding of Geometric Dimensioning and Tolerancing
- Dale Van Wyk for helping me understand statistical tolerancing

I am also grateful to the following people for their support and help in this effort.

- Bob Esposito and Linda Ludewig from McGraw-Hill for their faith in this work
- Sally Glover from McGraw-Hill for proofing the work
- Mike Tinker, Ted Moody, and Rita Casavantes for their management support
- Todd Flippin for his late-night help keeping the computers running
- Gene Mancias for the wealth of graphic support
- Kelli and Joe Mancuso (The Training Edge) for help with the layout and design
- Scott Peters for his help with the index and printing
- Douglas Winters III for his artistic talents, graphics, and cover design

I wish to thank the reviewers Tom Cheek, Gordon Cumming, Percy Mares, Jack Murphy, and Bob Wiles for their careful and thorough review of this material.

I am deeply indebted to Lowell Foster, for his review and endorsement of this work.

I especially want to thank my wife, Jane, for her patience, endless hours of editing, and perseverance. I could not have done this without her.

I wish to dedicate this book to God; my parents, Anne and Paul Drake; and my wife Jane and children Taylor, Ellen, and Madeline.

PART 1

HISTORY / LESSONS LEARNED

Chapter 1

Quality Thrust

Ron Randall
Ron Randall & Associates, Inc.
Dallas, Texas

Ron Randall is an independent consultant, and an associate of the Six Sigma Academy, specializing in applying the principles of Six Sigma quality. Since the 1980s, Ron has applied Statistical Process Control and Design of Experiments principles to engineering and manufacturing at Texas Instruments Defense Systems and Electronics Group. While at Texas Instruments, he served as chairman of the Statistical Process Control Council, a Six Sigma Champion, Six Sigma Master Black Belt, and a Senior Member of the Technical Staff. His graduate work has been in engineering and statistics with study at SMU, the University of Tennessee at Knoxville, and NYU's Stern School of Business under Dr. W. Edwards Deming. Ron is a Registered Professional Engineer in Texas, a Senior Member of the American Society for Quality, and a Certified Quality Engineer. Ron served two terms on the Board of Examiners for the Malcolm Baldrige National Quality Award.

1.1 Meaning of Quality

What do we mean by the word *quality*? The word quality has multiple meanings. Some very important meanings are:
- Quality consists of those product features that meet the needs of customers and thereby provide product satisfaction.
- Quality consists of freedom from deficiencies, or in other words, absence of defects. (Reference 5)

Most corporations manage the business by understanding the financials. They spend significant resources on financial planning, financial control, and financial improvement. Successful companies also spend significant effort on quality planning, quality control, and quality improvement.

1.2 The Evolution of Quality

The evolution of product quality and quality-of-service has received a great deal of attention by corporations, educational institutions, and health care providers especially in the last 15 years. (Reference 8) Some corporations have been very successful financially because the quality of the products and services is superior to anything offered by a competitor. The relationship of quality and financial success in the automotive industry in the 1980s is a familiar example.

The winners of the Deming Prize in Japan, the Malcolm Baldrige National Quality Award in the United States, and similar awards around the world all have something in common. They have proven the strong relationship of quality and customer satisfaction to business excellence and financial success.

1.3 Some Quality Gurus and Their Contributions

1.3.1 W. Edwards Deming

The most famous name in Japanese quality control is American.

Dr. W. Edwards Deming (1900–1993) was the quality control expert whose work in the 1950s led Japanese industry into new principles of management and revolutionized their quality and productivity.

In 1950, the Union of Japanese Scientists and Engineers (J.U.S.E.) invited Dr. Deming to lecture several times in Japan. These lectures turned out to be overwhelmingly successful. To commemorate Dr. Deming's visit and to further Japan's development of quality control, J.U.S.E. shortly thereafter established the Deming prizes to be presented each year to the Japanese companies with the most outstanding achievements in quality control. (Reference 6)

In 1985 Deming wrote:

> "For a long period after World War II, till around 1962, the world bought whatever American Industry produced. The only problem American management faced was lack of capacity to produce enough for the market. No ability was required for management under those circumstances. There was no way to lose.
>
> It is different now. Competition from Japan wrought challenges that Western industry was not prepared to meet. The change has been gradual and was, in fact, ignored and denied over a number of years. All the while, Western management generated explanations for decline of business that now can be described as creative. The plain fact is that management was caught off guard, unable to manage anything but an expanding market.
>
> People in management cannot learn on the job what the job of management is. Help must come from the outside.
>
> The statistician's job is to find sources of improvement and sources of trouble. This is done with the aid of the theory of probability, the characteristic that distinguishes statistical work from that of other professions. Sources of improvement, as well as sources of obstacles and inhibitors that afflict Western industry, lie in top management. Fighting fires and solving problems downstream is important, but relatively insignificant compared with the contributions that management must make. Examination of sources of improvement has brought the 14 points for management and an awareness of the necessity to eradicate the deadly diseases and obstacles that infest Western industry." (Reference 6)

In his book *Out of the Crisis* (Reference 2) published in 1982 and again in 1986, Deming illustrates his 14 points:

1. Create constancy of purpose for improvement of product and service.
2. Adopt the new philosophy.

3. Cease dependence on inspection to achieve quality.
4. End the practice of awarding business on the basis of price tag alone. Instead, minimize total cost by working with a single supplier.
5. Improve constantly and forever every process for planning, production, and service.
6. Institute training on the job.
7. Adopt and institute leadership.
8. Drive out fear.
9. Break down barriers between staff areas.
10. Eliminate slogans, exhortations, and targets for the work force.
11. Eliminate numerical quotas for the work force and numerical goals for management.
12. Remove barriers that rob people of pride of workmanship. Eliminate the annual rating or merit system.
13. Institute a vigorous program of education and self-improvement for everyone.
14. Put everybody in the company to work to accomplish the transformation.

Much of industry's Total Quality Management (TQM) practices stem from Deming's work. The turnaround of many U.S. companies is directly attributable to Deming. This author had the privilege of completing Deming's four-day course in 1987 and two subsequent courses at New York University in 1990 and 1991. He was a great man who completed great works.

1.3.2 Joseph Juran

Juran showed us how to organize for quality improvement.

Another pioneer and leader in the quality transformation is Dr. Joseph M. Juran (1904–), founder and chairman emeritus of the Juran Institute, Inc. in Wilton, Connecticut. Juran has authored several books on quality planning, and quality by design, and is the editor-in-chief of Juran's *Quality Control Handbook*, the fourth edition copyrighted in 1988. (Reference 5)

Juran was an especially important figure in the quality changes taking place in American industry in the 1980s. Through the Juran Institute, Juran taught industry that work is accomplished by processes. Processes can be improved, products can be improved, and important financial gains can be accomplished by making these improvements. Juran showed us how to organize for quality improvement, that the language of management is money, and promoted the concept of project teams to improve quality. Juran introduced the Pareto principle to American industry. The Italian economist, Wilfredo Pareto, demonstrated that a small fraction of the people held most of the wealth. As applied to the cost of poor quality, the Pareto principle states that a few contributors to the cost are responsible for most of the cost. From this came the 80-20 rule, which states 20% of all the contributors to cost, account for 80% of the total cost.

Juran taught us how to manage for quality, organize for quality, and design for quality. In his 1992 book, *Juran on Quality by Design* (Reference 4), he tells us that poor quality is usually planned that way and quality planning in the past has been done by amateurs.

Juran discussed the need for unity of language with respect to quality and defined key words and phrases that are widely accepted today: (Reference 4)

"A product is the output of a process. Economists define products as goods and services.

A product feature is a property possessed by a product that is intended to meet certain customer needs and thereby provide customer satisfaction.

Customer satisfaction is a result achieved when product features respond to customer needs. It is generally synonymous with product satisfaction. Product satisfaction is a stimulus to product salability. *The major impact is on share of market, and thereby on sales income.*

A product deficiency is a product failure that results in *product dissatisfaction. The major impact is on the costs* incurred to redo prior work, to respond to customer complaints, and so on.

Product deficiencies are, in all cases, sources of customer *dissatisfaction.*

Product satisfaction and product dissatisfaction are not opposites. Satisfaction has its origins in product features and is why clients buy the product. Dissatisfaction has its origin in non-conformances and is why customers complain. There are products that give no dissatisfaction; they do what the supplier said they would do. Yet, the customer is dissatisfied with the product if there is some competing product providing greater satisfaction.

A customer is anyone who is impacted by the product or process. Customers may be internal or external."

This author has had the honor and privilege to work with Dr. Juran on company and national quality efforts in the 1980s and 1990s. Dr. Juran showed us how to manage for quality. He is a great teacher, leader, and mentor.

1.3.3 Philip B. Crosby

Doing things right the first time adds nothing to the cost of your product of service. Doing things wrong is what costs money.

In his book, *Quality is Free—The Art of Making Quality Certain* (Reference 1) Crosby introduced valuable quality-building tools that caught the attention of Western Management in the early 1980s. Crosby developed many of these ideas and methods during his industrial career at International Telephone and Telegraph Corporation. Crosby went on to teach these methods to managers at the Crosby Quality College in Florida.

- Quality Management Maturity Grid—An entire objective system for measuring your present quality system. Easy to use, it pinpoints areas in your operation for potential improvement.
- Quality Improvement Program—A proven 14-step procedure to turn your business around.
- Make Certain Program—The first defect prevention program ever for white-collar and nonmanufacturing employees.
- Management Style Evaluation—A self-examination process for managers that shows how personal qualities may be influencing product quality.

Crosby demonstrated that the typical American corporation spends 15% to 20% of its sales dollars on inspection, tests, warranties, and other quality-related costs. Crosby's work went on to define the elements of the cost of poor quality that are in use today at many corporations. Prevention costs, appraisal costs, and failure costs are well defined, and a system for periodic accounting is demonstrated.

In this author's experience with many large corporations, there is a direct correlation between the number of defects produced and the cost of poor quality. Crosby was the leader who showed how to qualitatively correlate defects with money, which Juran showed us, is the language of management.

1.3.4 Genichi Taguchi

Monetary losses occur with any deviation from the nominal.

Dr. Genichi Taguchi is the Japanese engineer that understood and quantified the effects of variation on the final product quality. (Reference 11) He understood and quantified the fact that any deviation from the nominal will cause a quantifiable cost, or loss. Most of Western management thinking today still believes that loss occurs only when a specification has been violated, which usually results in scrap or rework. The truth is that any design works best when all elements are at their target value.

Taguchi quantified the cost of variation and set forth this important mathematical relationship. Taguchi quantified what Juran, Crosby and others continue to teach. The language of management is money, and deviations from standard are losses. These losses are in performance, customer satisfaction, and supplier and manufacturing efficiency. These losses are real and can be quantified in terms of money.

Taguchi's Loss Function (Fig. 1-1) is defined as follows:

Monetary loss is a function of each product feature (x), and its difference from the best (target) value.

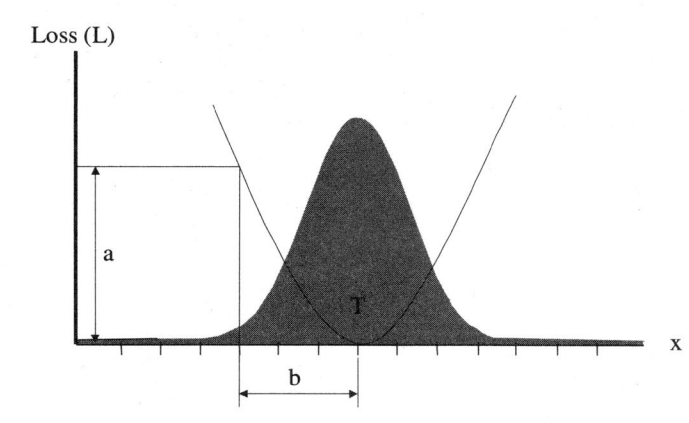

Figure 1-1 Taguchi's loss function and a normal distribution

x is a measure of a product characteristic
T is the target value of x
a = amount of loss when x is not on target T
b = amount that x is away from the target T

In this illustration, T = $\bar{x}$, where $\bar{x}$ is the mean of the sample of x's
In the simple case for one value of x, the loss is:
$L = k(x - T)^2$, where $k = a/b^2$

This simple quadratic equation is a good model for estimating the cost of not being on target.

The more general case can be expressed using knowledge of how the product characteristic (x) varies. The following model assumes a normal distribution, which is symmetrical about the average $\bar{x}$.

$L(x) = k[(\bar{x} - T)^2 + s^2]$, where s = the standard deviation of the sample of x's

The principles of Taguchi's Loss Function are fundamental to modern manufacturability and systems engineering analyses. Each function and each feature of a product can be analyzed individually. The summation of the estimated losses can lead an integrated design and manufacturing team to make tradeoffs quantitatively and early in the design process. (Reference 12)

1.4 The Six Sigma Approach to Quality

An aggressive campaign to boost profitability, increase market share, and improve customer satisfaction that has been launched by a select group of leaders in American Industry. (Reference 3)

1.4.1 The History of Six Sigma (Reference 10)

"In 1981, Bob Galvin, then chairman of Motorola, challenged his company to achieve a tenfold improvement in performance over a five-year period. While Motorola executives were looking for ways to cut waste, an engineer by the name of Bill Smith was studying the correlation between a product's field life and how often that product had been repaired during the manufacturing process. In 1985, Smith presented a paper concluding that if a product were found defective and corrected during the production process, other defects were bound to be missed and found later by the customer during the early use by the consumer. Additionally, Motorola was finding that best-in-class manufacturers were making products that required no repair or rework during the manufacturing process. (These were Six Sigma products.)

In 1988, Motorola won the Malcolm Baldrige National Quality Award, which set the standard for other companies to emulate.

(This author had the opportunity to examine some of Motorola's processes and products that were very near Six Sigma. These were nearly 2,000 times better than any products or processes that we at Texas Instruments (TI) Defense Systems and Electronics Group (DSEG) had ever seen. This benchmark caused DSEG to re-examine its product design and product production processes. Six Sigma was a very important element in Motorola's award winning application. TI's DSEG continued to make formal applications to the MBNQA office and won the award in 1992. Six Sigma was a very important part of the winning application.)

As other companies studied its success, Motorola realized its strategy to attain Six Sigma could be further extended." (Reference 3)

Galvin requested that Mikel J. Harry, then employed at Motorola's Government Electronics Group in Phoenix, Arizona, start the Six Sigma Research Institute (SSRI), circa 1990, at Motorola's Schaumburg, Illinois campus. With the financial support and participation of IBM, TI's DSEG, Digital Equipment Corporation (DEC), Asea Brown Boveri Ltd. (ABB), and Kodak, the SSRI began developing deployment strategies, and advanced applications of statistical methods for use by engineers and scientists.

Six Sigma Academy President, Richard Schroeder, and Harry joined forces at ABB to deploy Six Sigma and refined the breakthrough strategy by focusing on the relationship between net profits and product quality, productivity, and costs. The strategy resulted in a 68% reduction in defect levels and a 30% reduction in product costs, leading to $898 million in savings/cost reductions each year for two years. (Reference 13)

Schroeder and Harry established the Six Sigma Academy in 1994. Its client list includes companies such as Allied Signal, General Electric, Sony, Texas Instruments DSEG (now part of Raytheon), Bombardier, Crane Co., Lockheed Martin, and Polaroid. These companies correlate quality to the bottom line.

1.4.2 Six Sigma Success Stories

There are thousands of black belts working at companies worldwide. A blackbelt is an expert that can apply and deploy the Six Sigma Methods. (Reference 13)

Jennifer Pokrzywinski, an analyst with Morgan Stanley, Dean Witter, Discover & Co., writes "Six Sigma companies typically achieve faster working capital turns; lower capital spending as capacity is freed up; more productive R&D spending; faster new product development; and greater customer satisfaction." Pokrzywinski estimates that by the year 2000, GE's gross annual benefit from Six Sigma could be $6.6 billion, or 5.5% of sales. (Reference 7)

General Electric alone has trained about 6,000 people in the Six Sigma methods. The other companies mentioned above have trained thousands more. Each black belt typically completes three or four projects per year that save about $150,000 each. The savings are huge, and customers and shareholders are happier.

1.4.3 Six Sigma Basics

"The philosophy of Six Sigma recognizes that there is a direct correlation between the number of product defects, wasted operating costs, and the level of customer satisfaction. The Six Sigma statistic measures the capability of the process to perform defect-free work....

With Six Sigma, the common measurement index is defects per unit and can include anything from a component, piece of material, or line of code, to an administrative form, time frame, or distance. The sigma value indicates how often defects are likely to occur. The higher the sigma value, the less likely a process will produce defects.

Consequently, as sigma increases, product reliability improves, the need for testing and inspection diminishes, work in progress declines, costs go down, cycle time goes down, and customer satisfaction goes up.

Fig. 1-2 displays the short-term understanding of Six Sigma for a single critical-to-quality (CTQ) characteristic; in other words, when the process is centered. Fig. 1-3 illustrates the long-term perspective after the influence of process factors, which tend to affect process centering. From these figures, one can readily see that the short-term definition will produce 0.002 parts per million (ppm) defective. However, the long-term perspective reveals a defect rate of 3.4 ppm.

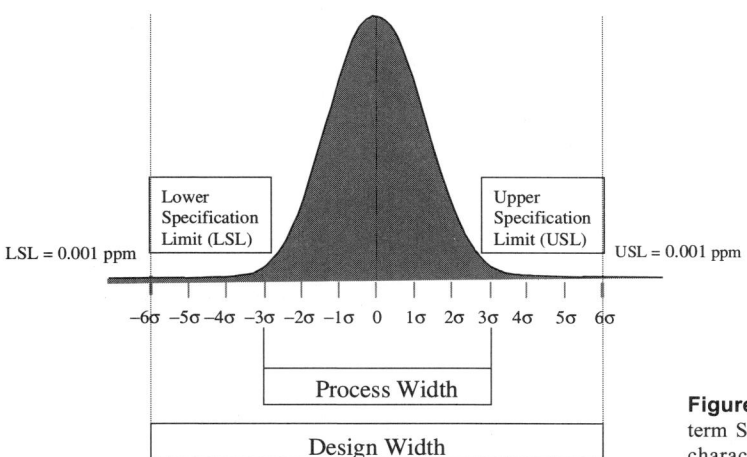

Figure 1-2 Graphical definition of short-term Six Sigma performance for a single characteristic

(This degradation in the short-term performance of the process is largely due to the adverse effect of long-term influences such as tool wear, material changes, and machine setup, just to mention a few. It is these types of factors that tend to upset process centering over many cycles of manufacturing. In fact, research has shown that a typical process is likely to deviate from its natural centered condition by approximately ±1.5 standard deviations at any given moment in time. With this principle in hand, one can make a rational estimate of the long-term process capability with knowledge of only the short-term performance. For example, if the capability of a CTQ characteristic is ±6.0 sigma in the short term, the long-term capability may be approximated as 6.0 sigma – 1.5 sigma = 4.5 sigma, or 3.4 ppm in terms of a defect rate.)" (Reference 3)

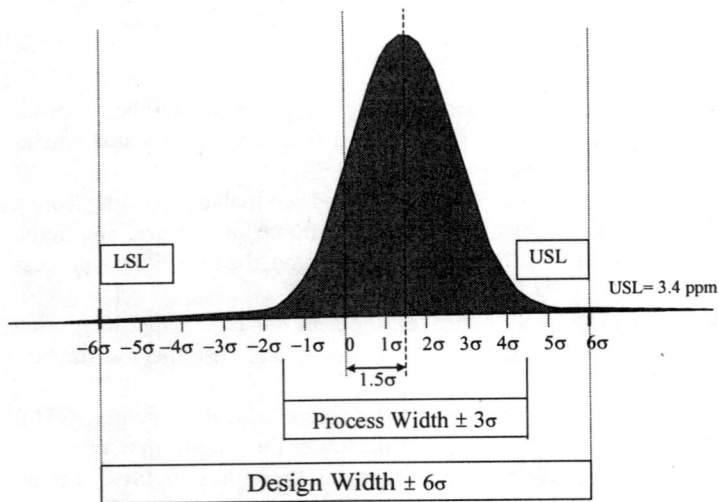

Figure 1-3 Graphical definition of long-term Six Sigma performance for a single characteristic (distribution shifted 1.5σ)

For designers of products, it is vitally important to know the capability of the process that will be used to manufacture a particular product feature. With this knowledge for each CTQ characteristic, an estimate of the number of defects that are likely to happen during manufacturing can be made. Extending this idea to the product level, a sigma value for the product design can be estimated. Products that are truly world-class have values around 6.0 sigma before manufacturing begins. Products that are extremely complex, like a large passenger jetliner, require sigma values greater than 6.0. Project managers and designers should know the sigma value of their design before production begins. The sigma value is a measure of the inherent manufacturability of the product.

Table 1-1 presents various levels of capability (manufacturability) and the implications to quality and costs.

Table 1-1 Practical impact of process capability

Sigma	Parts per Million	Cost of Poor Quality	
6 Sigma	3.4 defects per million	< 10% of sales	World class
5 Sigma	233 defects per million	10-15% of sales	
4 Sigma	6210 defects per million	15-20% of sales	Industry average
3 Sigma	66,807 defects per million	20-30% of sales	
2 Sigma	308,537 defects per million	30-40% of sales	Noncompetitive
1 Sigma	690,000 defects per million		

1.5 The Malcolm Baldrige National Quality Award (MBNQA)

Describe how new products are designed.

The criteria for the MBNQA asks companies to describe how new products are designed, and to describe how production processes are designed, implemented, and improved. Regarding design processes, the criteria further asks *"how design and production processes are coordinated to ensure trouble-free introduction and delivery of products."*

The winners of the MBNQA and other world-class companies have very specific processes for product design and product production. Most have an integrated product and process design process that requires early estimates of manufacturability. Following the Six Sigma methodology will enable design teams to estimate the quantitative measure of manufacturability.

What is the Malcolm Baldrige National Quality Award?

Congress established the award program in 1987 to recognize U.S. companies for their achievements in quality and business performance and to raise awareness about the importance of quality and performance excellence as a competitive edge. The award is not given for specific products or services. Two awards may be given annually in each of three categories: manufacturing, service, and small business.

While the Baldrige Award and the Baldrige winners are the very visible centerpiece of the U.S. quality movement, a broader national quality program has evolved around the award and its criteria. A report, *Building on Baldrige: American Quality for the 21st Century,* by the private Council on Competitiveness, states, "More than any other program, the Baldrige Quality Award is responsible for making quality a national priority and disseminating best practices across the United States."

The U.S. Commerce Department's National Institute of Standards and Technology (NIST) manages the award in close cooperation with the private sector.

Why was the award established?

In the early and mid-1980s, many industry and government leaders saw that a renewed emphasis on quality was no longer an option for American companies but a necessity for doing business in an ever expanding, and more demanding, competitive world market. But many American businesses either did not believe quality mattered for them or did not know where to begin. The Baldrige Award was envisioned as a standard of excellence that would help U.S. companies achieve world-class quality.

How is the Baldrige Award achieving its goals?

The criteria for the Baldrige Award have played a major role in achieving the goals established by Congress. They now are accepted widely, not only in the United States but also around the world, as the standard for performance excellence. The criteria are designed to help companies enhance their competitiveness by focusing on two goals: delivering ever improving value to customers and improving overall company performance.

The award program has proven to be a remarkably successful government and industry team effort. The annual government investment of about $3 million is leveraged by more than $100 million of private-sector contributions. This includes more than $10 million raised by private industry to help launch the program, plus the time and efforts of hundreds of largely private-sector volunteers.

The cooperative nature of this joint government/private-sector team is perhaps best captured by the award's Board of Examiners. Each year, more than 300 experts from industry, as well as universities, governments at all levels, and non-profit organizations, volunteer many hours reviewing applications for the award, conducting site visits, and providing each applicant with an extensive feedback report citing strengths and opportunities to improve. In addition, board members have given thousands of presentations on quality management, performance improvement, and the Baldrige Award.

The award-winning companies also have taken seriously their charge to be quality advocates. Their efforts to educate and inform other companies and organizations on the benefits of using the Baldrige Award framework and criteria have far exceeded expectations. To date, the winners have given approximately 30,000 presentations reaching thousands of organizations.

How does the Baldrige Award differ from ISO 9000?

The purpose, content, and focus of the Baldrige Award and ISO 9000 are very different. Congress created the Baldrige Award in 1987 to enhance U.S. competitiveness. The award program promotes quality awareness, recognizes quality achievements of U.S. companies, and provides a vehicle for sharing successful strategies. The Baldrige Award criteria focus on results and continuous improvement. They provide a framework for designing, implementing, and assessing a process for managing all business operations.

ISO 9000 is a series of five international standards published in 1987 by the International Organization for Standardization (ISO), Geneva, Switzerland. Companies can use the standards to help determine what is needed to maintain an efficient quality conformance system. For example, the standards describe the need for an effective quality system, for ensuring that measuring and testing equipment is calibrated regularly, and for maintaining an adequate record-keeping system. ISO 9000 registration determines whether a company complies with its own quality system.

Overall, ISO 9000 registration covers less than 10 percent of the Baldrige Award criteria. (Reference 9)

1.6 References

1. Crosby, Philip B. 1979. *Quality is Free—The Art of Making Quality Certain.* New York, NY: McGraw-Hill.
2. Deming, W. Edwards. 1982, 1986. *Out of the Crisis.* Cambridge, MA: Massachusetts Institute of Technology Center for Advanced Engineering Study.
3. Harry, Mikel J. 1998. Six Sigma: A Breakthrough Strategy for Profitability. *Quality Progress,* May, 60–64.
4. Juran, J.M. 1992. *Juran on Quality by Design.* New York: The Free Press.
5. Juran, J.M. 1988. *Quality Control Handbook.* 4th ed. New York, NY: McGraw-Hill.
6. Mann, Nancy R. 1985, 1987. *The Keys to Excellence.* Los Angeles: Prestwick Books.
7. Morgan Stanley, Dean Witter, Discover & Co. June 6, 1996. *Company Update.*
8. National Institute of Standards and Technology. 1998. U.S. Department of Commerce.
9. National Institute of Standards and Technology. U.S. Department of Commerce. 1998. Excerpt from "Frequently Asked Questions and Answers about the Malcolm Baldrige National Quality Award." Malcolm Baldrige National Quality Award Office, A537 Administration Building, NIST, Gaithersburg, Maryland 20899-0001.
10. Six Sigma is a federally registered trademark of Motorola.
11. Taguchi, Genichi. 1970. Quality Assurance and Design of Inspection During Production. *Reports of Statistical Applications and Research* 17(1). Japanese Union of Scientists and Engineers.
12. Taguchi, Genichi. 1985. *System of Experimental Design.* Vols. 1 and 2. White Plains, NY: Kraus International Publications.
13. The terms Breakthrough Strategy, Champion, Master Black Belt, Black Belt, and Green Belt are federally registered trademarks of Sigma Consultants, L.L.C., doing business as Six Sigma Academy.

Chapter 2

Dimensional Management

Robert H. Nickolaisen, P.E.
Dimensional Engineering Services
Joplin, Missouri

Robert H. Nickolaisen is president of Dimensional Engineering Services (Joplin, MO), which provides customized training and consulting in the field of Geometric Dimensioning and Tolerancing and related technologies. He also is a professor emeritus of mechanical engineering technology at Pittsburg State University (Pittsburg, Kansas). Professional memberships include senior membership in the Society of Manufacturing Engineers (SME) and the American Society of Mechanical Engineers (ASME). He is an ASME certified Senior Level Geometric Dimensioning and Tolerancing Professional (Senior GDTP), a certified manufacturing engineer (CMfgE), and a licensed professional engineer. Current standards activities include membership on the following national and international standards committees: US TAG ISO/TC 213 (Dimensional and Geometrical Product Specification and Verification), ASME Y14.5 (Dimensioning and Tolerancing), and ASME Y14.5.2 (Certification of GD&T Professionals).

2.1 Traditional Approaches to Dimensioning and Tolerancing

Engineering, as a science and a philosophy, has gone through a series of changes that explain and justify the need for a new system for managing dimensioning and tolerancing activities. The evolution of a system to control the dimensional variation of manufactured products closely follows the growth of the quality control movement.

Men like Sir Ronald Fisher, Frank Yates, and Walter Shewhart were introducing early forms of modern quality control in the 1920s and 1930s. This was also a period when engineering and manufacturing personnel were usually housed in adjacent facilities. This made it possible for the designer and fabricator to work together on a daily basis to solve problems relating to fit and function.

The importance of assigning and controlling tolerances that would consistently produce interchangeable parts and a quality product increased in importance during the 1940s and 1950s. Genichi Taguchi and W. Edwards Deming began to teach industries worldwide (beginning in Japan) that quality should be addressed before a product was released to production.

The space race and cold war of the 1960s had a profound impact on modern engineering education. During the 1960s and 1970s, the trend in engineering education in the United States shifted away from a design-oriented curriculum toward a more theoretical and mathematical approach. Concurrent with this change in educational philosophy was the practice of issuing contracts between customers and suppliers that increased the physical separation of engineering personnel from the manufacturing process. These two changes, education and contracts, encouraged the development of several different product design philosophies. The philosophies include engineering driven design, process driven design, and inspection driven design.

2.1.1   Engineering Driven Design

An engineering driven design is based on the premise that the engineering designer can specify any tolerance values deemed necessary to ensure the perceived functional requirements of a product. Traditionally, the design engineer assigns dimensional tolerances on component parts just before the drawings are released. These tolerance values are based on past experience, best guess, anticipated manufacturing capability, or build-test-fix methods during product development. When the tolerances are determined, there is usually little or no communication between the engineering and the manufacturing or inspection departments.

This method is sometimes called the "over-the-wall" approach to engineering design because once the drawings are released to production, the manufacturing and inspection personnel must live with whatever dimensional tolerance values are specified. The weakness of the approach is that problems are always discovered during or after part processing has begun, when manufacturing costs are highest. It also encourages disputes between engineering, manufacturing and quality personnel. These disputes in turn tend to increase manufacturing cycle times, engineering change orders, and overall costs.

2.1.2   Process Driven Design

A process driven design establishes the dimensional tolerances that are placed on a drawing based entirely on the capability of the manufacturing process, not on the requirements of the fit and function between mating parts. When the manufactured parts are inspected and meet the tolerance requirements of the drawings, they are accepted as good parts. However, they may or may not assemble properly. This condition occurs because the inspection process is only able to verify the tolerance specifications for the manufacturing process rather than the requirement for design fit and function for mating parts. This method is used in organizations where manufacturing "dictates" design requirements to engineering.

2.1.3   Inspection Driven Design

An inspection driven design derives dimensional tolerances from the expected measurement technique and equipment that will be used to inspect the manufactured parts. Inspection driven design does not use the functional limits as the assigned values for the tolerances that are placed on the drawing. The functional limits of a dimensional tolerance are the limits that a feature has to be within for the part to assemble and perform correctly.

One inspection driven design method assigns tolerances based on the measurement uncertainty of the measurement system that will be used to inspect finished parts. When this method is used, the tolerance values that are indicated on the drawing are derived by subtracting one-half of the measurement uncertainty from each end of the functional limits. This smaller tolerance value then becomes the basis for part acceptance or rejection.

Inspection driven design can be effective when the designer and metrologist work very closely together during the development stage of the product. However, the system breaks down when the designer has no knowledge of metrology, if the proposed measurement technique is not known, or if the measurements are not made as originally conceived.

2.2 A Need for Change

The need to change from the traditional approaches to dimensioning and tolerancing was not universally recognized in the United States until the 1980s. Prior to that time, tolerances were generally assigned as an afterthought of the build-test-fix product design process. The catalyst for change was that American industry began to learn and practice some of the techniques taught by Deming, Taguichi, Juran, and others (see Chapter 1).

The 1980s also saw the introduction of the Six Sigma Quality Method by a U.S. company (Motorola), adoption of the Malcolm Baldrige National Quality Award, and publication of the ISO 9000 Quality Systems Standards. The entire decade was filled with a renewed interest in a quality movement that emphasized statistical techniques, teams, and management commitment. These conditions provided the ideal setting for the birth of "dimensional management."

2.2.1 Dimensional Management

Dimensional management is a process by which the design, fabrication, and inspection of a product are systematically defined and monitored to meet predetermined dimensional quality goals. It is an engineering process that is combined with a set of tools that make it possible to understand and design for variation. Its purpose is to improve first-time quality, performance, service life, and associated costs. Dimensional management is sometimes called dimensional control, dimensional variation management or dimensional engineering.

2.2.2 Dimensional Management Systems

Inherent in the dimensional management process is the systematic implementation of dimensional management tools. A typical dimensional management system uses the following tools (see Fig. 2-1):

- Simultaneous engineering teams
- Written goals and objectives
- Design for manufacturability and design for assembly
- Geometric dimensioning and tolerancing

Simultaneous Engineering Teams

Written Goals and Objectives

Design for Manufacturability and Assembly

Geometric Dimensioning and Tolerancing

Key Characteristics

Statistical Process Control

Variation Measurement and Reduction

Variation Simulation Tolerance Analysis

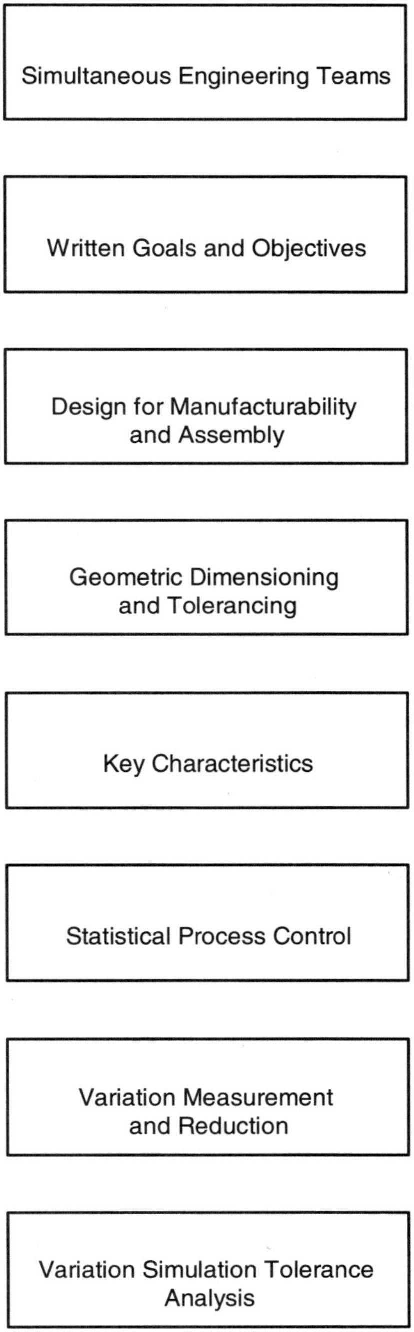

Figure 2-1 Dimensional management tools

- Key characteristics
- Statistical process control
- Variation measurement and reduction
- Variation simulation tolerance analysis

2.2.2.1 Simultaneous Engineering Teams

Simultaneous engineering teams are crucial to the success of any dimensional management system. They are organized early in the design process and are retained from design concept to project completion. Membership is typically composed of engineering design, manufacturing, quality personnel, and additional members with specialized knowledge or experience. Many teams also include customer representatives. Depending on the industry, they may be referred to as product development teams (PDT), integrated product teams (IPT), integrated process and product development (IPPD) teams, and design build teams (DBT).

The major purpose of a dimensional management team is to identify, document, and monitor the dimensional management process for a specific product. They are also responsible for establishing specific goals and objectives that define the amount of product dimensional variation that can be allowed for proper part fit, function, and assembly based on customer requirements and are empowered to ensure that these goals and objectives are accomplished. The overall role of any dimensional management team is to do the following:

- Participate in the identification, documentation, implementation, and monitoring of dimensional goals and objectives.
- Identify part candidates for design for manufacturability and assembly (DFMA).
- Establish key characteristics.
- Implement and monitor statistical process controls.
- Participate in variation simulation studies.
- Conduct variation measurement and reduction activities.
- Provide overall direction for dimensional management activities.

The most effective dimensional management teams are composed of individuals who have broad experience in all aspects of design, manufacturing, and quality assurance. A design engineer willing and able to understand and accept manufacturing and quality issues is a definite asset. A statistician with a firm foundation in process control and a dimensional engineer specializing in geometric dimensioning and tolerancing and variation simulation analysis add considerable strength to any dimensional management team. All members should be knowledgeable, experienced, and willing to adjust to the new dimensional management paradigm. Therefore, care should be taken in selecting members of a dimensional management team because the ultimate success or failure of any project depends directly on the support for the team and the individual team member's commitment and leadership.

2.2.2.2 Written Goals and Objectives

Using overall dimensional design criteria, a dimensional management team writes down the dimensional goals and objectives for a specific product. Those writing the goals and objectives also consider the capability of the manufacturing and measurement processes that will be used to produce and inspect the finished product. In all cases, the goals and objectives are based on the customer requirements for fit, function, and durability with quantifiable and measurable values.

In practice, dimensional management objectives are described in product data sheets. The purpose of these data sheets is to establish interface requirements early so that any future engineering changes related to the subject part are minimal. The data sheets typically include a drawing of the individual part or subassembly that identifies interface datums, dimensions, tolerance requirements, key characteristics, tooling locators, and the assembly sequence.

2.2.2.3 Design for Manufacturability (DFM) and Design for Assembly (DFA)

A design for manufacturability (DFM) program attempts to provide compatibility between the definition of the product and the proposed manufacturing process. The overall objective is for the manufacturing capabilities and process to achieve the design intent. This objective is not easy to accomplish and must be guided by an overall strategy. One such strategy that has been developed by Motorola Inc. involves six fundamental steps summarized below in the context of dimensional management team activities.

Step 1: Identify the key characteristics.

Step 2: Identify the product elements that influence the key characteristics defined in Step 1.

Step 3: Define the process elements that influence the key characteristics defined in Step 2.

Step 4: Establish maximum tolerances for each product and process element defined in Steps 2 and 3.

Step 5: Determine the actual capability of the elements presented in Steps 2 and 3.

Step 6: Assure $C_p \geq 2$; $C_{pk} \geq 1.5$. See Chapters 8, 10, and 11 for more discussion on C_p and C_{pk}.

Design for assembly (DFA) is a method that focuses on simplifying an assembly. A major objective of DFA is to reduce the number of individual parts in the assembly and to eliminate as many fasteners as possible. The results of applying DFA are that there are fewer parts to design, plan, fabricate, tool, inventory, and control. DFA will also lower cost and weight, and improve quality.

Some critical questions that are asked during a DFA study are as follows:

- Do the parts move relative to each other?
- Do the parts need to be made from different material?
- Do the parts need to be removable?

If the answer to all of these questions is no, then combining the parts should be considered. The general guidelines for conducting a DFA study should include a decision to:

- Minimize the overall number of parts.
- Eliminate adjustments and reorientation.
- Design parts that are easy to insert and align.
- Design the assembly process in a layered fashion.
- Reduce the number of fasteners.
- Attempt to use a common fastener and fastener system.
- Avoid expensive fastener operations.
- Improve part handling.
- Simplify service and packaging.

2.2.2.4 Geometric Dimensioning and Tolerancing (GD&T)

Geometric dimensioning and tolerancing is an international engineering drawing system that offers a practical method for specifying 3-D design dimensions and tolerances on an engineering drawing. Based on a universally accepted graphic language, as published in national and international standards, it improves communication, product design, and quality. Therefore, geometric dimensioning and tolerancing is accepted as the language of dimensional management and must be understood by all members of the dimensional management team. Some of the advantages of using GD&T on engineering drawings and product data sheets are that it:

- Removes ambiguity by applying universally accepted symbols and syntax.
- Uses datums and datum systems to define dimensional requirements with respect to part interfaces.
- Specifies dimensions and related tolerances based on functional relationships.
- Expresses dimensional tolerance requirements using methods that decrease tolerance accumulation.
- Provides information that can be used to control tooling and assembly interfaces.

See Chapters 3 and 5 for more discussion of the advantages of GD&T.

2.2.2.5 Key Characteristics

A key characteristic is a feature of an installation, assembly, or detail part with a dimensional variation having the greatest impact on fit, performance, or service life. The identification of key characteristics for a specific product is the responsibility of the dimensional management team working very closely with the customer.

Key characteristic identification is a tool for facilitating assembly that will reduce variability within the specification limits. This can be accomplished by using key characteristics to identify features where variation from nominal is critical to fit and function between mating parts or assemblies. Those features identified as key characteristics are indicated on the product drawing and product data sheets using a unique symbol and some method of codification. Features designated as "key" undergo variation reduction efforts. However, key characteristic identification does not diminish the importance of other nonkey features that still must comply with the quality requirements defined on the drawing.

The implementation of a key characteristic system has been shown to be most effective when the key characteristics are:

- Selected from interfacing control features and dimensions.
- Indicated on the drawings using a unique symbol.
- Established in a team environment.
- Few in number.
- Viewed as changeable over time.
- Measurable, preferably using variable data.
- Determined and documented using a standard method.

2.2.2.6 Statistical Process Control (SPC)

Statistical process control is a tool that uses statistical techniques and control charts to monitor a process output over time. Control charts are line graphs that are commonly used to identify sources of variation in a key characteristic or process. They can be used to reveal a problem, quantify the problem, help to solve the problem, and confirm that corrective action has eliminated the problem.

A standard deviation is a unit of measure used to describe the natural variation above an average or mean value. A normal distribution of a process output results in 68% of the measured data falling within ±1 standard deviation, 95% falling within ±2 standard deviations, and 99.7% falling within ±3 standard deviations.

The natural variation in a key characteristic or process defines its process capability. Capability refers to the total variation within the process compared to a six standard deviation spread. This capability is the amount of variation that is inherent in the process.

Process capability is expressed as a common ratio of "Cp" or "Cpk." Cp is the width of the engineering tolerance divided by the spread in the output of the process. The higher the Cp value, the less variance there is in the process for a given tolerance. A Cp ≥ 2.0 is usually a desired minimum value.

Cpk is a ratio that compares the average of the process to the tolerance in relation to the variation of the process. Cpk can be used to measure the performance of a process. It does not assume that the process is centered. The higher the Cpk value the less loss is associated with the variation. A Cpk ≥ 1.5 is usually a desired minimum value.

Cp and Cpk values are simply indicators of progress in the effort to refine a process and should be continuously improved. To reduce rework, the process spread should be centered between the specification limits and the width of the process spread should be reduced. See Chapters 8 and 10 for more discussion of Cp and Cpk.

2.2.2.7 Variation Measurement and Reduction

After key characteristics have been defined and process and tooling plans have been developed, parts must be measured to verify conformance with their dimensional specifications. This measurement data must be collected and presented in a format that is concise and direct in order to identify actual part variation. Therefore, measurement plans and procedures must be able to meet the following criteria:

- The measurement system must provide real-time feedback.
- The measurement process should be simple, direct, and correct.
- Measurements must be consistent from part to part; detail to assembly, etc.
- Data must be taken from fixed measurement points.
- Measurements must be repeatable and reproducible.
- Measurement data display and storage must be readable, meaningful, and retrievable.

A continuous program of gage and tooling verification and certification must also be integrated within the framework of the dimensional measurement plan. Gage repeatability and reproducibility (GR&R) studies and reports must be a standard practice. Assembly tooling must be designed so that their locators are coordinated with the datums established on the product drawings and product data sheets. This will ensure that the proper fit and function between mating parts has been obtained. The actual location of these tooling points must then be periodically checked and validated to ensure that they have not moved and are not introducing errors into the product. See Chapter 24 for more discussion of gage repeatability and reproducibility (GRER).

2.2.2.8 Variation Simulation Tolerance Analysis

Dimensional management tools have been successfully incorporated within commercial 3-D simulation software (see Chapter 15). The typical steps in performing a simulation study using simulation software are listed below (see Fig. 2-2):

Step 1: A conceptual design is created within an existing computer aided engineering (CAE) software program as a 3-D solid model.

Step 2: The functional features that are critical to fit and function for each component of an assembly are defined and relationships established using GD&T symbology and datum referencing.

Step 3: Dimensioning schemes are created in the CAE and are verified and analyzed by the simulation software for correctness to appropriate standards.

Step 4: Using information from the CAE database, a functional assembly model is mathematically defined and a definition of assembly sequence, methods, and measurements is created.

Step 5: Using the functional assembly model, a 3-D assembly tolerance analysis is statistically performed to identify, rank, and correct critical fit and functional relationships between the mating parts that make up the assembly.

The advantages of using simulation software are that it can be integrated directly with existing CAE software to provide a seamless communication tool from conceptual design to final assembly simulation without the expense of building traditional prototypes. The results also represent reality because the simulations are based on statistical concepts taking into account the relationship between functional requirements as well as the expected process and measurement capabilities.

2.3 The Dimensional Management Process

The dimensional management process can be divided into four general stages: concept, design, prototype, and production. These stages integrated with the various dimensional management tools can be represented by a flow diagram (see Fig. 2-3).

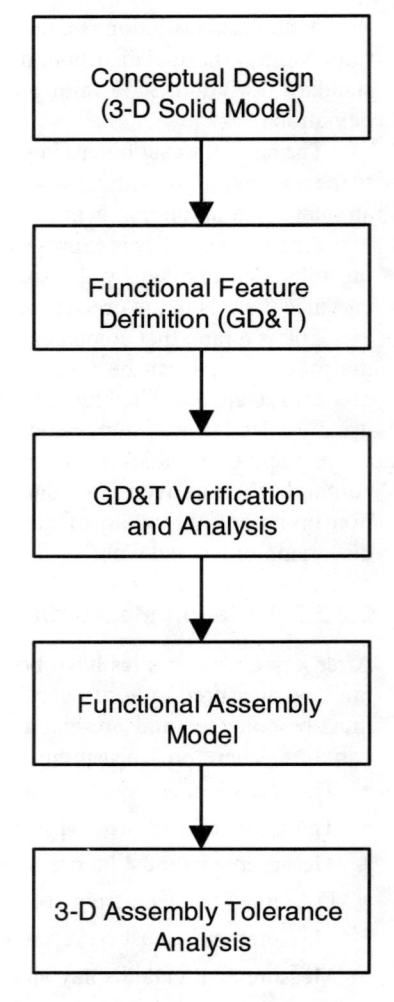

Figure 2-2 Variation simulation analysis

The key factor in the success of a dimensional management program is the commitment and support provided by upper management. Implementing and sustaining the dimensional management process requires a major investment in time, personnel, and money at the early stages of a design. If top management is not willing to make and sustain its commitment to the program throughout its life cycle, the program will fail. Therefore, no dimensional management program should begin until program directives from upper management clearly declare that sufficient personnel, budget, and other resources will be guaranteed throughout the duration of the project.

It is imperative that the product dimensional requirements are clearly defined in written objectives by the dimensional management team at the beginning of the design cycle. These written objectives must be based on the customer's requirements for the design and the process and measurement capabilities of the manufacturing system. If the objectives cannot be agreed upon by a consensus of the dimensional management team, the program cannot proceed to defining the design concept.

Dimensional Management 2-9

The design concept is defined by developing a 3-D solid model using a modern computer-aided engineering system. The 3-D model provides a product definition and is the basis for all future work.

Key characteristics are identified on individual features based on the functional requirements of the mating parts that make up assemblies and sub-assemblies. Features that are chosen as key characteristics will facilitate assembly and assist in reducing variability during processing and assembly.

Geometric dimensioning and tolerancing schemes are developed on the basis of the key characteristics that are chosen. Other requirements for correct fit and function between mating parts are also considered. A major objective for this GD&T activity is to establish datums and datum reference frames that will

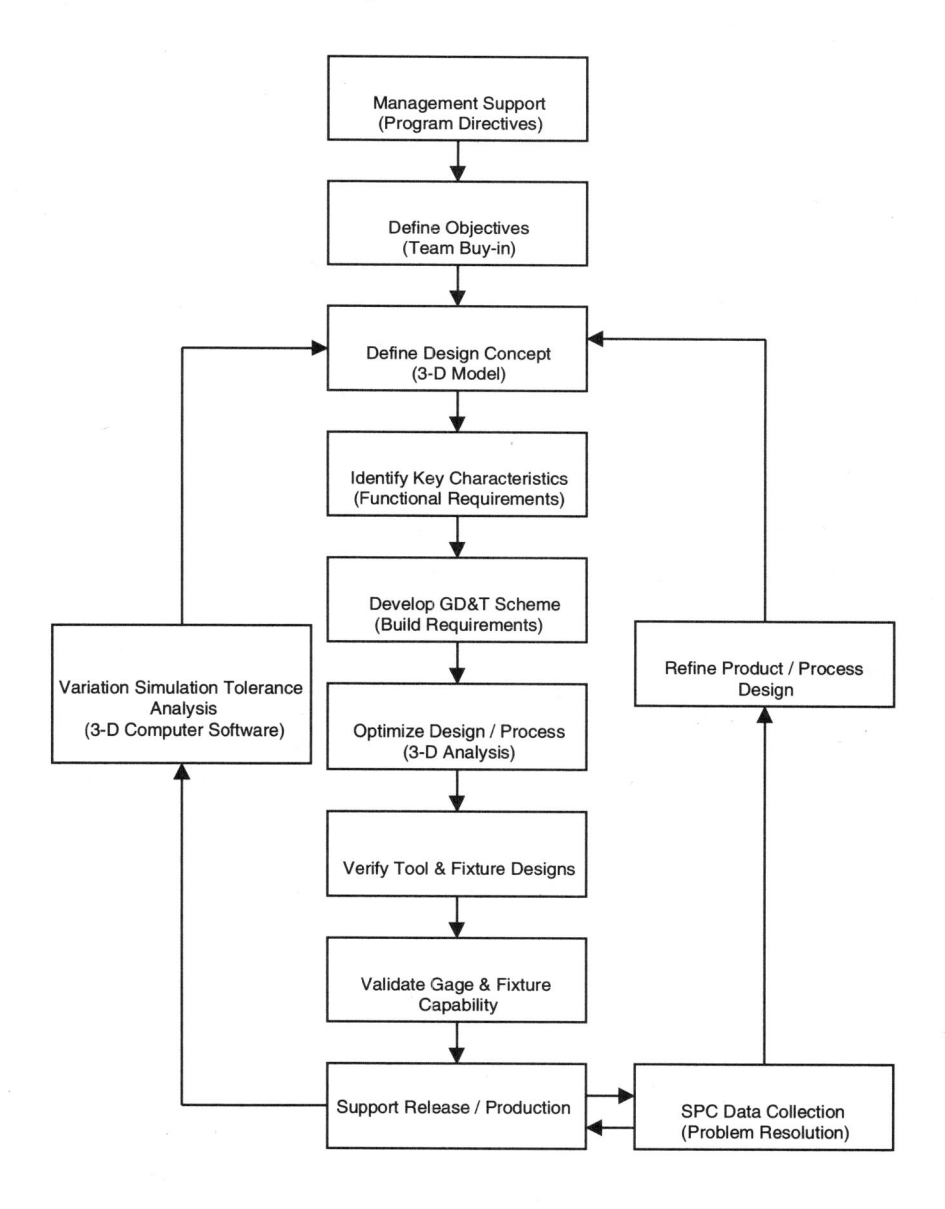

Figure 2-3 The dimensional management process

maintain correct interface between critical features during assembly. The datum system expressed by GD&T symbology also becomes the basis for determining build requirements that will influence processing, tooling, and inspection operations.

The product and process designs are optimized using variation simulation software that creates a functional assembly model. A mathematical definition of the assembly sequence, methods, and measurements that are based on the design concept, key characteristics, and GD&T scheme established in earlier stages of the program is created. This definition is used to statistically perform simulations based on known or assumed Cp and Cpk values, and to identify, rank, and correct critical fit and functional relationships between mating parts. These simulation tools are also used for the verification of the design of the tools and fixtures. This is done so that datums are correctly coordinated among part features, and the surfaces of tool and fixture locators are correctly positioned to reduce variation.

Measurement data is collected from gages and fixtures before production to verify their capability and compatibility with the product design. When the measurement data indicates that the tooling is not creating significant errors and meets the defined dimensional objectives, the product is released for production. If any problems are discovered that need a solution, further simulation and refinement is initiated.

During production statistical process control data is collected and analyzed to continually refine and improve the process. This in turn produces a product that has dimensional limits that will continue to approach their nominal values.

The dimensional management process can substantially improve dimensional quality for the following reasons:

- The product dimensional requirements are defined at the beginning of the design cycle.
- The design, manufacturing, and assembly processes all meet the product requirements.
- Product documentation is maintained and correct.
- A measurement plan is implemented that validates product requirements.
- Manufacturing capabilities achieve design intent.
- A feedback loop exists that ensures continuous improvement.

2.4 References

1. Craig, Mark. 1995. Using Dimensional Management. *Mechanical Engineering*, September, 986–988.
2. Creveling, C.M. 1997. *Tolerance Design*. Reading, MA: Addison-Wesley Longman Inc.
3. Harry, Mikel J. 1997. *The Nature of Six Sigma Quality*. Schaumburg, IL: Motorola University Press.
4. Larson, Curt, 1995. *Basics of Dimensional Management*. Troy, MI: Dimensional Control Systems Inc.
5. Liggett, John V. 1993. *Dimensional Variation Management Handbook*. Englewood Cliffs, NJ: Prentice-Hall Inc.
6. Nielsen, Henrik S. 1992. Uncertainty and Dimensional Tolerances. *Quality*, May, 25–28.

2.5 Glossary

Dimensional management - A process by which the design, fabrication, and inspection of a product is systematically defined and monitored to meet predetermined dimensional quality goals.

Dimensional management process - The integration of specific dimensional management tools into the concept, design, prototype, and production stages of a product life cycle.

Dimensional management system - A systematic implementation of dimensional management tools.

Key characteristics - A feature of an installation, assembly, or detail part with a dimensional variation having the greatest impact on fit, performance, or service life.

Variation measurement and reduction - Those activities relating to the measurement of fabricated parts to verify conformance with their dimensional specifications and give continuous dimensional improvement.

Variation simulation tolerance analysis - The use of 3-D simulation software in the early stages of a design to perform simulation studies in order to reduce dimensional variation before actual parts are fabricated.

Chapter 3

Tolerancing Optimization Strategies

Gregory A. Hetland, Ph.D.
Hutchinson Technology Inc.
Hutchinson, Minnesota

Dr. Hetland is the manager of corporate standards and measurement sciences at Hutchinson Technology Inc. With more than 25 years of industrial experience, he is actively involved with national, international, and industrial standards research and development efforts in the areas of global tolerancing of mechanical parts and supporting metrology. Dr. Hetland's research has focused on "tolerancing optimization strategies and methods analysis in a sub-micrometer regime."

3.1 Tolerancing Methodologies

This chapter will give a few examples to show the technical advantages of transitioning from linear dimensioning and tolerancing methodologies to geometric dimensioning and tolerancing methodologies. The key hypothesis is that geometric dimensioning and tolerancing strategies are far superior for clearly and unambiguously representing design intent, as well as allow the greatest amount of tolerance.

Geometric definitions can have only one clear technical interpretation. If there is more than one interpretation of a technical requirement, it causes problems not only at the design level, but also through manufacturing and quality. This problem not only adds confusion within an organization, but also adversely affects the supplier and customer base. This is not to say that utilization of geometric dimensioning and tolerancing will always make the drawing clear, because any language not used correctly can be misunderstood and can reflect design intent poorly.

3.2 Tolerancing Progression (Example #1)

Figs. 3-1 to 3-3 show three different dimensioning and tolerancing strategies that are "intended" to reflect designer's intent, and the supporting figures are intended to show the degree of variation allowed by the defined strategy. These three strategies reflect a progression of attempts to accomplish this goal.

3-2 Chapter Three

Fig. 3-3 depicts the optimum dimensioning and tolerancing strategy reflecting the greatest allowable flexibility for the designer and manufacturer. Note: Each of the drawings/figures is complete only to the degree necessary to discuss the features in question.

Prior to elaborating on each of the strategies, it is critical to understand what the designer was attempting to allow on the initial design. In this case, the designer intends to have the external boundary utilize a space of 6.35 mm ±0.025 mm "square," and to have the hub (inside diameter) on "center" of the square within ±0.025 mm. With this being the designer's goal, consider the following three strategies of dimensioning and tolerancing.

3.2.1 Strategy #1 (Linear)

Fig. 3-1a represents the original dimensioning and tolerancing strategy that is strictly linear. In this figure, the outside shape in the vertical and horizontal directions is 6.35 mm ±0.025 mm, while the hub is located at half the distance of the nominal width from the center of the part. Section A-A shows the allowable variation for the inside diameter.

Based on the defined goal of the designer, there are a number of problems that arise based on interpretation of any given national or international standard that exists today or in the past. All comments in this section will be limited to interpretation of the ASME Y14.5M-1994 (Y14.5) standard. It is critical to note that no industrial or company specification existed that would state anything different (related to reducing the ambiguities based on utilizing linear tolerancing methodologies) from the Y14.5 standard.

Paragraph 2.7.3 of Y14.5 addresses the "relationship between individual features," and states:

> *The limits of size do not control the orientation or location relationship between individual features. Features shown perpendicular, coaxial, or symmetrical to each other must be controlled for location or orientation to avoid incomplete drawing requirements.*

Based on the above-noted paragraph, it clearly indicates Fig. 3-1a to be lacking at least some geometric controls or at a minimum some notes to identify the degree of orientation and locational control. Figs. 3-1b to 3-1g show a few of the possible combinations of part variability (represented by dashed lines) that are allowed by the current "linear" callouts.

Fig. 3-1b shows a part perfectly square and made to its maximum size based on the tolerance specification (6.375 mm), which would be an acceptable part for size. Assuming the hub was exactly in the center where the designer would like it to be, this feature would measure 0.0125 mm off its ideal location based on this part's large size. Ideal nominal was 3.175 mm, and the actual value measured was 3.1875 mm, which would be a displacement of 0.0125 mm. It meets intended ideal, but fails specified ideal.

Like Fig. 3-1b, Fig. 3-1c shows a part that is perfectly square but is now made to its minimum allowable size based on specification (6.325 mm), which is again acceptable for size. Assuming the hub was exactly in the center where the designer would like it to be, this part also would measure 0.0125 mm off its ideal location based now on the part's small size. The ideal nominal was 3.175 mm, and the actual value measured was 3.1625 mm, which also shows a displacement of 0.0125 mm. Again, it meets intended ideal, but fails specified ideal.

Paragraph 2.7.3 of Y14.5 stated that "the limits of size do not control the orientation." Fig. 3-1d describes the condition that can occur based on the lack of geometric control for orientation. In this example, the part is restricted to the shape of a parallelogram, and the degree allowed is questionable. This particular example clearly shows the designer's intent would not be met if this condition was accepted. Based on the drawing callouts currently defined, it could not be rejected.

Fig. 3-1e shows a combination of Figs. 3-1b and 3-1c where it allows the shape to be small at one end and large at the other. Fig. 3-1f takes this one step further and shows a part that is, for the most part, large, except all the variability (0.05 mm) shows up on one edge.

Tolerancing Optimization Strategies 3-3

Fig. 3-1g is showing a part made to its large size (like Fig. 3-1b), and the hub shifted off the "designer's ideal" center, so it is centered on its nominal dimension. This figure also shows the effect this would have on its opposing corner which would be a displacement out to its worst-case tolerance of +0.025 mm (3.2 mm). The more challenging part would be to determine which edge is being measured, from one part to the next. This is somewhat difficult to do on a part that is designed perfectly symmetrical.

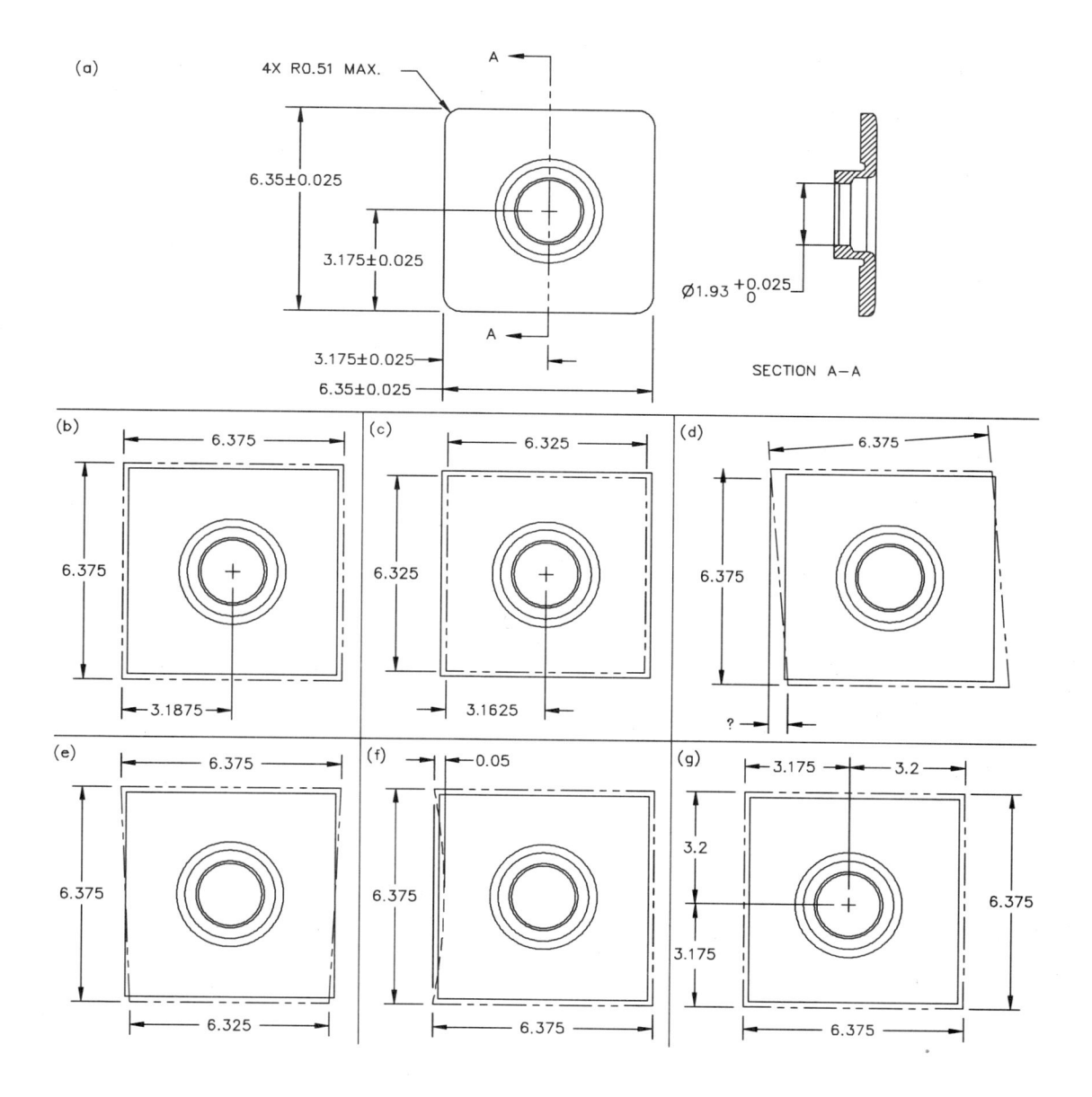

Figure 3-1 Linear dimensioning and tolerancing boundary example

3-4   Chapter Three

The above comments are not intended to identify all the potential problems, or even to touch on the probability of occurrence. These comments should identify a few obvious problems with this particular dimensioning and tolerancing strategy. It did not take long for the designer to realize this particular drawing was missing requirements to state what was intended to be allowed. Based on some initial training in geometric dimensioning and tolerancing, the designer modified the drawing as shown in Fig. 3-2a. This leads into strategy #2 which is a combination of linear and geometric tolerancing.

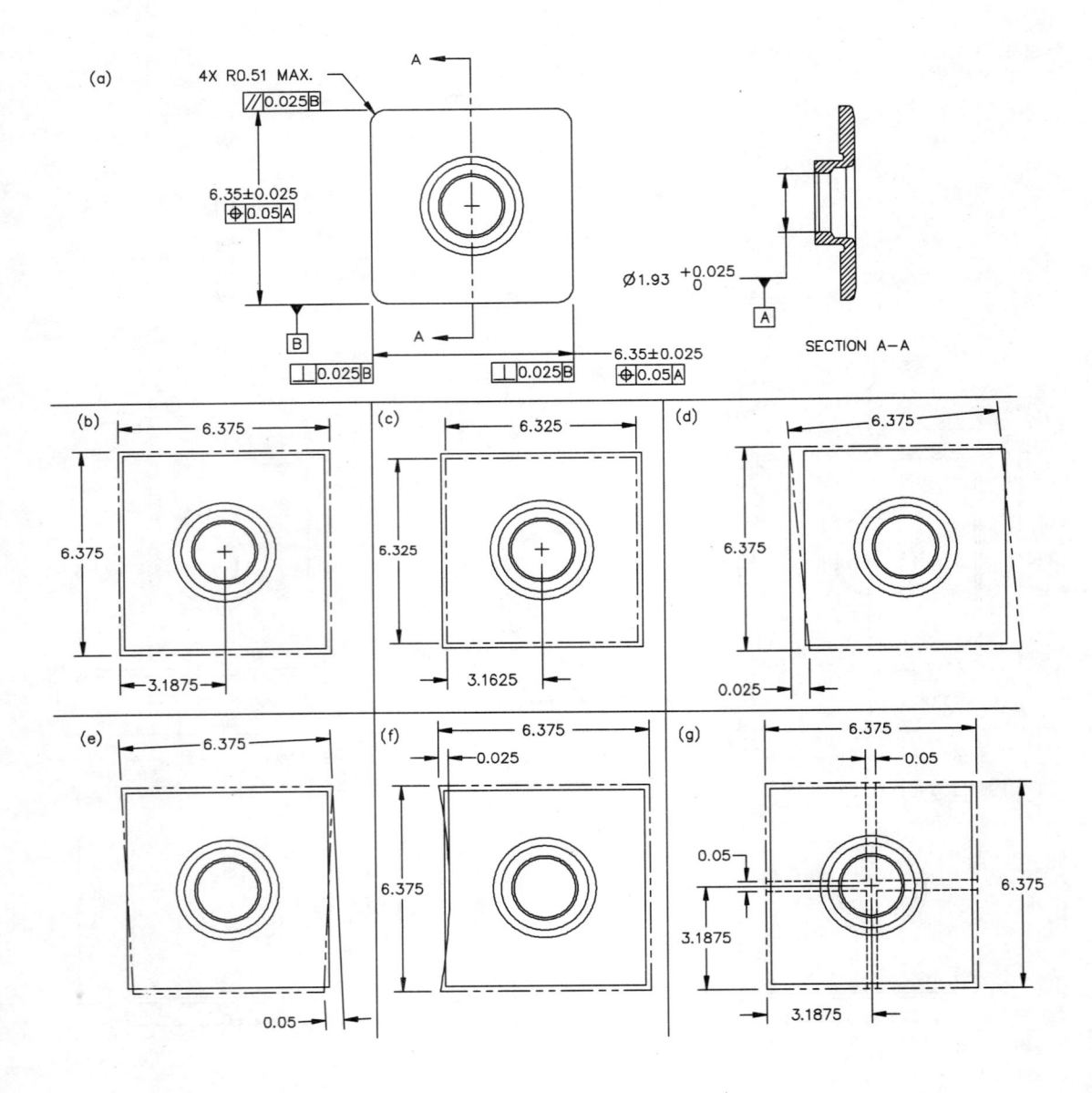

Figure 3-2 Linear and geometric dimensioning and tolerancing boundary example

3.2.2　Strategy #2 (Combination of Linear and Geometric)

Fig. 3-2a is a combination of linear and geometric callouts, and clearly adds controls for orientation of one surface to another. This is achieved with perpendicularity callouts on the left and right sides of the part in relationship to datum -B-, along with a parallelism callout on the top of the part, also to datum -B-. In addition, position callouts were added to each of the size dimensions (6.35 mm ±0.025 mm) and were controlled in relationship to datum -A-, which is the "axis" of the inside diameter (1.93 mm +0.025 mm / –0 mm). Figs. 3-2b to 3-2g define some of the conditions allowed by these drawing callouts.

Fig. 3-2b shows a part perfectly square and made to its maximum size based on the specification (6.375 mm), which would be an acceptable part for size. Assuming the hub was exactly in the center where the designer would like it to be, this part would measure 3.1875 mm. Unlike the negative impact mentioned in regards to Fig. 3-1b, this measurement adds no negative impact to specifications because the "center plane" is now being located from the "center" of the inside diameter.

Like Fig. 3-2b, Fig. 3-2c shows a part that is perfectly square and made to its minimum allowable size based on the specifications (6.325 mm), which is again acceptable for size. Again, assuming the hub was exactly in the center where the designer would like it to be, the 3.1625 mm measurement has no negative impact on specifications.

Fig. 3-2d (like Fig. 3-1d) shows a part on the large side of the tolerance allowed, with its orientation skewed to the shape of a parallelogram. In this example, however, the perpendicularity callouts added in Fig. 3-2a control the amount this condition can vary. In this case it is 0.025 mm. The problem that stands out here is that the designer's original intent stated: <u>to have the external boundary utilize a space of 6.35 mm ±0.025 mm "square."</u> Based on this requirement, it's clear this objective was not met. Granted, it is controlled tighter than the requirements defined in Fig. 3-1a, but it still does not meet the designer's expectations.

Fig. 3-2e shows a combination of Figs. 3-2b and 3-2c (like Figs. 3-1b and 3-1c), in that it allows the shape to be small at one end and large at the other. Unlike Figs. 3-1b and 3-1c, Fig. 3-2e restricts the magnitude of change from one end to the other by the parallelism and perpendicularity callouts shown in Fig. 3-2a.

Because this part is symmetrical, a unique problem surfaces in this example. Using Fig. 3-2e, assuming the bottom surface is datum -B-, the top surface is shown to be perfectly parallel. Due to the part being symmetrical, it is impossible to determine which surface is truly datum -B-. So, if we assume the left-hand edge of the part as shown in Fig. 3-2e was the datum, the opposite surface (based on the shape shown) would show to be out of parallel by 0.05 mm. This clearly shows that problems in the geometric callouts are not only in the design area, but also in the ability to measure consistently. Like-type parts could measure good or bad, depending on the surface identified as datum -B-.

Fig. 3-2f again shows displacement in shape allowed. In this case it shows a part that is for the most part large, except all the variability (0.025 mm) shows up on one edge. The limiting factor (depending on which surface is "chosen" as datum -B-) is the perpendicularity or parallelism callouts.

Fig. 3-2g is showing a part made to its large size (like Fig. 3-1b), and the 0.05 mm zone allowed by the position callout. Unlike Fig. 3-1g, the larger or smaller size of the square shape has no impact on the position. Based on the callout in Fig. 3-2a, the center planes (mid-planes) in both directions must fall inside the dashed boundaries.

The above comments concerning Fig. 3-2a are intended to show a tolerancing strategy that encompasses both liner and geometric callouts but still does not meet the designer's intended expectations. Based on this, the designer modified the drawing again, as shown by Fig. 3-3a, which led to strategy #3.

3.2.3 Strategy #3 (Fully Geometric)

Fig. 3-3a is the optimum dimensioning and tolerancing strategy for this design example. In this case, the outside shape is defined clearly as a square shape that is 6.35 mm "basic," and is controlled with two profile callouts. The 0.05 mm tolerance is shown in relationship to datums -B- and -A-, controlling primarily the "location" of the hub in relation to the outside shape (depicted by Fig. 3-3b). The 0.025 mm tolerance is shown in relationship to datum -B- and controls the total variation of "shape" (depicted by Fig. 3-3c). This tolerancing strategy clearly defines the designer's intent.

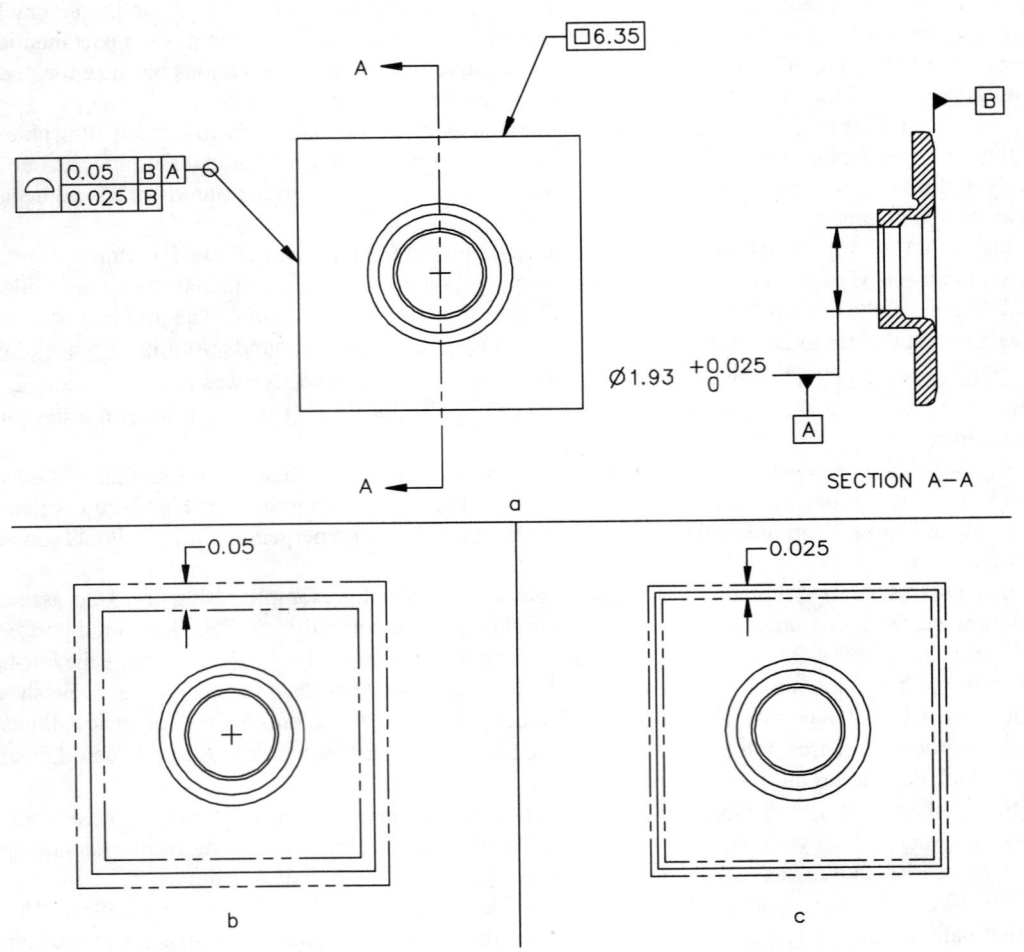

Figure 3-3 Fully geometric dimensioned and toleranced boundary example

3.3 Tolerancing Progression (Example #2)

This second example is intended to show the tolerancing progression for locating two mating plates (one plate with four holes and the other with four pins). Design intent requires both plates to be located within a size and location tolerance that will allow them to fit together, with a worst-case fit to be no tighter than a "line-to-line" fit. In addition, the relationship of the holes to the outside edges of the part is critical.

Tolerancing Optimization Strategies 3-7

The tolerance progression will start with linear dimensioning methodologies and will progress to using geometric symbology, which in this case will be position. This progression will conclude with the optimum tolerancing method for this design application, which will be a positional tolerance using zero tolerance at maximum material condition (MMC). All examples will follow the same "design intent" and use the same two plate configurations.

Initially, each figure showing a tolerancing progression will be displayed showing a "front and main view" for each part, along with a "tolerance stack-up graph" at the bottom of the figure (see Fig. 3-4 as an example). The component on the left will always show the part with four inside diameter holes, while the component on the right will always show the part with four pins. The tolerance stack-up graph will show the allowable location versus allowable size as they relate to the applicable component on their respective sides.

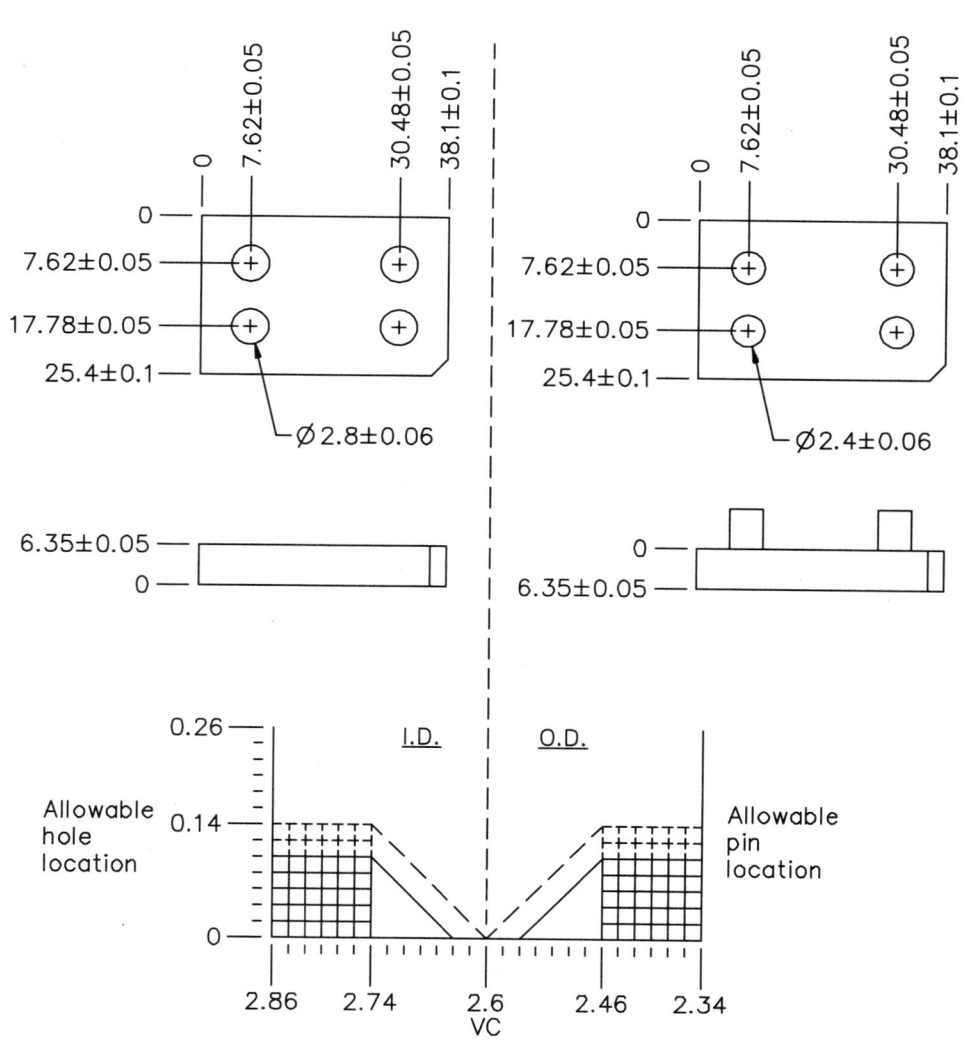

Figure 3-4 Tolerance stack-up graph (linear tolerancing)

3-8   Chapter Three

The critical items to follow in this example (as well as subsequent examples) are the dimensioning and tolerancing controls and the associative "tolerance stack-up" that occurs. Common practice for designers is to identify the worst-case condition that each component will allow, to ensure the components will assemble. This tolerance stack-up will be displayed graphically within each of the figures, such as the one shown at the bottom of Fig. 3-4.

Each component will be specified showing nominal size and tolerance for the inside diameter 2.8 mm ±?? mm) and outside diameter (2.4 mm ±?? mm "pins"). The size tolerance will change in some of the progressions, and the positional requirements will change in "each" of the progressions, both of which will be variables to monitor in the tolerance stack-up graph. The tolerance stack-up graph is the primary visual tool that monitors primary differences in the callouts. More filled-in graph area indicates that more tolerance is allowed by the dimensioning and tolerancing strategy.

To clarify the components of the graph so they are interpreted correctly, continue to follow along in Fig. 3-4. The horizontal scale of the graph shows size variation allowed by the size tolerance, while the vertical scale shows locational variation allowed by the feature's locational tolerance. Each square in the grid equals 0.02 mm for convenience. The center of the horizontal scale represents (in these examples) the "virtual condition" (VC), which is the worst case stack-up allowed by both components as the size and locational tolerances are combined. This condition tests for the line-to-line fit required by the designer.

Based on the above classifications, the reader should be able to follow along more easily with the differences in the following figures.

3.3.1   Strategy #1 (Linear)

Fig. 3-4 represents the original dimensioning and tolerancing strategy that is strictly "linear." The left side of the graph shows the allowable tolerance for the "inside diameter" to range from 2.74 mm to 2.86 mm, reflected by the numbers on the horizontal scale. The positional tolerance allowed in this example is 0.05 mm from its targeted (defined) nominal, or a total tolerance of 0.1 mm, reflected by the numbers on the vertical scale. The grid (solid line portion) indicates the combined size and locational variation "initially perceived" to be allowed as the drawing is currently defined.

The solid line that extends from the upper right corner of the "solid grid" pattern (intersection of 0.1 on the vertical scale and 2.74 on the horizontal scale) down to the 2.64 mark on the horizontal scale, represents the perceived virtual condition based on the noted tolerances. This area does not show up as a grid pattern (in this figure), because the actual space is not being used by either the size or positional tolerance.

The normal calculation for determining the virtual condition boundary is to take the MMC of the feature and subtract or add the allowable positional tolerance. This depends on whether it is an inside or outside diameter feature (subtract if it's an inside diameter, and add if it's an outside diameter). In this case, the MMC of the inside diameter is 2.74 mm and subtracting the allowable positional tolerance of 0.1 mm would derive a virtual condition of 2.64 mm.

This is where the first concern arises, which is depicted by the dashed grid area on the graph. Prior to detailed discussion on this dashed grid area, an explanation of the problem is necessary.

Fig. 3-5 reflects a tolerance zone comparison between a square tolerance zone and a diametral tolerance zone shown to be centered on the noted cross-hair. At the center of the figure is a cross-hair intended to depict the center axis of any one of the holes or pins, defined by the nominal location. In this example, use the upper-left hole shown in Fig. 3-4, which is equally located from the noted (zero) surfaces by 7.62 mm "nominal" in the x and y axes. In the center of this hole (as well as all others) there is a small cross-hair depicting the theoretically exact nominal. Based on the nominals noted, there is an allowable tolerance of 0.05 mm in the x and y axes.

Tolerance Zone Comparison

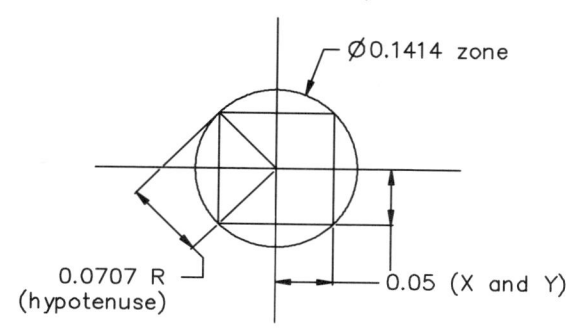

$$2\sqrt{\Delta X^2 + \Delta Y^2} = \varnothing \text{ tolerance zone}$$
$$2\sqrt{0.05^2 + 0.05^2} = 2(0.0707) = 0.1414 \; \varnothing \text{ tolerance zone}$$

Note: ±0.05 square zone = ∅ zone of 0.1414

Note: Benefit of changing square zone to diameter tolerance zone is:
  *Increases position tolerance zone by 57 %
  *Allows use of MMC principle

Figure 3-5 Plus/minus versus diametral tolerance zone comparison

The square shape shown in Fig. 3-5 represents the ±0.05 mm location tolerance. In evaluating the square tolerance zone, it becomes evident that from the center of the cross-hair, the axis of the hole can be further off (radially) in the corner than it can in the x and y axes. Calculating the magnitude of radial change shows a significant difference (0.05 mm to 0.0707 mm). The calculations at the bottom of Fig. 3-5 show a total conversion from a square to a diametral tolerance zone, which in this case yields a diametral tolerance boundary of 0.1414 mm (rounded to 0.14 mm for convenience of discussion).

Now, looking back at the graph in Fig. 3-4, the dashed grid area should now start to make some sense. The square (0.05 mm) tolerance boundary actually creates an awkward shaped boundary that under certain conditions can utilize a positional boundary of 0.14 mm. Based on this, the following is a recalculation of the virtual condition boundary. In this case, the MMC of the inside diameter is still 2.74 mm, and now subtracting the "potentially" allowable positional tolerance of 0.14 mm derives a virtual condition of 2.6 mm, which is what the second line (dashed) is intended to represent.

It should become very obvious that it makes little sense to tolerance the location of a round hole or pin with a square tolerance zone. Going on this premise, the two parts would, in fact, assemble if the location of a given hole (or pin) was produced at its maximum x and y tolerance. It would make sense to identify the tolerance boundary as diametral (cylindrical). The parts in fact will assemble based on this condition, which is why geometric tolerancing in Y14.5 progressed in this fashion. It needed some methodology to represent the tolerance boundary for the axes of the holes. A diametral boundary is one reason for the position symbol.

Up to this point, in referring to Fig. 3-4, comments have been limited to the part on the left side with the through holes. All comments apply in the same fashion to the part on the right side, except for the minor change in calculating the virtual condition. In this case, the maximum material condition of the pin is a diameter of 2.46 mm, so "adding" the allowable positional tolerance of 0.14 mm would result in a virtual condition boundary of 2.6 mm.

3-10   Chapter Three

Additional problems surface when utilizing linear tolerancing methodologies to locate individual holes or hole patterns, such as the ability to determine which surfaces should be considered as primary, secondary, and tertiary datums or if there is a need to distinguish a difference at all.

This ambiguity has the potential of resulting in a pattern of holes shaped like a parallelogram and/or being out of perpendicular to the primary datum or to the wrong primary datum. At a minimum, inconsistent inspection methodologies are natural by-products of drawings that are prone to multiple interpretations.

The above comments and the progression of Y14.5 leads to the utilization of geometric tolerancing using a feature control frame, and in this case specifically, the utilization of the position symbol, as shown in Fig. 3-6.

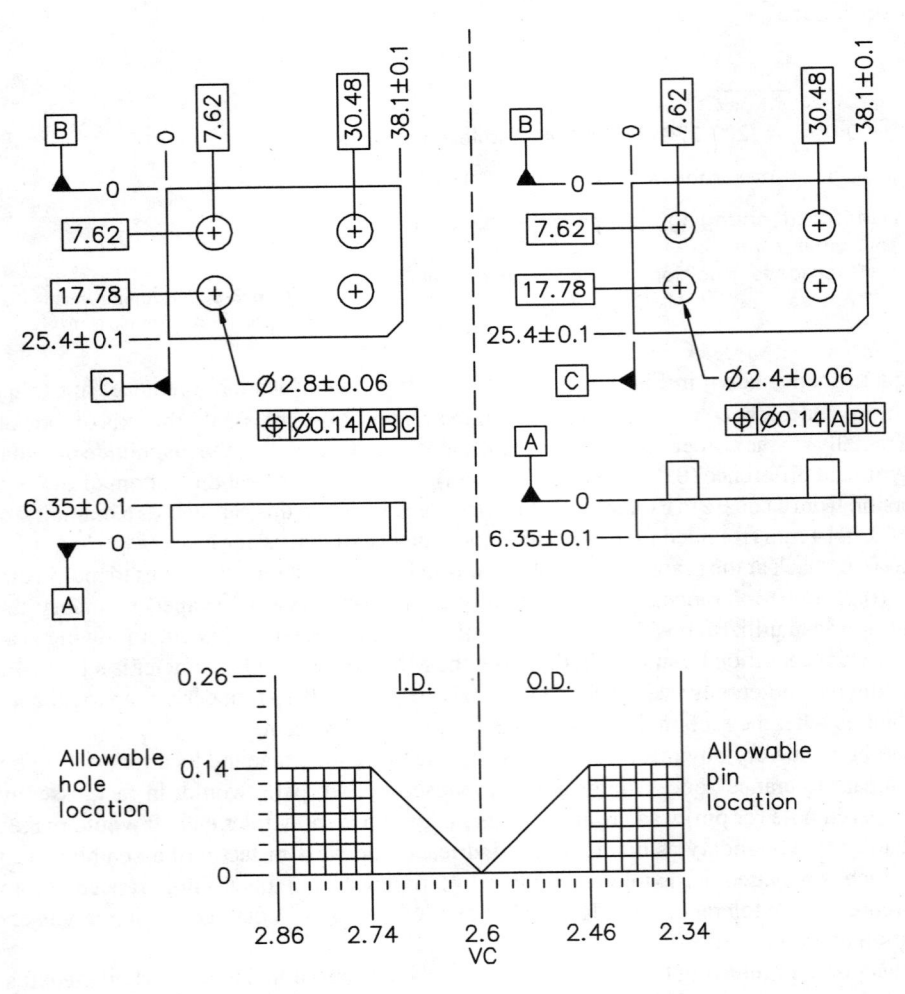

Figure 3-6   Tolerance stack-up graph (position at RFS)

3.3.2 Strategy #2 Geometric Tolerancing (⌖) Regardless of Feature Size

Fig. 3-6 shows the next progression using geometric tolerancing strategies. Tolerances for size are identical to Fig. 3-4. The only change is limited to the locational tolerances. In this example, the tolerance has been removed from the nominal locations and a box around the nominal location depicts it as being a "basic" (theoretically exact) dimension. The locational tolerance that relates to these basic dimensions is now located in the feature control frames, shown under the related features of size.

The diametral/cylindrical tolerance of 0.14 mm should look familiar at this point, as it was discussed earlier in relation to Figs. 3-4 and 3-5. This is a geometrically correct callout that is clear in its interpretation. The datums are clearly defined along with their order of precedence, and the tolerance zone is descriptive for the type of features being controlled.

The feature control frame would read as follows: The 2.8 mm holes (or 2.4 mm pins) are to be positioned within a cylindrical tolerance of 0.14 mm, regardless of their feature sizes, in relationship to primary datum -A-, secondary datum -B-, and tertiary datum -C-.

The graph at the bottom of Fig. 3-6 clearly describes the size and positional boundaries, along with associative lines depicting the virtual condition boundary, as noted in Fig. 3-4. Based on all the issues discussed in relation to Fig. 3-4, this would seem to be a very good example for positive utilization of geometric tolerances. There is, however, an opportunity that was missed by the designer in this example. It restricted flexibility in manufacturing as well as inspection and possibly added cost to each of the components.

Now a re-evaluation of the initial design criteria: Design intent required both plates to be dimensioned and located within a size and location tolerance that is adequate to allow them to fit together, with a worst-case fit to be no tighter than a "line-to-line" fit. In addition, the relationship of the holes to the outside edges of the part is critical.

Based on this, re-evaluate the feature control frame and the graph. It states the axis of the holes or pins are allowed to move around anywhere within the noted cylindrical tolerance of 0.14 mm, "regardless of the features size." This means that it does not matter whether the size is at its low or high limit of its noted tolerance and that the positional tolerance of 0.14 mm does not change.

It would make sense that if the hole on a given part was made to its smallest size (2.74 mm) and the pin on a given mating part was made to its largest size (2.46 mm), that the worst case allowable variation that could be allowed for position would each be 0.14 mm (2.74 mm - (minus) 2.46 mm = 0.28 mm total variation allowed between the two parts). The graph clearly shows this condition to reflect the worst case line-to-line fit.

If, however, the size of the hole on a given part was made to its largest size (2.86 mm) and the pin on a given mating part was made to its smallest size (2.34 mm), it would make sense that the worst case allowable positional variation could be larger than 0.14. Evaluating this further as was done above to determine a line-to-line fit would be as follows: 2.86 mm - 2.34 mm = 0.52 mm total variation allowed between the two parts.

The graph clearly indicates this condition. It would seem natural, due to the combined efforts of size and positional tolerance being used to determine the worst-case virtual condition boundary, that there should be some means of taking advantage of the two conditions. Fig. 3-7 depicts the flexibility to allow for this condition, which is the next step in this tolerance progression.

3.3.3 Strategy #3 (Geometric Tolerancing Progression at Maximum Material Condition)

Fig. 3-7 shows the next progression of enhancing the geometric strategy shown in Fig. 3-6. All tolerances are identical to Fig. 3-6. The only difference is the regardless of feature size condition noted in the feature control frame is changed to maximum material condition. Again, this would be considered a clean callout.

The feature control frame would now read as follows: The 2.8 mm holes (or 2.4 mm pins) are to be positioned within a cylindrical tolerance of 0.14 mm, at its maximum material condition, in relationship to primary datum -A-, secondary datum -B-, and tertiary datum -C-.

The graph at the bottom of Fig. 3-7 clearly describes the size and positional boundaries along with associative lines depicting the virtual condition boundary. Unlike Figs. 3-4 and 3-6, the grid area is no

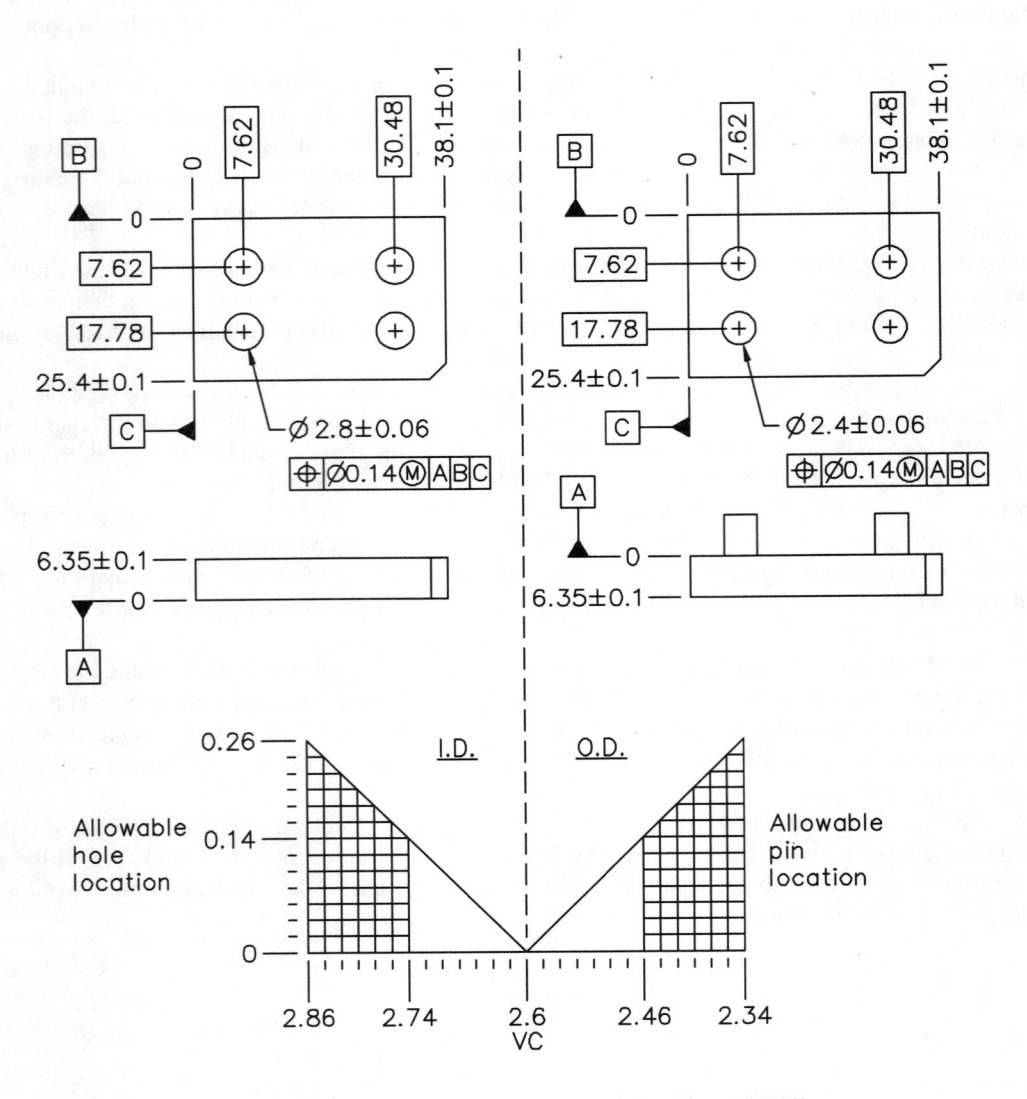

Figure 3-7 Tolerance stack-up graph (position at MMC)

longer rectangular. The range of the size boundary has not changed, but the range of the allowable positional boundary has changed significantly, due solely to the additional area above 0.14 mm being a function of size.

Evaluation of the feature control frame and graph depict the axis of the holes or pins, allowed to move around anywhere within the noted cylindrical tolerance of 0.14 mm when the feature is produced at its maximum material condition. The twist here is that as the feature departs from its maximum material condition, the displacement is additive one-for-one to the already defined positional tolerance. This supports the previous comments very well. Table 3-1 identifies the bonus tolerance gained to position as the feature's size is displaced from its maximum material condition and can be visually followed on the graph in Fig. 3-7.

Table 3-1 Bonus tolerance gained as the feature's size is displaced from its MMC

Feature Size	Displacement from MMC	Allowable Position Tolerance
2.74	0.00	0.14
2.76	0.02	0.16
2.78	0.04	0.18
2.80	0.06	0.20
2.82	0.08	0.22
2.84	0.10	0.24
2.86	0.12	0.26

The combined efforts of size and positional tolerance utilized in this fashion is a clean way of taking advantage of the two conditions. Individuals involved with the Y14.5 committee recognize this. There is, however, an opportunity here that still restricts "optimum" flexibility in many aspects. Fig. 3-8 depicts the flexibility to allow for this condition, which is the final step in this tolerance progression.

3.3.4 Strategy #4 (Tolerancing Progression "Optimized")

Fig. 3-8 shows the final/optimum strategy of this tolerancing progression. Both size and positional tolerances have been changed to reflect the spectrum of design, manufacturing, and measurement flexibility. Nominals for size were kept the same only for consistency in the graphs.

This tolerancing strategy is an extension of the concept shown in Fig. 3-7 that allowed bonus tolerancing for the locational tolerance to be gained as the feature departed from its maximum material condition. In similar fashion, the function of this part allows the flexibility to also add tolerance in the direction of size. In this case, when less locational tolerance is used, more tolerance is available for size.

The feature control frame now reads as follows: The 2.8 mm holes (or 2.4 mm pins) are to be positioned within a cylindrical tolerance of "0" (zero) at its maximum material condition in relationship to primary datum -A-, secondary datum -B-, and tertiary datum -C-.

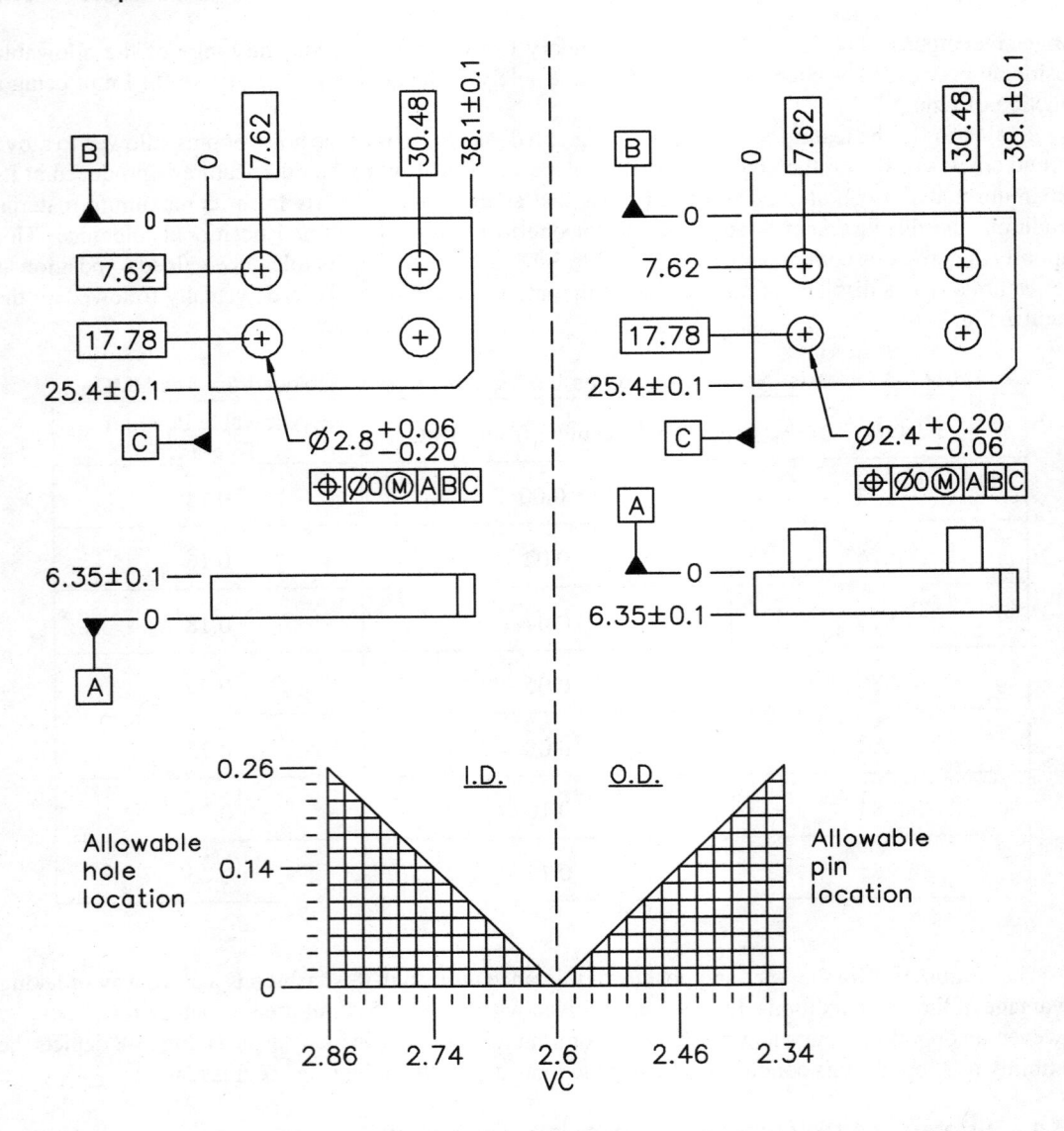

Figure 3-8 Tolerance stack-up graph (zero position at MMC)

According to the graph, when the feature is produced at its maximum material condition, there is no tolerance. But as the feature departs from it maximum material condition, its displacement is equal to the allowable tolerance for position. This supports the comments considered before very well. The same type of matrix as shown before could be developed to identify bonus tolerance gained to position as the feature's size is displaced from its maximum material condition. It can naturally be followed on the graph.

The virtual condition boundary still creates a worst case condition of 2.6 mm. The maximum material condition of both components now equals a cylindrical boundary of 2.6 mm, which means there is nothing left over for positional tolerance to be split between the two components.

3.4 Summary

Fig. 3-9 shows a summary of the boundaries each of the geometric progressions allowed. Each of these progressions is allowed by the current Y14.5 standard, but the flexibilities are not clearly understood. The intent of outlining these optimization strategies is to highlight the types of opportunities and strengths this engineering language makes available to industry in a sequential/graphical methodology.

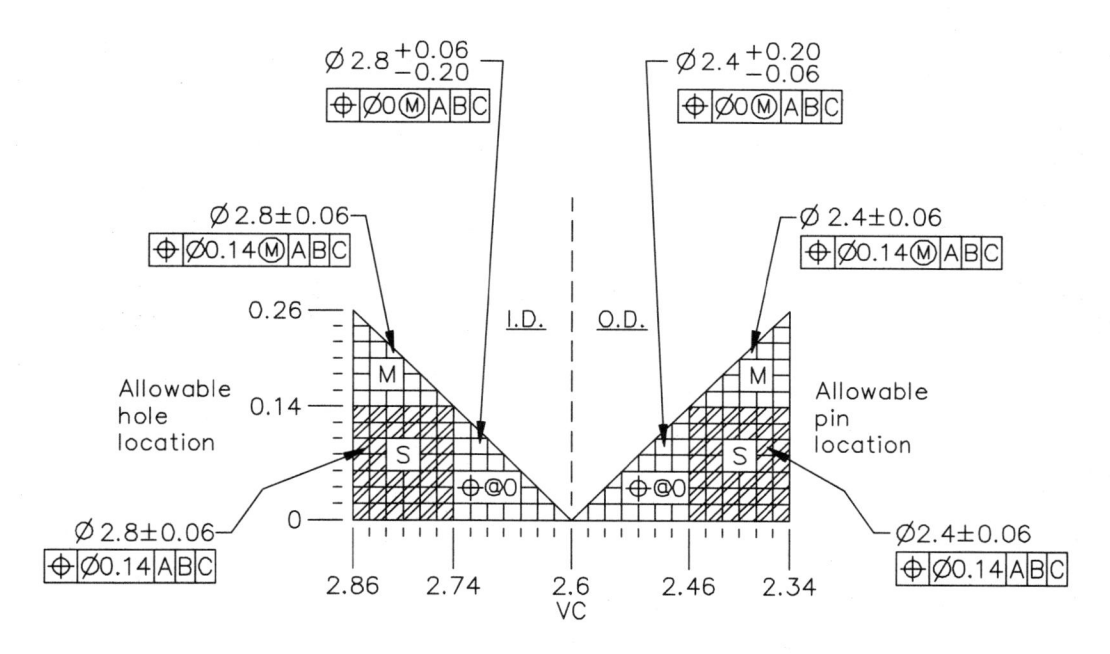

Figure 3-9 Summary graph

3.5 References

1. Hetland, Gregory A. 1995. Tolerancing Optimization Strategies and Methods Analysis in a Sub-Micrometer Regime. Ph.D. dissertation.
2. The American Society of Mechanical Engineers. 1995. ASME Y14.5M-1994, Engineering Drawings and Related Documentation Practices. New York, New York: The American Society of Mechanical Engineers.

P · A · R · T · 2

STANDARDS

Chapter 4

Drawing Interpretation

Patrick J. McCuistion, Ph.D
Ohio University
Athens, Ohio

Patrick J. McCuistion, Ph.D., Senior GDTP, is an associate professor of Industrial Technology at Ohio University. Dr. McCuistion taught for three years at Texas A&M University and previously worked in various engineering design, drafting, and checking positions at several manufacturing industries. He has provided instruction in geometric dimensioning and tolerancing and dimensional analysis to many industry, military, and educational institutions. He also has published one book, several articles, and given several academic presentations on those topics and dimensional management. Dr. McCuistion is an active member of several ASME/ANSI codes and standards subcommittees, including Y14 Main Committee, Y14.3 Multiview and Sectional View Drawings, Y14.5 Dimensioning and Tolerancing, Y14.11 Molded Part Drawings, Y14.35 Drawing Revisions, Y14.36 Surface Texture, and B89.3.6 Functional Gages.

4.1 Introduction

The engineering drawing is one of the most important communication tools that a company can possess. Drawings are not only art, but also legal documents. Engineering drawings are regularly used to prove the negligence of one party or another in a court of law. Their creation and maintenance are expensive and time consuming. For these reasons, the effort made in fully understanding them cannot be taken for granted.

Engineering drawings require extensive thought and time to produce. Many companies are using three-dimensional (3-D) computer aided design databases to produce parts and are bypassing the traditional two-dimensional (2-D) drawings. In many ways, creating an engineering drawing is the same as a part production activity. The main *difference* between drawing production and part production is that the drawing serves many different functions in a company. Pricing uses it to calculate product costs. Purchasing uses it to order raw materials. Routing uses it to determine the sequence of machine tools used to produce the part. Tooling uses it to make production, inspection, and assembly fixtures. Production uses the drawing information to make the parts. Inspection uses it to verify the parts have met the specifications. Assembly uses it to make sure the parts fit as specified.

This chapter provides a short drawing history and then covers the main components of mechanical engineering drawings.

4.2 Drawing History

The earliest known technical drawing was created about 4000 BC. It is an etching of the plan view of a fortress. The first written evidence of technical drawings dates to 30 BC. It is an architectural treatise stating the need for architects to be skillful as they create drawings.

The practice of drawing views of an object on projection planes (orthographic projection) was developed in the early part of the fifteenth century. Although none of Leonardo da Vinci's surviving drawings show orthographic views, it is likely that he used the technique. His treatise on painting used the perspective projection theory.

As a result of the industrial revolution, the number of people working for companies increased. This also increased the need for multiple copies of drawings. In 1876, the blueprinting machine was displayed at the bicentennial exposition in Philadelphia, PA. Although it was a messy process at first, it made multiple copies of large drawings possible. As drawings changed from an art form to a communication system, their creation also changed to a production activity.

From about 1750, when Gaspard Monge developed descriptive geometry practices, to about 1900, most drawings were created using first-angle projection. Starting in the late nineteenth century, most companies in the United States switched to third-angle projection. Third-angle projection is considered a more logical or natural positioning of views.

While it is common practice for many companies to create parts using a 3-D definition of the part, 2-D drawings are still the most widely used communication tool for part production. The main reason for this is, if a product breaks down in a remote location, a replacement part could be made on location from a 2-D drawing. The same probably would not be true from a 3-D computer definition.

4.3 Standards

If a machinist in a machine shop in a remote location is required to make a part for a US-built commercial aircraft, he or she must understand the drawings. This requires worldwide, standardized drafting practices. Many countries support a national standards development effort in addition to international participation. In the United States, the two groups of standards that are most influential are developed by the standards development bodies administered by the American National Standards Institute (ANSI) and the International Organization for Standardization (ISO). See Chapter 6 for a comparison of US and ISO standards.

4.3.1 ANSI

The ANSI administers the guidelines for standards creation in the United States. The American Society of Mechanical Engineers sponsors the development of the Y14 series of standards. The 26 standards in the series cover most facets of engineering drawings and related documents. Many of the concepts about how to read an engineering drawing presented in this chapter come from these standards. In addition to the Y14 series of standards, the complete library should also possess the B89 Dimensional Measurement standards series and the B46 Surface Texture standard.

4.3.2 ISO

The ISO, created in 1946, helped provide a structure to rebuild the world economy (primarily Europe) after World War II. Even though the United States has only one vote in international standards development, the US continues to propose many of the concepts presented in the ISO drafting standards.

4.4 Drawing Types

Of the many different types of drawings a manufacturing company might require, the three most common are note, detail, and assembly.

4.4.1 Note

Commonly used parts such as washers, nuts and bolts, fittings, bearings, tubing, and many others, may be identified on a note drawing. As the name implies, note drawings do not contain graphics. They are usually small drawings (A or A4 size) that contain a written description of the part. See Fig. 4-1.

4.4.2 Detail

The detail drawing should show all the specifications for one unique part. Examples of different types of detail drawings follow.

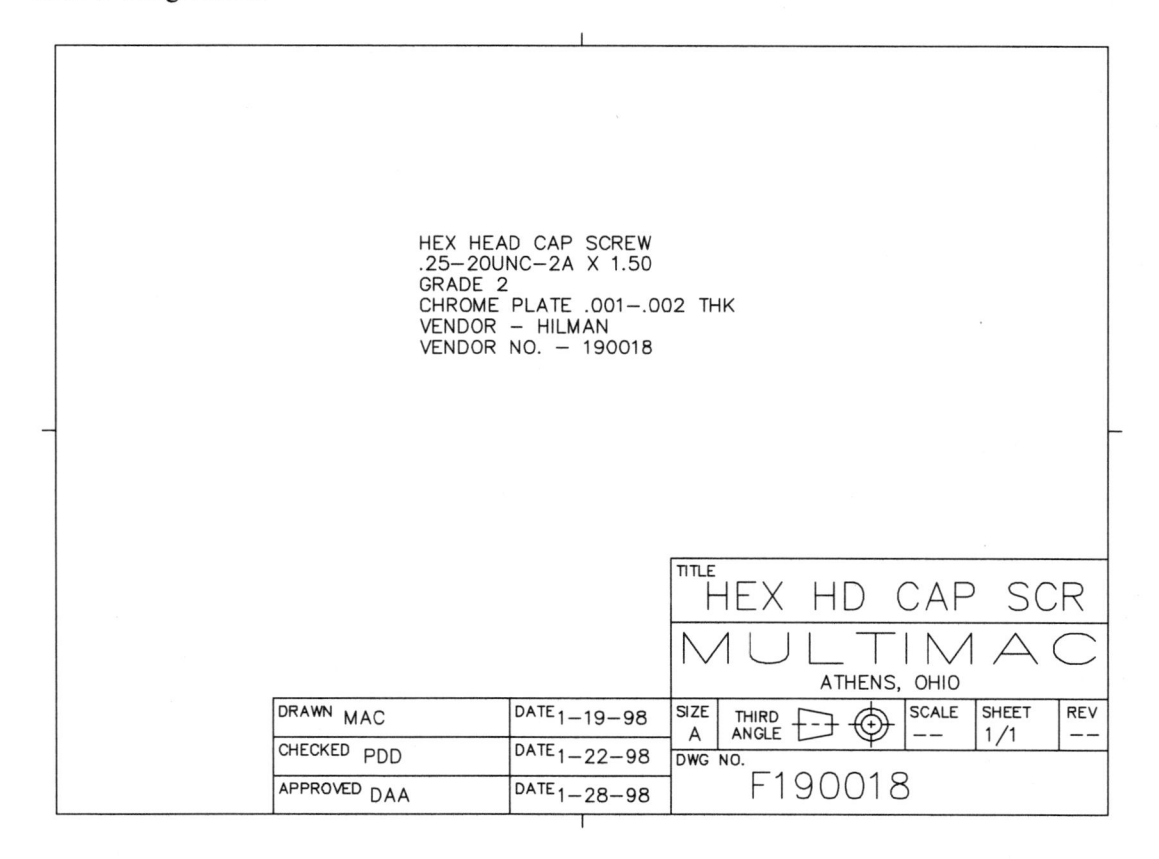

Figure 4-1 Note drawing

4.4.2.1 Cast or Forged Part

Along with normal dimensions, the detail drawing of a cast or forged part should show parting lines, draft angles, and any other unique features of the part prior to processing. See Fig. 4-2.

This drawing does not show any finished dimensions. Many companies combine cast or forged drawings with machined part drawings. Phantom lines are commonly used to show the cast or forged outline.

4.4.2.2 Machined Part

Finished dimensions are the main features of a machined part drawing. A machined part drawing usually does not specify how to achieve the dimensions. Fig. 4-3 shows a machined part made from a casting. Fig. 4-4 shows a machined part made from round bar stock.

4.4.2.3 Sheet Stock Part

Because there are different methods of forming sheet stock, drawings of these types of parts may look quite different. Fig. 4-5 shows a drawing of a structural component for an automobile frame. The part is illustrated primarily in 3-D with one 2-D view used to show detail. In these cases, the part geometry is stored in a computer database and is used throughout the company to produce the part. Fig. 4-6 shows a very different type of drawing. It is a flat pattern layout of a transition.

4.4.3 Assembly

Assembly drawings are categorized as subassembly or final assembly. Both show the relative positions of parts. They differ only in where they fit in the assembly sequence.

Assembly drawings are usually drawn in one of two forms: exploded pictorial view (see Fig. 4-7) or 2-D sectioned view (see Fig. 4-8). Two common elements of assembly drawings are identification balloons and parts lists. The item numbers in the balloons (circles with leaders pointing to individual parts) relate to the numbers in the parts list.

4.5 Border

The border is drawn around the perimeter of the drawing. It is a thick line with zone identification marks and centering marks. See Fig. 4-9.

4.5.1 Zones and Center Marks

The short marks around the rectangular border help to identify the location of points of interest on the drawing (similar to a road map). When discussing the details of a drawing over the telephone, the zone of the detail (A, 1 would be the location of the title block) is provided so the listener can find the same detail. This is particularly important for very detailed large drawings. The center marks, often denoted by arrows, are used to align the drawing on a photographic staging table when making microfilm negatives.

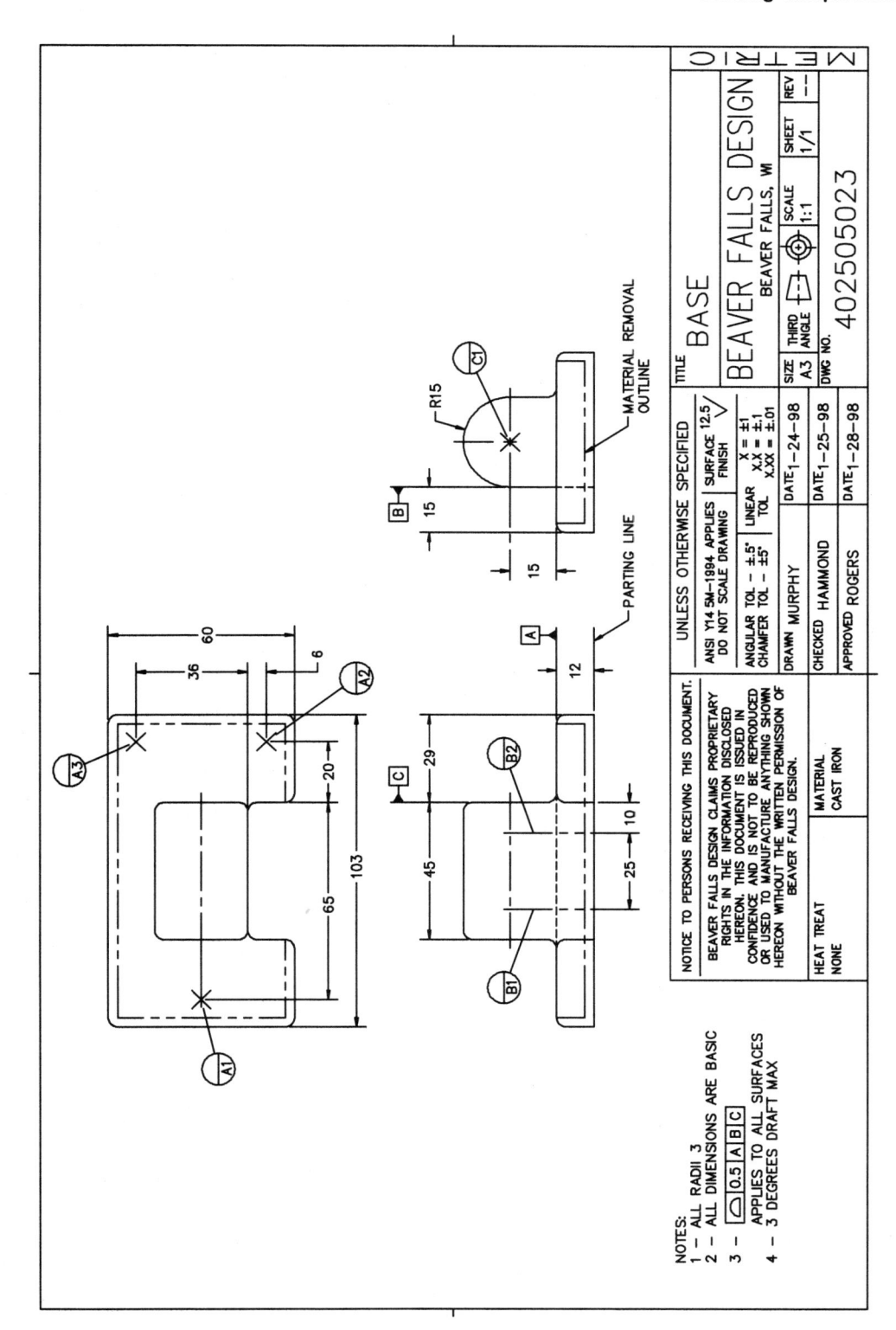

Figure 4-2 Casting drawing

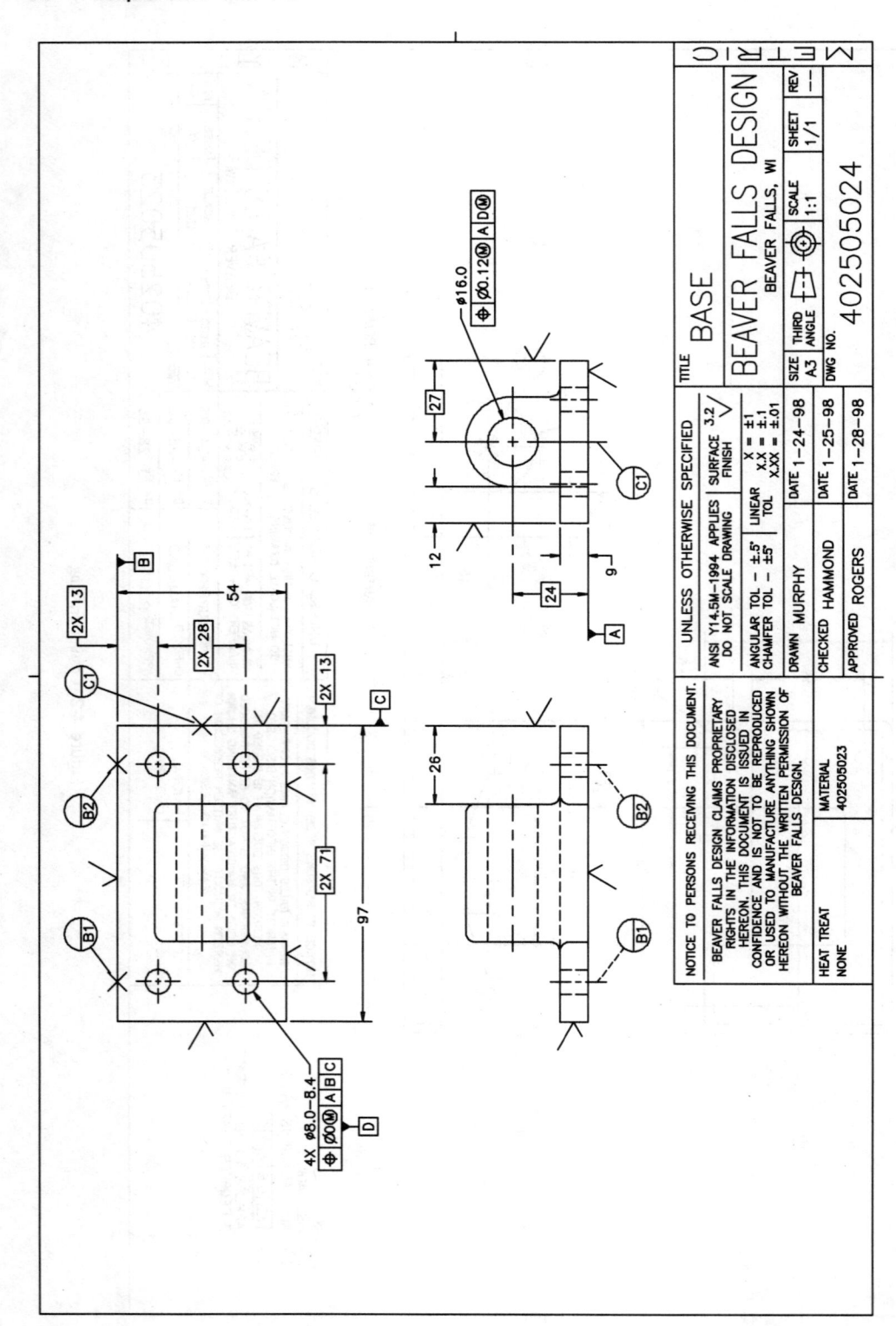

Figure 4-3 Machined part made from casting

Drawing Interpretation   4-7

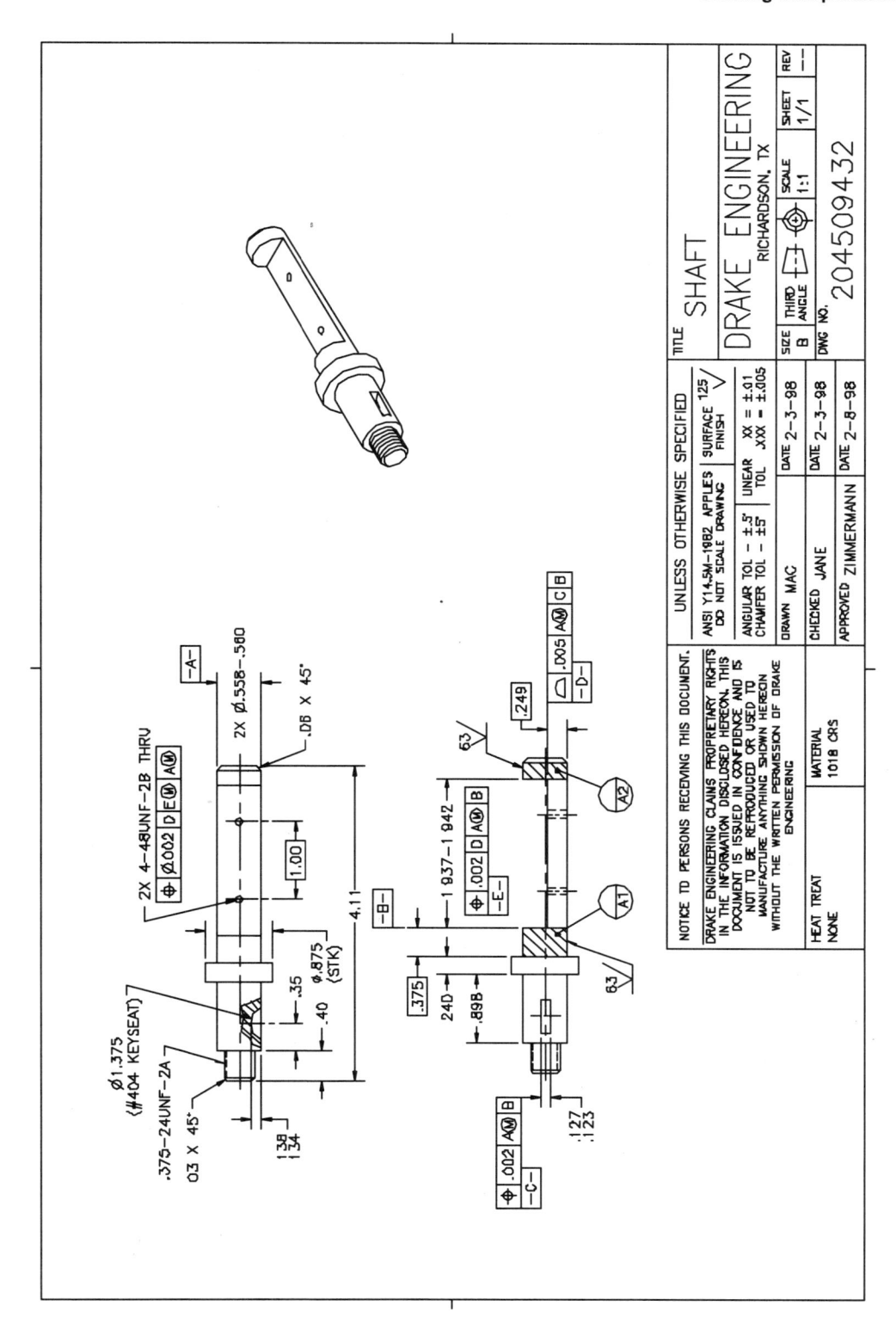

Figure 4-4 Machined part made from bar stock

4-8   Chapter Four

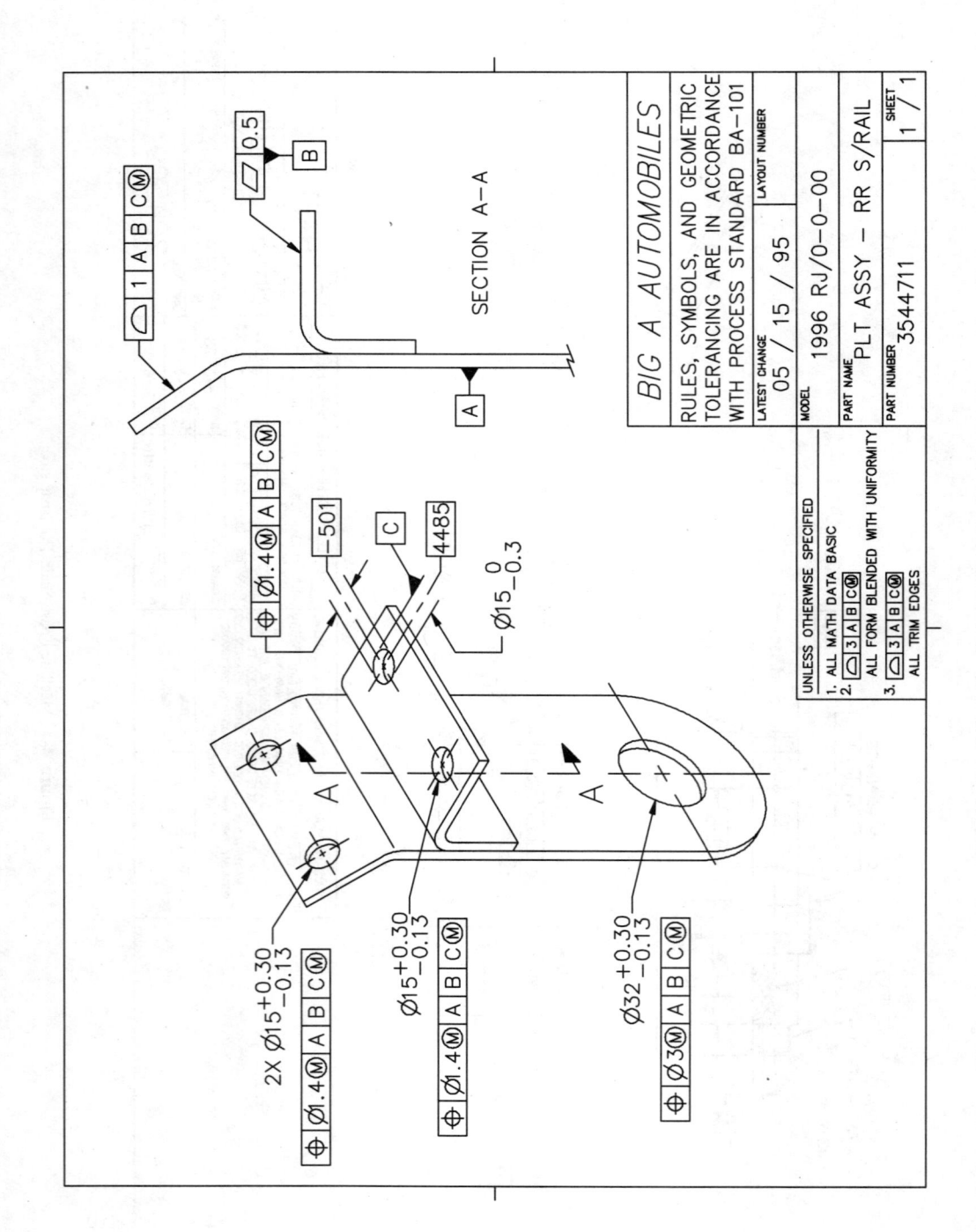

Figure 4-5 Stamped sheet metal part drawing

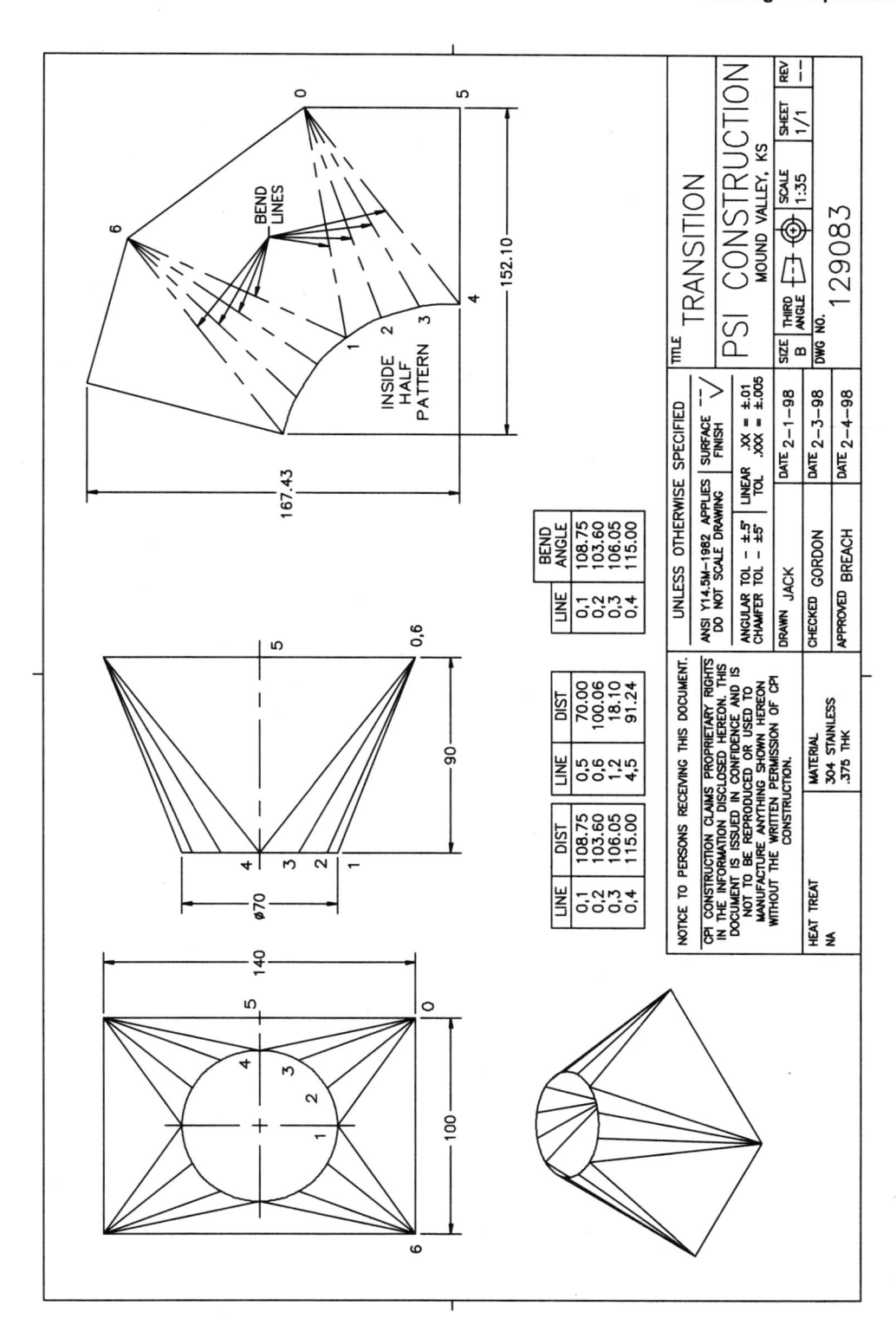

Figure 4-6 Flat pattern layout drawing

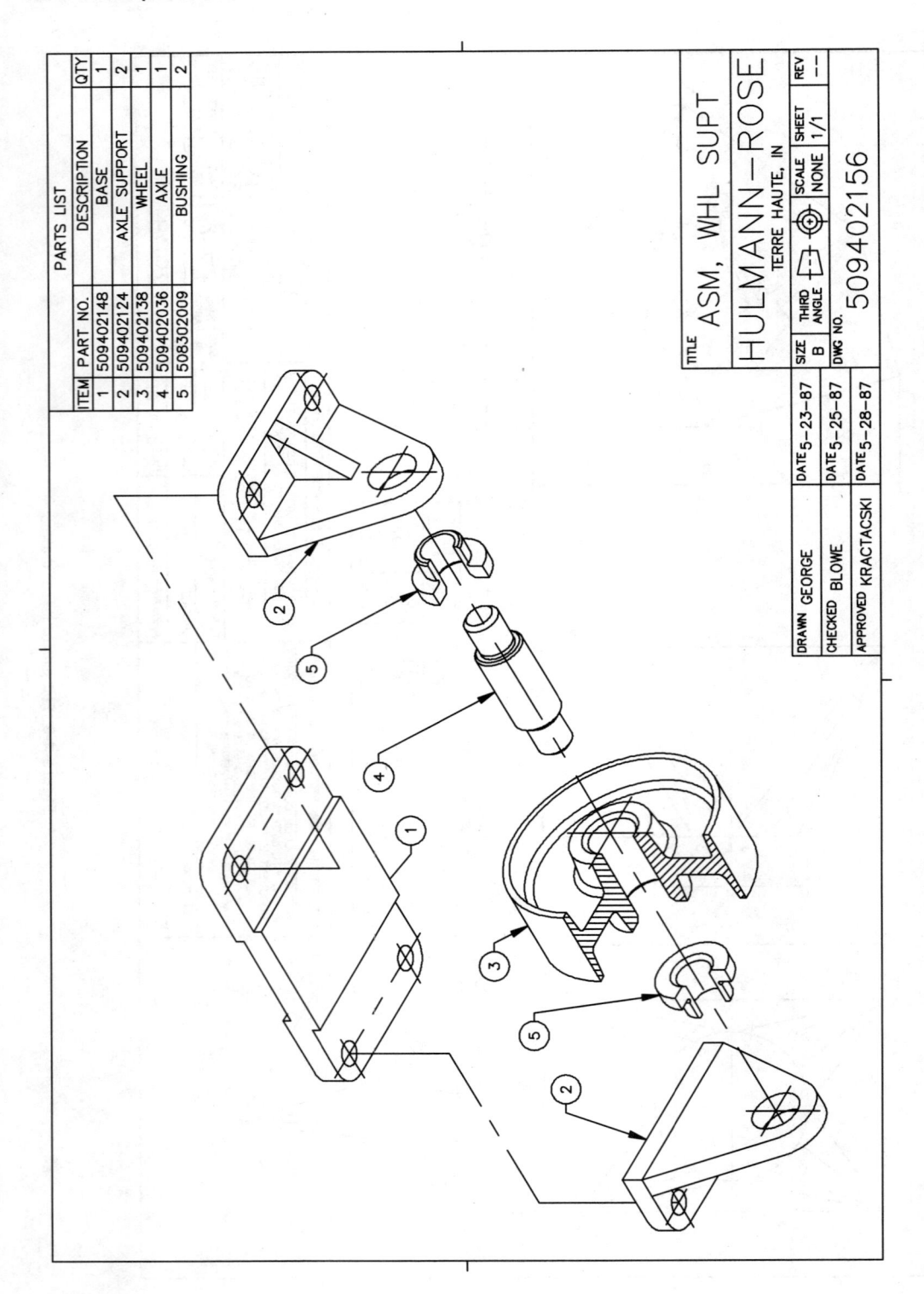

Figure 4-7 Exploded pictorial assembly drawing

Drawing Interpretation   4-11

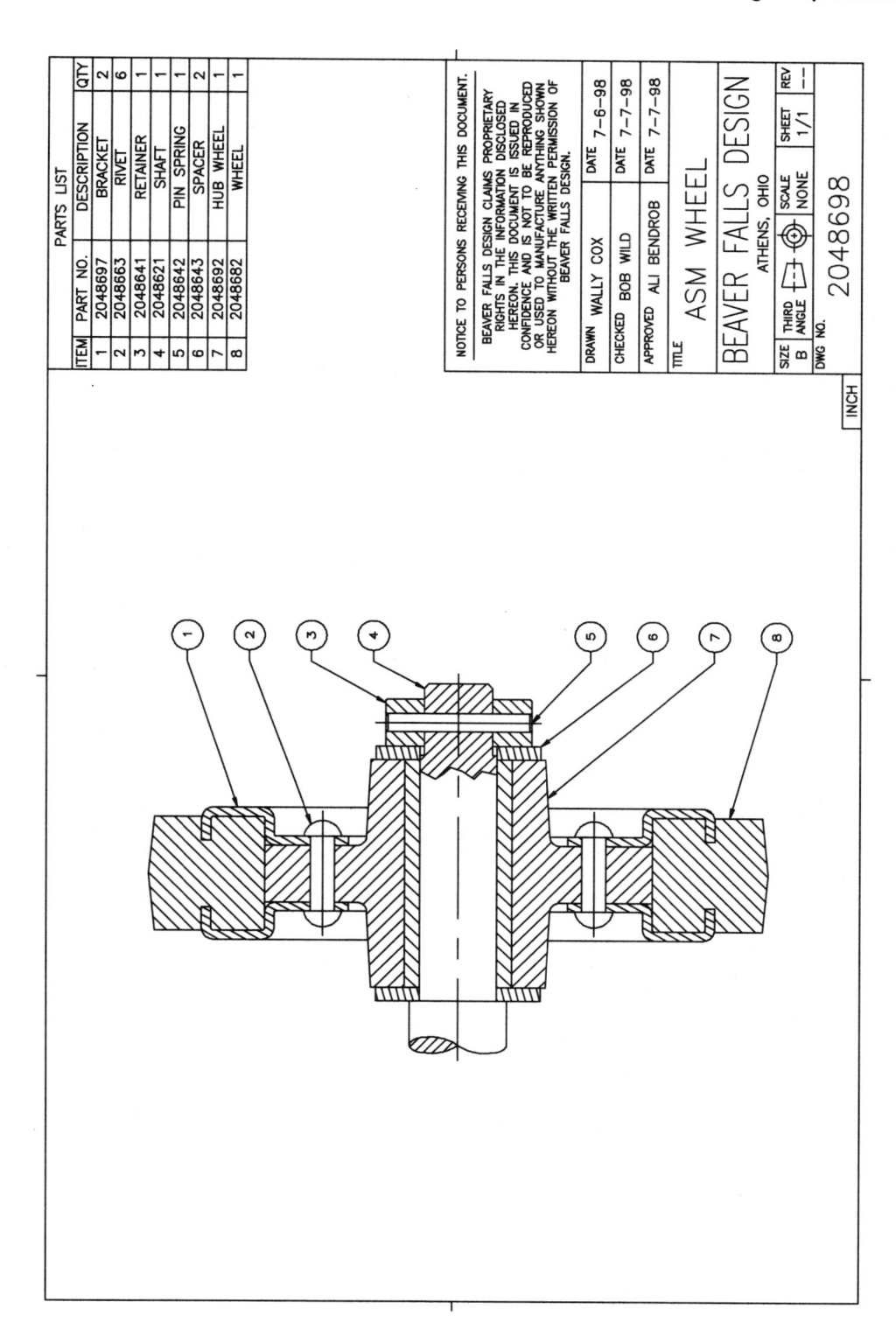

Figure 4-8 2-D sectioned assembly drawing

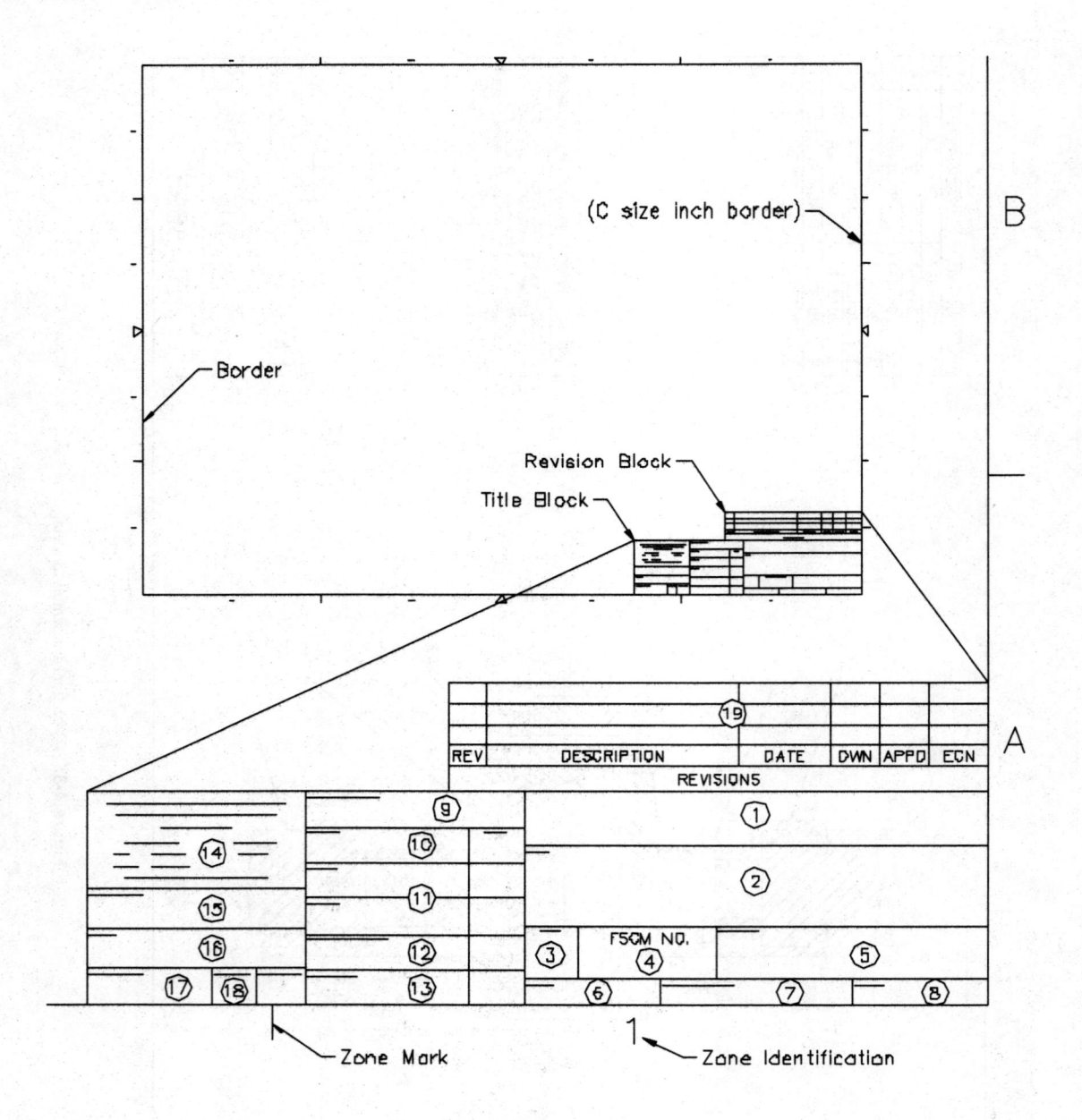

Figure 4-9 Border, title block, and revision block

4.5.2 Size Conventions

Most drawings conform to one of the sheet sizes listed below. If the drawing is larger than these sizes, it is generally referred to as a "roll size" drawing.

INCH		METRIC	
Code	Size	Code	Size
A	8.5 X 11	A4	210 X 297
B	11 X 17	A3	297 X 420
C	17 X 22	A2	420 X 594
D	22 X 34	A1	594 X 841
E	34 X 44	A0	841 X 1189

4.6 Title Blocks

The part of a drawing that has the highest concentration of information is usually the title block (see Fig. 4-9). It is the door to understanding the drawing and the company. Although there are many different arrangements possible, a good title block has the following characteristics.

- It is appropriate for the drawing type.
- It is intelligently constructed.
- It is filled in completely.
- All the signatures can be signed off within a short time frame.

Some drawing types will not use all of the following title block elements. For example: an assembly drawing may not require dimensional tolerances, surface finish, or next assembly. Although title block sizes and configurations have been standardized in ASME Y14.2, most companies will maintain the standard information but modify the configuration to suit their needs.

Reference Fig. 4-9 for the following standard title block items:

4.6.1 Company Name and Address

Many companies include their logo in addition to their name and address.

4.6.2 Drawing Title

When the drawing title is more than one word, it is often presented as the noun first and the adjective second. For example, SPRING PIN is written PIN, SPRING. This makes it easier to search all the titles when the first word is the key word in the title. There is no standard length for a title although many companies use about 15 character spaces. Abbreviations should not be used except for the words "assembly," "subassembly," and "installation," and trademarked names.

4.6.3 Size

The code letter for the sheet size is noted here. See Section 4.5.2 for common sheet sizes.

4.6.4 FSCM/CAGE

If your business deals with the federal government, you have a Federal Supply Code for Manufacturer's number. This number is the design activity code identification number.

4.6.5 Drawing Number

The drawing number is used for part identification and to ease storage and retrieval of the drawing and the produced parts. While there is no set way to assign part numbers, common systems are nonsignificant, significant, or some combination of the two previous systems.

Nonsignificant numbering systems are most preferred because no prior knowledge of significance is required.

Significant numbering systems could be used for commonly purchased items like fasteners. For example, the part number for a washer could include the inside diameter, outside diameters, thickness, material, and plating.

A combination of nonsignificant and significant numbering systems may use sections of the numbers in a hierarchical manner. For example, the last three digits could be the number assigned to the part (001, 002, 003, etc.). This would be nonsignificant. The remaining numbers could be significant: two numbers could be the model variation, the next two numbers could be the model number, and the next two could be the series number while the last two could be the project number. Many other possibilities exist.

4.6.6 Scale

There is no standard method of specifying the scale of a drawing. Scale examples for an object drawn at half its normal size are 1:2, 1=2, ½ or, HALF. They all mean the same thing. The first two examples are the easiest to use. If the one (1) is always on the left, the number on the right is the multiplication factor. For example, measure a distance on the drawing with a 1=1 scale and multiply that number by the number on the right (in this example, 2).

4.6.7 Release Date

This is the date the drawing was officially released for production.

4.6.8 Sheet Number

The sheet number shows how many individual sheets are required to completely describe a part. For many small parts, only one sheet is required. When parts are large, complicated, or both, multiple sheets are required. The number 4/12 would indicate the fourth (4) sheet of a twelve (12)-sheet drawing.

4.6.9 Contract Number

If this drawing was created as a part of a specific contract, the contract number is placed here. Other examples of drawing codes may be used to track the time spent on a project.

4.6.10 Drawn and Date

Some companies require the drafter to sign their name or initials. Other companies have the drafter type this information on the drawing. The date the drawing was started must be included.

4.6.11 Check, Design, and Dates

A drawing may be reviewed by more than one checker. For example, the drawing may go to a drafting checker first, then to a design checker, and maybe others. The checkers use the same method of identification as the drafters.

4.6.12 Design Activity and Date

As with checking, there may be multiple levels of approval before a document is released. The design activity is a representative of the area responsible for the design. All those approving the drawing use the same method of identification as the drafters.

4.6.13 Customer and Date

If the customer is required to approve the drawing, that name and date is placed here.

4.6.14 Tolerances

The items in this section apply unless it is stated differently on the field of the drawing. In addition to the general tolerance block that is shown in Fig. 4-9, other tolerance blocks might be used for sand casting, die casting, forging, and injection-molded parts.

Linear – Linear tolerances are presented in an equal format (±). It is also common to show multiple examples to indicate default numbers of decimal places.

Angular – Angular tolerances are also presented in an equal bilateral format (±). It is common to give one tolerance for general angles and a different tolerance for chamfers.

4.6.15 Treatment

Treatment might include manufacturing specifications, heat-treat notes, or plating specifications. Longer messages about processing are placed in a note. See Section 4.16.

4.6.16 Finish

The finish reveals the condition of part surfaces. It consists of roughness, waviness, and lay. The general surface roughness average is given in this space. See Section 4.15.

4.6.17 Similar To

Some companies prefer to have numbers of similar parts on the drawing in case the drawn part may be made from a like part.

4.6.18 Act Wt and Calc Wt

Providing the part weight on the drawing may help the personnel in the Routing area move the parts more efficiently.

4.6.19 Other Title Block Items

The part material must be stated on the drawing. The material is specified using codes provided by the Society of Automotive Engineers (SAE) or the American Society for Testing and Materials (ASTM).

The drawing number of the next assembly is often placed in the title block. Many standard parts have many different next assemblies. Each time a part is added to another assembly the drawing must be revised to add the next assembly number. The money spent maintaining these numbers causes some to question their value.

4.7 Revision Blocks

It is common for drawings to be revised several times for parts that are used for many years. During the life of a product, it may be revised to improve performance or reduce cost. After a drawing change request is made and accepted, the drawing is modified. Engineering change notices (ECN) are created to document the actual changes. The revision letter, description, date, drafter and approver identification, and ECN number are recorded in the revision block. See Fig. 4-9.

4.8 Parts Lists

A parts list names all the parts in an assembly. It lists the item number, description, part number, and quantity for each part in the assembly. The item number is placed in a circle (balloon) close to the part in the assembly view. A leader is drawn from the balloon pointing to the part. See Figs. 4-7 and 4-8.

4.9 View Projection

With the advent of orthographic (right-angle drawing) projection in the eighteenth century, battle fortifications could be visually described accurately and faster than mathematical methods. This contributed so much to Napoleon's success that it was kept secret during his time in power. Orthographic projection is a technique that uses parallel lines of sight intersecting mutually perpendicular planes of projection to create accurate 2-D views. The two variations most commonly used are first-angle and third-angle. As illustrated below, the names *first* and *third* relate into which 3-D quadrant the object is placed.

4.9.1 First-Angle Projection

The first-angle projection system is used primarily in Europe and other countries that only use ISO standards. When viewing a 2-D multiview drawing, the top view is placed below the front view and the right side view is placed on the left side of the front view. See Fig. 4-10.

4.9.2 Third-Angle Projection

The third-angle projection system is used primarily in the Americas. When viewing a 2-D multiview drawing, the top view is placed above the front view and the right side view is placed on the right side of the front view. See Fig. 4-11.

4.9.3 Auxiliary Views

Auxiliary views are those views drawn on projection planes other than the principal projection planes (see Figs. 4-12 and 4-19). Primary auxiliary views are drawn on projection planes constructed perpendicular to one of the principal projection planes. Successive auxiliary views are drawn on projection planes constructed perpendicular to any auxiliary projection plane.

4.10 Section Views

Section views show internal features of parts. Thin lines depict where solid material was cut. One of the opposing views will often have a cutting plane line showing the path of the cut. If the cutting plane in an assembly drawing passes through items that do not have internal voids, they should not be sectioned. Some of the items not usually sectioned are shafts, fasteners, rivets, keys, ribs, webs, and spokes. The following are standard types of sections.

Drawing Interpretation   **4-17**

Two-dimensional views are projected onto projection planes in the 1st quadrant of a three-dimensional Cartesian coordinate system. The planes are rotated 90° onto one two-dimensional plane.

Orthographic Views

Right Side   Front

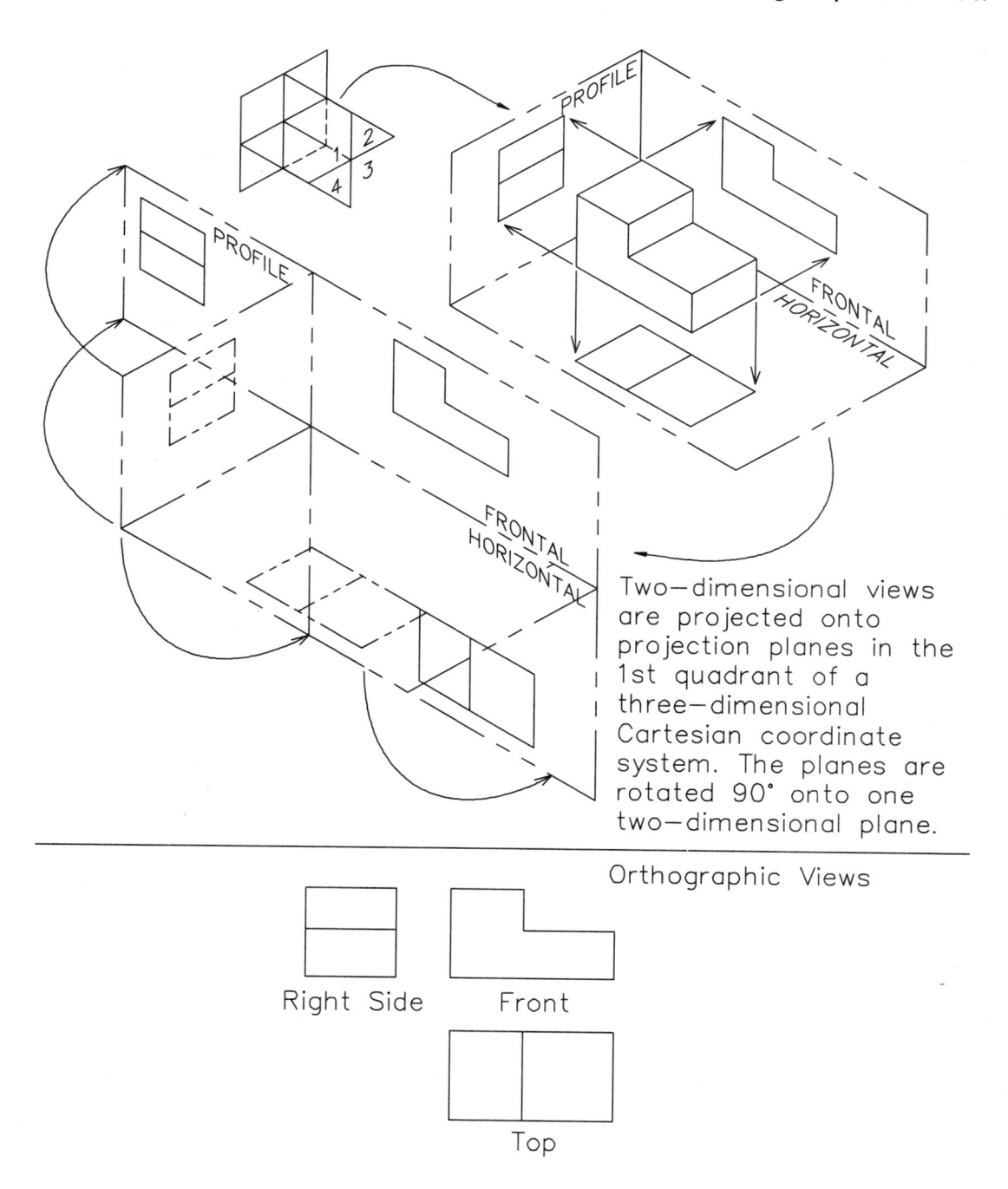

Top

Figure 4-10 First-angle projection

4-18   Chapter Four

Two-dimensional views are projected onto projection planes in the 3rd quadrant of a three-dimensional Cartesian coordinate system. The planes are rotated 90° onto one two-dimensional plane.

Orthographic Views

Top

Front   Right Side

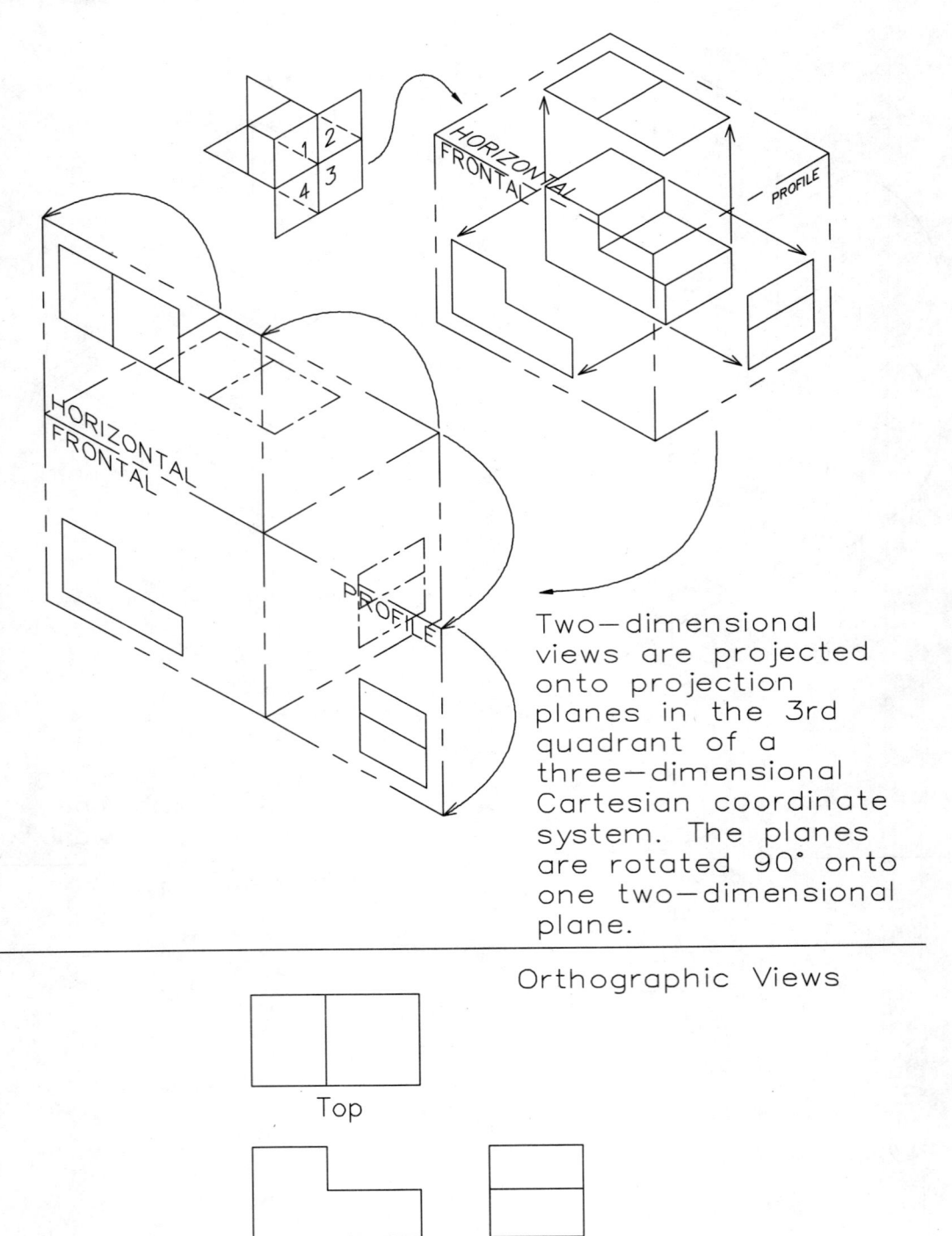

Figure 4-11 Third-angle projection

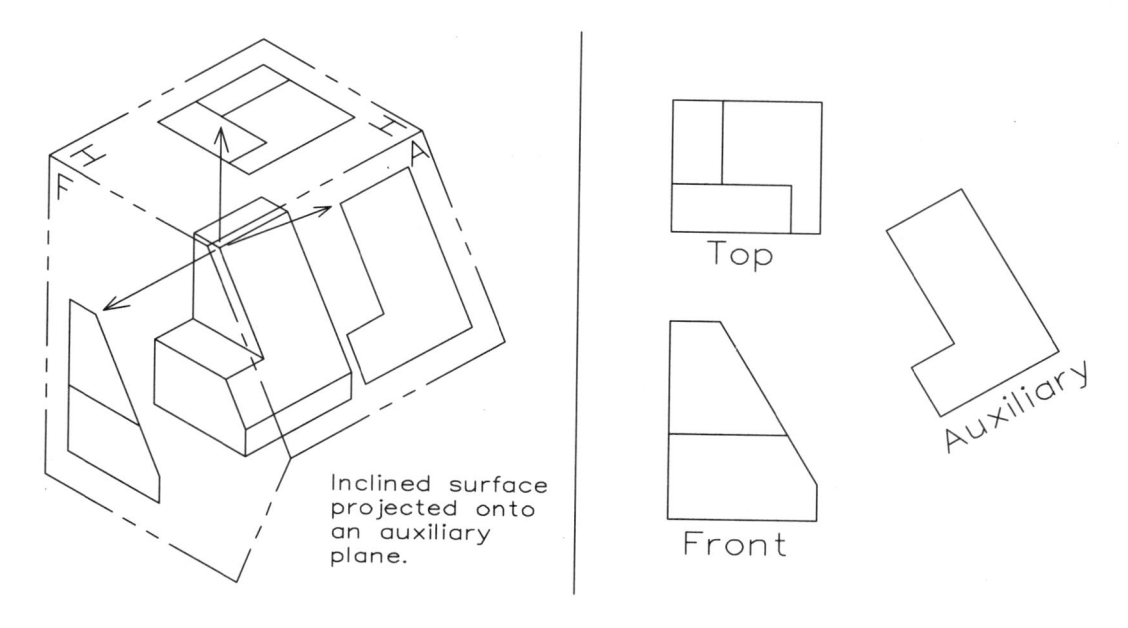

Figure 4-12 Auxiliary view development and arrangement

4.10.1 Full Sections

The view in full section appears to be cut fully from side to side. See Fig. 4-13. The cutting plane is one continuous plane with no offsets. If the location of the plane is obvious, it is not shown in an opposing view.

4.10.2 Half Sections

Half sections appear cut from one side to the middle of the part. See Fig. 4-14. In a half section, the side not in section does not show hidden lines. If the location of the plane is obvious, it is not shown in an opposing view.

4.10.3 Offset Sections

This type of sectioned view appears to be a full section, but when looking at the view where the section was taken, a cutting plane line will always show the direction of the cut through the part. See Fig. 4-15. The cutting plane changes direction to cut through the features of interest.

4.10.4 Broken-Out Section

The broken-out section of a view has the appearance of having been hit with a hammer to break a small part from the object. Rather than create a section through the entire part, only a localized portion of the object is sectioned. See Fig. 4-16.

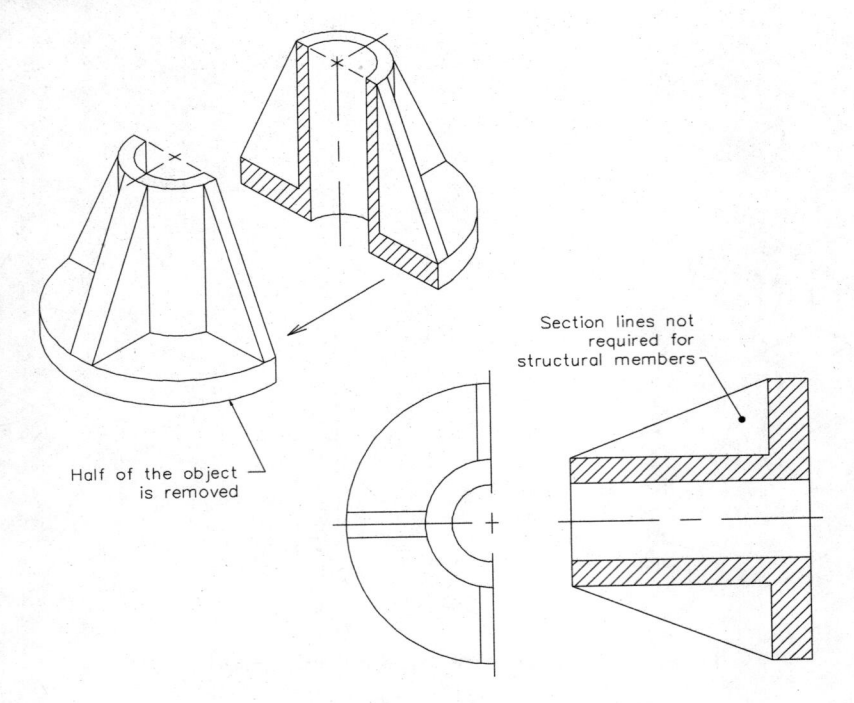

Figure 4-13 Full section

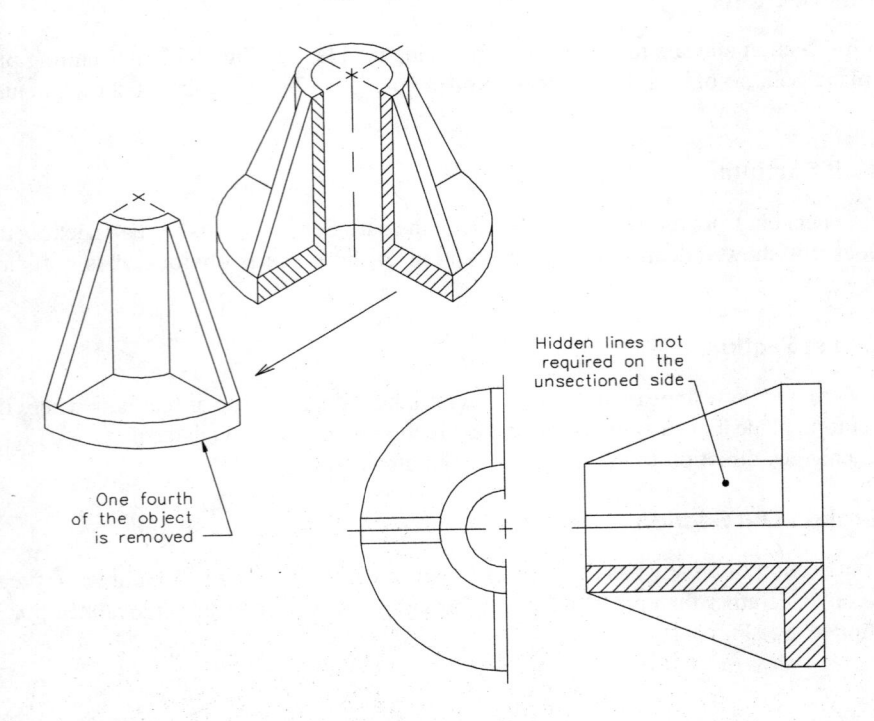

Figure 4-14 Half section

Drawing Interpretation 4-21

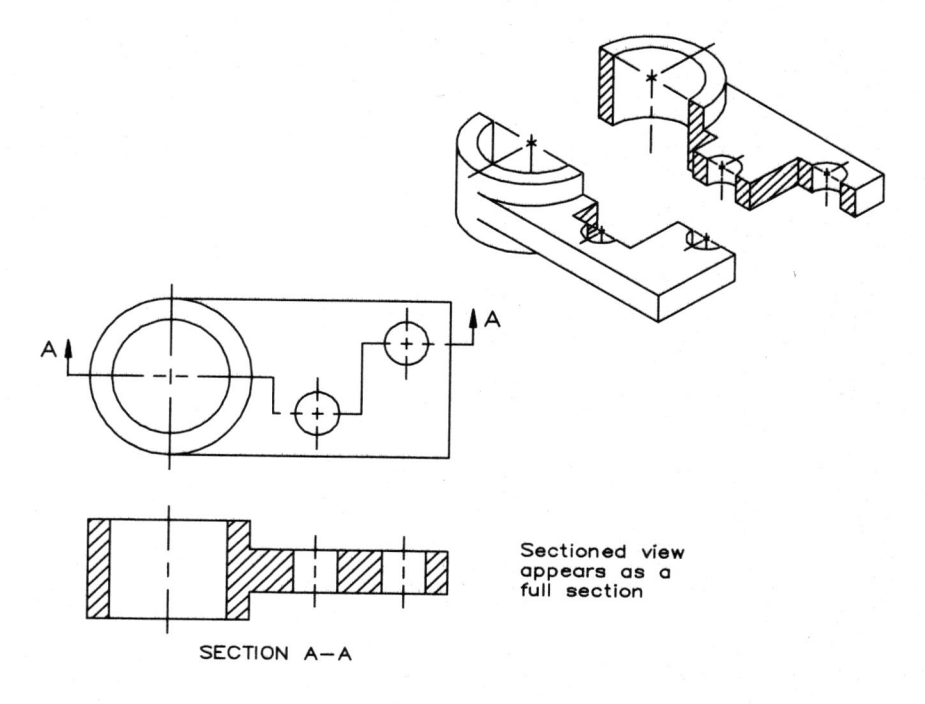

Figure 4-15 Offset section

Sectioned view appears as a full section

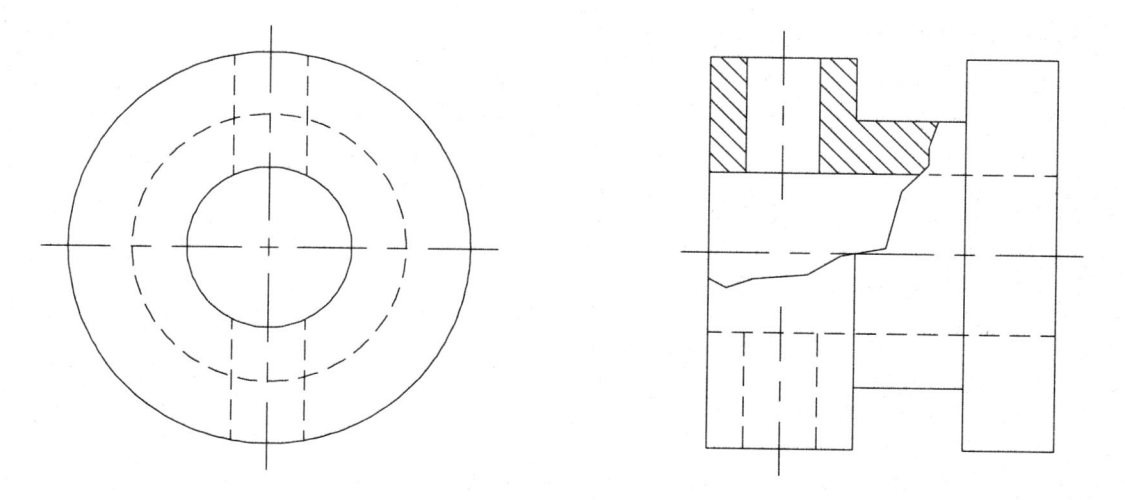

SECTION A—A

Figure 4-16 Broken-out section

4.10.5 Revolved and Removed Sections

The revolved and removed sections are developed in the same way. See Fig. 4-17. The concept is that a thin slice of an object is cut and rotated 90°. The section appears in the same view from where it was taken. The difference is the location of the sectioned view. The revolved view is placed at the point of revolution while the removed view is relocated to another more convenient location.

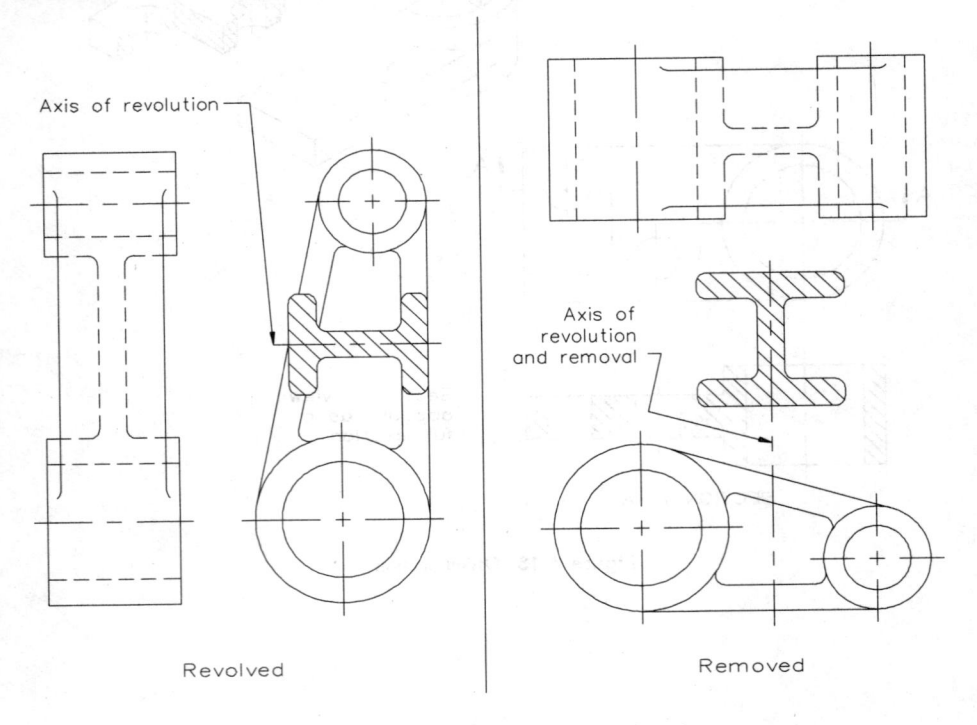

Figure 4-17 Revolved and removed section

4.10.6 Conventional Breaks

A conventional break is used to shorten a long consistent section length of material. See Fig. 4-18. There are conventional breaks for rods, bars, tubing, and woods.

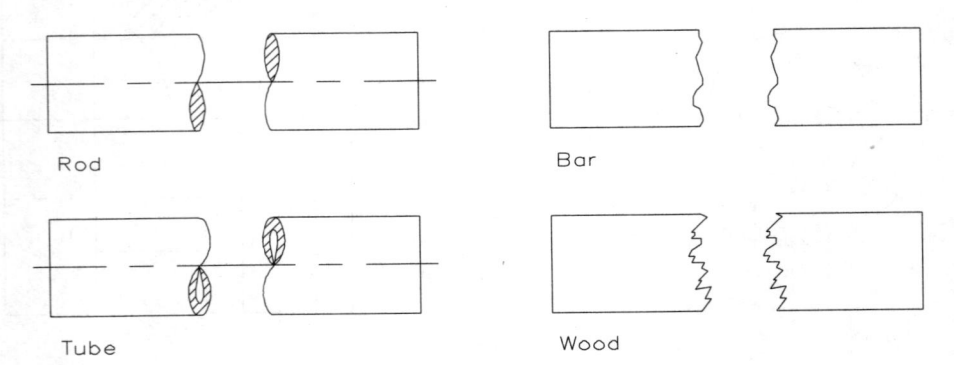

Figure 4-18 Conventional breaks

4.11 Partial Views

Partial views are regular views of an object with some lines missing. When it is confusing to show all the possible lines in any one view, some of the lines may be removed for clarity. See Fig. 4-19.

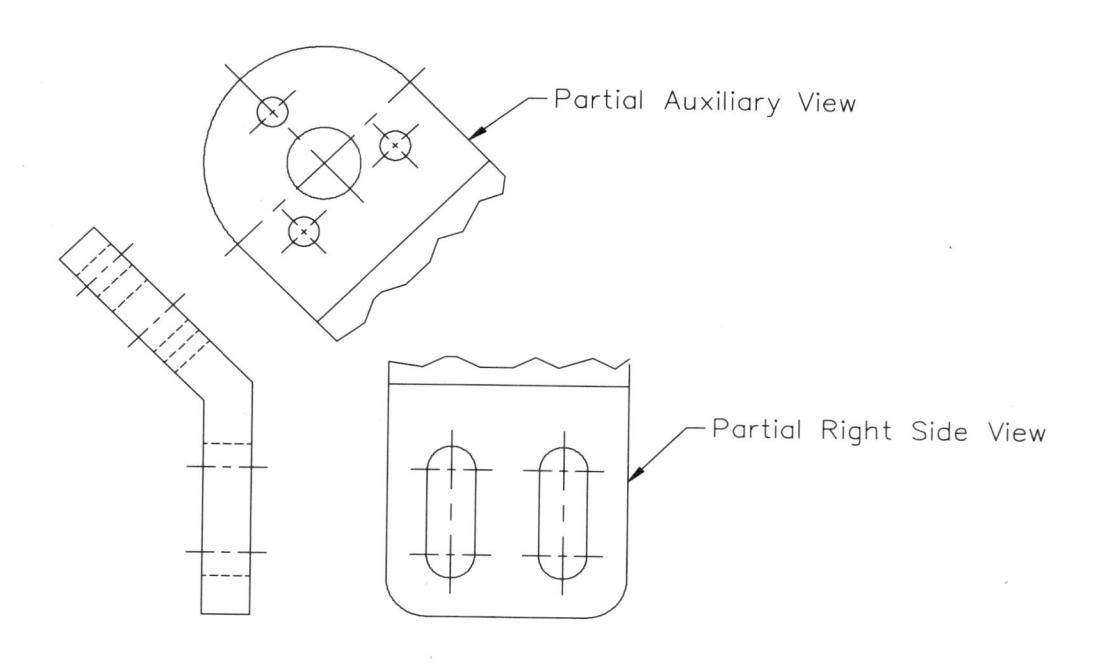

Figure 4-19 Partial views

4.12 Conventional Practices

It is not always practical to illustrate an object in its most correct projection. There are many occasions when altering the rules of orthographic projection is accepted. The following types of views represent common conventional practices.

4.12.1 Feature Rotation

Feature rotation is the practice of conceptually revolving features into positions that allow them to be viewed easily in an opposing view. For internal viewing, features may be rotated into a cutting plane. See Fig. 4-20. For external viewing, features may be rotated into a principal projection plane. This is often done to show the feature full size.

4.12.2 Line Precedence

When lines of different types occupy the same 2-D space, the lines are shown in the following order: object line, hidden line, cutting plane line, centerline, and phantom line.

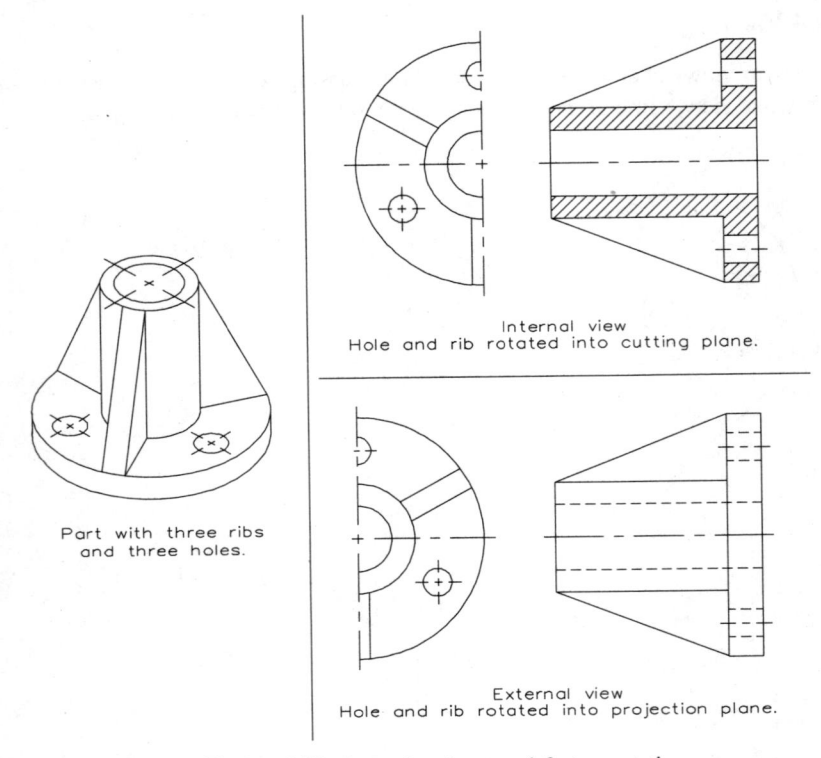

Figure 4-20 Internal and external feature rotation

4.13 Isometric Views

While many different methods may be used to show a pictorial view of a part, the isometric projection method is most common. To create an isometric projection, an object is rotated 45° in the top view then rotated 35°16' in the right side view. The resulting view appears 3-D. See Fig. 4-21. Fold line between the principal projection planes will measure 120° apart—hence, the name isometric or equal measures.

Companies that use 3-D computer programs to create part geometry may provide a 3-D view of the object along with conventional 2-D views. See Fig. 4-4. Some companies use 3-D views as their primary

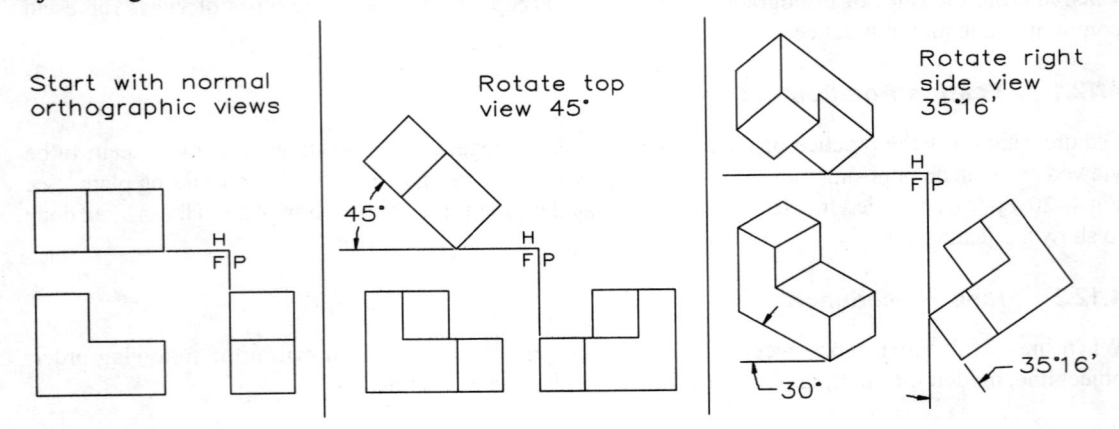

Figure 4-21 Isometric projection

view and 2-D views for sections. The object in Fig. 4-5 only shows critical size and geometric dimensioning. All other dimensions must be obtained from the computer database.

4.14 Dimensions

The role of the dimension on an engineering drawing has changed drastically for some companies. When dealing with traditional, manually created, 2-D drawings, the dimensions are the most important part of the drawing. The views are only a foundation for the dimensions. They could be quite inaccurate because the part is made from the dimensions and not the views.

When working with drawings created as a 3-D computer database, the geometry is most important. It must be created accurately because the computer database can be translated by another computer program into a language a machine tool can understand. In this scenario, the dimensions serve as a dimensional analysis tool and a reference document for inspection. See Chapter 16.

Dimensions may be of three different types: general dimensions, geometric dimensions, and surface texture. This section provides a brief introduction to general dimensioning and surface texture. Due to the extensive nature of geometric dimensioning, it is covered in Chapter 5. Prior to any discussion of dimensioning, the following underlying concepts must be understood.

4.14.1 Feature Types

Dimensions relate to features of parts. Features may be plane features, size features, or irregular features. A plane feature is considered nominally flat with a 2-D area. Size features are composed of two opposing surfaces like tabs and slots and surfaces with a constant radius like cylinders and spheres. Irregular features are free-form surfaces with defined undulations like the wing of an airplane or the outside surface of the hood of an automobile. Due to the nature of irregular surfaces, they are not usually defined only with general dimensions.

4.14.2 Taylor Principle / Envelope Principle

In 1905, an Englishman, William Taylor, was awarded the first patent for a full-form gage (GO-NOGO Gage) to inspect parts. His concept was that there is a space between the smallest size a feature can be and the largest size a feature can be and that all the surface elements must lie in that space. See Fig. 4-22.

A GO-NOGO gage is used to check the maximum and least material conditions of part features. The maximum material condition of a feature will make the part weigh more. The least material condition of a feature will make the part weigh less. Taylor's idea was to make a device that would reject a part whose form would exceed the maximum size of an external size feature or the minimum size of an internal size feature. For external size features, the device would be of two parallel plates separated by the maximum dimension for a tab or a largest sized hole for a shaft. For internal size features, the device would be two parallel plates at minimum separation for a slot or the smallest sized pin for a hole. See Chapter 19 for more information on gaging.

This idea was generally adopted by companies in the United States and was commonly known as the Taylor Principle. Product design uses a similar concept called the Envelope Principle. The Envelope Principle was adopted in the US because it unites the form of a feature with its 2-D size. It allows the allowance and maximum clearance to be calculated. Separate statements controlling the form of size features are not required.

The default condition adopted by the ISO is the Principle of Independency. This concept does not unite the form with the 2-D size of a feature—they are independent. If a form control is required, it must be stated. See Chapter 6 for the differences between the US and ISO standards.

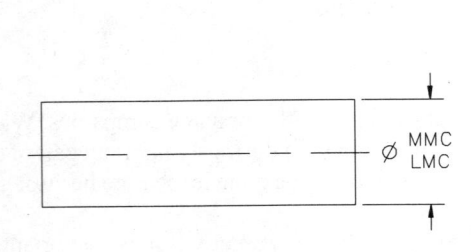

For features of size controlled only with size dimensions the allowance (AL) and maximum clearance (CL) may be calculated.

AL = MMC Hole − MMC Shaft
CL = LMC Hole − LMC Shaft

When produced at MMC, the parts must have perfect form.

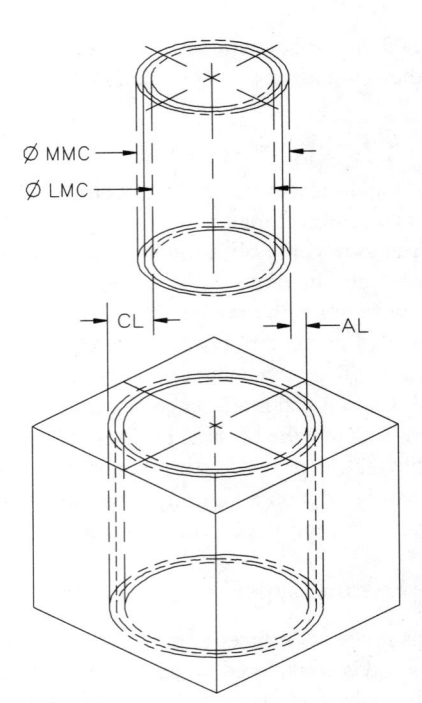

Figure 4-22 Envelope principle

4.14.3 General Dimensions

General dimensions provide size and location information. They can be classified with the names shown in Fig. 4-23.

All other dimensions use the UOS bilateral tolerances.

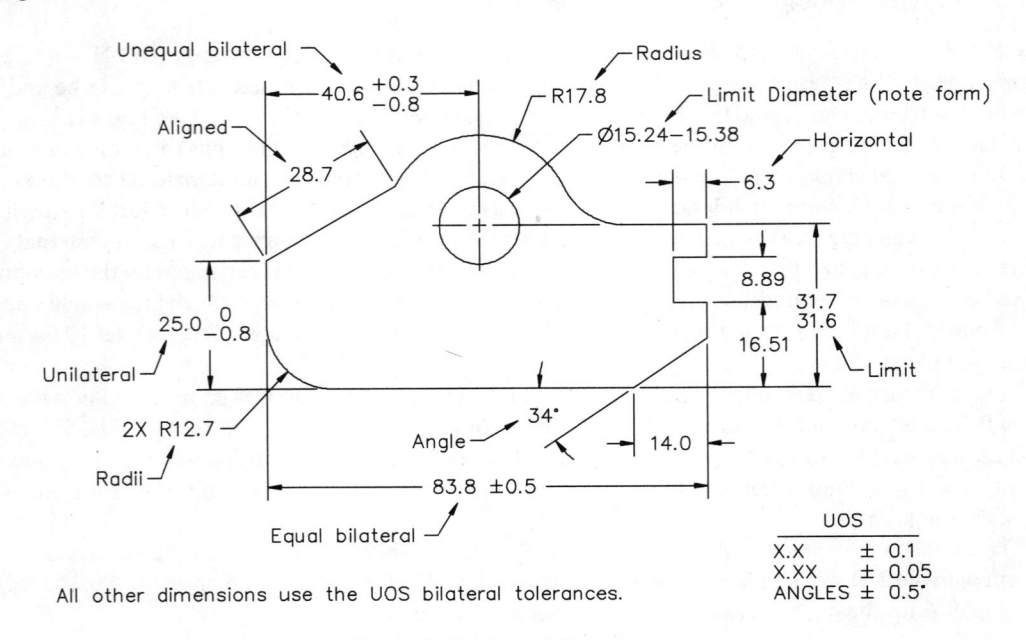

Figure 4-23 General dimension types

General dimensions have tolerances and, in the case of size features (in the US), conform to the Envelope Principle. They are most often placed on the drawing with dimension lines, dimension values, arrows, and leaders as shown on the left side of Fig. 4-24. Dimensions may be stated in a note, or the features can be coded with letters and the dimensions placed in a table in situations where there is not enough space to use extension lines and dimension lines.

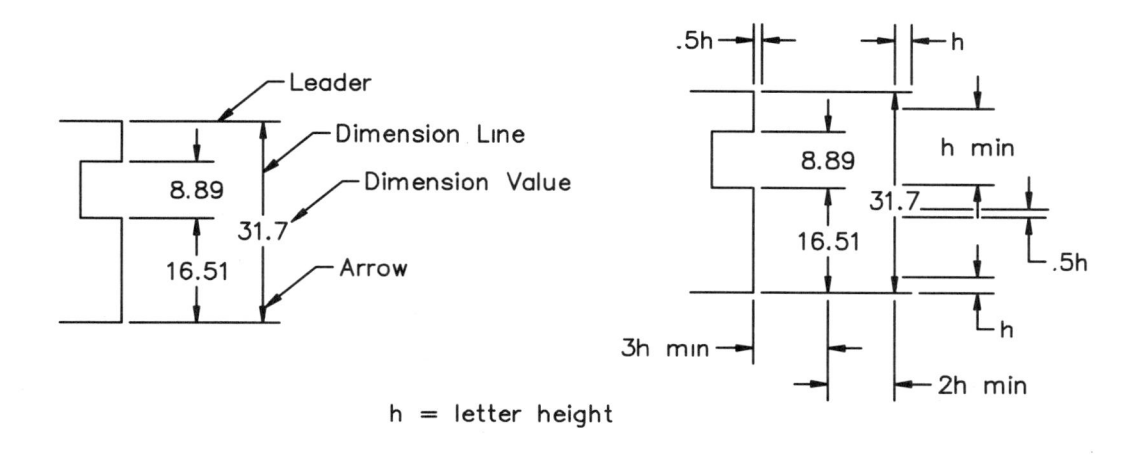

Figure 4-24 Dimension elements and measurements

4.14.4   Technique

Dimensioning techniques refer to the rudimentary details of arrow size, gap from the extension line to the object outline, length of the extension line past the dimension line, gap from the dimension line to the dimension value, and dimensioning symbols. The sizes shown on the right side of Fig. 4-24 are commonly used. Most computer aided drafting software will allow some or all of theses elements to be adjusted to the letter height, as shown, or some other constant. Additional dimensioning symbols are shown in Chapter 5.

4.14.5   Placement

Whereas dimensioning techniques are fairly common from drawing to drawing and company to company, dimension placement can vary. It may be based on view arrangement, part contour, function, size, or simple convenience. Some common dimension placement examples are shown in Figs. 4-2, 4-3, 4-4, 4-23, and dimensioned in Fig. 4-24.

The most important element to good placement is consistent spacing. This translates to easy readability and fewer mistakes. Some other placement techniques are:

- Provide a minimum of 10 mm from the object outline to the first dimension line
- Provide a minimum of 6 mm between dimension lines
- Place shorter dimensions inside longer dimensions
- Avoid crossing dimension lines with extension lines or other dimension lines
- Dimension where the true size contour of the object is shown
- Place dimensions that apply to two views between the views
- Dimension the size and location of size features in the same view

4.14.6 Choice

There are usually several different ways to dimension an assembly and its detail parts. Making the best dimensional choices involves understanding many different areas. Knowledge of the requirements of the design should be the most important. Other knowledge areas should include the type and use of tooling fixtures, manufacturing procedures and capabilities, inspection techniques, assembly methods, and dimensional management policies and procedures. Many other areas like pricing control or part routing may also influence the dimensioning activity. Due to the vast body of knowledge required and legal implications of incorrect dimensioning practices, the dimensioning activity should be carefully considered, thoroughly executed, and cautiously checked. Depending on the complexity of the product, it may be prudent to assign a team of dimensional control engineers to perform this activity.

4.14.7 Tolerance Representation

All dimensions must have a tolerance associated with them. Six different methods of expressing toleranced dimension are presented in Fig. 4-23.

1. The 31.6-31.7 dimension is an example of the limit type—it shows the extreme size possibilities (the large number is always on top).
2. The 15.24-15.38 dimension is the same as the limit dimension but is presented in note form (the small number is written first and the numbers are separated by a dash).
3. The 83.8 dimension is an example of the equal bilateral form—the dimension is allowed to vary from nominal by an equal amount.
4. The 40.6 dimension is an example of the unequal bilateral form—the dimension is allowed to vary more in one direction than another.
5. The 25.0 dimension is an example of the unilateral form—the dimension is only allowed to vary in one direction from nominal.
6. The dimensions with only one number are actually equal bilateral dimensions that show the nominal dimension while the tolerance appears in the Unless Otherwise Specified (UOS) part of the title block.

4.15 Surface Texture

Surface texture symbols specify the limits on surface roughness, surface waviness, lay, and flaws. A machined surface may be compared to the ocean surface in that the ocean surface is composed of small ripples on larger waves. See Fig. 4-25. Basic surface texture symbols are used on the drawing shown in Fig. 4-3.

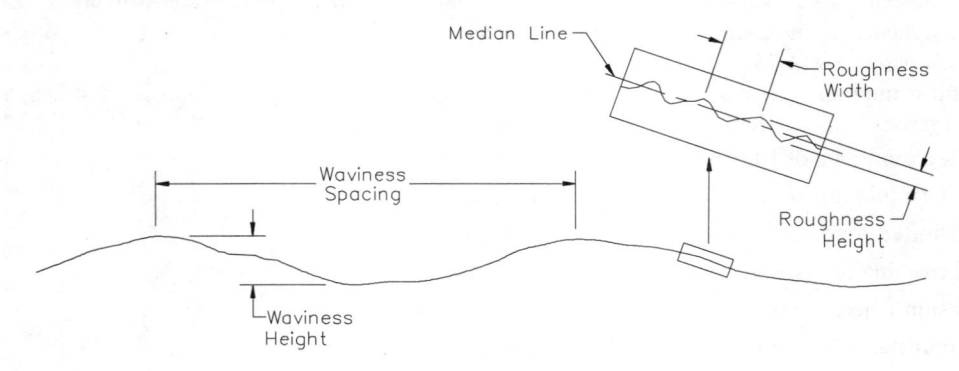

Figure 4-25 Surface characteristics

4.15.1 Roughness

The variability allowed for the small ripples on a surface is specified in micrometers or microinches. If only one number is given for the roughness average as shown in Fig. 4-26 (a) and (b), the measured values must be in a range between the stated number and 0. If two numbers are written one above the other as shown in example (c), the measured values must be within that range. Other roughness measures may be specified as shown in example (d).

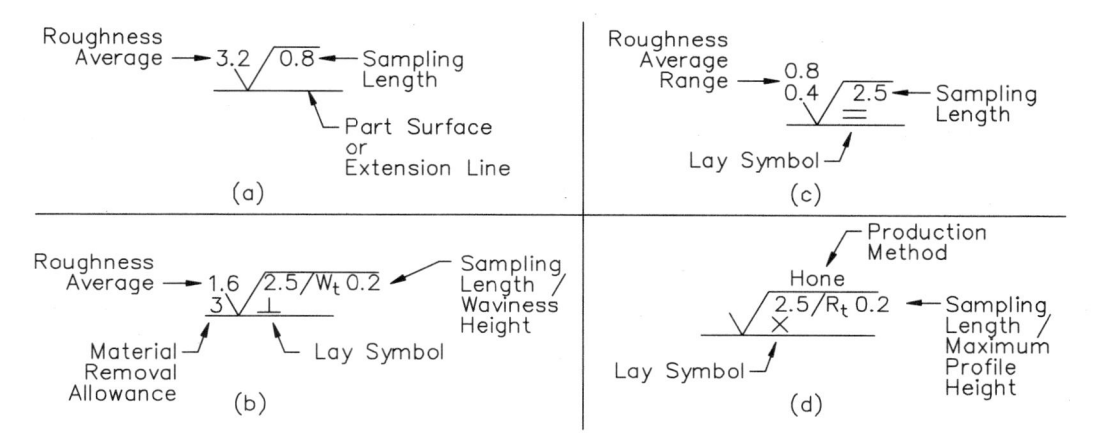

Figure 4-26 Surface texture examples and attributes

4.15.2 Waviness

The large waves are controlled by specifying the height (W_t) in millimeters. The placement of this parameter is shown in Fig. 4-26 (b).

4.15.3 Lay

The lay indicates the direction of the tool marks. See Fig. 4-26. Symbols or single letters are used to indicate perpendicular (b), parallel (c), crossed (d), multidirectional, circular, radial, particulate, nondirectional, or protuberant.

4.15.4 Flaws

Flaws are air pockets in the material that were exposed during production, scratches left by production or handling methods, or other nonintended surface irregularities. Flaw specifications are placed in the note section of the drawing.

4.16 Notes

Some information can be better stated in note form rather than in a dimension. See Fig. 4-2. Other information can *only* be stated in note form. Common notes specify default chamfer and radius values, information for plating or heat-treating, specific manufacturing operations, and many other pieces of information. Most companies group notes in one common location such as the upper left corner or to the left of the title block.

4.17　Drawing Status

The drawing life cycle may have several different stages. It may start as a sketch, progress to an experimental drawing, reach active status, and then be marked obsolete. Whatever their status, drawings require an accounting system to follow their changes in status. An engineering function, the data processing area, or a separate group may control this accounting system.

4.17.1　Sketch

A drawing often starts with a sketch of an assembly. From that sketch additional sketches may show interior parts and details of those parts. If the ideas seem worth the additional effort, the sketches may be transferred to formal detail and assembly drawings. Even though sketches may seem trivial at the time they are created, they should all be dated, signed, and stored for reference.

4.17.2　Configuration Layout

There may be different names for this type of drawing, but its main function is for analysis of geometric and dimensional details of an assembly. This activity has changed with the advent of computer simulations. Assemblies are built using 3-D digital models.

4.17.3　Experimental

Many ideas make the transition from sketches to experimental drawings. Parts made from these drawings may be tested and revised several times prior to being formally released as active production drawings.

4.17.4　Active

As the name implies, an active part drawing has gone through a formal release process. It will be released as any other drawing and, with good reason, should be accessible by any employee.

4.17.5　Obsolete

When a part is no longer sold, the drawing has reached the end of its life cycle. This does not mean a part could not be produced, but only that its status has changed to "Obsolete." Drawings are never destroyed. Drawings may be classified obsolete for production but retained for service, or obsolete for service but retained for production. If necessary, the drawing may be reactivated for production, service, or both.

4.18　Conclusion

With all the benefits realized by using a common drawing communication system, it is imperative that all personnel who deal with engineering drawings understand them completely. All the methods detailed in this chapter can be found in the appropriate standards. However, the standards covering this communication system are only guidelines. A company may choose to communicate their product specifications in different ways or to specify requirements not covered in the national standards. If this is the case, company-specific standards must be created and maintained.

4.19 References

1. The American Society of Mechanical Engineers. 1980. *ASME Y14.1-1980, Drawing Sheet Size and Format.* New York, New York: The American Society of Mechanical Engineers.
2. The American Society of Mechanical Engineers. 1995. *ASME B46.1-1995, Surface Texture (Surface Roughness, Waviness, and Lay).* New York, New York: The American Society of Mechanical Engineers.
3. The American Society of Mechanical Engineers. 1992. *ASME Y14.2M-1992, Line Conventions and Lettering.* New York, New York: The American Society of Mechanical Engineers.
4. The American Society of Mechanical Engineers. 1994. *ASME Y14.3-1994, Multiview and Sectional View Drawings.* New York, New York: The American Society of Mechanical Engineers.
5. The American Society of Mechanical Engineers. 1995. *ASME Y14.5M-1994, Dimensioning and Tolerancing.* New York, New York: The American Society of Mechanical Engineers.
6. The American Society of Mechanical Engineers. 1996. *ASME Y14.8M-1996, Castings and Forgings.* New York, New York: The American Society of Mechanical Engineers.
7. The American Society of Mechanical Engineers. 1996. *ASME Y14.36M-1996, Surface Texture and Symbols.* New York, New York: The American Society of Mechanical Engineers.

Chapter 5

Geometric Dimensioning and Tolerancing

Walter M. Stites
Paul Drake

Walter M. Stites
AccraTronics Seals Corp.
Burbank, California

Walter M. Stites is a graduate of California State University, Northridge. His 20-year tenure at AccraTronics Seals Corp began with six years in the machine shop, where he performed every task from operating a hand drill press to making tools and fixtures. Trained in coordinate measuring machine (CMM) programming in 1983, he has since written more than 1,000 CMM programs. He also performs product design, manufacturing engineering, and drafting. In 12 years of computer-assisted drafting, he's generated more than 800 engineering drawings, most employing GD&T. He has written various manuals, technical reports, and articles for journals. Mr. Stites is currently secretary of the ASME Y14.5 subcommittee and a key player in the ongoing development of national drafting standards.

5.1 Introducing Geometric Dimensioning and Tolerancing (GD&T)

When a hobbyist needs a simple part for a project, he might go straight to the little lathe or milling machine in his garage and produce it in a matter of minutes. Since he is designer, manufacturer, and inspector all in one, he doesn't need a drawing. In most commercial manufacturing, however, the designer(s), manufacturer(s), and inspector(s) are rarely the same person, and may even work at different companies, performing their respective tasks weeks or even years apart.

A designer often starts by creating an ideal assembly, where all the parts fit together with optimal tightnesses and clearances. He will have to convey to each part's manufacturer the ideal sizes and shapes, or *nominal dimensions* of all the part's surfaces. If multiple copies of a part will be made, the designer must recognize it's impossible to make them all identical. Every manufacturing process has unavoidable variations that impart corresponding variations to the manufactured parts. The designer must analyze his entire assembly and assess for each surface of each part how much variation can be allowed in size, form, orientation, and location. Then, in addition to the ideal part geometry, he must communicate to the manufacturer the calculated magnitude of variation or *tolerance* each characteristic can have and still contribute to a workable assembly.

For all this needed communication, words are usually inadequate. For example, a note on the drawing saying, "Make this surface real flat," only has meaning where all concerned parties can do the following:

- Understand English
- Understand to which surface the note applies, and the extent of the surface
- Agree on what "flat" means
- Agree on exactly how flat is "real flat"

Throughout the twentieth century, a specialized language based on graphical representations and math has evolved to improve communication. In its current form, the language is recognized throughout the world as *Geometric Dimensioning and Tolerancing (GD&T)*.

5.1.1 What Is GD&T?

Geometric Dimensioning and Tolerancing (GD&T) is a language for communicating engineering design specifications. GD&T includes all the symbols, definitions, mathematical formulae, and application rules necessary to embody a viable engineering language. As its name implies, it conveys both the nominal dimensions (ideal geometry), and the tolerances for a part. Since GD&T is expressed using line drawings, symbols, and Arabic numerals, people everywhere can read, write, and understand it regardless of their native tongues. It's now the predominant language used worldwide as well as the standard language approved by the American Society of Mechanical Engineers (ASME), the American National Standards Institute (ANSI), and the United States Department of Defense (DoD).

It's equally important to understand what GD&T is not. It is not a creative design tool; it cannot suggest how certain part surfaces should be controlled. It cannot communicate design intent or any information about a part's intended function. For example, a designer may intend that a particular bore function as a hydraulic cylinder bore. He may intend for a piston to be inserted, sealed with two Buna-N O-rings having .010" squeeze. He may be worried that his cylinder wall is too thin for the 15,000-psi pressure. GD&T conveys none of this. Instead, it's the designer's responsibility to translate his hopes and fears for his bore—his intentions—into unambiguous and measurable specifications. Such specifications may address the size, form, orientation, location, and/or smoothness of this cylindrical part surface as he deems necessary, based on stress and fit calculations and his experience. It's these objective specifications that GD&T codifies. Far from revealing what the designer has in mind, GD&T cannot even convey that the bore is a hydraulic cylinder, which gives rise to the Machinist's Motto.

> *Mine is not to reason why;*
> *Mine is but to tool and die.*

Finally, GD&T can only express what a surface shall be. It's incapable of specifying manufacturing processes for making it so. Likewise, there is no vocabulary in GD&T for specifying inspection or gaging methods. To summarize, GD&T is the language that designers use to translate design requirements into measurable specifications.

5.1.2 Where Does GD&T Come From?—References

The following American National Standards define GD&T's vocabulary and provide its grammatical rules.

- ASME Y14.5M-1994, Dimensioning and Tolerancing
- ASME Y14.5.1M-1994, Mathematical Definition of Dimensioning and Tolerancing Principles

Hereafter, to avoid confusion, we'll refer to these as "Y14.5" and "the Math Standard," respectively (and respectfully). The more familiar document, Y14.5, presents the entire GD&T language in relatively plain English with illustrated examples. Throughout this chapter, direct quotations from Y14.5 will appear in boldface. The supplemental Math Standard expresses most of GD&T's principles in more precise math terminology and algebraic notation—a tough read for most laymen. For help with it, see Chapter 7. Internationally, the multiple equivalent ISO standards for GD&T reveal only slight differences between ISO GD&T and the US dialect. For details, see Chapter 6.

Unfortunately, ASME offers no 800 number or hotline for Y14.5 technical assistance. Unlike computer software, the American National and ISO Standards are strictly rulebooks. Thus, in many cases, for ASME to issue an interpretation would be to arbitrate a dispute. This could have far-reaching legal consequences. Your best source for answers and advice are textbooks and handbooks such as this. As members of various ASME and ISO standards committees, the authors of this handbook are brimming with insights, experiences, interpretations, preferences, and opinions. We'll try to sort out the few useful ones and share them with you. In shadowboxes throughout this chapter, we'll concoct FAQs (frequently asked questions) to ourselves. Bear in mind, our answers reflect our own opinions, not necessarily those of ASME or any of its committees.

In this chapter, we've taken a very progressive approach toward restructuring the explanations and even the concepts of GD&T. We have solidified terminology, and stripped away redundancy. We've tried to take each principle to its logical conclusion, filling holes along the way and leaving no ambiguities. As you become more familiar with the standards and this chapter, you'll become more aware of our emphasis on practices and methodologies consistent with state-of-the-art manufacturing and high-resolution metrology.

FAQ: *I notice Y14.5 explains one type of tolerance in a single paragraph, but devotes pages and pages to another type. Does that suggest how frequently each should be used?*

A: No. There are some exotic principles that Y14.5 tries to downplay with scant coverage, but mostly, budgeting is based on a principle's complexity. That's particularly true of this handbook. We couldn't get by with a brief and vague explanation of a difficult concept just because it doesn't come up very often. Other supposed indicators, such as what questions show up on the Certification of GD&T Professionals exam, might be equally unreliable. Throughout this chapter, we'll share our preferences for which types of feature controls to use in various applications.

FAQ: *A drawing checker rejected one of my drawings because I used a composite feature control frame having three stacked segments. Is it OK to create GD&T applications not shown in Y14.5?*

A: Yes. Since the standards can neither discuss nor illustrate every imaginable application of GD&T, questions often arise as to whether or not a particular application, such as that shown in Fig. 5-127, is proper. Just as in matters of law, some of these questions can confound the experts. Clearly, if an illustration in the standard bears an uncanny resemblance to your own part, you'll be on pretty solid ground in copying that application. Just as often, however, the standard makes no mention of your specific application. You are allowed to take the explicit rules and principles and extend them to your application in any way that's consistent with all the rules and principles stated in the standard. Or, more simply, any application that doesn't

5-4   Chapter Five

> violate anything in the standard is acceptable. That's good news for a master practitioner who's familiar with the whole standard. Throughout this chapter we'll try to help novices by including "extension of principle" advice where it's appropriate.
>
> FAQ: *I've found what seem to be discrepancies between Y14.5 and the Math Standard. How can that be? Which standard supersedes?*
>
> A: You're right. There are a couple of direct contradictions between the two standards. Like any contemporary "living" language, GD&T is constantly evolving to keep pace with our modern world and is consequently imperfect. For instance, Y14.5 has 232 pages while the Math Standard has just 82. You could scarcely expect them to cover the same material in perfect harmony. Yet there's no clue in either document as to which one supersedes (they were issued only eight days apart). Where such questions arise, we'll discuss the issues and offer our preference.

5.1.3   Why Do We Use GD&T?

When several people work with a part, it's important they all reckon part dimensions the same. In Fig. 5-1, the designer specifies the distance to a hole's ideal location; the manufacturer measures off this distance and ("X marks the spot") drills a hole; then an inspector measures the actual distance to that hole. All three parties must be in perfect agreement about three things: from where to start the measurement, what direction to go, and where the measurement ends.

As illustrated in Chapter 3, when measurements must be precise to the thousandth of an inch, the slightest difference in the origin or direction can spell the difference between a usable part and an expensive paperweight. Moreover, even if all parties agree to measure to the hole's center, a crooked, bowed, or egg-shaped hole presents a variety of "centers." Each center is defensible based on a different design consideration. GD&T provides the tools and rules to assure that all users will reckon each dimension the same, with perfect agreement as to origin, direction, and destination.

It's customary for GD&T textbooks to spin long-winded yarns explaining how GD&T affords more tolerance for manufacturing. By itself, it doesn't. GD&T affords however much or little tolerance the designer specifies. Just as ubiquitous is the claim that using GD&T saves money, but these claims are never accompanied by cost or Return on Investment (ROI) analyses. A much more fundamental reason for

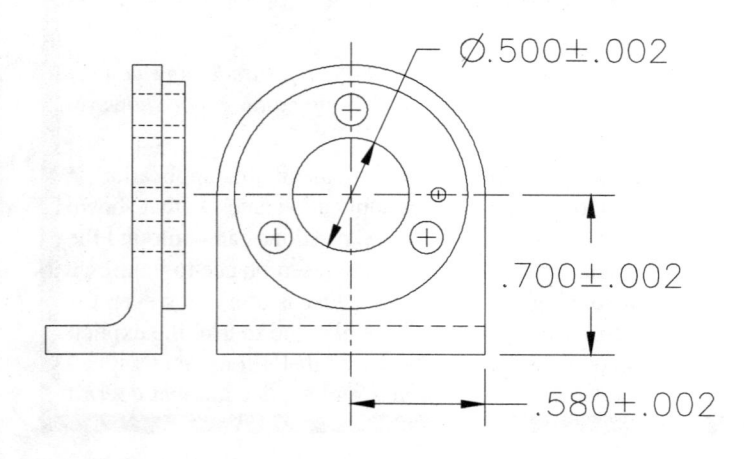

Figure 5-1 Drawing showing distance to ideal hole location

using GD&T is revealed in the following study of how two very different builders approach constructing a house.

A primitive builder might start by walking around the perimeter of the house, dragging a stick in the dirt to mark where walls will be. Next, he'll lay some long boards along the lines on the uneven ground. Then, he'll attach some vertical boards of varying lengths to the foundation. Before long, he'll have a framework erected, but it will be uneven, crooked, and wavy. Next, he'll start tying or tacking palm branches, pieces of corrugated aluminum, or discarded pieces of plywood to the crude frame. He'll overlap the edges of these flexible sidings 1-6 inches and everything will fit just fine. Before long, he'll have the serviceable shanty shown in Fig. 5-2, but with some definite limitations: no amenities such as windows, plumbing, electricity, heating, or air conditioning.

Figure 5-2 House built without all of the appropriate tools

A house having such modern conveniences as glass windows and satisfying safety codes requires more careful planning. Materials will have to be stronger and more rigid. Spaces inside walls will have to be provided to fit structural members, pipes, and ducts.

To build a house like the one shown in Fig. 5-3, a modern contractor begins by leveling the ground where the house will stand. Then a concrete slab or foundation is poured. The contractor will make the slab as level and flat as possible, with straight, parallel sides and square corners. He will select the straightest wooden plates, studs, headers, and joists available for framing and cut them to precisely uniform lengths. Then he'll use a large carpenter's square, level, and plumb bob to make each frame member parallel or perpendicular to the slab.

Why are such precision and squareness so important? Because it allows him to make accurate measurements of his work. Only by making accurate measurements can he assure that such prefabricated

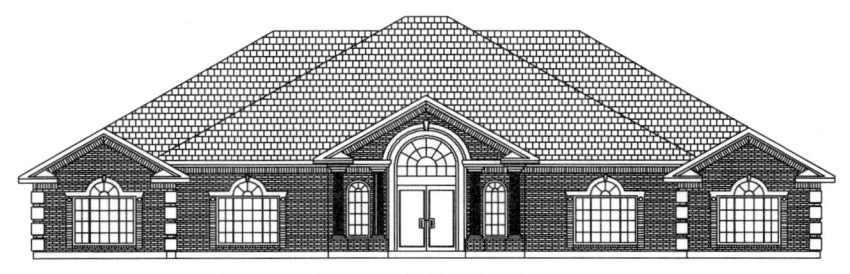

Figure 5-3 House built using the correct tools

5-6　Chapter Five

items as Sheetrock, windows, bathtubs, and air conditioning ducts will fit in the spaces between his frame members. Good fits are important to conserve space and money. It also means that when electrical outlet boxes are nailed to the studs 12" up from the slab, they will all appear parallel and neatly aligned. Remember that it all derives from the flatness and squareness of the slab.

By now, readers with some prior knowledge of GD&T have made the connection: The house's concrete slab is its "primary datum." The slab's edges complete the "datum reference frame." The wooden framing corresponds to "tolerance zones" and "boundaries" that must contain "features" such as pipes, ducts, and windows.

Clearly, the need for precise form and orientation in the slab and framing of a house is driven by the fixtures to be used and how precisely they must fit into the framing. Likewise, the need for GD&T on a part is driven by the types and functions of its features, and how precisely they must relate to each other and/ or fit with mating features of other parts in the assembly. The more complex the assembly and the tighter the fits, the greater are the role and advantages of GD&T.

Fig. 5-4 shows a non-GD&T drawing of an automobile wheel rotor. Despite its neat and uniform appearance, the drawing leaves many relationships between part features totally out of control. For example, what if it were important that the Ø5.50 bore be perpendicular to the mounting face? Nothing on the drawing addresses that. What if it were critical that the Ø5.50 bore and the Ø11.00 OD be on the same axis? Nothing on the drawing requires that either. In fact, Fig. 5-5 shows the "shanty" that could be built. Although all its dimensions are within their tolerances, it seems improbable that any "fixtures" could fit it.

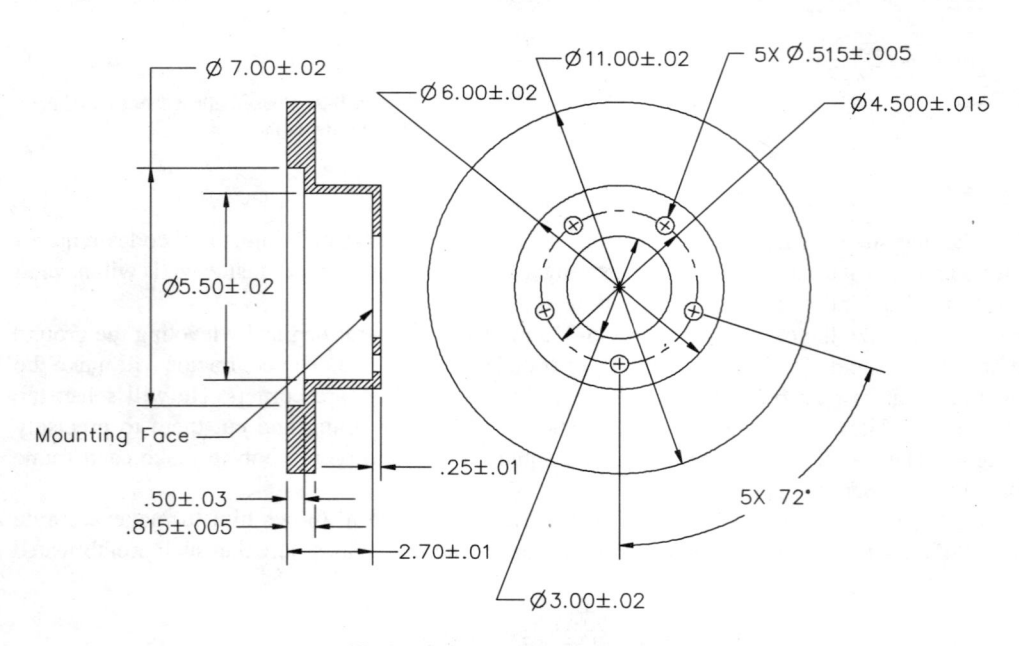

Figure 5-4 Drawing that does not use GD&T

In Fig. 5-6, we've applied GD&T controls to the same design. We've required the mounting face to be flat within .005 and then labeled it datum feature A. That makes it an excellent "slab" from which we can launch the rest of the part. Another critical face is explicitly required to be parallel to A within .003. The perpendicularity of the Ø5.50 bore is directly controlled to our foundation, A. Now the Ø5.50 bore can be labeled datum feature B and provide an unambiguous origin—a sturdy "center post"—from which the Ø.515 bolt holes and other round features are located. Datum features A and B provide a very uniform and well-aligned framework from which a variety of relationships and fits can be precisely controlled. Just as

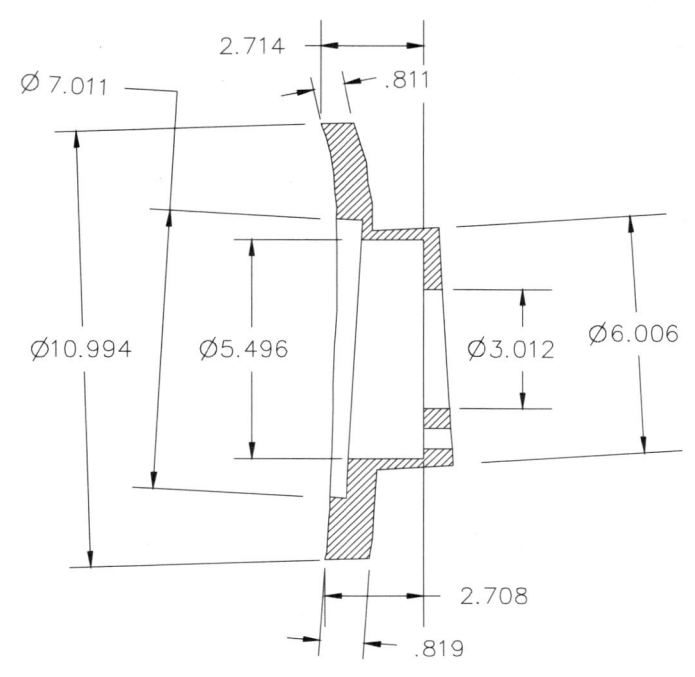

Figure 5-5 Manufactured part that conforms to the drawing without GD&T (Fig. 5-4)

importantly, GD&T provides unique, unambiguous meanings for each control, precluding each person's having his own competing interpretation. GD&T, then, is simply a means of controlling surfaces more precisely and unambiguously.

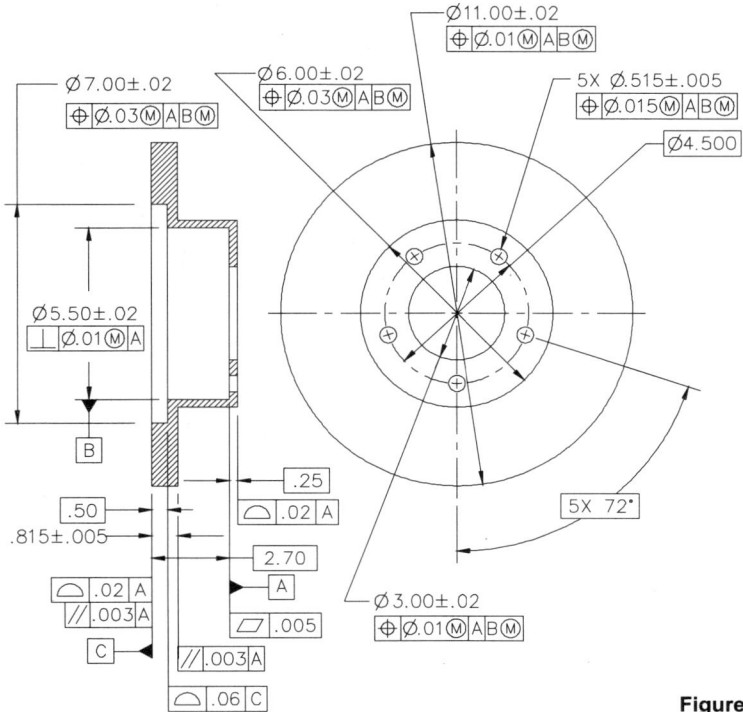

Figure 5-6 Drawing that uses GD&T

And that's the fundamental reason for using GD&T. It's the universal language throughout the world for communicating engineering design specifications. Clear communication assures that manufactured parts will function and that functional parts won't later be rejected due to some misunderstanding. Fewer arguments. Less waste.

As far as that ROI analysis, most of the costs GD&T reduces are hidden, including the following:

- Programmers wasting time trying to interpret drawings and questioning the designers
- Rework of manufactured parts due to misunderstandings
- Inspectors spinning their wheels, deriving meaningless data from parts while failing to check critical relationships
- Handling and documentation of functional parts that are rejected
- Sorting, reworking, filing, shimming, etc., of parts in assembly, often in added operations
- Assemblies failing to operate, failure analysis, quality problems, customer complaints, loss of market share and customer loyalty
- The meetings, corrective actions, debates, drawing changes, and interdepartmental vendettas that result from each of the above failures

It all adds up to an enormous, yet unaccounted cost. Bottom line: use GD&T because it's the right thing to do, it's what people all over the world understand, and it saves money.

5.1.4 When Do We Use GD&T?

In the absence of GD&T specifications, a part's ability to satisfy design requirements depends largely on the following four "laws."

1. Pride in workmanship. Every industry has unwritten customary standards of product quality, and most workers strive to achieve them. But these standards are mainly minimal requirements, usually pertaining to cosmetic attributes. Further, workmanship customs of precision aerospace machinists are probably not shared by ironworkers.
2. Common sense. Experienced manufacturers develop a fairly reliable sense for what a part is supposed to do. Even without adequate specifications, a manufacturer will try to make a bore very straight and smooth, for example, if he suspects it's for a hydraulic cylinder.
3. Probability. Sales literature for modern machining centers often specifies repeatability within 2 microns (.00008"). Thus, the running gag in precision manufacturing is that part dimensions should never vary more than that. While the performance of a process can usually be predicted statistically, there are always "special causes" that introduce surprise variations. Further, there's no way to predict what processes might be used, how many, and in what sequence to manufacture a part.
4. Title block, workmanship, or contractual ("boiler plate") standards. Sometimes these provide clarification, but often, they're World War II vintage and inadequate for modern high-precision designs. An example is the common title block note, "All diameters to be concentric within .005."

Dependence on these four "laws" carries obvious risks. Where a designer deems the risks too high, specifications should be rigorously spelled out with GD&T.

FAQ: *Should I use GD&T on every drawing?*

A: Some very simple parts, such as a straight dowel, flat washer, or hex nut may not need GD&T. For such simple parts, Rule #1 (explained in section 5.6.3.1), which pertains to size limits, may provide adequate control by itself. However, some practitioners always use GD&T positional tolerancing for holes and width-type features (slots and tabs). It depends primarily on how much risk there is of a part being made, such as that shown in Fig. 5-5, which conforms to all the non-GD&T tolerances but is nevertheless unusable.

FAQ: *Can I use GD&T for just one or two selected surfaces on a drawing, or is it "all or nothing?"*

A: On any single drawing you can mix and match all the dimensioning and tolerancing methods in Y14.5. For example, one pattern of holes may be controlled with composite positional tolerance while other patterns may be shown using coordinate dimensions with plus and minus tolerances. Again, it depends on the level of control needed. But, if you choose GD&T for any individual feature or pattern of features, you must give that feature the full treatment. For example, you shouldn't dimension a hole with positional tolerance in the X-axis, and plus and minus tolerance in the Y-axis. Be consistent. Also, it's a good idea to control the form and orientational relationships of surfaces you're using as datum features.

FAQ: *Could GD&T be used on the drawings for a house?*

A: Hmmm. Which do you need, shanty or chateau?

5.1.5 How Does GD&T Work?—Overview

In the foregoing paragraphs, we alluded to the goal of GD&T: to guide all parties toward reckoning part dimensions the same, including the origin, direction, and destination for each measurement. GD&T achieves this goal through four simple and obvious steps.

1. Identify part surfaces to serve as origins and provide specific rules explaining how these surfaces establish the starting point and direction for measurements.
2. Convey the nominal (ideal) distances and orientations from origins to other surfaces.
3. Establish boundaries and/or tolerance zones for specific attributes of each surface along with specific rules for conformance.
4. Allow dynamic interaction between tolerances (simulating actual assembly possibilities) where appropriate to maximize tolerances.

5.2 Part Features

Up to this point, we've used the terms *surface* and *feature* loosely and almost interchangeably. To speak GD&T, however, we must begin to use the vocabulary as Y14.5 does.

> *Feature* is **the general term applied to a physical portion of a part, such as a surface, pin, tab, hole, or slot.**

Usually, a part feature is a single surface (or a pair of opposed parallel plane surfaces) having uniform shape. You can establish datums from, and apply GD&T controls to features only. The definition implies that no feature exists until a part is actually produced. There are two general types of features: those that have a built-in dimension of "size," and those that don't.

5-10 Chapter Five

> FAQ: *Is a center line a feature?*
>
> A: No, since a center line or center plane can never be a physical portion of a part.
>
> FAQ: *Well, what about a nick or a burr? They're "physical portions of a part," right?*
>
> A: True, but Y14.5 doesn't mean to include nicks and burrs as features. That's why we've added "having uniform shape" to our own description.
>
> FAQ: *With transitions at tangent radii or slight angles, how can I tell exactly where one feature ends and the adjacent feature begins?*
>
> A: You can't. The Math Standard points out, "Generally, features are well defined only in drawings and computer models." Therefore, you are free to reckon the border between features at any single location that satisfies all pertinent tolerances.

5.2.1 Nonsize Features

A *nonsize feature* is a surface having no unique or intrinsic size (diameter or width) dimension to measure. Nonsize features include the following:

- A nominally flat planar surface
- An irregular or "warped" planar surface, such as the face of a windshield or airfoil
- A *radius*—a portion of a cylindrical surface encompassing less than 180° of arc length
- A *spherical radius*—a portion of a spherical surface encompassing less than 180° of arc length
- A *revolute*—a surface, such as a cone, generated by revolving a spine about an axis

5.2.2 Features of Size

A *feature of size* is **one cylindrical or spherical surface, or a set of two opposed elements or opposed parallel surfaces, associated with a size dimension.**

A feature of size has opposing points that partly or completely enclose a space, giving the feature an intrinsic dimension—size—that can be measured apart from other features. Holes are "internal" features of size and pins are "external" features of size. Features of size are subject to the principles of material condition modifiers, as we'll explain in section 5.6.2.1.

"Opposed parallel surfaces" means the surfaces are designed to be parallel to each other. To qualify as "opposed," it must be possible to construct a perpendicular line intersecting both surfaces. Only then, can we make a meaningful measurement of the size between them. From now on, we'll call this type of feature a *width-type feature*.

> FAQ: *Where a bore is bisected by a groove, is the bore still considered a single feature of size, or are there two distinct bores?*
>
> A: A similar question arises wherever a boss, slot, groove, flange, or step separates any two otherwise continuous surfaces. A specification preceded by **2X** clearly denotes two distinct features. Conversely, Y14.5 provides no symbol for linking interrupted surfaces. For example, an extension line that connects two surfaces by bridging across an interruption has no standardized meaning. Where a single feature control shall apply to all portions of an interrupted surface, a note, such as **TWO SURFACES AS A SINGLE FEATURE,** should accompany the specification.

5.2.2.1 Screw Threads

A screw thread is a group of complex helical surfaces that can't directly be reckoned with as a feature of size. However, the abstract *pitch cylinder* derived from the thread's flanks best represents the thread's functional axis in most assemblies. Therefore, by default, the pitch cylinder "stands in" for the thread as a datum feature of size and/or as a feature of size to be controlled with an orientation or positional tolerance. The designer may add a notation specifying a different abstract feature of the thread (such as MAJOR DIA, or MINOR DIA). This notation is placed beneath the feature control frame or beneath or adjacent to the "datum feature" symbol, as applicable.

> FAQ: *For a tapped hole, isn't it simpler just to specify the minor diameter?*
>
> A: Simpler, yes. But it's usually a mistake, because the pitch cylinder can be quite skewed to the minor diameter. The fastener, of course, will tend to align itself to the pitch cylinder. We've seen projected tolerance zone applications where parts would not assemble despite the minor diameters easily conforming to the applicable positional tolerances.

5.2.2.2 Gears and Splines

Gears and splines, like screw threads, need a "stand in" feature of size. But because their configurations and applications are so varied, there's no default for gears and splines. In every case, the designer shall add a notation specifying an abstract feature of the gear or spline (such as MAJOR DIA, PITCH DIA, or MINOR DIA). This notation is placed beneath the feature control frame or beneath the "datum feature" symbol, as applicable.

5.2.3 Bounded Features

There is a type of feature that's neither a sphere, cylinder, nor width-type feature, yet clearly has "a set of two opposed elements." The D-hole shown in Fig. 5-70, for example, is called an "irregular feature of size" by some drafting manuals, while Y14.5's own coverage for this type of feature is very limited. Although the feature has obvious MMC and LMC boundaries, it's arguable whether the feature is "associated with a size dimension." We'll call this type of feature a *bounded feature*, and consider it a nonsize feature for our purposes. However, like features of size, bounded features are also subject to the principles of material condition modifiers, as we'll explain in section 5.6.2.1.

5.3 Symbols

In section 5.1, we touched on some of the shortcomings of English as a design specification language. Fig. 5-7 shows an attempt to control part features using mostly English. Compare that with Fig. 5-6, where GD&T symbols are used instead. Symbols are better, because of the following reasons:

- Anyone, regardless of his or her native tongue, can read and write symbols.
- Symbols mean exactly the same thing to everyone.
- Symbols are so compact they can be placed close to where they apply, and they reduce clutter.
- Symbols are quicker to draw and easier for computers to draw automatically.
- Symbols are easier to spot visually. For example, in Figs. 5-6 and 5-7, find all the positional callouts.

5-12   Chapter Five

- Feature has a cylindrical tolerance zone whose size is ⌀.03 at maximum material condition and is positioned to datum A primary and datum B secondary at maximum material condition.

- Perpendicular to datum A within a cylindrical tolerance zone of ⌀.01 at maximum material condition.   Datum feature B.

- These features are located within a cylindrical tolerance zone of ⌀.01 at maximum material condition to datum A primary and datum B secondary at maximum material condition.

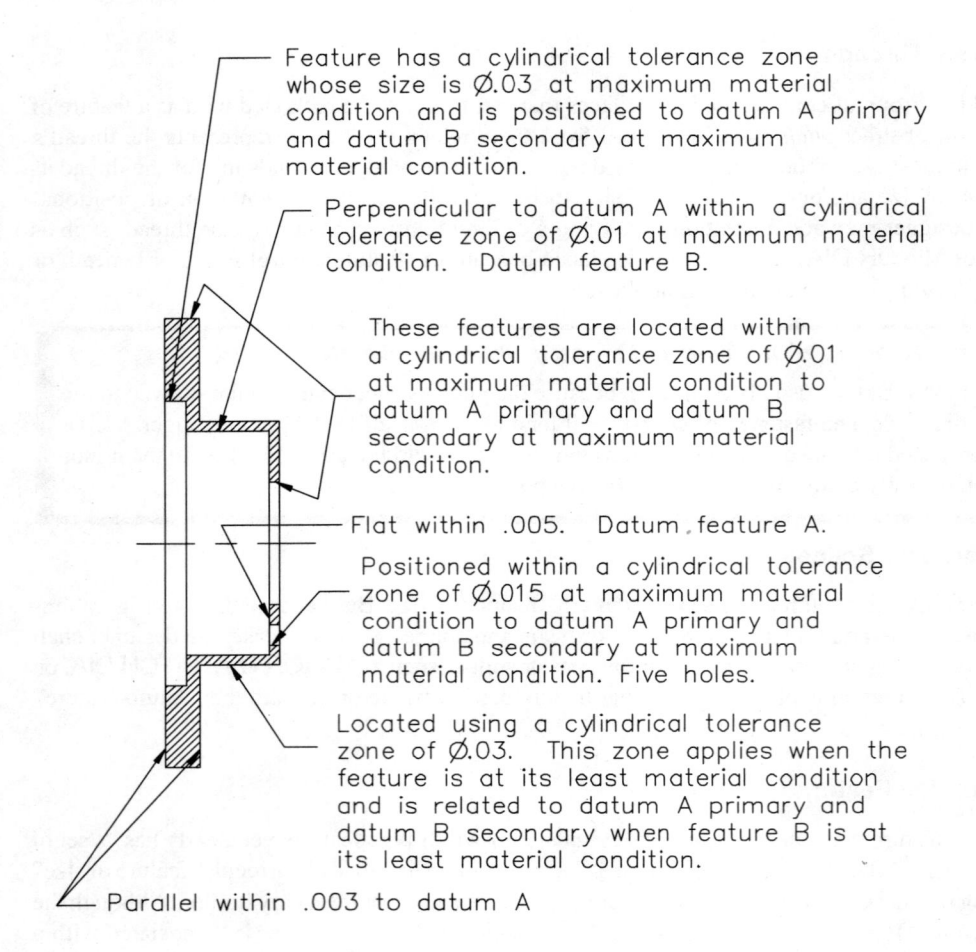

- Flat within .005.   Datum feature A.

- Positioned within a cylindrical tolerance zone of ⌀.015 at maximum material condition to datum A primary and datum B secondary at maximum material condition. Five holes.

- Located using a cylindrical tolerance zone of ⌀.03.   This zone applies when the feature is at its least material condition and is related to datum A primary and datum B secondary when feature B is at its least material condition.

- Parallel within .003 to datum A

Figure 5-7 Using English to control part features

In the following sections, we'll explain the applications and meanings for each GD&T symbol. Unfortunately, the process of replacing traditional words with symbols is ongoing and complicated, requiring coordination among various national and international committees. In several contexts, Y14.5 suggests adding various English-language notes to a drawing to clarify design requirements. However, a designer should avoid notes specifying methods for manufacture or inspection.

5.3.1   Form and Proportions of Symbols

Fig. 5-8 shows each of the symbols used in dimensioning and tolerancing. We have added dimensions to the symbols themselves, to show how they are properly drawn. Each linear dimension is expressed as a multiple of h, a variable equal to the letter height used on the drawing. For example, if letters are drawn .12" high, then $h = .12"$ and $2h = .24"$. It's important to draw the symbols correctly, because to many drawing users, that attention to detail indicates the draftsman's (or programmer's) overall command of the language.

Geometric Dimensioning and Tolerancing   5-13

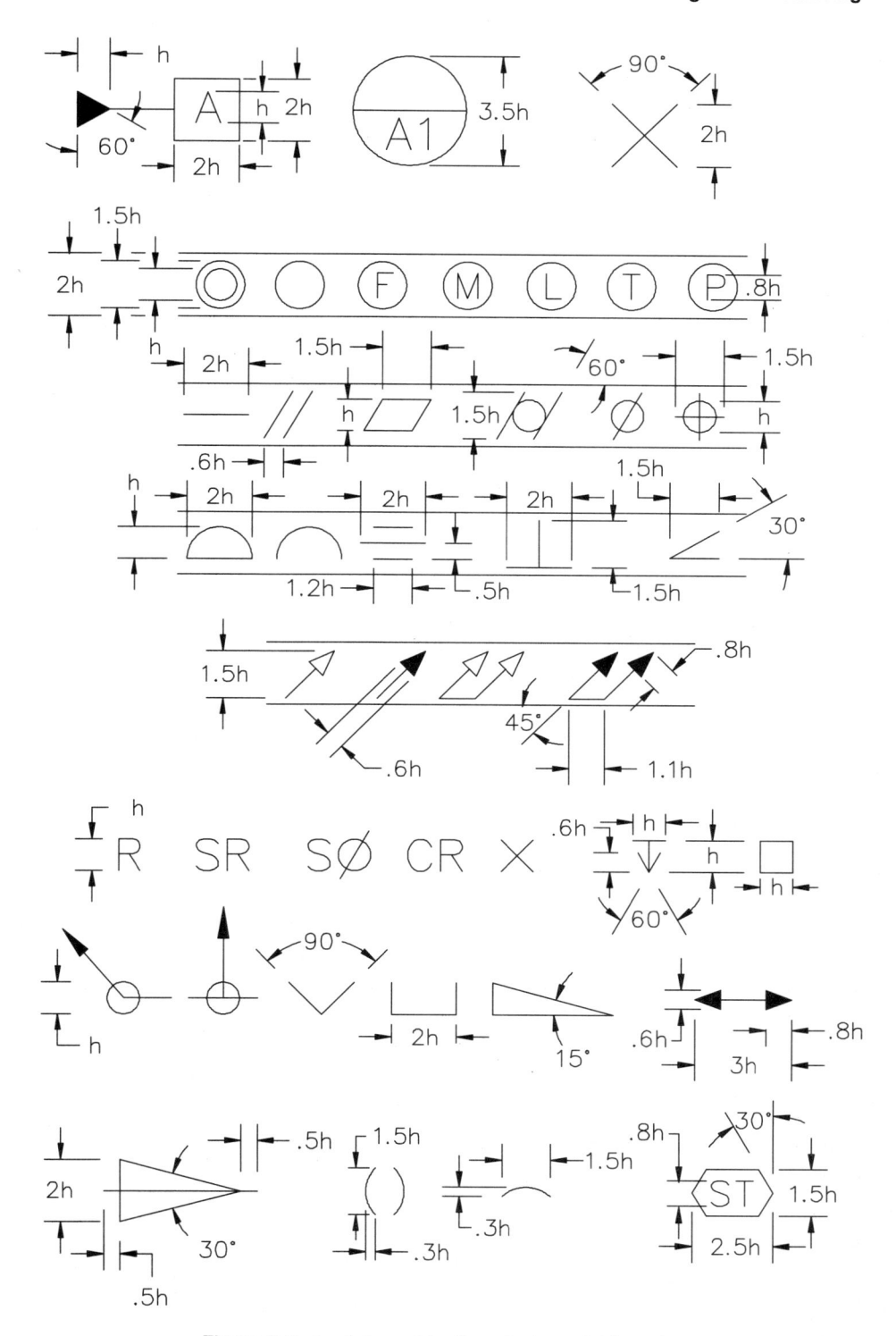

Figure 5-8 Symbols used in dimensioning and tolerancing

5.3.2 Feature Control Frame

Each geometric control for a feature is conveyed on the drawing by a rectangular sign called a *feature control frame*. As Fig. 5-9 shows, the feature control frame is divided into compartments expressing the following, sequentially from left to right.

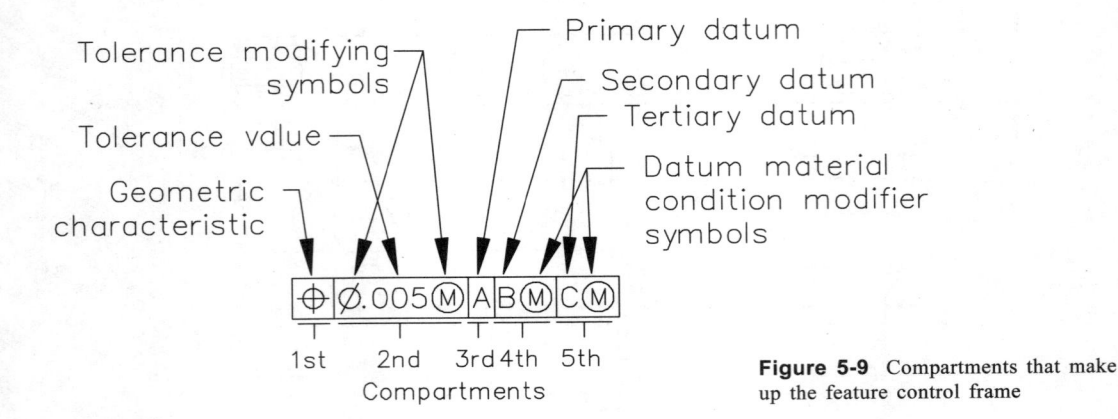

Figure 5-9 Compartments that make up the feature control frame

The **1**st compartment contains a *geometric characteristic symbol* specifying the type of geometric control. Table 5-1 shows the 14 available symbols.

The **2**nd compartment contains the geometric tolerance value. Many of the *modifying symbols* in Table 5-2 can appear in this compartment with the tolerance value, adding special attributes to the geometric control. For instance, where the tolerance boundary or zone is cylindrical, the tolerance value is preceded by the "diameter" symbol, Ø. Preceding the tolerance value with the "SØ" symbol denotes a spherical boundary or zone. Other optional modifying symbols, such as the "statistical tolerance" symbol, may follow the tolerance value.

The **3**rd, **4**th, and **5**th compartments are each added only as needed to contain (sequentially) the primary, secondary, and tertiary datum references, each of which may be followed by a material condition modifier symbol as appropriate.

Thus, each feature control frame displays most of the information necessary to control a single geometric characteristic of the subject feature. Only basic dimensions (described in section 5.3.3) are left out of the feature control frame.

5.3.2.1 Feature Control Frame Placement

Fig. 5-10(a) through (d) shows four different methods for attaching a feature control frame to its feature.
(a) Place the frame below or attached to a leader-directed callout or dimension pertaining to the feature.
(b) Run a leader from the frame to the feature.
(c) Attach either side or either end of the frame to an extension line from the feature, provided it is a plane surface.
(d) Attach either side or either end of the frame to an extension of the dimension line pertaining to a feature of size.

Geometric Dimensioning and Tolerancing   5-15

Table 5-1 summarizes the application options and rules for each of the 14 types of geometric tolerances. For each type of tolerance applied to each type of feature, the table lists the allowable "feature control frame placement options." Multiple options, such as "a" and "d," appearing in the same box yield identical results. Notice, however, that for some tolerances, the type of control depends on the feature control frame placement. For a straightness tolerance applied to a cylindrical feature, for instance, placement "b" controls surface elements, while placements "a" or "d" control the derived median line.

Table 5-1 Geometric characteristics and their attributes

CHARACTERISTIC	SYMBOL	TYPE OF FEATURE CONTROLLED	FEATURE CONTROL FRAME PLACEMENT OPTIONS (SEE LEGEND)	BOUNDARY/TOL ZONE SHAPE MODIFIER	TOLERANCE MODIFIABLE TO MMC OR LMC	NUMBER OF DATUM REFERENCES ALLOWED	MMC/LMC ALLOWED FOR DATUM REFERENCE(S)	BASIC DIMENSIONS REQD
STRAIGHTNESS	—	CYL–SURFACE ELEMENTS	b			0		
		CYL–DERIVED MEDIAN LINE	a, d	Ø	✓	0		
		PLANE–LINE ELEMENTS	b, c			0, 1		
FLATNESS	▱	PLANE	b, c			0		
		WIDTH–DERIVED MEDIAN PLANE	a, d		✓	0		
CIRCULARITY	○	REVOLUTE, SPHERE	a, b, d			0		
CYLINDRICITY	⌭	CYLINDER	a, b, d			0		
PROFILE OF A LINE	⌒	ALL	b			0–3	✓	✓
PROFILE OF A SURFACE	⌓	REVOLUTE	b			0–3	✓	✓
		OTHER (NON-REVOLUTE)	b			0–3	✓	✓
		COPLANARITY OF PLANES	b			0		
PERPENDICULARITY	⊥	PLANE (INCL LINE ELEMENTS)	b, c			1–3	✓	
PARALLELISM	∥	CYLINDER	a, d	Ø	✓	1–3	✓	
		WIDTH	a, d		✓	1–3	✓	
		REVOLUTE–RADIAL ELEMENT	b, c			1–3	✓	
ANGULARITY	∠	PLANE (INCL LINE ELEMENTS)	b, c			1–3	✓	✓
		CYLINDER	a, d	Ø	✓	1–3	✓	✓
		WIDTH	a, d		✓	1–3	✓	✓
		REVOLUTE–RADIAL ELEMENT	b, c			1–3	✓	✓
POSITION	⌖	CYLINDER	a, d	Ø	✓	1–3	✓	✓
		WIDTH	a, d		✓	1–3	✓	✓
		SPHERE	a, d	SØ	✓	1–3	✓	✓
CONCENTRICITY	◎	ALL NON–SPHERICAL	a, b, d	Ø		1–3		
		SPHERE	a, b, d	SØ		1–3		
SYMMETRY	⌯	OPPOSED POINTS	a, d			1–3		
CIRCULAR RUNOUT	↗	REVOLUTE	a, b, d			1–2		
TOTAL RUNOUT	↗↗	CYLINDER	a, b, d			1–2		
		PLANE PERP TO AXIS	b, c			1–2		

FEATURE CONTROL FRAME PLACEMENT OPTIONS (LEGEND)

(a) Place the frame below or attached to a leader-directed callout or dimension pertaining to the feature.

(b) Run a leader from the frame to the feature.

(c) Attach either side or either end of the frame to an extension line from the feature, provided it is a plane surface.

(d) Attach either side or either end of the frame to an extension of the dimension line pertaining to a feature of size.

Table 5-2 Modifying symbols

Characteristic	Symbol
At maximum material condition	Ⓜ
At least material condition	Ⓛ
Projected tolerance zone	Ⓟ
Free state	Ⓕ
Tangent plane	Ⓣ
Diameter	⌀
Spherical diameter	S⌀
Radius	R
Spherical radius	SR
Controlled radius	CR
Reference	( )
Arc length	⌒
Statistical tolerance	⟨ST⟩
Between	↔

5.3.2.2 Reading a Feature Control Frame

It's easy to translate a feature control frame into English and read it aloud from left to right. Tables 5-1 and 5-2 show equivalent English words to the left of each symbol. Then, we just add the following English-language preface for each compartment:

    1st compartment—"*The...*"
    2nd compartment—"*...of this feature shall be within...*"
    3rd compartment—"*...to primary datum...*"
    4th compartment—"*...and to secondary datum...*"
    5th compartment—"*...and to tertiary datum...*"

Now, read along with us Fig. 5-9's feature control frame. "*The* position *of this feature shall be within* diameter .005 at maximum material condition *to primary datum* A *and to secondary datum* B at maximum material condition *and to tertiary datum* C at maximum material condition." Easy.

Geometric Dimensioning and Tolerancing 5-17

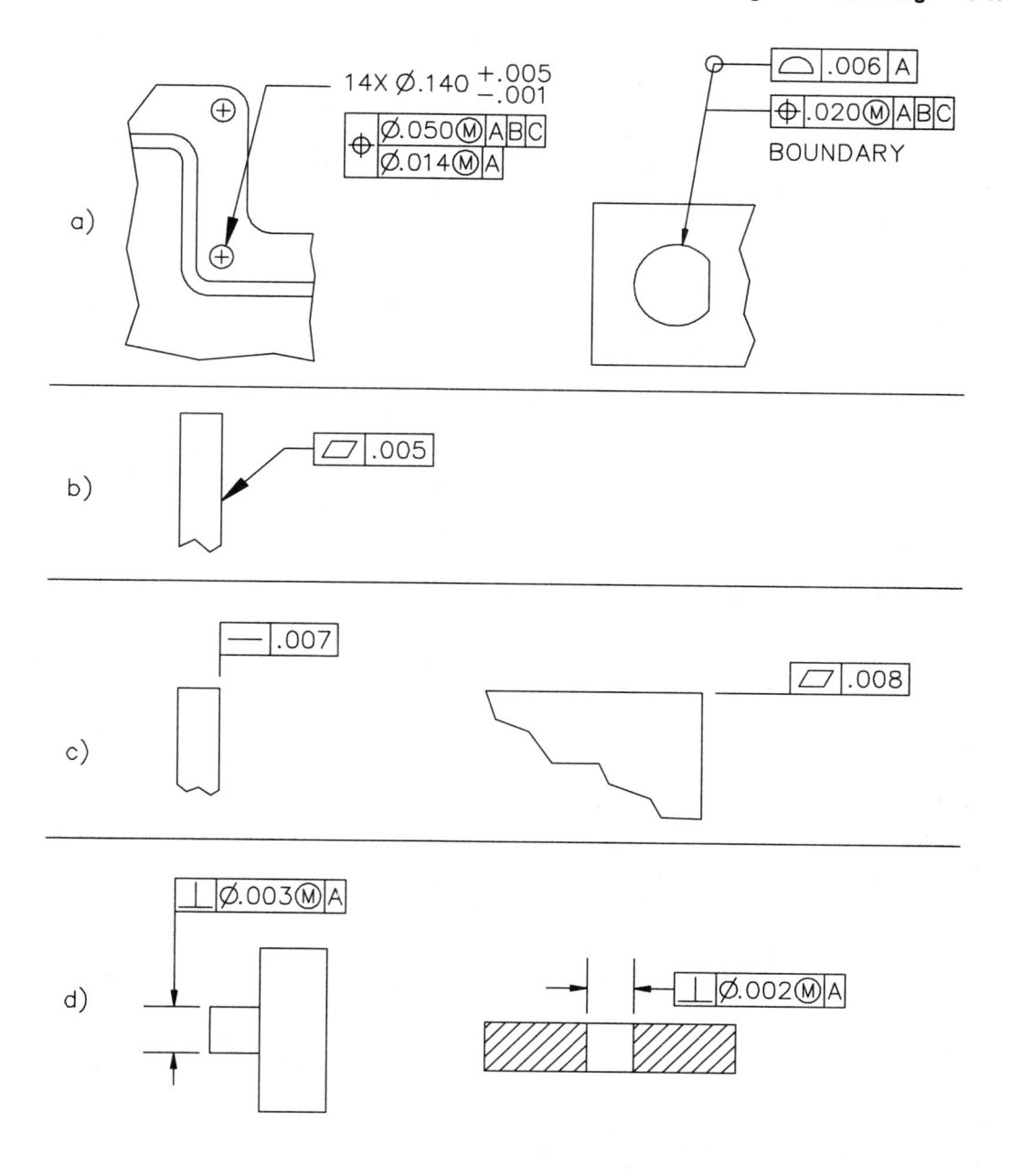

Figure 5-10 Methods of attaching feature control frames

5.3.3 Basic Dimensions

A *basic dimension* is **a numerical value used to describe the theoretically exact size, profile, orientation, or location of a feature or datum target.** The value is usually enclosed in a rectangular frame, as shown in

Figure 5-11 Method of identifying a basic .875 dimension

Fig. 5-11. Permissible variation from the basic value is specified in feature control frames, notes, or in other toleranced dimensions.

5.3.4 Reference Dimensions and Data

A *reference dimension* is a dimension, usually without tolerance, used for information only. On a drawing, a dimension (or other data) is designated as "reference" by enclosing it in parentheses. In written notes, however, parentheses retain their more common grammatical interpretation unless otherwise specified. Where a basic dimension is shown as a reference, enclosure in the "basic dimension frame" is optional. Although superfluous data and advice should be minimized on a drawing, a well-placed reference dimension can prevent confusion and time wasted by a user trying to decipher a relationship between features. Reference data shall either repeat or derive from specifications expressed elsewhere on the drawing or in a related document. However, the reference data itself shall have no bearing on part conformance.

5.3.5 "Square" Symbol

A square shape can be dimensioned using a single dimension preceded (with no space) by the "square" symbol shown in Fig. 5-47. The symbol imposes size limits and Rule #1 between each pair of opposite sides. (See section 5.6.3.1.) However, perpendicularity between adjacent sides is merely implied. Thus, the "square" symbol yields no more constraint than if 2X preceded the dimension.

5.3.6 Tabulated Tolerances

Where the tolerance in a feature control frame is tabulated either elsewhere on the drawing or in a related document, a representative letter is substituted in the feature control frame, preceded by the abbreviation TOL. See Figs. 5-116 and 5-117.

5.3.7 "Statistical Tolerance" Symbol

Chapters 8 and 10 explain how a *statistical tolerance* can be calculated using statistical process control (SPC) methods. Each tolerance value so calculated shall be followed by the "statistical tolerance" symbol shown in Fig. 5-12. In a feature control frame, the symbol follows the tolerance value and any applicable modifier(s). In addition, a note shall be placed on the drawing requiring statistical control of all such tolerances. Chapter 11 explains the note in greater detail and Chapter 24 shows several applications.

Figure 5-12 "Statistical tolerance" symbol

5.4 Fundamental Rules

Before we delve into the detailed applications and meanings for geometric tolerances, we need to understand a few fundamental ground rules that apply to every engineering drawing, regardless of the types of tolerances used.

(a) Each dimension shall have a tolerance, except for those dimensions specifically identified as reference, maximum, minimum, or stock (commercial stock size). The tolerance may be applied directly to the dimension (or indirectly in the case of basic dimensions), indicated by a general note, or located in a supplementary block of the drawing format. See ANSI Y14.1.

(b) Dimensioning and tolerancing shall be complete so there is full understanding of the characteristics of each feature. Neither scaling (measuring the size of a feature directly from an engineering drawing) nor assumption of a distance or size is permitted, except as follows: Undimensioned drawings, such as loft, printed wiring, templates, and master layouts prepared on stable material, are excluded provided the necessary control dimensions are specified.

(c) Each necessary dimension of an end product shall be shown. No more dimensions than those necessary for complete definition shall be given. The use of reference dimensions on a drawing should be minimized.

(d) Dimensions shall be selected and arranged to suit the function and mating relationship of a part and shall not be subject to more than one interpretation.

(e) The drawing should define a part without specifying manufacturing methods. Thus, only the diameter of a hole is given without indicating whether it is to be drilled, reamed, punched, or made by any other operation. However, in those instances where manufacturing, processing, quality assurance, or environmental information is essential to the definition of engineering requirements, it shall be specified on the drawing or in a document referenced on the drawing.

(f) It is permissible to identify as nonmandatory certain processing dimensions that provide for finish allowance, shrink allowance, and other requirements, provided the final dimensions are given on the drawing. Nonmandatory processing dimensions shall be identified by an appropriate note, such as **NONMANDATORY (MFG DATA)**.

(g) Dimensions should be arranged to provide required information for optimum readability. Dimensions should be shown in true profile views and refer to visible outlines.

(h) Wires, cables, sheets, rods, and other materials manufactured to gage or code numbers shall be specified by linear dimensions indicating the diameter or thickness. Gage or code numbers may be shown in parentheses following the dimension.

(i) A 90° angle applies where center lines and lines depicting features are shown on a drawing at right angles and no angle is specified.

(j) A 90° basic angle applies where center lines of features in a pattern or surfaces shown at right angles on the drawing are located or defined by basic dimensions and no angle is specified.

(k) Unless otherwise specified, all dimensions are applicable at 20°C (68°F). Compensation may be made for measurements made at other temperatures.

(l) All dimensions and tolerances apply in a free state condition. This principle does not apply to nonrigid parts as defined in section 5.5.

(m) Unless otherwise specified, all geometric tolerances apply for full depth, length, and width of the feature.

(n) Dimensions and tolerances apply only at the drawing level where they are specified. A dimension specified for a given feature on one level of drawing, (for example, a detail drawing) is not mandatory for that feature at any other level (for example, an assembly drawing).

5.5 Nonrigid Parts

A *nonrigid part* is a part that can have different dimensions while restrained in assembly than while relaxed in its "free state." Rubber, plastic, or thin-wall parts may be obviously nonrigid. Other parts might reveal themselves as nonrigid only after assembly or functioning forces are applied. That's why the exemption of "nonrigid parts" from Fundamental Rule *(l)* is meaningless. Instead, the rule must be interpreted as applying to all parts and meaning, "Unless otherwise specified, all dimensions and tolerances apply in a free state condition." Thus, a designer must take extra care to assure that a suspected nonrigid part will have proper dimensions while assembled and functioning. To do so, one or more tolerances may be designated to apply while the part is restrained in a way that simulates, as closely as practicable, the restraining forces exerted in the part's assembly and/or functioning.

5.5.1 Specifying Restraint

A nonrigid part might conform to all tolerances only in the free state, only in the restrained state, in both states, or in neither state. Where a part, such as a rubber grommet, may or may not need the help of restraint for conformance, the designer may specify optional restraint. This allows all samples to be inspected in their free states. Parts that pass are accepted. Those that fail may be reinspected—this time, while restrained. Where there is a risk that restraint could introduce unacceptable distortion, the designer should specify mandatory restraint instead.

Restraint may be specified by a note such as UNLESS OTHERWISE SPECIFIED, ALL DIMENSIONS AND TOLERANCES MAY (or SHALL) APPLY IN A RESTRAINED CONDITION. Alternatively, the note may be directed only to certain dimensions with flags and modified accordingly. The note shall always include (or reference a document that includes) detailed instructions for restraining the part. A typical note, like that shown in Fig. 5-134, identifies one or two functional datum features (themselves nonrigid) to be clamped into some type of gage or fixture. The note should spell out any specific clamps, fasteners, torques, and other forces deemed necessary to simulate expected assembly conditions.

5.5.2 Singling Out a Free State Tolerance

Even where restraint is specified globally on a drawing, a geometric tolerance can be singled out to apply only in the free state. Where the "free state" symbol follows a tolerance (and its modifiers), the tolerance shall be verified with no external restraining forces applied. See section 5.8.7 and Fig. 5-45 for an example.

5.6 Features of Size—The Four Fundamental Levels of Control

Four different levels of GD&T control can apply to a feature of size. Each higher-level tolerance adds a degree of constraint demanded by the feature's functional requirements. However, all lower-level controls remain in effect. Thus, a single feature can be subject to many tolerances simultaneously.

Level 1: Controls size and (for cylinders or spheres) circularity at each cross section only.
Level 2: Adds overall form control.
Level 3: Adds orientation control.
Level 4: Adds location control.

5.6.1 Level 1—Size Limit Boundaries

For every feature of size, the designer shall specify the largest and the smallest the feature can be. In section 5.7, we discuss three different ways the designer can express these *size limits* (also called "limits of size") on the drawing. Here, we're concerned with the exact requirements these size limits impose on a feature. The Math Standard explains how specified size limits establish small and large *size limit boundaries* for the feature. The method may seem complicated at first, but it's really very simple.

It starts with a geometric element called a *spine*. The spine for a cylindrical feature is a simple (nonselfintersecting) curve in space. Think of it as a line that may be straight or wavy. Next, we take an imaginary solid ball whose diameter equals the small size limit of the cylindrical feature, and sweep its center along the spine. This generates a "wormlike" 3-dimensional (3-D) boundary for the feature's smallest size.

Fig. 5-13 illustrates the spine, the ball, and the 3-D boundary. Likewise, we may create a second spine, and sweep another ball whose diameter equals the large size limit of the cylindrical feature. This generates a second 3-D boundary, this time for the feature's largest size.

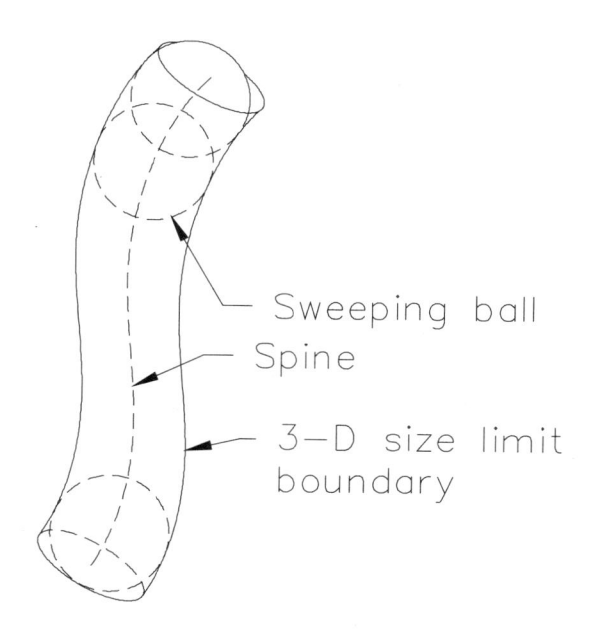

Figure 5-13 Generating a size limit boundary

As Fig. 5-14 shows, a cylindrical feature of size conforms to its size limits when its surface can contain the smaller boundary and be contained within the larger boundary. (The figure shows a hole, but the requirement applies to external features as well.) Under Level 1 control, the curvatures and relative locations of each spine may be adjusted as necessary to achieve the hierarchy of containments, except that the small size limit boundary shall be entirely contained within the large size limit boundary.

For a width-type feature (slot or tab), a spine is a simple (nonself-intersecting) surface. Think of it as a plane that may be flat or warped. The appropriate size ball shall be swept all over the spine, generating a 3-D boundary resembling a thick blanket. Fig. 5-15 illustrates the spines, balls, and 3-D boundaries for both size limits. Again, whether an internal or external feature, both feature surfaces shall contain the smaller boundary and be contained within the larger boundary.

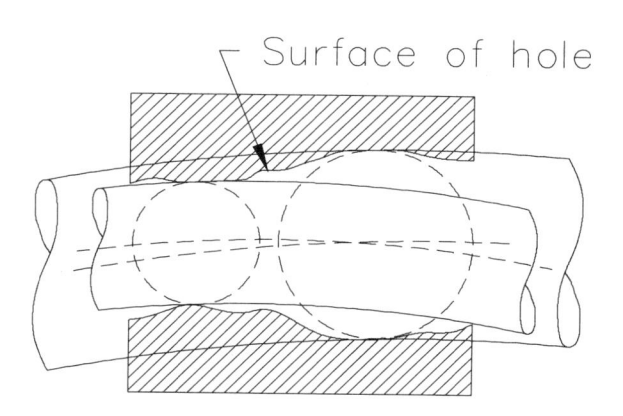

Figure 5-14 Conformance to limits of size for a cylindrical feature

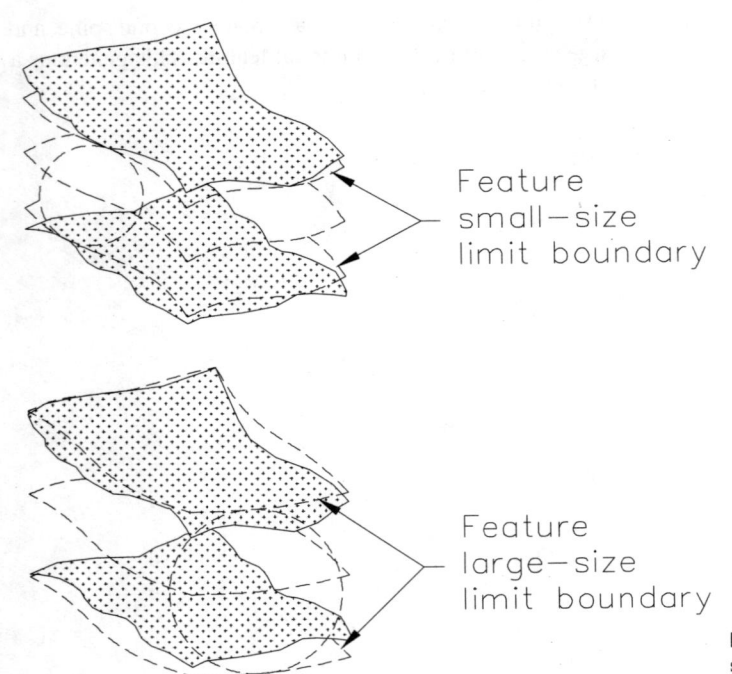

Figure 5-15 Conformance to limits of size for a width-type feature

The boundaries for a spherical feature of size are simply a small size limit sphere and a large size limit sphere. The rules for containment are the same and the boundaries need not be concentric.

In addition to limiting the largest and smallest a feature can be at any cross section, the two size limit boundaries control the *circularity* (roundness) at each cross section of a cylindrical or spherical feature of size. Fig. 5-16 shows a single cross section through a cylindrical feature and its small and large size limit boundaries. Notice that even though the small boundary is offset within the large boundary, the difference between the feature's widest and narrowest diameters cannot exceed the total size tolerance without violating a boundary. This Level 1 control of size and circularity at each cross section is adequate for most nonmating features of size. If necessary, circularity may be further refined with a separate circularity tolerance as described in section 5.8.5.

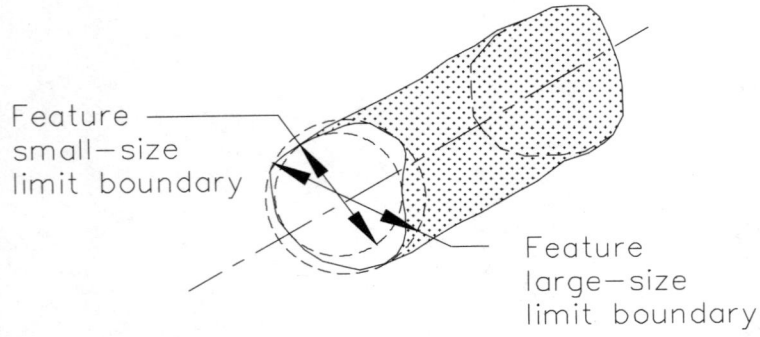

Figure 5-16 Size limit boundaries control circularity at each cross section

Obviously, the sweeping ball method is an ideal that cannot be realized with hard gages, but can be modeled by a computer to varying degrees of accuracy approaching the ideal. Since metrology (measuring) will always be an inexact science, inspectors are obliged to use the available tools to try to approximate the ideals. If the tool at hand is a pair of dial calipers or a micrometer, the inspector can only make "two-point" measurements across the width or diameter of a feature. But the inspector should make many such measurements and every measured value shall be between the low and high size limits. The inspector should also visually inspect the surface(s) for high or low regions that might violate a size limit boundary without being detected by the two-point measurements.

Before publication of the Math Standard, size limits were interpreted as applying to the smallest and largest two-point measurements obtainable at any cross section. However, with no spine linking the cross sections, there's no requirement for continuity. A cylindrical boss could resemble coins carelessly stacked. It was agreed that such abrupt offsets in a feature are unsatisfactory for most applications. The new "sweeping ball" method expands GD&T beyond the confines of customary gaging methods, creating a mathematically perfect requirement equal to any technology that might evolve.

5.6.2 Material Condition

Material condition is another way of thinking about the size of an object taking into account the object's nature. For example, the nature of a mountain is that it's a pile of rock material. If you pile on more material, its "material condition" increases and the mountain gets bigger. The nature of a canyon is that it's a void. As erosion decreases its "material condition," the canyon gets bigger.

If a mating feature of size is as small as it can be, will it fit tighter or sloppier? Of course, you can't answer until you know whether we're talking about an internal feature of size, such as a hole, or an external feature of size, such as a pin. But, if we tell you a feature of size has less material, you know it will fit more loosely regardless of its type. *Material condition*, then, is simply a shorthand description of a feature's size in the context of its intended function.

Maximum material condition (abbreviated MMC) is **the condition in which a feature of size contains the maximum amount of material within the stated limits of size.**

You can think of MMC as the condition where the most part material is present at the surface of a feature, or where the part weighs the most (all else being equal). This equates to the smallest allowable hole or the largest allowable pin, relative to the stated size limits.

Least material condition (abbreviated LMC) is **the condition in which a feature of size contains the least amount of material within the stated limits of size.**

You can think of LMC as the condition where the least part material is present at the surface of a feature, or where the part weighs the least (all else being equal). This equates to the largest allowable hole or the smallest allowable pin, relative to the stated size limits.

It follows then, that for every feature of size, one of the size limit boundaries is an *MMC boundary* corresponding to an *MMC limit*, and the other is an *LMC boundary* corresponding to an *LMC limit*. Depending on the type of feature and its function, the MMC boundary might ensure matability or removal of enough stock in a manufacturing process; the LMC boundary may ensure structural integrity and strength or ensure that the feature has enough stock for removal in a subsequent manufacturing process.

5.6.2.1 Modifier Symbols

Each geometric tolerance for a feature of size applies in one of the following three *contexts*:
- Regardless of Feature Size (RFS), the default
- modified to Maximum Material Condition (MMC)
- modified to Least Material Condition (LMC)

Table 5-1 shows which types of tolerances may be optionally "modified" to MMC or LMC. As we'll detail in the following paragraphs, such modification causes a tolerance to establish a new and useful fixed-size boundary based on the geometric tolerance and the corresponding size limit boundary. Placing a material condition modifier symbol, either a circled M or a circled L, immediately following the tolerance value in the feature control frame modifies a tolerance. As we'll explain in section 5.9.8.4, either symbol may also appear following the datum reference letter for each datum feature of size. In notes outside a feature control frame, use the abbreviation "MMC" or "LMC."

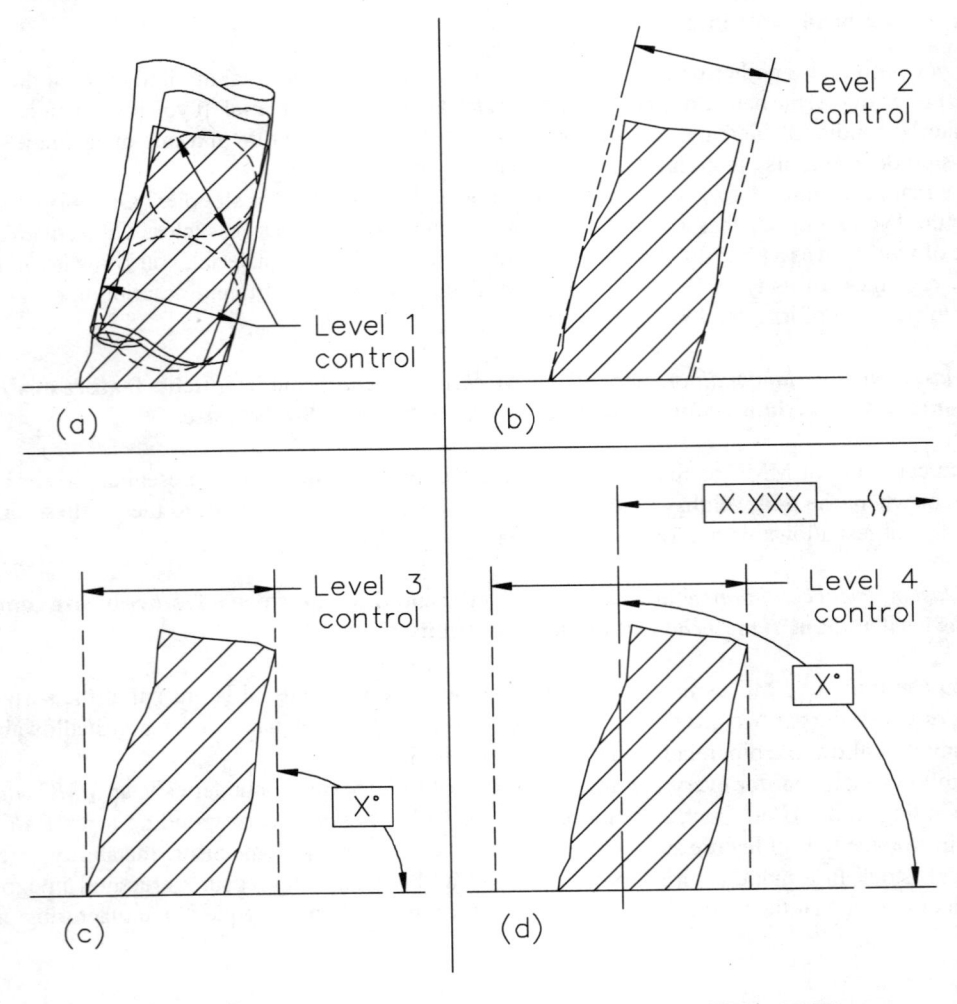

Figure 5-17 Levels of control for geometric tolerances modified to MMC

Geometric Dimensioning and Tolerancing 5-25

A geometric tolerance applied to a feature of size with no modifying symbol applies RFS. A few types of tolerances can only apply in an RFS context. As we'll explain in section 5.6.4, a Level 2, 3, or 4 tolerance works differently in an RFS context. Rather than a fixed-size boundary, the tolerance establishes a central tolerance zone.

5.6.3 Method for MMC or LMC

Geometric tolerances modified to MMC or LMC extend the system of boundaries for direct control of the feature surface(s). At each level of control, the applied tolerances establish a unique boundary, shown in Fig. 5-17(a) through (d) and Fig. 5-18(a) through (d), beyond which the feature surface(s) shall not encroach. Each higher-level tolerance creates a new boundary with an added constraint demanded by the feature's functional (usually mating) requirements. However, all lower-level controls remain in effect, regardless of their material condition contexts. Thus, a single feature can be subject to many boundaries simultaneously. The various boundaries are used in establishing datums (see Section 9), calculating tolerance stackups (see Chapters 9 and 11), and functional gaging (see Chapter 19).

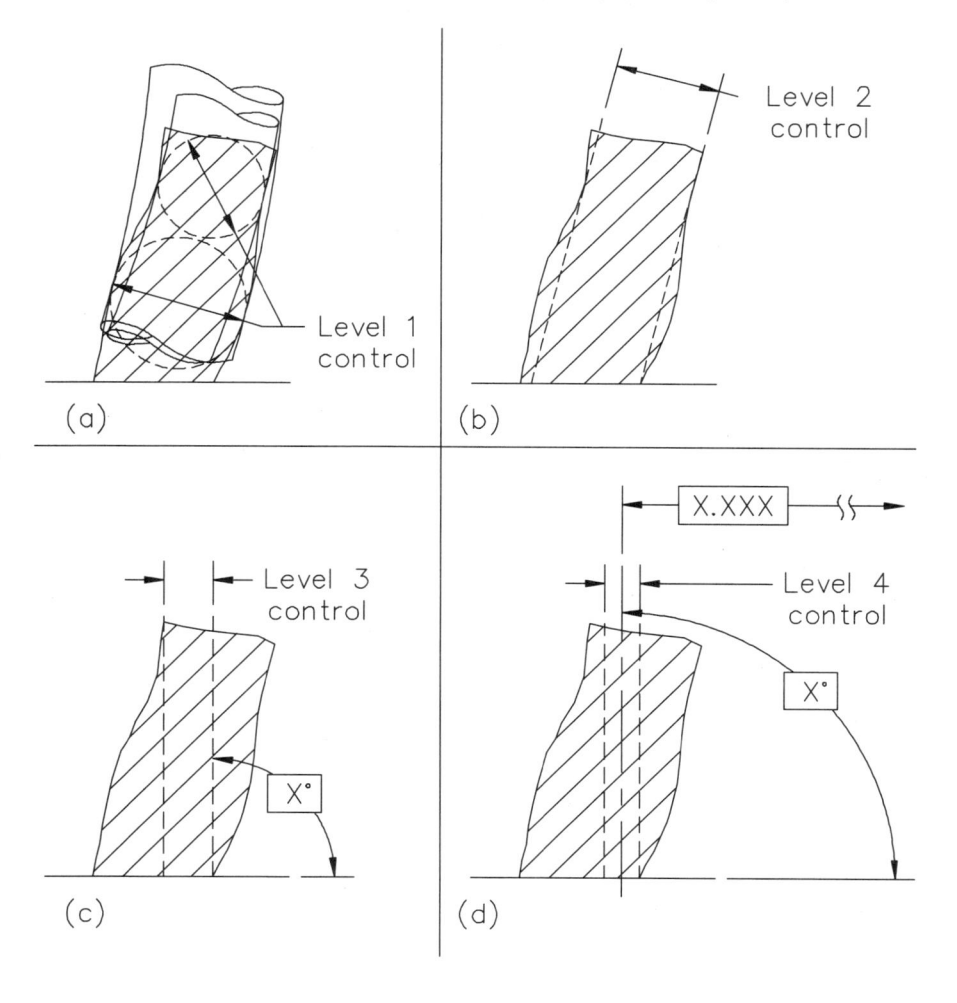

Figure 5-18 Levels of control for geometric tolerances modified to LMC

5-26 Chapter Five

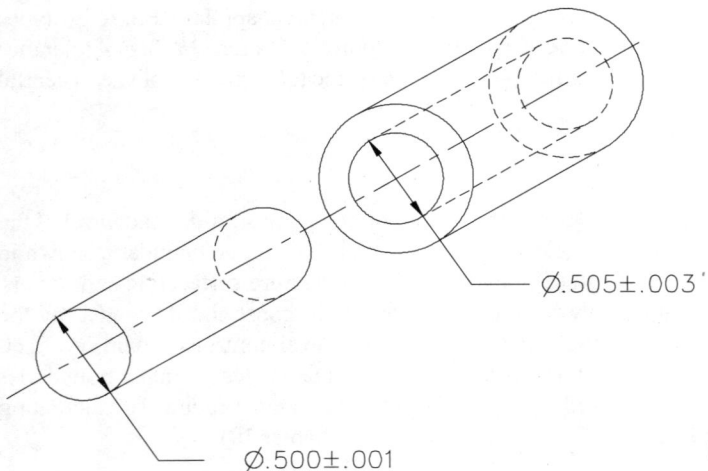

Figure 5-19 Cylindrical features of size that must fit in assembly

5.6.3.1 Level 2—Overall Feature Form

For features of size that must achieve a clearance fit in assembly, such as those shown in Fig. 5-19, the designer calculates the size tolerances based on the assumption that each feature, internal and external, is straight. For example, the designer knows that a ∅.501 maximum pin will fit in a ∅.502 minimum hole if both are straight. If one is banana shaped and the other is a lazy "S," as shown in Fig. 5-20, they usually won't

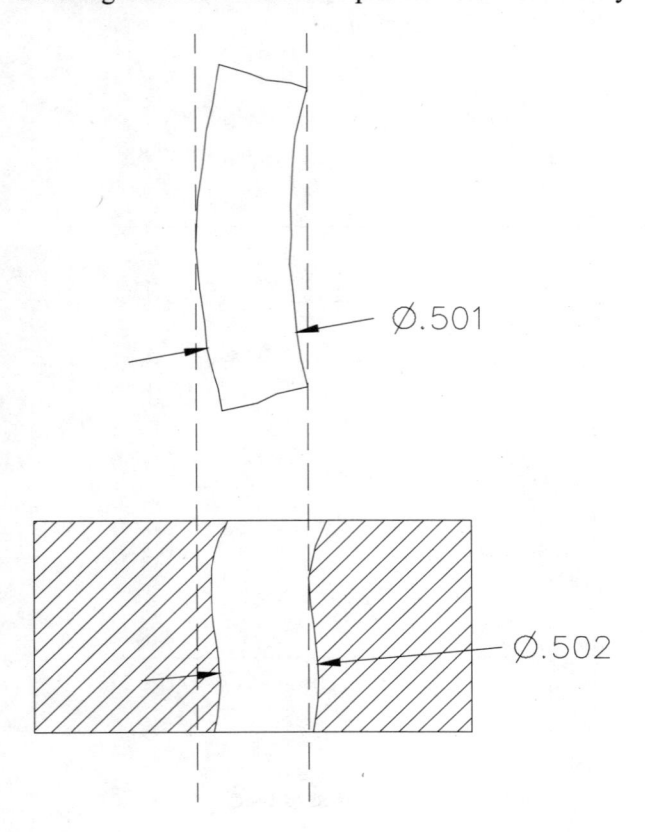

Figure 5-20 Level 1's size limit boundaries will not assure assemblability

go together. Because Level 1's size limit boundaries can be curved, they can't assure assemblability. Level 2 adds control of the overall geometric shape or *form* of a feature of size by establishing a perfectly formed boundary beyond which the feature's surface(s) shall not encroach.

Boundaries of Perfect Form—A size limit spine can be required to be perfectly formed (straight or flat, depending on its type). Then, the sweeping ball generates a *boundary of perfect form*, either a perfect cylinder or pair of parallel planes. The feature surface(s) must then achieve some degree of straightness or flatness to avoid violating the boundary of perfect form. Boundaries of perfect form have no bearing on the orientational, locational, or coaxial relationships between features. However, this Level 2 control is usually adequate for a feature of size that relates to another feature in the absence of any orientation or location restraint between the two features—that is, where the features are free-floating relative to each other. Where necessary, overall form control may be adjusted with a separate straightness, flatness, or cylindricity tolerance, described in sections 5.8.2, 5.8.4, and 5.8.6, respectively.

For an individual feature of size, the MMC and LMC size limit boundaries can be required to have perfect form in four possible combinations: MMC only, LMC only, both, or neither. Each combination is invoked by different rules which, unfortunately, are scattered throughout Y14.5. We've brought them together in the following paragraphs. (Only the first rule is numbered.)

At MMC (Only)—Rule #1—Based on the assumption that most features of size must achieve a clearance fit, Y14.5 established a default rule for perfect form. Y14.5's *Rule #1* decrees that, unless otherwise specified or overridden by another rule, a feature's MMC size limit spine shall be perfectly formed (straight or flat, depending on its type). This invokes a boundary of perfect form at MMC (also called an *envelope*). Rule #1 doesn't require the LMC boundary to have perfect form.

In our example, Fig. 5-21 shows how Rule #1 establishes a ⌀.501 boundary of perfect form at MMC (envelope) for the pin. Likewise, Rule #1 mandates a ⌀.502 boundary of perfect form at MMC (envelope)

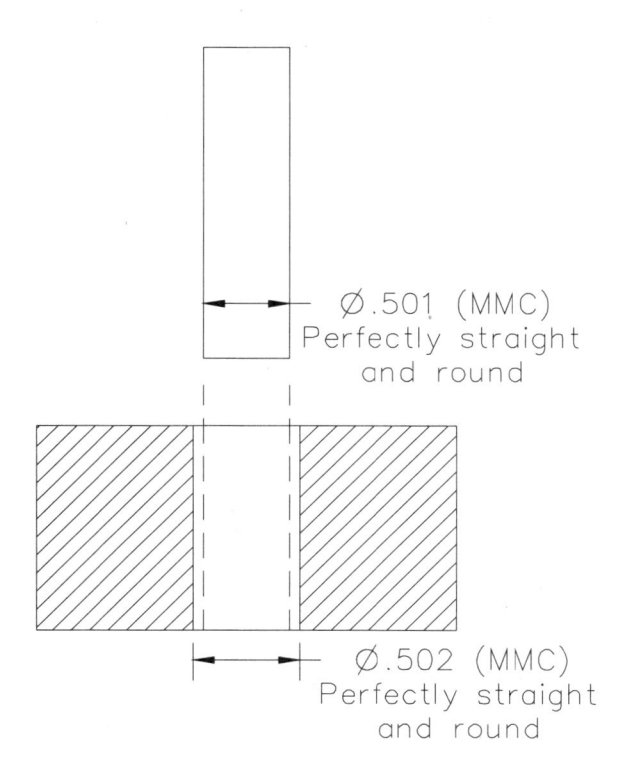

Figure 5-21 Rule #1 specifies a boundary of perfect form at MMC

5-28 Chapter Five

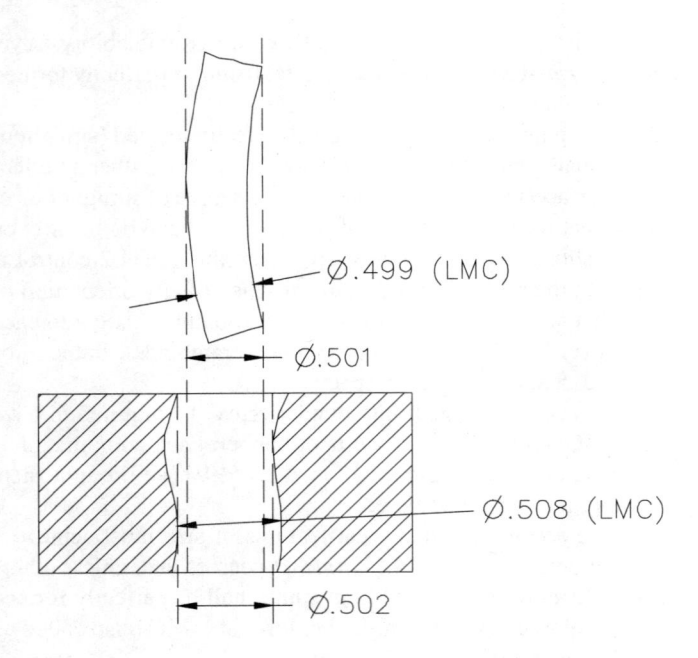

Figure 5-22 Rule #1 assures matability

for the hole. Fig. 5-22 shows how matability is assured for any pin that can fit inside its ⌀.501 envelope and any hole that can contain its ⌀.502 envelope. This simple hierarchy of fits is called the *envelope principle*.

At LMC (Only)—(Y14.5 section 5.3.5)—Fig. 5-23 illustrates a case where a geometric tolerance is necessary to assure an adequate "skin" of part material in or on a feature of size, rather than a clearance fit. In such an application, the feature of size at LMC represents the worst case. An LMC modifier applied to the geometric tolerance overrides Rule #1 for the controlled feature of size. Instead, the feature's LMC spine shall be perfectly formed (straight or flat, depending on its type). This invokes a boundary of perfect form at LMC. The MMC boundary need not have perfect form. The same is true for a datum feature of size referenced at LMC.

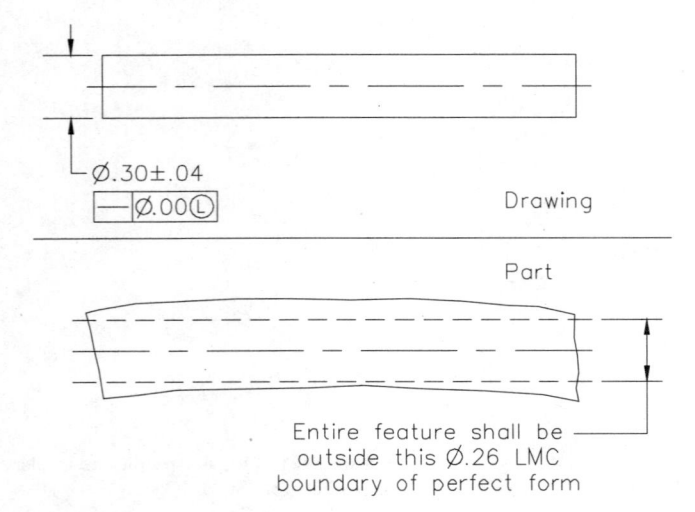

Figure 5-23 Using an LMC modifier to assure adequate part material

At both MMC and LMC—There are rare cases where a feature of size is associated with an MMC modifier in one context, and an LMC modifier in another context. For example, in Fig. 5-24, the datum B bore is controlled with a perpendicularity tolerance at MMC, then referenced as a datum feature at LMC. Each modifier for this feature, MMC and LMC, invokes perfect form for the feature's corresponding size limit boundary.

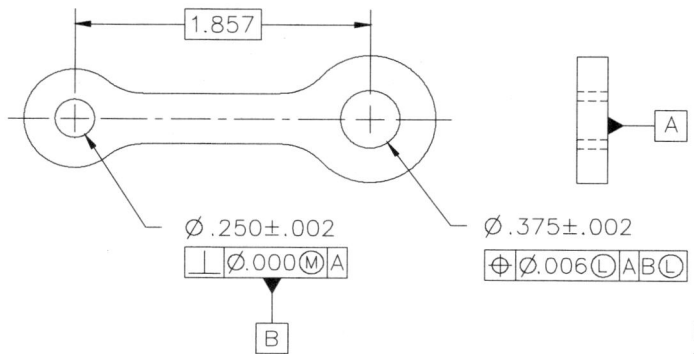

Figure 5-24 Feature of size associated with an MMC modifier and an LMC modifier

At neither MMC nor LMC—the Independency Principle—Y14.5 exempts the following from Rule #1.
- **Stock, such as bars, sheets, tubing, structural shapes, and other items produced to established industry or government standards that prescribe limits for straightness, flatness, and other geometric characteristics. Unless geometric tolerances are specified on the drawing of a part made from these items, standards for these items govern the surfaces that remain in the as-furnished condition on the finished part.**
- Dimensions for which restrained verification is specified in accordance with section 5.5.1
- A cylindrical feature of size having a straightness tolerance associated with its diameter dimension (as described in section 5.8.2)
- A width-type feature of size having a straightness or (by extension of principle) flatness tolerance associated with its width dimension (as described in section 5.8.4)

In these cases, feature form is either noncritical or controlled by a straightness or flatness tolerance separate from the size limits. Since Rule #1 doesn't apply, the size limits by themselves impose neither an MMC nor an LMC boundary of perfect form.

Fig. 5-25 is a drawing for an electrical bus bar. The cross-sectional dimensions have relatively close tolerances, not because the bar fits closely inside anything, but rather because of a need to assure a

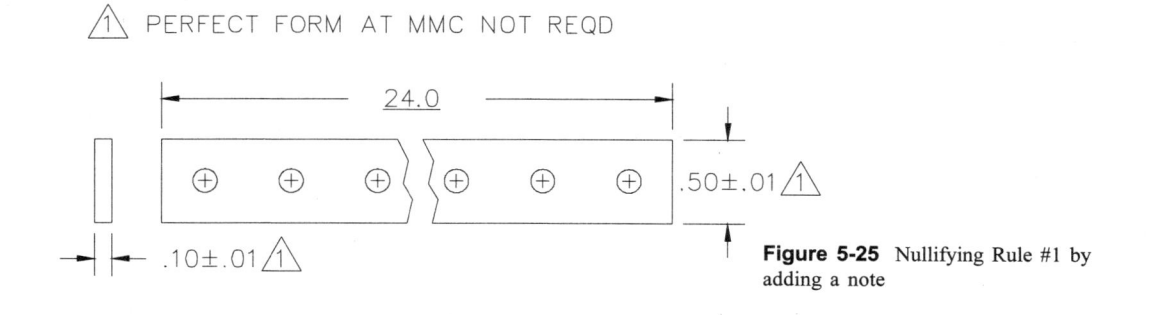

Figure 5-25 Nullifying Rule #1 by adding a note

minimum current-carrying capacity without squandering expensive copper. Neither the MMC nor the LMC boundary need be perfectly straight. However, if the bus bar is custom rolled, sliced from a plate, or machined at all, it won't automatically be exempted from Rule #1. In such a case, Rule #1 shall be explicitly nullified by adding the note PERFECT FORM AT MMC NOT REQD adjacent to each of the bus bar's size dimensions.

Many experts argue that Rule #1 is actually the "exception," that fewer than half of all features of size need any boundary of perfect form. Thus, for the majority of features of size, Rule #1's perfect form at MMC requirement accomplishes nothing except to drive up costs. The rebuttal is that Y14.5 prescribes the "perfect form not required" note and designers simply fail to apply it often enough. Interestingly, ISO defaults to "perfect form not required" (sometimes called the *independency principle*) and requires application of a special symbol to invoke the "envelope" (boundary) of perfect form at MMC. This is one of the few substantial differences between the US and ISO standards.

Regardless of whether the majority of features of size are mating or nonmating, regardless of which principle, envelope or independency, is the default, <u>every</u> designer should consider for <u>every</u> feature of size whether a boundary of perfect form is a necessity or a waste.

Virtual Condition Boundary for Overall Form—There are cases where a perfect form boundary is needed, but at a different size than MMC. Fig. 5-26 shows a drawn pin that will mate with a very flexible socket in a mating connector. The pin has a high aspect (length-to-diameter) ratio and a close diameter tolerance. It would be extremely difficult to manufacture pins satisfying both Rule #1's boundary of perfect form at MMC (Ø.063) and the LMC (Ø.062) size limit. And since the mating socket has a flared lead-in, such near-perfect straightness isn't functionally necessary.

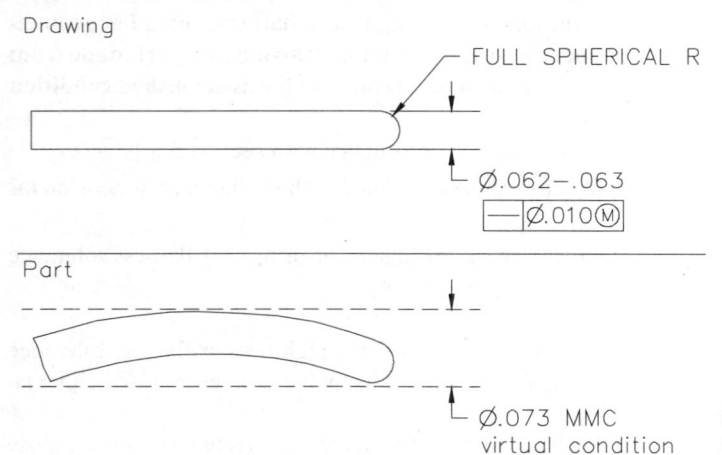

Figure 5-26 MMC virtual condition of a cylindrical feature

Fig. 5-27 shows a flat washer to be stamped out of sheet stock. The thickness (in effect, of the sheet stock) has a close tolerance because excessive variation could cause a motor shaft to be misaligned. Here again, for the tolerance and aspect ratio, Rule #1 would be unnecessarily restrictive. Nevertheless, an envelope is needed to prevent badly warped washers from jamming in automated assembly equipment.

In either example, the note PERFECT FORM AT MMC NOT REQD could be added, but would then allow pins as curly as a pig's tail or washers as warped as a potato chip. A better solution is to control the pin's overall form with a separate straightness tolerance modified to MMC. This replaces Rule #1's boundary of perfect form at MMC with a new perfect form boundary, called a *virtual condition boundary*, at some size other than MMC. Likewise, the washer's overall flatness can be controlled with a separate flatness tolerance modified to MMC. For details on how to apply these tolerances, see sections 5.8.2 and 5.8.4.

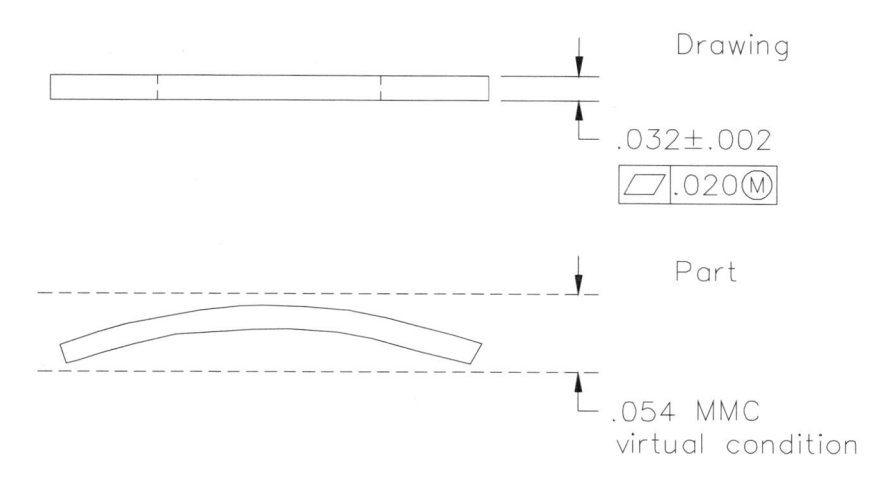

Figure 5-27 MMC virtual condition of a width-type feature

Any geometric tolerance applied to a feature of size and modified to MMC establishes a virtual condition boundary in the air adjacent to the feature surface(s). The boundary constitutes a restricted air space into which the feature shall not encroach. A geometric tolerance applied to a feature of size and modified to LMC likewise establishes a virtual condition boundary. However, in the LMC case, the boundary is embedded in part material, just beneath the feature surface(s). This boundary constitutes a restricted core or shell of part material into which the feature shall not encroach. The perfect geometric shape of any virtual condition boundary is a counterpart to the nominal shape of the controlled feature and is usually expressed with the form tolerance value, as follows.

Straightness Tolerance for a Cylindrical Feature—The "∅" symbol precedes the straightness tolerance value. The tolerance specifies a virtual condition boundary that is a cylinder. The boundary cylinder extends over the entire length of the actual feature.

Flatness Tolerance for a Width-Type Feature—No modifying symbol precedes the flatness tolerance value. The tolerance specifies a virtual condition boundary of two parallel planes. The boundary planes extend over the entire length and breadth of the actual feature.

Whether the form tolerance is modified to MMC or LMC determines the size of the virtual condition boundary relative to the feature's specified size limits.

Modified to MMC—The MMC virtual condition boundary represents a restricted air space reserved for the mating part feature. In such a mating interface, the internal feature's MMC virtual condition boundary must be at least as large as that for the external feature. *MMC virtual condition* (the boundary's fixed size) is determined by three factors: 1) the feature's type (internal or external); 2) the feature's MMC size limit; and 3) the specified geometric tolerance value.

For an internal feature of size:
    MMC virtual condition = MMC size limit − geometric tolerance

For an external feature of size:
    MMC virtual condition = MMC size limit + geometric tolerance

5-32 Chapter Five

Four notes regarding these formulae:
1. For the pin in Fig. 5-26, the diameter of the virtual condition boundary equals the pin's MMC size plus the straightness tolerance value: ∅.063 + ∅.010 = ∅.073. This boundary can be simulated with a simple ∅.073 ring gage.
2. A Level 2 (straightness or flatness) tolerance value of zero at MMC is the exact equivalent of Rule #1 and therefore redundant.
3. For an internal feature, a geometric tolerance greater than the MMC size limit yields a negative virtual condition. This is no problem for computerized analysis, but it precludes functional gaging.
4. For a screw thread, an MMC virtual condition can be calculated easily based on the MMC pitch diameter. The boundary, however, has limited usefulness in evaluating an actual thread.

Modified to LMC—The LMC virtual condition boundary assures a protected core of part material within a pin, boss, or tab, or a protected case of part material around a hole or slot. *LMC virtual condition* (the boundary's fixed size) is determined by three factors: 1) the feature's type (internal or external); 2) the feature's LMC size limit; and 3) the specified geometric tolerance value.

For an internal feature of size:
    LMC virtual condition = LMC size limit + geometric tolerance

For an external feature of size:
    LMC virtual condition = LMC size limit − geometric tolerance

Fig. 5-28 shows a part where straightness of datum feature A is necessary to protect the wall thickness. Here, the straightness tolerance modified to LMC supplants the boundary of perfect form at LMC. The tolerance establishes a virtual condition boundary embedded in the part material beyond which the feature surface shall not encroach. For datum feature A in Fig. 5-28, the diameter of this boundary equals the LMC size minus the straightness tolerance value: ∅.247 − ∅.005 = ∅.242. Bear in mind the difficulties of verifying conformance where the virtual condition boundary is embedded in part material and can't be simulated with tangible gages.

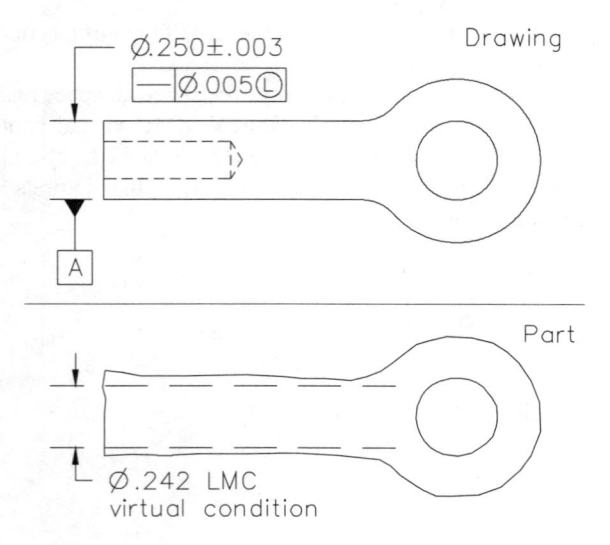

Figure 5-28 LMC virtual condition of a cylindrical feature

5.6.3.2 Level 3—Virtual Condition Boundary for Orientation

For two mating features of size, Level 2's perfect form boundaries can only assure assemblability in the absence of any orientation or location restraint between the two features—that is, the features are free-floating relative to each other. In Fig. 5-29, we've taken our simple example of a pin fitting into a hole, and added a large flange around each part. We've also stipulated that the two flanges shall bolt together and make full contact. This introduces an orientation restraint between the two mating features. When the flange faces are bolted together tightly, the pin and the hole must each be very square to their respective flange faces. Though the pin and the hole might each respect their MMC boundaries of perfect form, nothing prevents those boundaries from being badly skewed to each other.

We can solve that by taking the envelope principle one step further to Level 3. An orientation tolerance applied to a feature of size, modified to MMC or LMC, establishes a virtual condition boundary beyond which the feature's surface(s) shall not encroach. For details on how to apply an orientation tolerance, see section 5.10.1. In addition to perfect form, this new boundary has perfect orientation in all applicable degrees of freedom relative to any datum feature(s) we select (see section 5.9.7). The shape and size of the virtual condition boundary for orientation are governed by the same rules as for form at Level 2. A single feature of size can be subject to multiple virtual condition boundaries.

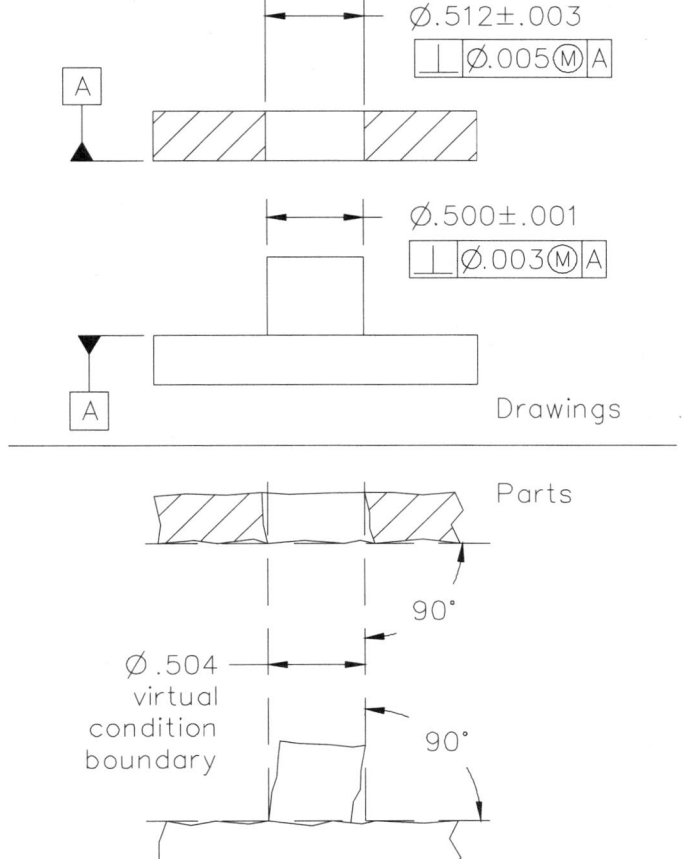

Figure 5-29 Using virtual condition boundaries to restrain orientation between mating features

For each example part in Fig. 5-29, we've restrained the virtual condition boundary perpendicular to the flange face. The lower portion of Fig. 5-29 shows how matability is assured for any part having a pin that can fit inside its ⌀.504 MMC virtual condition boundary and any part having a hole that can contain its ⌀.504 MMC virtual condition boundary.

5.6.3.3 Level 4—Virtual Condition Boundary for Location

For two mating features of size, Level 3's virtual condition boundary for orientation can only assure assemblability in the absence of any location restraint between the two features, for example, where no other mating features impede optimal location alignment between our pin and hole. In Fig. 5-30, we've

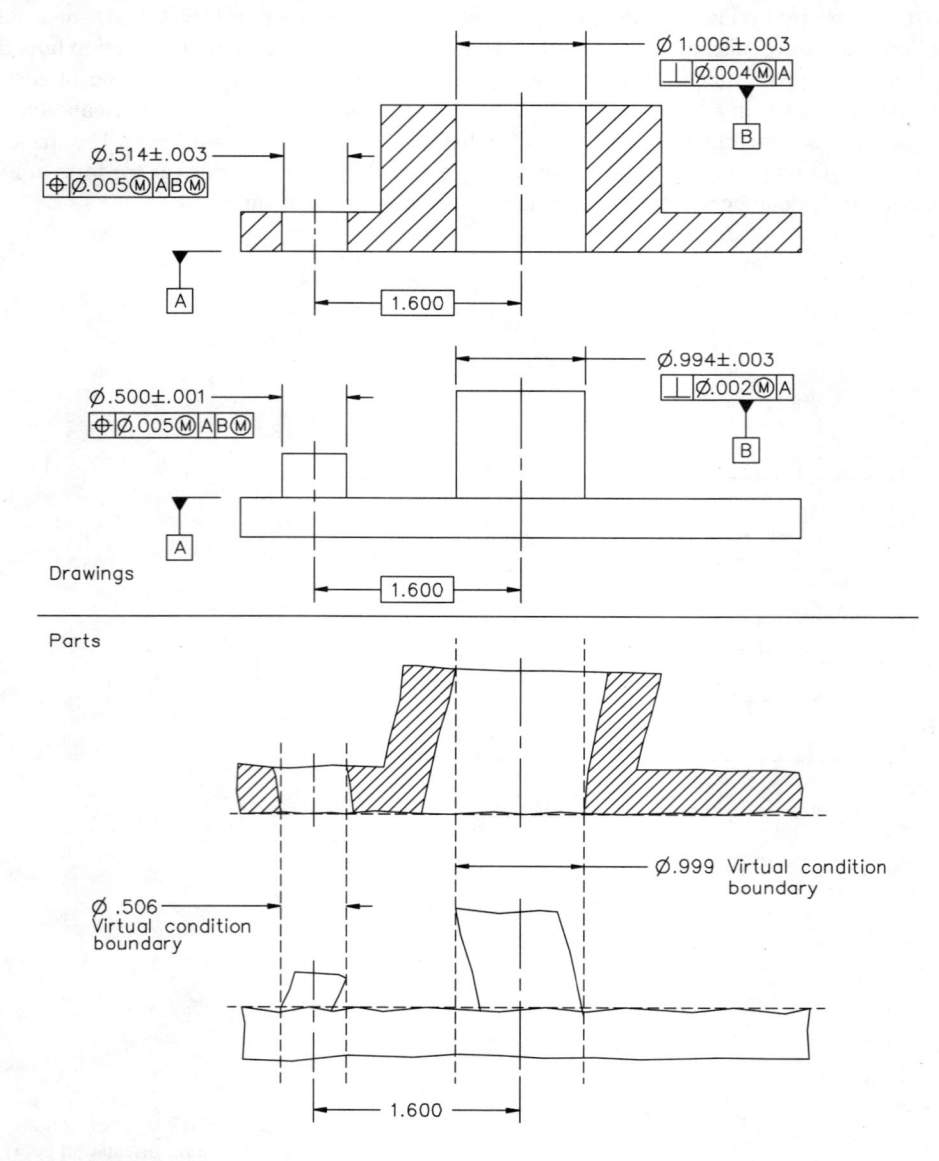

Figure 5-30 Using virtual condition boundaries to restrain location (and orientation) between mating features

moved the pin and hole close to the edges of the flanges and added a larger bore and boss mating interface at the center of the flanges. When the flange faces are bolted together tightly and the boss and bore are fitted together, the pin and the hole must each still be very square to their respective flange faces. However, the parts can no longer slide freely to optimize the location alignment between the pin and the hole. Thus, the pin and the hole must each additionally be accurately located relative to its respective boss or bore.

A positional tolerance applied to a feature of size, modified to MMC or LMC, takes the virtual condition boundary one step further to Level 4. For details on how to apply a positional tolerance, see section 5.11.2. In addition to perfect form and perfect orientation, the new boundary shall have perfect location in all applicable degrees of freedom relative to any datum feature(s) we select (see section 5.9.7). The shape and size of the virtual condition boundary for location are governed by the same rules as for form at Level 2 and orientation at Level 3, with one addition. For a spherical feature, the tolerance is preceded by the "S$\varnothing$" symbol and specifies a virtual condition boundary that is a sphere. A single feature of size can be subject to multiple virtual condition boundaries—one boundary for each form, orientation, and location tolerance applied.

In Fig. 5-30, we've identified four datums and added dimensions and tolerances for our example assembly. The central boss has an MMC size limit of $\varnothing$.997 and a perpendicularity tolerance of $\varnothing$.002 at MMC. Since it's an external feature of size, its virtual condition is $\varnothing$.997 + $\varnothing$.002 = $\varnothing$.999. The bore has an MMC size limit of $\varnothing$1.003 and a perpendicularity tolerance of $\varnothing$.004 at MMC. Since it's an internal feature of size, its virtual condition is $\varnothing$1.003 − $\varnothing$.004 = $\varnothing$.999. Notice that for each perpendicularity tolerance, the datum feature is the flange face. Each virtual condition boundary for orientation is restrained perfectly perpendicular to its referenced datum, derived from the flange face. As the lower portion of Fig. 5-30 shows, the boss and bore will mate every time.

The pin and hole combination requires MMC virtual condition boundaries with location restraint added. Notice that for each positional tolerance, the primary datum feature is the flange face and the secondary datum feature is the central boss or bore. Each virtual condition boundary for location is restrained perfectly perpendicular to its referenced primary datum, derived from the flange face. Each boundary is additionally restrained perfectly located relative to its referenced secondary datum, derived from the boss or bore. This restraint of both orientation and location on each part is crucial to assuring perfect alignment between the boundaries on both parts, and thus, assemblability. The pin has an MMC size limit of $\varnothing$.501 and a positional tolerance of $\varnothing$.005 at MMC. Since it's an external feature of size, its virtual condition is $\varnothing$.501 + $\varnothing$.005 = $\varnothing$.506. The hole has an MMC size limit of $\varnothing$.511 and a positional tolerance of $\varnothing$.005 at MMC. Since it's an internal feature of size, its virtual condition is $\varnothing$.511 − $\varnothing$.005 = $\varnothing$.506. Any pin contained within its $\varnothing$.506 boundary can assemble with any hole containing its $\varnothing$.506 boundary. Try *that* without GD&T!

5.6.3.4 Level 3 or 4 Virtual Condition Equal to Size Limit (Zero Tolerance)

All the tolerances in our example assembly were chosen to control the fit between the two parts. Subsequent chapters deal with the myriad considerations involved in determining fits. To simplify our example, we matched virtual condition sizes for each pair of mating features. All our intermediate values, however, were chosen arbitrarily.

For example, in Fig. 5-30, the boss's functional extremes are at $\varnothing$.991 and $\varnothing$.999. Between them, the total tolerance is $\varnothing$.008. Based on our own assumptions about process variation, we arbitrarily divided this into $\varnothing$.006 for size and $\varnothing$.002 for orientation. Thus, the $\varnothing$.997 MMC size limit has no functional significance. We might just as well have divided the $\varnothing$.008 total into $\varnothing$.004 + $\varnothing$.004, $\varnothing$.006 + $\varnothing$.002, or even $\varnothing$.008 + $\varnothing$.000.

5-36 Chapter Five

In a case such as this, where the only MMC design consideration is a clearance fit, it's not necessary for the designer to apportion the fit tolerance. Why not give it all to the manufacturing process and let the process divvy it up as needed? This is accomplished by stretching the MMC size limit to equal the MMC virtual condition size and reducing the orientation or positional tolerance to zero.

Fig. 5-31 shows our example assembly with orientation and positional tolerances of zero. Notice that now, the central boss has an MMC size limit of ⌀.999 and a perpendicularity tolerance of ⌀.000 at MMC.

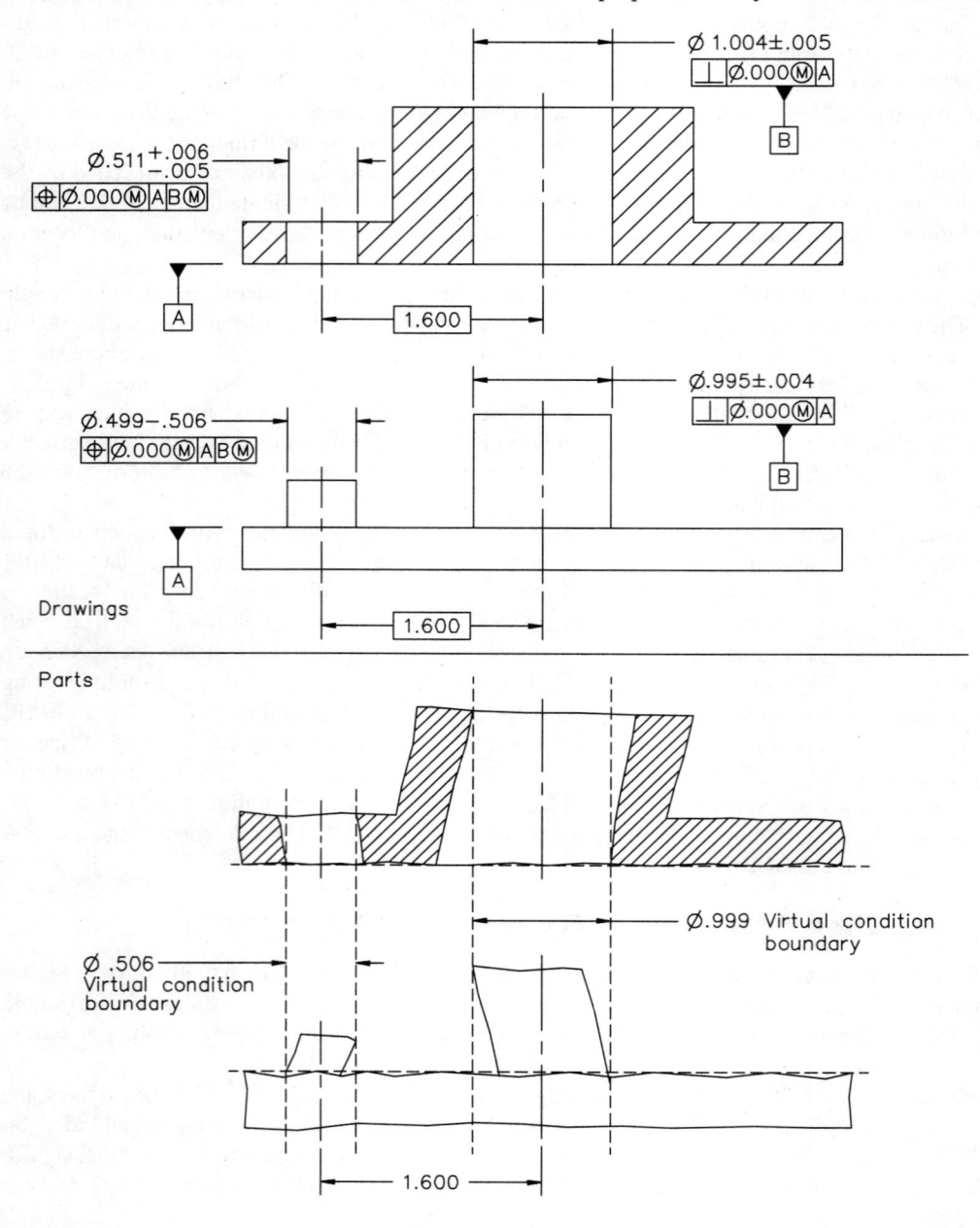

Figure 5-31 Zero orientation tolerance at MMC and zero positional tolerance at MMC

Since it's an external feature of size, its virtual condition is $\varnothing.999 + \varnothing.000 = \varnothing.999$.

Compare the lower portions of Figs. 5-30 and 5-31. The conversion to zero orientation and positional tolerances made no change to any of the virtual condition boundaries, and therefore, no change in assemblability and functionality. However, manufacturability improved significantly for both parts. Allowing the process to apportion tolerances opens up more tooling choices. In addition, a perfectly usable part having a boss measuring $\varnothing.998$ with perpendicularity measuring $\varnothing.0006$ will no longer be rejected.

The same rationale may be applied where a Level 3 or 4 LMC virtual condition exists. Unless there's a functional reason for the feature's LMC size limit to differ from its LMC virtual condition, make them equal by specifying a zero orientation or positional tolerance at LMC, as appropriate.

Some novices may be alarmed at the sight of a zero tolerance. "How can anything be made perfect?" they ask. Of course, a zero tolerance doesn't require perfection; it merely allows parity between two different levels of control. The feature shall be manufactured with size and orientation adequate to clear the virtual condition boundary. In addition, the feature shall nowhere encroach beyond its opposite size limit boundary.

5.6.3.5 Resultant Condition Boundary

For the $\varnothing.514$ hole in Fig. 5-30, we have primary and secondary design requirements. Since the hole must clear the $\varnothing.500$ pin in the mating part, we control the hole's orientation and location with a positional tolerance modified to MMC. This creates an MMC virtual condition boundary that guarantees air space for the mating pin. But now, we're worried that the wall might get too thin between the hole and the part's edge.

To address this secondary concern, we need to determine the farthest any point around the hole can range from "true position" (the ideal center). That distance constitutes a worst-case perimeter for the hole shown in Fig. 5-32 and called the *resultant condition boundary*. We can then compare the resultant condition boundary with that for the flange diameter and calculate the worst-case thin wall. We may then need to adjust the positional tolerance and/or the size limits for the hole and/or the flange.

Resultant condition is defined as a variable value obtained by adding the total allowable geometric tolerance to (or subtracting it from) the feature's actual mating size. Tables in Y14.5 show resultant condition values for feature sizes between the size limits. However, the only resultant condition value that anyone cares about is the single worst-case value defined below, as determined by three factors: 1) the feature's type (internal or external); 2) the feature's size limits; and 3) the specified geometric tolerance value.

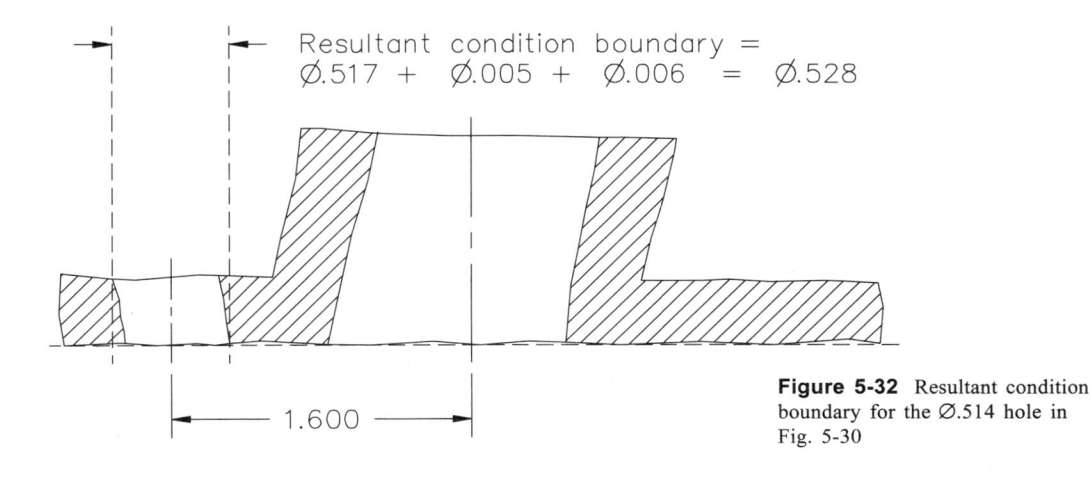

Resultant condition boundary = $\varnothing.517 + \varnothing.005 + \varnothing.006 = \varnothing.528$

1.600

Figure 5-32 Resultant condition boundary for the $\varnothing.514$ hole in Fig. 5-30

For an internal feature of size controlled at MMC:
   Resultant condition = LMC size limit + geometric tolerance + size tolerance

For an external feature of size controlled at MMC:
   Resultant condition = LMC size limit − geometric tolerance − size tolerance

For an internal feature of size controlled at LMC:
   Resultant condition = MMC size limit − geometric tolerance − size tolerance

For an external feature of size controlled at LMC:
   Resultant condition = MMC size limit + geometric tolerance + size tolerance

5.6.4 Method for RFS

A geometric tolerance applied to a feature of size with no modifying symbol applies RFS. A few types of tolerances can only apply in an RFS context. Instead of a boundary, a Level 2, 3, or 4 tolerance RFS establishes a central tolerance zone, within which a geometric element derived from the feature shall be contained. Each higher-level tolerance adds a degree of constraint demanded by the feature's functional requirements, as shown in Fig. 5-33(a) through (d). However, all lower-level controls remain in effect, regardless of their material condition contexts. Thus, a single feature can be subject to many tolerance zones and boundaries simultaneously. Unfortunately, tolerance zones established by RFS controls cannot be simulated by tangible gages. This often becomes an important design consideration.

5.6.4.1 Tolerance Zone Shape

The geometrical shape of the RFS tolerance zone usually corresponds to the shape of the controlled feature and is expressed with the tolerance value, as follows.

   For a Width-Type Feature—Where no modifying symbol precedes the tolerance value, the tolerance specifies a tolerance zone bounded by two parallel planes separated by a distance equal to the specified tolerance. The tolerance planes extend over the entire length and breadth of the actual feature.

   For a Cylindrical Feature—The tolerance value is preceded by the "⌀" symbol and specifies a tolerance zone bounded by a cylinder having a diameter equal to the specified tolerance. The tolerance cylinder extends over the entire length of the actual feature.

   For a Spherical Feature—The tolerance is preceded by the "S⌀" symbol and specifies a tolerance zone bounded by a sphere having a diameter equal to the specified tolerance.

5.6.4.2 Derived Elements

A multitude of geometric elements can be derived from any feature. A geometric tolerance RFS applied to a feature of size controls one of these five:

- Derived median line (from a cylindrical feature)
- Derived median plane (from a width-type feature)
- Feature center point (from a spherical feature)
- Feature axis (from a cylindrical feature)
- Feature center plane (from a width-type feature)

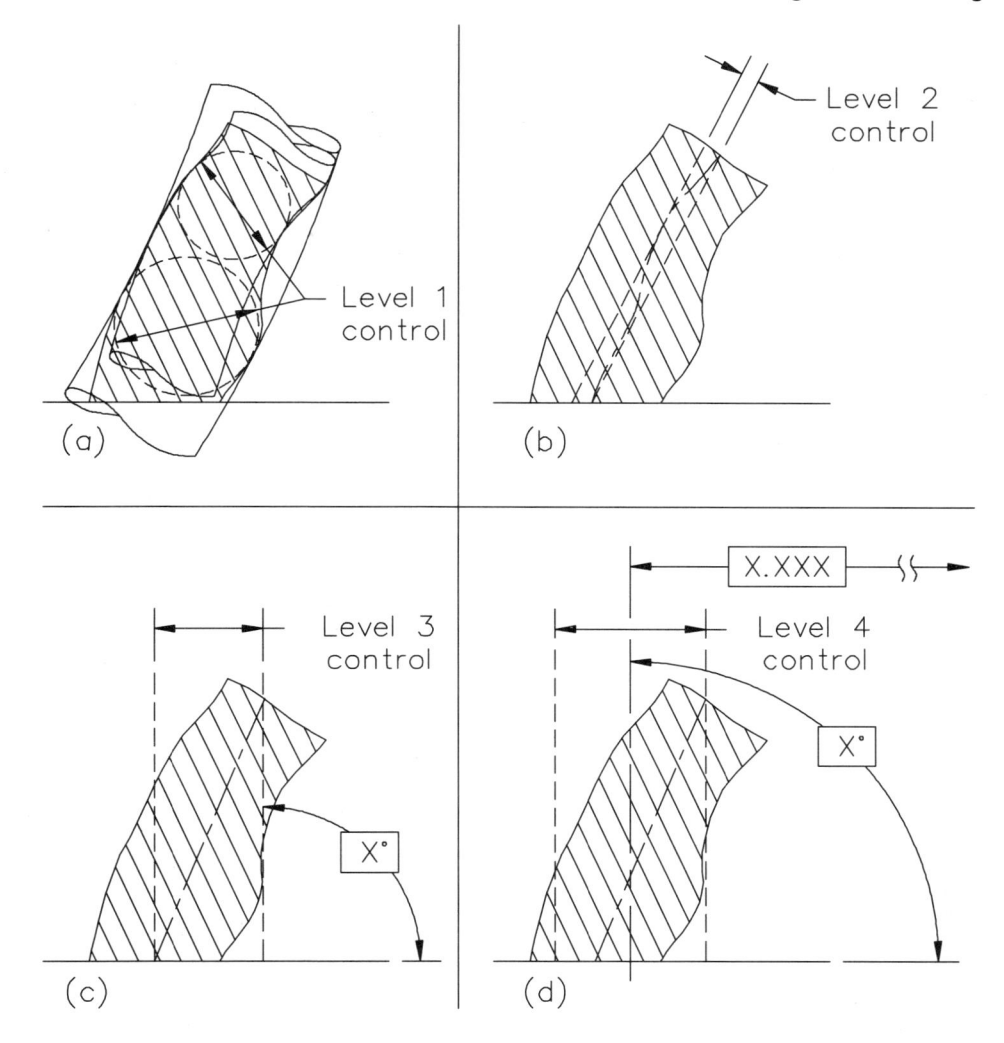

Figure 5-33 Levels of control for geometric tolerances applied RFS

A Level 2 (straightness or flatness) tolerance nullifies Rule #1's boundary of perfect form at MMC. Instead, the separate tolerance controls overall feature form by constraining the derived median line or derived median plane, according to the type of feature.

A cylindrical feature's *derived median line* is **an imperfect line (abstract) that passes through the center points of all cross sections of the feature. These cross sections are normal to the axis of the actual mating envelope. The cross section center points are determined as per ANSI B89.3.1.**

A width-type feature's *derived median plane* is **an imperfect plane (abstract) that passes through the center points of all line segments bounded by the feature. These line segments are normal to the actual mating envelope.**

5-40 Chapter Five

In Fig. 5-34, the absence of a material condition modifier symbol means the straightness tolerance applies RFS by default. This specifies a tolerance zone bounded by a cylinder having a diameter equal to the tolerance value, within which the derived median line shall be contained. In Fig. 5-35, the flatness tolerance applies RFS by default. This specifies a tolerance zone bounded by two parallel planes separated by a distance equal to the tolerance value, within which the entire derived median plane shall be contained. Both size limits are still in force, but neither the spine for the MMC size boundary nor the spine for the LMC size boundary need be perfectly formed. A straightness or flatness tolerance value may be less than, equal to, or greater than the size tolerance.

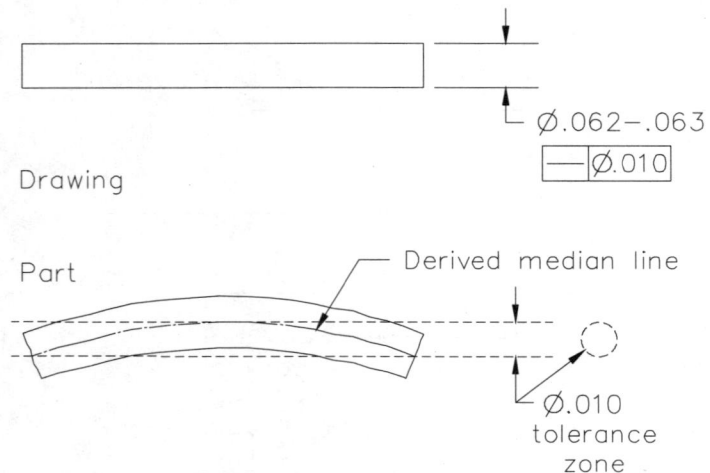

Figure 5-34 Tolerance zone for straightness control RFS

As you can imagine, deriving a median line or plane is a complex procedure that's extremely difficult without the help of a microprocessor-based machine. But where it's necessary to control overall form with a tolerance that remains constant, regardless of feature size, there are no simpler options. However, once we've assured overall form with Rule #1 or a separate form tolerance, we can apply Level 3 and 4 tolerances to geometric elements that are more easily derived: a center point, perfectly straight axis, or perfectly flat center plane. These elements must be defined and derived to represent the features' worst-case functionality.

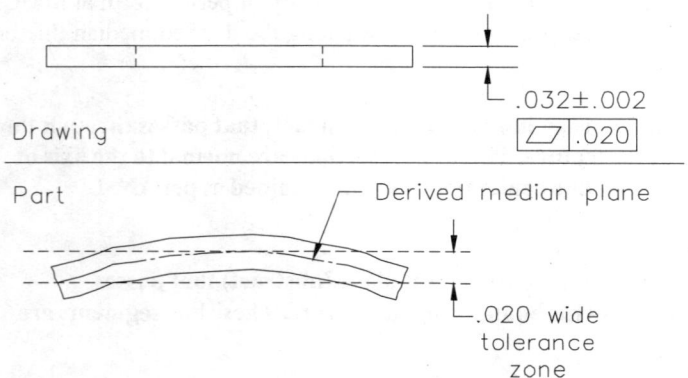

Figure 5-35 Tolerance zone for flatness control RFS

In an RFS context, the *feature center point*, *feature axis*, or *feature center plane* is the center of the feature's *actual mating envelope*. In all cases, a feature's axis or center plane extends for the full length and/or breadth of the feature.

The *actual mating envelope* is a surface, or pair of parallel-plane surfaces, of perfect form, which correspond to a part feature of size as follows:

(a) For an External Feature. A similar perfect feature counterpart of smallest size, which can be circumscribed about the feature so that it just contacts the feature surface(s). For example, a smallest cylinder of perfect form or two parallel planes of perfect form at minimum separation that just contact(s) the surface(s).

(b) For an Internal Feature. A similar perfect feature counterpart of largest size, which can be inscribed within the feature so that it just contacts the feature surface(s). For example, a largest cylinder of perfect form or two parallel planes of perfect form at maximum separation that just contact(s) the surface(s).

In certain cases, the orientation, or the orientation and location of an actual mating envelope shall be restrained to one or two datums (see Fig. 5-36 and Table 5-3). In Fig. 5-37, for example, the true geometric counterpart of datum feature B is the actual mating envelope (smallest perfect cylinder) restrained perpendicular to datum plane A.

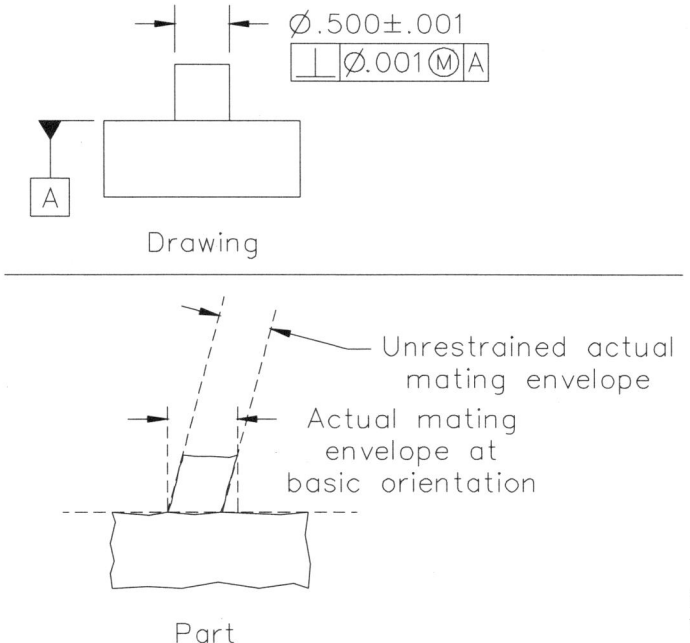

Figure 5-36 Example of restrained and unrestrained actual mating envelopes

Be careful not to confuse the actual mating envelope with the boundary of perfect form at MMC "envelope." Our above definitions are cobbled together from both Y14.5 and the Math Standard, since the standards differ slightly. Table 5-3 shows that in most cases, the actual mating envelope is unrestrained— that is, allowed to achieve any orientation and location when fitted to the feature. As we'll discuss later, when simulating a secondary or tertiary datum feature RFS, the actual mating envelope shall be oriented (held square) to the higher precedence datum(s). Obviously, that restraint will produce a different fit.

Table 5-3 Actual mating envelope restraint

	APPROPRIATE RESTRAINT	
PURPOSE OF ENVELOPE	Unrestrained	Restrained to higher datum(s)
Evaluate conformance to:		
Rule #1	X	
orientation tolerance	X	
positional tolerance	X	
Establish True Geometric Counterpart RFS for a datum feature:		
primary	X	
secondary, tertiary		X
Actual mating size of datum feature for DRF displacement		
primary	X	
secondary, tertiary		X

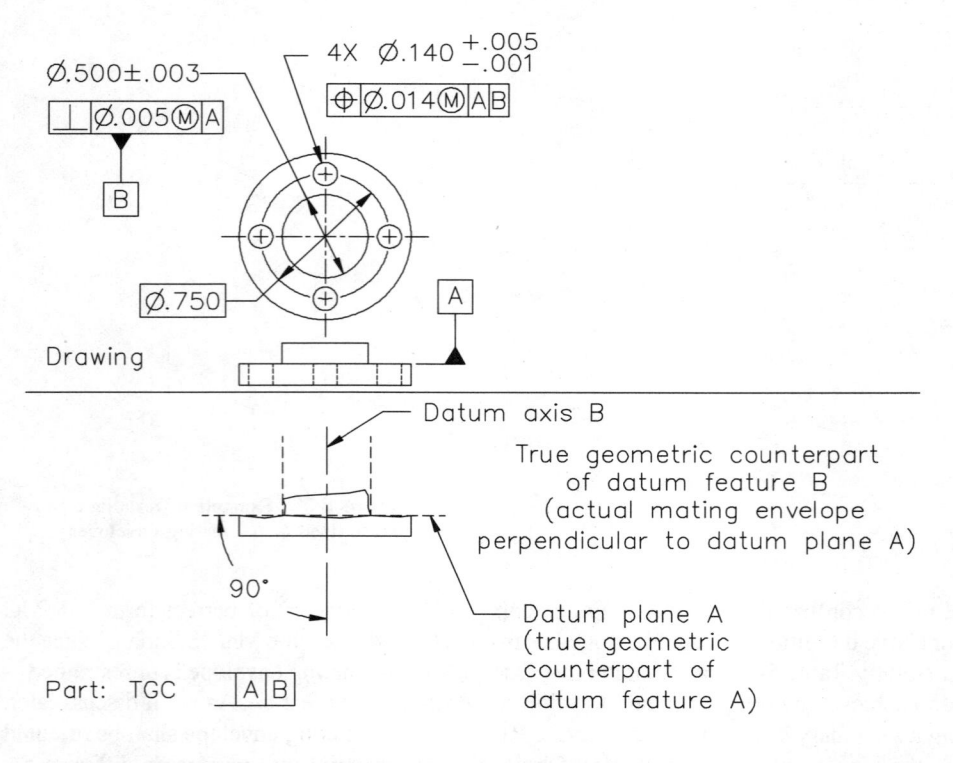

Figure 5-37 The true geometric counterpart of datum feature B is a restrained actual mating envelope

There are even some cases where the actual mating envelope's location shall be held stationary relative to the higher precedence datum(s). In addition, when calculating positional tolerance deviations, there are circumstances where a "restrained" actual mating envelope shall be used. We'll explain these applications in greater detail in later sections.

In practice, the largest cylindrical gage pin that can fit in a hole can often simulate the hole's actual mating envelope. The actual mating envelope for a slot can sometimes be approximated by the largest stack of Webber (or "Jo") blocks that can fit. External features are a little tougher, but their actual mating envelopes might be simulated with cylindrical ring gages or Webber block sandwiches.

Cases calling for a restrained actual mating envelope really challenge hard gaging methods. Traditionally, inspectors have fixtured parts to coordinate measuring machine (CMM) tables (on their datum feature surfaces) and held cylindrical gage pins in a drill chuck in the CMM's ram. This practice is only marginally satisfactory, even where relatively large tolerances are involved.

5.6.5 Alternative "Center Method" for MMC or LMC

As we explained in section 5.6.3, Level 2, 3, and 4 geometric tolerances applied to features of size and modified to MMC or LMC establish virtual condition boundaries for the features. Chapter 19 explains how functional gages use pins, holes, slots, tabs, and other physical shapes to simulate the MMC virtual condition boundaries, emulating worst-case features on the mating part as if each mating feature were manufactured at its MMC with its worst allowable orientation and location. However, without a functional gage or sophisticated CMM software, it might be very difficult to determine whether or not a feature encroaches beyond its virtual condition boundary. Therefore, the standards provide an alternative method that circumvents virtual boundaries, enabling more elementary inspection techniques. We call this alternative the *center method*.

Where a Level 2, 3, or 4 geometric tolerance is applied to a feature of size in an MMC or LMC context, the tolerance may optionally be interpreted as in an RFS context—that is, it establishes a central tolerance zone, within which a geometric element derived from the feature shall be contained. However, unlike in the RFS context, the MMC or LMC tolerance zone shall provide control approximating that of the virtual condition boundary. To accomplish this, the size of the tolerance zone shall adjust according to the feature's actual size.

5.6.5.1 Level 3 and 4 Adjustment—Actual Mating/Minimum Material Sizes

The adjustment for Level 3 and 4 tolerances is very simple: The tolerance zone is uniformly enlarged by *bonus tolerance*—a unit value to be added to the specified geometric tolerance.

At MMC—Bonus tolerance equals the arithmetic difference between the feature's actual mating size and its specified MMC size limit.

> *Actual mating size* is **the dimensional value of the actual mating envelope** (defined in section 5.6.4.2), and represents the worst-case mating potential for a feature of size. See Fig. 5-38.

Thus, actual mating size is the most suitable measure of actual size in clearance-fit applications or for most features having a boundary of perfect form at MMC. For a hole having an actual mating size ∅.001 larger than its MMC, ∅.001 of bonus tolerance is added to the specified geometric tolerance. Likewise, for a tab .002 smaller than its MMC, .002 is added to the specified tolerance value.

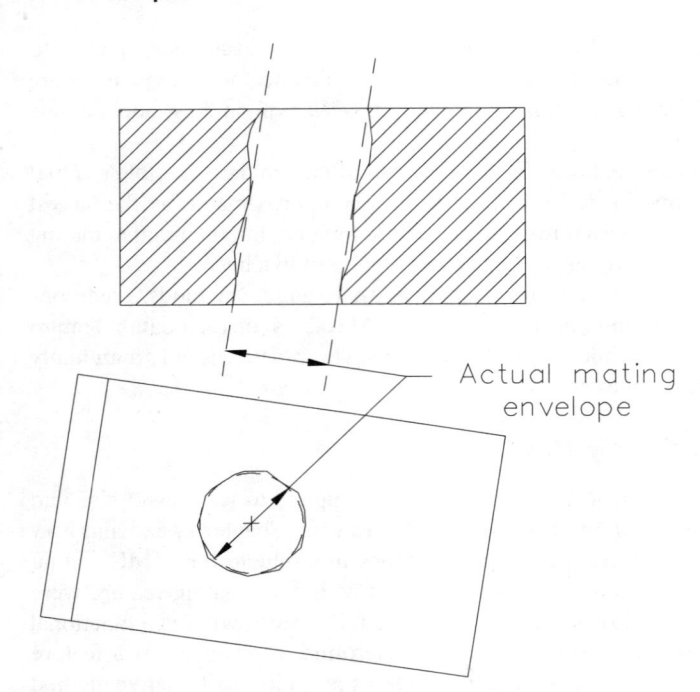

Figure 5-38 Actual mating envelope of an imperfect hole

At LMC—Bonus tolerance equals the arithmetic difference between the feature's actual minimum material size and its specified LMC size limit.

Actual minimum material size is the dimension of the actual minimum material envelope.
Actual minimum material envelope is defined according to the type of feature, as follows:
(a) *For an External Feature.* A similar perfect feature counterpart of largest size, which can be inscribed within the feature so that it just contacts the surface(s).
(b) *For an Internal Feature.* A similar perfect feature counterpart of smallest size, which can be circumscribed about the feature so that it just contacts the surface(s).
In certain cases, the orientation, or the orientation and location of an actual minimum material envelope shall be restrained to one or two datums.

Notice from Fig. 5-39 that the actual minimum material envelope is the inverse of the actual mating envelope. While the actual mating envelope resides in the "air" at the surface of a feature, the actual minimum material envelope is embedded in part material. That makes it impossible to simulate with tangible gages. The actual minimum material envelope can only be approximated by scanning point data into a computer and modeling the surface—a process called *virtual gaging* or *softgaging*.

Let's consider a cast boss that must have an adequate "shell" of part material all around for cleanup in a machining operation. If its LMC size limit is ∅.387 and its actual minimum material size is ∅.390, a "bonus" of ∅.003 shall be added to the specified geometric tolerance.

In section 5.6.3.1, we described some rare features having boundaries of perfect form at both MMC and LMC. Those features have an actual mating envelope and actual mating size that's used in the context of the geometric tolerance and/or datum reference at MMC. For the LMC context, the same feature additionally has an actual minimum material envelope and actual minimum material size. As might be apparent from Fig. 5-39, the greater the feature's form deviation (and orientation deviation, as applicable), the greater is the difference between the two envelopes and sizes.

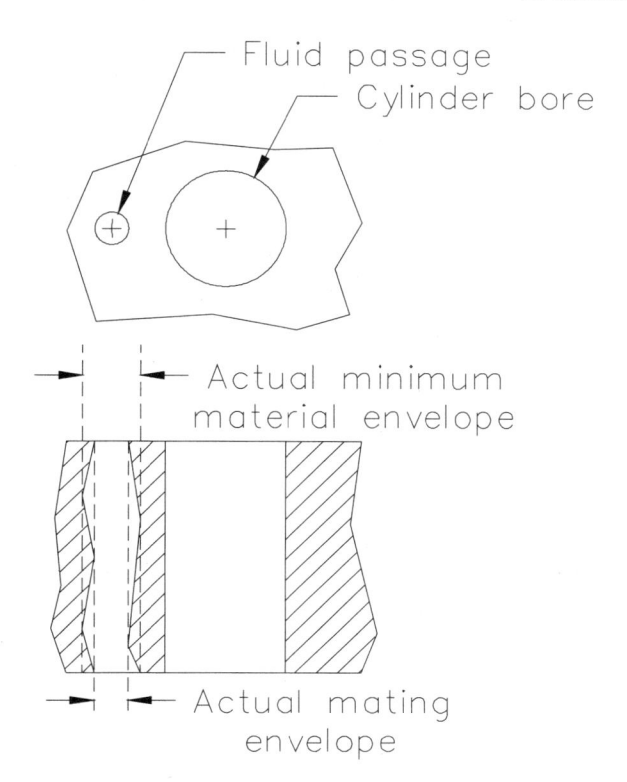

Figure 5-39 Actual minimum material envelope of an imperfect hole

5.6.5.2 Level 2 Adjustment—Actual Local Sizes

Since Level 3 and 4 tolerances impose no additional form controls, the "center method" permits use of a uniform tolerance zone and an all-encompassing envelope size. Level 2 tolerances, however, are intended to control feature form. Thus, the tolerance zone must interact with actual feature size independently at each cross section of the feature. Though the effective control is reduced from 3-D down to 2-D, inspection is paradoxically more complicated. Perhaps because there's rarely any reason to use the alternative "center method" for Level 2 tolerances, neither Y14.5 nor the Math Standard defines it thoroughly. In our own following explanations, we've extended actual mating/minimum material envelope principles to emulate accurately the controls imposed by Level 2 virtual condition boundaries.

Straightness of a Cylindrical Feature at MMC—The central tolerance zone is bounded by a revolute, within which the derived median line shall be contained. At each cross-sectional slice, the diameter of the tolerance zone varies according to the actual mating local size. Within any plane perpendicular to the axis of the actual mating envelope, *actual mating local size* is the diameter of the largest perfect circle that can be inscribed within an internal feature, or the smallest that can be circumscribed about an external feature, so that it just contacts the feature surface. The straightness tolerance zone local diameter equals the stated straightness tolerance value plus the diametral difference between the actual mating local size and the feature's MMC limit size.

At any cross section of the pin shown in Fig. 5-26, as the pin's actual mating local size approaches MMC ($\varnothing$.063), the straightness tolerance zone shrinks to the specified diameter ($\varnothing$.010). Conversely, as the pin's actual mating local size approaches LMC ($\varnothing$.062), the tolerance zone expands to $\varnothing$.011. Either way, for any pin satisfying both its size limits and its straightness tolerance, the surface of the pin will nowhere encroach beyond its $\varnothing$.073 virtual condition boundary.

Straightness of a Cylindrical Feature at LMC—The central tolerance zone is bounded by a revolute, within which the derived median line shall be contained. At each cross-sectional slice, the diameter of the tolerance zone varies according to the actual minimum material local size. Within any plane perpendicular to the axis of the actual minimum material envelope, *actual minimum material local size* is the diameter of the smallest perfect circle that can be circumscribed about an internal feature, or the largest that can be inscribed within an external feature, so that it just contacts the feature surface. The straightness tolerance zone local diameter equals the stated straightness tolerance value plus the diametral difference between the actual minimum material local size and the feature's LMC limit size.

Flatness of a Width-Type Feature at MMC or LMC—The central tolerance zone is bounded by two mirror image imperfect planes, within which the derived median plane shall be contained. At each point on the derived median plane, the corresponding local width of the tolerance zone equals the stated flatness tolerance value plus the difference between the feature's actual local size and the feature's MMC (in an MMC context) or LMC (in an LMC context) limit size. *Actual local size* is the distance between two opposite surface points intersected by any line perpendicular to the center plane of the actual mating envelope (MMC context), or of the actual minimum material envelope (LMC context).

At any cross section of the washer shown in Fig. 5-27, as the washer's actual local size approaches MMC (.034), the flatness tolerance zone shrinks to the specified width (.020). Conversely, as the washer's actual local size approaches LMC (.030), the tolerance zone expands to .024. Either way, for any washer satisfying both its size limits and its flatness tolerance, neither surface of the washer will anywhere encroach beyond the .054 virtual condition boundary.

5.6.5.3 Disadvantages of Alternative "Center Method"

By making the geometric tolerance interact with the feature's actual size, the "center method" closely emulates the preferred (virtual condition) *boundary method*. For a hypothetical perfectly formed and perfectly oriented feature, the two methods yield identical conformance results. For imperfect features, however, the Math Standard offers a detailed explanation of how the "center method" might reject a barely conforming feature, or worse, accept a slightly out-of-tolerance feature. Be very careful with older CMMs and surface plate techniques roughly employing the "center method." Generally, the boundary method will be more forgiving of marginal features, but will never accept a nonfunctional one.

The Math Standard uses actual mating size for all actual envelope size applications in RFS and MMC contexts, and applies actual minimum material size in all LMC contexts. Y14.5 does not yet recognize actual minimum material size and uses actual mating size in all contexts. In an LMC context, local voids between the feature surface and the actual mating envelope represent portions of the feature at risk for violating the LMC virtual condition boundary. Since actual mating size is unaffected by such voids, it can't provide accurate emulation of the LMC virtual condition boundary. This discrepancy causes some subtle contradictions in Y14.5's LMC coverage, which this chapter circumvents by harmonizing with the Math Standard.

5.6.6 Inner and Outer Boundaries

Many types of geometric tolerances applied to a feature of size, for example, runout tolerances, establish an *inner boundary* and/or *outer boundary* beyond which the feature surface(s) shall not encroach. Since the standards don't define feature controls in terms of these inner and outer boundaries, the boundaries are considered the result of other principles at work. See section 5.12.9. They're sometimes useful in tolerance calculations. See Chapter 9, section 9.3.3.3.

5.6.7 When Do We Use a Material Condition Modifier?

The functional differences between RFS, MMC, and LMC contexts should now be clear. Obviously, an MMC or LMC modifier can only be associated with a feature of size or a bounded feature. A modifier can only apply to a datum reference in a feature control frame, or to a straightness, flatness, orientation, or positional tolerance in a feature control frame. In all such places, we recommend designers use a modifier, either MMC or LMC, unless there is a specific requirement for the unique properties of RFS.

MMC for clearance fits—Use MMC for any feature of size that assembles with another feature of size on a mating part and the foremost concern is that the two mating features clear (not interfere with) each other. Use MMC on any datum reference where the datum feature of size itself makes a clearance fit, and the features controlled to it likewise make clearance fits. Because clearance fits are so common, and because MMC permits functional gaging, many designers have wisely adopted MMC as a default. (Previously, Y14.5 made it the default.) Where a screw thread must be controlled with GD&T or referenced as a datum, try to use MMC.

LMC for minimum stock protection—Use LMC where you must guarantee a minimum "shell" of material all over the surface of any feature of size, for example:
- For a cast, forged, or rough-machined feature to assure stock for cleanup in a subsequent finishing operation
- For a nonmating bore, fluid passage, etc., to protect minimum wall thickness for strength
- For a nonmating boss around a hole, to protect minimum wall thickness for strength
- For the gaging features of a functional gage to assure the gage won't clear a nonconforming part
- For a boss that shall completely cover a hole in the mating part

Where a fluid passage is drilled next to a cylinder bore, as shown in Fig. 5-39, the designer may be far more concerned with the thinnest wall between them than with the largest pin that can fit into the fluid passage. An MMC virtual condition boundary can't prevent a void deep down inside the hole created by an errant drill. In cases such as this, where we're more concerned with presence of material than with a clearance fit, LMC is preferred.

You don't often see LMC applied to datum features, but consider an assembly where datum features of size pilot two mating parts that must be well centered to each other. LMC applied to both datum features guarantees a minimal offset between the two parts regardless of how loose the fit. This is a valuable technique for protecting other mating interfaces in the assembly. And on functional gages, LMC is an excellent choice for datum references.

Compared to MMC, LMC has some disadvantages in gaging and evaluation. It's difficult to assess the actual minimum material size. Functional gages cannot be used.

RFS for centering—RFS is obsessed with a feature's center to the point of ignorance of the feature's actual size. In fact, RFS allows no dynamic interaction between size and location or between size and orientation of a feature. However, this apparent limitation of RFS actually makes it an excellent choice for self-centering mating interfaces where the mating features always fit together snugly and center on each other regardless of their actual mating sizes. Examples of self-centering mating interfaces include the following:
- Press fits
- Tapers, such as Morse tapers and countersinks for flat-head screws
- Elastic parts or elastic intermediate parts, such as O-rings
- An adjustable interface where an adjusting screw, shim, sleeve, etc., will be used in assembly to center a mating part
- Glued or potted assemblies

In such interfaces, it's obvious to the designer that the actual sizes of the mating features have no relevance to the allowable orientation or positional tolerance for those features. In the case of an external O-ring groove, for example, MMC would be counterproductive, allowing eccentricity to increase as diameter size gets smaller. Here, RFS is the wiser choice.

There are certain geometric characteristics, such as runout and concentricity, where MMC and LMC are so utterly inappropriate that the rules prohibit material condition modifiers. For these types of tolerances, RFS always applies.

Y14.5 allows RFS to be applied to any tolerance and any datum reference in conjunction with any feature of size having a defined center. In fact, RFS principles now apply by default in the absence of any material condition modifier. (Note that's different from earlier editions of Y14.5.) But RFS is versatile like a monkey wrench. You can use it on everything, but for most of your choices, there is a more suitable tool (MMC or LMC) that will fit the work better and cost less. For example, RFS is a poor choice in clearance-fit mating interfaces because it doesn't allow dynamic tolerance interaction. That means smaller tolerances, usable parts rejected, and higher costs.

Remember that RFS principles are based on a feature's center. To verify most RFS controls, the inspector must derive the center(s) of the involved feature(s). Functional gages with fixed-size elements cannot be used with RFS. RFS applied to a feature pattern referenced as a datum, or to any type of feature for which Y14.5 doesn't define a center, is sure to provoke a debate somewhere and waste more money.

FAQ: *Should I use RFS instead of MMC whenever I need greater precision?*

A: Not always. A tolerance applied RFS is more restrictive than an equal tolerance modified to MMC. That fact leads to the common misconception that RFS is therefore a more precise tool. This is like comparing the precision of a saw and a hammer. We've tried to emphasize the differences between MMC, LMC, and RFS. Each tool is the most precise for its intended function. RFS works differently from MMC, often with different rules and different results. As a broadly general statement based on drawings we've seen, MMC is hugely underused, LMC is somewhat underused, and RFS is hugely overused.

FAQ: *Why, then, is RFS now the default?*

A: For what it's worth, the default now agrees with the ISO 8015 standard. It's like "training wheels" for users who might fail to comprehend properly and apply RFS where it's genuinely needed.

5.7 Size Limits (Level 1 Control)

For every feature of size, the designer shall specify the largest and the smallest the feature can be. In section 5.6.1, we discussed the exact requirements these size limits impose on the feature. The standards provide three options for specifying size limits on the drawing: symbols for limits and fits, *limit dimensioning*, and *plus and minus tolerancing*. Where tolerances directly accompany a dimension, it's important to coordinate the number of decimal places expressed for each value to prevent confusion. The rules depend on whether the dimension and tolerance values are expressed in inches or millimeters.

5.7.1 Symbols for Limits and Fits

Inch or metric size limits may be indicated using a standardized system of preferred sizes and fits. Using this system, standard feature sizes are found in tables in ANSI B4.1 (inch) or ANSI B4.2 (metric), then expressed on the drawing as a basic size followed by a tolerance symbol, for example, ⌀.625 LC5 or 30 f7.

For other fit conditions, limits must be calculated using tables in the standard's appendix that list deviations from the basic size for each tolerance zone symbol (alphanumeric designation). When introducing this system in an organization, it's a good idea to show as reference either the basic size and tolerance symbol, or the actual MMC and LMC limits.

5.7.2   Limit Dimensioning

The minimum and maximum limits may be specified directly. Place the high limit (maximum value) above the low limit (minimum value). When expressed in a single line, place the low limit preceding the high limit with a dash separating the two values.

$$\varnothing {.500 \atop .495} \qquad \text{or} \qquad \varnothing .495\text{--}.500$$

5.7.3   Plus and Minus Tolerancing

The nominal size may be specified, followed by plus and minus tolerance values.

$$.497 {+.003 \atop -.002} \qquad \text{or} \qquad .500 \pm .005$$

5.7.4   Inch Values

In all dimensions and tolerances associated with a feature, the number of decimal places shall match. It may be necessary to add one or more trailing zeros to some values. Express each plus and minus tolerance with the appropriate plus or minus sign.

$$.500 {+.005 \atop -.000} \qquad \text{not} \qquad .500 {+.005 \atop 0}$$

$$.500 \pm .005 \qquad \text{not} \qquad .50 \pm .005$$

$$ {.750 \atop .748} \qquad \text{not} \qquad {.75 \atop .748}$$

.310		.31
with	not	with
⊕ ⌀.008 Ⓜ A B C		⊕ ⌀.008 Ⓜ A B C

5.7.5   Millimeter Values

For any value less than one millimeter, precede the decimal point with a zero.

$$0.9 \qquad \text{not} \qquad .9$$

Eliminate unnecessary trailing zeros.

25.1	not	25.10
12	not	12.0

32		32.00
with	not	with
⌖ ⌀0.25Ⓜ A B C		⌖ ⌀0.25Ⓜ A B C

The exceptions are limit dimensions and bilateral (plus and minus) tolerances, where the number of decimal places shall match. It may be necessary to add a decimal point and one or more trailing zeros to some values. Plus and minus tolerances are each expressed with the appropriate plus or minus sign.

$$\begin{array}{ccc} 25.45 & & 25.45 \\ 25.00 & \text{not} & 25 \end{array}$$

$$32 {}^{+0.25}_{-0.10} \quad \text{not} \quad 32 {}^{+0.25}_{-0.1}$$

For unilateral tolerances, express the nil value as a single zero digit with no plus or minus sign.

$$32 {}^{\;\;0}_{-0.02} \quad \text{or} \quad 32 {}^{+0.02}_{\;\;0}$$

5.8 Form (Only) Tolerances (Level 2 Control)

In section 5.6.1, we described how imaginary balls define for a feature of size MMC and LMC size limit boundaries. For a cylindrical or spherical feature, these boundaries control to some degree the circularity of the feature at each cross section. In section 5.6.3.1, we described how Rule #1 imposes on a feature of size a default boundary of perfect form at MMC. This perfect-form boundary controls to some degree the straightness of a cylindrical feature's surface or the flatness of a width-type feature's surfaces. A boundary of perfect form at LMC imposes similar restraint. The level of form control provided by size limits and default boundaries of perfect form is adequate for most functional purposes. However, there are cases where a generous tolerance for overall feature size is desirable, but would allow too much surface undulation. Rather than reduce the size tolerance, a separate form (only) tolerance may be added. For most features of size, such a separate form tolerance must be less than the size tolerance to have any effect.

A form (only) tolerance is specified on the drawing using a feature control frame displaying one of the four form (only) characteristic symbols, followed by the tolerance value. Only two types of form tolerance may be meaningfully modified to MMC or LMC. Since form tolerances have no bearing on orientation or location relationships between features, datum references are meaningless and prohibited. Each type of form tolerance works differently and has different application rules.

5.8.1 Straightness Tolerance for Line Elements

Where a straightness tolerance feature control frame is placed according to option (b) in Table 5-1 (leader-directed to a feature surface or attached to an extension line of a feature surface), the tolerance controls only line elements of that surface. The feature control frame may only appear in a view where the controlled surface is represented by a straight line. The tolerance specifies a tolerance zone plane containing a tolerance zone bounded by two parallel lines separated by a distance equal to the tolerance value. As the tolerance zone plane sweeps the entire feature surface, the surface's intersection with the plane shall everywhere be contained within the tolerance zone (between the two lines). Within the plane, the orientation and location of the tolerance zone may adjust continuously to the part surface while sweeping. See Fig. 5-40.

Of a Cylindrical or Conical Feature—The straightness tolerance zone plane shall be swept radially about the feature's axis, always containing that axis. (Note that the axis of a cone isn't explicitly defined.) Within the rotating tolerance zone plane, the tolerance zone's orientation relative to the feature axis may adjust continuously. Since Rule #1 already controls a cylinder's surface straightness within size limits, a separate straightness tolerance applied to a cylindrical feature must be less than the size tolerance to be meaningful.

Of a Planar Feature—The orientation and sweep of the tolerance zone plane is not explicitly related to any other part feature. The plane is merely implied to be parallel to the view plane and swept perpendicular to the view plane (toward and away from the viewer). Again, the zone itself may tilt and shift within the tolerance zone plane to accommodate gross surface undulations. See Fig. 5-40. Where it's important to relate the tolerance zone plane to datums, specify instead a profile of a line tolerance, as described in section 5.13.8.

For a width-type feature of size, Rule #1 automatically limits the flatness and straightness deviation of each surface—no extra charge. Thus, to have any meaning, a separate straightness tolerance applied to either single surface must be less than the total size tolerance.

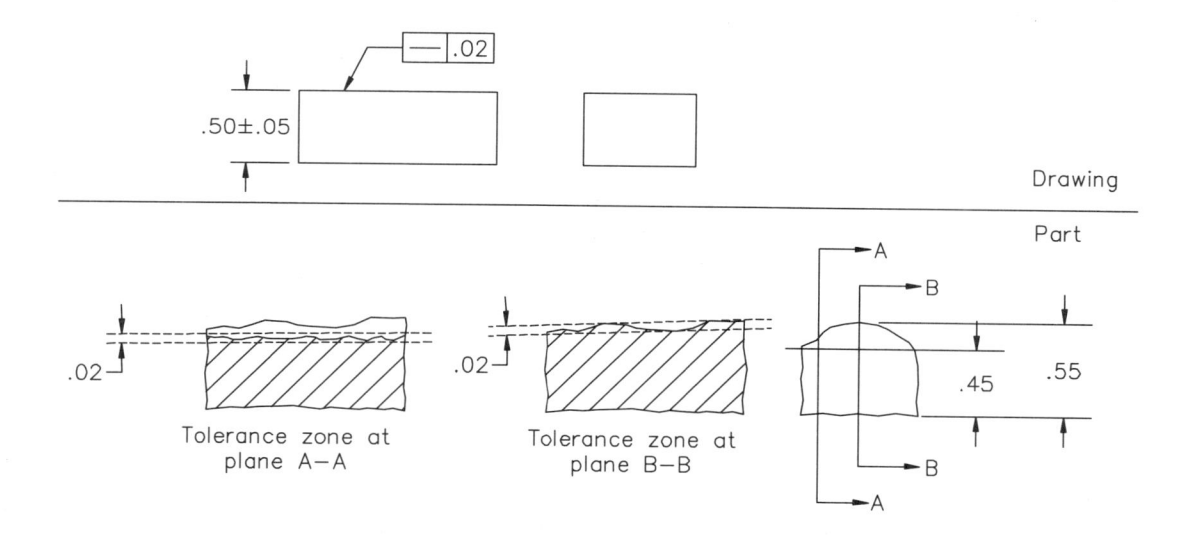

Figure 5-40 Straightness tolerance for line elements of a planar feature

5.8.2 Straightness Tolerance for a Cylindrical Feature

A straightness tolerance feature control frame placed according to options (a) or (d) in Table 5-1 (associated with a diameter dimension) replaces Rule #1's requirement for perfect form at MMC with a separate tolerance controlling the overall straightness of the cylindrical feature. Where the tolerance is modified to MMC or LMC, it establishes a Level 2 virtual condition boundary as described in section 5.6.3.1 and Figs. 5-17(b) and 5-18(b). Alternatively, the "center method" described in section 5.6.5.2 may be applied to a straightness tolerance at MMC or LMC, but there's rarely any benefit to offset the added complexity. Unmodified, the tolerance applies RFS and establishes a central tolerance zone as described in section 5.6.4.1, within which the feature's derived median line shall be contained.

5.8.3 Flatness Tolerance for a Single Planar Feature

Where a flatness tolerance feature control frame is placed according to options (b) or (c) in Table 5-1 (leader-directed to a feature or attached to an extension line from the feature), the tolerance applies to a single nominally flat feature. The flatness feature control frame may be applied only in a view where the element to be controlled is represented by a straight line. This specifies a tolerance zone bounded by two parallel planes separated by a distance equal to the tolerance value, within which the entire feature surface shall be contained. The orientation and location of the tolerance zone may adjust to the part surface. See Fig. 5-41. A flatness tolerance cannot control whether the surface is fundamentally concave, convex, or stepped; just the maximum range between its highest and lowest undulations.

For a width-type feature of size, Rule #1 automatically limits the flatness deviation of each surface. Thus, to have any meaning, a separate flatness tolerance applied to either single surface must be less than the total size tolerance.

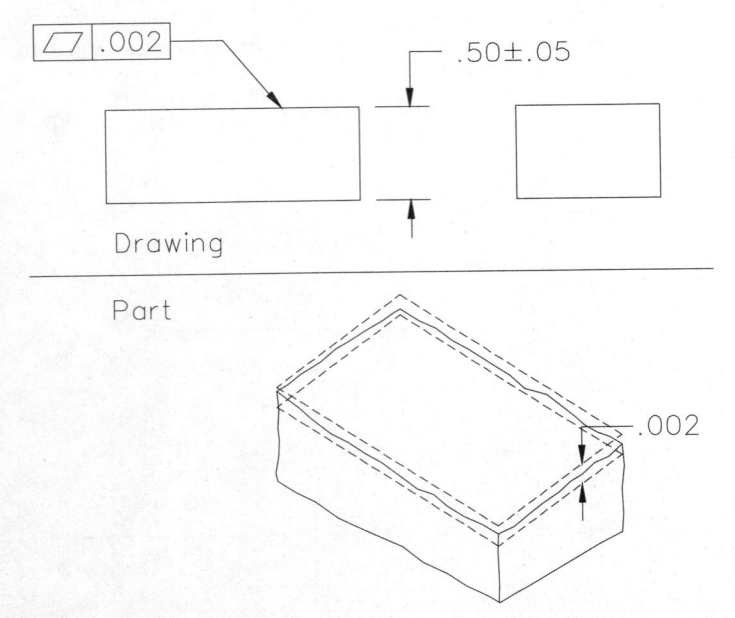

Figure 5-41 Flatness tolerance for a single planar feature

5.8.4 Flatness Tolerance for a Width-Type Feature

A flatness tolerance feature control frame placed according to options (a) or (d) in Table 5-1 (associated with a width dimension) replaces Rule #1's requirement for perfect form at MMC with a separate tolerance controlling the overall flatness of the width-type feature. Where the tolerance is modified to MMC or

LMC, it establishes a Level 2 virtual condition boundary as described in section 5.6.3.1 and Figs. 5-17(b) and 5-18(b). Alternatively, the "center method" described in section 5.6.5.2 may be applied to a flatness tolerance at MMC or LMC, but there's rarely any benefit to offset the added complexity. Unmodified, the tolerance applies RFS and establishes a central tolerance zone as described in section 5.6.4.1, within which the feature's derived median plane shall be contained.

This application of a flatness tolerance is an extension of the principles of section 5.8.2. Y14.5 suggests an equivalent control using the "straightness" characteristic symbol. We think it's inappropriate to establish a parallel plane tolerance zone using the straightness symbol. However, where strict adherence to Y14.5 is needed, then straightness symbol should be used.

5.8.5 Circularity Tolerance

A circularity tolerance controls a feature's *circularity* (roundness) at individual cross sections. Thus, a circularity tolerance may be applied to any type of feature having uniformly circular cross sections, including spheres, cylinders, revolutes (such as cones), tori (doughnut shapes), and bent rod and tubular shapes.

Where applied to a nonspherical feature, the tolerance specifies a tolerance zone plane containing an annular (ring-shaped) tolerance zone bounded by two concentric circles whose radii differ by an amount equal to the tolerance value. See Fig. 5-42. The tolerance zone plane shall be swept along a simple, nonself-

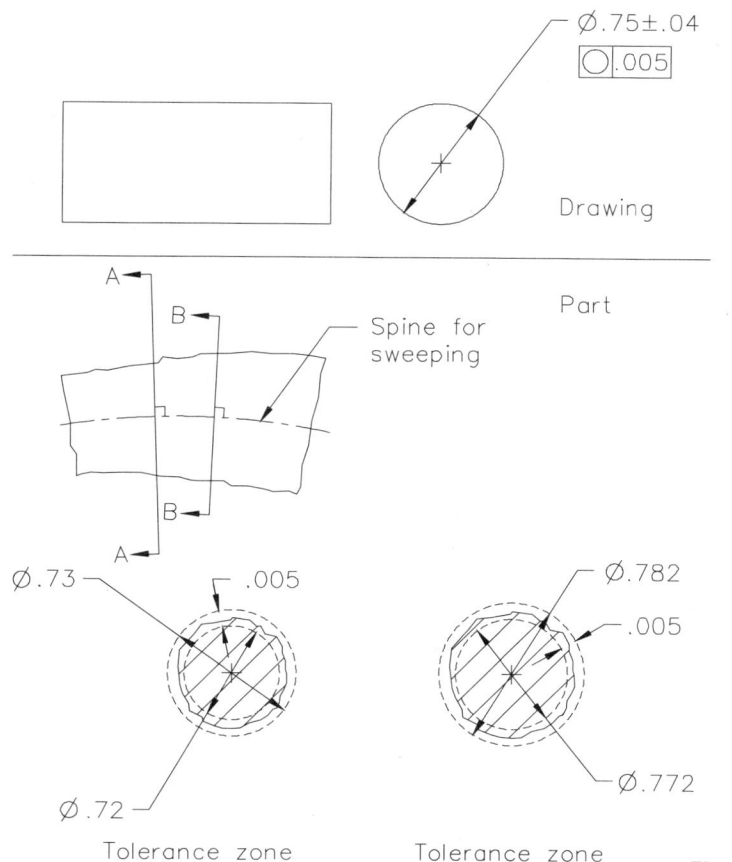

Figure 5-42 Circularity tolerance (for nonspherical features)

5-54   Chapter Five

intersecting, tangent-continuous curve (spine). At each point along the spine, the tolerance zone plane shall be perpendicular to the spine and the tolerance zone centered on the spine. As the tolerance zone plane sweeps the entire feature surface, the surface's intersection with the plane shall everywhere be contained within the annular tolerance zone (between the two circles). While sweeping, the tolerance zone may continually adjust in overall size, but shall maintain the specified radial width. This effectively removes diametral taper from circularity control. Additionally, the spine's orientation and curvature may be adjusted within the aforementioned constraints. This effectively removes axial straightness from circularity control. The circularity tolerance zone need not be concentric with either size limit boundary.

A circularity tolerance greater than the total size tolerance has no effect. A circularity tolerance between the full size tolerance and one-half the size tolerance limits only single-lobed (such as D-shaped and egg-shaped) deviations. A circularity tolerance must be less than half the size tolerance to limit multi-lobed (such as elliptical and tri-lobed) deviations.

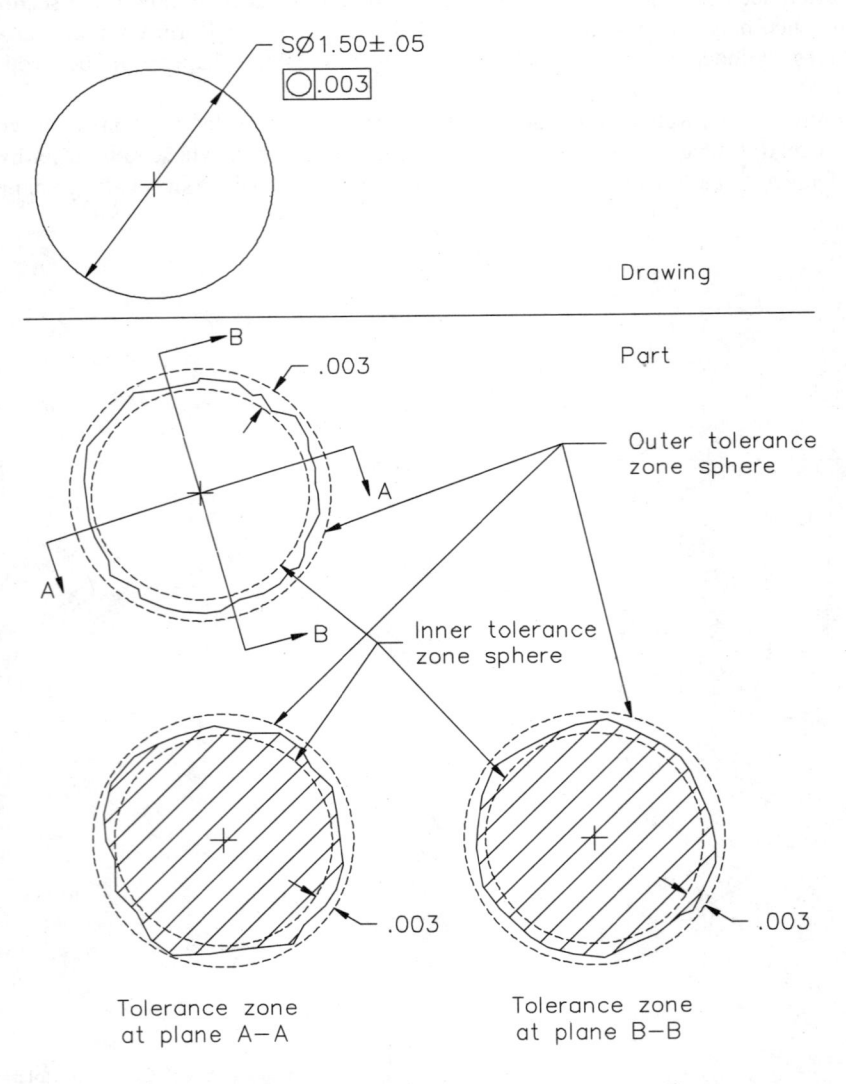

Figure 5-43 Circularity tolerance applied to a spherical feature

Note that Y14.5's explanation refers to an "axis," which could be interpreted as precluding curvature of the spine. Either way, most measuring equipment can only inspect circularity relative to a straight line.

5.8.5.1 Circularity Tolerance Applied to a Spherical Feature

The standards also use a tolerance zone plane to explain a circularity tolerance applied to a spherical feature. Since any pair of surface points can be included in such a plane, their respective distances from a common center shall not differ by more than the circularity tolerance. Therefore, the explanation can be simplified as follows: The tolerance specifies a tolerance zone bounded by two concentric spheres whose radii differ by an amount equal to the tolerance value. The tolerance zone may adjust in overall size, but shall maintain the specified radial width. All points on the considered spherical feature shall be contained within the tolerance zone (between the two spheres). See Fig. 5-43. Since the tolerance zone need not be concentric with either size limit boundary, a circularity tolerance must be less than half the size tolerance to limit multi-lobed form deviations.

5.8.6 Cylindricity Tolerance

A cylindricity tolerance is a composite control of form that includes circularity, straightness, and taper of a cylindrical feature. A cylindricity tolerance specifies a tolerance zone bounded by two concentric cylinders whose radii differ by an amount equal to the tolerance value. See Fig. 5-44. The entire feature surface shall be contained within the tolerance zone (between the two cylinders). The tolerance zone cylinders may adjust to any diameter, provided their radial separation remains equal to the tolerance value. This effectively removes feature size from cylindricity control. As with circularity tolerances, a cylindricity tolerance must be less than half the size tolerance to limit multi-lobed form deviations. Since neither a cylindricity nor a circularity tolerance can nullify size limits for a feature, there's nothing to be gained by modifying either tolerance to MMC or LMC.

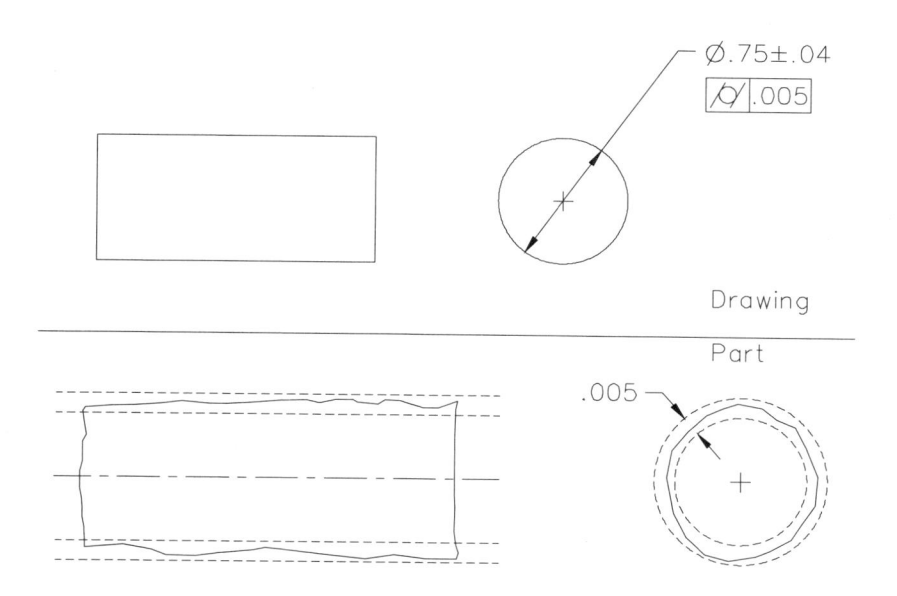

Figure 5-44 Cylindricity tolerance

5.8.7 Circularity or Cylindricity Tolerance with Average Diameter

The thin-wall nylon bushing shown in Fig. 5-45 is typical of a nonrigid part having diameters that fit rather closely with other parts in assembly. If customary diameter size limits were specified, no matter how liberal, their inherent circularity control would be overly restrictive for the bushing in its free state (unassembled). The part's diameters in the free state cannot and need not stay as round as they'll be once restrained in assembly. We need a different way to control size-in-assembly, while at the same time guarding against collapsed or grotesquely out-of-round bushings that might require excessive assembly force or jam in automated assembly equipment.

The solution is to specify limits for the feature's *average diameter* along with a generous circularity tolerance. Where a diameter tolerance is followed by the note AVG, the size limit boundaries described in section 5.6.1 do not apply. Instead, the tolerance specifies limits for the feature's average diameter. *Average diameter* is defined somewhat nebulously as the average of at least four two-point diameter measurements. A contact-type gage may deflect the part, yielding an unacceptable measurement. Where practicable, average diameter may be found by dividing a peripheral tape measurement by π. When the part is restrained in assembly, its effective mating diameter should correspond closely to its average diameter in the free state.

Though we told you our nylon bushing is a nonrigid part, the drawing itself (Fig. 5-45) gives no indication of the part's rigidity. In particular, there's no mention of restraint for verification as described in section 5.5.1. Therefore, according to Fundamental Rule *(l)*, a drawing user shall interpret all dimensions and tolerances, including the circularity tolerance, as applying in the free state. The standard

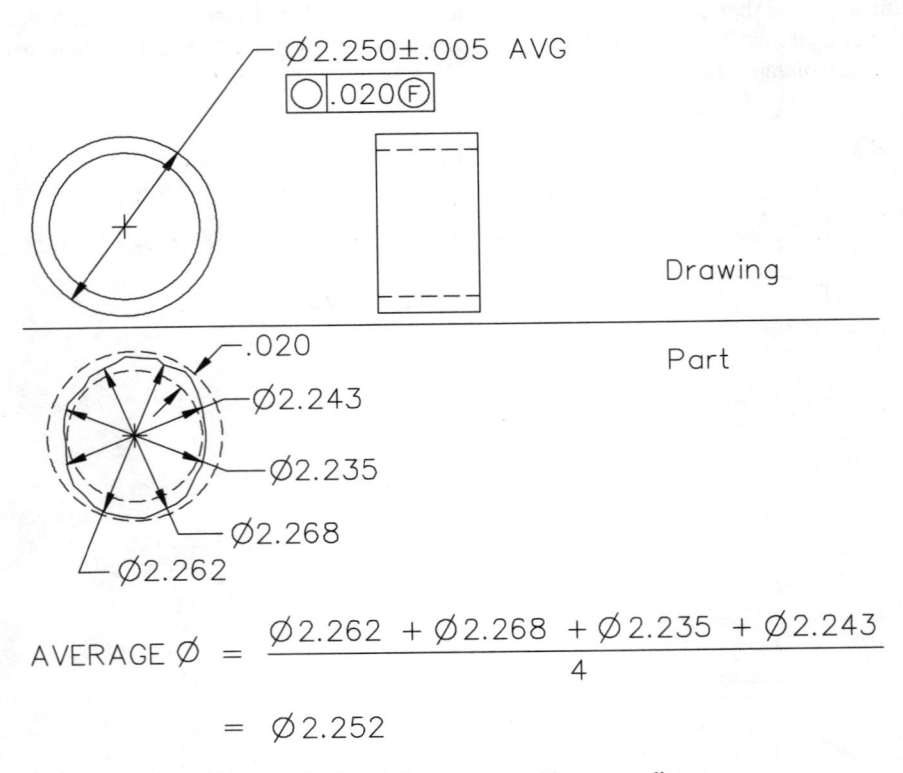

$$\text{AVERAGE } \emptyset = \frac{\emptyset 2.262 + \emptyset 2.268 + \emptyset 2.235 + \emptyset 2.243}{4}$$

$$= \emptyset 2.252$$

Figure 5-45 Circularity tolerance with average diameter

implies average diameter can only be used in conjunction with the "free state" symbol. For that reason only, we've added the "free state" symbol after the circularity tolerance value. A feature's conformance to both tolerances shall be evaluated in the free state—that is, with no external forces applied to affect its size or form.

The same method may be applied to a longer nonrigid cylindrical feature, such as a short length of vinyl tubing. Simply specify a relatively liberal cylindricity tolerance modified to "free state," along with limits for the tube's average diameter.

5.8.8 Application Over a Limited Length or Area

Some designs require form control over a limited length or area of the surface, rather than the entire surface. In such cases, draw a heavy chain line adjacent to the surface, basically dimensioned for length and location as necessary. See Fig. 5-46. The form tolerance applies only within the limits indicated by the chain line.

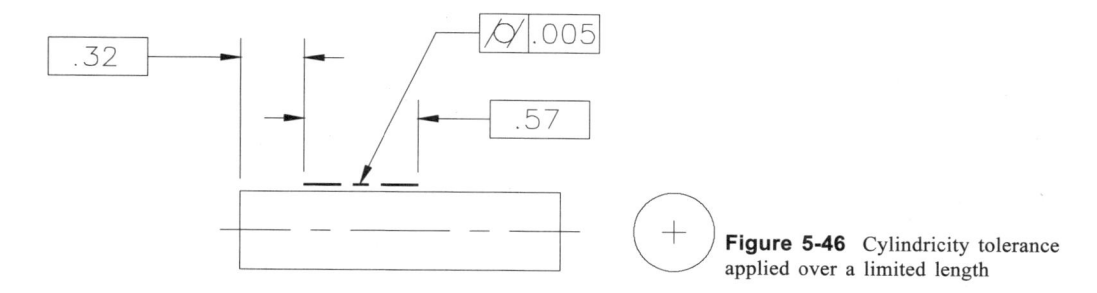

Figure 5-46 Cylindricity tolerance applied over a limited length

5.8.9 Application on a Unit Basis

There are many features for which the design could tolerate a generous amount of form deviation, provided that deviation is evenly distributed over the total length and/or breadth of the feature. This is usually the case with parts that are especially long or broad in proportion to their cross-sectional areas.

The 6' piece of bar stock shown in Fig. 5-47 could be severely bowed after heat-treating. But if the bar is then sawed into 6" lengths, we're only concerned with how straight each 6" length is. The laminated honeycomb panel shown in Fig. 5-48 is an airfoil surface. Gross flatness of the entire surface can reach .25". However, any abrupt surface variation within a relatively small area, such as a dent or wrinkle, could disturb airflow over the surface, degrading performance.

These special form requirements can be addressed by specifying a form (only) tolerance on a unit basis. The size of the unit length or area, for example 6.00 or 3.00 X 3.00, is specified to the right of the form tolerance value, separated by a slash. This establishes a virtual condition boundary or tolerance zone as usual, except limited in length or area to the specified dimension(s). As the limited boundary or tolerance zone sweeps the entire length or area of the controlled feature, the feature's surface or derived element (as applicable) shall conform at every location.

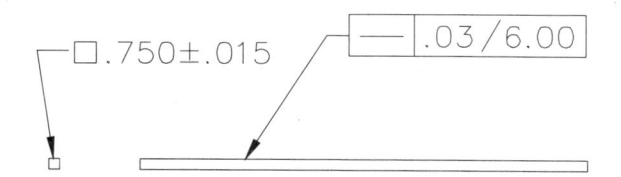

Figure 5-47 Straightness tolerance applied on a unit basis

5-58   Chapter Five

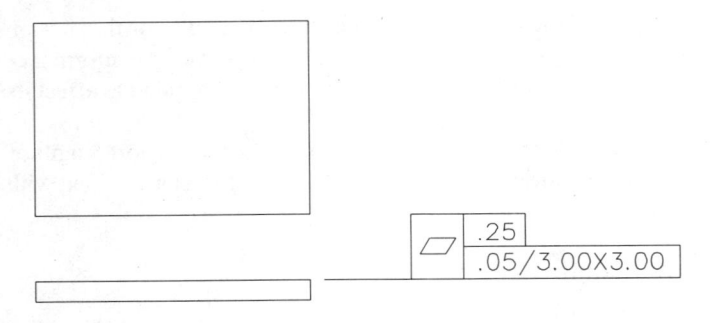

Figure 5-48 Flatness tolerance applied on a unit basis

Since the bar stock in Fig. 5-47 may be bowed no more than .03" in any 6" length, its accumulated bow over 6' cannot exceed 4.38". The automated saw can handle that. In contrast, the airfoil in Fig. 5-48 may be warped as much as .05" in any 3 x 3" square. Its maximum accumulated warp over 36" is 6.83". A panel that bowed won't fit into the assembly fixture. Thus, for the airfoil, a compound feature control frame is used, containing a single "flatness" symbol with two stacked segments. The upper segment specifies a flatness tolerance of .25" applicable to the entire surface. The lower segment specifies flatness per unit area, not to exceed .05" in any 3 x 3" square. Obviously, the per-unit tolerance value must be less than the total-feature tolerance.

5.8.10   Radius Tolerance

A *radius* (plural, *radii*) is a portion of a cylindrical surface encompassing less than 180° of arc length. A radius tolerance, denoted by the symbol R, establishes a zone bounded by a minimum radius arc and a maximum radius arc, within which the entire feature surface shall be contained. As a default, each arc shall be tangent to the adjacent part surfaces. See Fig. 5-49. Where a center is drawn for the radius, as in Fig. 5-50, two concentric arcs of minimum and maximum radius bound the tolerance zone. Within the tolerance zone, the feature's contour may be further refined with a "controlled radius" tolerance, as described in the following paragraph.

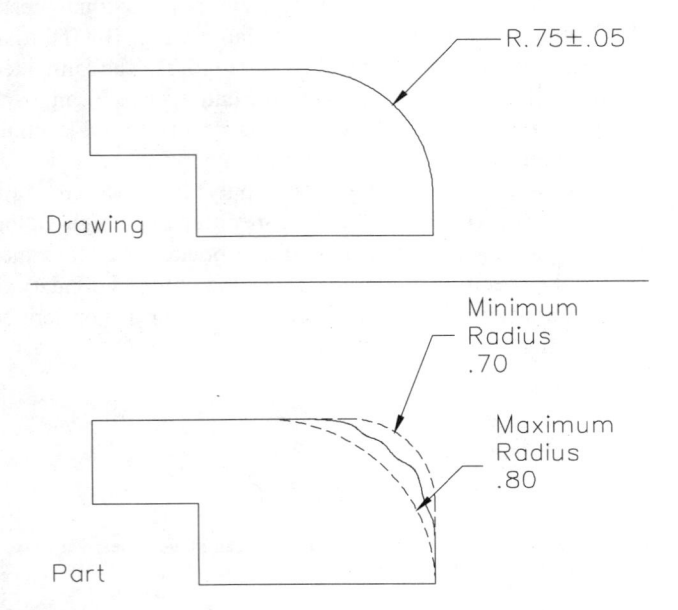

Figure 5-49 Radius tolerance zone (where no center is drawn)

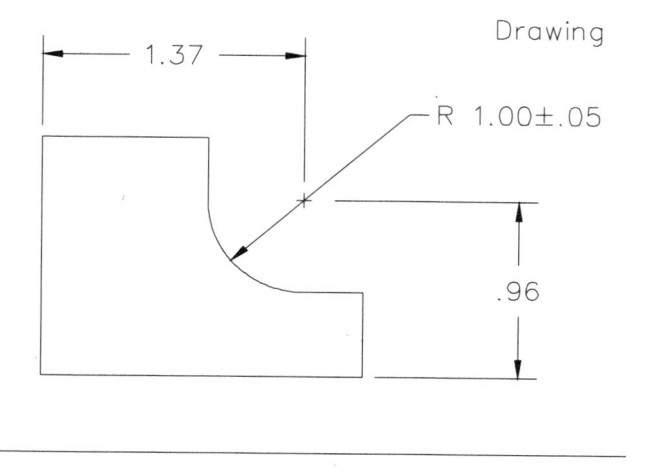

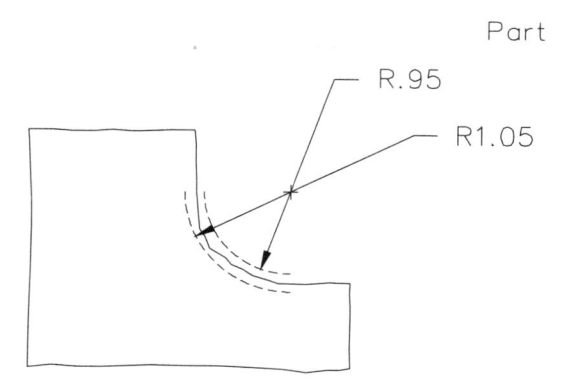

Figure 5-50 Radius tolerance zone where a center is drawn

5.8.10.1 Controlled Radius Tolerance

Where the symbol **CR** is applied to a radius, the tolerance zone is as described in section 5.8.10, but there are additional requirements for the surface. The surface contour shall be **a fair curve without reversals**. We interpret this to mean a tangent-continuous curve that is everywhere concave or convex, as shown in Fig. 5-51. Before the 1994 Revision of Y14.5, there was no **CR** symbol, and these additional controls applied to every radius tolerance. The standard implies that **CR** can only apply to a tangent radius, but we feel that by extension of principle, the refinement can apply to a "centered" radius as well.

5.8.11 Spherical Radius Tolerance

A *spherical radius* is a portion of a spherical surface encompassing less than 180° of arc length. A spherical radius tolerance, denoted by the symbol SR, establishes a zone bounded by a minimum radius arc and a maximum radius arc, within which the entire feature surface shall be contained. As a default, each arc shall be tangent to the adjacent part surfaces. Where a center is drawn for the radius, two concentric spheres of minimum and maximum radius bound the tolerance zone. The standards don't address "controlled radius" refinement for a spherical radius.

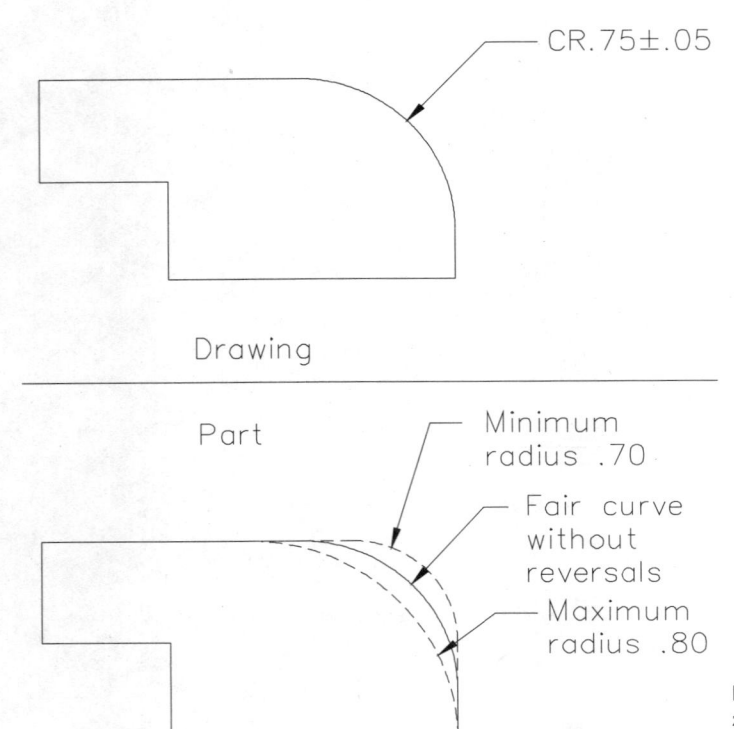

Figure 5-51 Controlled radius tolerance zone

5.8.12 When Do We Use a Form Tolerance?

As we explain in the next section, datum simulation methods can accommodate warped and/or out-of-round datum features. However, datum simulation will usually be more repeatable and error free with well-formed datum features. We discuss this further in section 5.9.12.

As a general rule, apply a form (only) tolerance to a nondatum feature only where there is some risk that the surface will be manufactured with form deviations severe enough to cause problems in subsequent manufacturing operations, inspection, assembly, or function of the part. For example, a flatness tolerance might be appropriate for a surface that seals with a gasket or conducts heat to a heat sink. A roller bearing might be controlled with a cylindricity tolerance. A conical bearing race might have both a straightness of surface elements tolerance and a circularity tolerance. However, such a conical surface might be better controlled with profile tolerancing as explained in section 5.13.11.

FAQ: *If feature form can be controlled with profile tolerances, why do we need all the form tolerance symbols?*

A: In section 5.13.11, we explain how profile tolerances may be used to control straightness or flatness of features. While such applications are a viable option, most drawing users prefer to see the "straightness" or "flatness" characteristic symbols because those symbols convey more information at a glance.

5.9 Datuming

5.9.1 What Is a Datum?

According to the dictionary, a *datum* is a single piece of information. In logic, a datum may be a given starting point from which conclusions may be drawn. In surveying, a datum is any level surface, line, or point used as a reference in measuring. Y14.5's definition embraces all these meanings.

A *datum* is **a theoretically exact point, axis, or plane derived from the true geometric counterpart of a specified datum feature. A datum is the origin from which the location or geometric characteristics of features of a part are established.**

A *datum feature* is **an actual feature of a part that is used to establish a datum.**

A *datum reference* is an alpha letter appearing in a compartment following the geometric tolerance in a feature control frame. It specifies a datum to which the tolerance zone or acceptance boundary is basically related. A feature control frame may have zero, one, two, or three datum references.

The diagram in Fig. 5-52 shows that a "datum feature" begets a "true geometric counterpart," which begets a "datum," which is the building block of a "datum reference frame," which is the basis for tolerance zones for other features. Even experts get confused by all this, but keep referring to Fig. 5-52 and we'll sort it out one step at a time.

5.9.2 Datum Feature

In section 5.1.5, we said the first step in GD&T is to "identify part surfaces to serve as origins and provide specific rules explaining how these surfaces establish the starting point and direction for measurements." Such a part surface is called a *datum feature*.

According to the Bible, about five thousand years ago, God delivered some design specifications for a huge water craft to a nice guy named Noah. "Make thee an ark of gopher wood… The length of the ark shall be three hundred cubits, the breadth of it fifty cubits, and the height of it thirty cubits." Modern scholars are still puzzling over the ark's material, but considering the vessel would be half again bigger than a football field, Noah likely had to order material repeatedly, each time telling his sons, "Go fer wood." For the "height of thirty cubits" dimension, Noah's sons, Shem and Ham, made the final measurement from the level ground up to the top of the "poop" deck, declaring the measured size conformed to the Holy Specification "close enough." Proudly looking on from the ground, Noah was unaware he was standing on the world's first datum feature!

Our point is that builders have long understood the need for a consistent and uniform origin from which to base their measurements. For the ancients, it was a patch of leveled ground; for modern manufacturers, it's a flat surface or a straight and round diameter on a precision machine part. Although any type of part feature can be a datum feature, selecting one is a bit like hiring a sheriff who will provide a strong moral center and direction for the townsfolk. What qualifications should we look for?

5.9.2.1 Datum Feature Selection

The most important quality you want in a datum feature (or a sheriff) is leadership. A good datum feature is a surface that most strongly influences the orientation and/or location of the part in its assembly. We call that a "functional" datum feature. Rather than being a slender little wisp, a good datum feature, such

5-62 Chapter Five

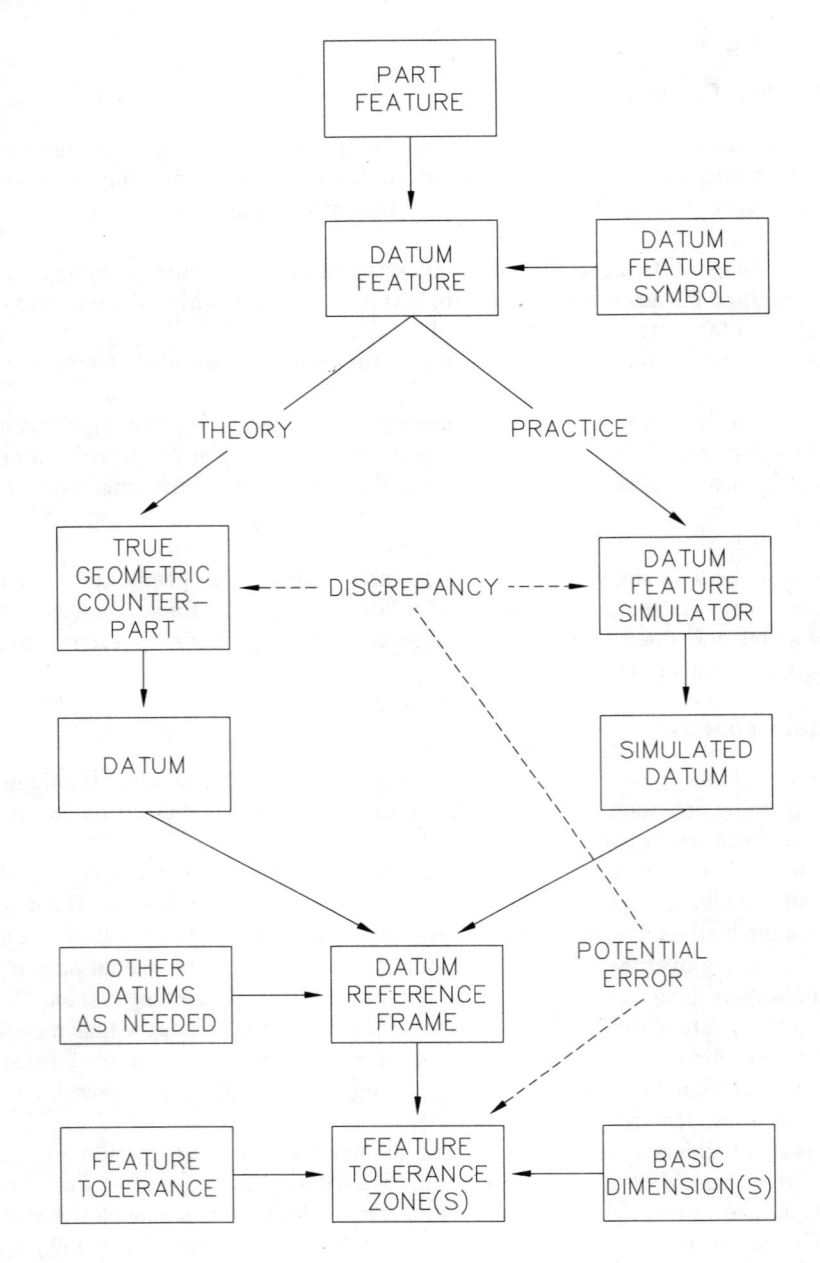

Figure 5-52 Establishing datum reference frames from part features

as that shown in Fig. 5-53, should have "broad shoulders" able to take on the weight of the part and provide stability. Look for a "straight arrow" with an even "temperament" and avoid "moody" and unfinished surfaces with high and low spots. Just as you want a highly visible sheriff, choose a datum feature that's likewise always accessible for fixturing during manufacturing, or for inspection probing at various stages of completion.

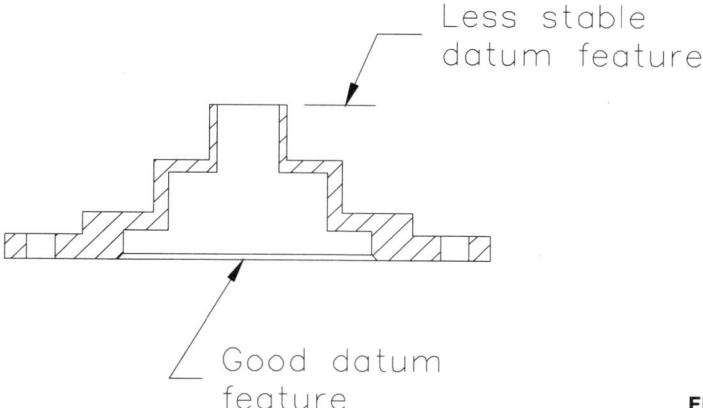

Figure 5-53 Selection of datum features

5.9.2.2 Functional Hierarchy

It's tough to judge leadership in a vacuum, but you can spot it intuitively when you see how a prospect relates to others. Fig. 5-54 shows three parts of a car engine: engine block, cylinder head, and rocker arm cover. Intuitively, we rank the dependencies of the pieces: The engine block is our foundation to which we bolt on the cylinder head, to which we in turn bolt on the rocker arm cover. And in fact, that's the

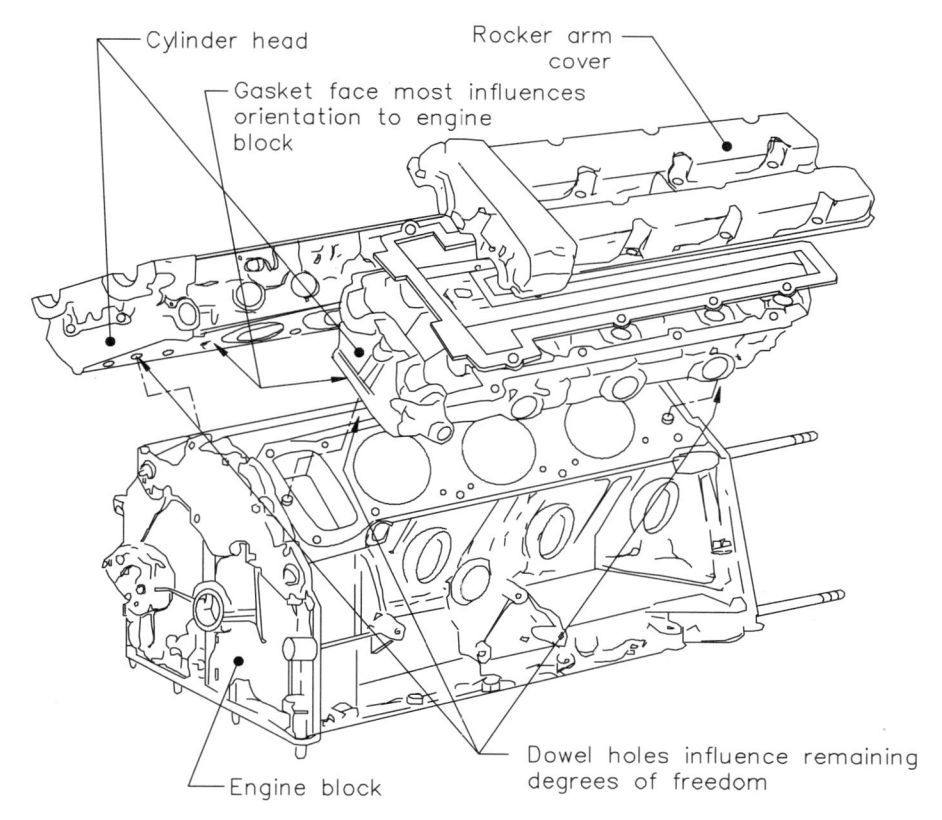

Figure 5-54 Establishing datums on an engine cylinder head

typical assembly sequence. Thus, in "interviewing" candidates for datum feature on the cylinder head, we want the feature that most influences the head's orientation to the engine block. A clear choice would be the bottom (head gasket) face. The two dowel holes are the other key players, influencing the remaining degree of orientation as well as the location of the head on the block. These datum features, the bottom face and the dowel holes, satisfy all our requirements for good, functional datum features. To select the upper surface of the cylinder head (where the rocker cover mounts) as a datum feature for the head seems backwards—counterintuitive.

In our simple car engine example, functional hierarchy is based on assembly sequence. In other types of devices, the hierarchy may be influenced or dominated by conflicting needs such as optical alignment. Thus, datum feature selection can sometimes be as much art as science. In a complicated assembly, two experts might choose different datum features.

5.9.2.3 Surrogate and Temporary Datum Features

Often, a promising candidate for datum feature has all the leadership, breadth, and character we could ever hope for and would get sworn in on the spot if only it weren't so reclusive or inaccessible. There are plenty of other factors that can render a functional datum feature useless to us. Perhaps it's an O-ring groove diameter or a screw thread—those are really tough to work with. In such cases, it may be wiser to select a nonfunctional *surrogate datum feature*, as we've done in Fig. 5-55. A prudent designer might choose a broad flange face and a convenient outside diameter for surrogate datum features even though in assembly they contact nothing but air.

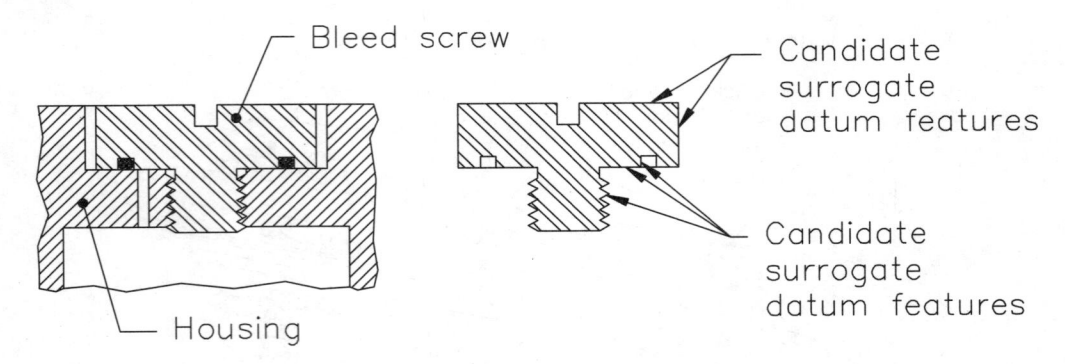

Figure 5-55 Selecting nonfunctional datum features

Many parts require multiple steps, or *operations*, in multiple machines for their manufacture. Such parts, especially castings and forgings, may need to be fixtured or inspected even before the functional datum features are finished. A thoughtful designer will anticipate these manufacturing needs and identify some *temporary datum features* either on an intermediate operation drawing or on the finished part drawing.

The use of surrogate and temporary datum features often requires extra precautions. These nonfunctional surfaces may have to be made straighter, rounder, and/or smoother than otherwise necessary. Also, the relationship between these features and the real, functional features may have to be closely controlled to prevent tolerances from stacking up excessively. There is a cost tradeoff in passing over functional datum features that may be more expensive to work with in favor of nonfunctional datum features that may be more expensive to manufacture.

5.9.2.4 Identifying Datum Features

Once a designer has "sworn in" a datum feature, he needs to put a "badge" on it to denote its authority. Instead of a star, we use the "datum feature" symbol shown in Fig. 5-56. The symbol consists of a capital letter enclosed in a square frame, a leader line extending from the frame to the datum feature, and a terminating triangle. The triangle may optionally be solid filled, making it easier to spot on a busy drawing.

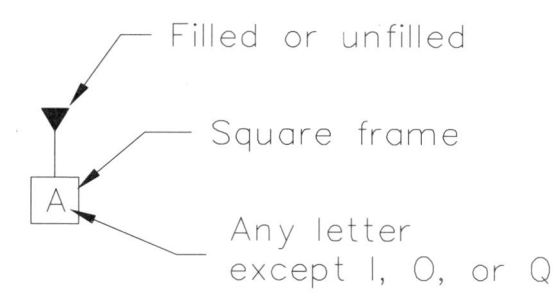

Figure 5-56 Datum feature symbol

Each datum feature shall be identified with a different letter of the alphabet (except I, O, or Q). When the alphabet is exhausted, double letters (AA through AZ, BA through BZ, etc.) are used and the frame is elongated to fit. Datum identifying letters have no meaning except to differentiate datum features. Though letters need not be assigned sequentially, or starting with A, there are advantages and disadvantages to doing both. In a complicated assembly, it may be desirable to coordinate letters among various drawings, so that the same feature isn't B on the detail part drawing, and C on the assembly drawing. It can be confusing when two different parts in an assembly both have a datum feature G and those features don't mate. On the other hand, someone reading one of the detail part drawings can be frustrated looking for nonexistent datums where letters are skipped. Such letter choices are usually left to company policy, and may be based on the typical complexity of the company's drawings.

The datum feature symbol is applied to the concerned feature surface outline, extension line, dimension line, or feature control frame as follows:

(a) **placed on the outline of a feature surface, or on an extension line of the feature outline, clearly separated from the dimension line, when the datum feature is the surface itself.** See Fig. 5-57(a).

(b) **placed on an extension of the dimension line of a feature of size when the datum is the axis or center plane. If there is insufficient space for the two arrows, one of them may be replaced by the datum feature triangle.** See Fig. 5-57(b).

(c) **placed on the outline of a cylindrical feature surface or an extension line of the feature outline, separated from the size dimension, when the datum is the axis.** The triangle may be drawn tangent to the feature. See Fig. 5-57(c).

(d) **placed on a dimension leader line to the feature size dimension where no geometrical tolerance and feature control frame are used.** See Fig. 5-57(d).

(e) **placed on the planes established by datum targets on complex or irregular datum features** (see section 5.9.13.6)**, or to reidentify previously established datum axes or planes on repeated or multisheet drawing requirements. Where the same datum feature symbol is repeated to identify the same feature in other locations of a drawing, it need not be identified as reference.**

(f) **placed above or below and attached to the feature control frame when the feature (or group of features) controlled is the datum axis or datum center plane.** See Fig. 5-57(e).

(g) placed on a chain line that indicates a partial datum feature.

Formerly, the "datum feature" symbol consisted of a rectangular frame containing the datum-identifying letter preceded and followed by a dash. Because the symbol had no terminating triangle, it was placed differently in some cases.

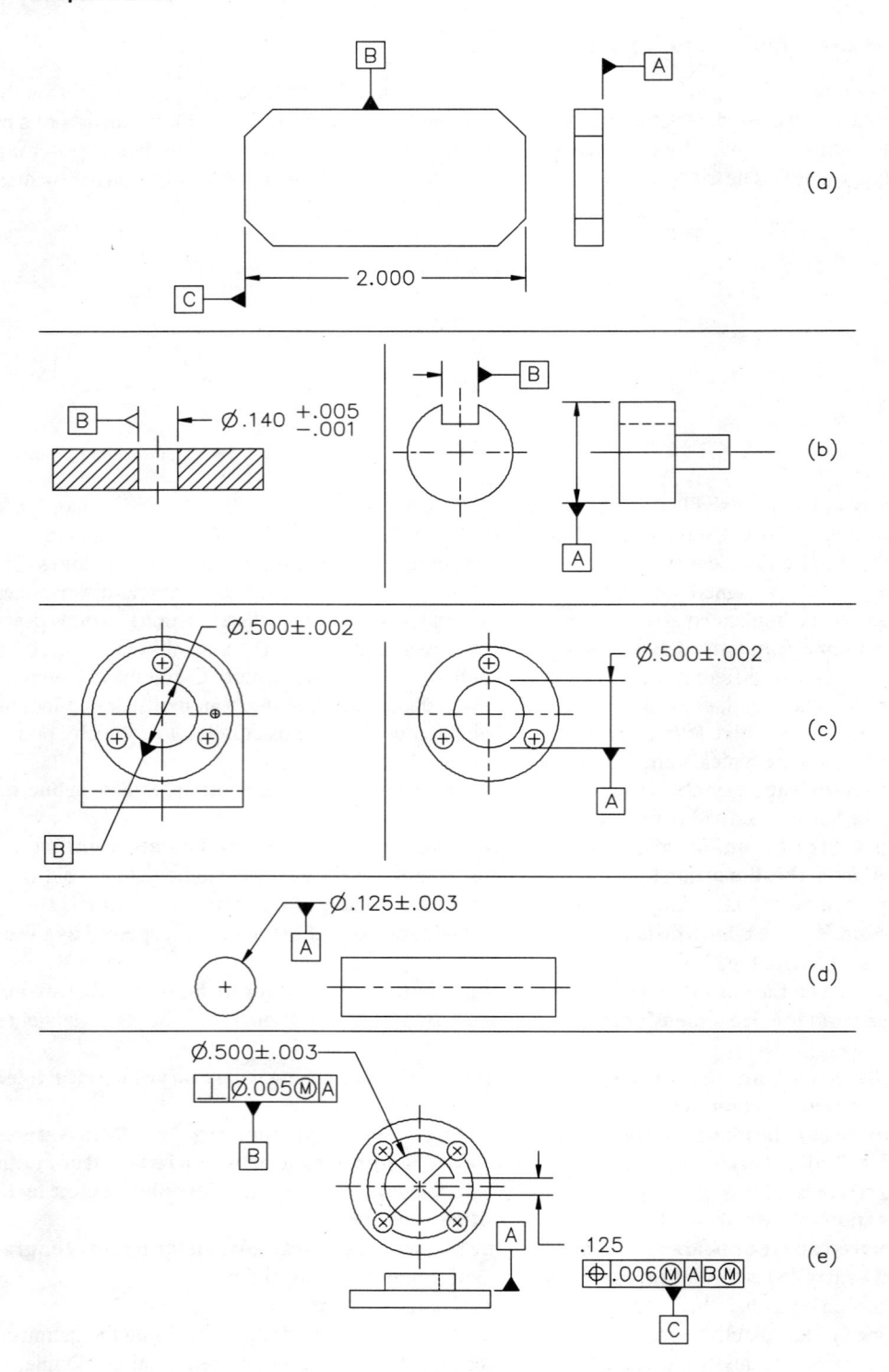

Figure 5-57 Methods of applying datum feature symbols

5.9.3 True Geometric Counterpart (TGC)—Introduction

Simply deputizing a part surface as a "datum feature" still doesn't give us the uniform origin necessary for highly precise measurements. As straight, flat, and/or round as that feature may be, it still has slight irregularities in its shape that could cause differences in repeated attempts to reckon from it. To eliminate such measurement variation, we need to reckon from a geometric shape that's, well, perfect. Such a perfect shape is called a *true geometric counterpart (TGC)*.

If we look very closely at how parts fit together in Fig. 5-58, we see they contact each other only at a few microscopic points. Due to infinitesimal variations and irregularities in the manufacturing process, these few peaks or *high points* stand out from the surrounding part surface. Now, we realize that when parts are clamped together with bolts and other fastening forces, sometimes at thousands of pounds per square inch, surface points that were once the elite "high" get brutally mashed down with the rank and file. Flanges warp and bores distort. Flat head screws stretch and bend tortuously as their cones squash into countersinks. We hope these plastic deformations and realignments are negligible in proportion to assembly tolerances. In any event, we lack the technology to account for them. Thus, GD&T's datum principles are based on the following assumptions: 1) The foremost design criterion is matability; and 2) high points adequately represent a part feature's matability. Thus, like it or not, all datum methods are based on surface high points.

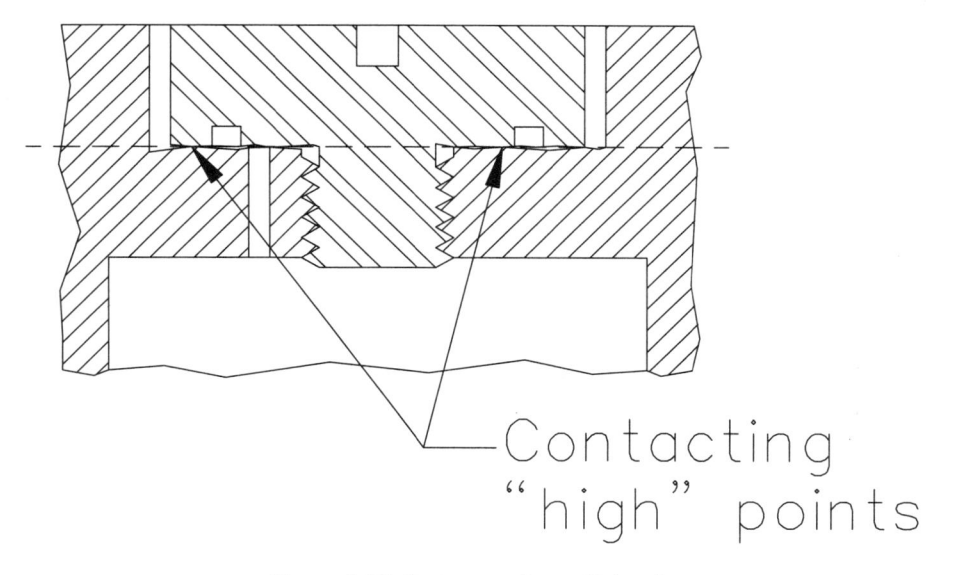

Figure 5-58 Parts contacting at high points

From Table 5-4, you'll notice for every datum feature, there's at least one TGC (perfect shape) that's related to its surface high points. In many cases, the TGC and the datum feature surface are conceptually brought together in space to where they contact each other at one, two, or three high points on the datum feature surface. In some cases, the TGC is custom fitted to the datum feature's high points. In yet other cases, the TGC and datum feature surface are meant to clear each other. We'll explain the table and the three types of relationships in the following sections.

Table 5-4 Datum feature types and their TGCs

Datum Feature Type	Datum Precedence	True Geometric Counterpart (TGC)	Restraint of TGC*	Contact Points	Typical Datum Simulator(s)
nominally flat plane	primary	tangent plane	none	1-3	surface plate or other flat base
	secondary or tertiary	tangent plane	O	1-2	restrained square or fence
math-defined (contoured) plane	primary	tangent math-defined contour	none	1-6	contoured fixture
	secondary or tertiary	tangent math-defined contour	O	1-2	restrained contoured fixture
feature of size, RFS	primary	actual mating envelope	none	3-4	adjustable-size chuck, collet, or mandrel; fitted gage pin, ring, or Jo blocks
	secondary or tertiary	actual mating envelope	O	2-3	same as for primary (above), but restrained
feature of size, MMC	primary	boundary of perfect form at MMC	none	0-4	gage pin, ring, or Jo blocks, at MMC size
	primary w/straightness or flatness tol at MMC	MMC virtual condition boundary	none	0-4	gage pin, ring, or Jo blocks, at MMC virtual condition size
	secondary or tertiary	MMC virtual condition boundary	O,L	0-2	restrained pin, hole, block, or slot, at MMC virtual condition size
feature of size, LMC	primary	boundary of perfect form at LMC	none	0-4	computer model at LMC size
	primary w/straightness or flatness tol at LMC	LMC virtual condition boundary	none	0-4	computer model at LMC virtual condition size
	secondary or tertiary	LMC virtual condition boundary	O,L	0-2	computer model at LMC virtual condition size
bounded feature, MMC	primary	MMC profile boundary	none	0-5	fixture or computer model
	secondary or tertiary	MMC virtual condition boundary	O,L	0-3	fixture or computer model
bounded feature, LMC	primary	LMC profile boundary	none	0-5	computer model
	secondary or tertiary	LMC virtual condition boundary	O,L	0-3	computer model

* to higher-precedence datum(s)   O = restrained in orientation, L = restrained in location

5.9.4 Datum

Remember the definition: A *datum* is **a theoretically exact point, axis, or plane derived from the true geometric counterpart of a specified datum feature.** Once we have a TGC for a feature, it's simple to derive the datum from it based on the TGC's shape. This is shown in Table 5-5.

Table 5-5 TGC shape and the derived datum

TGC SHAPE	DERIVED DATUM
tangent plane	identical plane
math-defined contour	3 mutually perpendicular planes (complete DRF)
sphere	(center) point
cylinder	axis (straight line)
opposed parallel planes	(center) plane
revolute	axis and point along axis
bounded feature	2 perpendicular planes

5.9.5 Datum Reference Frame (DRF) and Three Mutually Perpendicular Planes

Datums can be thought of as building blocks used to build a dimensioning grid called a *datum reference frame (DRF)*. The simplest DRFs can be built from a single datum. For example, Fig. 5-59(a) shows how a datum plane provides a single dimensioning axis with a unique orientation (perpendicular to the plane) and an origin. This DRF, though limited, is often sufficient for controlling the orientation and/or location of other features. Fig. 5-59(b) shows how a datum axis provides one dimensioning axis having an orientation with no origin, and two other dimensioning axes having an origin with incomplete orientation. This DRF is adequate for controlling the coaxiality of other features.

Simple datums may be combined to build a 2-D Cartesian coordinate system consisting of two perpendicular axes. This type of DRF may be needed for controlling the location of a hole. Fig. 5-60 shows the ultimate: a 3-D Cartesian coordinate system having a dimensioning axis for height, width, and depth. This top-of-the-line DRF has three mutually perpendicular planes and three mutually perpendicular axes. Each of the three planes is perpendicular to each of the other two. The line of intersection of each pair of planes is a dimensioning axis having its origin at the point where all three axes intersect. Using this DRF, the orientation and location of any type of feature can be controlled to any attitude, anywhere in space. Usually, it takes two or three datums to build this complete DRF.

Since each type of datum has different abilities, it's not very obvious which ones can be combined, nor is it obvious how to build the DRF needed for a particular application. In the following sections, we'll help you select datums for each type of tolerance. In the meantime, we'll give you an idea of what each datum can do.

5.9.6 Datum Precedence

Where datums are combined to build a DRF, they shall always be basically (perfectly) oriented to each other. In some cases, two datums shall also be basically located, one to the other. Without that perfect alignment, the datums won't define a unique and unambiguous set of mutually perpendicular planes or axes.

5-70 Chapter Five

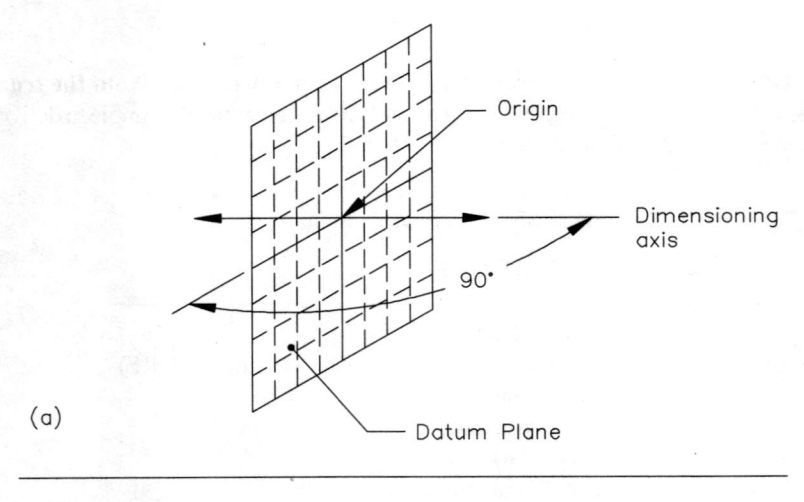

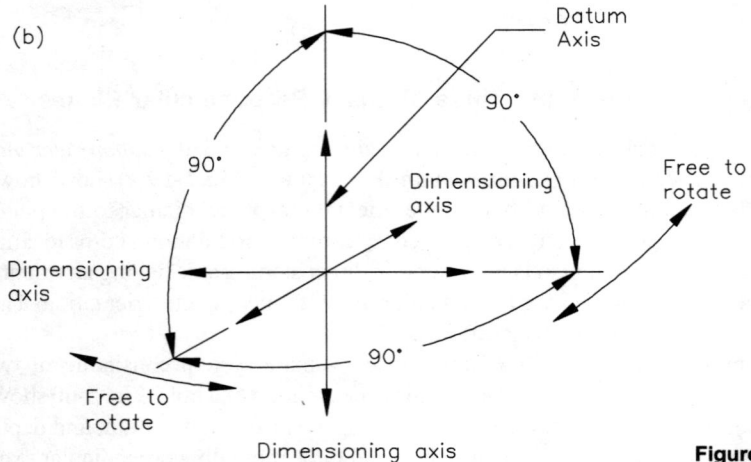

(a)

(b)

Figure 5-59 Building a simple DRF from a single datum

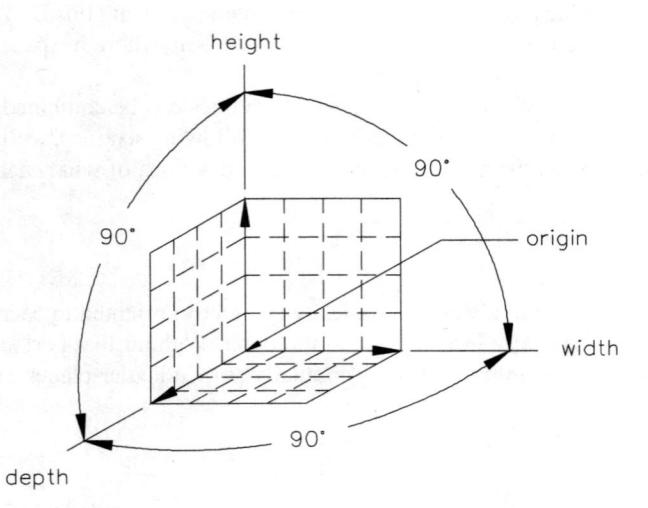

Figure 5-60 3-D Cartesian coordinate system

In functional hierarchy, Fig. 5-61's "cover" is a part that will be mounted onto a "base." The cover's broad face will be placed against the base, slid up against the fences on the base, then spot welded in place. Using our selection criteria for functional datum features, we've identified the cover's three planar mounting features as datum features A, B, and C. Considered individually, the TGC for each datum feature is a full-contact tangent plane. Since the datum feature surfaces are slightly out-of-square to each other, their full-contact TGCs would likewise be out-of-square to each other, as would be the three datum planes derived from them. Together, three out-of-square datum planes cannot yield a unique DRF. We need the three datum (and TGC) planes to be mutually perpendicular. The only way to achieve that is to excuse at least two of the TGC planes from having to make full contact with the cover's datum features.

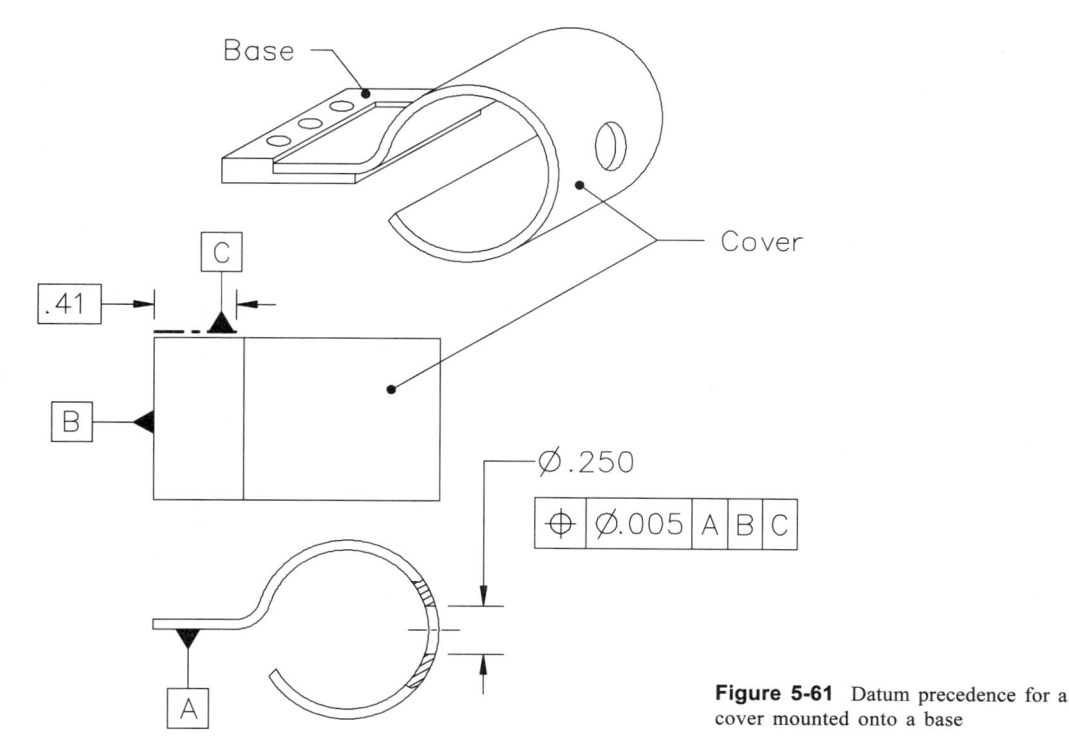

Figure 5-61 Datum precedence for a cover mounted onto a base

On the other hand, if we allow each of the three TGCs to contact only a single high point on its respective datum feature, we permit a wide variety of alignment relationships between the cover and its TGCs. Intuitively, we wouldn't expect the cover to assemble by making only one-point contact with the base. And certainly, this scheme is no good if we want repeatability in establishing DRFs. Instead, we should try to maximize contact between our datum feature surfaces and their TGC planes. Realizing we can't have full contact on all three surfaces, we'll have to prioritize the three datum features, assigning each a different requirement for completeness of contact.

Using the same criteria by which we selected datum features A, B, and C in the first place, we examine the leadership each has over the cover's orientation and location in the assembly. We conclude that datum feature A, being the broad face that will be clamped against the base, is the most influential. The datum feature B and C edges will be pushed up against fences on the base. Datum feature B, being longer, will tend to overpower datum feature C in establishing the cover's rotation in assembly. However, datum feature C will establish a unique location for the cover, stopping against its corresponding fence on the base.

Thus, we establish *datum precedence* for the cover, identifying datum A as the *primary datum*, datum B as the *secondary datum*, and datum C as the *tertiary datum*. We denote datum precedence by placing the datum references sequentially in individual compartments of the feature control frame. The tolerance compartment is followed by the primary datum compartment, followed by the secondary datum compartment, followed by the tertiary datum compartment. In text, we can express the same precedence A|B|C. The specified datum precedence tells us how to prioritize establishment of TGCs, allowing us to fit three mutually perpendicular TGC planes to our out-of-square cover. Here's how it works.

5.9.7 Degrees of Freedom

Let's start with a system of three mutually perpendicular TGC planes as shown in Fig. 5-62(a). For discussion purposes, let's label one plane "A," one "B," and one "C." The lines of intersection between each pair of planes can be thought of as axes, "AB," "BC," and "CA." Remember, this is a system of TGC planes, not a DRF (yet).

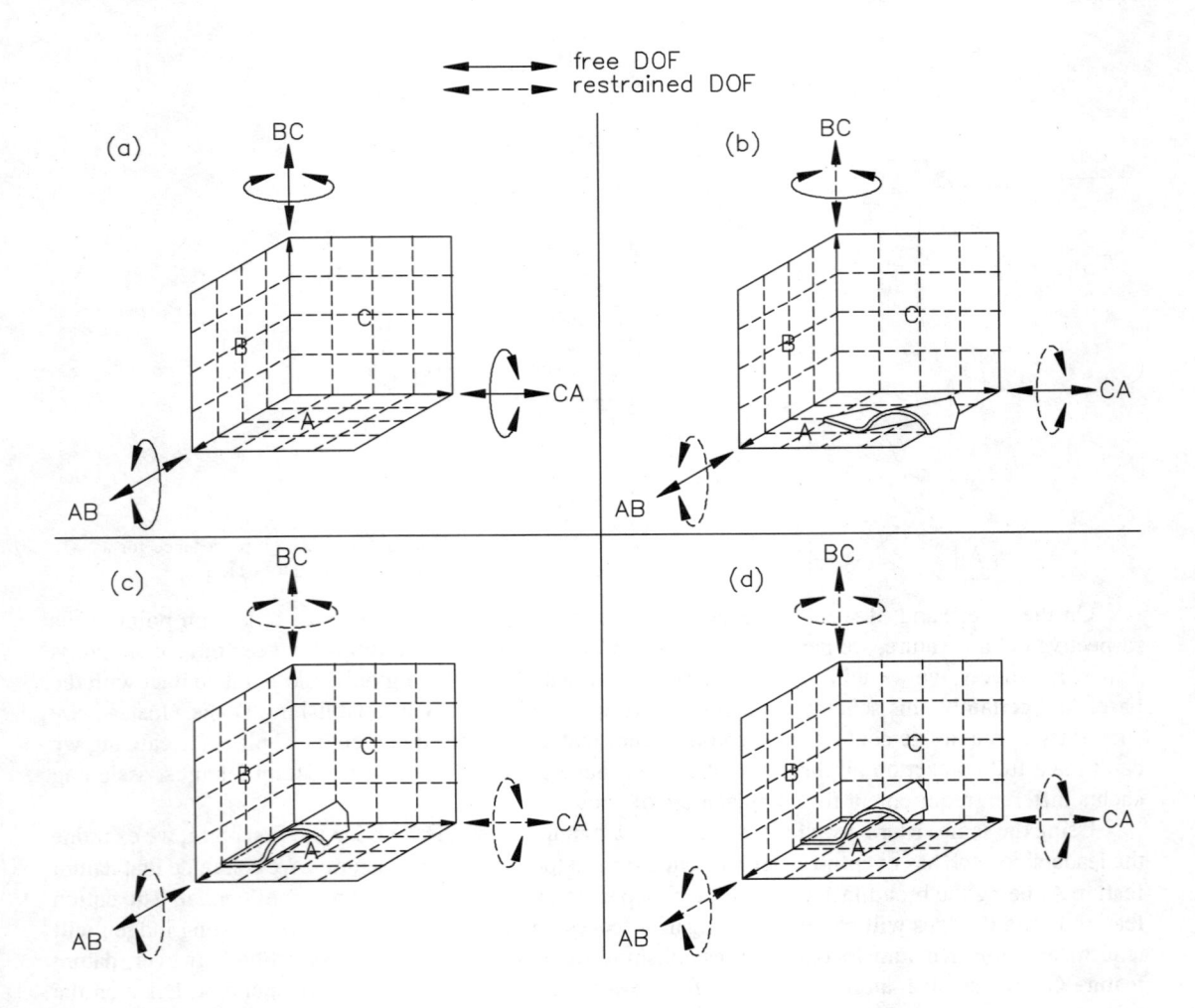

Figure 5-62 Arresting six degrees of freedom between the cover and the TGC system

Imagine the cover floating in space, tumbling all about, and drifting in a randomly winding motion relative to our TGC system. (The CMM users among you can imagine the cover fixed in space, and the TGC system floating freely about—Albert Einstein taught us it makes no difference.) We can describe all the relative free-floating motion between the cover and the TGC system as a combination of *rotation* and *translation* (linear movement) parallel to each of the three TGC axes, AB, BC, and CA. These total *six degrees of freedom*. In each portion of Fig. 5-62, we represent each degree of freedom with a double-headed arrow. To achieve our goal of fixing the TGC system and cover together, we must arrest each one of the six degrees of relative motion between them. Watch the arrows; as we restrain each degree of freedom, its corresponding arrow will become dashed.

Each datum reference in the feature control frame demands a level of congruence (in this case, contact) between the datum feature and its TGC plane. The broad face of the cover is labeled datum feature A, the primary datum feature. That demands maximum congruence between datum feature A and TGC plane A. Fig. 5-62(b) shows the cover slamming up tight against TGC plane A and held there, as if magnetically. Suddenly, the cover can no longer rotate about the AB axis, nor can it rotate about the CA axis. It can no longer translate along the BC axis. Three degrees of freedom arrested, just like that. (Notice the arrows.) However, the cover is still able to twist parallel to the BC axis and translate at will along the AB and CA axes. We'll have to put a stop to that.

The long edge of the cover is labeled datum feature B, the secondary datum feature. Fig. 5-62(c) illustrates the cover sliding along plane A, slamming up tight against plane B and held there. However, this time the maximum congruence possible is limited. As the cover slides, all three degrees of freedom arrested by any higher precedence datum feature—datum feature A in this case—shall remain arrested. Thus, datum feature B can only arrest degrees of freedom left over from datum feature A. This means the cover can't rotate about the BC axis anymore, nor can it translate along the CA axis. Two more degrees of freedom are now arrested. We've reduced the cover to sliding to and fro in a perfectly straight line parallel to axis AB. One more datum reference should finish it off.

The short edge of the cover is labeled datum feature C, the tertiary datum feature. Fig. 5-62(d) now shows the cover sliding along axis AB, slamming up tight against plane C and held there. Again, the maximum congruence possible is even more limited. As the cover slides, all degrees of freedom arrested by higher precedence datum features—three by datum feature A and two by datum feature B—shall remain arrested. Thus, datum feature C can only arrest the last remaining degree of freedom, translation along axis AB. Finally, all six degrees of freedom have been arrested; the cover and its three TGC planes are now totally stuck together.

The next steps are to derive the datum from each TGC, then construct the DRF from the three datums. Since we used such a simple example, in this case, the datums are the same planes as the TGCs, and the three mutually perpendicular planes of the DRF are the very same datum planes. Sometimes, it's just that simple!

Because we were so careful in selecting and prioritizing the cover's datum features according to their assembly functions, the planes of the resulting DRF correspond as closely as possible to the mating surfaces of the base. That's important because it allows us to maximize tolerances for other features controlled to our DRF. Just as importantly, we can unstick the cover, set it toppling and careening all over again, then repeat the above three alignment steps. No matter who tells it, no matter who performs it, no matter which moves, TGCs or cover, the cover's three datum features and their TGC planes will always slam together exactly the same. We'll always get the same useful DRF time after time.

"Always," that is, when datum precedence remains the same, A|B|C. Note that in Fig. 5-63(a), the DRF's orientation was optimized for the primary datum feature, A, first and foremost. The orientation was only partly optimized for the secondary datum feature, B. Orientation was not optimized at all for the tertiary datum feature, C. If we transpose datum precedence to A|C|B, as in Fig. 5-63(b), our first alignment step remains the same. We still optimize orientation of the TGC system to datum feature A. However, now

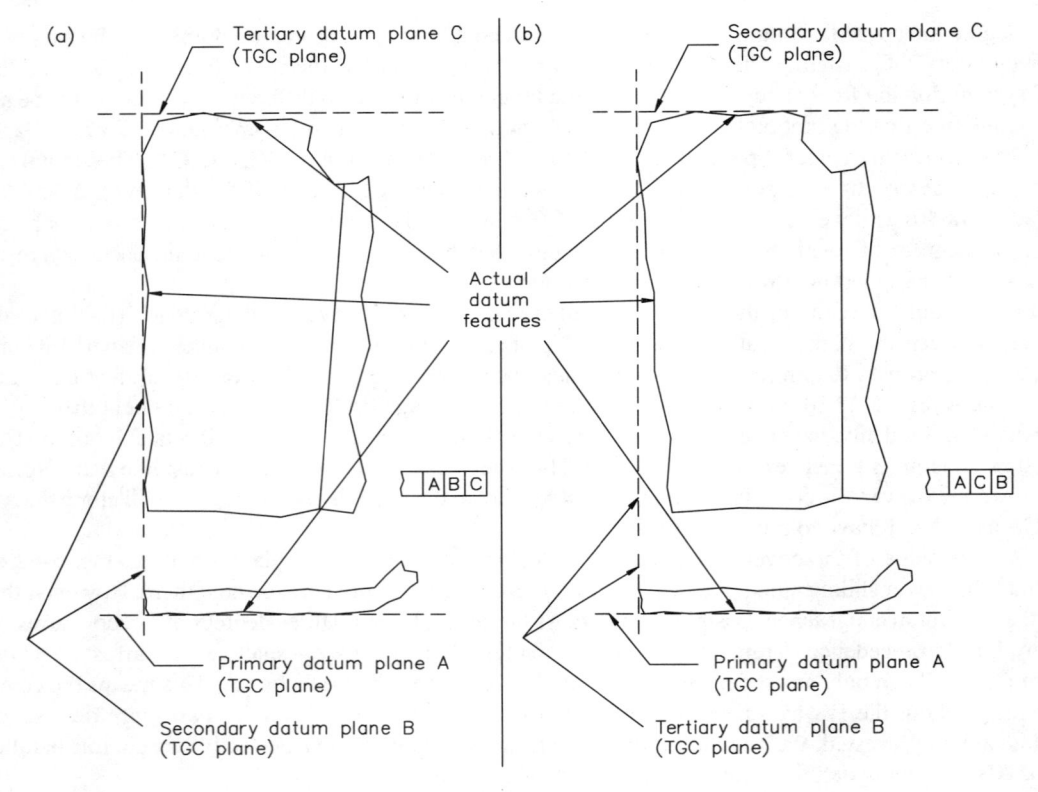

Figure 5-63 Comparison of datum precedence

our second step is to optimize orientation partly for secondary datum C. Datum feature B now has no influence over orientation. Thus, changing datum precedence yields a different DRF. The greater the out-of-squareness between the datum features, the greater the difference between the DRFs.

Our example part needs three datums to arrest all six degrees of freedom. On other parts, all six degrees can be arrested by various pairings of datums, including two nonparallel lines, or by certain types of math-defined contours. Further, it's not always necessary to arrest all six degrees of freedom. Many types of feature control, such as coaxiality, require no more than three or four degrees arrested.

> FAQ: *Is there any harm in adding more datum references than necessary in a feature control frame—just to be on the safe side?*
>
> A: Superfluous datum references should be avoided to prevent confusion. A designer must fully understand every datum reference, including the appropriate TGC, the type of datum derived, the degrees of freedom arrested based on its precedence, and that datum's role in constructing the DRF. Doubt is unacceptable.

5.9.8 TGC Types

Table 5-4 shows that each type of datum feature has a corresponding TGC. Each TGC either has no size, adjustable size, or fixed size, depending on the type of datum feature and the referenced material condition. Also, a TGC is either restrained or unrestrained, depending on the datum precedence.

5.9.8.1 Restrained Versus Unrestrained TGC

We saw in our cover example how all the degrees of freedom arrested by higher precedence datum features flowed down to impose limitations, or *restraint*, on the level of congruence achievable between each lower-precedence datum feature and its TGC. As we mentioned, such restraint is necessary in all DRFs to establish mutually perpendicular DRF planes. In the case of a primary datum feature, there is no higher precedence datum, and therefore, no restraint. However, where a secondary TGC exists, it's restrained relative to the primary TGC in all three or four degrees arrested by the primary datum feature. Likewise, where a tertiary TGC exists, it's restrained relative to the primary and secondary TGCs in all five degrees arrested by the primary and secondary datum features.

In our simple cover example, secondary TGC plane B is restrained perpendicular to TGC plane A. The translation arrested by plane A has no effect on the location of plane B. Tertiary TGC plane C is first restrained perpendicular to TGC plane A, then perpendicular to TGC plane B as well. The two degrees of translation arrested by planes A and B have no effect on the location of plane C.

In all cases, the orientation of secondary and tertiary TGCs is restrained. Where a secondary or tertiary datum feature is nominally angled (neither parallel nor perpendicular) to a higher precedence datum, its TGC shall be restrained at the basic angle expressed on the drawing. The planes of the DRF remain normal to the higher precedence datums. If the angled datum arrests a degree of translation, the origin is where the angled datum (not the feature itself) intersects the higher precedence datum. As we'll explain in section 5.9.8.4, there are cases where the location of a TGC is also restrained relative to higher-precedence datums.

5.9.8.2 Nonsize TGC

Look at the "Datum Feature Type" column of Table 5-4. Notice that for a nominally flat plane, the TGC is a tangent plane. For a math-defined (contoured) plane, the TGC is a perfect, tangent, math-defined contour. These TGC planes, whether flat or contoured, have no intrinsic size. As we saw in Fig. 5-62(b), the TGC plane and the datum feature surface are brought together in space to where they just contact at as many high points on the datum feature surface as possible (as many as three for a flat plane, or up to six for a contoured plane). "Tangent" means the TGC shall contact, but not encroach beyond the datum feature surface. In other words, all noncontacting points of the datum feature surface shall lie on the same side of the TGC plane.

Notice under the "Restraint of TGC" column, for a primary flat or contoured tangent plane TGC, no restraint is possible. For a secondary or tertiary tangent plane TGC, orientation is always restrained and location is never restrained to the higher-precedence datum(s). If location were restrained, it might be impossible to achieve contact between the datum feature surface and its TGC.

5.9.8.3 Adjustable-size TGC

Looking again at Table 5-4, we notice that for a feature of size referenced as a datum RFS, the TGC is an actual mating envelope as defined in section 5.6.4.2. An actual mating envelope is either a perfect sphere, cylinder, or pair of parallel planes, depending on the type of datum feature of size. See Fig. 5-64. The actual mating envelope's size shall be adjusted to make contact at two to four high points on the datum feature surface(s) without encroaching beyond it.

According to the Math Standard, for a secondary or tertiary actual mating envelope TGC, orientation is always restrained and location is never restrained to the higher-precedence datum(s). See Fig. 5-65.

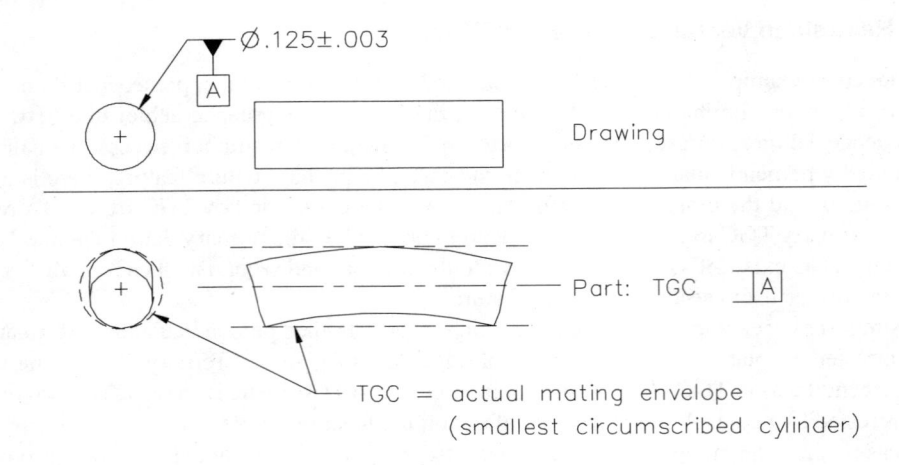

5-76   Chapter Five

Drawing

⌀.125±.003 [A]

Part: TGC [A]

TGC = actual mating envelope
(smallest circumscribed cylinder)

Figure 5-64 Feature of size referenced as a primary datum RFS

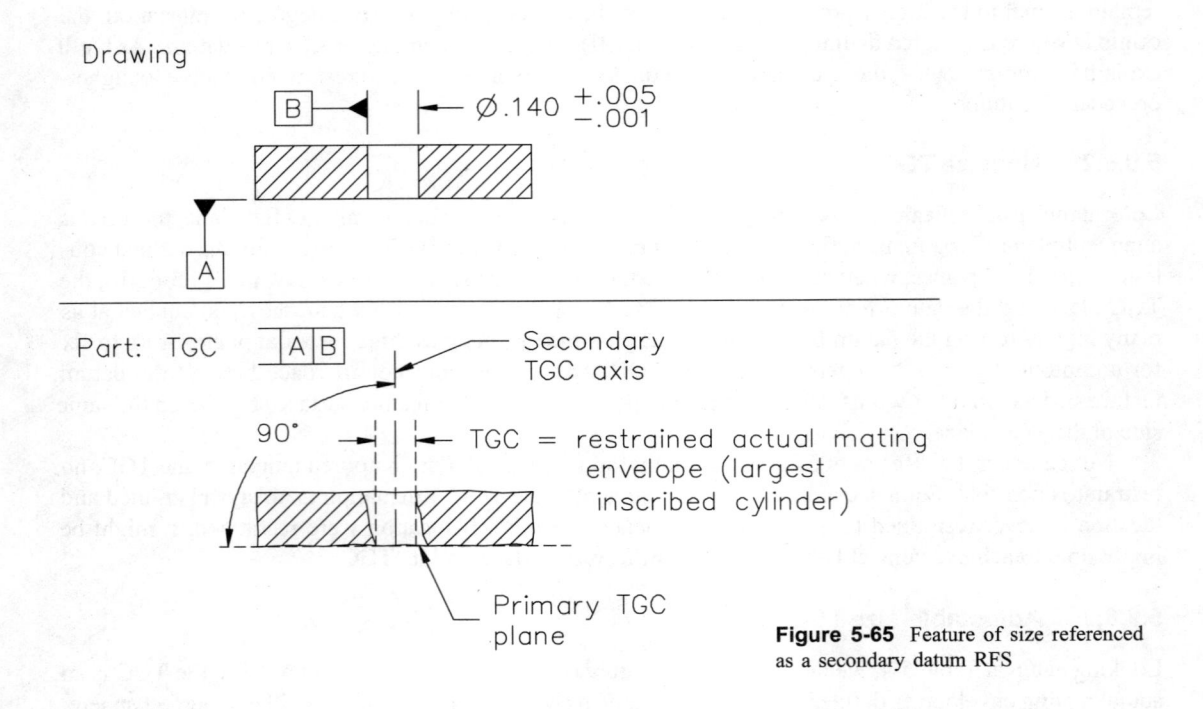

Drawing

[B] ⌀.140 +.005/−.001

[A]

Part: TGC [A|B]

90°

Secondary TGC axis

TGC = restrained actual mating envelope (largest inscribed cylinder)

Primary TGC plane

Figure 5-65 Feature of size referenced as a secondary datum RFS

FAQ: *But, if I have a shaft (primary datum A) with a shallow radial anti-rotation hole (secondary datum B), how can the hole arrest the DRF's rotation if its TGC isn't fixed (located) on center with the shaft?*

A: In this example, datum feature B, by itself, can't arrest the rotational degree of freedom satisfactorily. It must work jointly with datum feature A. Both A and B should be referenced as secondary co-datum features, as described in section 5.9.14.2. The DRF would be A|A-B.

5.9.8.4 Fixed-size TGC

According to Table 5-4, for features of size and bounded features referenced as datums at MMC or LMC, the TGCs include MMC and LMC boundaries of perfect form, MMC and LMC virtual condition boundaries, and MMC and LMC profile boundaries. See Figs. 5-66 through 5-71. Each of these TGCs has a fixed size and/or fixed shape. For an MMC or LMC boundary of perfect form, the size and shape are defined by size limits (see section 5.6.3.1 and Figs. 5-66 and 5-68). A virtual condition boundary is defined by a

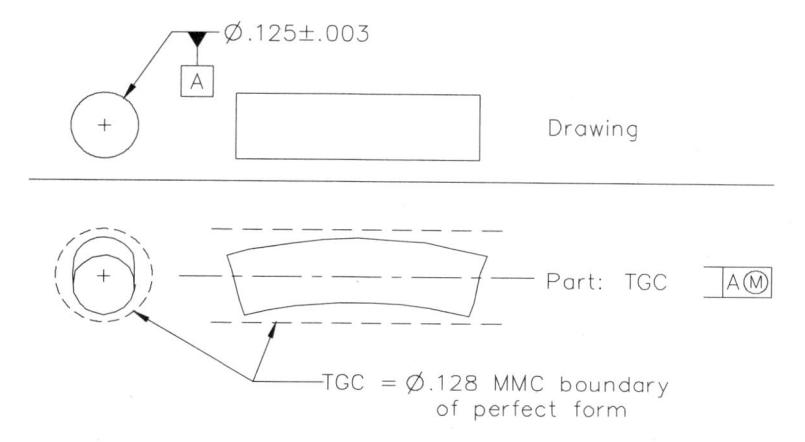

Figure 5-66 Feature of size referenced as a primary datum at MMC

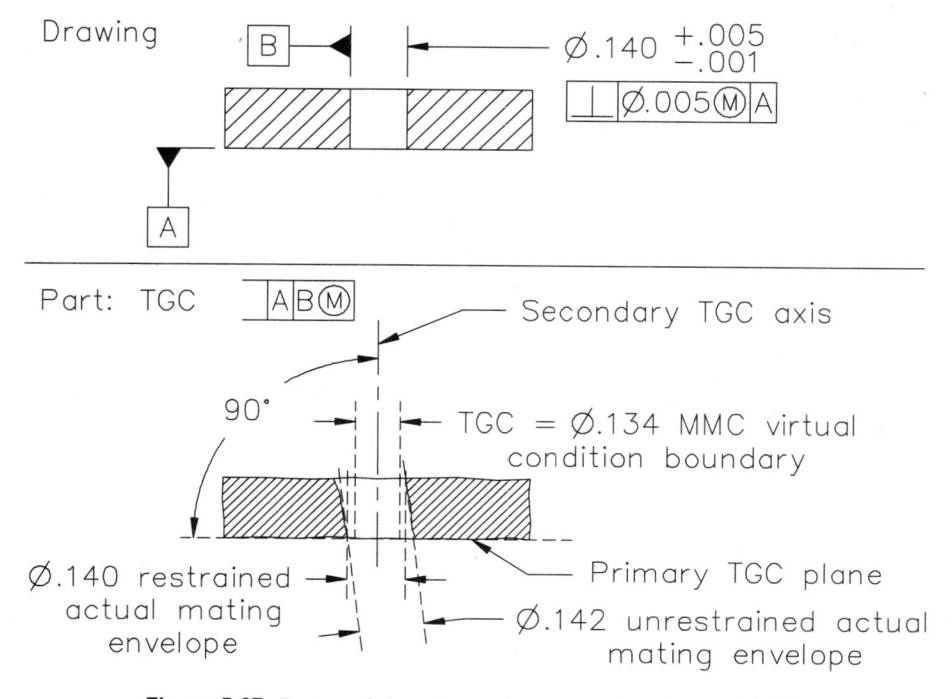

Figure 5-67 Feature of size referenced as a secondary datum at MMC

5-78 Chapter Five

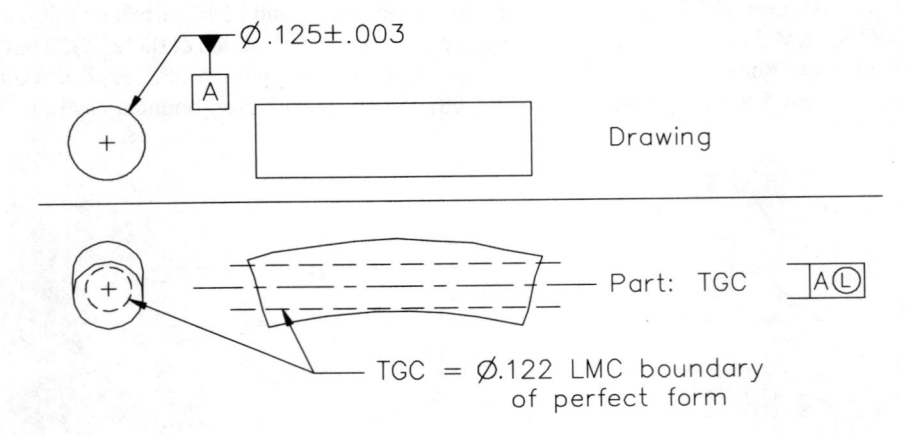

Figure 5-68 Feature of size referenced as a primary datum at LMC

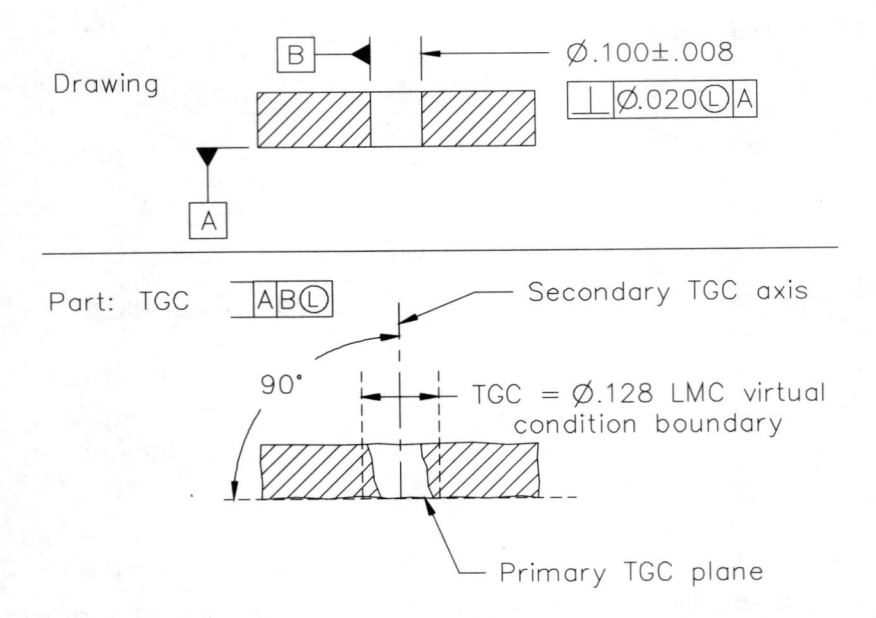

Figure 5-69 Feature of size referenced as a secondary datum at LMC

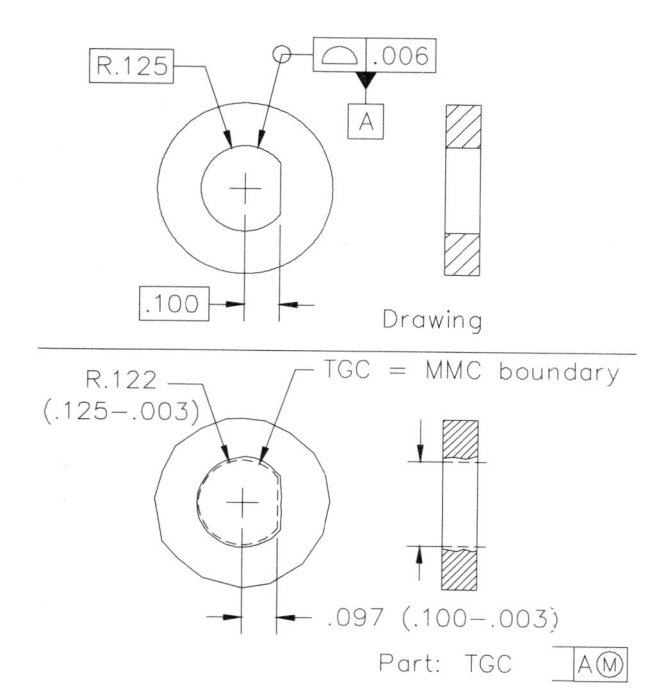

Figure 5-70 Bounded feature referenced as a primary datum at MMC

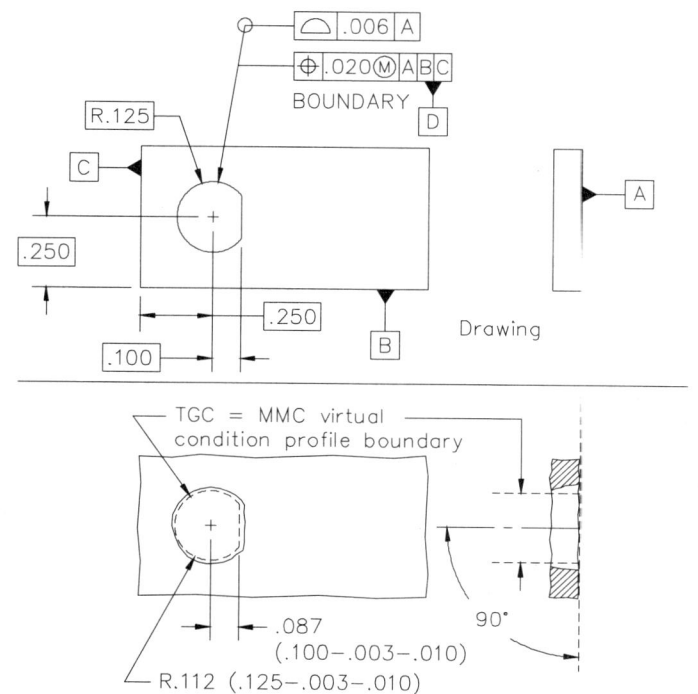

Figure 5-71 Bounded feature referenced as a secondary datum at MMC

combination of size limits and a geometric tolerance (see section 5.6.3.2 and Figs. 5-67 and 5-69). A profile boundary is defined by a profile tolerance (see section 5.13.4 and Figs. 5-70 and 5-71). Thus, none of these boundaries are generated by referencing the feature as a datum feature. It's just that when the feature is referenced, its appropriate preexisting boundary becomes its TGC.

A straightness tolerance at MMC or LMC applied to a primary datum feature cylinder, or a straightness or flatness tolerance at MMC or LMC applied to a primary datum feature width establishes a Level 2 virtual condition boundary for that primary datum feature. See Fig. 5-72. This unrestrained virtual condition boundary becomes the TGC for the datum feature.

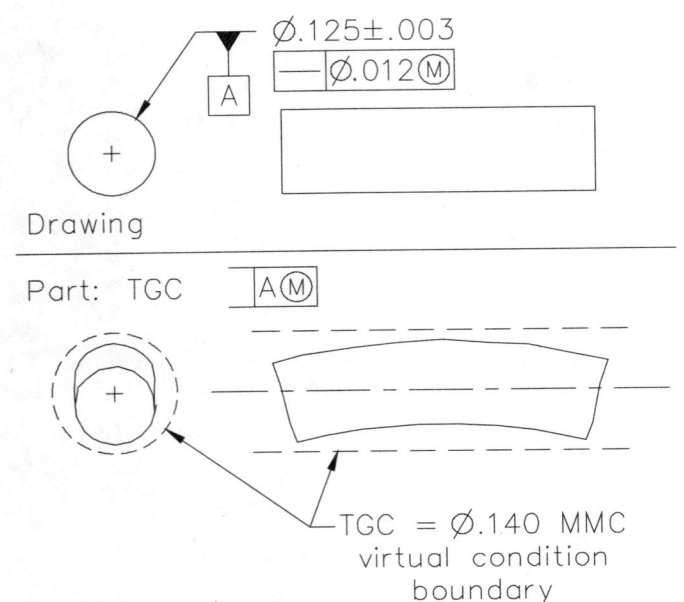

Figure 5-72 Cylindrical feature of size, with straightness tolerance at MMC, referenced as a primary datum at MMC

For a secondary or tertiary datum feature of size or bounded feature referenced at MMC or LMC, the TGC is an MMC or LMC virtual condition boundary. For this virtual condition boundary TGC, orientation is always restrained at the basic angle to the higher-precedence datum(s). Where the virtual condition boundary is also basically located relative to higher precedence datum(s), the TGC's location is always restrained at the basic location as well. In Fig. 5-24, the datum B bore is controlled with a perpendicularity tolerance at MMC, then referenced as a datum at LMC. Such applications should be avoided because the standards don't clearly define the TGC for datum B.

A fixed-size TGC is meant to emulate an assembly interface with a fixed-size feature on the mating part. Since contact may or may not occur between the two mating features, contact is likewise permitted but not required between the datum feature surface and its fixed-size TGC.

5.9.9 Datum Reference Frame (DRF) Displacement

The requirement for maximum contact between a planar surface and its nonsize TGC should yield a unique fit. Likewise, an actual mating envelope's maximum expansion within an internal feature of size or its contraction about an external feature of size ought to assure a repeatable fit. Each of those types of TGC should always achieve a unique and repeatable orientation and location relative to its datum feature. Conversely, a fixed-size TGC is not fitted to the datum feature, and need not even contact the datum

Geometric Dimensioning and Tolerancing 5-81

feature surface(s). Rather than achieving a unique and repeatable fit, the fixed-size TGC can achieve a variety of orientations and/or locations relative to its datum feature, as shown in Fig. 5-73. This effect, called *datum reference frame (DRF) displacement*, is considered a virtue, not a bug, since it emulates the variety of assembly relationships achievable between potential mating parts.

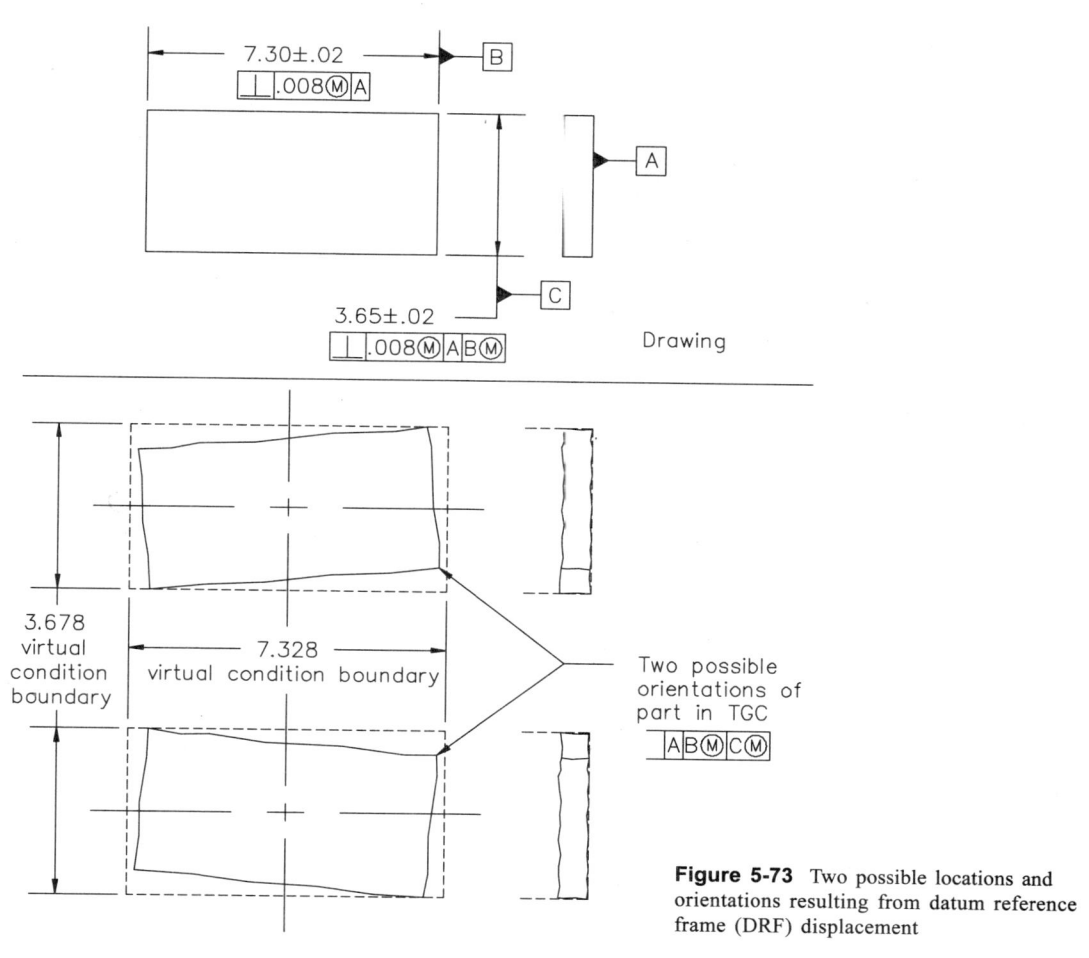

Figure 5-73 Two possible locations and orientations resulting from datum reference frame (DRF) displacement

Usually, a looser fit between two mating parts eases assembly. You may have experienced situations where screws can't seem to find their holes until you jiggle the parts around a little, then the screws drop right through. Where a designer can maximize the assembly clearances between piloting features, those clearances can be exploited to allow greater tolerances for such secondary features as screw holes. This may reduce manufacturing costs without harming assemblability.

5.9.9.1 Relative to a Boundary of Perfect Form TGC

In Fig. 5-74, we have three parts, shaft, collar, and pin. Let's assume our only design concern is that the pin can fit through both the collar and the shaft. We've identified as datum features the shaft's diameter and the collar's inside diameter. Notice that the smaller the shaft is made, the farther its cross-hole can stray from center and the pin will still assemble. Likewise, the larger the collar's inside diameter, the farther off-center its cross-hole can be and the pin will still assemble. On the shaft or the collar, we can make the hole's

5-82 Chapter Five

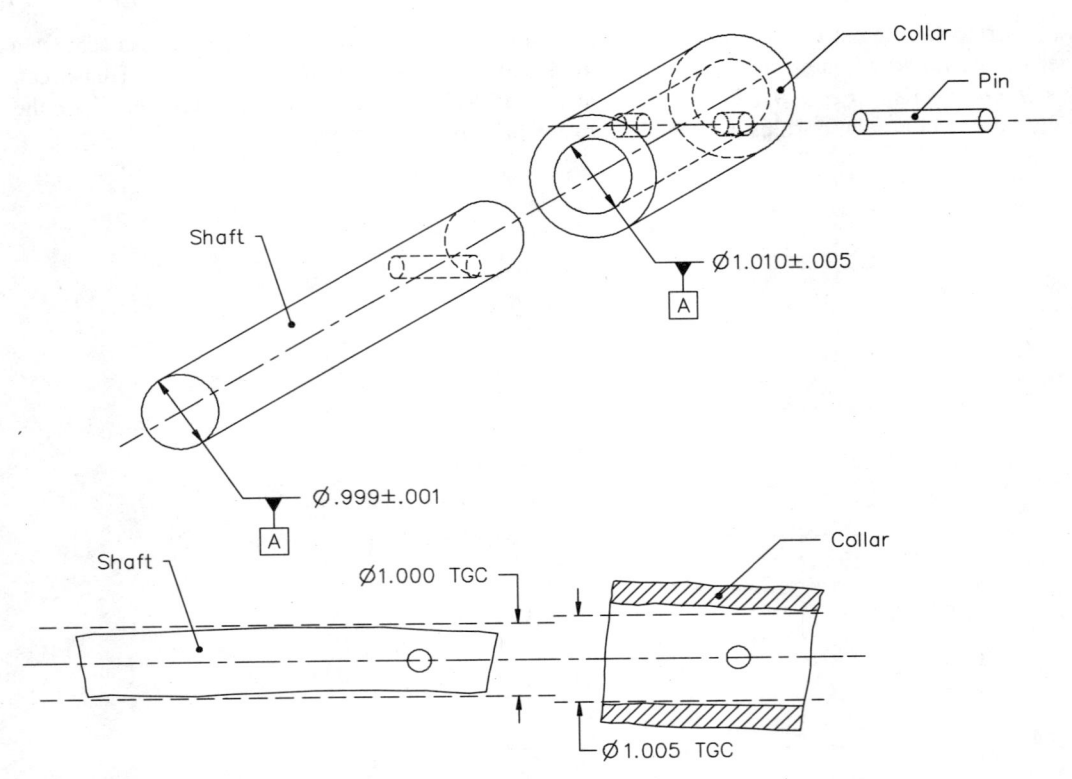

Figure 5-74 DRF displacement relative to a boundary of perfect form TGC

positional tolerance interact with the actual size of the respective datum feature, always permitting the maximum positional tolerance. We'll explain the tolerance calculations in Chapter 22, but right now, we're concerned with how to establish the DRFs for the shaft and the collar.

The shaft's datum feature is a feature of size. According to Table 5-4, if we reference that feature as a primary datum at MMC, its boundary of perfect form at MMC also becomes its TGC. That's a perfect ⌀1.000 cylinder. Any shaft satisfying its size limits will be smaller than ⌀1.000 (MMC) and able to rattle around, to some extent, within the ⌀1.000 TGC cylinder. (Remember, the datum feature surface need not contact the TGC anywhere.) This rattle, or DRF displacement, is relative motion permitted between the datum feature surface and its TGC. You can think of either one (or neither one) as being fixed in space. In the case of the shaft's primary datum, DRF displacement may include any combination of shifting and tilting. In fact, of the six degrees of freedom, none are absolutely restrained. Instead, rotation about two axes, and translation along two axes are merely limited. The limitations are that the TGC may not encroach beyond the datum feature surface. Obviously, the greater the clearance between the datum feature surface and its TGC, the greater the magnitude of allowable DRF displacement.

Similarly, the collar's datum feature is a feature of size. Referenced as a primary datum feature at MMC, its TGC is its ⌀1.005 boundary of perfect form at MMC. Any collar satisfying its size limits will be larger than ⌀1.005 (MMC) and able to rattle around about the ⌀1.005 TGC cylinder.

By extension of principle, an entire bounded feature may be referenced as a datum feature at MMC or LMC. Where the bounded feature is established by a profile tolerance, as in Fig. 5-70, the appropriate MMC or LMC profile boundary also becomes the TGC. As with simpler shapes, DRF displacement derives from clearances between the datum bounded feature surface and the TGC. As always, the TGC may not encroach beyond the datum feature surface.

5.9.9.2 Relative to a Virtual Condition Boundary TGC

A primary datum diameter or width may have a straightness tolerance at MMC, or a feature of size may be referenced as a secondary or tertiary datum at MMC. In these cases, DRF displacement occurs between the datum feature surface and the TGC that is the MMC virtual condition boundary. Table 5-4 reminds us that for a secondary or tertiary datum feature of size at MMC, degrees of rotation (orientation) and/or translation (location) already restrained by higher precedence datums shall remain restrained. Thus, DRF displacement may be further limited to translation along one or two axes and/or rotation about just one axis.

5.9.9.3 Benefits of DRF Displacement

As Fig. 5-52 shows, a TGC defines a datum, which, in turn, defines or helps define a DRF. This DRF, in turn, defines a framework of tolerance zones and/or acceptance boundaries for controlled features. Thus, allowable displacement between a datum feature surface and its TGC equates to identical displacement between the datum feature surface and the framework of tolerance zones. DRF displacement thereby allows freedom and flexibility in manufacturing, commensurate with what will occur in actual assembly. Because DRF displacement is a dynamic interaction, it's often confused with the other type of interaction, "bonus tolerance," described in section 5.6.5.1. Despite what anyone tells you:

Unlike "bonus tolerance," allowable DRF displacement never increases any tolerances. All virtual condition boundaries and/or tolerance zones remain the same size.

5.9.9.4 Effects of All Datums of the DRF

Allowable displacement of the entire DRF is governed by all the datums of that DRF acting in concert. In Fig. 5-75, datum boss B, acting alone as a primary datum, could allow DRF displacement including translation along three axes and rotation about three axes. Where datum A is primary and B is secondary (as shown), DRF displacement is limited to translation in two axes, and rotation only about the axis of B. Addition of tertiary datum C still permits some DRF displacement, but the potential for translation is not equal in all directions. Rotation of the DRF lessens the magnitude of allowable translation, and conversely, translation of the DRF lessens the magnitude of allowable rotation.

5.9.9.5 Effects of Form, Location, and Orientation

The actual form, location, and orientation of each datum feature in a DRF may allow unequal magnitudes for displacement in various directions. In Fig. 5-76, the datum shaft is out-of-round, but is still within its size limits. In Fig. 5-77, the tertiary datum boss deviates from true position, yet conforms to its positional tolerance. In both examples, the potential for DRF translation in the X-axis is significantly greater than in the Y-axis.

5.9.9.6 Accommodating DRF Displacement

In any DRF, the effects described above in sections 5.9.9.4 and 5.9.9.5 may combine to produce a potential for displacement with complex and interactive magnitudes that vary in each direction. As we said, the allowable displacement has no effect on the sizes of any virtual condition boundaries or tolerance zones for controlled features. DRF displacement may be completely and correctly accommodated by softgaging or (in MMC applications) by a functional gage. (See Chapter 19.) (The best way to learn about DRF displacement is to feel with your hands the clearances or "rattle" between a part and its functional gage.)

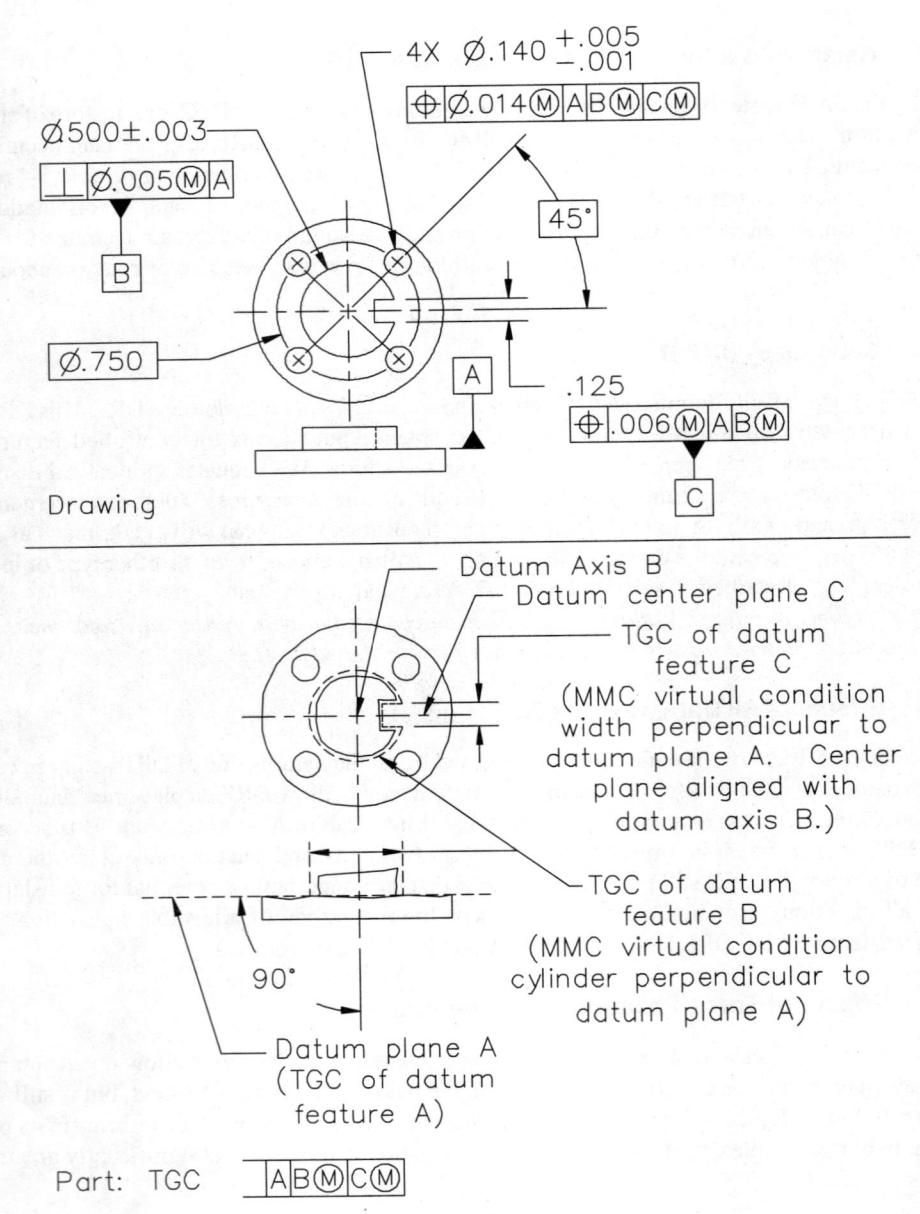

Figure 5-75 DRF displacement allowed by all the datums of the DRF

In DRFs having a single datum feature of size referenced at MMC, allowable displacement may be approximated by calculating the size difference between the datum feature's TGC and its actual mating envelope. Find the appropriate entities to use in Tables 5-3 and 5-4. For a primary datum feature, both the TGC and the actual mating envelope are unrestrained. For a secondary or tertiary datum feature, both entities must be restrained identically for proper results.

For example, in Fig. 5-67, secondary datum feature B's TGC is a cylindrical virtual condition boundary restrained perpendicular to datum A. To calculate allowable DRF displacement, we compare the size of this

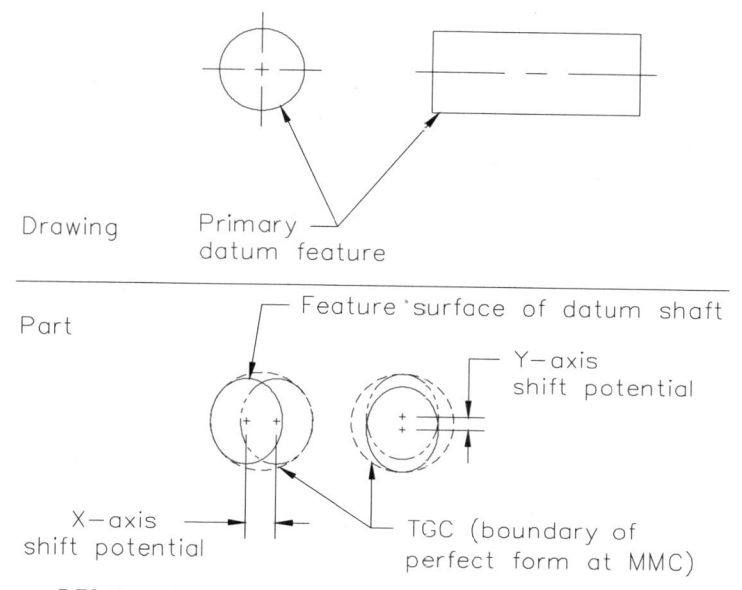

Figure 5-76 Unequal X and Y DRF displacement allowed by datum feature form variation

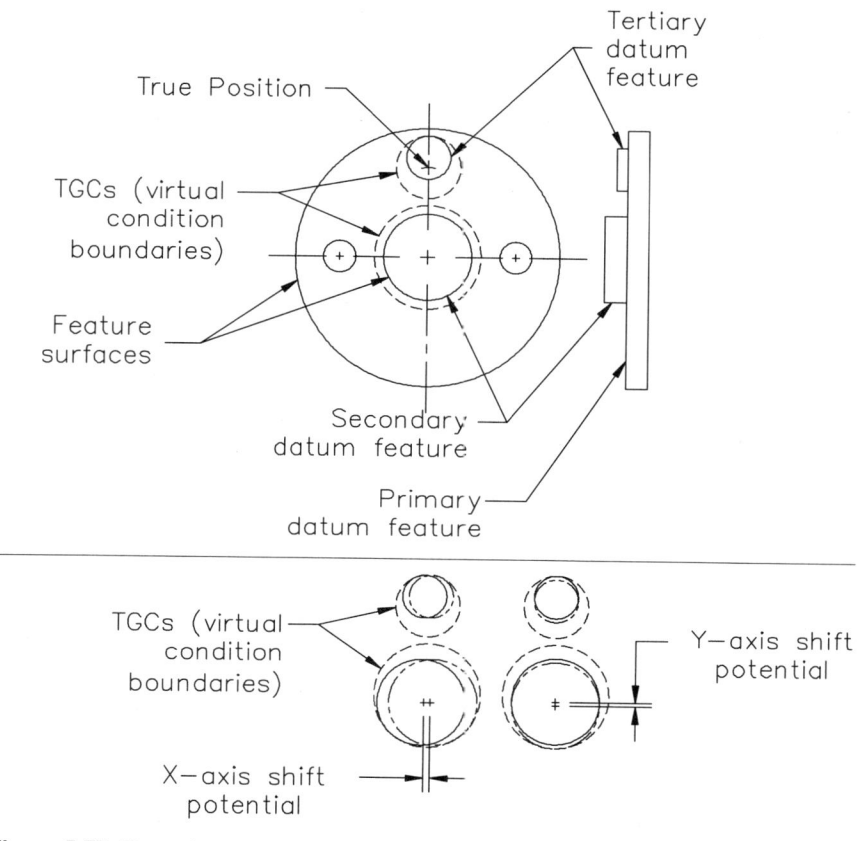

Figure 5-77 Unequal X and Y DRF displacement allowed by datum feature location variation

boundary (⌀.134) with datum feature B's actual mating size (⌀.140), derived from the actual mating envelope that is likewise restrained perpendicular to datum A. The calculated size difference (⌀.006) approximates the total clearance. With the actual mating envelope centered about the virtual condition boundary as shown, the clearance all around is uniform and equal to one-half the calculated size difference (⌀.006 ÷ 2 = .003). Thus, the DRF may translate up to that amount (.003) in any direction before the mating envelope and the TGC interfere. In our example, the ⌀.142 unrestrained actual mating envelope is larger than the ⌀.140 restrained envelope. Calculations erroneously based on the larger unrestrained envelope will overestimate the clearance all around, perhaps allowing acceptance of a part that won't assemble.

In using fitted envelopes, this simple approximation method is like the alternative center method described in section 5.6.5 and has similar limitations: It's awkward for LMC contexts, it doesn't accommodate allowable tilting, and the least magnitude for translation in any direction is applied uniformly in all directions. Consequently, it will reject some marginal parts that a proper functional gage will accept. Where used properly, however, this method will never accept a nonconforming part.

5.9.10 Simultaneous Requirements

We mentioned that DRF displacement emulates the variety of orientation and/or location relationships possible between two parts in assembly. In most cases, however, the parts will be fastened together at just one of those possible relationships. Thus, there shall be at least one relationship where all the holes line up, tab A fits cleanly into slot B, and everything works smoothly without binding. Stated more formally, there shall be a single DRF to which all functionally related features simultaneously satisfy all their tolerances. This rule is called *simultaneous requirements*.

By default, the "simultaneous requirements" rule applies to multiple features or patterns of features controlled to a "common" DRF having allowable DRF displacement. Obviously, DRF displacement can only occur where one or more of the datum features is a feature of size or bounded feature referenced at MMC or LMC. Fig. 5-78 demonstrates why "common DRF" must be interpreted as "identical DRF."

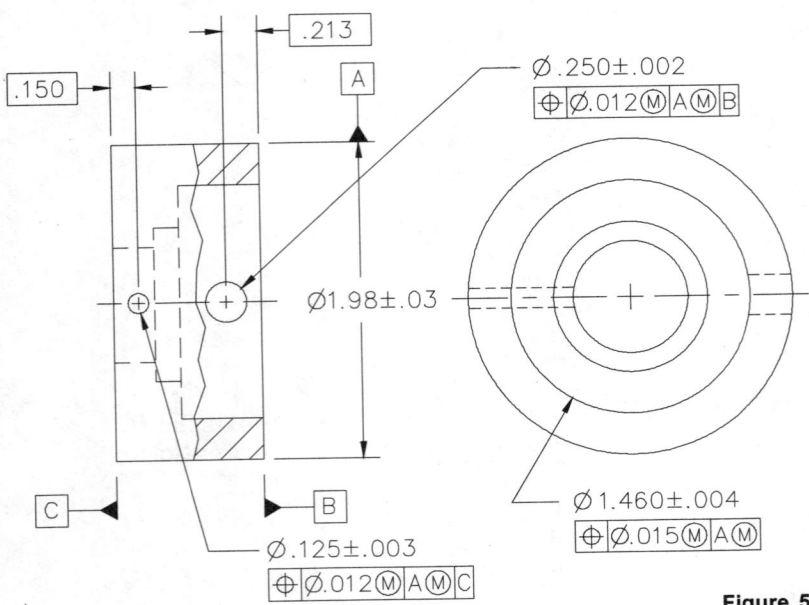

Figure 5-78 "Common DRF" means "identical DRF"

Though primary datum A is "common" to all three feature control frames, we can't determine whether the DRF of datum A alone should share simultaneous requirements with A|B or with A|C. Thus, no simultaneous requirements exist unless there is a one-to-one match of datum references, in the same order of precedence, and with the same modifiers, as applicable.

The part in Fig. 5-79 will assemble into a body where all the features will mate with fixed counterparts. The designer must assure that all five geometrically controlled features will fit at a single assembly relationship. Rather than identifying the slot or one of the holes as a clocking datum, we have controlled all five features to a single DRF. The angular relationships among the .125 slot and the holes are fixed by 90° and 180° basic angles implied by the crossing center lines, according to Fundamental Rule *(j)*. As a result, all five features share simultaneous requirements, and all five geometric tolerances can be inspected with a single functional gage in just a few seconds.

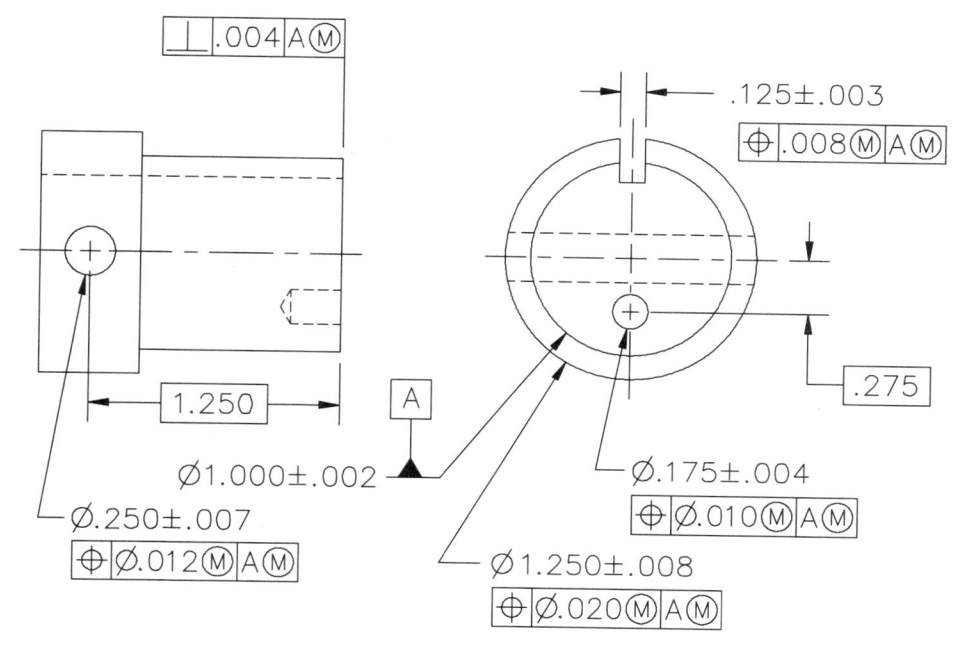

Figure 5-79 Using simultaneous requirements rule to tie together the boundaries of five features

Without such a gage, simultaneous requirements can become a curse. An inspector may be required to make multiple surface plate setups, struggling to reconstruct each time the identical DRF. Older CMMs generally establish all datums as if they were RFS, simply ignoring allowable DRF displacement. That's fine if all simultaneous requirement features conform to that fixed DRF. More sophisticated CMM software can try various displacements of the DRF until it finds a legitimate one to which all the controlled features conform.

Given the hardships it can impose, designers should nullify the "simultaneous requirements" rule wherever it would apply without functional benefit. Do this by placing the note **SEP REQT** adjacent to each applicable feature control frame, as demonstrated in Fig. 5-80. Where separate requirements are allowed, a part may still be accepted using a common setup or gage. But a "SEP REQT" feature (or pattern) cannot be deemed discrepant until it has been evaluated separately. For details on how simultaneous or separate requirements apply among composite and stacked feature control frames, see section 5.11.7.3 and Table 5-7.

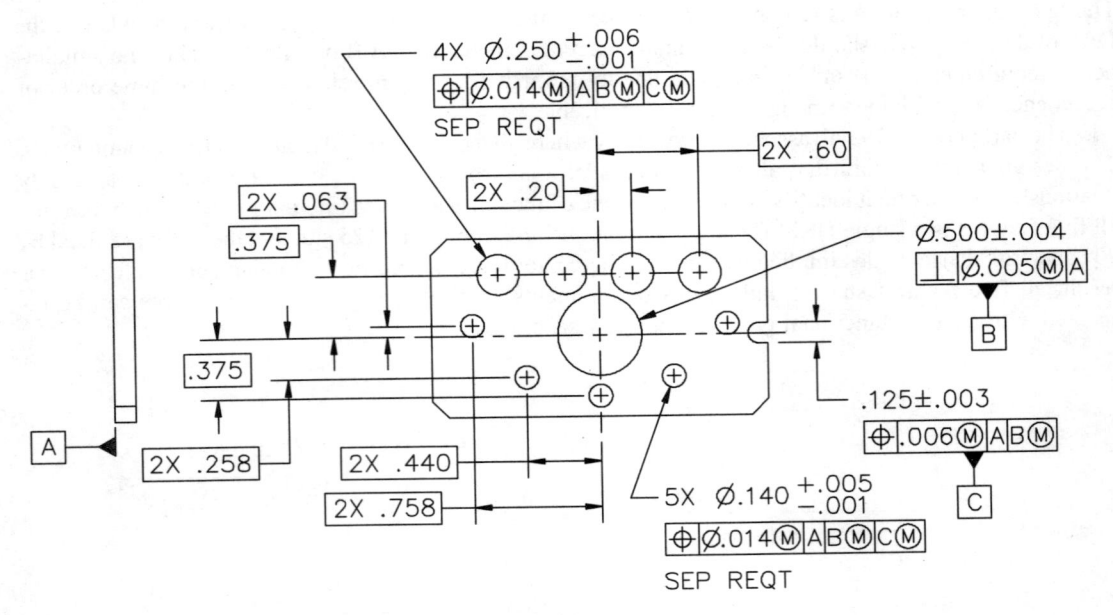

Figure 5-80 Specifying separate requirements

FAQ: *Do simultaneous requirements include profile and orientation tolerances?*

A: Y14.5 shows an example where simultaneous requirements include a profile tolerance, but neither standard mentions the rule applying to orientation tolerances. We feel that, by extension of principle, orientation tolerances are also included automatically, but a designer might be wise to add the note **SIM REQT** adjacent to each orientation feature control frame that should be included, as we have in Fig. 5-81.

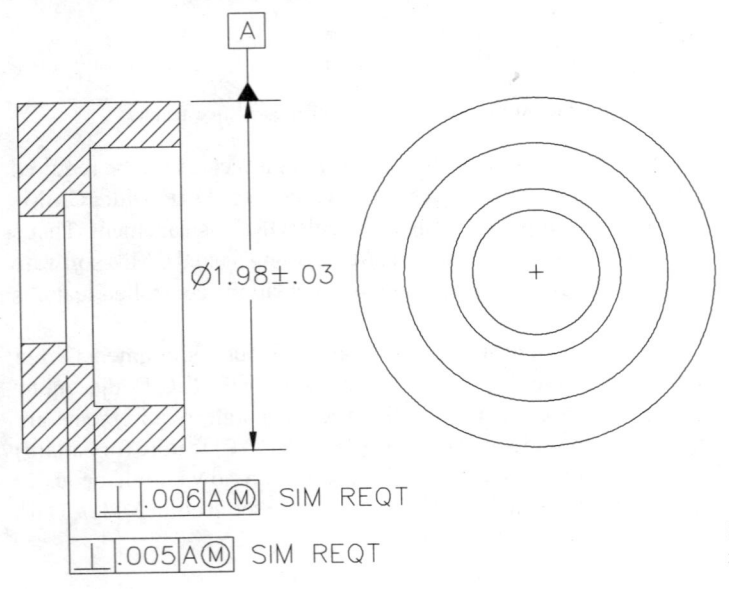

Figure 5-81 Imposing simultaneous requirements by adding a note

5.9.11 Datum Simulation

In sections 5.9.8.1 through 5.9.8.4, we discussed how perfectly shaped TGCs are theoretically aligned, fitted, or otherwise related to their datum features. The theory is important to designers, because it helps them analyze their designs and apply proper geometric controls. But an inspector facing a produced part has no imaginary perfect shapes in his toolbox. What he has instead include the following:

- Machine tables and surface plates (for planar datum features)
- Plug and ring gages (for cylindrical datum features)
- Chucks, collets, and mandrels (also for cylindrical datum features)
- Contoured or offset fixtures (for mathematically defined datum features)

Inspectors must use such high quality, but imperfect tools to derive datums and establish DRFs. The process is called *datum simulation* because it can only simulate the true datums with varying degrees of faithfulness. The tools used, called *datum feature simulators*, though imperfect, are assumed to have a unique tangent plane, axis, center plane, or center point, called the *simulated datum*, that functions the same as a theoretical datum in establishing a DRF.

Fig. 5-52 shows the relationship between the terms Y14.5 uses to describe the theory and practice of establishing datums. Errors in the form, orientation, and/or location of datum simulators create a discrepancy between the simulated datum and the true datum, so we always seek to minimize the magnitude of such errors. "Dedicated" tools, such as those listed above, are preferred as simulators, because they automatically find and contact the surface high points. Alternatively, flexible processing equipment, such as CMMs may be used, but particular care must be taken to seek out and use the correct surface points. The objective is to simulate, as nearly as possible, the theoretical contact or clearance between the TGC and the datum feature's high or tangent points. Table 5-4 includes examples of appropriate datum feature simulators for each type of datum feature.

5.9.12 Unstable Datums, Rocking Datums, Candidate Datums

Cast and forged faces tend to be bowed and warped. An out-of-tram milling machine will generate milled faces that aren't flat, perhaps with steps in them. Sometimes, part features distort during machining and heat treating processes. Fig. 5-82 shows a datum feature surface that's convex relative to its tangent TGC

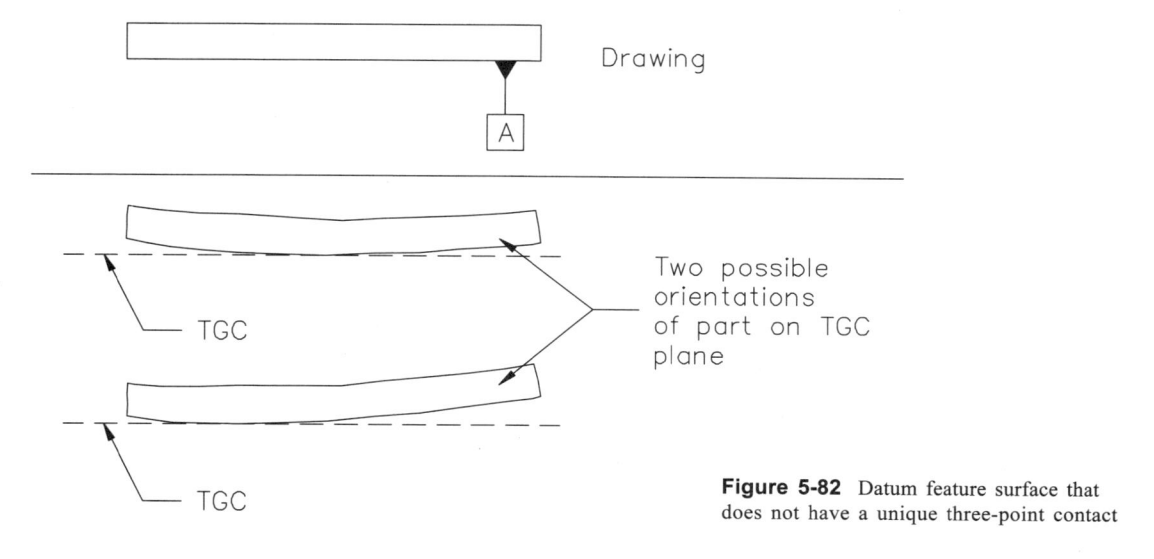

Figure 5-82 Datum feature surface that does not have a unique three-point contact

plane, and can't achieve a unique three-point contact relationship. In fact, contact may occur at just one or two high points. This is considered an "unstable" condition and produces what's called a *rocking datum*. In other words, there are a variety of tangent contact relationships possible, each yielding a different *candidate datum* and resulting *candidate datum reference frame*. These terms derive from the fact that each "candidate" is qualified to serve as the actual datum or DRF. The standards allow a user to elect any single expedient candidate datum.

Let's suppose an inspector places a part's primary datum face down on a surface plate (a datum simulator) and the part teeters under its own weight. The inspector needs the part to hold still during the inspection. Y14.5 states the inspector may "adjust" the part "to an optimum position," presumably a position where all features that reference that DRF conform to their tolerances. The prescribed "adjustment" usually involves placing some shims or clay strategically between the part and the surface plate.

The only way a CMM can properly establish a usable candidate datum from a rocking surface is by collecting hundreds or even thousands of discrete points from the surface and then modeling the surface in its processor. It must also have data from all features that reference the subject DRF. Then, the processor must evaluate the conformance of the controlled features to various candidate DRFs until it finds a candidate DRF to which all those features conform.

We mentioned an example part that "teeters under its own weight," but really, neither standard cites gravity as a criterion for candidate datums. A part such as that shown in Fig. 5-83 may be stable under its own weight, but may rock on the surface plate when downward force is applied away from the center of gravity. In fact, one side of any part could be lifted to a ludicrous angle while the opposite edge still makes one- or two-point contact with the simulator. Recognizing this, the Math Standard added a restriction saying (roughly simplified) that for a qualified candidate datum, the TGC's contact point(s) cannot all lie on one "side" of the surface, less than one-third of the way in from the edge. (One-third is the default; the drawing can specify any fraction.) This restriction eliminates, at least in most cases, "optimizations," such as shown at the bottom of Fig. 5-83, that might be functionally absurd.

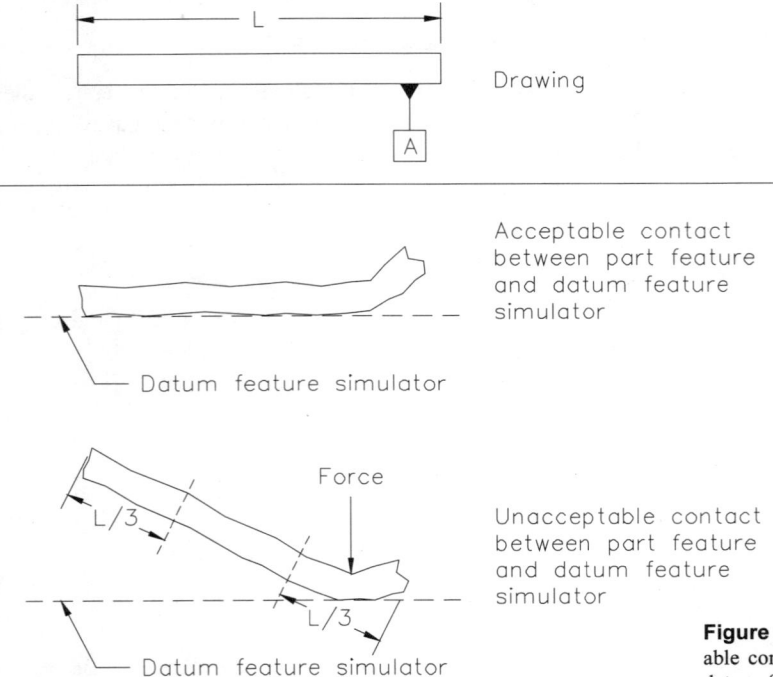

Figure 5-83 Acceptable and unacceptable contact between datum feature and datum feature simulator

This entire "adjusting to an optimum position" scheme is fraught with pitfalls and controversy. Depending on the inspection method, the optimization may not be repeatable. Certainly, the part will not achieve the same artificially optimized orientation in actual assembly. For example, a warped mounting flange might flatten out when bolted down, not only invalidating the DRF to which the part conformed in inspection, but possibly physically distorting adjacent features as well. It's fairly certain the designer didn't account for a rocking datum in his tolerance calculations.

> **FAQ:** *Can't we come up with a standard method for deriving a unique and repeatable datum from a rocker?*
>
> **A:** A variety of methods have been proposed, each based on different assumptions about the form, roughness, rigidity, and function of typical features. But this debate tends to eclipse a larger issue. A rocking datum feature betrays a failure in the design and/or manufacturing process, and may portend an even larger disaster in the making. Rather than quarrel over how to deal with rocking datums, we believe engineers should direct their energies toward preventing them. Designers must adequately control the form of datum features. They should consider datum targets (explained below) for cast, forged, sawed, and other surfaces that might reasonably be expected to rock. Manufacturing engineers must specify processes that will not produce stepped or tottering datum features. Production people must be sure they produce surfaces of adequate quality. Inspectors finding unstable parts should report to production and help correct the problem.

5.9.13 Datum Targets

So far, we've discussed how a datum is derived from an entire datum feature. TGC (full-feature) datum simulation demands either a fixture capable of contacting any high points on the datum feature, or sampling the entire datum feature with a probe. These methods are only practicable, however, where the datum feature is relatively small and well formed with simple and uniform geometry. Few very large datum features, such as an automobile hood or the outside diameter of a rocket motor, mate with other parts over their entire length and breadth. More often, the assembly interface is limited to one or more points, lines, or small areas. Likewise, **non-planar or uneven surfaces produced by casting, forging, or molding; surfaces of weldments; and thin-section surfaces subject to bowing, warping, or other inherent or induced distortions** rarely mate or function on a full-feature basis. More than just being impracticable and cost prohibitive in such cases, full-feature simulation could yield erroneous results. The obvious solution is to isolate only those pertinent points, lines, and/or limited areas, called *datum targets*, to be used for simulation. The datum thus derived can be used the same as a datum derived from a TGC. It can be referenced alone, or combined with other datums to construct a DRF.

5.9.13.1 Datum Target Selection

For each "targeted" datum feature, the type of target used should correspond to the type of mating feature or to the desired simulator and the necessary degree of contact, according to the following table.

Multiple target types may be combined to establish a single datum. However, the type(s), quantity, and placement of datum targets on a feature shall be coordinated to restrain the same degrees of freedom as would a full-feature simulator. For example, a targeted primary datum plane requires a minimum of three noncolinear points, or a line and a noncolinear point, or a single area of sufficient length and breadth. While the number of targets should be minimized, additional targets may be added as needed to simulate assembly, and/or to support heavy or nonrigid parts. For example, the bottom side of an automobile hood

Table 5-6 Datum target types

MATING FEATURE OR SIMULATOR TYPE	TARGET TYPE
spherical or pointed	POINT (0-dimensional contact)
"side" of a cylinder or "knife" edge	LINE (1-dimensional contact)
flat or elastic "pad" area	AREA (2-dimensional contact)

may need six or more small target areas. Unless target locations correspond to mating interfaces, multiple targets for a single datum should be spread as far apart as practicable to provide maximum stability.

5.9.13.2 Identifying Datum Targets

First, wherever practicable, the datum feature itself should be identified in the usual way with a "datum feature" symbol to clarify the DRF origin. As detailed in the following paragraphs, each datum target is shown on or within the part outline(s) in one or more views. Outside the part outline(s), one "datum target" symbol is leader directed to each target point, line, and area. Where the target is hidden in the view, perhaps on the far side of the part, the leader line shall be dashed. The "datum target" symbol is a circle divided horizontally into halves. See Figs. 5-8 and 5-84. The lower half always contains the target label, consisting of the datum feature letter, followed by the target number, assigned sequentially starting with 1 for each datum feature. The upper half is either left blank, or used for defining the size of a target area, as described below.

Datum Target Point—A datum target point is indicated by the "target point" symbol, dimensionally located on a direct view of the surface or on two adjacent views if there's no direct view. See Fig. 5-85.

Datum Target Line—A datum target line is indicated by the "target point" symbol on an edge view of the surface, a phantom line on the direct view, or both. See Fig. 5-85. The location (in one or two axes) and length of the datum target line shall be directly dimensioned as necessary.

Datum Target Area—A datum target area is indicated on a direct view of the surface by a phantom outline of the desired shape with section lines inside. The location (in one or two axes) and size of the datum target area shall be dimensioned as necessary. See Fig. 5-84(a) and (b). Notice that the diameter value of the target area is either contained within the upper half of the "datum target" symbol (space permitting) or leader directed there. Where it's not practicable to draw a circular phantom outline, the "target point" symbol may be substituted, as in Fig. 5-84(c).

> FAQ: *Can the upper half of the "datum target" symbol be used to specify a noncircular area?*
>
> A: Nothing in the standard forbids it. A size value could be preceded by the "square" symbol instead of the "diameter" symbol. A rectangular area, such as .25 X .50, could also be specified. The phantom outline shall clearly show the orientation of any noncircular target area.

Geometric Dimensioning and Tolerancing   5-93

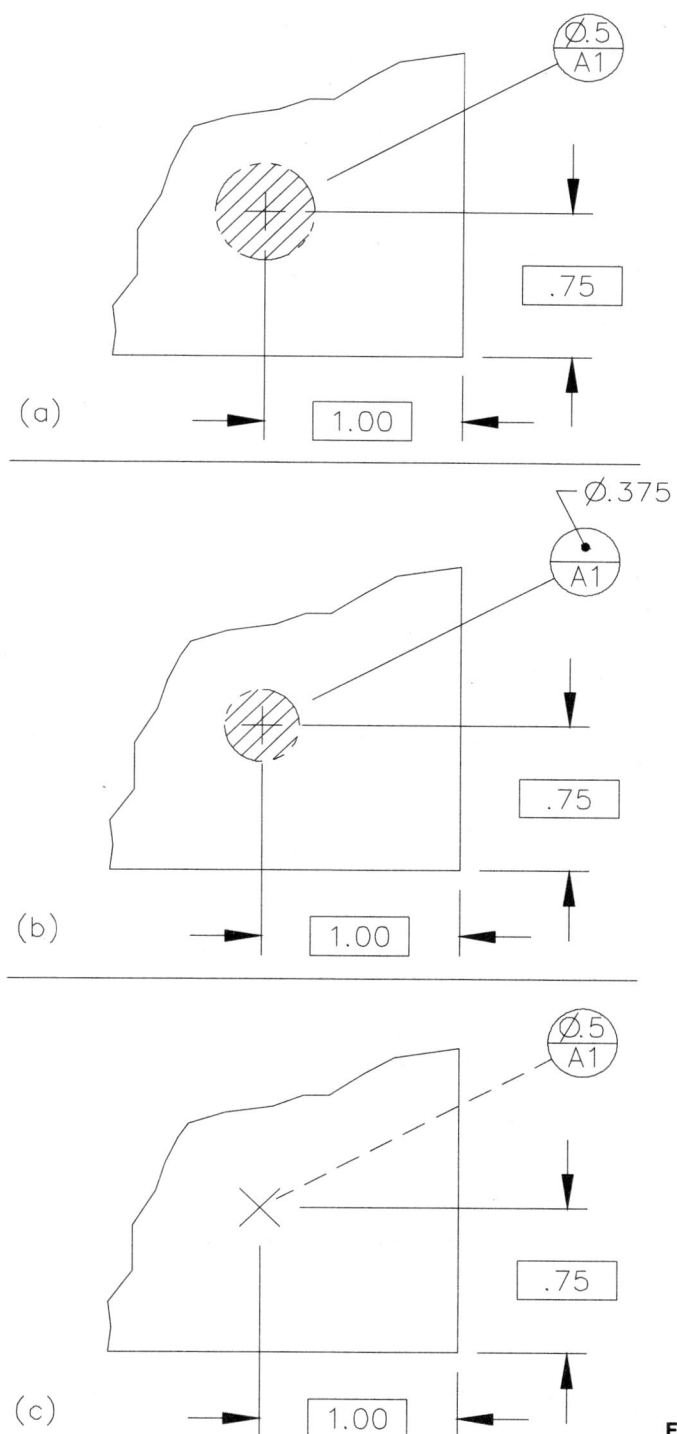

Figure 5-84 Datum target identification

5-94 Chapter Five

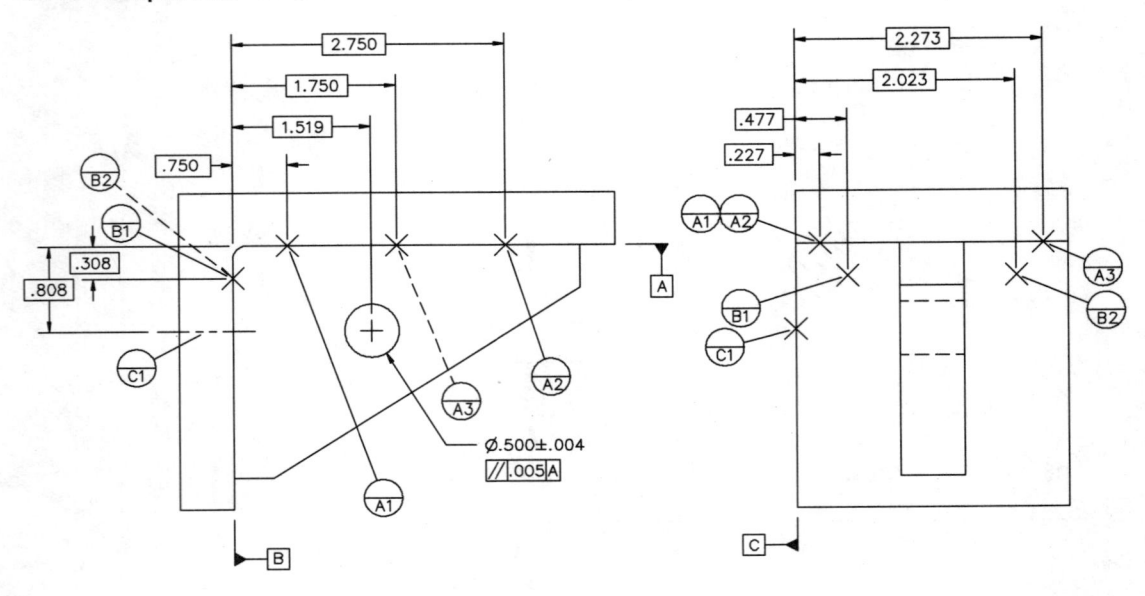

Figure 5-85 Datum target application on a rectangular part

5.9.13.3 Datum Target Dimensions

The location and size, where applicable, of datum targets are defined with either basic or toleranced dimensions. If defined with basic dimensions, established tooling or gaging tolerances apply. Such dimensions are unconventional in that they don't pertain to any measurable attribute of the part. They are instead specifications for the process of datum simulation, in effect saying, "Simulation for this datum feature shall occur here."

On any sample part, the datum simulation process may be repeated many times with a variety of tools. For example, the part could be made in multiple machines, each having its own fixture using the datum targets. The part might then be partially inspected with a CMM that probes the datum feature only at the datum targets. Final inspection may employ a functional gage that uses the datum targets. Thus, dimensions and tolerances for a datum target actually apply directly to the location (and perhaps, size) of the simulator (contacting feature) on each tool, including CMM probe touches. Variations within the applicable tolerances contribute to discrepancies between the DRFs derived by different tools.

> **FAQ:** *Where can I look up "established tooling or gaging tolerances" for locating simulators?*
>
> **A:** We're not aware of any national or military standard and it's unlikely one will emerge. The traditional rule of thumb—5% or 10% of the feature tolerance—is quite an oversimplification in this context. (And to which feature would it refer?) While tolerances of controlled features are certainly a factor in determining target tolerances, there are usually many other factors, including the form and surface roughness of the datum feature, and the type and size of the simulator. For example, on a forged surface, the point of contact of a $\varnothing$1mm spherical simulator is usually more critical than that of a $\varnothing$4mm simulator. (Both are common CMM styli.)

5.9.13.4 Interdependency of Datum Target Locations

In Fig. 5-85, three targeted datum features establish a DRF. Notice that targets A1, A2, and A3 are located relative to datums B and C. Targets B1 and B2 are located relative to datums A and C. Likewise, target C1 is located relative to datums A and B. This interdependency creates no problem for hard tooling that simulates all three datums simultaneously. However, methods that simulate the datums sequentially encounter a paradox: The targets for any one datum cannot be accurately found until the other two datums have been properly established. A CMM, for example, may require two or three iterations of DRF construction to achieve the needed accuracy in probing the targets. Even for the simple parallelism callout that references only datum A, all three datums must be simulated and the entire A|B|C DRF properly constructed.

> **FAQ:** *Should the parallelism callout in Fig. 5-85 reference all three datums, then, A|B|C?*
>
> **A:** No. Referencing datum B would add an unnecessary degree of restraint to the parallelism tolerance. An excellent solution is to extend positional tolerancing principles (RFS) to datum targets. See section 11. A feature control frame complete with datum references may be placed beneath the #1 "datum target" symbol for each datum (for example, A1, B1, and C1). This method overcomes all the shortcomings of plus and minus coordinate tolerancing, and unambiguously controls the locations of all six targets to a common and complete DRF. (In our example, A|B|C should be referenced for each of the three target sets.) The standard neither prohibits nor shows this method, so a drawing user might welcome guidance from a brief general note.

5.9.13.5 Applied to Features of Size

Datum targets may be applied to a datum feature of size for RFS simulation. The simulators shall be adjustable to contact the feature at all specified targets. Simulators on hard tools shall expand or contract uniformly while maintaining all other orientation and location relationships relative to each other and to other datums in the subject DRF.

Width-Type Feature—In the tertiary datum slot in Fig. 5-86, simulators C1 and C2 shall expand apart. Proper simulation is achieved when each simulator contacts the slot, each is equidistant from datum plane BY, and each is the specified distances from datum planes A and BX.

Cylindrical Feature—A datum target line or area may be wrapped around a cylindrical feature, specifying what amounts to a TGC of zero or limited length. Alternatively, datum target points or lines (longitudinal) may be equally spaced around the feature. For the secondary datum boss in Fig. 5-86, simulators B1, B2, and B3 shall contract inward to trap the feature. A hard tool, perhaps a precision chuck, shall have a set of three equally spaced simulators (jaws) capable of moving radially at an equal rate from a common axis. Proper simulation is achieved when each simulator contacts the boss and each is equidistant from the datum axis.

Poor feature form, orientation, or location may prevent one or more simulators from making contact, despite obeying all the rules. Where, for example, we need to derive a primary datum from a forged rod, we may specify target points A1, A2, and A3 around one end, and A4, A5, and A6 around the other end. This requires all six simulators to contract uniformly. The larger rod end will be trapped securely, while at the smaller end, never more than two simulators can touch. This yields a rocking datum. One solution is to relabel A4, A5, and A6 as B1, B2, and B3, and then establish co-datum A-B. This allows the two simulator sets, A and B, to contract independently of each other, thereby ensuring contact at all six targets.

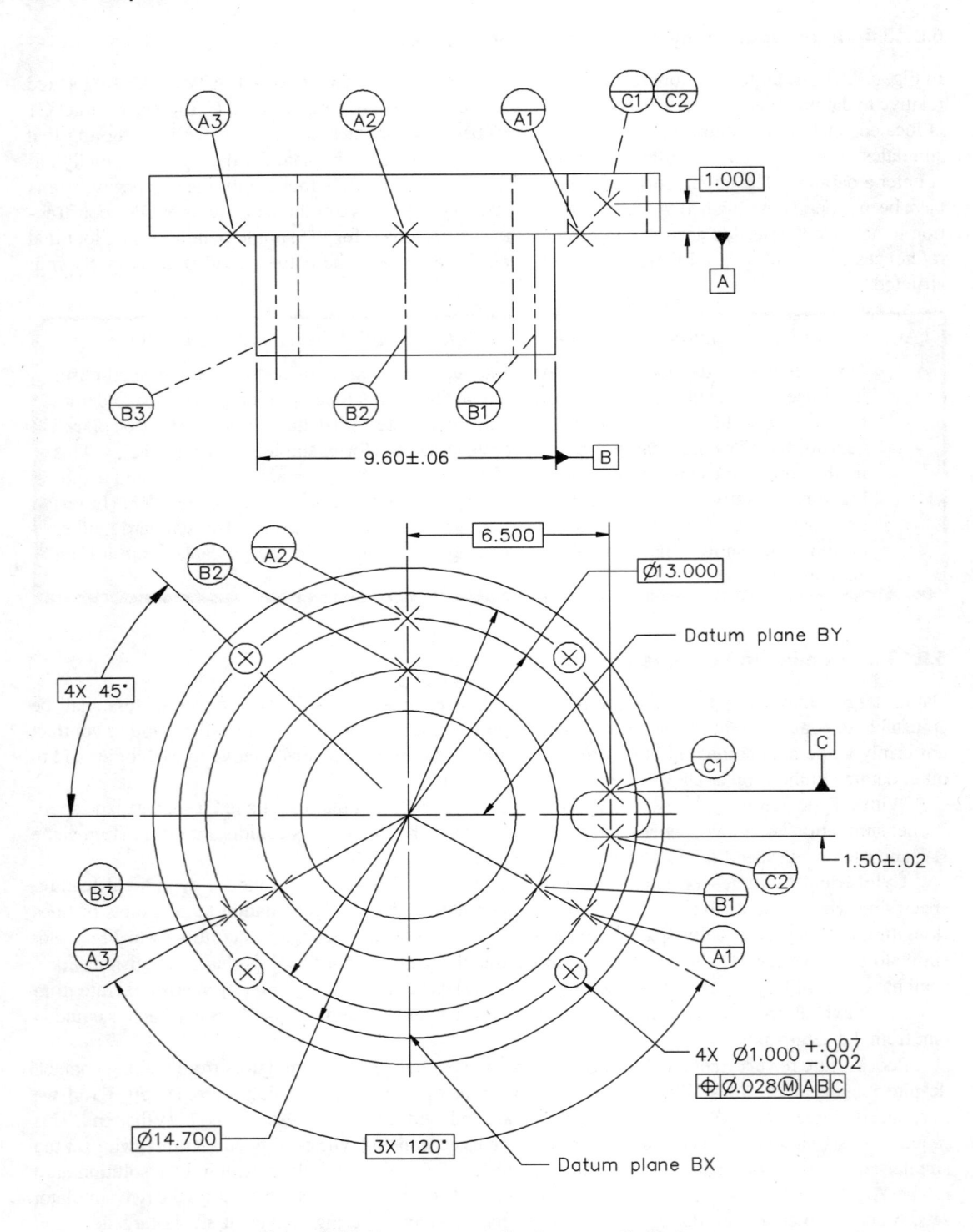

Figure 5-86 Datum target application on a cylindrical part

> FAQ: *Can datum targets be applied to a feature of size on an MMC basis?*
>
> A: Nothing in the standard precludes it. We've been careful to emphasize that datum targets are targets for simulation, not necessarily contact. For MMC, a typical hard tool would have simulators at a fixed diameter or width based on the datum feature's MMC. With advanced software, a CMM can easily accommodate MMC and LMC applications. All the DRF displacement principles of section 5.9.9 apply, except that the target set does not comprise a TGC.

5.9.13.6 Applied to Any Type of Feature

Datum targets provide the means for simulating a usable datum from any imaginable type and shape of feature. With irregular datum features, the designer must carefully assure that all nonadjustable relationships between targets are dimensioned, preferably using just one coordinate system. Any relationship between targets left undimensioned shall be considered adjustable.

Particularly with a complex drawing, a drawing user may have trouble identifying a datum plane or axis derived and offset from a stepped or irregularly shaped datum feature. In such cases only, it's permissible to attach a "datum feature" symbol to a center line representing the datum.

Stepped Plane—A datum plane can be simulated from multiple surfaces that are parallel but not coplanar. Datum targets should be defined such that at least one target lies in the datum plane. Offset distances of other targets are defined with dimensions normal to the datum plane. This also permits convenient application of profile tolerancing to the part surfaces.

Revolutes—A revolute is generated by revolving a 2-D spine (curve) about a coplanar axis. This can yield a cone (where the spine is a straight nonparallel line), a toroid (where the spine is a circular arc), or a vase or hourglass shape. It may be difficult or impossible to define TGCs for such shapes. Further, full-feature datum simulation based on nominal or basic dimensions may not achieve the desired fit or contact. Where a revolute must be referenced as a datum feature, it's a good idea to specify datum targets at one or two circular elements of the feature. At each circular element, a triad of equally spaced datum target points or lines, or a single circular target "line" may be used.

Fig. 5-87 shows a datum axis derived from a chicken egg. Targets A1, A2, and A3 are equally spaced on a fixed $\varnothing 1.250$ basic circle. These simulators neither expand nor contract relative to each other. Targets B1, B2, and B3 are likewise equally spaced on a fixed $\varnothing 1.000$ basic circle. The drawing implies basic coaxiality and clocking between the two target sets. However, the distance between the two sets is undimensioned and therefore, adjustable. This distance shall close until contact occurs at all six targets and the egg is immobilized. In the positional tolerance feature control frame for the egg's $\varnothing .250$ observation port (peephole), co-datum axis A-B is referenced RFS (see section 5.9.14.2). The .500 basic dimension for the observation hole originates from the plane of the datum A target set.

Fig. 5-88 shows one possible setup for drilling the observation hole. Despite the egg's frailty, we've chosen pointed simulators over spherical ones to assure that contact always occurs at the specified basic diameters. Simulators A1, A2, and A3 are affixed to the "stationary" jaw of a precision vise. Simulators B1, B2, and B3 are attached to the "movable" vise jaw. "Stationary" and "movable" are always relative terms. In this case, mobility is relative to the machine spindle.

To simulate the egg's datum axis at MMC, a basic or toleranced dimension shall be added for the distance between the two triads of targets. The targets are labeled *A*1 through *A*6 and establish datum axis *A* (where *A* is any legal identifying letter). Since none of the simulators would be adjustable in any direction, the egg can rattle around between them. (On a hard tool, one or more simulators would have to be removable to let the egg in and out.)

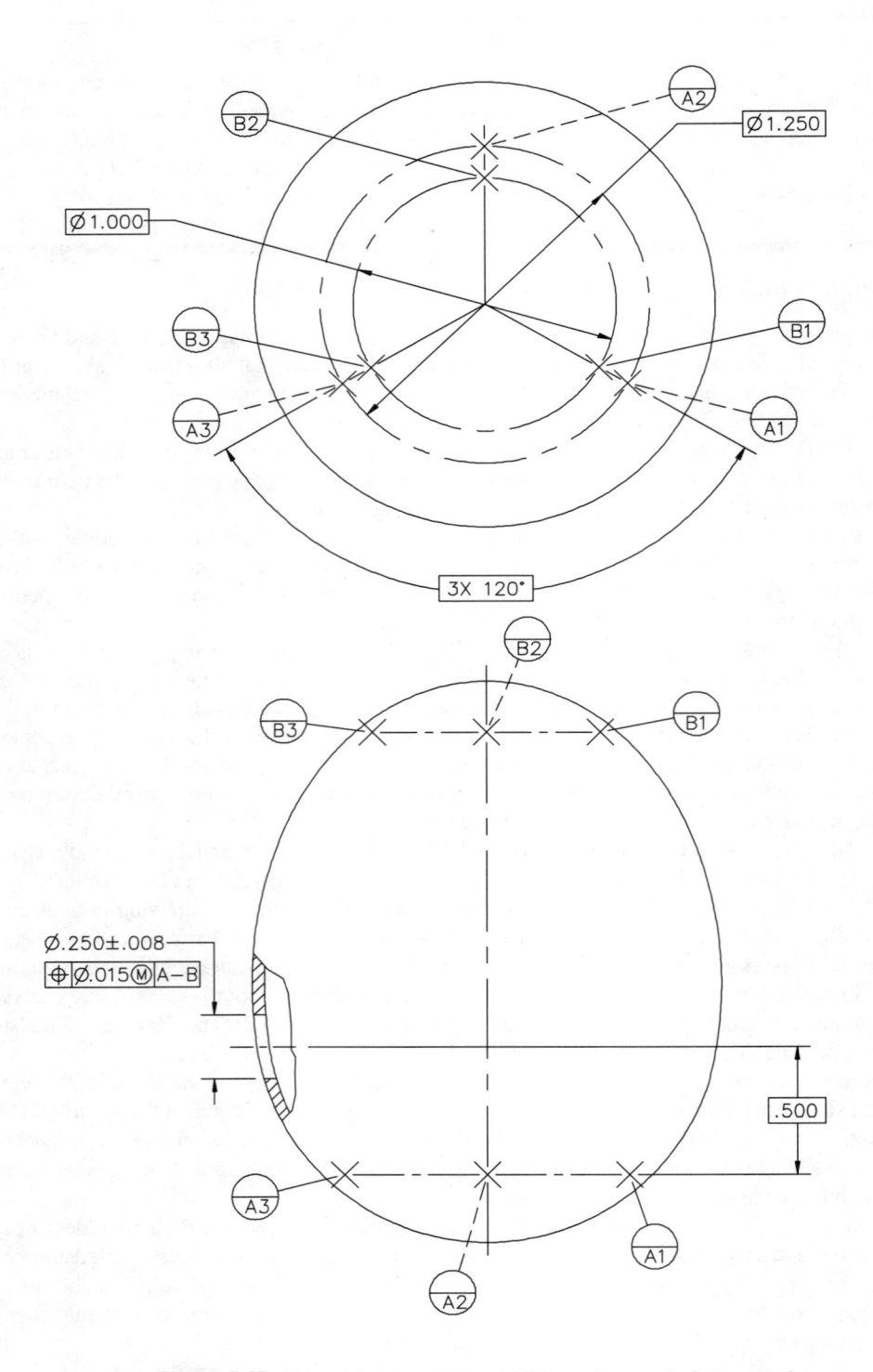

Figure 5-87 Using datum targets to establish a primary axis from a revolute

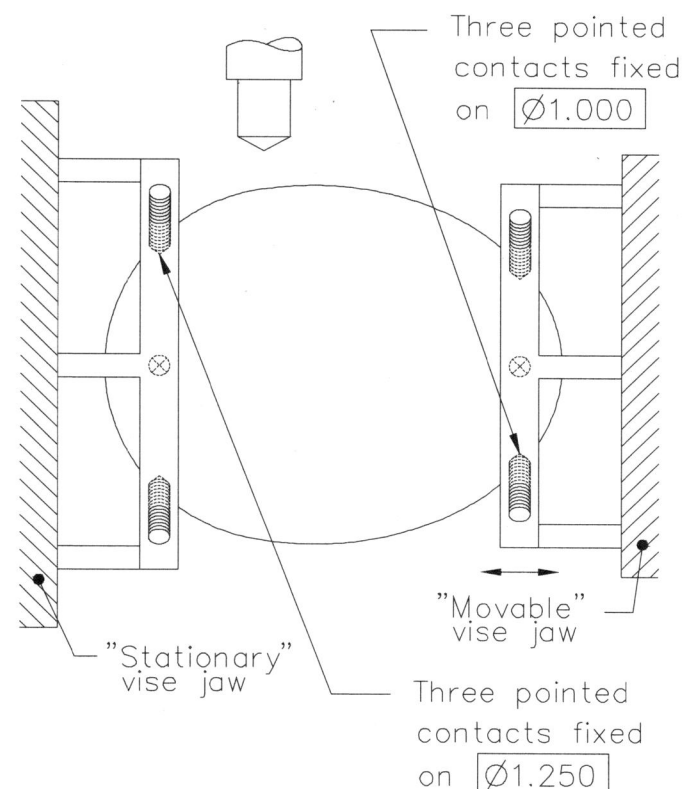

Figure 5-88 Setup for simulating the datum axis for Fig. 5-87

Math-Defined Feature—Datum targets can be placed on radii, spherical radii, and any type of nominally warped planar surface. The desired datum planes can establish a coordinate system for defining the location of each target in 3-D space. In some cases, it may be simpler if every target is offset from the datum planes.

Bounded Feature—All the above principles can apply.

5.9.13.7 Target Set with Switchable Precedence

In Fig. 5-89, datum B is the primary datum for a parallelism tolerance, so we've identified the minimum necessary target points, B1, B2, and B3. However, in the other DRF, A|B|C, datum B is the secondary datum. Here, we only need and want to use points B1 and B2. On a very simple drawing, such as ours, a note can be added, saying, "IN DATUM REFERENCE FRAME A|B|C, OMIT TARGET B3." On a more complex drawing, a table like the one below could be added. The right column can list either targets to use or targets to omit, whichever is simpler.

IN DATUM REFERENCE FRAME	OMIT TARGET(S)
A|B|C	B3
B|A	A3
D|E|F	E3, F2, F3

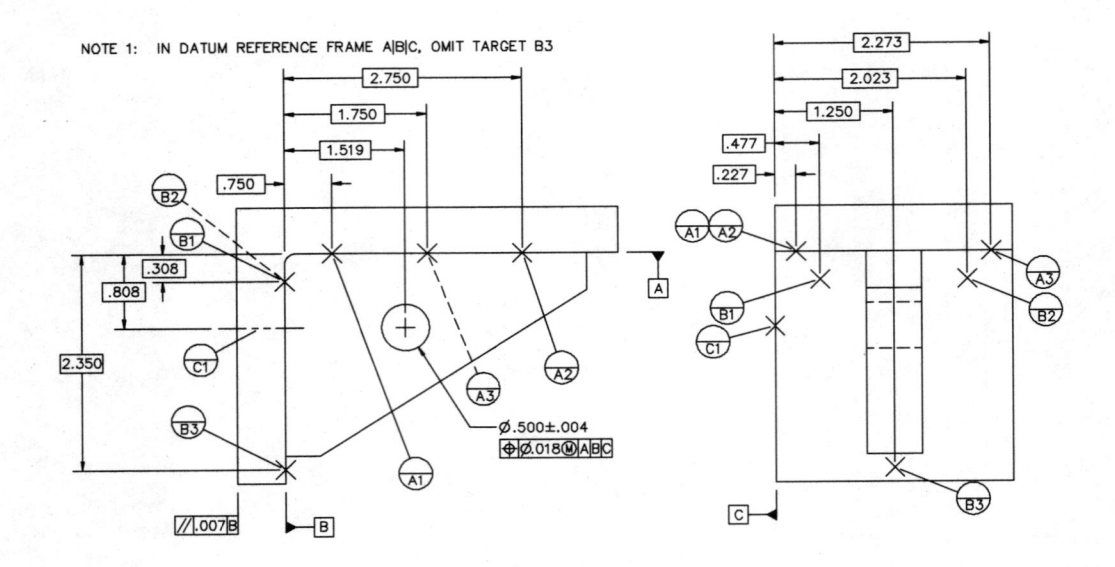

Figure 5-89 Target set with switchable datum precedence

5.9.14 Multiple Features Referenced as a Single Datum Feature

In some cases, multiple features can be teamed together and treated as a single datum feature. This is a frontier of datuming, not fully developed in the standards. When referencing multiple features in this way, designers must be extremely careful to understand the exact shapes, sizes (where applicable), and interrelationships of the TGC(s); simulation tools that might be used; and the exact degrees of freedom arrested. If any of these considerations won't be obvious to drawing users, the designer must explain them in a drawing note or auxiliary document.

5.9.14.1 Feature Patterns

While discussing Fig. 5-54, we said the cylinder head's bottom face is an obvious choice for the primary datum feature. The two dowel holes are crucial in orienting and locating the head on the block. One hole could be the secondary datum feature and the other tertiary, but the holes would then have unequal specifications requiring unequal treatment. Such datum precedence is counterintuitive, since both holes play exactly equal roles in assembly. This is an example where a pattern of features can and should be treated as a single datum feature. Rather than a single axis or plane, however, we can derive two perpendicular datum planes, both oriented and located relative to the holes.

Fig. 5-90 shows just three of many options for establishing the origin from our pattern of dowel holes. The designer must take extra care to clarify the relationship between a datum feature pattern and the origins of the coordinate system derived therefrom.

Fig. 5-91 shows a feature pattern referenced as a single datum feature at MMC. Rather than a single TGC, the datum B reference establishes a pattern or *framework* of multiple, identical, fixed-size TGCs. Within this framework, the orientation and location of all the TGCs are fixed relative to one another according to the basic dimensions expressed on the drawing. As the figure's lower portion shows, two perpendicular planes are derived, restricting all three remaining degrees of freedom. For discussion pur-

Geometric Dimensioning and Tolerancing 5-101

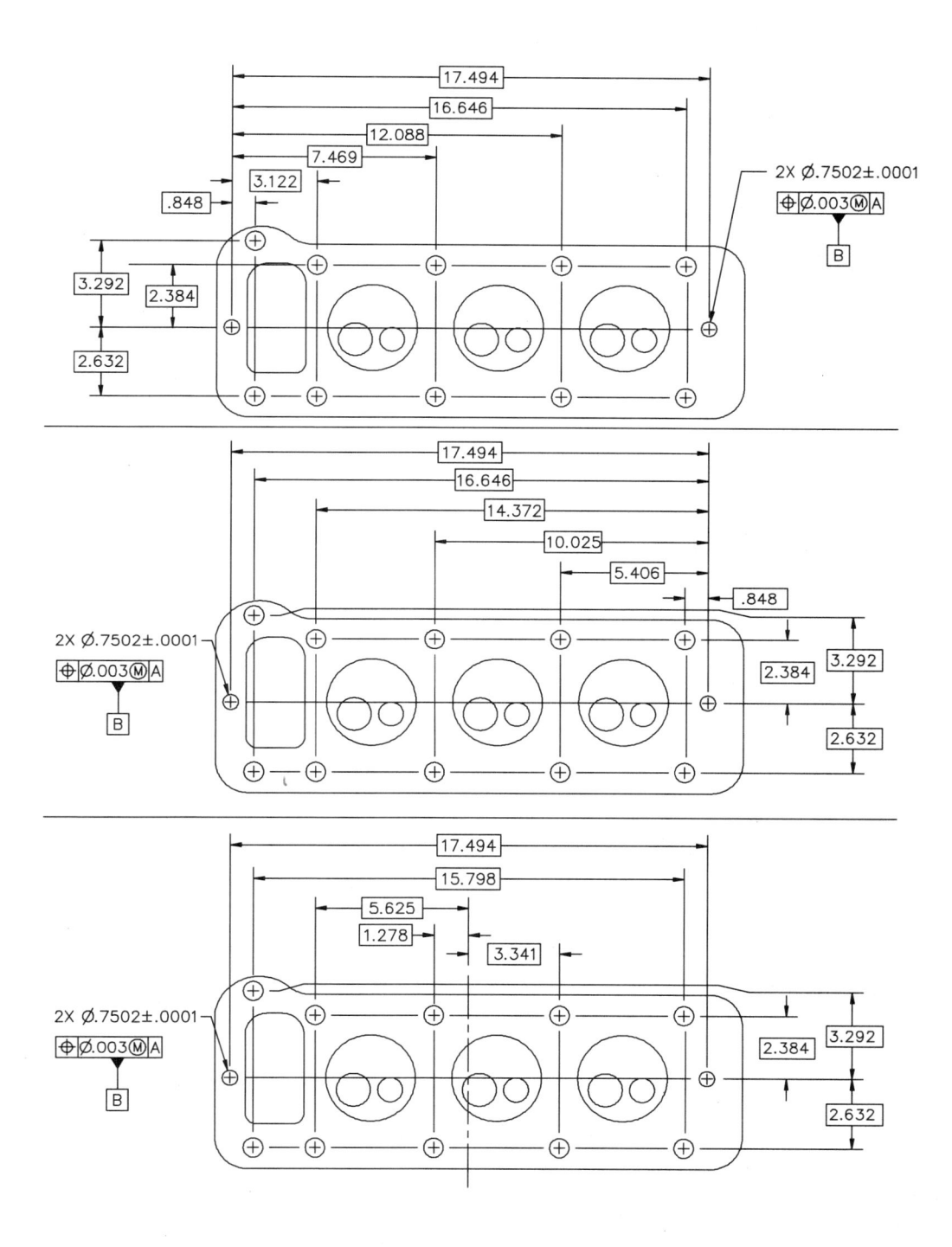

Figure 5-90 Three options for establishing the origin from a pattern of dowel holes

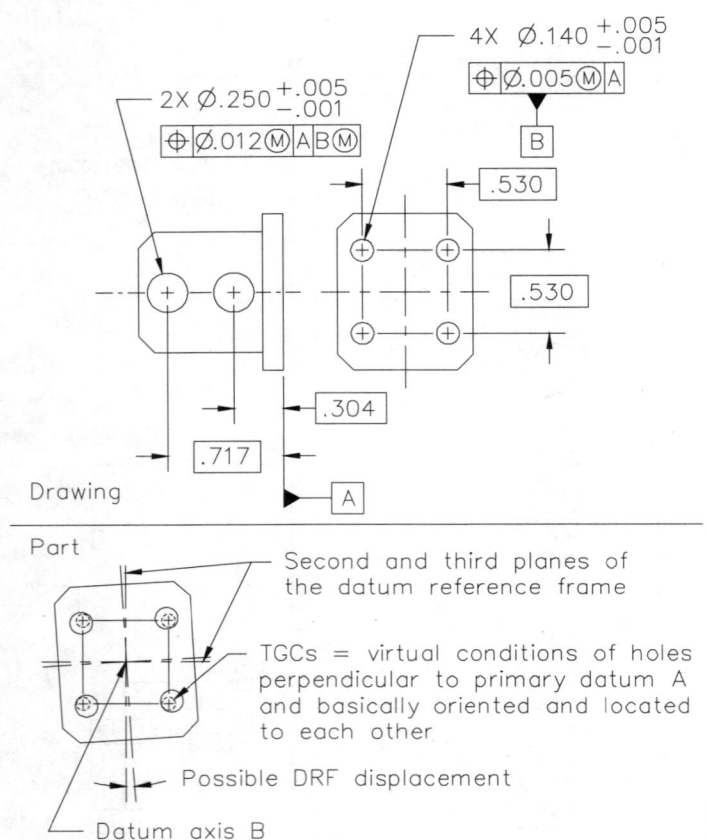

Figure 5-91 Pattern of holes referenced as a single datum at MMC

poses, we've labeled the intersection of these planes "datum axis B." Since each individual feature in the pattern clears its respective TGC, DRF displacement is possible, including rotation about datum axis B, and translation in any direction perpendicular to datum axis B. The rules for simultaneous requirements are the same as if datum feature B were a single feature.

> **FAQ:** *Can a datum feature pattern be referenced at LMC or RFS?*
>
> **A:** At LMC, yes, but this will require softgaging. The datum feature simulator is a set of virtual fixed-size TGCs. For RFS, the simulator should be a set of adjustable TGCs, each expanding or contracting to fit its individual feature. But differences among the size, form, orientation, and location of individual features raise questions the standards don't address. Must the TGCs adjust simultaneously and uniformly? Must they all end up the same size? In such a rare application, the designer must provide detailed instructions for datum simulation, because the standards don't.

5.9.14.2 Coaxial and Coplanar Features

Fig. 5-131 shows another example of separate features—this time, two bearing journals—that have exactly equal roles in orienting and locating the shaft in assembly. Again, to give one feature precedence over the other seems inappropriate. Here, however, the features are not the same size, and can't be considered a feature pattern.

The solution is to identify each datum feature separately, but include both identifying letters in a single datum reference, separated by a hyphen. It doesn't matter which letter appears first in the compartment, since neither datum feature has precedence over the other.

Rather than a single TGC, a hyphenated co-datum reference establishes a pair of perfectly coaxial or coplanar TGCs (depending on the feature types). In our example, datum features A and B are both referenced RFS. Their TGCs are coaxial actual mating envelopes that shall contract independently until each makes a minimum of two-point contact, jointly arresting four degrees of freedom. Hyphenated co-datum features are usually the same type of feature, with matching material conditions, and thus, matching TGC types. But not necessarily. The principle is equally applicable at MMC, LMC, or any pairing of material conditions.

> FAQ: *How can this simulation scheme work if the two datum features are badly eccentric?*
>
> A: The simulation will still work, but the part might not. Deriving meaningful datums (and DRFs, for that matter) from multiple features always demands careful control (using GD&T) of the orientation and location relationships between the individual datum features. For our example shaft, section 5.12.4 and Fig. 5-132 describe an elegant way to control coaxiality between the two bearing journals.

5.9.15 Multiple DRFs

On larger and/or more complicated parts, it may be impractical to control all features to a single DRF. Where features have separate functional relationships, relating them to the same DRF might be unnecessarily restrictive. Multiple DRFs may be used, but only with great care. Designers typically use too many datums and different DRFs, often without realizing it. Remember that any difference in datum references, their order of precedence, or their material conditions, constitutes a separate DRF. The tolerances connecting these DRFs start stacking up to where the designer quickly loses control of the part's overall integrity. A good way to prevent this and to unify the design is to structure multiple DRFs as a tree. That means controlling the datum features of each "branch" DRF to a common "trunk" DRF.

5.10 Orientation Tolerance (Level 3 Control)

Orientation is a feature's angular relationship to a DRF. An *orientation tolerance* controls this relationship without meddling in location control. Thus, an orientation tolerance is useful for relating one datum feature to another and for refining the orientation of a feature already controlled with a positional tolerance.

5.10.1 How to Apply It

An orientation tolerance is specified using a feature control frame displaying one of the three orientation characteristic symbols. See Fig. 5-92. The symbol used depends on the basic orientation angle, as follows.

0° or 180°—"parallelism" symbol
90° or 270°—"perpendicularity" symbol
any other angle—"angularity" symbol

All three symbols work exactly the same. The only difference is that where the "angularity" symbol is used, a basic angle shall be explicitly specified. Where the "parallelism" or "perpendicularity" symbol is used, the basic angle is implied by a drawing view that shows the parallel or perpendicular relationship. Though a single generic "orientation" symbol has been proposed repeatedly, most users prefer separate symbols for parallelism and perpendicularity because each tells the whole story at a glance. The feature control frame includes the orientation tolerance value followed by one or two datum references.

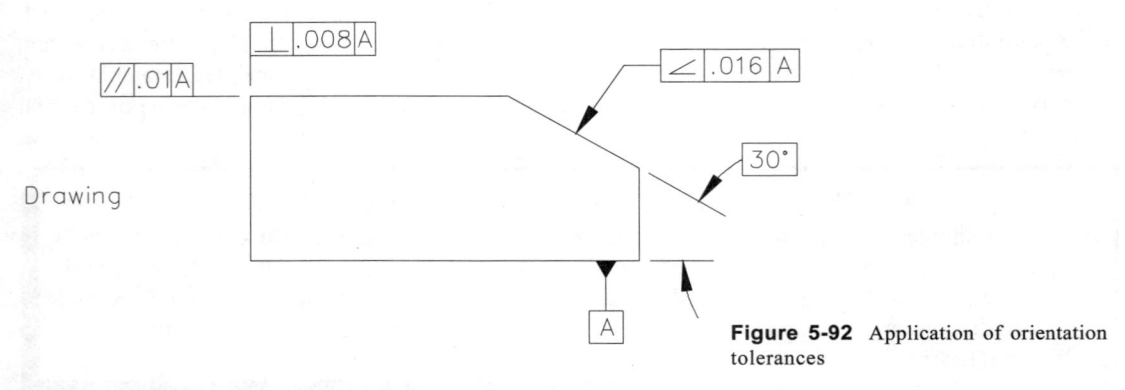

Figure 5-92 Application of orientation tolerances

5.10.2 Datums for Orientation Control

Orientation control requires a DRF. A primary datum plane or axis always establishes rotation about two axes of the DRF and is usually the only datum reference needed for orientation control. There are cases where it's necessary to establish rotation about the third axis as well and a secondary datum reference is needed. Sometimes, a secondary datum is needed to orient and/or locate a tolerance zone plane for controlling line elements of a feature. In other cases, hyphenated co-datums (see section 5.9.14.2) may be used to arrest rotation. Since all three rotational degrees of freedom can be arrested with just two datums, a tertiary datum is usually meaningless and confusing.

5.10.3 Applied to a Planar Feature (Including Tangent Plane Application)

Any nominally flat planar feature can be controlled with an orientation tolerance. Fig. 5-93 shows the tolerance zone bounded by two parallel planes separated by a distance equal to the tolerance value. The surface itself shall be contained between the two parallel planes of the tolerance zone. Form deviations including bumps, depressions, or waviness in the surface could prevent its containment. Thus, an orientation tolerance applied to a plane also controls flatness exactly the same as an equal flatness tolerance. In a mating interface, however, depressions in the surface may be inconsequential. After all, only the surface's three highest points are likely to contact the mating face (assuming the mating face is perfectly flat). Here, we may want to focus the orientation control on only the three highest or *tangent points*, excluding all other points on the surface from the tolerance. We do this by adding the "tangent plane" symbol (a circled T) after the tolerance value in the feature control frame. See Fig. 5-94. Now, only the perfect plane constructed tangent to the surface's three highest points shall be contained within the tolerance zone. Since it's acceptable for lower surface points to lie outside the zone, there's no flatness control.

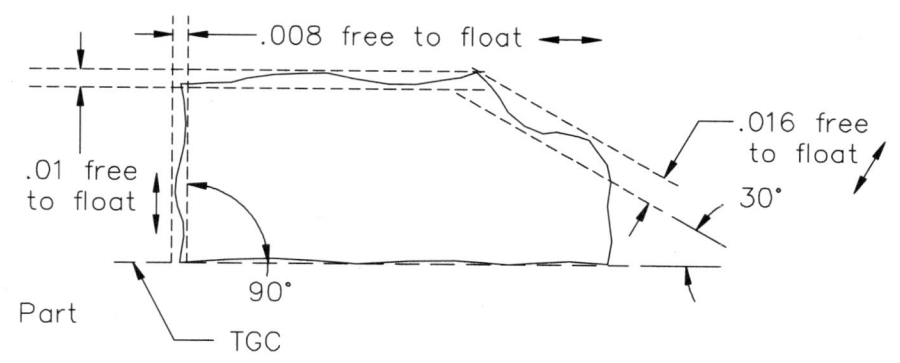

Figure 5-93 Tolerance zones for Fig. 5-92

The validity of "tangent plane" orientation control depends on the surface having exactly three noncolinear points that rise above the rest, allowing construction of exactly one tangent plane. Any other condition allows multiple candidate tangent planes to be constructed—a catastrophe not addressed by any standard. The method also assumes the mating face will be perfectly flat. If it too has three outstanding points, it's unlikely that contact will occur in either surface's tangent plane. Be careful with the "tangent plane" symbol.

For a width-type feature of size, Rule #1 automatically limits the parallelism of each surface to the other. Thus, a separate orientation tolerance meant to control parallelism between the two surfaces won't have any effect unless it's less than the total size tolerance.

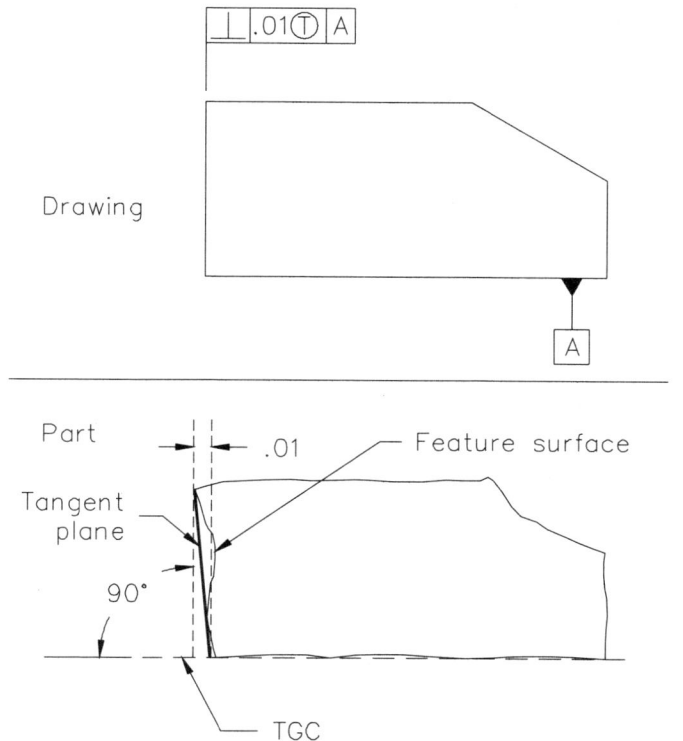

Figure 5-94 Application of tangent plane control

5.10.4 Applied to a Cylindrical or Width-Type Feature

Where an orientation tolerance feature control frame is placed according to options (a) or (d) in Table 5-1 (associated with a diameter or width dimension), the tolerance controls the orientation of the cylindrical or width-type feature. Where the tolerance is modified to MMC or LMC, it establishes a Level 3 virtual condition boundary as described in section 5.6.3.2 and Figs. 5-17(c) and 5-18(c). Alternatively, the "center method" described in section 5.6.5.1 may be applied to an orientation tolerance at MMC or LMC. Unmodified, the tolerance applies RFS and establishes a central tolerance zone as described in section 5.6.4.1, within which the feature's axis or center plane shall be contained. See Fig. 5-95. Applied to a feature of size, the orientation tolerance provides no form control beyond Level 2.

Fig. 5-95 shows the center plane of a slot contained within a central parallel-plane tolerance zone ("center method"). Y14.5 also allows the orientation of an axis to be controlled within a parallel-plane tolerance zone. Since this would not prevent the axis from revolving like a compass needle between the two parallel planes, such an application usually accompanies a larger positional tolerance. In Fig. 5-96, a "diameter" symbol precedes the angulation tolerance value. Here, the central tolerance zone is bounded by a cylinder having a diameter equal to the tolerance value. This control is more like a positional tolerance, except the orientation zone is not basically located from the datums.

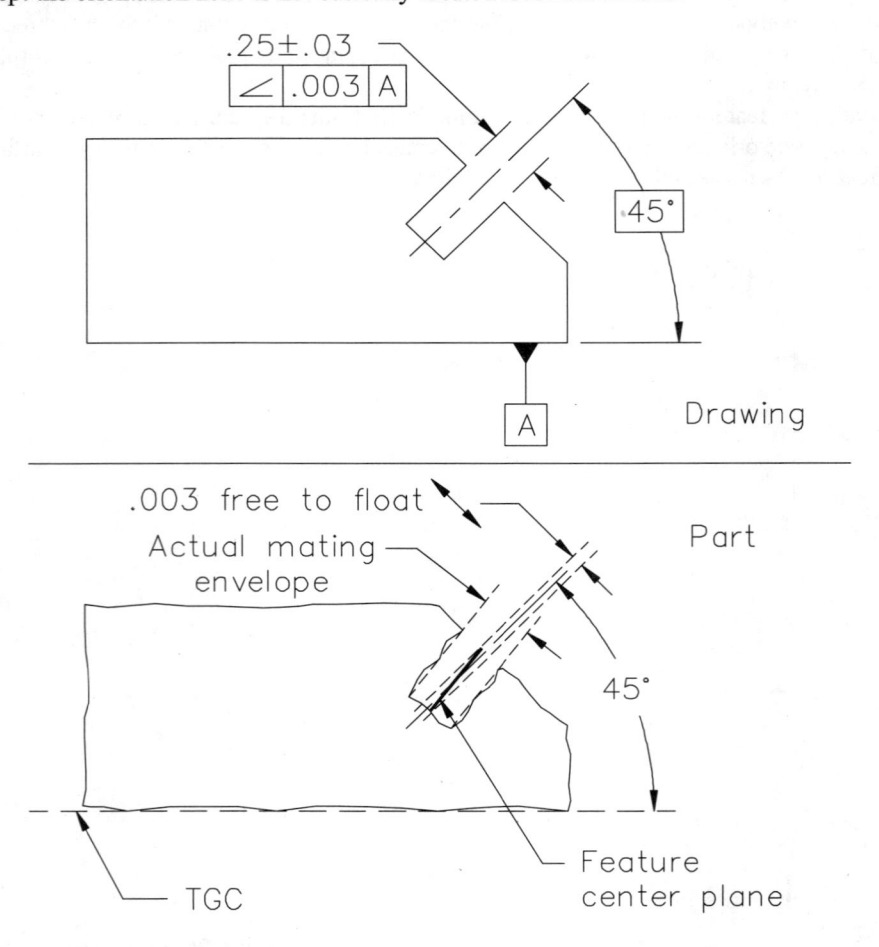

Figure 5-95 Applying an angularity tolerance to a width-type feature

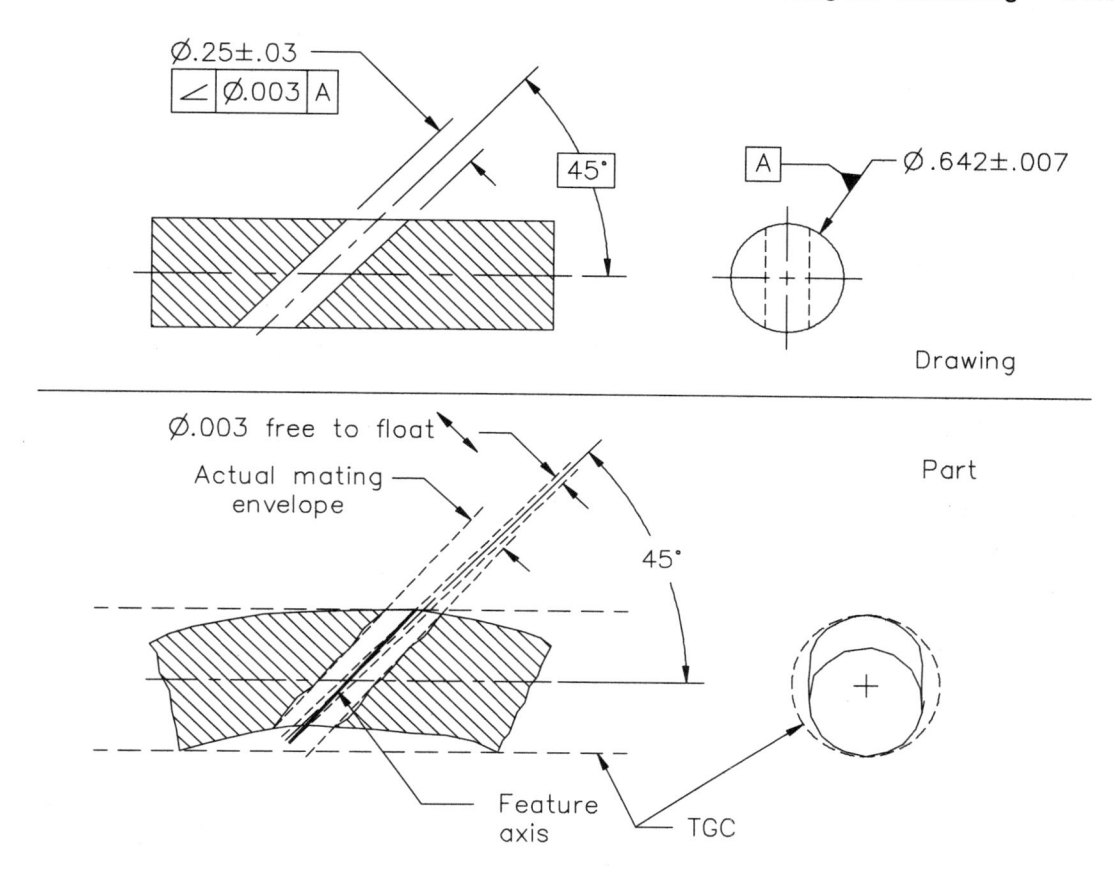

Figure 5-96 Applying an angularity tolerance to a cylindrical feature

A positional tolerance also controls orientation for a feature of size to the same degree as an equal orientation tolerance. Thus, for any feature of size, an orientation tolerance equal to or greater than its positional tolerance is meaningless. Conversely, where the designer needs to maximize positional tolerance while carefully protecting orientation, a generous positional tolerance can be teamed up with a more restrictive orientation tolerance.

5.10.4.1 Zero Orientation Tolerance at MMC or LMC

Where the only MMC design consideration is a clearance fit, there may be no reason for the feature's MMC size limit to differ from its Level 3 virtual condition. In such a case, we recommend stretching the MMC size limit to equal the MMC virtual condition size and reducing the orientation tolerance to zero as described in section 5.6.3.4. In LMC applications, as well, a zero orientation tolerance should be considered.

5.10.5 Applied to Line Elements

Where a profiled surface performs a critical function, it's sometimes necessary to control its orientation to a DRF. For the cam surface shown in Fig. 5-97, the 3-D control imposed by a parallel-planes tolerance zone is inappropriate because the surface isn't supposed to be flat. Here, we want to focus the orientation

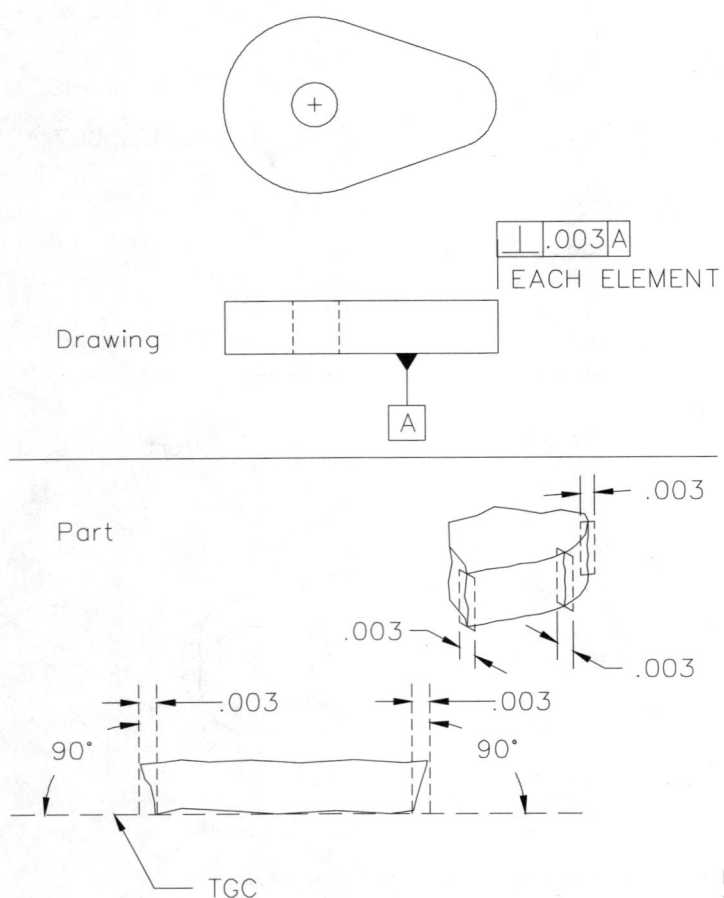

Figure 5-97 Controlling orientation of line elements of a surface

tolerance only on individual cross sections of the surface, one at a time. We do this by adding a note such as **EACH ELEMENT** or **EACH RADIAL ELEMENT** adjacent to the orientation feature control frame. This specifies a tolerance zone plane containing a tolerance zone bounded by two parallel lines separated by a distance equal to the tolerance value. As the tolerance zone plane sweeps the entire surface, the surface's intersection with the plane shall everywhere be contained within the tolerance zone (between the two lines). Within the plane, the tolerance zone's location may adjust continuously to the part surface while sweeping, but its orientation shall remain fixed at the basic angle relative to the DRF. This type of 2-D control allows unlimited surface undulation in only one direction.

Of a Surface Constructed About a Datum Axis—The note **EACH RADIAL ELEMENT** adjacent to the feature control frame means the tolerance zone plane shall sweep radially about a datum axis, always containing that axis. If the orienting (primary) datum doesn't provide an axis of revolution for the tolerance zone plane, a secondary datum axis shall be referenced. Note that within the rotating tolerance zone plane, the tolerance zone's location may adjust continuously.

Of a Profiled Surface—Where only a primary datum is referenced, as in Fig. 5-97, the tolerance zone plane shall sweep all around the part, always basically oriented to the datum, and always normal (perpendicular) to the controlled surface at each location. Where a secondary datum is referenced, the tolerance zone plane shall instead remain basically oriented to the complete DRF as it sweeps.

5.10.6 The 24 Cases

So far, in this section we've described the following:
- Four different types of orientation tolerance zone containments ("center method")
  - Plane (feature surface, tangent, or center) between two parallel planes
  - Axis between two parallel planes
  - Axis within a cylinder
  - Line element between two parallel lines
- Two types of primary datums for orientation
  - Plane
  - Axis
- Three orientation tolerance symbols
  - Parallelism (0° or 180°)
  - Perpendicularity (90° or 270°)
  - Angularity (any other angle)

These components can be combined to create 24 (4 x 2 x 3) different fundamental applications (or "cases") of orientation tolerance, illustrated in Fig. 5-98. In many cases, a secondary datum may be added for additional control. The illustrated parts are simplified abstracts, meant to show only the orientation control. On real parts, the orientation tolerances often accompany positional or profile tolerances.

5.10.7 Profile Tolerance for Orientation

As we'll see in Section 13, a single profile tolerance can control the size, form, orientation, and location of any feature, depending on the feature's type and the completeness of the referenced DRF. Where a profile tolerance already establishes the "size" and shape of a feature, incorporating orientation control may be as simple as adding another datum reference or expanding the feature control frame for composite profile control. Otherwise, it's better to use one of the dedicated orientation symbols.

5.10.8 When Do We Use an Orientation Tolerance?

Most drawings have a tolerance block or a general note that includes default plus and minus tolerances for angles. This default tolerance applies to any angle explicitly dimensioned without a tolerance. The angle between the depicted features shall be within the limits established by the angle dimension and the default angle tolerance. The default tolerance can be overridden by attaching a greater or lesser tolerance directly to an angle dimension. Either way, since neither feature establishes a datum for the other, the angular control between the features is reciprocal and balanced. The same level of control occurs where center lines and/or surfaces of part features are depicted on a drawing intersecting at right angles. Here, an implied 90° angle is understood to apply along with the default plus and minus angle tolerances. As before, there is no datum hierarchy, so all affected angular relationships are mutual.

The type of plus and minus angle tolerances just described does not establish a tolerance zone, wedge shaped or otherwise, to control the <u>angulation</u> of either feature. Be careful not to misinterpret Y14.5's Fig. 2-13, which shows a wedge-shaped zone controlling the <u>location</u> of a planar surface. Because it's still possible for the surface to be angled out of tolerance within the depicted zone, the "MEANS THIS" portion of the figure adds the note, **its angle shall not be less than 29°30' nor more than 30°30'.**

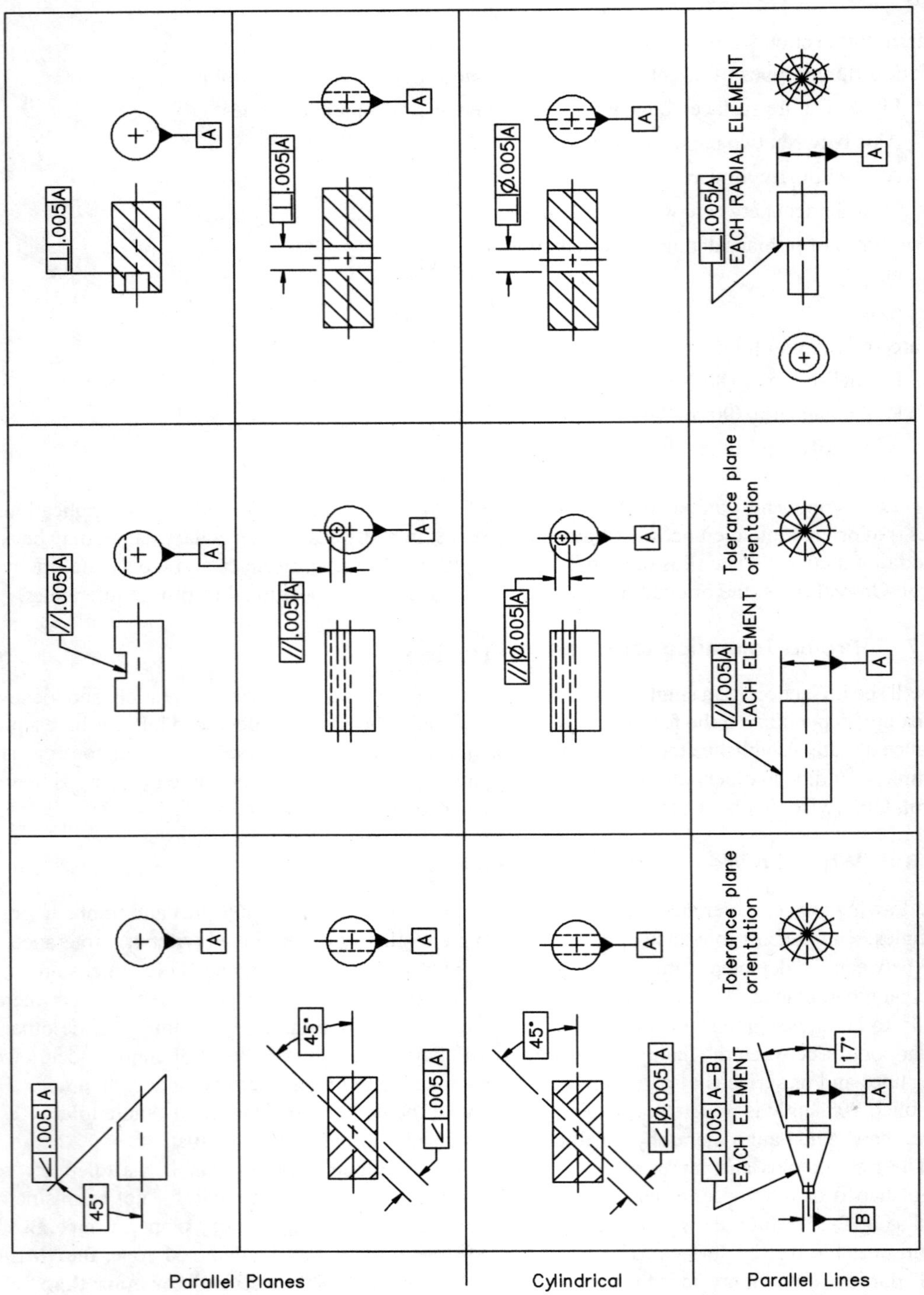

Figure 5-98 Applications of orientation tolerances

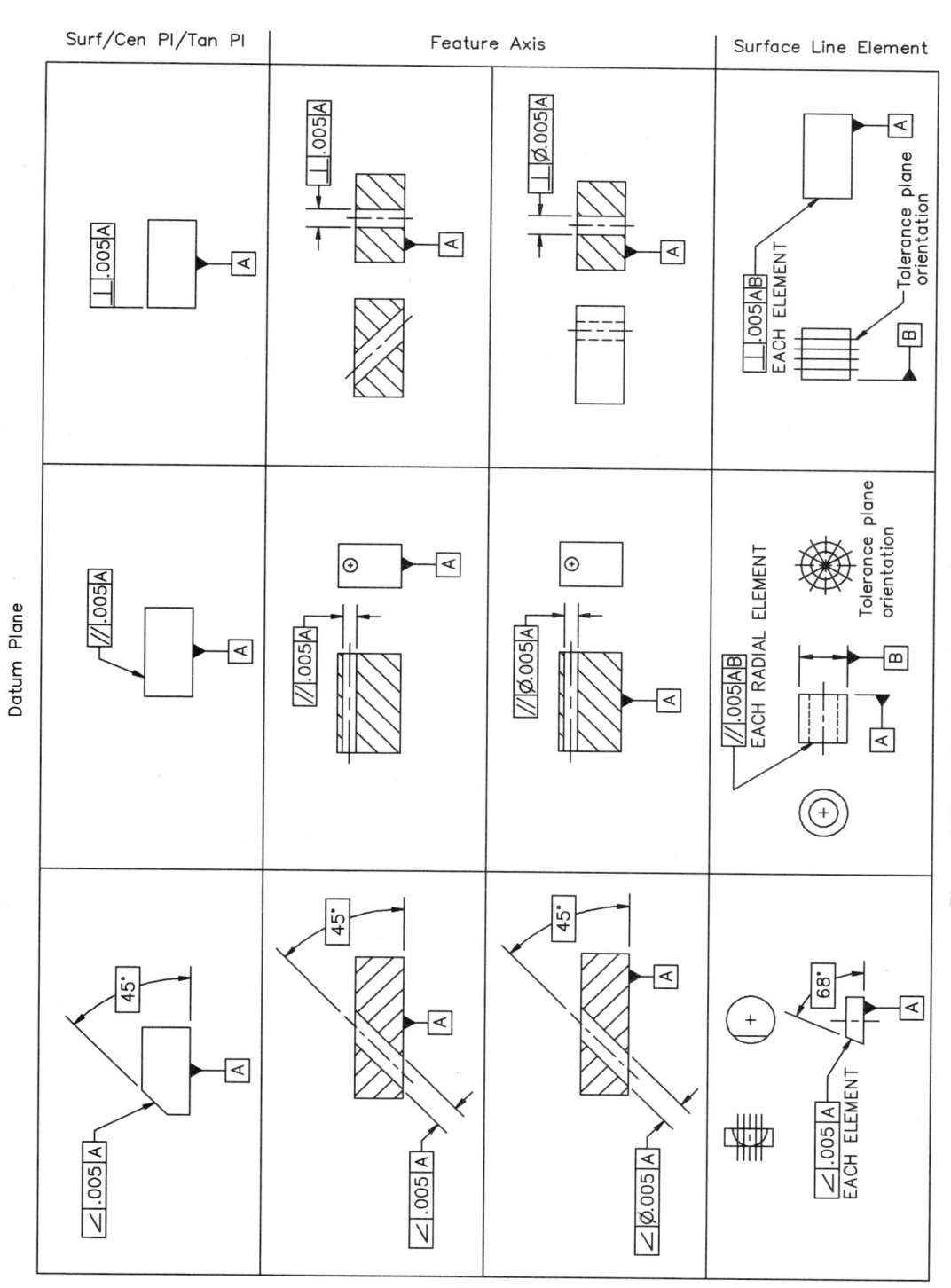

Figure 5-98 continued Applications of orientation tolerances

5-112 Chapter Five

Now, let's consider a different case, illustrated in Fig. 5-99, where two planar features intersect at an angle controlled with plus and minus tolerances and location is not an issue. For the sake of discussion, we'll attach the "dimension origin" symbol to the extension line for one surface, ostensibly making it a "quasi-datum" feature and the other a "controlled" feature. We'll suppose the "controlled" feature shall be contained within some wedge-shaped tolerance zone. Without a rule for locating its vertex (a line), such a zone would be meaningless. For example, if we could locate the vertex a mile away from the part, the zone could easily contain the "controlled" feature, the whole part, and probably the whole building! Since the standards are mute on all this, let's be reasonable and suppose the vertex can be located anywhere in our supposed "datum plane," as we've shown in the lower portion of the figure.

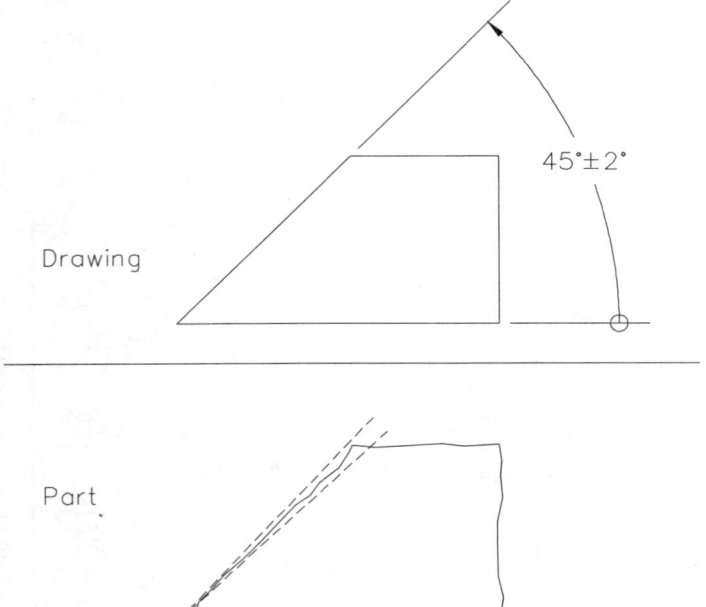

Figure 5-99 Erroneous wedge-shaped tolerance zone

Now here's the problem: Approaching the vertex, the width of our wedge-shaped tolerance zone approaches zero. Of course, even a razor edge has a minute radius. So we can assume that because of an edge radius, our "controlled" feature won't quite extend all the way to the vertex of the tolerance zone. But depending on the "size" of the radius and the angular tolerance, the zone could be only a few microns wide at the "controlled" feature's edge. Thus, the "controlled" feature's line elements parallel to the vertex shall be straight within those few microns, and angularity of the feature shall likewise approach perfection. Those restrictions are absurd.

Thus, even with a "dimension origin" symbol, a plus and minus angle tolerance establishes no defensible or usable tolerance zone for angulation. Instead, the tolerance applies to the angle measured between the two features. Imperfections in feature form complicate the measurement, and different alignments of the measuring scale yield different measurements. Unfortunately, the standards provide no guidance in either area. Despite these limitations, plus and minus angle tolerances are often sufficient for noncritical relationships where inspectors can be trusted to come up somehow with adequately repeatable and reproducible measurements.

Where a feature's orientation is more critical and the above methods are too ambiguous, an orientation tolerance feature control frame should be applied. In theory, datum simulation methods can accommodate out-of-squareness between datum features in a DRF. However, datum simulation will be more repeatable and error free where squareness of the secondary and tertiary datum features has been carefully and directly controlled to the higher-precedence datum(s).

As we'll see in the following sections, positional and profile tolerances automatically control feature orientation. But often, a generous positional or profile tolerance must be accompanied by a more strict orientation tolerance to assure functionality.

5.11 Positional Tolerance (Level 4 Control)

In the past, it was customary to control the location of a feature on a part by specifying for each direction a nominal dimension accompanied by plus and minus tolerances. In Fig. 5-100, the measured hole location shall be 1.625 ± .005 from the end of the shaft. Since the hole is drawn on the center line of the shaft, we know it must be well centered. But plus or minus how much? Let's assume the tolerance for centrality should match that for the 1.625 length. In effect, then, the axis of the hole shall lie within a .010" x .010" square box. Such a "square box" tolerance zone rarely represents the true functional requirements. Chapter 3 further elaborates on the shortcomings of plus and minus tolerances for location. The standards neither explain nor prohibit this method, but Y14.5 expresses a clear preference for its own brand of *positional tolerance* to control the orientation and location of one or more features of size, or in some cases, bounded features, relative to a DRF. A positional tolerance provides no form control beyond Level 2.

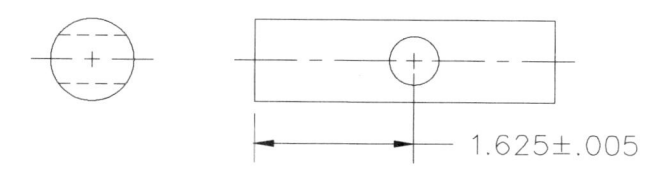

Figure 5-100 Controlling the location of a feature with a plus and minus tolerance

5.11.1 How Does It Work?

A positional tolerance may be specified in an RFS, MMC, or LMC context.

At MMC or LMC—Where modified to MMC or LMC, the tolerance establishes a Level 4 virtual condition boundary as described in section 5.6.3.3 and Figs. 5-17(d) and 5-18(d). Remember that the virtual condition boundary and the corresponding size limit boundary differ in size by an amount equal to the positional tolerance. In section 5.6.3.4, we discuss the advantages of unifying these boundaries by specifying a positional tolerance of zero. A designer should always consider this option, particularly in fastener applications.

At RFS—Unmodified, the tolerance applies RFS and establishes a central tolerance zone as described in section 5.6.4.1, within which the feature's center point, axis, or center plane shall be contained.

Alternative "Center Method" for MMC or LMC—Where the positional tolerance applies to a feature of size at MMC or LMC, the alternative "center method" described in section 5.6.5.1 may be applied.

For any feature of size, including cylindrical, spherical, and width-type features, a virtual condition boundary and/or derived center element is easily defined, and positional tolerancing is readily applicable.

Positional tolerancing can also be applied to a bounded feature for which an MMC or LMC virtual condition boundary can be defined relative to size limit and/or profile tolerance boundaries.

> **FAQ:** *Can positional tolerancing be applied to a radius?*
>
> **A:** No. Neither virtual condition boundaries nor central tolerance zones can be used to control the orientation or location of a radius or a spherical radius. There are no definitions for MMC, LMC, axis, or center point for these nonsize features.

5.11.2 How to Apply It

A positional tolerance is specified using a feature control frame displaying the "position" characteristic symbol followed by a compartment containing the positional tolerance value. See Fig. 5-9. Within the compartment, the positional tolerance value may be followed by an MMC or LMC modifying symbol. Any additional modifiers, such as "statistical tolerance," and/or "projected tolerance zone" follow that. The tolerance compartment is followed by one, two, or three separate compartments, each containing a datum reference letter. Within each compartment, each datum reference may be followed by an MMC or LMC modifying symbol, as appropriate to the type of datum feature and the design.

For each individual controlled feature, a unique *true position* shall be established with basic dimensions relative to a specified DRF. True position is the nominal or idal orientation and location of the feature and thus, the center of the virtual condition boundary or positional tolerance zone. The basic dimensions may be shown graphically on the drawing, or expressed in table form either on the drawing or in a document referenced by the drawing. Figs. 5-101 and 5-102 show five different methods for establishing true positions, explained in the following five paragraphs.

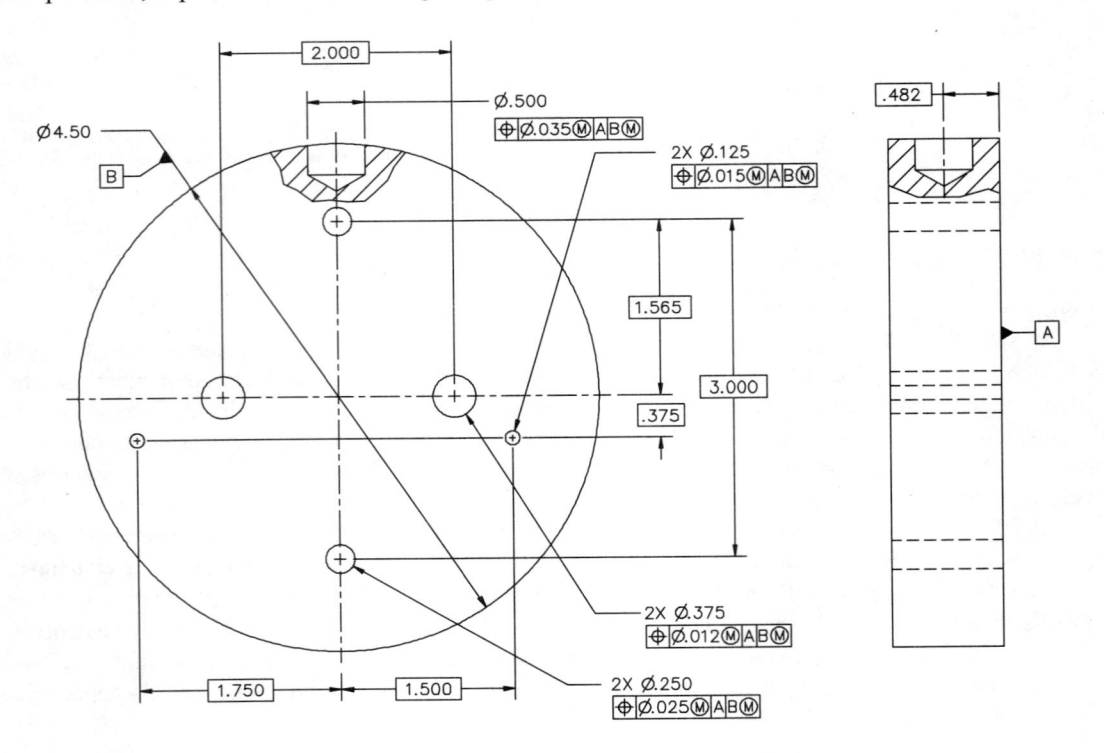

Figure 5-101 Methods for establishing true positions

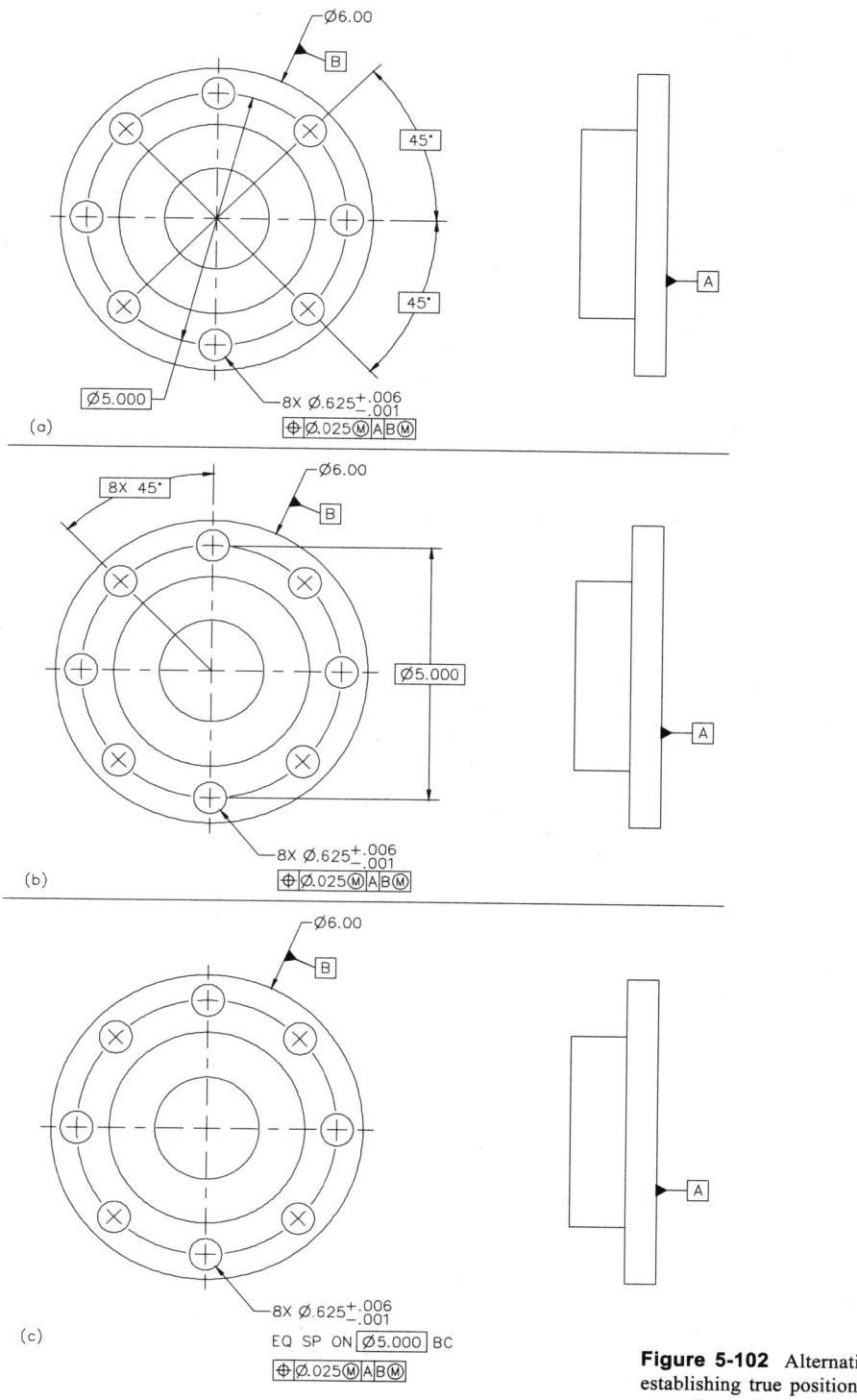

Figure 5-102 Alternative methods for establishing true positions using coordinate dimensioning

5-116   Chapter Five

Base line dimensioning—For each of the two ∅.125 holes shown in Fig. 5-101, a basic dimension originates from each plane of the DRF. Manufacturers prefer this method because it directly provides them the coordinates for each true position relative to the datum origin. CMM inspection is simplified, using a single 0,0 origin for both holes.

Chain dimensioning—In Fig. 5-101, a basic dimension of 1.565 locates the upper ∅.250 hole directly from the center plane. However, the lower ∅.250 hole is located with a 3.000 basic dimension from the true position of the upper hole. People often confuse the 3.000 basic as originating from the actual axis of the upper hole, rather than from its true position. A manufacturer needing the coordinate of the lower hole will have to calculate it: $1.565 - 3.000 = -1.445$. Or is it -1.435?

Implied symmetry dimensioning—In many cases, the applicable basic dimensions are implied by drawing views. In Fig. 5-101, the true positions of the two ∅.375 holes have a single 2.000 basic dimension between them, but no dimension that relates either hole to the planes of the DRF. Since the holes appear symmetrical about the center plane of the DRF, that symmetrical basic relationship is implied.

Implied zero-basic dimensions—The view implies the relationship of the ∅.500 hole to the planes of the DRF as represented by the view's center lines. Obviously, the hole's basic orientation is 0° and its basic offset from center is 0. These implied zero-basic values need not be explicated.

Polar coordinate dimensioning—Rather than by "rectangular coordinates" corresponding to two perpendicular axes of the DRF, the true positions of the eight ∅.625 holes shown in Fig. 5-102(a) are defined by *polar coordinates* for angle and diameter. The ∅5.000 "bolt circle" is basically centered at the intersection of the datum planes, and the two 45° basic angles originate from a plane of the DRF. Figs. 5-102(b) and (c) show alternative approaches that yield equivalent results, based on various methods and fundamental rules we've presented.

All the above methods are acceptable. Often, a designer can choose between base line and chain dimensioning. While both methods yield identical results, we prefer base line dimensioning even if the designer has to make some computations to express all the dimensions originating from the datum origin. Doing so once will preclude countless error-prone calculations down the road.

5.11.3 Datums for Positional Control

One of the chief advantages of a GD&T positional tolerance over plus and minus coordinate tolerances is its relationship to a specific DRF. Every positional tolerance shall reference one, two, or three datum features. The DRF need not restrain all six degrees of freedom, only those necessary to establish a unique orientation and location for true position. (Degrees of freedom are explained in section 5.9.7.) For example, the DRF established in Fig. 5-103 restrains only four degrees of freedom. The remaining two degrees, rotation about and translation along the datum axis, have no bearing on the controlled feature's true position. Thus, further datum references are meaningless and confusing.

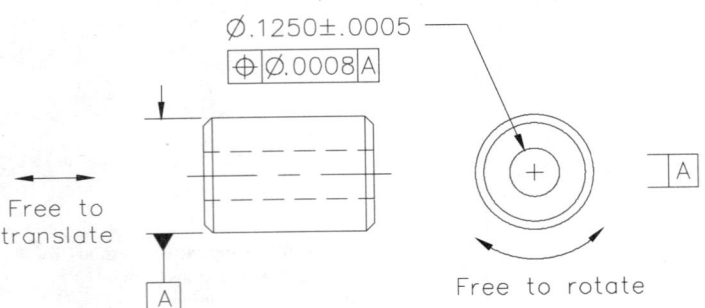

Figure 5-103 Restraining four degrees of freedom

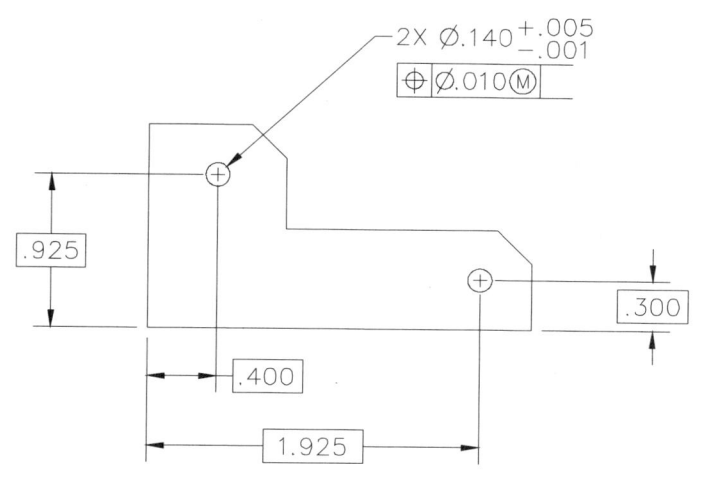

Figure 5-104 Implied datums are not allowed

For many positional tolerances, such as those shown in Fig. 5-104, the drawing view makes it quite obvious which part features are the origins, even if they weren't identified as datum features and referenced in the feature control frame. Before the 1982 revision of Y14.5, *implied datums* were recognized and not required to be explicitly referenced in such cases. In Fig. 5-104, although we all may agree the part's left and lower edges are clearly datum features, we might disagree on their precedence in establishing the orientation of the DRF. In another example, where a part has multiple coaxial diameters, it might be obvious to the designer, but very unclear to the reader, which diameter is supposed to be the datum feature. For these reasons, Y14.5 no longer allows implied datums; the savings in plotter ink aren't worth the confusion.

A datum feature of size can be referenced RFS (the default where no modifier symbol appears), at MMC, or at LMC. Section 5.6.7 discusses modifier choices. When MMC or LMC is selected, the DRF is not fixed to the part with a unique orientation and location. Instead, the DRF can achieve a variety of orientations and/or locations relative to the datum feature(s). The stimulating details of such allowable "DRF displacement" are bared in section 5.9.9.

5.11.4 Angled Features

Positional tolerancing is especially suited to angled features, such as those shown in Fig. 5-105. Notice how the true position for each angled feature is carefully defined with basic lengths and angles relative only to planes of the DRF. In contrast, Fig. 5-106 shows a common error: The designer provided a basic dimension to the point where the hole's true position axis intersects the surrounding face. Thus, the true position is established by a face that's not a datum feature. This is an example of an implied datum, which is no longer allowed.

5.11.5 Projected Tolerance Zone

A positional tolerance, by default, controls a feature over its entire length (or length and breadth). This presumes the feature has no functional interface beyond its own length and breadth. However, in Fig. 5-107, a pin is pressed into the controlled hole and expected to mate with another hole in a cover plate. The mating feature is not the pin hole itself, but rather the pin, which represents a projection of the hole. Likewise, the mating interface is not within the length of the pin hole, but above the hole, within the thickness of the cover plate.

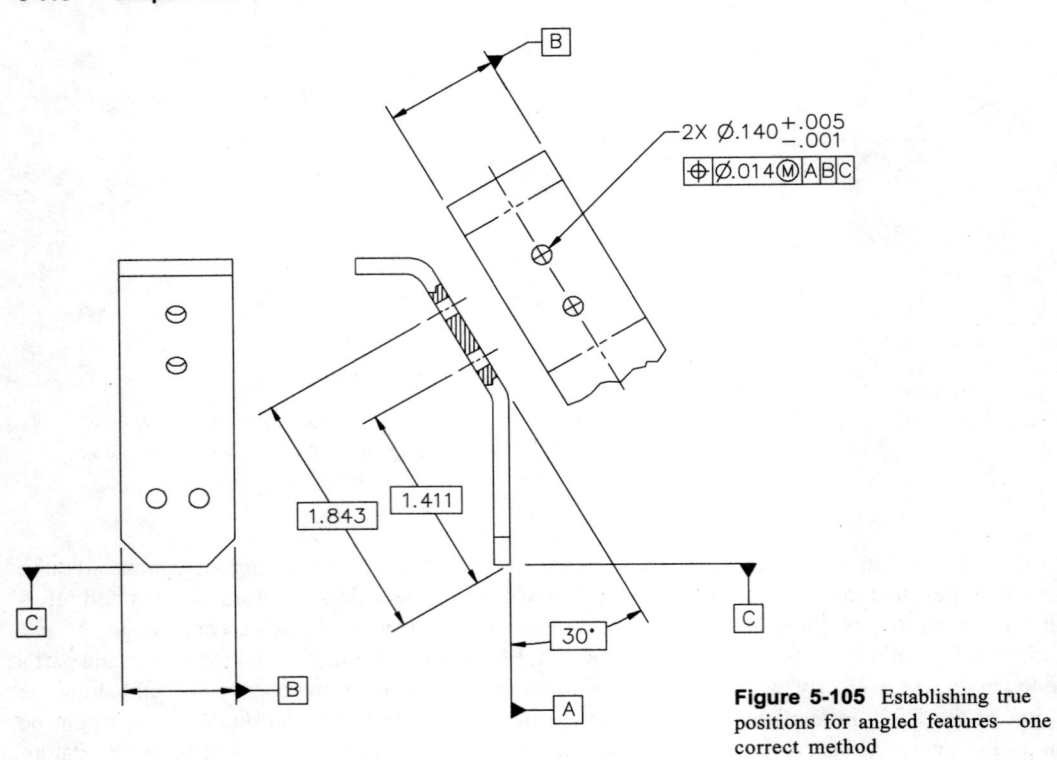

Figure 5-105 Establishing true positions for angled features—one correct method

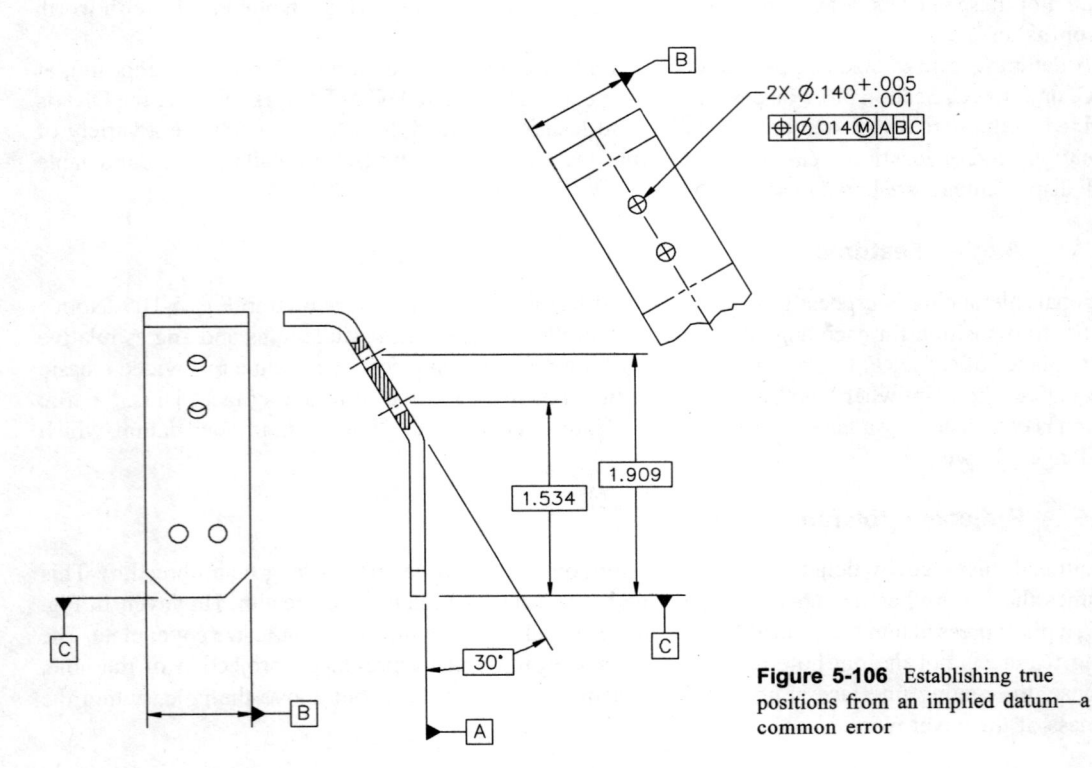

Figure 5-106 Establishing true positions from an implied datum—a common error

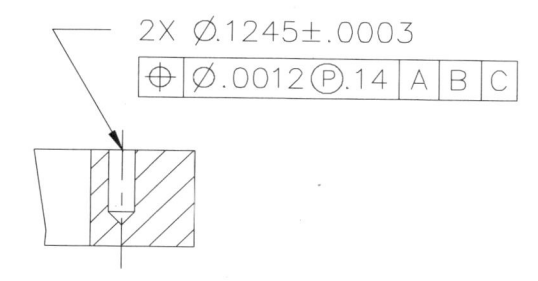

Figure 5-107 Specifying a projected tolerance zone

If the pin hole were perfectly perpendicular to the planar interface between the two parts, there would be no difference between the location of the hole and the pin. Any angulation, however, introduces a discrepancy in location. This discrepancy is proportional to the length of projection. Thus, directly controlling the location of the pin hole itself is inadequate to assure assemblability. Instead, we need to control the location of the hole's projection, which could be thought of as a phantom pin. This is accomplished with a positional tolerance modified with a projected tolerance zone.

A projected tolerance zone is specified by placing the "projected tolerance zone" symbol (a circled P) after the tolerance value in the position feature control frame. This establishes a constant-size central tolerance zone bounded either by two parallel planes separated by a distance equal to the specified tolerance, or by a cylinder having a diameter equal to the specified tolerance. For blind holes and other applications where the direction of projection is obvious, the length of projection may be specified after the symbol in the feature control frame. This means the projected tolerance zone terminates at the part face and at the specified distance from the part face (away from the part, and parallel to the true position axis or center plane). The projection length should equal the maximum extension of the mating interface. In our pin and cover plate example, the projection length must equal the cover plate's maximum thickness, .14. Where necessary, the extent and direction of the projected tolerance zone are shown in a drawing view as a dimensioned value with a heavy chain line drawn next to the center line of the feature, as in Fig. 5-108.

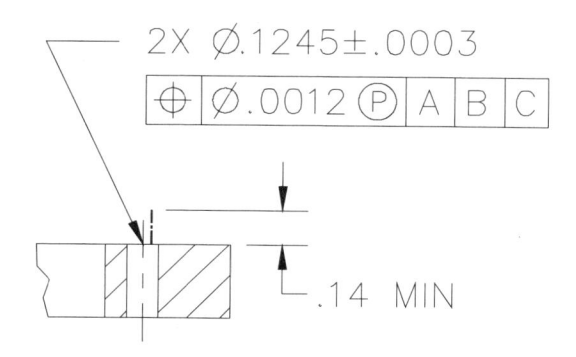

Figure 5-108 Showing extent and direction of projected tolerance zone

At RFS—The extended axis or center plane of the feature's actual mating envelope (as defined in section 5.6.4.2) shall be contained within the projected tolerance zone.

At MMC—The extended axis or center plane of the feature's applicable Level 2 MMC perfect form boundary (as defined in section 5.6.3.1) shall be contained within the projected tolerance zone. See Fig. 5-109. As the feature's size departs from MMC, the feature fits its MMC perfect form boundary more loosely. This permits greater deviation in the feature's orientation and/or location. A hole's departure from MMC permits assembly with a mating pin having its axis anywhere within a conical zone. The alternative

5-120   Chapter Five

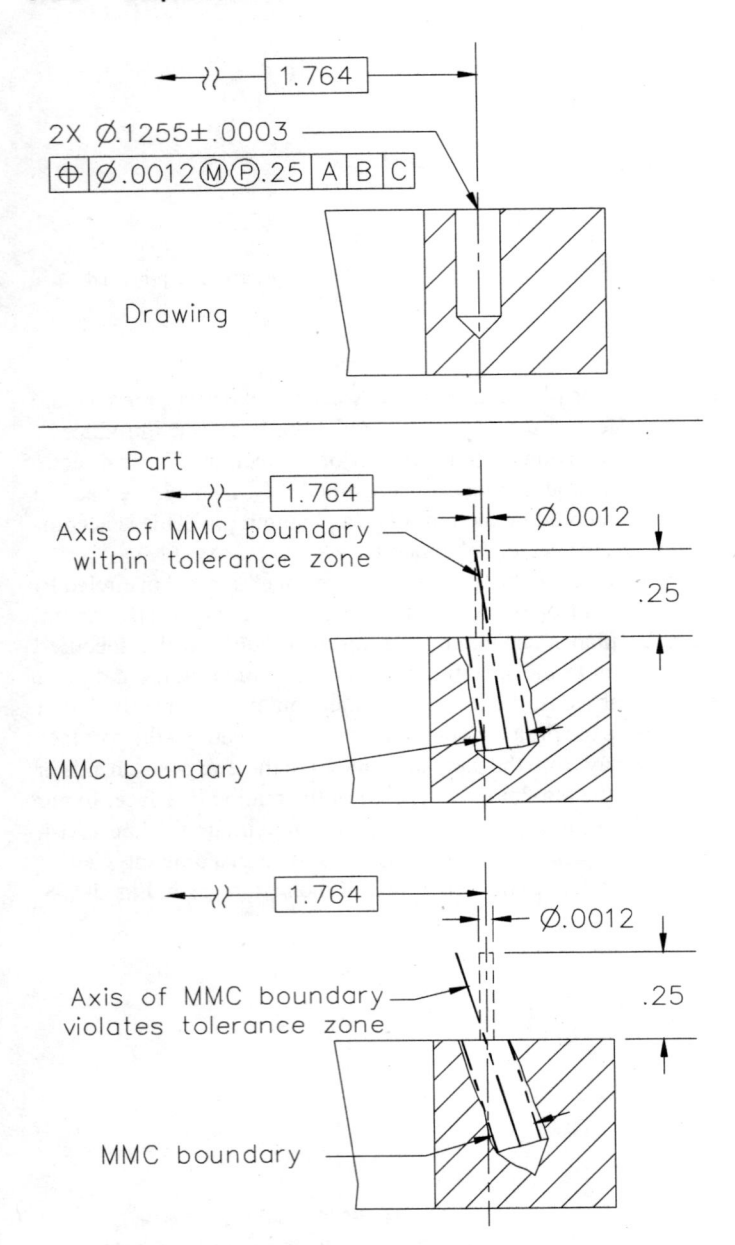

Figure 5-109 Projected tolerance zone at MMC

"center method" described in section 5.6.5.1 cannot be used for a projected tolerance zone. Its "bonus tolerance" would simply enlarge the projected tolerance zone uniformly along its projected length, failing to emulate the feature's true functional potential.

At LMC—The extended axis or center plane of the feature's Level 2 LMC perfect form boundary (as defined in section 5.6.3.1) shall be contained within the projected tolerance zone. As the feature's size departs from LMC, the feature fits its LMC perfect form boundary more loosely. This permits greater deviation in the feature's orientation and/or location. The alternative "center method" described in section 5.6.5.1 cannot be used for a projected tolerance zone.

5.11.6 Special-Shaped Zones/Boundaries

We stated that a "square box" tolerance zone rarely represents a feature's true functional requirements, and that the shape of a positional tolerance zone usually corresponds to the shape of the controlled feature. There are exceptions, however, and GD&T has been made flexible enough to accommodate them.

5.11.6.1 Tapered Zone/Boundary

Where a relatively long or broad feature of size has different location requirements at opposite extremities, a separate positional tolerance can be specified for each extremity. This permits maximization of both tolerances. "Extremities" are defined by nominal dimensions. Thus, for the blind hole shown in Fig. 5-110, the ⌀.010 tolerance applies at the intersection of the hole's true position axis with the surrounding part face (Surface C). The ⌀.020 tolerance applies .750 (interpreted as basic) below that.

At MMC or LMC—The tolerances together establish a Level 4 virtual condition boundary as described in section 5.6.3.3 and Figs. 5-17(d) and 5-18(d), except that in this case, the boundary is a frustum (a cone or wedge with the pointy end chopped off). The virtual condition size at each end derives from the regular applicable formula and applies at the defined extremity.

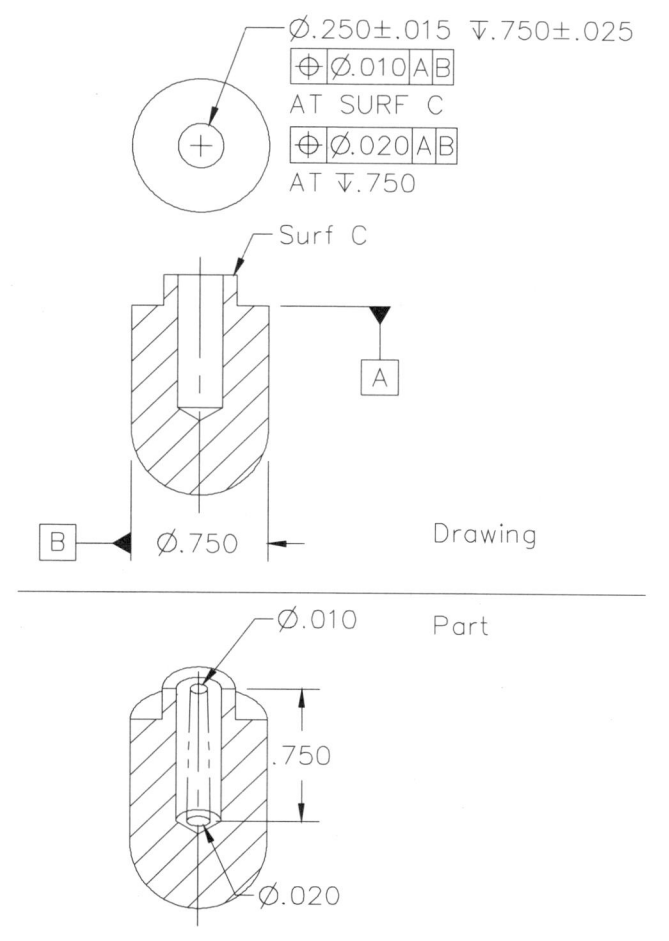

Figure 5-110 Different positional tolerances (RFS) at opposite extremities

At RFS—Unmodified, the tolerances apply RFS and establish a central tolerance zone bounded by a conical or wedge-shaped frustum, within which the feature's axis or center plane shall be contained. The specified tolerance zone sizes apply at the defined extremities. See Fig. 5-110.

Alternative "Center Method" for MMC or LMC—Where modified to MMC or LMC, the tolerances may optionally be interpreted as in an RFS context—that is, they establish a central tolerance zone bounded by a conical or wedge-shaped frustum, within which the feature's axis or center plane shall be contained. However, unlike in the RFS context, the size of the MMC or LMC tolerance zone shall be enlarged at each defined extremity by a single "bonus tolerance" value, derived according to section 5.6.5.1.

5.11.6.2 Bidirectional Tolerancing

A few features have different positional requirements relative to different planes of the DRF. Where these differences are slight, or where even the lesser tolerance is fairly generous, the more restrictive value can be used in an ordinary positional tolerance. In most cases, the manufacturing process will vary nearly equally in all directions, so an extra .001" of tolerance in just one direction isn't much help. However, where the difference is significant, a separate feature control frame can be specified for each direction. Y14.5 calls this practice *bidirectional tolerancing*. It can be used with a cylindrical feature of size located with two coordinates, or with a spherical feature of size located with three coordinates.

Each bidirectional feature control frame may be evaluated separately, just as if each controls a separate feature of size. However, as with separate features, rules for simultaneous or separate requirements apply (see section 5.9.10). By convention, the "diameter" symbol (Ø) is not used in any bidirectional feature control frames. The exact meanings of bidirectional tolerances are deceivingly complex. They depend on whether true position is defined in a rectangular or polar coordinate system, and on whether the tolerances apply in an RFS, MMC, or LMC context.

In a Rectangular Coordinate System—Fig. 5-111 shows a coupling ball located with rectangular coordinates in three axes. Each of the three separate feature control frames constrains the ball's location

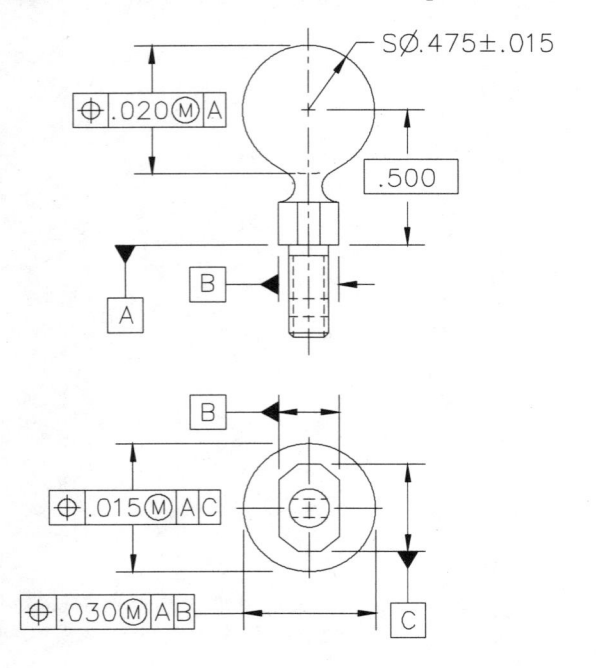

Figure 5-111 Bidirectional positional tolerancing, rectangular coordinate system

relative only to the DRF plane that is perpendicular to the dimension line. The .020 tolerance, for example, applies only to the .500 BASIC coordinate, relative to the horizontal plane of the DRF.

At MMC or LMC (Rectangular)—Each positional tolerance establishes a tolerance plane perpendicular to its dimension line. Each tolerance plane contains the center point (or axis, for a cylinder) of a Level 4 virtual condition boundary as described in section 5.6.3.3. However, within this plane, the location (and for a cylinder, orientation) of the boundary center is unconstrained. Thus, by itself, each tolerance would permit the controlled feature to spin and drift wildly within its tolerance plane. But, the combined restraints of three (or two, for a cylinder) perpendicular tolerance planes are usually adequate to control the feature's total location (and orientation, for a cylinder).

The virtual condition boundaries for a shaft at MMC are external to the shaft. As each cylindrical boundary spins and drifts within its tolerance plane, it generates an effective boundary of two parallel planes. The intersection of these parallel-plane boundaries is a fixed size rectangular box at true position. See Fig. 5-112. Thus, a single functional gage having a fixed rectangular cutout can gauge both bidirectional positional tolerances in a single pass. The same is not true where the virtual condition boundaries are internal to a hole at MMC, since a hole cannot contain parallel-plane boundaries.

At RFS (Rectangular)—Unmodified, each positional tolerance applies RFS and specifies a central tolerance zone bounded by two parallel planes separated by a distance equal to the specified tolerance. The intersection of these parallel-plane tolerance zones is a rectangular box centered at true position, within which the feature's axis or center point shall be contained. See Fig. 5-113.

Alternative "Center Method" for MMC or LMC (Rectangular)—Where modified to MMC or LMC, both tolerances may optionally be interpreted as in an RFS context—that is, each establishes a central

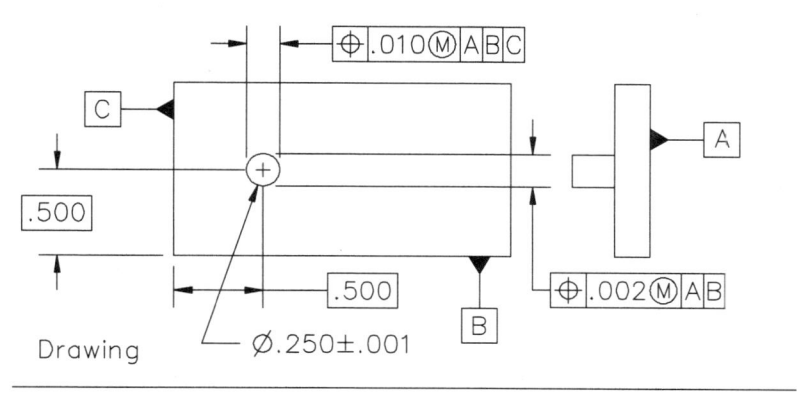

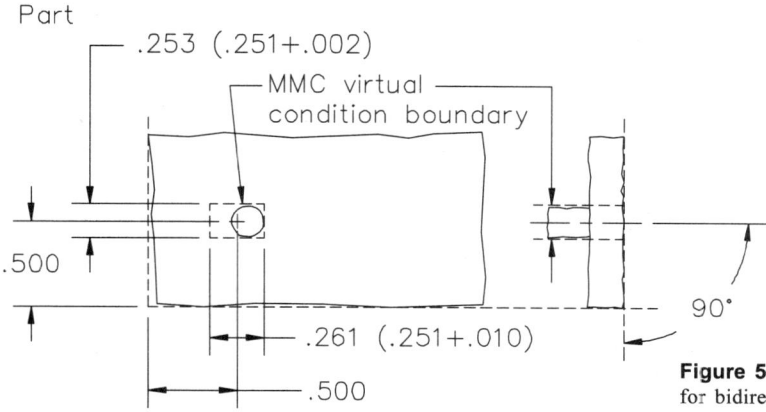

Figure 5-112 Virtual condition boundaries for bidirectional positional tolerancing at MMC, rectangular coordinate system

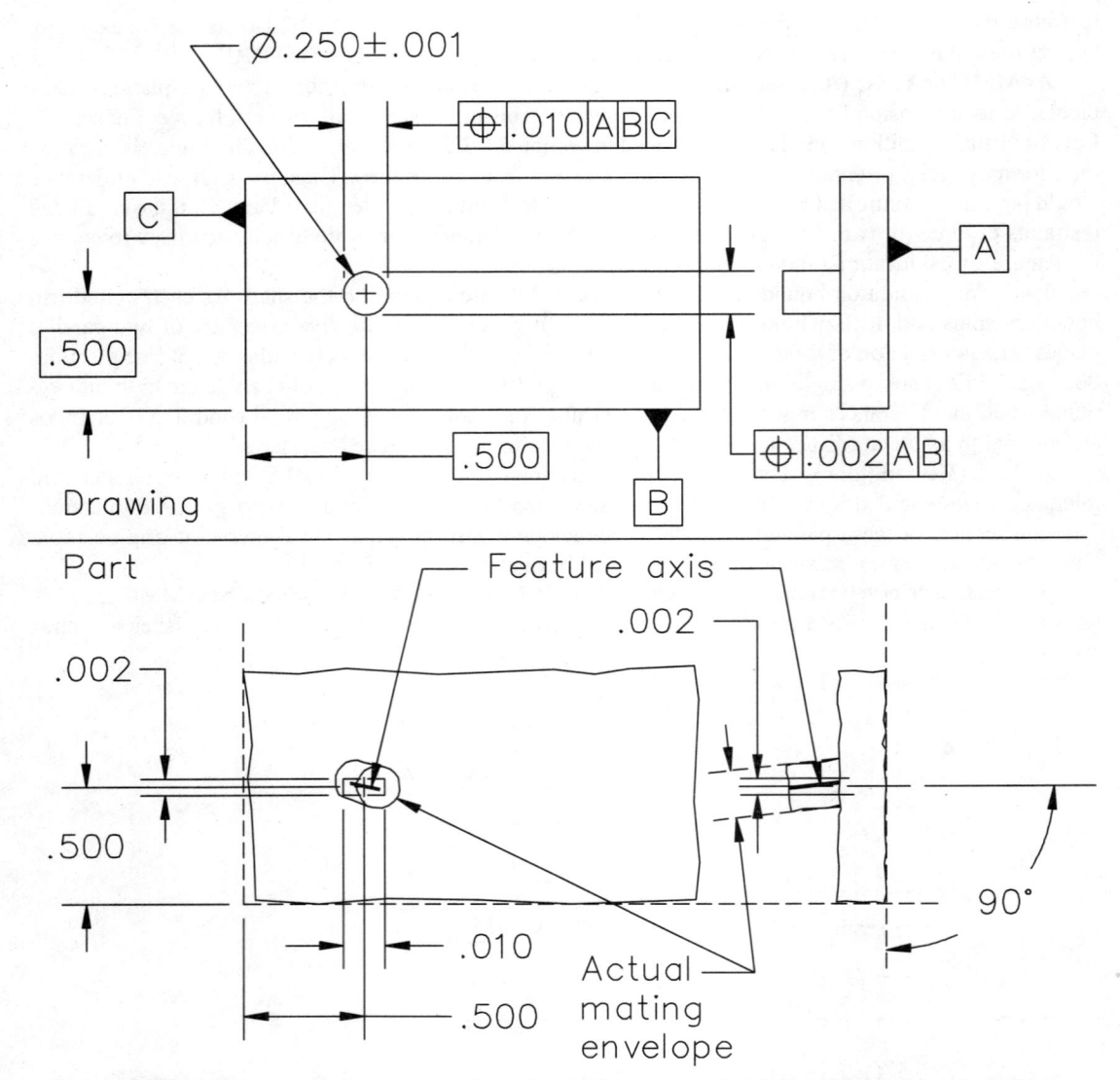

Figure 5-113 Tolerance zone for bidirectional positional tolerancing applied RFS, rectangular coordinate system

tolerance zone bounded by a pair of parallel planes, within which the feature's axis or center point shall be contained. However, unlike in the RFS context, the size of each MMC or LMC tolerance zone shall be enlarged by a single "bonus tolerance" value, derived according to section 5.6.5.1.

In a Polar Coordinate System—Fig. 5-114 shows a hole located with polar coordinates, one for radius and one for angle. The .020 tolerance constrains the hole's location relative only to the R.950 basic coordinate—in effect, its radial distance from the DRF origin point. The .010 tolerance constrains the hole relative only to a center plane rotated 47° basic relative to the DRF plane.

At MMC or LMC (Polar)—In this type of application, no virtual condition boundary is defined, due to problems in defining its restraint. The "center method," described on the next page, shall be used instead.

Figure 5-114 Bidirectional positional tolerancing, polar coordinate system

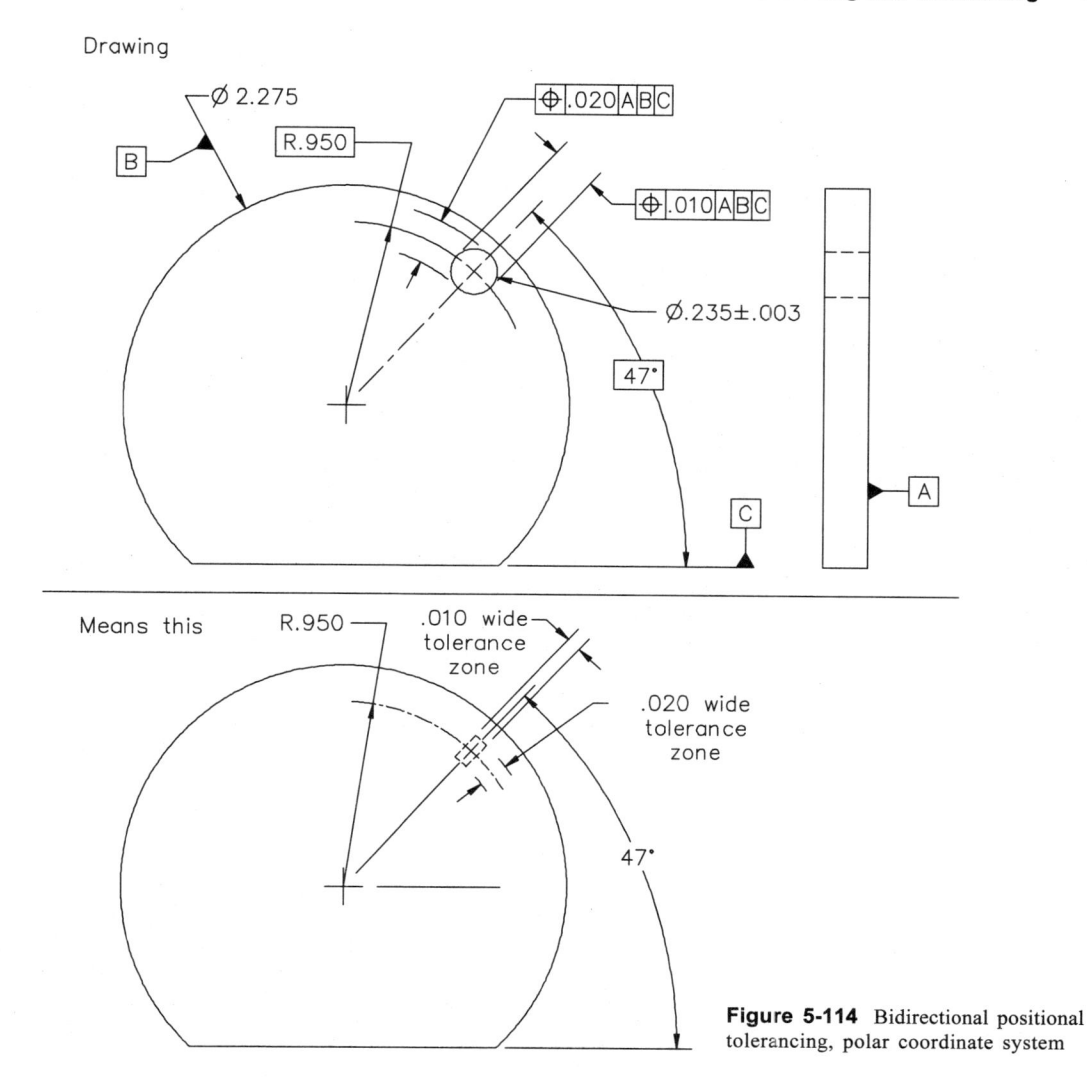

At RFS (Polar)—Unmodified, each positional tolerance applies RFS. One tolerance specifies a central tolerance zone bounded by two parallel planes separated by a distance equal to the specified tolerance. The other tolerance specifies a tolerance zone bounded by two concentric cylinders radially separated by a distance equal to the specified tolerance. The intersection of these tolerance zones is an arc-shaped space (shown in the lower portion of Fig. 5-114) centered at true position, within which the feature's axis or center point shall be contained.

"Center Method" for MMC or LMC (Polar)—Where modified to MMC or LMC, both tolerances shall be interpreted as in an RFS context—that is, each establishes a central tolerance zone bounded by a pair of parallel planes and a pair of concentric cylinders, within which the feature's axis or center point shall be contained. However, unlike in the RFS context, the size of each MMC or LMC tolerance zone shall be enlarged by a single "bonus tolerance" value, derived according to section 5.6.5.1.

5.11.6.3 Bounded Features

Positional tolerance can be applied judiciously to bounded features having opposing elements that partly or completely enclose a space.

At MMC or LMC—If the positional tolerance is modified to MMC, the bounded feature shall have a defined and discernible MMC size/form boundary. This can derive from multiple size dimensions or profile tolerance(s) (see Section 13). In an LMC context, an LMC size/form boundary shall be defined. The tolerance establishes a Level 4 virtual condition boundary uniformly offset from the applicable MMC or LMC size/form limit boundary by an amount equal to one-half the specified positional tolerance. For clarification, the term **BOUNDARY** is placed beneath the feature control frames.

At RFS—RFS is not applicable unless the designer specifies a detailed procedure for deriving unique and repeatable center elements. Then, the tolerance establishes one or more central tolerance zones within which the derived center element(s) shall be contained.

Fig. 5-115 shows a bounded feature controlled with two different positional tolerances. In this example, the concept is identical to that for bidirectional tolerancing described in section 5.11.6.2, except the controlled feature is noncircular with a separate size dimension corresponding to each positional tolerance. Where bidirectional control is not necessary, we recommend using instead composite profile tolerancing, as detailed in section 5.13.13.

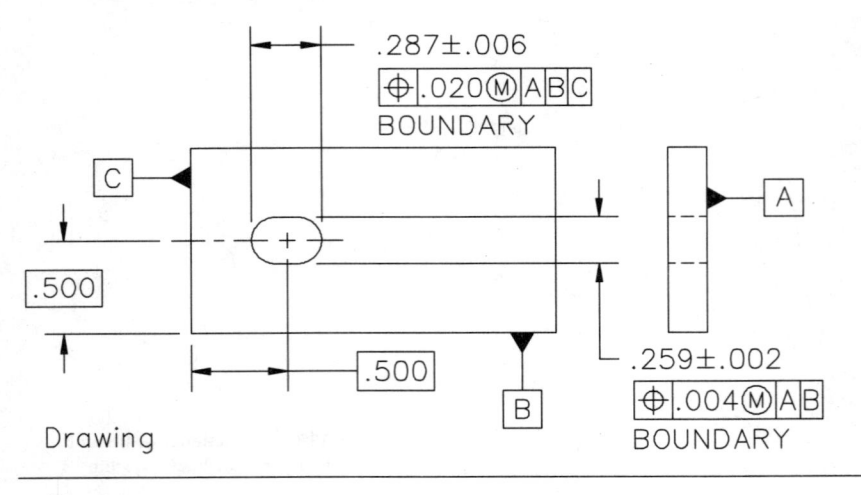

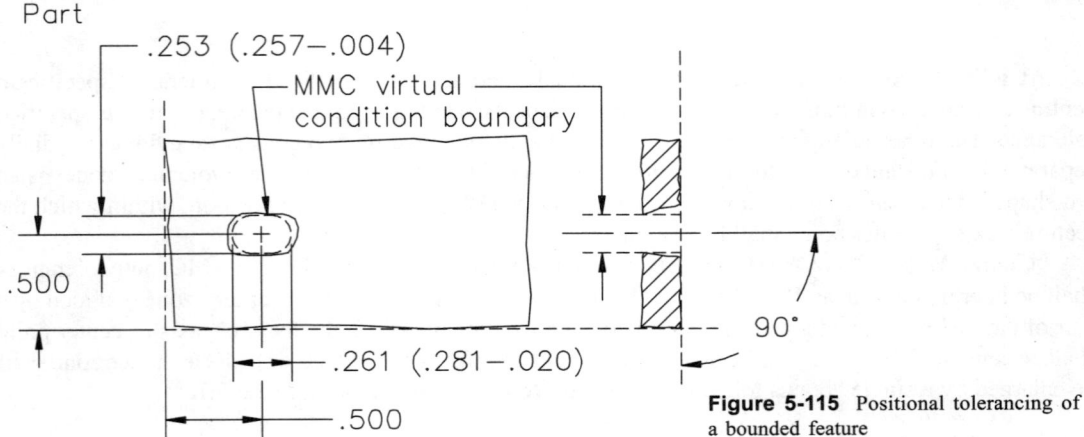

Figure 5-115 Positional tolerancing of a bounded feature

5.11.7 Patterns of Features

In many assemblies, two parts are attached to each other through a pattern of (multiple) features of size. For example, a closure cover may be bolted to a pump body with 24 3/8" bolts. A positional tolerance may be applied to the entire pattern, controlling the orientation and location of each individual feature relative to a DRF, and relative to every other feature in the pattern. Rather than a single boundary or tolerance zone, a positional tolerance applied to a feature pattern establishes a pattern (*framework*) of multiple boundaries or tolerance zones. Within this framework, the orientation and location of all the boundaries (or zones) are fixed relative to one another according to the basic dimensions expressed on the drawing.

At MMC or LMC—Where modified to MMC or LMC, the tolerance establishes a framework of Level 4 virtual condition boundaries as described in section 5.6.3.3.

At RFS—Unmodified, the tolerance applies RFS and establishes a framework of central tolerance zones as described in section 5.6.4.1.

Alternative "Center Method" for MMC or LMC—Where the positional tolerance applies to features of size at MMC or LMC, the alternative "center method" described in section 5.6.5.1 may be applied. The size of each tolerance zone adjusts independently according to the actual size of its corresponding feature.

In the following discussion, we're going to focus on cylindrical mating features and their Level 4 MMC virtual condition boundaries. However, pattern controls are equally effective for width-type features, and just as usable in LMC and RFS contexts. The few simplified calculations we'll be making are just to illustrate the concepts of pattern control. Subsequent chapters, particularly 22 and 24, present a more thorough discussion of positional tolerance calculations.

5.11.7.1 Single-Segment Feature Control Frame

The handle shown in Fig. 5-116 is for lifting an avionics "black box" out of a plane. It will be attached to a die-cast aluminum box using six 8-32 machine screws into blind tapped holes. The handle is a standard catalog item, chosen partly for its ready availability and low cost. Had it been a custom design, we might have specified tighter tolerances for the mounting holes. Nevertheless, through careful use of GD&T, we can still specify a pattern of tapped holes that will always allow hassle-free mounting of any sample handle.

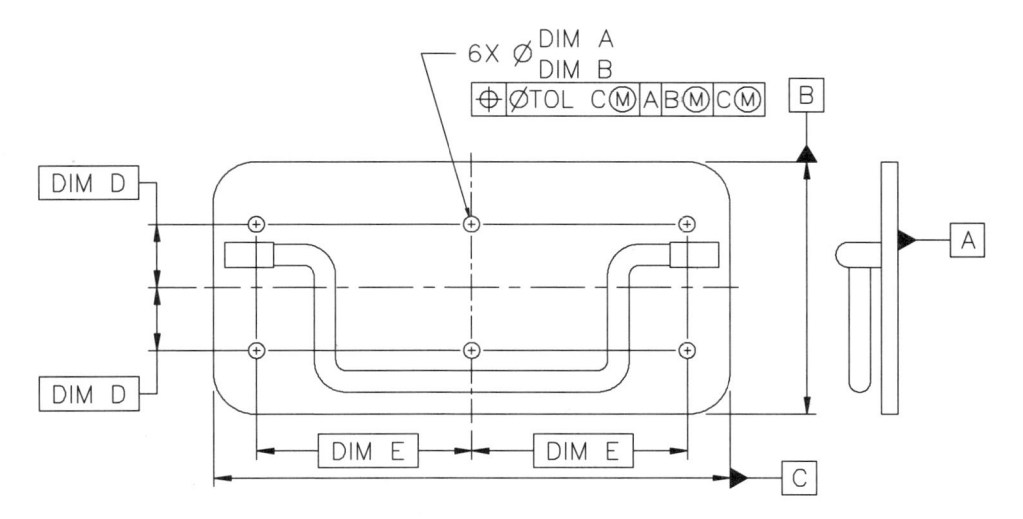

Figure 5-116 Standard catalog handle

5-128 Chapter Five

For ease of assembly, we primarily need to assure a clearance fit between each of the handle's holes and the major diameter of its corresponding 8-32 screw. Worst-case assemblability is therefore represented by the MMC virtual conditions of the holes and the MMC virtual conditions of the screws. The handle's Technical Bulletin (Fig. 5-117) tells us the mounting holes can be as small as ∅.186. At that MMC size, a hole's positional deviation can be as much as ∅.014 (likely a conversion from ±.005 coordinate tolerances). According to the formula in section 5.6.3.1, the MMC virtual condition for each hole (internal feature) is ∅.186 − ∅.014 = ∅.172.

Dash #	Screw Size	DIM A	DIM B	TOL C	DIM D	DIM E
−1	#4					
−2	#6					
−3	#8	.191	.186	.014	.750	2.500
−4	#10					

Figure 5-117 Handle Technical Bulletin

To assure a clearance fit, then, we must establish for each screw a Level 4 virtual condition boundary no larger than ∅.172. While we can't apply a positional tolerance directly to the screws, we can apply a tolerance to the pattern of tapped holes. The most difficult assembly would result from a screw with its pitch diameter at MMC and its major diameter at MMC (∅.1640), torqued into a tapped hole that's also at MMC. Functionally, this is only slightly more forgiving than a simple ∅.164 boss. For our tapped holes, then, if we model our virtual conditions on a substitute ∅.164 boss, our tolerances will be slightly conservative, which is fine.

For a ∅.164 boss, the maximum allowable positional tolerance is found by simply reversing our virtual condition formula—that is, by starting with the desired MMC virtual condition size and subtracting the feature's MMC size: ∅.172 − ∅.164 = ∅.008. In Fig. 5-118, we've specified a single positional tolerance of ∅.008 for the entire pattern of six tapped holes. The tolerance controls the location of each hole to the DRF A|B|C, and at the same time, the spacings between holes. Assemblability is assured. Problem solved.

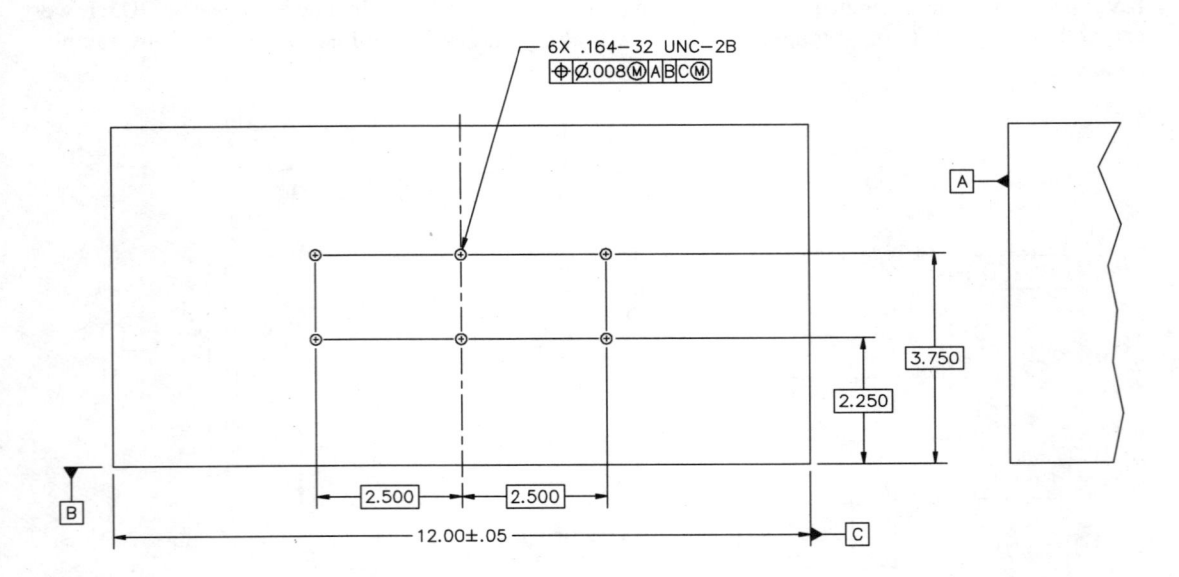

Figure 5-118 Avionics "black box" with single positional tolerance on pattern of holes

"Problem solved," that is, until we discover that about half the boxes made have one or more tapped holes exceeding their ⌀.008 positional tolerance. On closer analysis, we find the same problem on every rejected box: Though the hole-to-hole spacings are excellent and handles can assemble easily, the entire pattern of holes is shifted relative to the datum C width. We often find that processes can make hole-to-hole spacings more precise than the overall location of the pattern. Fortunately, most designs can afford a significantly greater tolerance for overall location. In our example, ⌀.008 is necessary for the hole-to-hole spacings, but we could actually allow the entire pattern (the handle itself) to shift around on the box 1/8" or so in any direction.

5.11.7.2 Composite Feature Control Frame

In Fig. 5-119, we've applied a *composite positional tolerance* feature control frame to our pattern of tapped holes. As does the more common single-segment frame already described, the composite frame has a single "position" symbol. Unlike the single-segment frame, the composite frame has two segments, upper and lower, each establishing a distinct framework of virtual condition boundaries or central tolerance zones. Notice the difference in tolerance values and datum references between the two segments. The intent of a composite feature control frame is for the upper segment to provide a complete overall location control, then for the lower segment to provide a specialized refinement within the constraints of the upper segment. Here's how it works.

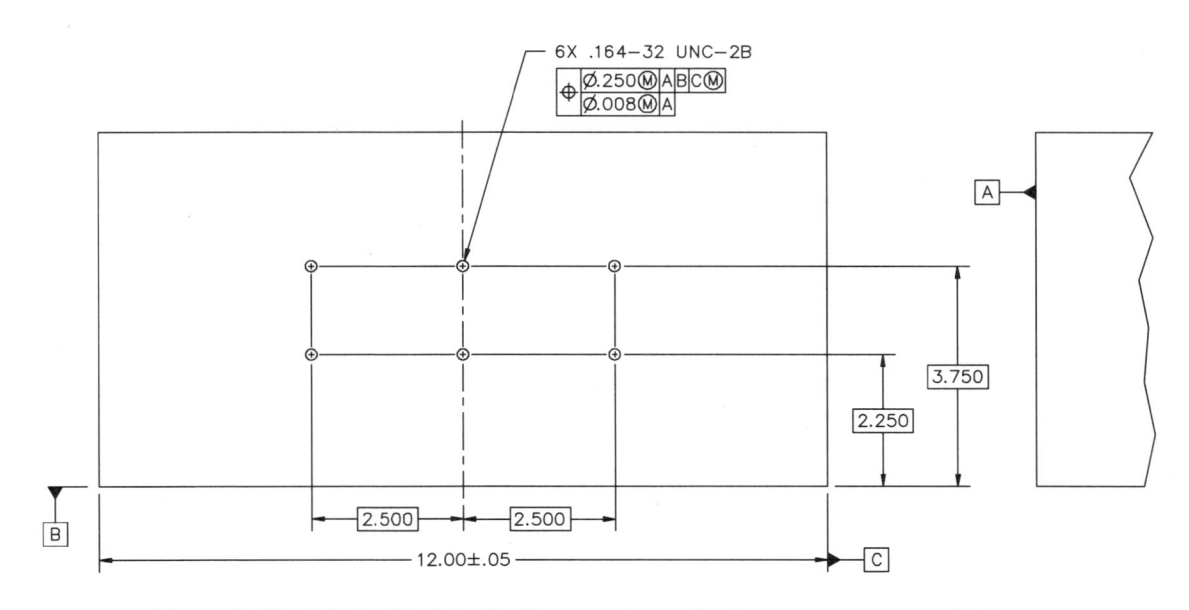

Figure 5-119 Avionics "black box" with composite positional tolerance on pattern of holes

The *upper segment* means the same as a single-segment positional tolerance feature control frame. In our Fig. 5-119 example, positional tolerance of ⌀.250 is permitted for each hole, relative to the DRF A|B|C. This establishes a *Pattern Locating Tolerance Zone Framework (PLTZF)* (pronounced "Plahtz") comprising six virtual condition boundaries for the holes, all basically parallel and basically located to each other. In addition, the orientation and location of the entire PLTZF is restrained relative to the referenced DRF A|B|C. In this case, the tapped holes would have negative virtual conditions. Fig. 5-120 shows instead the PLTZF virtual condition boundaries for our substitute ⌀.164 bosses.

5-130 Chapter Five

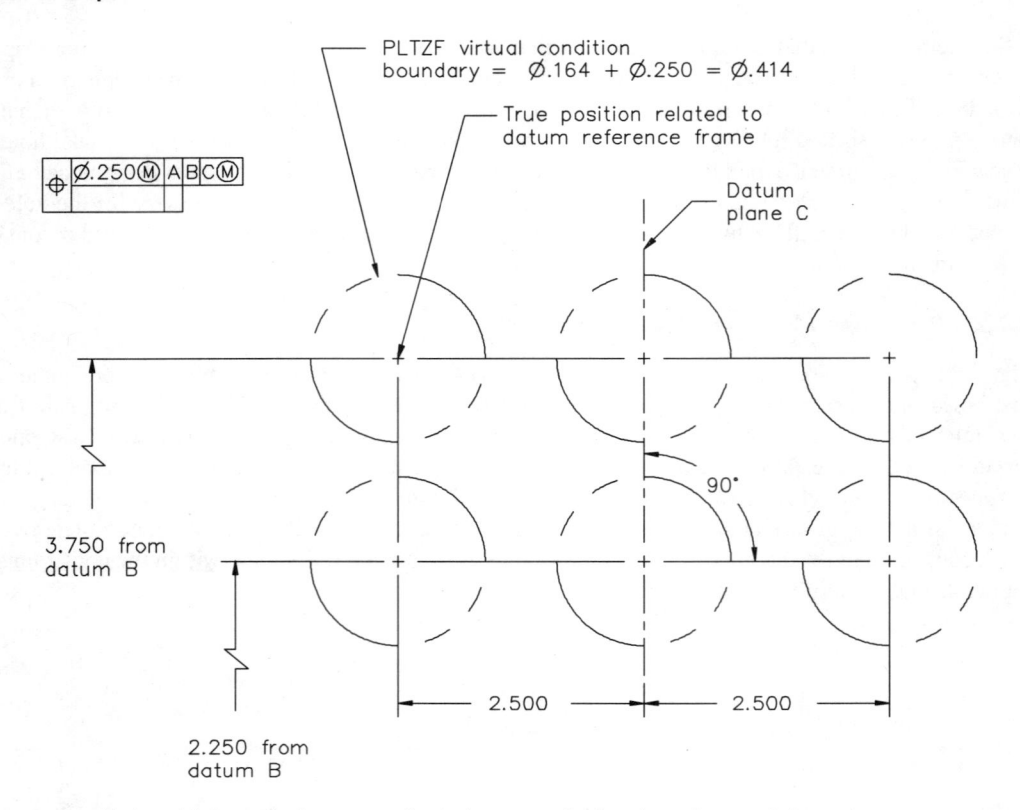

[Diagram with annotations:
- PLTZF virtual condition boundary = ⌀.164 + ⌀.250 = ⌀.414
- True position related to datum reference frame
- Datum plane C
- Feature control frame: ⌀.250 Ⓜ A B C Ⓜ
- 3.750 from datum B
- 2.250 from datum B
- 2.500, 2.500
- 90°]

Note: All boundaries are oriented perpendicular to primary datum A.

Figure 5-120 PLTZF virtual condition boundaries for Fig. 5-119

Compared with the single-segment positional tolerance of Fig. 5-118, the upper segment tolerance in our example affords much more freedom for the overall location of the handle on the box. However, ⌀.250 allows too much feature-to-feature variation to assure assemblability. That's where the lower segment kicks in.

The *lower segment* establishes the *Feature Relating Tolerance Zone Framework (FRTZF)* (pronounced "Fritz"). This segment may have zero, one, two, or three datum references. Where datums are referenced, they restrain only the orientation of the FRTZF, never its location. Fig. 5-121 shows the FRTZF virtual condition boundaries for our substitute bosses at work. Notice that datum A restrains the orientation of the FRTZF. This is crucial to the handle's fitting flush. However, datum A couldn't possibly restrain the location of the FRTZF, since the holes are perpendicular to datum A. In our example, then, the rule against location restraint is moot. In a moment, we'll show how the difference can become relevant.

Compared with the single-segment positional tolerance of Fig. 5-118, the lower segment tolerance in our example has the same tolerance value, and affords exactly the same feature-to-feature control. However, the lower segment's entire FRTZF is able to translate freely relative to the DRF, affording no restraint at all for the overall location of the handle on the box.

To summarize, we've solved our handle mounting problem with a composite positional tolerance that's really two tolerances in one: a larger tolerance to control the overall location of the handle on the box; and a smaller tolerance to control the orientation (perpendicularity) of the holes to the mounting face, as well as the hole-to-hole spacings. Assemblability is assured. Problem solved.

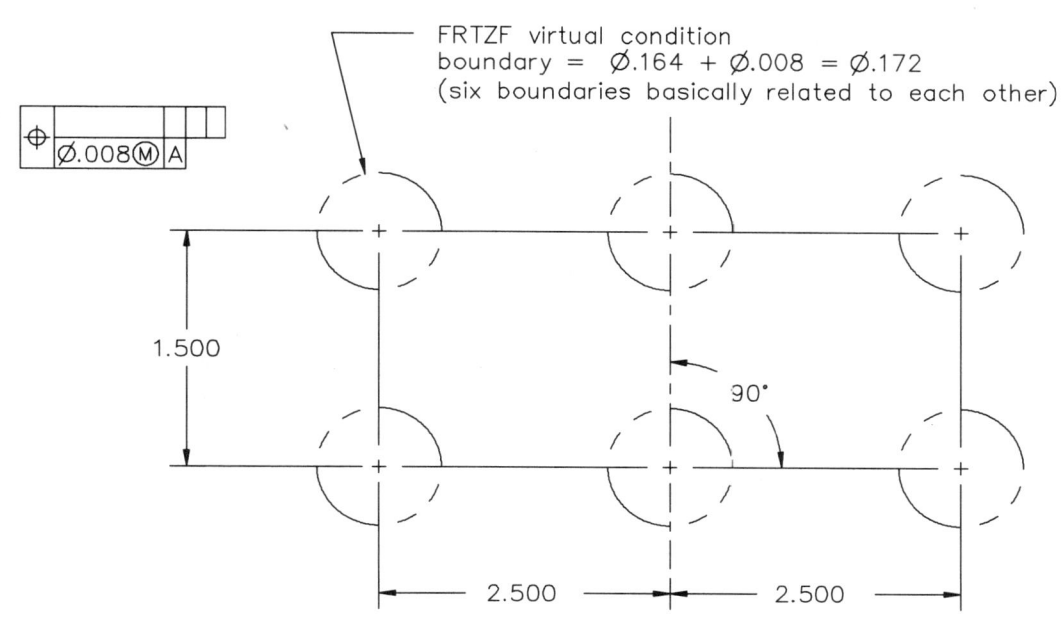

Figure 5-121 FRTZF virtual condition boundaries for Fig. 5-119

With a Secondary Datum in the Lower Segment—With composite control, there's no explicit congruence requirement between the PLTZF and the FRTZF. But, if features are to conform to both tolerances, the FRTZF will have to drift to where its virtual condition boundaries (or central tolerance zones) have enough overlap with those of the PLTZF. Fig. 5-122 shows for our example one possible valid relationship between the PLTZF and FRTZF. Again, the virtual condition boundaries are based on a substitute ⌀.164 boss. Notice that the PLTZF virtual conditions are so large, they allow considerable rotation of the pattern of tapped holes. The FRTZF offers no restraint at all of the pattern relative to datums B or C. This could allow a handle to be visibly crooked on the box.

In Fig. 5-123, we've corrected this limitation by simply referencing datum B as a secondary datum in the lower segment. Now, the orientation (rotation) of the FRTZF is restrained normal to the datum B plane. Although datum B could also restrain the basic location of the FRTZF, in a composite control such as this, it's not allowed to. Thus, while the pattern of tapped holes is now squared up, it can still shift around nearly as much as before.

5.11.7.3 Rules for Composite Control

Datum References—Since the lower segment provides specialized refinement only within the constraints of the upper segment, the lower segment may never reference any datum(s) that contradicts the DRF of the upper segment. Neither shall there be any mismatch of material condition modifier symbols. This leaves four options for referencing datums in the lower segment.

1. Reference no datums.
2. Copy only the primary datum and its modifier (if any).
3. Copy the primary and secondary datums and their modifiers, in order.
4. Copy the primary, secondary, and tertiary datums and their modifiers, in order.

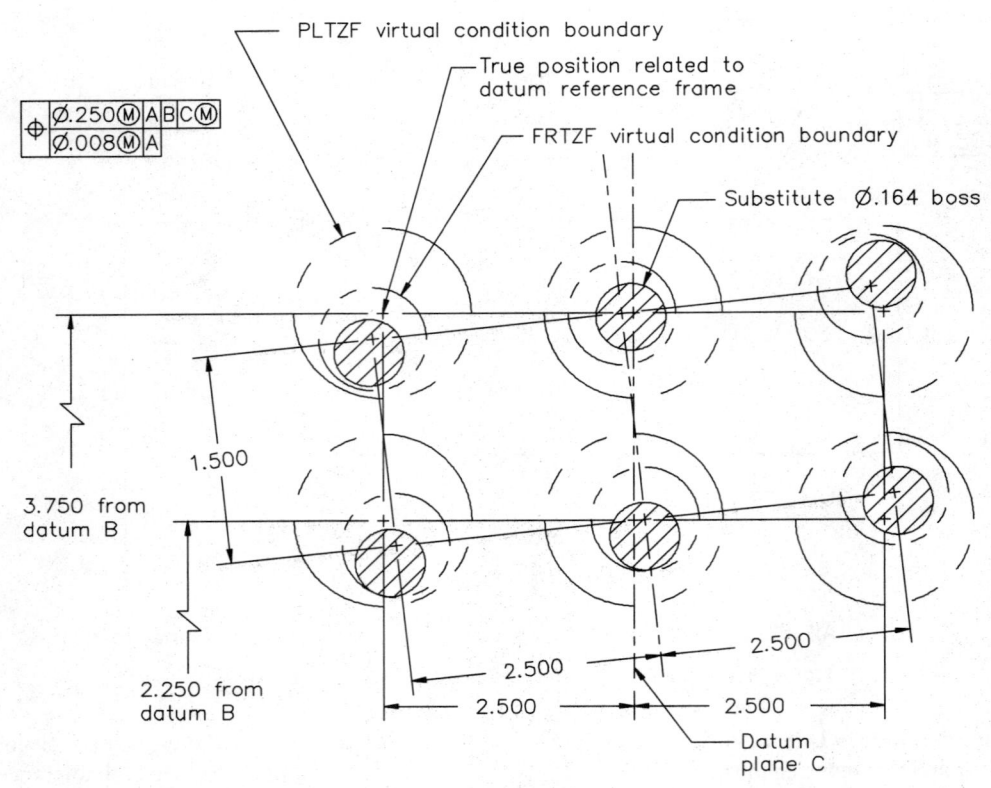

Figure 5-122 One possible relationship between the PLTZF and FRTZF for Fig. 5-119

Only datums needed to restrain the <u>orientation</u> of the FRTZF may be referenced. The need for two datum references in a lower segment is somewhat rare, and for three, even more uncommon.

Tolerance Values—The upper-segment tolerance shall be greater than the lower-segment tolerance. Generally, the difference should be enough to make the added complexity worthwhile.

Simultaneous Requirements—The upper and lower segments may be verified separately, perhaps using two different functional gages. Thus, where both upper and lower segments reference a datum feature of size at MMC or at LMC, each segment may use a different datum derived from that datum feature. Table 5-7 shows the defaults for simultaneous requirements associated with composite control. Simultaneous requirements are explained in section 5.9.10.

> FAQ: *The Table 5-7 defaults seem somewhat arbitrary. Can you explain the logic?*
> A: No, it escapes us too.

Notice that the lower segments of composite feature control frames default to separate requirements. Placing the note **SIM REQT** adjacent to a lower segment that references one or more datums overrides the default and imposes simultaneous requirements. If the lower segment references no datums, functionally related features of differing sizes should instead be grouped into a single pattern of features controlled

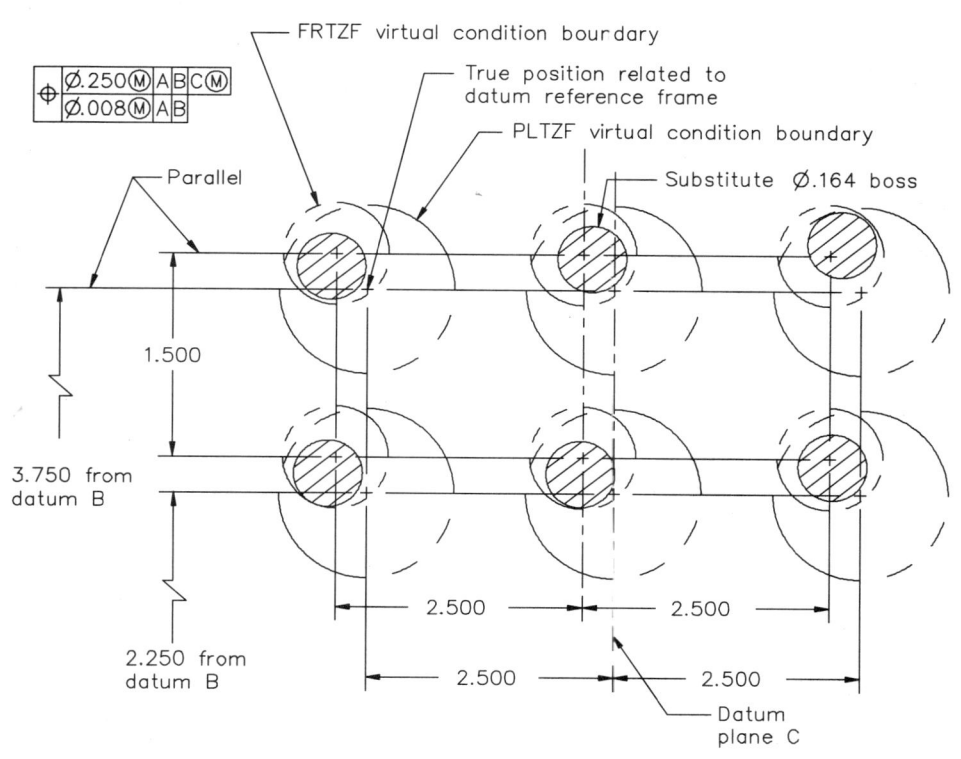

Figure 5-123 One possible relationship between the PLTZF and FRTZF with datum B referenced in the lower segment

Table 5-7 Simultaneous/separate requirement defaults

Between	Default	Modifiable?
Upper and lower segments within a single composite feature control frame	SEP REQTS	NO
Upper segments (only) of two or more composite feature control frames	SIM REQTS	YES
Lower segments (only) of two or more composite feature control frames	SEP REQTS	YES
Upper segment of a composite and a single-segment feature control frame	SIM REQTS	YES
Lower segment of a composite and a single-segment feature control frame	SEP REQTS	YES

with a single composite feature control frame. This can be done with a general note and flags, or with a note such as THREE SLOTS or TWO COAXIAL HOLES placed adjacent to the shared composite feature control frame.

5.11.7.4 Stacked Single-Segment Feature Control Frames

A composite positional tolerance cannot specify different location requirements for a pattern of features relative to different planes of the DRF. This is because the upper segment allows equal translation in all directions relative to the locating datum(s) and the lower segment has no effect at all on pattern translation. In section 5.11.6.2, we explained how bidirectional positional tolerancing could be used to specify different location requirements relative to different planes of the DRF. This works well for an individual feature of size, but applied to a pattern, the feature-to-feature spacings would likewise have a different tolerance for each direction.

Fig. 5-124 shows a sleeve with four radial holes. In this design, centrality of the holes to the datum A bore is critical. Less critical is the distance of the holes from the end of the sleeve, datum B. Look closely at the feature control frames. The appearance of <u>two</u> "position" symbols means this is <u>not</u> a composite positional feature control frame. What we have instead are simply two single-segment positional tolerance feature control frames stacked one on top of the other (with no space between). Each feature control frame, upper and lower, establishes a distinct framework of Level 4 virtual condition boundaries or central tolerance zones.

Fig. 5-125 shows the virtual condition boundaries for the upper frame. The boundaries are basically oriented and located to each other. In addition, the framework of boundaries is basically oriented and located relative to the referenced DRF A|B. The generous tolerance in the upper frame adequately locates the holes relative to datum B, but not closely enough to datum A.

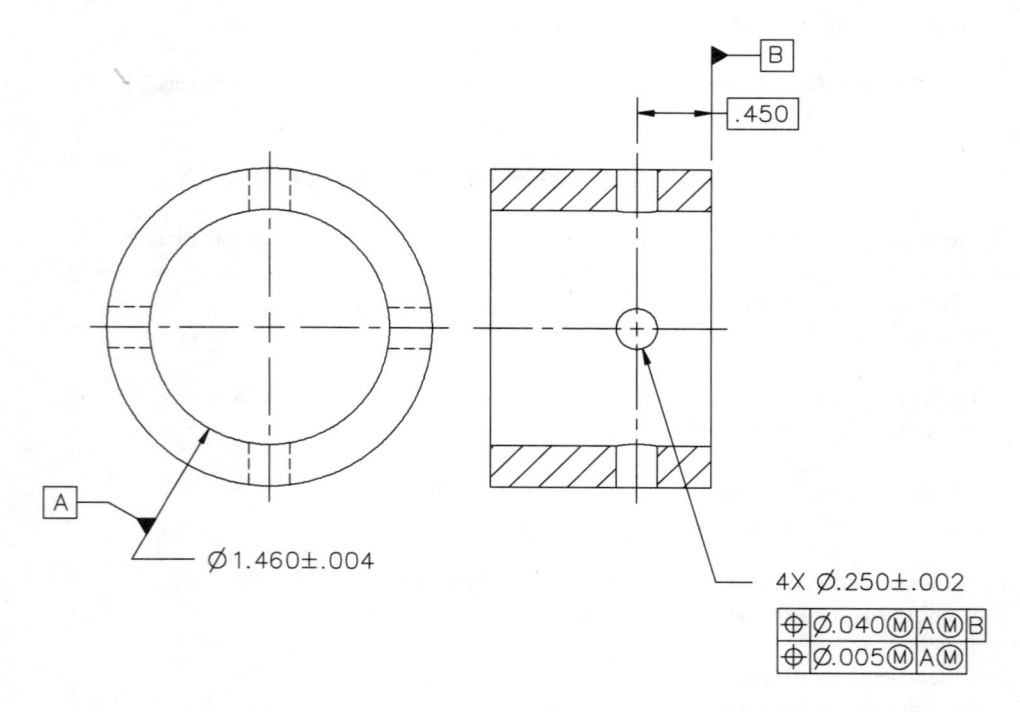

Figure 5-124 Two stacked single-segment feature control frames

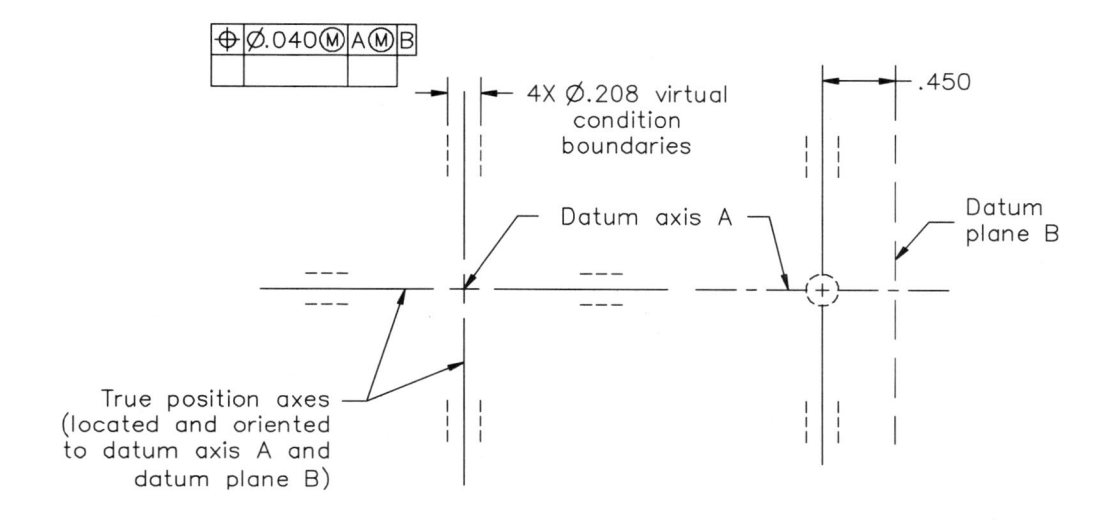

Figure 5-125 Virtual condition boundaries of the upper frame for Fig. 5-124

Fig. 5-126 shows the virtual condition boundaries for the lower frame. The boundaries are basically oriented and located to each other. In addition, the framework of boundaries is basically oriented and located relative to the referenced datum A. The comparatively close tolerance adequately centers the holes to the bore, but has no effect on location relative to datum B.

There is no explicit congruence requirement between the two frameworks. But, if features are to conform to both tolerances, virtual condition boundaries (or central tolerance zones) must overlap to some extent.

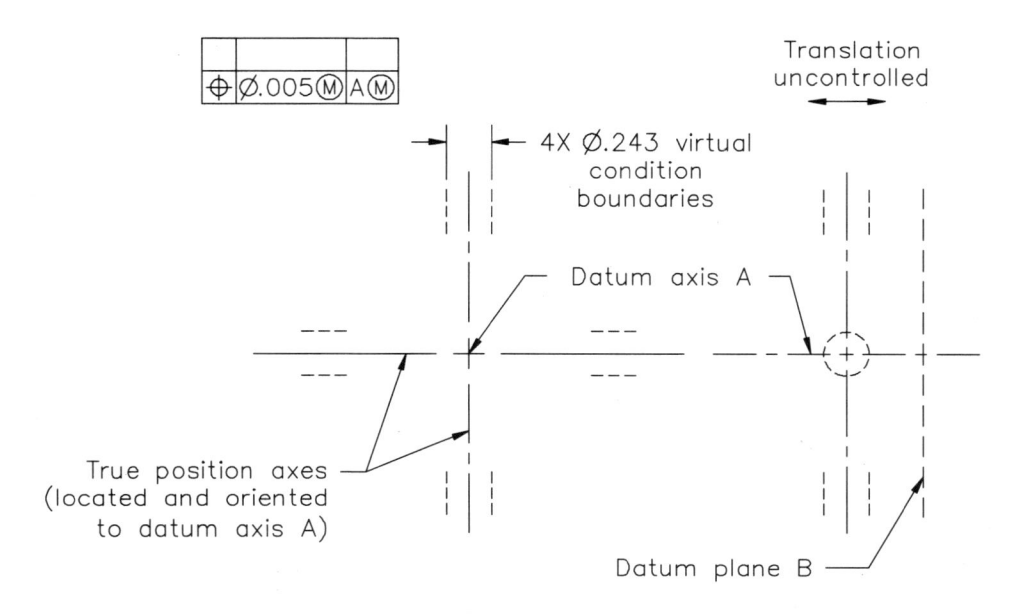

Figure 5-126 Virtual condition boundaries of the lower frame for Fig. 5-124

5.11.7.5 Rules for Stacked Single-Segment Feature Control Frames

Datum References—As with any pair of separate feature control frames, each may reference whatever datum(s), in whatever precedence, and with whatever modifiers are appropriate for the design, provided the DRFs are not identical (which would make the larger tolerance redundant). Since one frame's constraints may or may not be contained within the constraints of the other, the designer must carefully assure that the feature control frames together provide the necessary controls of feature orientation and location to the applicable datums.

Tolerance Values—Generally, the tolerances should differ enough to justify the added complexity. It's customary to place the frame with the greater tolerance on top.

Simultaneous Requirements—Since the two frames reference non-matching DRFs, they shall be evaluated separately, perhaps using two different functional gages. As explained in section 5.9.10, each feature control frame defaults to sharing simultaneous requirements with any other feature control frame(s) that references the identical DRF, as applicable.

FAQ: *I noticed that the 1994 revision of Y14.5 has much more coverage for pattern location than the 1982 revision. Is that just because the principles are so complicated, or does it mean I should make more use of composite and stacked feature control frames?*

A: Y14.5M-1982 was unclear about composite control as to whether the lower segment affects pattern location. Perhaps because most users assumed it did, Y14.5M-1994 includes dozens of figures meant to clarify that it does not and to introduce the method of using stacked frames. Don't interpret the glut of coverage as a sign that composite tolerancing is extremely complicated or that it's underused. The next revision might condense pattern location coverage.

FAQ: *How should I interpret composite tolerancing on drawings made before the 1994 revision? Does the lower segment control pattern location or not?*

A: That remains a huge controversy. Here's what ASME Y14.5M-1982 says (in section 5.4.1.4) about an example lower segment: "The axes of individual holes must also lie within 0.25 diameter feature-relating tolerance zones basically related to each other and basically oriented to datum axis A." Though it would have been very pertinent in the example, basic location to datum A is not mentioned. If we interpret this as an error of omission, we can likewise interpret anything left out of the standard as an error and do whatever we please. Thus, we feel the "not located" interpretation is more defensible. Where an "oriented and located" interpretation is needed on an older drawing, there's no prohibition against "retrofitting" stacked single-segment frames.

5.11.7.6 Coaxial and Coplanar Features

All the above principles for locating patterns of features apply as well to patterns of cylindrical features arranged in-line on a common axis, or width-type features arranged on a common center plane. Fig. 5-127 shows a pattern of two coaxial holes controlled with a composite positional tolerance. Though we've added a third segment to our composite feature control frame, the meaning is consistent with what we described in section 5.11.7.2. The upper segment's PLTZF controls the location and orientation of the pair of holes to the referenced DRF. The middle segment refines only the orientation (parallelism) of a FRTZF relative to datum A. The lower segment establishes a separate free-floating FRTZF that refines only the feature-to-feature coaxiality of the individual holes. Child's play. Different sizes of in-line features can share a common positional tolerance if their size specifications are stacked above a shared feature control frame.

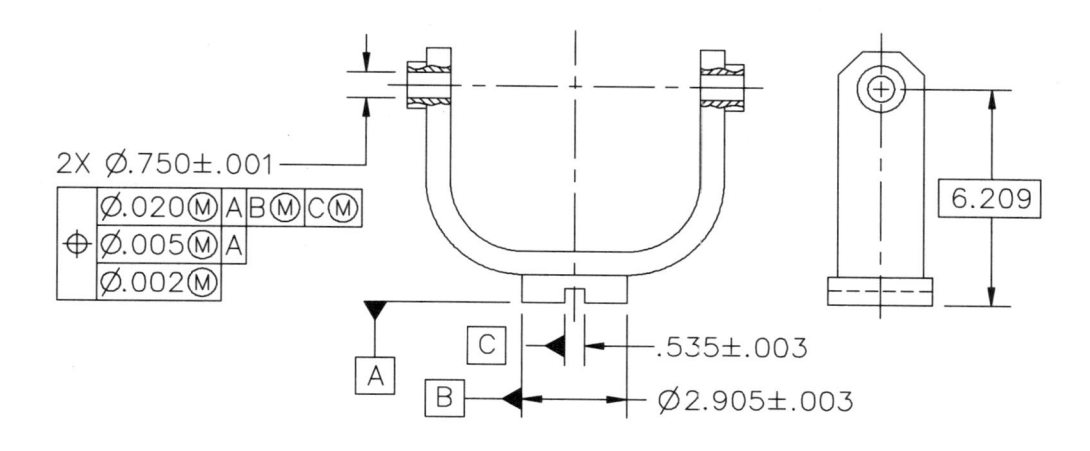

Figure 5-127 Three-segment composite feature control frame

5.11.8 Coaxiality and Coplanarity Control

Coaxiality is the relationship between multiple cylindrical or revolute features sharing a common axis. Coaxiality can be specified in several different ways, using a runout, concentricity, or positional tolerance. As Section 12 explains, a runout tolerance controls surface deviations directly, without regard for the feature's axis. A concentricity tolerance, explained in section 5.14.3, controls the midpoints of diametrically opposed points.

The standards don't have a name for the relationship between multiple width-type features sharing a common center plane. We will extend the term *coplanarity* to apply in this context. Coplanarity can be specified using either a symmetry or positional tolerance. A symmetry tolerance, explained in section 5.14.4, controls the midpoints of opposed surface points.

Where one of the coaxial or coplanar features is identified as a datum feature, the coaxiality or coplanarity of the other(s) can be controlled directly with a positional tolerance applied at RFS, MMC, or LMC. Likewise, the datum reference can apply at RFS, MMC, or LMC. For each controlled feature, the tolerance establishes either a Level 4 virtual condition boundary or a central tolerance zone (see section 5.11.1) located at true position. In this case, no basic dimensions are expressed, because true position is coincident with the referenced datum axis or datum center plane.

All the above principles can be extended to a pattern of coaxial feature groups. For a pattern of counterbored holes, the pattern of holes is located as usual. A single "datum feature" symbol is attached according to section 5.9.2.4. Coaxiality for the counterbores is specified with a separate feature control frame. In addition, a note such as **4X INDIVIDUALLY** is placed under the "datum feature" symbol and under the feature control frame for the counterbores, indicating the number of places each applies on an individual basis.

Where the coaxiality or coplanarity of two features is controlled with a positional tolerance of zero at MMC and the datum is also referenced at MMC, it makes no difference which of these features is the datum. For each feature, its TGC, its virtual condition, and its MMC size limit are identical. The same is true in an all-LMC context.

5-138   Chapter Five

> FAQ: *Where a piston's ring grooves interrupt the outside diameter (OD), do I need to control coaxiality among the three separate segments of the OD?*
>
> A: If it weren't for those pesky grooves, Rule #1 would impose a boundary of perfect form at MMC for the entire length of the piston's OD. Instead of using **3X** to specify multiple same-size ODs, place the note **THREE SURFACES AS A SINGLE FEATURE** adjacent to the diameter dimension. That forces Rule #1 to ignore the interruptions. A similar note can simplify orientation and/or location control of a pattern of coaxial or coplanar same-size features.

5.12   Runout Tolerance

Runout is one of the oldest and simplest concepts used in GD&T. Maybe as a child you stood your bicycle upside down on the ground and spun a wheel. If you fixed your stare on the shiny rim where it passed a certain part of the frame, you could see the rim wobble from side to side and undulate inward and outward. Instead of the rim running in a perfect circle, it, well—ran out. *Runout*, then, is the variation in the surface elements of a round feature relative to an axis.

5.12.1   Why Do We Use It?

In precision assemblies, runout causes misalignment and/or balance problems. In Fig. 5-128, runout of the ring groove diameters relative to the piston's diameter might cause the ring to squeeze unevenly around the piston or force the piston off center in its bore. A motor shaft that runs out relative to its bearing journals will cause the motor to run out-of-balance, shortening its working life. A designer can prevent such wobble and lopsidedness by specifying a runout tolerance. There are two levels of control, *circular runout* and *total runout*. Total runout adds further refinement to the requirements of circular runout.

5.12.2   How Does It Work?

For as long as piston ring grooves and motor shafts have been made, manufacturers have been finding ways to spin a part about its functional axis while probing its surface with a dial indicator. As the indicator's tip surfs up and down over the undulating surface, its dial swings gently back and forth, visually display-

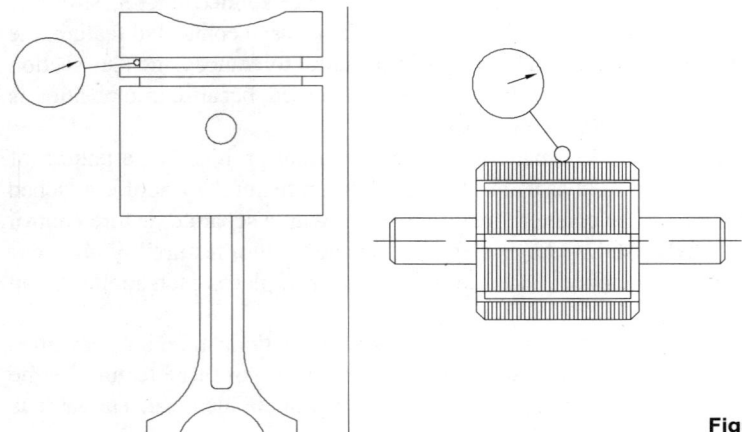

Figure 5-128 Design applications for runout control

ing the magnitude of runout. Thus, measuring runout can be very simple as long as we agree on three things:

- What surface(s) establish the functional axis for spinning—datums
- Where the indicator is to probe
- How much swing of the indicator's dial is acceptable

The whole concept of "indicator swing" is somewhat dated. Draftsmen used to annotate it on drawings as TIR for "Total Indicator Reading." Y14.5 briefly called it FIR for "Full Indicator Reading." Then, in 1973, Y14.5 adopted the international term, FIM for "Full Indicator Movement." *Full Indicator Movement (FIM)* is the difference (in millimeters or inches) between an indicator's most positive and most negative excursions. Thus, if the lowest reading is −.001" and the highest is +.002", the FIM (or TIR or FIR) is .003".

Just because runout tolerance is defined and discussed in terms of FIM doesn't mean runout tolerance can only be applied to parts that spin in assembly. Neither does it require the part to be rotated, nor use of an antique twentieth century, jewel-movement, dial indicator to verify conformance. The "indicator swing" standard is an ideal meant to describe the requirements for the surface. Conformance can be verified using a CMM, optical comparator, laser scanning with computer modeling, process qualification by SPC, or any other method that approximates the ideal.

5.12.3 How to Apply It

A runout tolerance is specified using a feature control frame displaying the characteristic symbol for either "circular runout" (a single arrow) or "total runout" (two side-by-side arrows). As illustrated in Fig. 5-129, the arrowheads may be drawn filled or unfilled. The feature control frame includes the runout tolerance value followed by one or two (but never three) datum references.

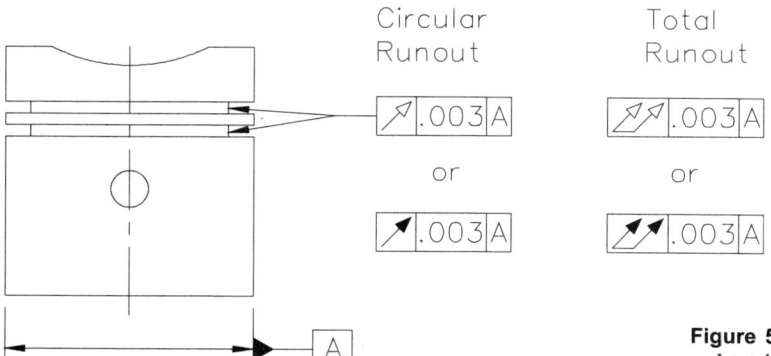

Figure 5-129 Symbols for circular runout and total runout

Considering the purpose for runout tolerance and the way it works, there's no interaction between a feature's size and its runout tolerance that makes any sense. In our piston ring groove diameter example, an MMC modifier would be counterproductive, allowing the groove diameter's eccentricity to increase as it gets smaller. That would only aggravate the squeeze and centering problems we're trying to correct. Thus, material condition modifier symbols, MMC and LMC, are prohibited for both circular and total runout tolerances and their datum references. If you find yourself wishing you could apply a runout tolerance at MMC, you're not looking at a genuine runout tolerance application; you probably want positional tolerance instead.

5-140　Chapter Five

5.12.4　Datums for Runout Control

A runout tolerance controls surface elements of a round feature relative to a datum axis. GD&T modernized runout tolerancing by applying the rigors and flexibility of the DRF. Every runout tolerance shall reference a datum axis. Fig. 5-130 shows three different methods for doing this.

Since a designer wishes to control the runout of a surface as directly as possible, it's important to select a functional feature(s) to establish the datum axis. During inspection of a part such as that shown in Fig. 5-130(a), the datum feature might be placed in a V-block or fixtured in a precision spindle so that the part can be spun about the axis of the datum feature's TGC. This requires that the datum feature be long enough and that its form be well controlled (perhaps by its own size limits or form tolerance). In addition, the datum feature must be easily accessible for such fixturing or probing.

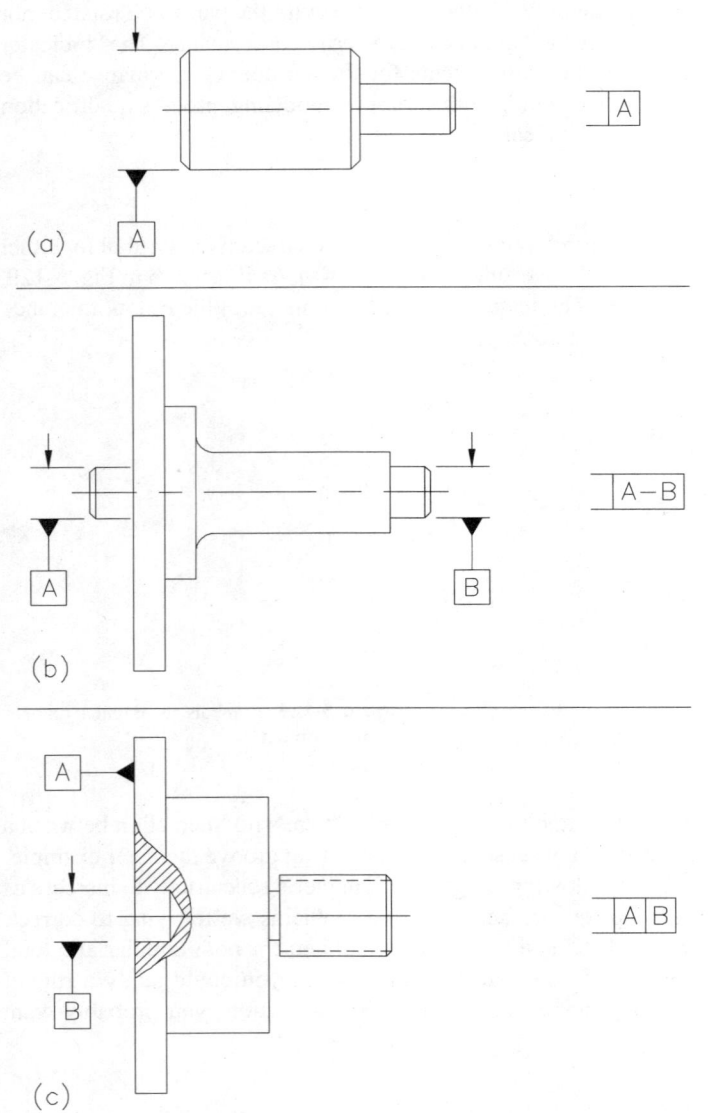

Figure 5-130　Datums for runout control

There are many cases where the part itself is a spindle or rotating shaft that, when assembled, will be restrained in two separate places by two bearings or two bushings. See Fig. 5-131. If the two bearing journals have ample axial separation, it's unrealistic to try to fixture on just one while ignoring the other. We could better stabilize the part by identifying each journal as a datum feature and referencing both as equal co-datum features. In the feature control frame, the datum reference letters are placed in a single box, separated by a hyphen. As we explained in section 5.9.14.2, hyphenated co-datum features work as a team. Neither co-datum feature has precedence over the other. We can't assume the two journals will be made perfectly coaxial. To get a decent datum axis from them, we should add a runout tolerance for each journal, referencing the common datum axis they establish. See Fig. 5-132. This is one of the few circumstances where referencing a feature as a datum in its own feature control frame is acceptable.

Where a single datum feature or co-datum feature pair establishes the axis, further datum references are meaningless and confusing. However, there are applications where a shoulder or end face exerts more leadership over the part's orientation in assembly while the diametral datum feature merely establishes the center of revolution. In Fig. 5-130(c), for example, the face is identified as primary datum feature A and the bore is labeled secondary datum feature B. In inspection, the part will be spun about datum axis B which, remember, is restrained perpendicular to datum plane A.

5.12.5 Circular Runout Tolerance

Circular runout is the lesser level of runout control. Its tolerance applies to the FIM while the indicator probes over a single circle on the part surface. That means the indicator's body is to remain stationary both axially and radially relative to the datum axis as the part is spun at least 360° about its datum axis. The tolerance applies at every possible circle on the feature's surface, but each circle may be evaluated separately from the others.

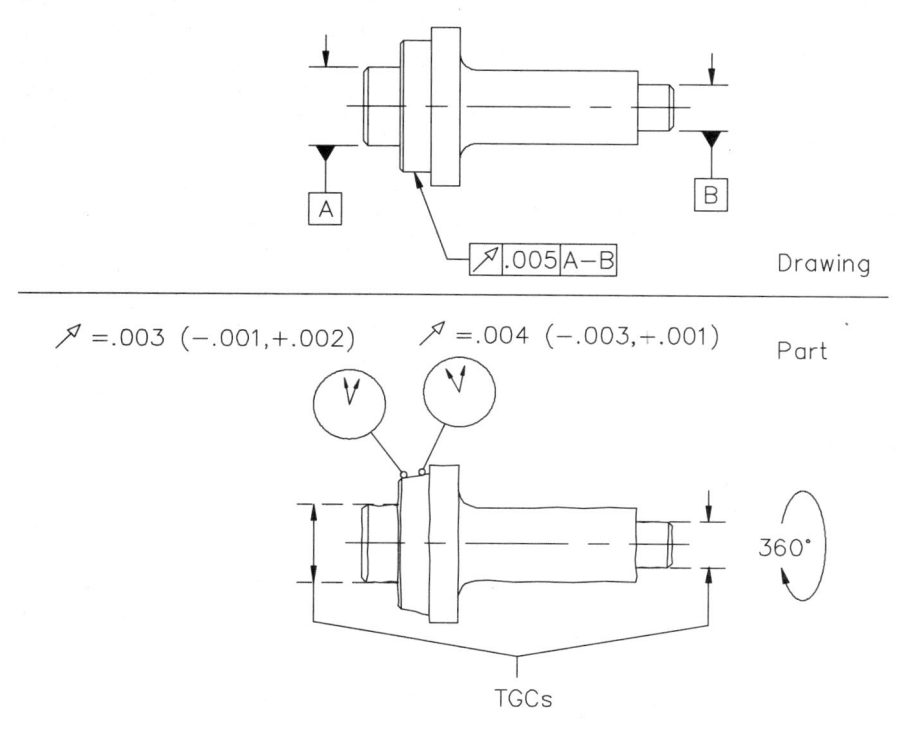

Figure 5-131 Two coaxial features establishing a datum axis for runout control

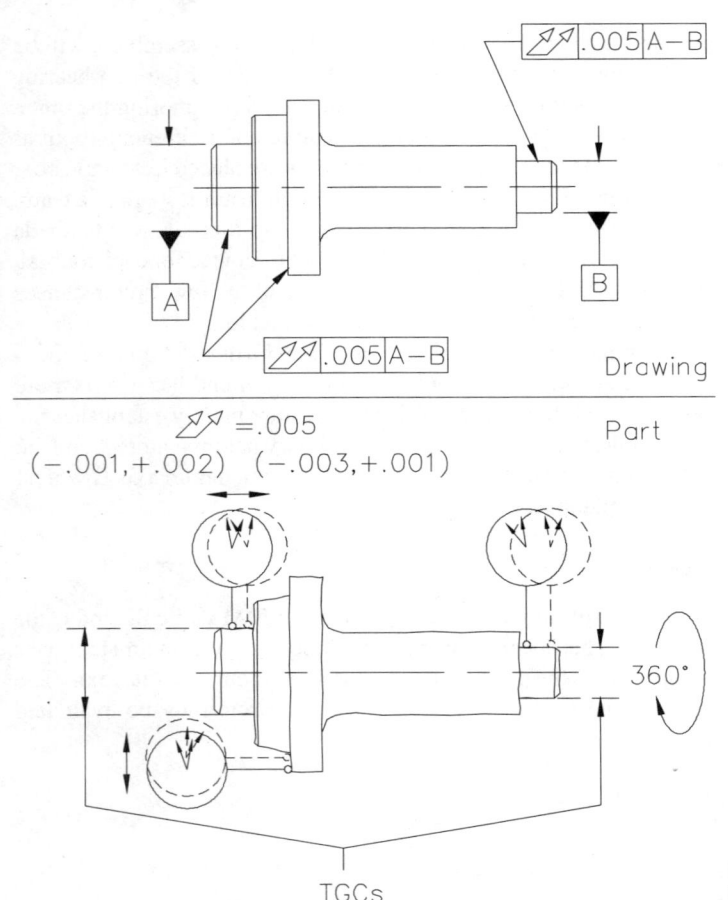

Figure 5-132 Runout control of hyphenated co-datum features

Let's evaluate the .005 circular runout tolerance of Fig. 5-131. We place an indicator near the left end of the controlled diameter and spin the part 360°. We see that the farthest counterclockwise excursion of the indicator dial reaches −.001" and the farthest clockwise excursion reaches +.002". The circular runout deviation at that circle is .003". We move the indicator to the right and probe another circle. Here, the indicator swings between −.003" and +.001". The difference, .004", is calculated without regard for the readings we got from the first circle. The FIM for each circle is compared with the .005" tolerance separately.

Obviously, we can't spend all day trying to measure infinitely many circles, but after probing at both ends of the feature and various places between, we become confident that no circle along the feature would yield an FIM greater than, perhaps, .004". Then, we can conclude the feature conforms to the .005" circular runout tolerance.

Circular runout can be applied to any feature that is nominally cylindrical, spherical, toroidal, conical, or any revolute having round cross sections (perpendicular to the datum axis). When evaluating noncylindrical features, the indicator shall be continually realigned so that its travel is always normal to the surface at the subject circle. See Fig. 5-133. Circular runout can also be applied to a face or face groove that is perpendicular to the datum axis. Here, the surface elements are circles of various diameters, each concentric to the datum axis and each evaluated separately from the others.

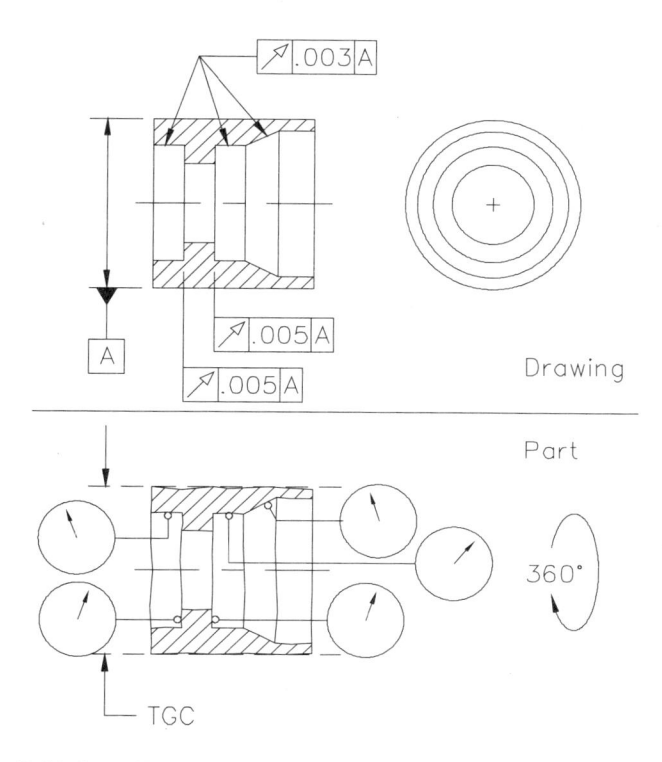

Figure 5-133 Application of circular runout

5.12.6 Total Runout Tolerance

Total runout is the greater level of runout control. Its tolerance applies to the FIM while the indicator sweeps over the entire controlled surface. Rather than each circular element being evaluated separately, the total runout FIM encompasses the highest and lowest of all readings obtained at all circles.

For a nominally cylindrical feature, the indicator's body shall be swept parallel to the datum axis, covering the entire length of the controlled feature, as the part is spun 360° about the datum axis. See Fig. 5-132. Any taper or hourglass shape in the controlled feature will increase the FIM.

For a nominally flat face perpendicular to the datum axis, the indicator's body shall be swept in a line perpendicular to the datum axis, covering the entire breadth of the controlled feature. Any conicity, wobble, or deviations from flatness in the controlled feature increase the FIM. The control imposed by this type of total runout tolerance is identical to that of an equal perpendicularity tolerance with an RFS datum reference.

> FAQ: *Can total runout tolerance be applied to a cone?*
>
> A: For any features other than cylinders or flat perpendicular faces, the indicator would have to be swept along a path neither parallel nor perpendicular to the datum axis. Since the standards have not adequately defined these paths, avoid such applications.

5.12.7 Application Over a Limited Length

Since a runout tolerance applies to surface elements, it sometimes makes sense to limit the control to a limited portion of a surface. A designer can do this easily by applying a chain line as described in section 5.8.8.

5.12.8 When Do We Use a Runout Tolerance?

Runout tolerances are especially suited to parts that revolve about a datum axis in assembly, and where alignments and dynamic balance are critical. Circular runout tolerance is often ideal for O-ring groove diameters, but watch out for surfaces inaccessible to an indicator tip. This might be an internal O-ring groove where the cylinder bore is the datum. How can an inspector spin the part about that bore and get his indicator tip into the groove at the same time? As we said, there are other inspection methods, but a designer should always keep one eye on practicality.

The following equations pertain to the controls imposed by circularity, cylindricity, concentricity, circular runout, and total runout when applied to a revolute or cylindrical feature.

$$\text{CIRCULARITY} + \text{CONCENTRICITY} = \text{CIRCULAR RUNOUT}$$

$$\text{CYLINDRICITY} + \text{CONCENTRICITY} = \text{TOTAL RUNOUT}$$

Remember that FIM is relatively simple to measure and reflects the combination of out-of-roundness and eccentricity. It's quite complex to differentiate between these two constituent variations. That means checking circularity or concentricity apart from the other requires more sophisticated and elaborate techniques. Of course, there are cases where the design requires tight control of one (say, circularity); to impose the same tolerance for the other (concentricity) would significantly complicate manufacturing. However, if this won't be a problem, use a runout tolerance.

A runout tolerance applies directly to surface elements. That distinguishes it from a positional tolerance RFS that controls only the coaxiality of the feature's actual mating envelope. Positional tolerancing provides no form control for the surface. While the positional tolerance coaxiality control is similar to that for runout tolerance, the positional tolerance is modifiable to MMC or LMC. Thus, where tolerance interaction is desirable and size limits will adequately control form, consider a positional tolerance instead of a runout tolerance.

FAQ: *Can I apply a runout tolerance to a gear or a screw thread?*

A: Avoid doing that. Remember that a runout tolerance applies to the FIM generated by surface elements. Some experts suggest modifying the runout tolerance by adding the note **PITCH CYLINDER**. We feel that subverts the purpose for runout tolerance and requires unique and complicated inspection methods. Consider a positional tolerance instead.

FAQ: *A feature's runout tolerance has to be less than its size tolerance, right?*

A: Wrong. A feature's size limits don't control its runout; neither does a runout tolerance control the feature's size. Depending on design considerations, a runout tolerance may be less than, equal to, or greater than the size tolerance. One can imagine scenarios justifying just about any ratio. That's why it's important to consider each runout tolerance independently and carefully.

FAQ: *Can I apply a runout tolerance "unless otherwise specified" in the tolerance block or by a general note?*

A: Yes, but identify a datum feature and reference it with the runout tolerance. A runout tolerance with no datum reference is meaningless and illegal. Many novice inspectors encountering a general runout tolerance with no datum reference start checking every possible pairing of features—for five diameters, that's 20 checks! Also, consider each feature to which the runout tolerance will apply and be careful not to rob any feature of usable and needed tolerance.

5.12.9 Worst Case Boundaries

Instead of troweling on feature control frames for form and location, a clever designer can often simplify requirements by using a few well-thought-out runout tolerances to control combinations of relationships.

A circular runout or total runout tolerance applied to an internal or external diameter feature yields a worst case inner boundary equal in size to the feature's small-limit size minus the value of its runout tolerance and a worst case outer boundary equal in size to the feature's large-limit size plus the value of its runout tolerance. The inner or outer boundary can be exploited to protect a secondary requirement for clearance without using a separate positional tolerance.

5.13 Profile Tolerance

In the previous sections, we've covered nearly all the principles needed to control planar features and simple features of size. In the old MIL-STD-8 drawing standards, that was as far as GD&T went. However, automobiles, airplanes, and ships are replete with parts having nonplanar, noncylindrical, nonspherical features. Such irregularly shaped *profiled features* couldn't be geometrically controlled until 1966 when the first edition of Y14.5 introduced "profile of a line" and "profile of a surface" characteristic symbols and feature control frames for controlling profiled features. The 1973 revision of Y14.5 introduced datum references in profile feature control frames. Finally, designers could apply all the power and precision of GD&T to nearly every imaginable type of part feature.

The 1982 and 1994 revisions of Y14.5 enhanced the flexibility of profile tolerancing to the extent that now just about every characteristic of just about every type of feature (including planes and simple features of size) can be controlled with a profile tolerance. Thus, some gurus prescribe profile tolerancing for everything, as if it's "the perfect food." (We address that notion in Section 17.)

The fundamental principles of profile tolerancing are so simple that the Math Standard covers them fully with just one column of text. However, the Math Standard only addresses the meaning of the tolerance. Profile tolerancing's multitude of application options and variations comprise quite a lot of material to learn.

5.13.1 How Does It Work?

Every profile tolerance relies on a *basic profile*. See Fig. 5-134. This is the profiled feature's nominal shape usually defined in a drawing view with basic dimensions. A profile tolerance zone is generated by offsetting each point on the basic profile in a direction normal to the basic profile at that point. This offsetting creates a "band" that follows the basic profile. The part feature (or 2-D element thereof) shall be contained within the profile tolerance zone. In addition, the surface (or 2-D element) shall "blend" everywhere. We interpret this to mean it shall be tangent-continuous.

There are two levels of profile tolerance control. The difference between the two levels is analogous to the difference between flatness and straightness tolerances. *Profile of a surface* provides complete 3-D control of a feature's total surface. *Profile of a line* provides 2-D control of a feature's individual cross-sectional elements. Either type of control may be related to a DRF.

5.13.2 How to Apply It

Application of a profile tolerance is a three-step process: 1) define the basic profile, 2) define the tolerance zone disposition relative to the basic profile, and 3) attach a profile feature control frame.

5-146 Chapter Five

Drawing

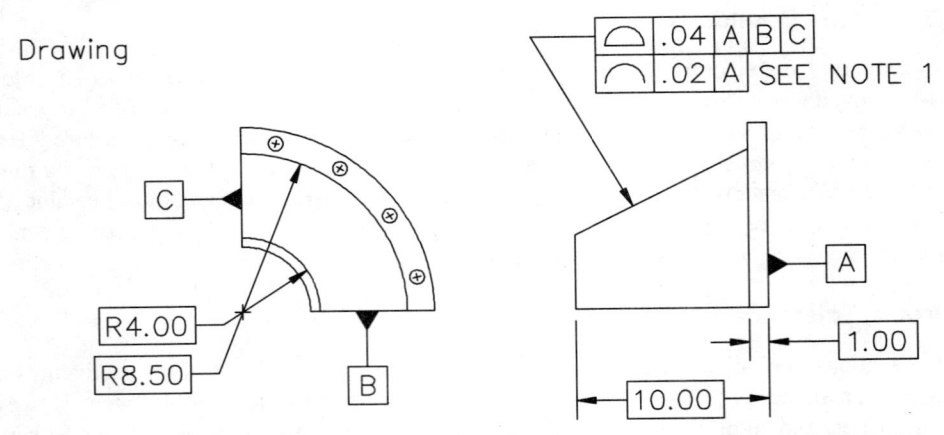

⌒	.04	A	B	C
⌒	.02	A	SEE NOTE 1

R4.00
R8.50
B
C
A
1.00
10.00

NOTE 1: THIS TOLERANCE APPLIES WHEN DATUM FEATURE A IS MOUNTED AGAINST A FLAT SURFACE USING FOUR .250-28 BOLTS TORQUED TO 10 FOOT POUNDS.

Part

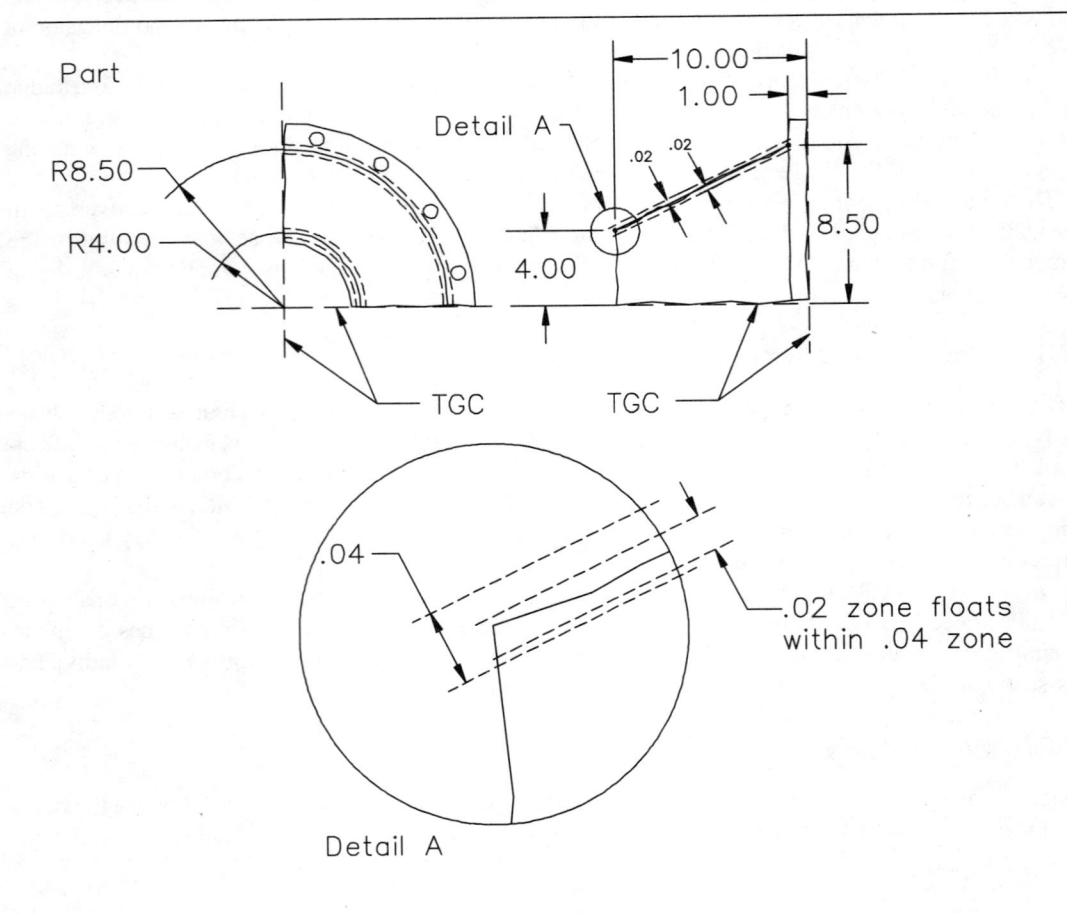

R8.50
R4.00
Detail A
.02 .02
4.00
10.00
1.00
8.50
TGC
TGC

.04
.02 zone floats within .04 zone

Detail A

Figure 5-134 Application of profile tolerances

5.13.3 The Basic Profile

You can specify the basic profile by any method that defines a unique and unambiguous shape for the controlled feature. The most common methods are projecting a 3-D figure onto a plane or taking cross sections through the figure. The resulting 2-D profile is shown in a drawing view. We call this 2-D graphical representation the *profile outline*. Basic dimensions are specified for the basic profile to define each of its elements. Such basic dimensions may include lengths, diameters, radii, and angles. Alternatively, a coordinate grid system might be established, with points or nodes on the basic profile listed in a table. Yet another method is to provide one or more mathematical formulas that define the elements of the basic profile, perhaps accompanied by one or more basically dimensioned nodes or end points.

A CAD/CAM model's digital representation of a basic profile also qualifies. It's not necessary to attach basic dimensions to the model since the computer already "understands" the ones and zeros that define it. In a paperless manufacturing environment, the "undimensioned" model along with a profile tolerance specification are all that's needed by automated equipment to make and inspect the profiled feature. This method accommodates truly 3-D–profiled features having varying cross sections, such as a turbine blade or an automobile windshield.

While any of these or other methods could be used, the designer must take into account the expected manufacturing methods and ensure that the basic profile specifications are accessible and usable. This consideration may prescribe multiple 2-D drawing views to show, for example, an airplane wing at several different cross sections.

5.13.4 The Profile Tolerance Zone

As depicted in Fig. 5-135, the profile tolerance zone is generated by offsetting each point on the basic profile in a direction normal to the basic profile at that point. This tolerance zone may be *unilateral* or *bilateral* relative to the basic profile. For a unilateral profile tolerance, the basic profile is offset totally in one direction or the other by an amount equal to the profile tolerance. See Figs. 5-135(b) and (c). For a bilateral profile tolerance, the basic profile is offset in both directions by a combined amount equal to the profile tolerance. Equal offsets of half the tolerance in each direction—*equal-bilateral tolerance*—is the default. See Fig. 5-135(a). Though the offsets need not be equal, they shall be uniform everywhere along the basic profile.

Regardless of the tolerance zone's disposition relative to the basic profile, it always represents the range of allowable variation for the feature. You could also think of this disposition as the basic profile running along one boundary of the tolerance band, or somewhere between the two boundaries. In any case, since the variations in most manufacturing processes tend to be equal/bidirectional, programmers typically program tool paths to target the mean of the tolerance zone. With an equal-bilateral tolerance, the basic profile runs right up the middle of the tolerance zone. That simplifies programming because the drawing's basic dimensions directly define the mean tool path without any additional calculations. Programmers love equal-bilateral tolerances, the default.

Of course, a unilateral tolerance is also acceptable. The drawing shall indicate the offset direction relative to the basic profile. Do this as shown in Fig. 5-135(b) and (c) by drawing a phantom line parallel to the basic profile on the tolerance zone side. Draw the phantom line (or curve) only long enough to show clearly. The distance between the profile outline and the phantom line is up to the draftsman, but should be no more than necessary for visibility after copying (don't forget photoreduction), and need not be related to the profile tolerance value.

A pair of short phantom lines can likewise be drawn to indicate a bilateral tolerance zone with unequal distribution. See Fig. 5-135(d). Draw one phantom line on each side of the profile outline with one visibly farther away to indicate the side having more offset. Then, show one basic dimension for the distance between the basic profile and one of the boundaries represented by a phantom line.

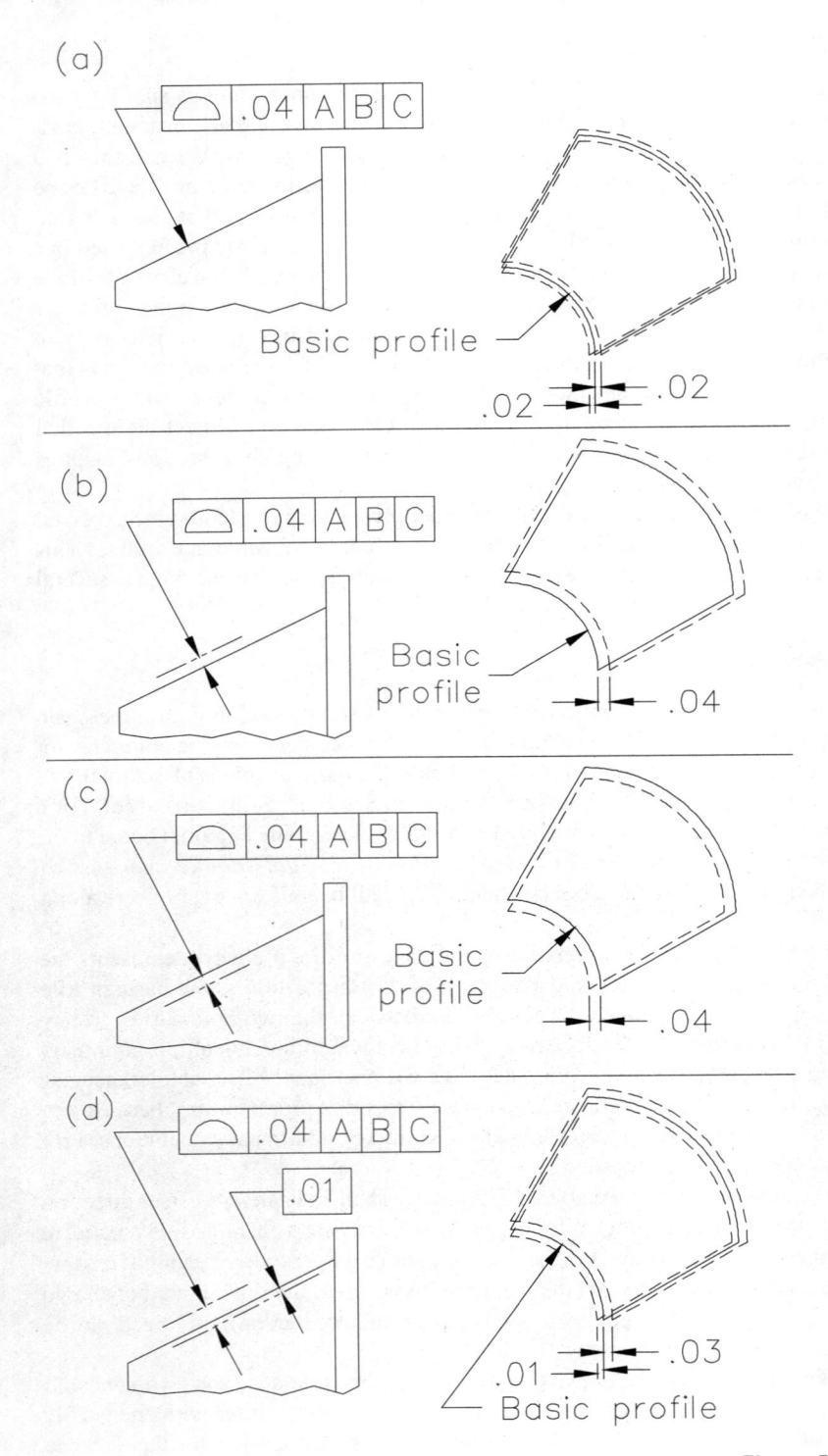

Figure 5-135 Profile tolerance zones

On complex and dense drawings, readers often fail to notice and comprehend such phantom lines, usually with disastrous consequences. Unequal-bilateral tolerancing is particularly confusing. If practicable, designers should spend a few extra minutes to convert the design for equal-bilateral tolerances. The designer will only have to make the computations once, precluding countless error-prone calculations down the road.

5.13.5 The Profile Feature Control Frame

A profile tolerance is specified using a feature control frame displaying the characteristic symbol for either "profile of a line" (an arc with no base line) or "profile of a surface" (same arc, with base line). The feature control frame includes the profile tolerance value followed by up to three datum references, if needed. Where the profile tolerance is equal-bilateral, the feature control frame is simply leader-directed to the profile outline, as in Fig. 5-135(a). Where the tolerance is unilateral or unequal-bilateral, dimension lines are drawn for the width of the tolerance zone, normal to the profile as in Fig. 5-135(b) through (d). One end of a dimension line is extended to the feature control frame.

5.13.6 Datums for Profile Control

Where a profile tolerance need only control a feature's shape, it's unnecessary to relate the profile tolerance zone to any DRF. Thus, there are many applications where the profile feature control frame should have no datum references. Where the tolerance must also control the orientation, or orientation and location of the profiled feature, the tolerance zone shall be related to a DRF. Depending on design requirements, the DRF may require one, two, or three datum references in the profile feature control frame.

5.13.7 Profile of a Surface Tolerance

A feature control frame bearing the "profile of a surface" symbol specifies a 3-D tolerance zone having a total width equal to the tolerance value. The entire feature surface shall everywhere be contained within the tolerance zone. If a DRF is referenced, it restrains the orientation, or orientation and location of the tolerance zone.

5.13.8 Profile of a Line Tolerance

A feature control frame bearing the "profile of a line" symbol specifies a tolerance zone plane containing a 2-D profile tolerance zone having a total width equal to the tolerance value. As the entire feature surface is swept by the tolerance zone plane, its intersection with the plane shall everywhere be contained within the tolerance zone.

Where no DRF is referenced, the tolerance plane's orientation and sweep shall be normal to the basic profile at each point along the profile. For a revolute, such as shown in Fig. 5-136, the plane shall sweep radially about an axis. Within the plane, the orientation and location of the tolerance zone may adjust continuously to the part surface while sweeping. Alternatively, one or two datums may be referenced as necessary to restrain the orientation of the tolerance plane as it sweeps. Depending on the datums chosen, the DRF might also restrain the orientation of the tolerance zone within the sweeping plane. Any basic dimensions that locate the zone relative to the referenced DRF will restrain the zone's location as well. Addition of a secondary or tertiary datum reference could arrest for the zone all three degrees of translation. For a nominally straight surface, the sweeping plane would then generate a 3-D zone identical to that specified by the "profile of a surface" symbol. To limit the control to 2-D, then, a designer must be careful not to overrestrain the tolerance plane and zone.

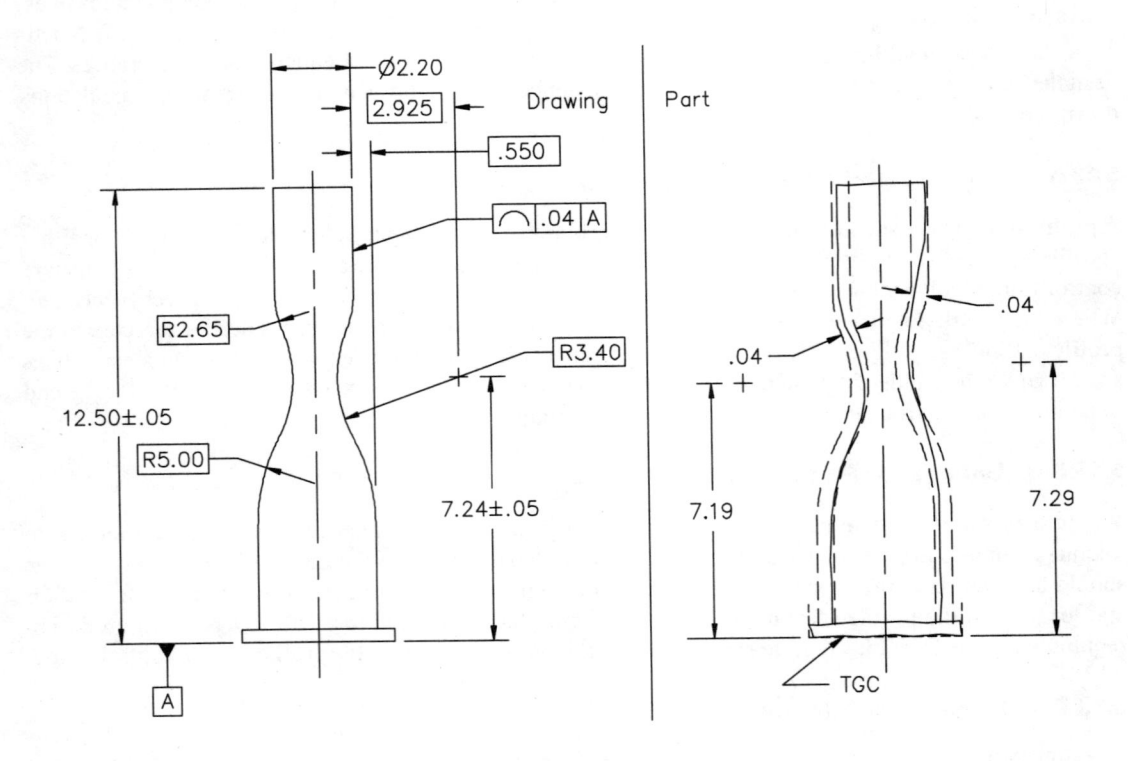

Figure 5-136 Profile of a line tolerance

> **FAQ:** *How can I get the orientation restraint I need from a DRF without getting location restraint I don't want?*
>
> **A:** Currently, there's no symbolic way to "switch off" a DRF's origins. In the rare case where basic dimensions define the basic profile, but you don't want the location restraint, you'll have to add a note to the drawing.

5.13.9 Controlling the Extent of a Profile Tolerance

By default, a single profile tolerance applies to a single tangent-continuous profiled feature. There are cases where a feature's tangency or continuity is interrupted, inconveniently dividing it into two or more features. We'd hate to plaster identical profile feature control frames all around a drawing view like playbills at a construction site. In other cases, different portions of a single feature should have different profile tolerances. An example is where only a portion of a feature is adjacent to a thin wall.

Y14.5 provides three tools for expanding or limiting the extent of a profile tolerance: the "all around" symbol, the **ALL OVER** note, and the "between" symbol. These allow the designer very precise control of profiled features. In our explanations for them, we'll be referring to the *subject view*—a single drawing view that shows a profile outline with a profile feature control frame.

Geometric Dimensioning and Tolerancing 5-151

The "all around" symbol (a circle) modifies a profile tolerance to apply all around the entire outline shown in the subject view regardless of breaks in tangency. As in Fig. 5-137, the symbol is drawn at the "elbow" in the leader line from the feature control frame. "All around" control does not extend to surfaces or edges parallel to the viewing plane or to any feature not shown in the subject view.

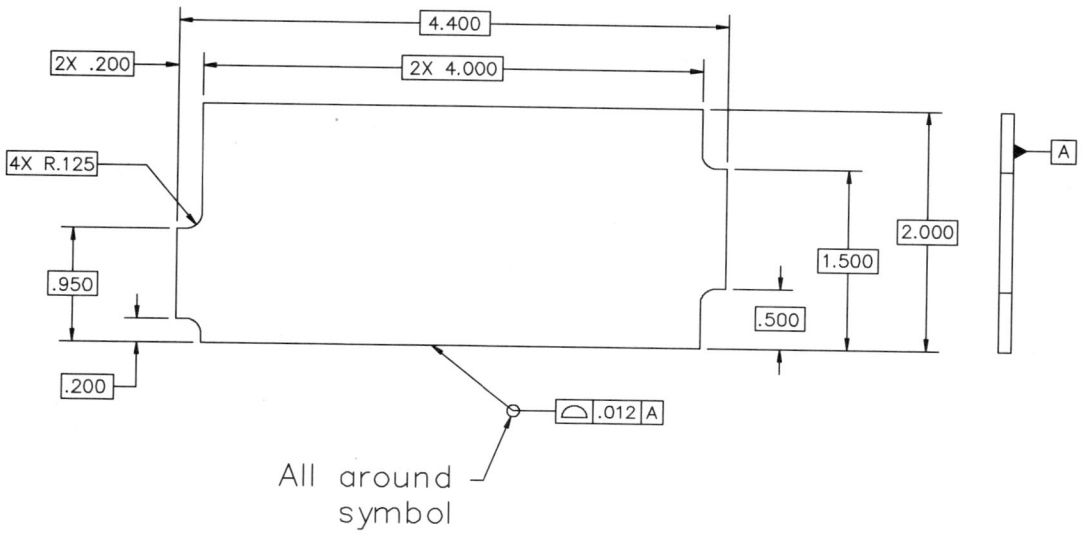

Figure 5-137 Profile "all around"

The note **ALL OVER** has not yet been replaced with a symbol. When the note appears below a profile feature control frame, as in Fig. 5-138, it modifies the profile tolerance to extend all over every surface of the part, including features or sections not shown in the subject view. (Any feature having its own specifications is exempt.) The few applications where this is appropriate include simple parts, castings, forgings, and truly 3-D profiled features. For example, we might specify an automobile door handle or the mold for a shampoo bottle with profile of a surface **ALL OVER**.

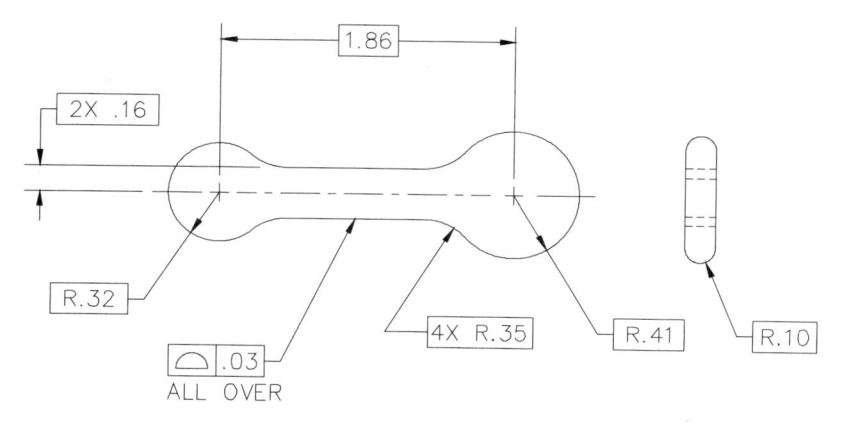

Figure 5-138 Profile "all over"

The third method is to indicate (in the subject view) two points along the basic profile as terminations for the subject tolerance zone. Each point is designated by directing a reference letter to the point with a leader. See Fig. 5-139. If a terminating point is not located at an obvious break in the continuity or tangency of the basic profile, it shall be located with basic dimensions. In addition, the same two reference letters are repeated adjacent to the profile feature control frame, separated by the "between" symbol (a two-headed arrow). The tolerance applies along the basic profile only between the designated terminating points. Neither the choice of reference letters, their relative placement in the subject view, nor their sequence before or after the "between" symbol have any bearing on which portion of the feature is concerned. Where the profile outline closes upon itself, as in Fig. 5-139, the terminating points divide the outline into two portions, both of which can be interpreted as "between" the pair of points. The tolerance applies only to the portion having a leader from the feature control frame. A more complex profile outline having multiple feature control frames with more than two terminating points might require more care in clarifying the extents of the zones.

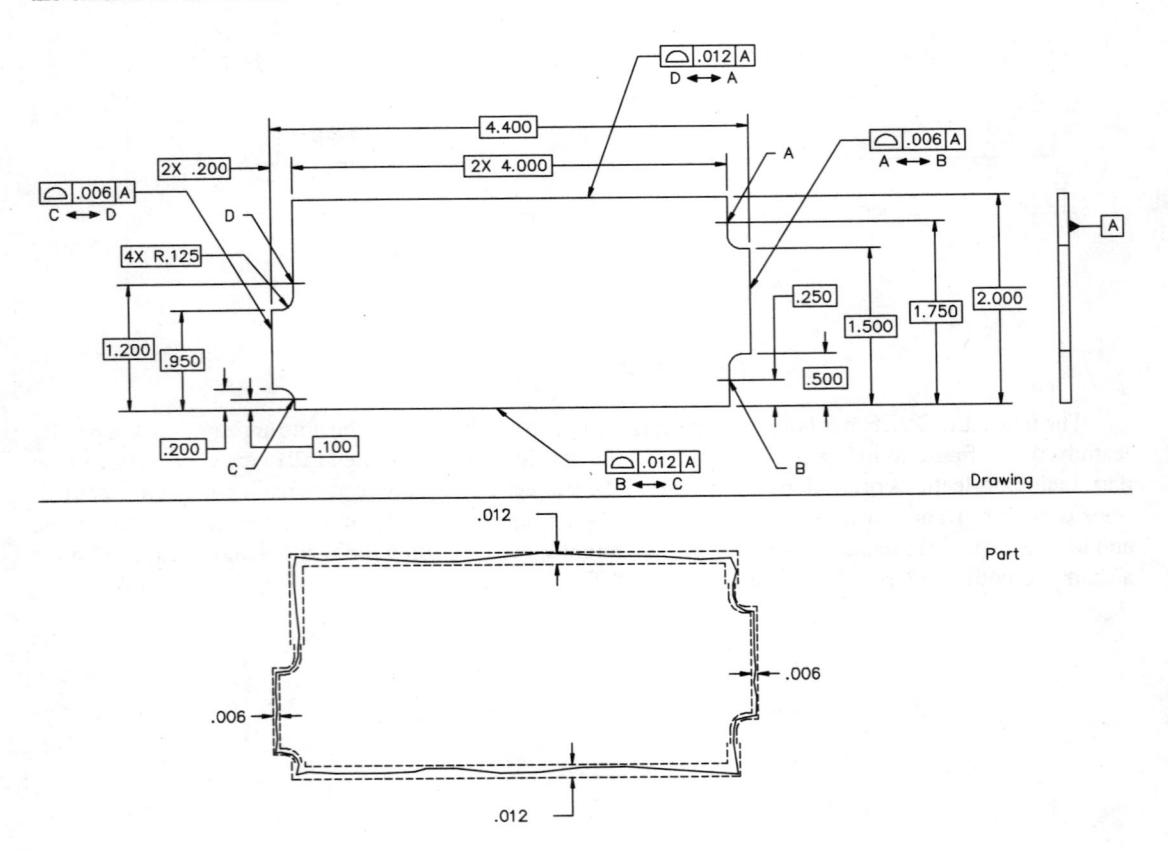

Figure 5-139 Profile "between" points

If, by using any of the above techniques, a profile tolerance is extended to include a sharp corner, the boundary lines for each adjacent surface are extended to intersect. In some designs, the intersection of the zones may not provide adequate control of the corner radius. A separate radius tolerance (as described in section 5.8.10) may be applied as a refinement of the profile control.

5.13.10 Abutting Zones

Abutting profile tolerance zones having boundaries with dissimilar offsets can impose weird or even impossible constraints on the surface. For example, if a zone unilaterally offset in one direction abuts a zone unilaterally offset in the other direction, the transition between zones has zero width. Where zones intersect at a corner, the surface radius could have concave, convex, and straight portions. A designer must carefully consider what the surface contour will be through the transition.

Remember that manufacturing variation tends to be equal/bidirectional, and that tool path programmers target the mean of the tolerance zone. Thus, where the designer makes a narrow unilateral zone abut a much wider unilateral zone, the tool path within the wider zone is "programmer's choice." The programmer might choose to do one of the following.

- Keep the tool path consistently close to the basic profile, discarding tolerance in the wider zone.
- Make an abrupt step in the surface to always follow the median.
- Make a tapered transition to the median.

Since none of the choices are completely satisfactory, we have one more reason to try to use equal-bilateral tolerance zones.

5.13.11 Profile Tolerance for Combinations of Characteristics

By skillfully manipulating tolerance values and datum references, an expert designer can use profile tolerancing to control a surface's form, orientation, and/or location. That's desirable where other types of tolerances, such as size limits, flatness, and angularity tolerances are inapplicable or awkward. For example, in Fig. 5-140, the profile tolerance controls the form of a conical taper. The reference to datum A additionally controls the cone's orientation, and the reference to datum B controls the axial location of the cone relative to the end face. In this case, size limits are useless, but a single profile tolerance provides simple and elegant control. In other cases where more specialized controls will work just fine, it's usually less confusing if the designer applies one or more of them instead.

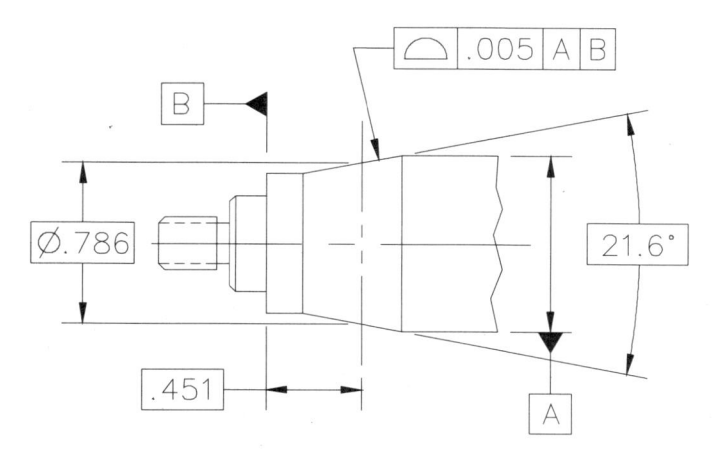

Figure 5-140 Profile tolerancing to control a combination of characteristics

5.13.11.1 With Positional Tolerancing for Bounded Features

Profile tolerancing can be teamed with positional tolerancing to control the orientation and location of bounded features having opposing elements that partly or completely enclose a space. See section 5.11.6.3.

5.13.12 Patterns of Profiled Features

The principles explained in sections 5.11.7 through 5.11.7.5 for controlling patterns of features of size can be extended to patterns of profiled features. Rather than a framework of Level 4 virtual condition boundaries, a profile tolerance applied to a feature pattern establishes a framework of multiple profile tolerance zones. Within this framework, the orientation and location of all the zones are fixed relative to one another according to the basic dimensions expressed on the drawing.

5.13.12.1 Single-Segment Feature Control Frame

Where feature "size," form, orientation, location, and feature-to-feature spacing can all share a single tolerance value, a single-segment profile feature control frame is recommended. Fig. 5-141 shows a pattern of three mounting feet controlled for coplanarity. All points on all three feet shall be contained between a pair of parallel plane boundaries. This effectively controls the flatness of each foot as well as the coplanarity of all three together to prevent rocking. (A flatness tolerance would apply to each foot only on an individual basis.)

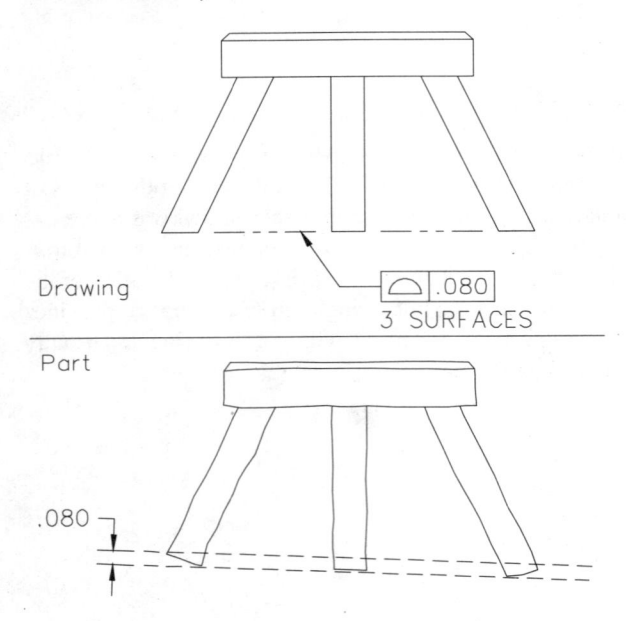

Figure 5-141 Profile tolerance to control coplanarity of three feet

5.13.12.2 Composite Feature Control Frame

A composite feature control frame can specify separate tolerances for overall pattern location and spacing. The few differences in symbology between composite positional and composite profile controls are obvious when comparing Fig. 5-119 with Fig. 5-142. The composite profile feature control frame contains a single entry of the "profile of a surface" symbol. The upper segment establishes a framework (PLTZF) of wider profile tolerance zones that are basically located and oriented relative to the referenced datums. The lower segment provides a specialized refinement within the constraints of the upper segment. It establishes a framework (FRTZF) of comparatively narrower zones that are basically oriented, but not located, relative to the referenced datums. All the rules given in section 5.11.7.3 governing datum references, tolerance values, and simultaneous requirements apply for composite profile tolerances as well.

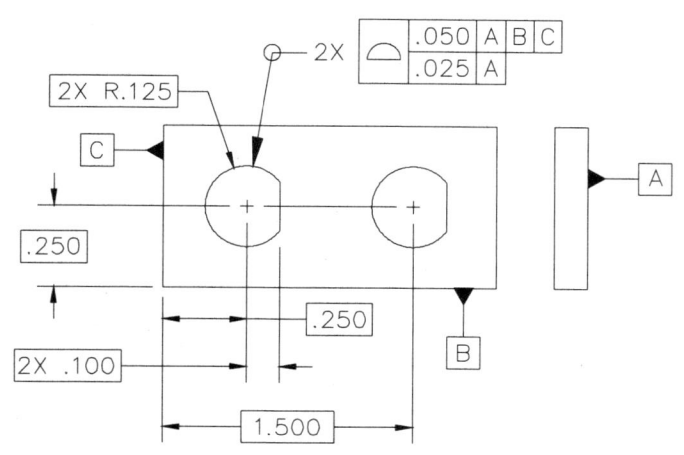

Figure 5-142 Composite profile for a pattern

5.13.12.3 Stacked Single-Segment Feature Control Frames

Where it's necessary to specify different location requirements for a pattern of profiled features relative to different planes of the DRF, stacked single-segment profile feature control frames may be applied as described in section 5.11.7.4. Each of the stacked feature control frames establishes a framework of profile tolerance zones that are basically located and oriented relative to the referenced datums. There is no explicit congruence requirement between the two frameworks. But, if features are to conform to both tolerances, tolerance zones must overlap to some extent. All the rules given in section 5.11.7.5 governing datum references, tolerance values, and simultaneous requirements apply for stacked single-segment profile tolerances as well.

5.13.12.4 Optional Level 2 Control

For features of size such as holes, size limits or tolerances and Rule #1 specify Level 2 form control. For profiled features, each profile tolerance zone provides a degree of Level 2 control (for feature "size" and form). However, where no pattern-controlling tolerance provides adequate Level 2 control, a separate profile tolerance may be added above and separated from the pattern-controlling frame(s). In Fig. 5-143,

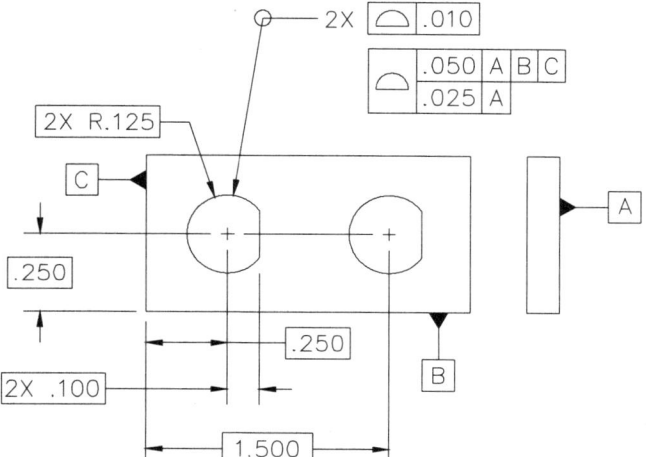

Figure 5-143 Composite profile tolerancing with separate Level 2 control

the profile tolerance of .010 establishes a discrete profile tolerance zone for each individual feature. As with the Level 2 size limit boundaries for holes in a pattern, there is no basic relationship between these Level 2 profile zones. They are all free to float relative to each other and relative to any datums. (Note: If the Level 2 feature control frame were added as a third segment of the composite control, the Level 2 profile zones would be basically related to each other.) Of course, the Level 2 tolerance must be less than any pattern-controlling tolerances to have any effect.

5.13.13 Composite Profile Tolerance for a Single Feature

For features of size, different characteristic symbols denote the four different levels of control. But, for irregularly shaped nonsize features, the same "profile of a surface" symbol is used for each level. In Fig. 5-144, for example, we want to refine a bounded feature's orientation within the constraints of its locating tolerance. Simply stacking two single-segment profile feature control frames would be confusing. Many people would question whether the .020 tolerance controls location relative to datum B. Instead, we've borrowed from pattern control the composite feature control frame containing a single entry of the "profile of a surface" symbol. Though our "pattern" has only one feature, the tolerances mean the same.

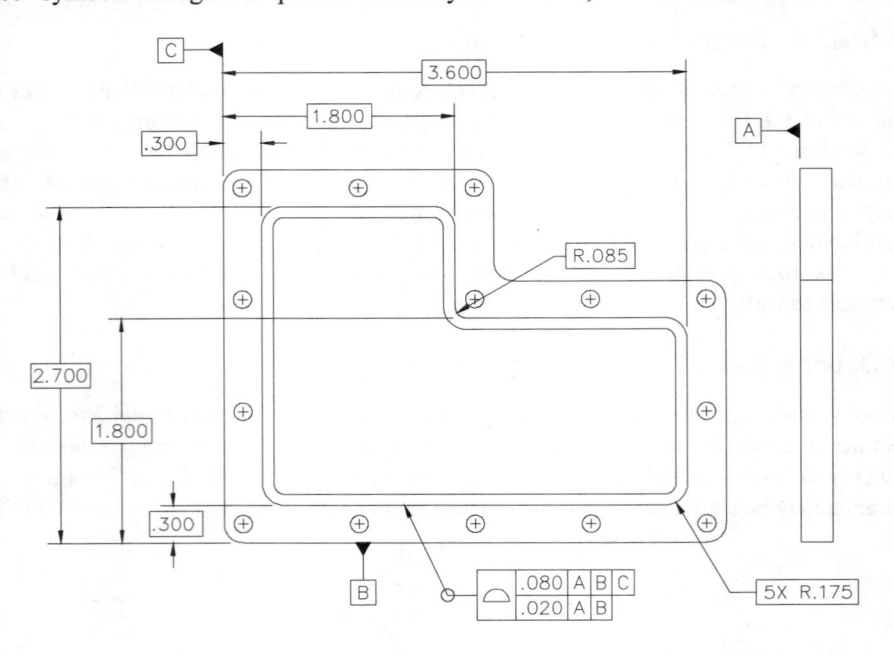

Figure 5-144 Composite profile tolerance for a single feature

In Fig. 5-144, the upper segment establishes a .080 wide profile tolerance zone basically located and oriented relative to the DRF A|B|C. The lower segment provides a specialized refinement within the constraints of the upper segment. It establishes a .020 wide zone basically oriented, but not located, relative to the DRF A|B. All the rules given in section 5.11.7.3 governing datum references, tolerance values, and simultaneous requirements apply for a composite profile "pattern of one."

5.14 Symmetry Tolerance

Symmetry is the correspondence in size, contour, and arrangement of part surface elements on opposite sides of a plane, line, or point. We usually think of symmetry as the twofold mirror-image sort of balance about a center plane shown in Fig. 5-145(a) and (b). There are other types as well. A three-lobe cam can have symmetry, both the obvious twofold kind about a plane as shown in Fig. 5-145(c), and a threefold kind about an axis as shown in Fig. 5-145(d). The pentagon shown in Fig. 5-145(e) has fivefold symmetry about an axis. GD&T's symmetry tolerances apply at the *lowest order of symmetry*—the lowest prime divisor of the number of sides, facets, blades, lobes, etc., that the feature is supposed to have. Thus, a 27-blade turbine would be controlled by threefold symmetry. For a hexagonal flange (six sides), twofold symmetry applies. By agreement, a nominally round shaft or sphere is subject to twofold symmetry as well.

5.14.1 How Does It Work?

The Math Standard describes in detail how symmetry tolerancing works. Generically, a symmetry tolerance prescribes that a datum plane or axis is extended all the way through the controlled feature. See Fig. 5-146. From any single point on that datum within the feature, vectors or *rays* perpendicular to the datum

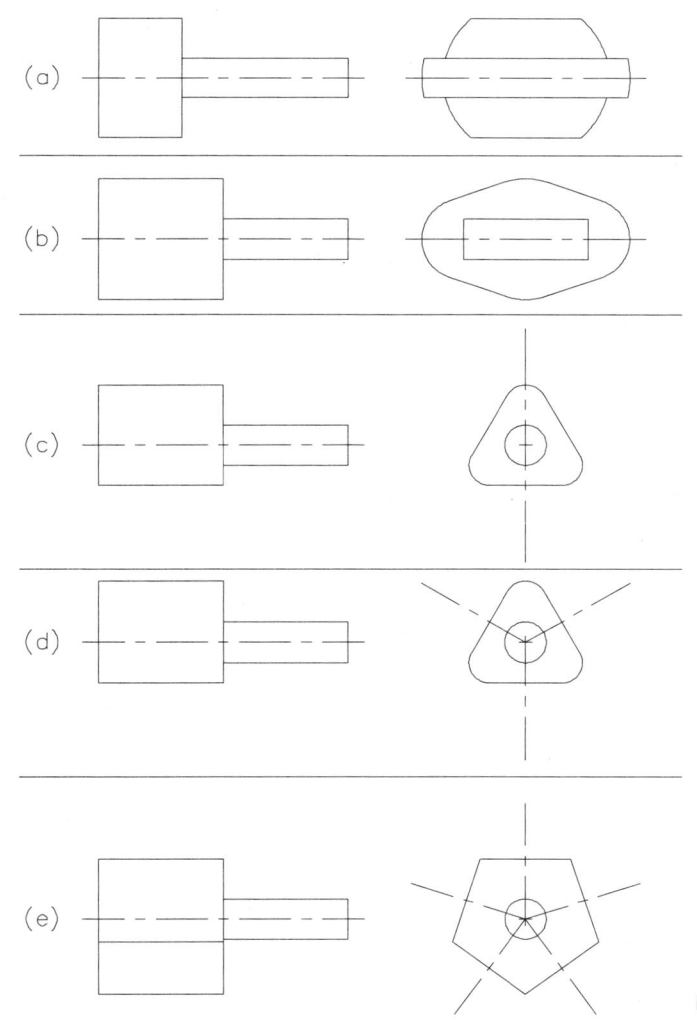

Figure 5-145 Types of symmetry

5-158 Chapter Five

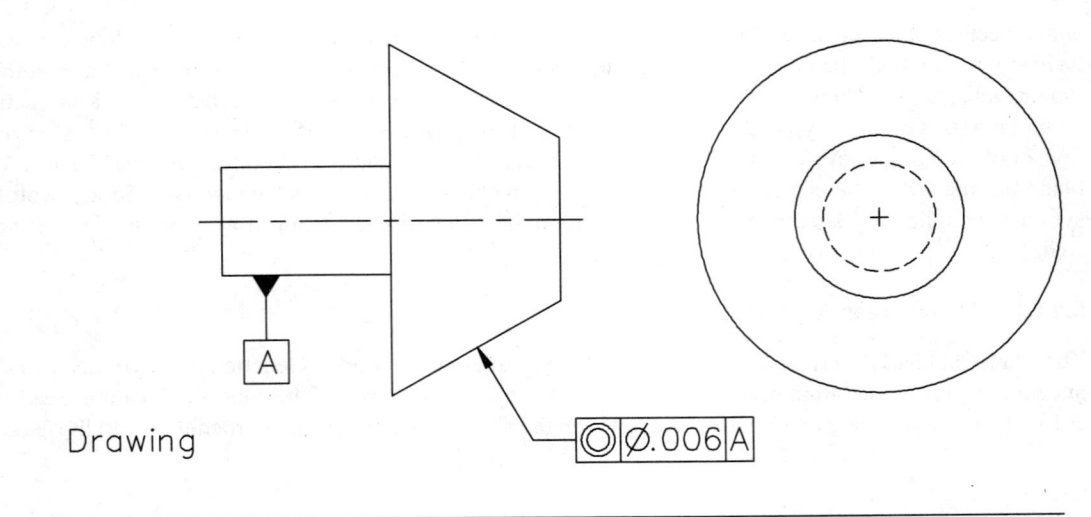

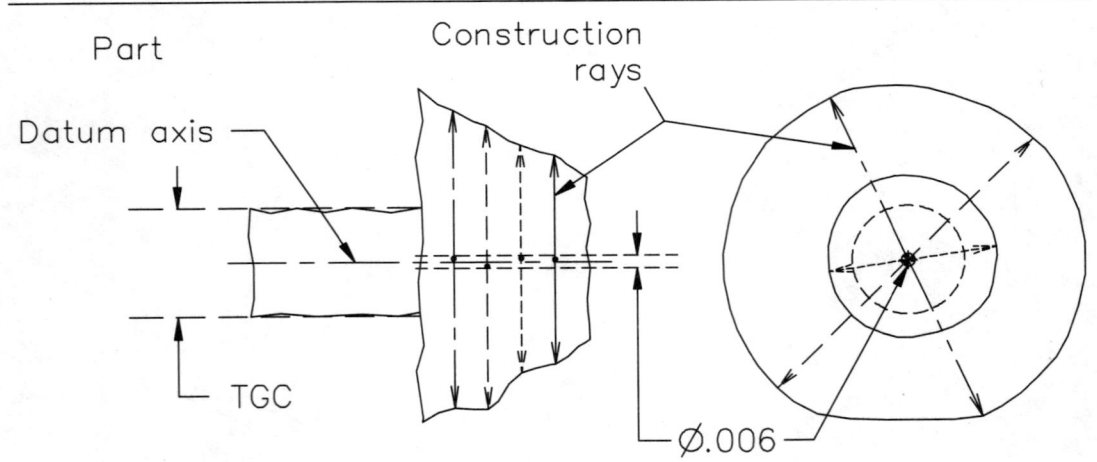

Figure 5-146 Symmetry construction rays

are projected to intersect the feature surface(s). For common twofold symmetry, two rays are projected, 180° apart. From those intersection points, a median point (centroid) is constructed. This median point shall lie within a tolerance zone that is uniformly distributed about the datum.

If one of the construction rays hits a small dent in the surface, but an opposite ray intersects a uniform portion of the surface, the median point might lie outside the tolerance zone. Thus, symmetry tolerancing demands that any local "low spot" in the feature surface be countered by another "low spot" opposite. Similarly, any "high spot" must have a corresponding "high spot" opposite it. Symmetry tolerancing primarily prevents "lopsidedness."

As you can imagine, inspecting a symmetry tolerance is no simple matter. Generally, a CMM with advanced software or a dedicated machine with a precision spindle should be used. For an entire feature to conform to its symmetry tolerance, all median points shall conform, for every possible ray pattern, for every possible origin point on the datum plane or axis within the feature. Although it's impossible to verify infinitely many median points, a sufficient sample (perhaps dozens or hundreds) should be constructed and evaluated.

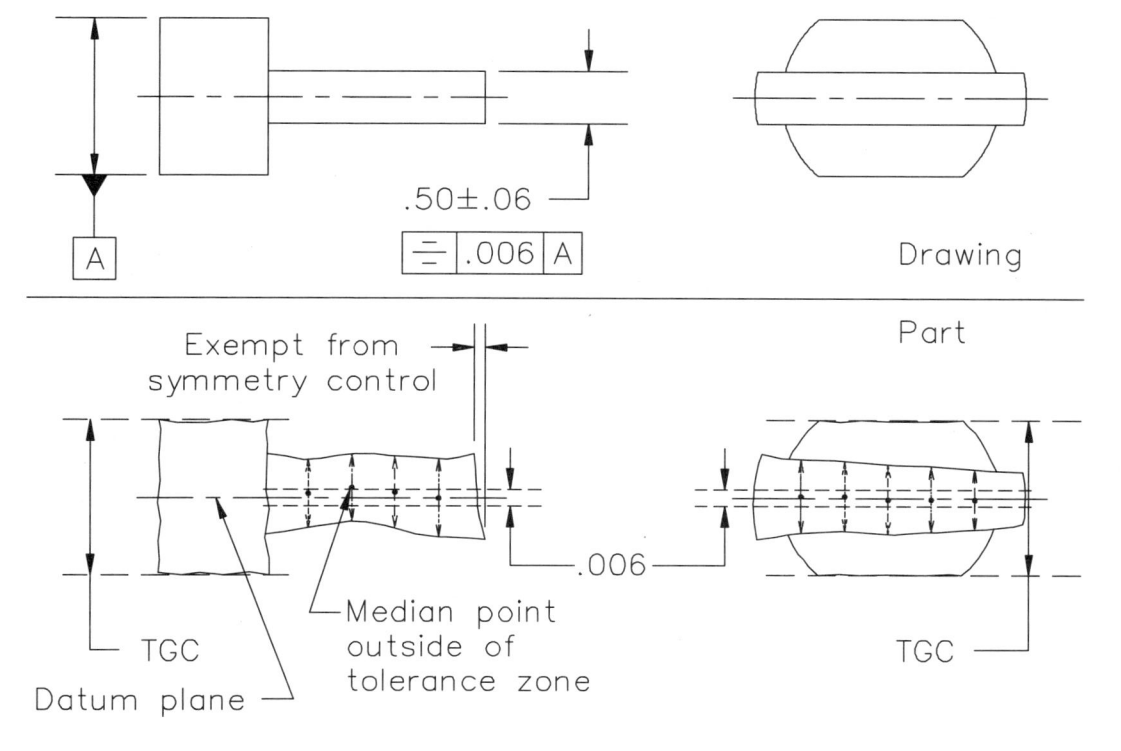

Figure 5-147 Symmetry tolerance about a datum plane

At the ends of every actual bore or shaft, and at the edges of every slot or tab, for example, the terminating faces will not be perfectly perpendicular to the symmetry datum. Though one ray might intersect a part surface at the extreme edge, the other ray(s) could just miss and shoot off into the air. This also happens at any cross-hole, flat, keyseat, or other interruption along the controlled feature(s). Obviously then, unopposed points on the surface(s), as depicted in Fig. 5-147, are exempt from symmetry control. Otherwise, it would be impossible for any feature to conform.

5.14.2 How to Apply It

A symmetry tolerance is specified using a feature control frame displaying the characteristic symbol for either "concentricity" (two concentric circles) or "symmetry about a plane" (three stacked horizontal bars). See Figs. 5-146 through 5-148. The feature control frame includes the symmetry tolerance value followed by one, two, or three datum references.

There's no practical interaction between a feature's size and the acceptable magnitude of lopsidedness. Thus, material condition modifier symbols, MMC and LMC, are prohibited for all symmetry tolerances and their datum references.

5.14.3 Datums for Symmetry Control

Symmetry control requires a DRF. A primary datum plane or axis usually arrests the three or four degrees of freedom needed for symmetry control. All datum references shall be RFS.

5.14.4 Concentricity Tolerance

Concentricity tolerancing of a revolute, as illustrated in Fig. 5-146, is one of the most common applications of symmetry tolerancing. It's specified by a feature control frame containing the "concentricity" symbol. In this special symmetry case, the datum is an axis. There are two rays 180° apart (colinear) perpendicular to the datum axis. The rays intersect the feature surface at two diametrically opposed points. The midpoint between those two surface points shall lie within a cylindrical tolerance zone coaxial to the datum and having a diameter equal to the concentricity tolerance value.

At each cross-sectional slice, the revolving rays generate a locus of distinct midpoints. As the rays sweep the length of the controlled feature, these 2-D loci of midpoints stack together, forming a 3-D "wormlike" locus of midpoints. The entire locus shall be contained within the concentricity tolerance cylinder. Don't confuse this 3-D locus with the 1D derived median line defined in section 5.6.4.2.

5.14.4.1 Concentricity Tolerance for Multifold Symmetry about a Datum Axis

The explanation of concentricity in Y14.5 is somewhat abstruse because it's also meant to support multifold symmetry about an axis. Any prime number of rays can be projected perpendicular from the datum axis, provided they are coplanar with equal angular spacing. For the 3-lobe cam in Fig. 5-148, there are three rays, 120° apart. A 25-blade impeller would require five rays spaced 72° apart, etc.

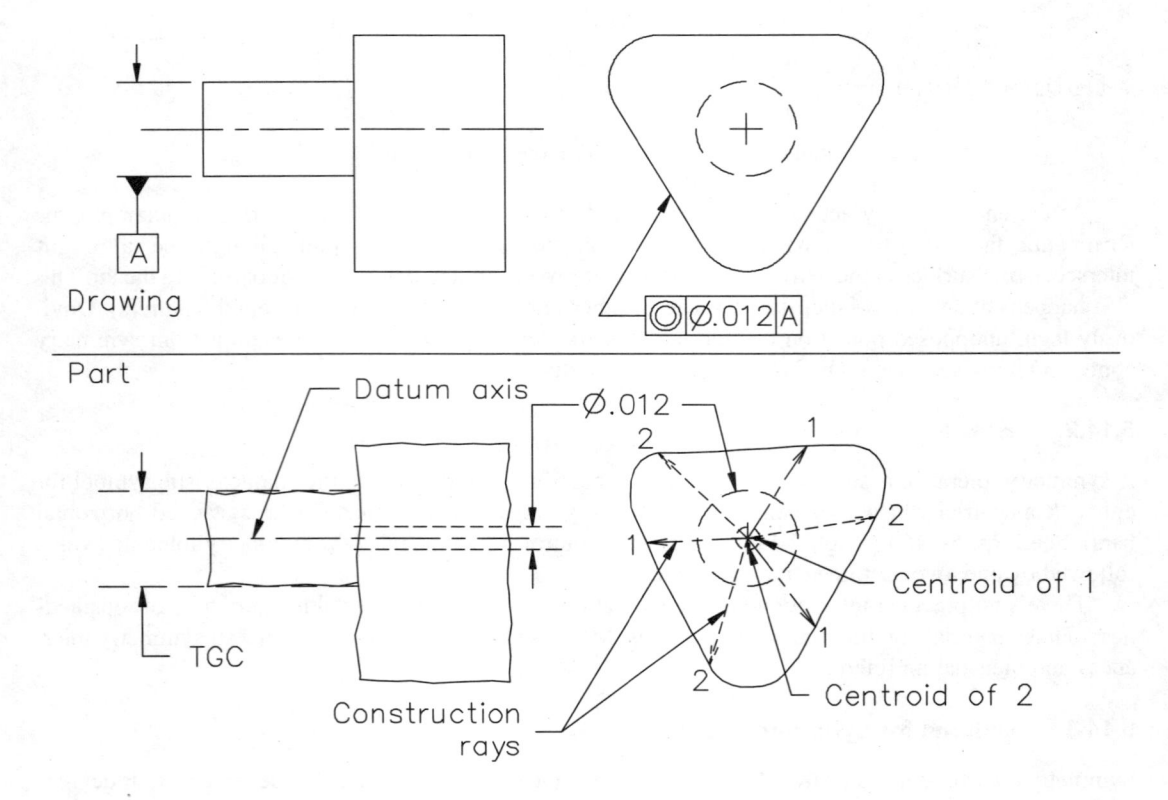

Figure 5-148 Multifold concentricity tolerance on a cam

From the multiple intersection points, a centroid is then constructed and checked for containment within the tolerance zone. The standards don't specify how to derive the centroid, but we recommend the Minimum Radial Separation (MRS) method described in ANSI B89.3.1-1972. Obviously, verification is well beyond the capability of an inspector using multiple indicators and a calculator. Notice that as the rays are revolved about the datum axis, they intersect the surface(s) at vastly different distances from center. Nevertheless, if the part is truly symmetrical, the centroid still remains within the tolerance cylinder.

5.14.4.2 Concentricity Tolerance about a Datum Point

The "concentricity" symbol can also be used to specify twofold or multifold symmetry about a datum point. This could apply to a sphere, tetrahedron, dodecahedron, etc. In all cases, the basic geometry defines the symmetry rays, and centroids are constructed and evaluated. The tolerance value is preceded by the symbol SØ, specifying a spherical tolerance zone.

5.14.5 Symmetry Tolerance about a Datum Plane

The other symmetry symbol, having three horizontal bars, designates symmetry about a plane. Y14.5 calls this application **Symmetry Tolerancing to Control the Median Points of Opposed or Correspondingly-Located Elements of Features.** Despite this ungainly and nondescriptive label, symmetry tolerancing about a plane works just like concentricity except for two differences: the symmetry datum is a plane instead of an axis; and the symmetry can only be twofold. See Fig. 5-147. From any point on the datum plane between the controlled surfaces, two rays are projected perpendicular to the datum, 180° apart (colinear). The rays intersect the surfaces on either side of the datum. The midpoint between those two surface points shall be contained between two parallel planes, separated by a distance equal to the symmetry tolerance value. The two tolerance zone planes are equally disposed about (thus, parallel to) the datum plane. All midpoints shall conform for every possible origin point on the datum plane between the controlled surfaces.

As the rays sweep, they generate a locus of midpoints subtly different from the derived median plane defined in section 5.6.4.2. The symmetry rays are perpendicular to the datum plane, while the derived median plane's construction lines are perpendicular to the feature's own center plane. It's not clear why the methods differ or whether the difference is ever significant.

Symmetry tolerancing about a plane does not limit feature size, surface flatness, parallelism, or straightness of surface line elements. Again, the objective is that the part's mass be equally distributed about the datum. Although a symmetry or concentricity tolerance provides little or no form control, it always accompanies a size dimension that provides some restriction on form deviation according to Rule #1.

5.14.6 Symmetry Tolerancing of Yore (Past Practice)

Until the 1994 edition, Y14.5 described concentricity tolerancing as an "axis" control, restraining a separate "axis" at each cross-section of the controlled feature. A definition was not provided for *axis*, nor was there any explanation of how a two-dimensional imperfect shape (a circular cross-section) could even have such a thing. As soon as the Y14.5 Subcommittee defined the term *feature axis*, it realized two things about the feature axis: it's what ordinary positional tolerance RFS controls, and it has nothing to do with lopsidedness (balance). From there, symmetry rays, median points, and worms evolved.

The "Symmetry Tolerance" of the 1973 edition was exactly the same as positional tolerance applied to a noncylindrical feature RFS. (See the note at the bottom of Fig. 140 in that edition.) The three-horizontal bars symbol was simply shorthand, saving draftsmen from having to draw circle-S symbols. Partly because of its redundancy, the "symmetry tolerance" symbol was cut from the 1982 edition.

5.14.7 When Do We Use a Symmetry Tolerance?

Under any symmetry tolerance, a surface element on one "side" of the datum can "do anything it wants" just as long as the opposing element(s) mirrors it. This would appear to be useful for a rotating part that must be dynamically balanced. However, there are few such assemblies where GD&T alone can adequately control balance. More often, the assembly includes setscrews, keyseats, welds, or other attachments that entail a balancing operation after assembly. And ironically, a centerless ground shaft might have near-perfect dynamic balance, yet fail the concentricity tolerance because its out-of-roundness is 3-lobed.

FAQ: Could a note be added to modify the concentricity tolerance for a cylinder to 3-fold symmetry?

A: Sure.

FAQ: Can I use a symmetry tolerance if the feature to be controlled is offset (not coaxial or coplanar) from the datum feature?

A: Nothing in the standard prohibits that, either. Be sure to add a basic dimension to specify the offset. You may also need two or even three datum references.

FAQ: Since a runout tolerance includes concentricity control and is easier to check, wouldn't it save money to replace every concentricity tolerance with an equal runout tolerance? We wouldn't need concentricity at all.

A: Though that is the policy at many companies, there's another way to look at it. Let's consider a design where significant out-of-roundness can be tolerated as long as it's symmetrical. A concentricity tolerance is carefully chosen. We can still use runout's FIM method to inspect a batch of parts. Of those conforming to the concentricity tolerance, all or most parts will pass the FIM test and be accepted quickly and cheaply. Those few parts that fail the FIM inspection may be re-inspected using the formal concentricity method. The concentricity check is more elaborate and expensive than the simple FIM method, but also more forgiving, and would likely accept many of the suspect parts. Alternatively, management may decide it's cheaper to reject the suspect parts without further inspection and to replace them. The waste is calculated and certainly no worse than if the well-conceived concentricity tolerance had been arbitrarily converted to a runout tolerance. The difference is this: If the suspect parts are truly usable, the more forgiving concentricity tolerance offers a chance to save them.

5.15 Combining Feature Control Frames

In section 5.6, we defined four different levels of GD&T control for features of size. In fact, the four levels apply for every feature.

        Level 1: 2-D form at individual cross sections
        Level 2: Adds third dimension for overall form control
        Level 3: Adds orientation control
        Level 4: Adds location control

For every feature of every part, a designer must consider all the design requirements, including function, strength, assemblability, life expectancy, manufacturability, verification, safety, and appearance. The designer must then adequately control each part feature, regardless of its type, at each applicable level of control, to assure satisfaction of all design requirements. For a nonsize feature, a single "profile" or "radius" tolerance will often suffice. Likewise, a feature of size might require nothing more than size limits and a single-segment positional tolerance.

In addition to the design requirements listed, many companies include cost considerations. In cost-sensitive designs, this often means maximizing a feature's tolerance at each level of control. The designer must understand the controls imposed at each level by a given tolerance. For example, where a Level 4 (location) tolerance has been maximized, it might not adequately restrict orientation. Thus, a separate lesser Level 3 (orientation) tolerance must be added. Even that tolerance, if properly maximized, might not adequately control 3-D form, etc. That's why it's not uncommon to see two, or even three feature control frames stacked for one feature, each maximizing the tolerance at a different level.

5.16 "Instant" GD&T

Y14.5 supports several general quasi-GD&T practices as alternatives to the more rigorous methods we've covered. To be fair, they're older practices that evolved as enhancements to classical tolerancing methods. However, despite the refinement and proliferation of more formal methods, the quasi-GD&T practices are slow to die and you'll still see them used on drawings. Designers might be tempted to use one or two of them to save time, energy, and plotter ink. We'll explain why, for each such practice, we feel that's false economy.

5.16.1 The "Dimension Origin" Symbol

The "dimension origin" symbol, shown in Fig. 5-149, is not associated with any datum feature or any feature control frame. It's meant to **indicate that a dimension between two features shall originate from one of these features and not the other.** The specified treatment for the originating surface is exactly the same as if it were a primary datum feature. But for some unfathomable reason, Y14.5 adds, **This concept does not establish a datum reference frame...** The treatment for the other surface is exactly the same as if it were controlled with a profile of a surface tolerance. We explained in section 5.10.8 why this practice is meaningless for many angle dimensions. Prevent confusion; instead of the "dimension origin" symbol, use a proper profile or positional tolerance.

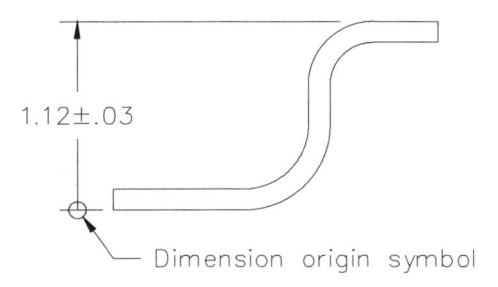

Figure 5-149 Dimension origin symbol

5.16.2 General Note to Establish Basic Dimensions

Instead of drawing the "basic dimension" frame around each basic dimension, a designer may designate dimensions as basic by **specifying on the drawing (or in a document referenced on the drawing) the general note: UNTOLERANCED DIMENSIONS LOCATING TRUE POSITION ARE BASIC.** This could be extremely confusing where other untoleranced dimensions are not basic, but instead default to tolerances expressed in a tolerance block. Basic dimensions for angularity and profile tolerances, datum targets, and more would still have to be framed unless the note were modified. Either way, the savings in ink are negligible compared to the confusion created. Just draw the frames.

5.16.3 General Note in Lieu of Feature Control Frames

Y14.5 states that linear and angular dimensions may be related to a DRF without drawing a feature control frame for each feature. **[T]he desired order of precedence may be indicated by a note such as: UNLESS OTHERWISE SPECIFIED, DIMENSIONS ARE RELATED TO DATUM A (PRIMARY), DATUM B (SECONDARY), AND DATUM C (TERTIARY).** However, applicable datum references shall be included in any feature control frames used. It's not clear whether or not this practice establishes virtual condition boundaries or central tolerance zones for the affected features, and if so, of what sizes and shapes. As we explained in section 5.10.8, for some angle dimensions a wedge-shaped zone is absurd.

The hat trick of "instant" GD&T is to combine the above two "instant basic dimensions" and "instant datum references" notes with an "instant feature control" note, such as **PERFECT ORIENTATION (or COAXIALITY or LOCATION OF SYMMETRICAL FEATURES) AT MMC REQUIRED FOR RELATED FEATURES.** This should somehow provide cylindrical or parallel-plane tolerance zones equivalent to zero positional or zero orientation tolerances at MMC for all "related features" of size.

Throughout this chapter, we've emphasized how important it is for designers to consider carefully and individually each feature to maximize manufacturing tolerances. Certainly, troweling on GD&T with general notes does not require such consideration, although, neither does the practice preclude it. And while there may be drawings that would benefit from consolidation and unification of feature controls, we prefer to see individual, complete, and well-thought-out feature control frames.

5.17 The Future of GD&T

GD&T's destiny is clearly hitched to that of manufacturing technology. You wouldn't expect to go below deck on *Star Trek's USS Enterprise* and find a machine room with a small engine lathe and a Bridgeport mill. You might find instead some mind-bogglingly precise process that somehow causes a replacement "Support, Dilithium Crystal" to just "materialize" out of a dust cloud or a slurry. Would Scotty need to measure such a part?

Right now, the rapid-prototyping industry is making money with technology that's only a couple of generations away from being able to "materialize" high-strength parts in just that way. If such a process were capable of producing parts having precision at least an order of magnitude more than what's needed, the practice of measuring parts would indeed become obsolete, as would the language for specifying dimensional tolerances. Parts might instead be specified with only the basic geometry (CAD model) and a process capability requirement.

History teaches us that new technology comes faster than we ever expected. Regardless of our apprehension about that, history also reveals that old technology lingers on longer than we expected. In fact, the better the technology, the slower it dies. An excellent example is the audio Compact Cassette, introduced to the world by Philips in 1963. Even though Compact Discs have been available in every music store since 1983, about one-fourth of all recorded music is still sold on cassette tapes. We can likewise expect material removal processes and some form of GD&T to enjoy widespread use for at least another two decades, regardless of new technology.

In its current form, GD&T reflects its heritage as much as its aspirations. It evolved in relatively small increments from widespread, time-tested, and work-hardened practices. As great as it is, GD&T still has much room for improvement. There have been countless proposals to revamp it, ranging from moderate streamlining to total replacement. Don't suppose for one second that all such schemes have been harebrained. One plan, for example, would define part geometry just as a coordinate measuring machine sees it—vectorially. Such a system could expedite automated inspection, and be simpler to learn. But does it preclude measurements with simple tools and disenfranchise manufacturers not having access to a CMM? What about training? Will everyone have to be fluent in two totally different dimensioning and tolerancing languages?

As of this writing, the international community is much more receptive to radical change than the US. Europe is a hotbed of revolutionary thought; any daring new schemes will likely surface there first. Americans can no longer play isolationism as they could decades ago. Many US companies are engaged in multinational deals where a common international drawing standard is mandatory. Those companies are scarcely able to insist that standard be Y14.5. There are always comments about "the tail wagging the dog," but the US delegation remains very influential in ISO TC 213 activity pertaining to GD&T. Thus, in the international standards community, it's never quite clear where the tail ends and the dog begins.

Meanwhile, Americans are always looking for ways to simplify GD&T, to make their own Y14.5 Standard thinner (or at least to slow its weight gain). You needn't study GD&T long to realize that a few characteristic symbols are capable of controlling many more attributes than some others control. For example, a surface profile tolerance can replace an equal flatness tolerance. Why do we need the "flatness" symbol? And if the only difference between parallelism, perpendicularity, and angularity is the basic angle invoked, why do we need three different orientation symbols? In fact, couldn't the profile of a surface characteristic be modified slightly to control orientation?

These are all valid arguments, and taken to the next logical step, GD&T could be consolidated down to perhaps four characteristic symbols. And following in the same logic, down to three or two symbols, then down to one symbol. For that matter, not even one symbol would be needed if it were understood that each feature has default tolerance boundaries according to its type. The document that defines such tolerance zones might have only thirty pages. This would be GD&T at its leanest and meanest! OK, so why don't we do it?

That argument assumes that the complexity of a dimensioning and tolerancing system is proportional to the number of symbols used. Imagine if English had only 100 words, but the meanings of those words change depending on the context and the facial expression of the speaker. Would that be simpler? Easier to learn? No, because instead of learning words, a novice would have to learn all the rules and meanings for each word just to say "Hello." There's a lot to be gained from simplification, but there's also a huge cost.

In fact, GD&T's evolution could be described as a gradual shift from simplicity toward flexibility. As users become more numerous and more sophisticated, they request that standards add coverage for increasingly complex and esoteric applications. Consequently, most issues faced by the Y14.5 committee boil down to a struggle to balance simplicity with flexibility.

It's impossible to predict accurately where GD&T is headed, but it seems reasonable to expect the Y14.5 committee will continue to fine-tune a system that is rather highly developed, mature, and in widespread international use. Radical changes cannot be ruled out, but they would likely follow ISO activity. Be assured, GD&T's custodial committees deeply contemplate the future of dimensioning and tolerancing.

Standards committee work is an eye-opening experience. Each volunteer meets dozens of colleagues representing every sector of the industry, from the mainstream Fortune 500 giants to the tiniest outpost ma-and-pa machine shops. GD&T belongs equally to all these constituents. Often, what seemed a brilliant inspiration to one volunteer withers under the hot light of committee scrutiny. That doesn't mean that nothing can get through committee; it means there are very few clearly superior and fresh ideas under the sun. Perhaps, though, you've got one. If so, we encourage you to pass it along to this address.

The American Society of Mechanical Engineers
Attention: Secretary, Y14 Main Committee
345 East 47th Street
New York, NY 10017

5.18 References

1. The American Society of Mechanical Engineers. 1972. *ANSI B89.3.1-1972. Measurement of Out-Of-Roundness.* New York, New York: The American Society of Mechanical Engineers.
2. The American Society of Mechanical Engineers. 1972. *ANSI B4.1-1967. Preferred Limits and Fits for Cylindrical Parts.* New York, New York: The American Society of Mechanical Engineers.
3. The American Society of Mechanical Engineers. 1978. *ANSI B4.2-1978. Preferred Metric Limits and Fits.* New York, New York: The American Society of Mechanical Engineers.
4. The American Society of Mechanical Engineers. 1982. *ANSI Y14.5M-1982, Dimensioning and Tolerancing.* New York, New York: The American Society of Mechanical Engineers.
5. The American Society of Mechanical Engineers. 1995. *ASME Y14.5M-1994, Dimensioning and Tolerancing.* New York, New York: The American Society of Mechanical Engineers.
6. The American Society of Mechanical Engineers. 1994. *ASME Y14.5.1-Mathematical Definition of Dimensioning and Tolerancing Principles.* New York, New York: The American Society of Mechanical Engineers.
7. International Standards Organization. 1985. *ISO8015. Technical Drawings -- Fundamental Tolerancing Principle.* International Standards Organization: Switzerland.

Chapter 6

Differences Between US Standards and Other Standards

Alex Krulikowski
Scott DeRaad

Alex Krulikowski
General Motors Corporation
Westland, Michigan

A Standards manager at General Motors and a member of SME and AQC, Mr. Krulikowski has written articles for several magazines and speaks frequently at public seminars and in-house training programs. He has written 12 books on dimensioning and tolerancing, produced videotapes, computer based training, and other instructional materials. He serves on several corporate and national committees on dimensioning and tolerancing.

Scott DeRaad
General Motors Corporation
Ann Arbor, Michigan

A co-author of Quick Comparison of Dimensioning Standards - 1997 Edition, Mr. DeRaad is an instructor of the ASME Y14.5M-1994 GD&T standard with international teaching experience. He is an automotive automatic transmission design and development engineer for GM Powertrain. Mr. DeRaad is a cum laude graduate of the University of Michigan holding a B.S.E. Engineering-Physics.

6.1   Dimensioning Standards

Dimensioning standards play a critical role in the creation and interpretation of engineering drawings. They provide a uniform set of symbols, definitions, rules, and conventions for dimensioning. Without standards, drawings would not be able to consistently communicate the design intent. A symbol or note could be interpreted differently by each person reading the drawing. It is very important that the drawing user understands which standards apply to a drawing before interpreting the drawing.

Most dimensioning standards used in industry are based on either the American Society of Mechanical Engineers (ASME) or International Organization for Standardization (ISO) standards. Although these two standards have emerged as the primary dimensioning standards, there are also several other standards worldwide that are in use to a lesser degree. There is increasing pressure to migrate toward a common international standard as the world evolves toward a global marketplace. (Reference 5)

This chapter introduces the various standards, briefly describes their contents, provides an overview of the originating bodies, and compares the Y14.5M-1994 and ISO dimensioning standards.

6.1.1 US Standards

In the United States, the most common standard for dimensioning is ASME Y14.5M - 1994. The ASME standards are established by the American Society of Mechanical Engineers, which publishes hundreds of standards on various topics. A list of the ASME standards that are related to dimensioning is shown in Table 6-1.

Table 6-1 ASME standards that are related to dimensioning

STD Number	Title	STD Date
Y14.5M	Dimensioning and Tolerancing	1994
Y14.5.1M	Mathematical Definition of Dimensioning and Tolerancing Principles	1994
Y14.8M	Castings and Forgings	1996
Y14.32.1	Chassis Dimensioning Practices	1994

The ASME Y14.5M - 1994 Dimensioning and Tolerancing Standard covers all the topics of dimensioning and tolerancing. The Y14.5 standard is 227 pages long and is updated about once every ten years. The other Y14 standards in Table 6-1 are ASME standards that provide terminology and examples for the interpretation of dimensioning and tolerancing of specific applications.

Subcommittees of ASME create ASME standards. Each subcommittee consists of representatives from industry, government organizations, academia, and consultants. There are typically 8 to 25 members on a subcommittee. Once the subcommittee creates a draft of a standard, it goes through an approval process that includes a public review. (Reference 5)

6.1.2 International Standards

Outside the United States, the most common standards for dimensioning are established by the International Organization for Standardization (ISO). ISO is a worldwide federation of 40 to 50 national standards bodies (ISO member countries). The ISO federation publishes hundreds of standards on various topics. A list of the ISO standards that are related to dimensioning is shown in Table 6-2.

Table 6-2 ISO standards that are related to dimensioning

STD Number	Title	STD Date
128	Technical Drawings - General principles of presentation	1982
129	Technical Drawings - Dimensioning - General principles, definitions, methods of execution and special indications	1985
406	Technical Drawings - Tolerancing of linear and angular dimensions	1987
1101	Technical drawings - Geometrical tolerancing - Tolerances of form, orientation, location and runout - Generalities, definitions, symbols, indications on drawings	1983
1660	Technical drawings - Dimensioning and tolerancing of profiles	1987
2692	Technical drawings - Geometrical tolerancing - Maximum material principle	1988
2768-1	General tolerances - Part 1: Tolerances for linear and angular dimensions without individual tolerance indications	1989
2768-2	General tolerances - Part 2: Tolerances for features without individual tolerance indications	1989
2692	Amendment 1: Least material requirement	1992
3040	Technical drawings - Dimensioning and tolerancing - Cones	1990
5458	Technical drawings - Geometrical tolerancing - Positional tolerancing	1987
5459	Technical drawings - Geometrical tolerancing - Datums and datum system for geometrical tolerances	1981
7083	Technical drawings - Symbols for geometrical tolerancing - Proportions and dimensions	1983
8015	Technical drawings - Fundamental tolerancing principle	1985
10209-1	Technical product documentation vocabulary - Part 1: Terms relating to technical drawings - General and types of drawings	1992
10578	Technical drawings - Tolerancing of orientation and location - Projected tolerance zone	1992
10579	Technical drawings - Dimensioning and tolerancing - Non-rigid parts	1993
13715	Technical drawings - Corners of undefined shape - Vocabulary and indication on drawings	1997

The ISO standards divide dimensioning and tolerancing into topic subsets. A separate ISO standard covers each dimensioning topic. The standards are typically short, approximately 10 to 20 pages in length. When using the ISO standards for dimensioning and tolerancing, it takes 15 to 20 standards to cover all the topics involved.

The work of preparing international standards is normally carried out through ISO technical committees. Each country interested in a subject for which a technical committee has been established has the right to be represented on that committee. International organizations, governmental and nongovernmental, in liaison with ISO, also take part in the work. The ISO standards are an agreement of major points among countries. Many companies (or countries) that use the ISO dimensioning standards also have additional dimensioning standards to supplement the ISO standards.

A Draft International Standard is prepared by the technical committee and circulated to the member countries for approval before acceptance as an international standard by the ISO Council. Draft Standards are approved in accordance with ISO procedures requiring at least 75% approval by the member countries voting. Each member country has one vote. (Reference 5)

6.1.2.1 ISO Geometrical Product Specification Masterplan

Many of the ISO standards that are related to dimensioning contain duplications, contradictions and gaps in the definition of particular topics. For instance, Tolerance of Position is described in at least four ISO standards (#1101, 2692, 5458, 10578).

The ISO technical report (#TR 14638), *Geometrical Product Specification (GPS) - Masterplan*, was published in 1995 as a guideline for the organization of the ISO standards and the proper usage of the standards at the appropriate stage in product development. The report contains a matrix model that defines the relationship among standards for particular geometric characteristics (e.g., size, distance, datums, and orientation) in the context of the product development process. The product development process is defined as a chain of six links (Chain Link 1-6) that progresses through design, manufacturing, inspection and quality assurance for each geometric characteristic. The intent of the matrix model is to ensure a common understanding and eliminate any ambiguity between standards. The general organization of the matrix model is shown in Table 6-3. (Reference 3)

Table 6-3 Organization of the matrix model from ISO technical report (#TR 14638)

The Fundamental GPS Standards	**The Global GPS Standards** GPS standards or related standards that deal with or influence several or all General GPS chains of standards.
	General GPS Matrix 18 General GPS Chains of Standards
	Complementary GPS Matrix Complementary GPS Chains of Standards A. Process Specific Tolerance Standards B. Machine Element Geometry Standards

6.2 Comparison of ASME and ISO Standards

Most worldwide dimensioning standards used in industry are based on either the ASME or ISO dimensioning standards. These two standards have emerged as the primary dimensioning standards. In the United States, the ASME standard is used in an estimated 90% of major corporations.

The ASME and ISO standards organizations are continually making revisions that bring the two standards closer together. Currently the ASME and ISO dimensioning standards are 60 to 70% common. It is predicted that in the next five years the two standards will be 80 to 90% common. Some industry experts predict that the two dimensioning standards will be merged into a single common standard sometime in the future. (Reference 5)

6.2.1 Organization and Logistics

An area of difference between ASME and ISO standards is in the organization and logistics of documentation. With regards to the approach to dimensioning in the ASME and ISO standards, the ASME standard uses product function as the primary basis for establishing tolerances. This is supported with numerous illustrated examples of tolerancing applications throughout the ASME standard. The ISO dimensioning standard is more theoretical in its explanation of tolerancing. It contains a limited number of generic examples that explain the interpretation of tolerances, with functional application a lesser consideration. Table 6-4 summarizes the differences between standards. (Reference 5)

Table 6-4 Differences between ASME and ISO standards

Item	ASME Y14.5M - 1994	ISO
Approach to dimensioning	Functional	Theoretical
Level of explanation	Thorough explanation and complementary illustrations	Minimal explanations, select examples
Number of standards	Single standard	Multiple Standards (15-20 separate publications)
Revision frequency	About every ten years	Select individual standards change yearly
Cost of standards	Less than $100 USD	$700 - $1000 USD

6.2.2 Number of Standards

The ASME and ISO organizations have a significantly different approach to documenting dimensioning and tolerancing standards. ASME publishes a single standard that explains all dimensioning and tolerancing topics. ISO publishes multiple standards on subsets of dimensioning and tolerancing topics. The relative advantages and disadvantages of each approach are presented in Table 6-5. (Reference 5)

6.2.3 Interpretation and Application

The differences in drawing interpretation and application as defined by the ASME and ISO standards are important to the user of dimensioning and tolerancing standards. Differences between the two standards, summarized in Tables 6-6 through 6-13, are organized into the following eight categories:
1. General: Tables 6-6 A through 6-6 F
2. Form: Tables 6-7 A through 6-7 B
3. Datums: Tables 6-8 A through 6-8 D
4. Orientation: Tables 6-9
5. Tolerance of Position: Tables 6-10 A through 6-10 D
6. Symmetry: Table 6-11
7. Concentricity: Table 6-12
8. Profile: Tables 6-13 A through 6-13 B

Table 6-5 Advantages and disadvantages of the number of ASME and ISO standards

Standard	Advantages	Disadvantages
ASME Y14.5M - 1994 *Single Standard*	All the information on dimensioning and tolerancing is contained in one document.	A larger document takes more time to create and revise than does a shorter document.
	Relatively infrequent revisions allow industry to thoroughly integrate the standard into the workforce.	If an error is in the document, it will be around for a long time.
	Ensures that the terms and concepts are at the same revision level at the time of publication.	
	Easy to specify and understand which standards apply to a drawing for dimensioning and tolerancing.	
ISO *Multiple Standards*	Shorter documents can be created and revised in less time than a longer document.	Industry needs adequate time to integrate new standards into the workforce. Training, software development, and multiple standards all require time to address.
	Additional topics can be added without revising all the existing standards.	New or revised standards may introduce terms or concepts that conflict with other existing standards.
		Multiple standards have multiple revision dates.
		Can be difficult to determine which standards apply to a drawing.
		One belief is the ISO standards that are in effect on the date of the drawing are the versions that apply to the drawing. This method is indirect, and many drawing users do not know which standards are in effect for a given date.

Differences include those of interpretation, items or allowances in one standard that are not allowed in the other, differences in terminology and drawing conventions.

6.2.3.1 ASME

The ASME standard referenced in Tables 6-6 through 6-13 is ASME Y14.5M-1994. The number in the parentheses represents the paragraph number from Y14.5M-1994. For example, (3.3.11) refers to paragraph 3.3.11 in ASME Y14.5M-1994.

6.2.3.2 ISO

The ISO standards referenced in Tables 6-6 through 6-13 are:

ISO 1101-1983	ISO 8015-1985	ISO 10578-1992	ISO 1660-1987
ISO 5458-1987	ISO 10579-1993	ISO 2692-1988	ISO 5460-1985
ISO 129-1985	ISO 2768-1989	ISO 5459-1981	

The numbers in the parentheses represent the standard and paragraph number. For example, (#1101.14.6) refers to ISO 1101, paragraph 14.6.

Differences Between US Standards and Other Standards 6-7

Table 6-6A General

General

Reprinted by permission of Effective Training Inc.

Concept / Term	ASME Y14.5M-1994		ISO	
	SYMBOL OR EXAMPLE		SYMBOL OR EXAMPLE	
All around	⊙—Symbol	Symbolic means of indicating that a tolerance applies to surfaces all around the part in the view shown. (3.3.18)	None	Use a note
Basic dimension	▢	Basic dimension (1.3.9)	▢	Theoretically exact dimension (#1101, 10)
Between	↔	Symbolic means of indicating that a tolerance applies to a limited segment of a surface between designated extremities. (3.3.11)	None	Use a note
Controlled radius	CR	Tolerance zone defined by two arcs (the minimum and maximum radii) that are tangent to the adjacent surfaces. The part contour must be a fair curve without reversals. Radii taken at all points on the part contour must be within size limits. (2.15.2)	None	Use a note
Counterbore / Spotface	⌴	Symbolic means of indicating a counterbore or spotface. The symbol precedes, with no space, the dimension of the counterbore or spotface. (3.3.12)	None	Use a note
Countersink	⌵	Symbolic means of indicating a countersink. The symbol precedes, with no space, the dimension of the countersink. (3.3.13)	None	Dimensioned by showing either the required diametral dimension at the surface and the included angle, or the depth and the included angle. (#129, 6.4.2)

Table 6-6B General

General

Reprinted by permission of Effective Training Inc.

Concept / Term	ASME Y14.5M-1994		ISO	
	SYMBOL OR EXAMPLE		SYMBOL OR EXAMPLE	
Depth / Deep	▽	Symbolic means of indicating that a dimension applies to the depth of a feature. (3.3.14)	None	Use a note
Diameter symbol usage	⌀	Diameter symbol precedes all diametral values. (1.8.1)	⌀	Diameter symbol may be omitted where the shape is clearly defined. (#129, 4.4.4)
Extension (Projection) lines		Extension lines start with a short visible gap from the outline of the part (1.7.2)		Extension lines start from the outline of the part without any gap. (#129,4.2)
Feature control frame		Feature control frame (3.4)		Tolerance frame (#1101, 5.1)
Feature control frame placement		Feature: leader line drawn to the surface of the toleranced feature. (6.4.1.1.1) Feature of size: (To control axis or median plane) feature control frame is associated with the feature of size dimension. (6.4.1.1.2)		Feature: Tolerance frame connection to the toleranced feature by a leader line drawn to toleranced feature or extension of the feature outline. (#1101,6) Feature of size: (To control axis or median plane) Tolerance frame connection to the toleranced feature as an extension of a dimension line. (#1101,6)
Feature control frame placement		Common nominal axis or median plane: Each individual feature of size is toleranced separately. Note: Direction of arrow of leader line is not important.		Common nominal axis or median plane: Tolerance applies to the axis or median plane of all features common to the toleranced axis or median plane. Note: Direction of arrow of leader line defines the direction of the tolerance zone width. (#1101, 7)

Differences Between US Standards and Other Standards   6-9

Table 6-6C   General

General

Reprinted by permission of Effective Training Inc.

Concept / Term	ASME Y14.5M-1994		ISO	
	SYMBOL OR EXAMPLE		SYMBOL OR EXAMPLE	
General tolerances		General tolerances are not covered in Y14.5.	ISO 2768-(*)	When the note "ISO-2768-(*)" appears on a drawing a set of general tolerances are invoked for linear and angular dimensions without individual tolerances shown.(#2768,4,5)
				Unless otherwise stated, workpieces exceeding the general tolerance shall not lead to automatic rejection provided that the ability of the workpiece to function is not impaired.(#2768,6)
			* M, F, C, V	* A letter is shown to denote which set of tolerances apply from the standard.
Non-rigid part	None	Non-rigid parts do not require a designation. (6.8)		Non-rigid parts shall include the following indications as appropriate:
		Restraint note may be used for measurement of tolerances. (6.8.2)		A. "ISO 10579-NR" designation in or near the title block.
		"AVG" denotes average diameter for a form control verified in the free state. (6.8.3)	ISO 10579-NR	B. In a note, the conditions under which the part shall be restrained to meet the drawing requirements.
		A Ⓕ free state symbol may be used to denote a tolerance is checked in the free state (6.8.1)		C. Geometric variations allowed in the free state (by using Ⓕ)
				D. The conditions under which the geometric tolerances in this free state are achieved, such as direction of gravity, orientation of the part, etc.
Numerical notation	X.X	Decimal point (.) separates the whole number from the decimal fraction (1.6.3)	X,X	Comma (,) separates the whole number from the decimal fraction.

6-10   Chapter Six

Table 6-6D General

General

Reprinted by permission of Effective Training Inc.

Concept / Term	ASME Y14.5M-1994		ISO	
	SYMBOL OR EXAMPLE		SYMBOL OR EXAMPLE	
Radius	R	A radius is any straight line extending from the center to the periphery of a circle or sphere (2.15) Flats and reversals are allowed on the surface of a radius.	R	No formal definition in ISO standards.
Reference dimension	( )	Reference dimension (1.3.10)	( )	Auxiliary dimension (#129.3.1.1.3)
Regardless of feature size (RFS)	None Default per Rule #1 (S)	Rule #2, All applicable geometric tolerances: RFS applies, with respect to the individual tolerance, datum reference, or both, where no modifying symbol is specified. (by default) (2.8) Rule #2a, For a tolerance of position, RFS may be specified on the drawing with respect to the individual tolerance, datum reference, or both, as applicable. (2.8)	None Default	RFS by default (no exceptions) (#8015.5.2)
Screw threads	None	Pitch diameter rule: Each tolerance of orientation or position and datum reference specified for a screw thread applies to the axis of the thread derived from the pitch cylinder. (2.9, 2.10, 4.5.9)	None	None

Differences Between US Standards and Other Standards   6-11

Table 6-6E   General

General

Reprinted by permission of Effective Training Inc.

Concept / Term	ASME Y14.5M-1994		ISO	
	SYMBOL OR EXAMPLE		SYMBOL OR EXAMPLE	
Size / form control	None Default per Rule #1	Rule #1 (Taylor Principle): Controls both size and form simultaneously. The surface or surfaces of a feature shall not extend beyond a boundary (envelope) of perfect form at MMC. Exceptions: stock, such as bars, sheets, tubing, etc. produced to established standards; parts subject to free state variation in the unrestrained condition. Rule #1 holds for all engineering drawings specifying ANSI/ASME standards unless explicitly stated that Rule #1 is not required (2.7.1 - 2.7.2)	(E)	Principle of Independency: (ISO Default) Size control only - no form control. Form tolerance is additive to size tolerance. (#8015,4) Envelope Principle: Optional ISO specification with note/symbol equals ASME Rule #1. Envelope principle can be invoked for entire engineering drawings by stating such in a general note or title block; envelope principle can be applied to individual dimensions with the application of the appropriate symbol: an encircled capital letter (E). (#8015,6)
Square symbol usage	□	Symbol precedes the dimension with no space. (3.3.15)	□	Square symbol may be omitted where the shape is clearly defined. (#129, 4.4.4)
Statistical tolerance	⬡ST	Assigning of tolerances to related components of an assembly on the basis of sound statistics. (2.16) Symbolic means of indicating that a tolerance is based on statistical tolerancing. An additional note is required on the drawing referencing SPC. (3.3.10)	None	None

Table 6-6F General

General

Reprinted by permission of Effective Training Inc.

Concept / Term	ASME Y14.5M-1994		ISO	
	SYMBOL OR EXAMPLE		SYMBOL OR EXAMPLE	
Tangent plane modifier	⊤	Where it is desired to control a feature surface established by the contacting points of that surface, the tangent plane symbol is added in the feature control frame after the stated tolerance. (6.6.1.3)	None	None
Tolerance zones	(example with 0.1 A, Part, Tolerance zone)	The direction of the width of the tolerance zone is always normal to the nominal geometry of the part.	(example with 0.1 A, 80°)	The width of the tolerance zone is in the direction of the arrow of the leader line joining the tolerance frame to the toleranced feature, unless the tolerance value is preceded by the sign Ø. (#1101, 7.1) The default direction of the width of the tolerance zone is always normal to the nominal geometry of the part. The direction and width of the tolerance zone can be specified (#1101, 7.2-7.3)
View projection	⊕	Third angle projection (1.2)	⊖	First angle projection (#128)

Differences Between US Standards and Other Standards 6-13

Table 6-7A Form

Form

Reprinted by permission of Effective Training Inc.

Concept / Term	ASME Y14.5M-1994		ISO	
	SYMBOL OR EXAMPLE		SYMBOL OR EXAMPLE	
Flatness	(Example showing two surfaces with 0.08 flatness callout labeled "TWO SURFACES")	Flatness can only be applied to a single surface. (6.4.2) (Profile is used to control flatness / coplanarity of multiple surfaces (6.5.6.1))	(Example with 0.1 flatness on single surface) (Example with ABC 0.1 applied to surfaces A, B, C) (Example with COMMON ZONE 0.1 on features A, B) (Example with COMMON ZONE applied to features A, A)	Flatness can be applied to a single surface or flatness can have a single tolerance frame applied to multiple surfaces simultaneously. (#1101, 7.4) Flatness can have a single tolerance frame with toleranced feature indicators. (#1101, 7.4) Use of COMMON ZONE above the tolerance frame is used to indicate that a common tolerance zone is applied to several separate features. (#1101, 7.5)

Table 6-7B Form

Form

Reprinted by permission of Effective Training Inc.

Concept / Term	ASME Y14.5M-1994		ISO	
	SYMBOL OR EXAMPLE		SYMBOL OR EXAMPLE	
Form qualifying notes	No examples shown		⌭ 0,1 NOT CONVEX ⌭ 0,1 ⟵ NOT CONVEX	NOT CONVEX / NOT CONCAVE: Indications qualifying the form of the feature within the tolerance zone shall be written near the tolerance frame and may be connected by a leader line (#1101, 5.3)
Restrictive tolerance	Only allowed for geometrical tolerances without datum references. Straightness (6.4.1.1.4) Flatness (6.4.2.1.1) — | 0.4 — | 0.1/25		// | 0,1 | A | 0,05/200 |	If a smaller tolerance of the same type is added to the tolerance on the whole feature, but restricted over a limited length, the restrictive tolerance shall be indicated in the lower compartment. (#1101,9.2) Restricitve tolerances are allowed for geometrical toelrances with datum references.
Straightness applied to a planar feature of size	Straightness can be applied to a planar feature of size. The tolerance zone is two parallel planes. Each line element of the centerplane of the toleranced feature of size must lie within the tolerance zone. (6.4.1.1)		None	None

Differences Between US Standards and Other Standards 6-15

Datums

Reprinted by permission of Effective Training Inc.

Concept / Term	ASME Y14.5M-1994		ISO	
	SYMBOL OR EXAMPLE		SYMBOL OR EXAMPLE	
Centerpoint of a circle as a datum	None			A line element of the cylinder is used as the datum. (#5460, 5.3.1)
Common axis formed by two features		A single datum axis may be established by two coaxial diameters. Each diameter is designated as a datum feature and the datum axis applies when they are referenced as co-datums (A-B). (4.5.7.2)		A common axis can be formed by two features by placing the datum symbol on the centerline of the features.(#1101,8.2) (The Y14.5 method shown may also be used.)

Table 6-8A Datums

6-16   Chapter Six

Table 6-8B   Datums

Datums

Reprinted by permission of Effective Training Inc.

Concept / Term	ASME Y14.5M-1994		ISO	
	SYMBOL OR EXAMPLE		SYMBOL OR EXAMPLE	
Datum axis	*(symbols/examples)*	Datum symbol is placed on the extension line of a feature of size. OR Placed on the outline of a cylindrical feature surface or an extension line of the feature outline, separated from the size dimension. OR Placed on a dimension leader line to the feature of size dimension where no geometrical tolerance is used. OR Attached above or below the feature control frame for a feature or group of features.	*(symbols/examples)*	Datum symbol is placed on the centerline of a feature of size. OR Placed on the outline of a cylindrical feature surface or an extension line of the feature outline, separated from the size dimension.

Differences Between US Standards and Other Standards   6-17

Table 6-8C   Datums

Datums

Reprinted by permission of Effective Training Inc.

Concept / Term	ASME Y14.5M-1994		ISO	
	SYMBOL OR EXAMPLE		SYMBOL OR EXAMPLE	
Datum letter specified / implied	[// | 0.2 | A] with leader to [A]	Datum letter must be specified. (3.3.2)	[// | 0.2] with leader to box	If the tolerance frame can be directly connected with the datum feature by a leader line, the datum letter may be omitted. (#1101, 8.3)
Datum sequence	[A | B | C]	Primary, Secondary, or Tertiary must be specified. (4.4)	[A B C]	Primary, Secondary, Tertiary Ambiguous order allowed when datum sequence not important. (#1101, 8.4)
Datum target line	─ ─ ─ ─ ─ ─	Phantom line on direct view. Target point symbol on edge view. Both applications can be used in conjunction for clarity. (4.6.1.2)	✕ ─── ✕	Target point symbol on edge view. Two crosses connected by a thin continuous line (direct view). (#5459, 7.1.2)
Generating line as a datum	None	None	[A] with cylinder symbol	A line element of the cylinder is used as the datum. (#5460, 5.3.1)
Mathematically defined surface as a datum	None	Any compound geometry that can be mathematically defined and related to a three plane datum reference frame. (4.5.10.1)	None	None

Table 6-8D Datums

Datums

Reprinted by permission of Effective Training Inc.

Concept / Term	ASME Y14.5M-1994		ISO	
	SYMBOL OR EXAMPLE		SYMBOL OR EXAMPLE	
Median plane	(figure: datum symbol A on extension line of feature of size)	Datum symbol placed on the extension line of a feature of size.	(figure: datum symbol A on median plane)	Datum symbol is placed on the median plane. (#1101, 8.2)
		OR		OR
	(figure: datum symbol A on dimension leader line with XX XX)	Placed on a dimension leader line to the feature of size dimension where no geometrical tolerance is used.	(figure: datum symbol A on extension line of feature of size)	Placed on the extension line of a feature of size. (#1101, 8.2)
		OR		
	(figure: feature control frame ⌖ ⌀0.1 Ⓜ A B with datum attached)	Attached above or below the feature control frame for a feature or group of features. (3.3.2)	(figure: 4 Holes with ⌖ ⌀0.05 Ⓜ A B D and C datum)	Attached to the tolerance frame for a group of features as the datum. (#5459, 9)
Virtual condition datum		In Y14.5, the virtual condition of the datum axes includes the geometrical tolerance at MMC by default even though the MMC symbol is not explicitly applied. (4.5.4)		ISO practices that the datum axes should be interpreted as specified. Therefore if the virtual condition of the datum axes is to include the affect of the geometrical tolerance at MMC, the symbol must be explicitly applied to the tolerance.

Table 6-9   Orientation

Orientation

Reprinted by permission of Effective Training Inc.

Concept / Term	ASME Y14.5M-1994	ISO
Angular tolerances	**SYMBOL OR EXAMPLE** Angular tolerance controls both the general orientation of lines or line elements of surfaces and their form. All points of the actual lines or surface must lie within the tolerance zone defined by the angular tolerance. (2.12) All surface elements must be within the tolerance zone. (2.12)	**SYMBOL OR EXAMPLE** Angular tolerance controls only the general orientation of line elements of surfaces but not their form. The general orientation of the line derived from the actual surface is the orientation of the contacting line of ideal geometrical form. The maximum distance between the contacting line and the actual line shall be the least possible. (#8015, 5.1.2) Plane formed by the high points of the surface must be within the tolerance zone. (#8015, 5.1.2)

Differences Between US Standards and Other Standards   6-19

6-20 Chapter Six

Table 6-10A Tolerance of Position

Tolerance of Position
Reprinted by permission of Effective Training Inc.

Concept / Term	ASME Y14.5M-1994		ISO	
	SYMBOL OR EXAMPLE		SYMBOL OR EXAMPLE	
Composite positional tolerance	⊕ | ⌀0.5 Ⓜ | A | B | C | ⊕ | ⌀0.1 Ⓜ | A | B |	A composite application of positional tolerancing for the location of feature patterns as well as the interrelation (position and orientation) of features within these patterns. (5.4.1) The upper segment controls the location of the toleranced pattern. The lower segment controls the orientation and spacing within the pattern.	⊕ | ⌀0.5 Ⓜ | A | B | C | ⊕ | ⌀0.1 Ⓜ | A | B |	When a tolerance frame is as shown, it is interpreted as two separate requirements.
Extremities of long holes	8X ⌀ 12.8 12.5 ⊕ | ⌀0.5 Ⓜ | A | B | C | AT SURFACE C ⊕ | ⌀1 Ⓜ | A | B | C | AT SURFACE D	Different positional tolerances may be specified for the extremities of long holes; this establishes a conical rather than a cylindrical tolerance zone.	None	
Flat surface	None		[A] [B] 35 105° ⊕ | 0,05 | A | B |	Tolerance zone is limited by two parallel planes 0,05 apart and disposed symmetrically with respect to the theoretically exact position of the considered surface. (#1101, 14.10)

Differences Between US Standards and Other Standards   6-21

Table 6-10B Tolerance of Position

Tolerance of Position
Reprinted by permission of Effective Training Inc.

Concept / Term	ASME Y14.5M-1994	ISO
Line	**SYMBOL OR EXAMPLE** None	**SYMBOL OR EXAMPLE** [example with ⌖ 0,05 A] Tolerance zone is limited by two parallel straight lines 0,05 apart and disposed symmetrically with respect to the theoretically exact position of the considered line if the tolerance is specified only in one direction (#1101, 14.10)
Point	[⌖ SØ 0.8 A B] Only when applied to control a spherical feature. (5.2) Spherical tolerance zone. (5.15)	[example with ⌖ Ø 0,3] Tolerance zone is limited by two parallel straight lines 0,3 apart and disposed symmetrically with respect to the theoretically exact position of the considered line if the tolerance is specified only in one direction (#1101, 14.10)
Projected tolerance zone	[6X M20 X2-6H, ⌖ Ø 0.4 Ⓜ Ⓟ 35 A B C] The projected tolerance zone symbol is placed in the feature control frame along with the dimension indicating the minimum height of the tolerance zone. (3.4.7) [6X M20 X2-6H, ⌖ Ø 0.4 Ⓜ Ⓟ A B C] For clarification, the projected tolerance zone symbol may be shown in the feature control frame and a zone height dimension indicated with a chain line on a drawing view. The height dimension may then be omitted from the feature control frame. (2.4.7)	[example with 8X Ø25, ⌖ Ø 0.02 Ⓟ A B] The projected tolerance zone is indicated on a drawing view with the Ⓟ symbol followed by the projected dimension: represented by a chain thin double-dashed line in the corresponding drawing view, and indicated in the tolerance frame by the symbol Ⓟ placed after the tolerance value. (#1101,11;#10578,4)

6-22   Chapter Six

Table 6-10C Tolerance of Position

Tolerance of Position

Reprinted by permission of Effective Training Inc.

Concept / Term	ASME Y14.5M-1994	ISO
Simultaneous gaging requirement	SYMBOL OR EXAMPLE Where two or more features or patterns of features are located by basic dimensions related to common datum features referenced in the same order of precedence and the same material condition, as applicable, they are considered as a composite pattern with the geometric tolerances applied simultaneously (4.5.12)	Groups of features shown on same axis to be a single pattern (example has same datum references) (#5458, 3.4) SYMBOL OR EXAMPLE Unless otherwise stated by an appropriate instruction. (#5458, 3.4)

Table 6-10D Tolerance of Position

Tolerance of Position
Reprinted by permission of Effective Training Inc.

Concept / Term	ASME Y14.5M-1994		ISO	
	SYMBOL OR EXAMPLE		SYMBOL OR EXAMPLE	
Requirements for application	⊕	Basic dimensions to specified datums, position symbol, tolerance value, applicable material condition modifiers, applicable datum references (5.2)	⊕	Theoretically exact dimensions locate features in relation to each other or in relation to one or more datums. (#5458, 3.2) (No chain basic of dimensions necessary to datums.)
	None	None	4X Ø 6, ⊕ Ø 0.2, 16±0.5, 20	When the group of features is individually located by positional tolerancing and the pattern location by coordinate tolerances, each requirement shall be met independently. (#5458, 4.1)
Tolerance of position for a group of features	⊕ Ø 0.2 Y Z / ⊕ Ø 0.2 A	Separately-specified feature-relating tolerance, using a second single-segment feature control frame is used when each requirement is to be met independently. (5.4.1) Do not use composite positional tolerancing method for independent requirements.	⊕ Ø 0.2 Y Z / ⊕ Ø 0.2 A, 15, 30, A	When the group of features is individually located by positional tolerancing and the pattern location by positional tolerancing, each requirement shall be met independently. (#5458, 4.2)
True position	None	True position (1.3.36)	None	Theoretical exact position (#5458, 3.2)

Table 6-11 Symmetry

Symmetry

Reprinted by permission of Effective Training Inc.

Concept / Term	ASME Y14.5M-1994		ISO	
	SYMBOL OR EXAMPLE		SYMBOL OR EXAMPLE	
Symmetry	⌯	Can be applied to planar features of size. The tolerance zone is two parallel planes that control median points of opposed or correspondingly-located elements of two or more feature surfaces. (5.14) Symmetry tolerance and the datum reference can only apply RFS.	⌯	Can be applied to planar or diametrical features of size. (#1101, 14.12) The tolerance zone is two parallel planes. Controls the median plane of the toleranced feature. (#1101 14.12.1) (Equivalent to Y14.5 tolerance of position RFS) OR The tolerance zone is two parallel straight lines (when symmetry is applied to a diameter in only one direction) (#1101, 14.12.2) OR The tolerance zone is a parallelepiped (when symmetry is applied to a diameter in two directions) (#1101, 14.12.2) Can be applied at MMC, LMC, or RFS.

Differences Between US Standards and Other Standards   6-25

Table 6-12 Concentricity

Concentricity

Reprinted by permission of Effective Training Inc.

Concept / Term	ASME Y14.5M-1994		ISO	
	SYMBOL OR EXAMPLE		SYMBOL OR EXAMPLE	
Concentricity (Y14.5) **Coaxiality (ISO)**	◎	Can be applied to a surface of revolution about a datum axis. (5.12) Controls median points of the toleranced feature. (5.12) Can only apply RFS	◎	Can be applied to a surface of revolution or circular elements about a datum axis. Controls the axis or centerpoint of the toleranced feature. (#1101, 14.11.1) Can apply at RFS, MMC, or LMC. (#1101, 14.11.2, #2692, 8.2, #2692 Amd. 1, 4, fig B.4)

Table 6-13A Profile

Profile

Reprinted by permission of Effective Training Inc.

Concept / Term	ASME Y14.5M-1994		ISO	
	SYMBOL OR EXAMPLE		SYMBOL OR EXAMPLE	
Composite profile tolerance	⌓ | 0.8 | A | B | C | | 0.1 | A |	Application to control location of a profile feature as well as the requirement of form, orientation, and in some instances, the size of the feature within the larger profile location tolerance zone. (6.5.9.1)	None	Use a note
Direction of profile tolerance zone		The tolerance zone is always normal to the true profile (6.5.3)		The default direction of the width of the tolerance zone is normal to the true profile, however the direction can be specified. (#1101, 7.2 - 7.3 see *General: tolerance zones*, p.7)

6-26   Chapter Six

Table 6-13B   Profile

Profile

Reprinted by permission of Effective Training Inc.

Concept / Term	ASME Y14.5M-1994		ISO	
	SYMBOL OR EXAMPLE		**SYMBOL OR EXAMPLE**	
Profile tolerance zone	[R 50, 0.4 A example] Bilateral tolerance zone equal distribution [R 50, 0.1 example] Bilateral tolerance zone unequal distribution [0.4 A example] Unilateral tolerance zone	For profile of a surface and line, the tolerance value represents the distance between two boundaries equally or unequally disposed about the true profile or entirely disposed on one side of the profile (6.5.3)	[R 30, 0.03 example] [R 10, S Ø 0.03 example] [0.03 example with graph] [Ø 0.03 example]	For profile of a surface - the tolerance zone is limited by two surfaces enveloping spheres of diameter *t*, the centers of which are situated on a surface having the true geometric form (#1101, 14.6) For profile of a line - the tolerance zone is limited by two lines enveloping circles of diameter *t*, the centers of which are situated on a line having the true geometric form (#1101, 14.5) In both cases the zone is equally disposed on either side of the true profile of the surface (#1660, 4.2)

The information contained in Figs. 6-6 through 6-13 is intended to be a quick reference for drawing interpretation. Many of the tables are incomplete by intent and should not be used as a basis for design criteria or part acceptance. (References 2,3,4,5,7)

6.3 Other Standards

Although most dimensioning standards used in industry are based on either ASME or ISO standards, there are several other dimensioning and tolerancing standards in use worldwide. These include national standards based on ISO or ASME, US government standards, and corporate standards.

6.3.1 National Standards Based on ISO or ASME Standards

There are more than 20 national standards bodies (Table 6-14) and three international standardizing organizations (Table 6-15) that publish technical standards. (Reference 6) Many of these groups have developed geometrical standards based on the ISO standards. For example, the German Standards (DIN) have adopted several ISO standards directly (ISO 1101, ISO 5458, ISO 5459, ISO 3040, ISO 2692, and ISO 8015), in addition to creating their own standards such as DIN 7167. (Reference 2)

Table 6-14 A sample of the national standards bodies that exist

Country	National Standards Body
Australia	Standards Australia (SAA)
Canada	Standards Council of Canada (SCC)
Finland	Finnish Standards Association (SFS)
France	Association Française de Normalisation (AFNOR)
Germany	Deutches Institut fur Normung (DIN)
Greece	Hellenic Organization for Standardization (ELOT)
Ireland	National Standards Authority of Ireland (NSAI)
Iceland	Icelandic Council for Standardization (STRI)
Italy	Ente Nazionale Italiano di Unificazione (UNI)
Japan	Japanese Industrial Standards Committee (JISC)
Malaysia	Standards and Industrial Research of Malaysia (SIRIM)
Netherlands	Nederlands Nomalisatie-instituut (NNI)
New Zealand	Standards New Zealand
Norway	Norges Standardiseringsforbund (NSF)
Portugal	Instituto Portugues da Qualidade (IPQ)
Saudi Arabia	Saudi Arabian Standards Organization (SASO)
Slovenia	Standards and Metrology Institute (SMIS)
Sweden	SIS - Standardiseringen i Svergie (SIS)
United Kingdom	British Standards Institute (BSI)
United States	American Society of Mechanical Engineers (ASME)

Table 6-15 International standardizing organizations

Abbreviation	Organization Name
ISO	International Organization for Standardization
IEC	International Electrotechnical Commission
ITU	International Telecommunication Union

6.3.2 US Government Standards

The United States government is a very large organization with many suppliers. Therefore, using common standards is a critical part of being able to conduct business. The United States government creates and maintains standards for use with companies supplying parts to the government.

The Department of Defense Standard is approved for use by departments and agencies of the Department of Defense (DoD). The *Department of Defense Standard Practice for Engineering Drawing Practices* is created and maintained by the US Army Armament Research Group in Picatinny Arsenal, New Jersey. This standard is called MIL-STD-100G. The "G" is the revision level. This revision was issued on June 9, 1997. The standard is used on all government projects.

The *Department of Defense Standard Practice for Engineering Drawings Practices* (MIL-STD-100G) references ASME and other national standards to cover a topic wherever possible. The ASME Y14.5M - 1994 standard is referenced for dimensioning and tolerancing of engineering drawings that reference MIL-STD-100G. (Reference 5)

The MIL-STD-100G contains a number of topics in addition to dimensioning and tolerancing:
- Standard practices of the preparation of engineering drawings, drawing format and media for delivery
- Requirements for drawings derived from or maintained by Computer Aided Design (CAD)
- Definitions and examples of types of engineering drawings to be prepared for the DoD
- Procedures for the creation of titles for engineering drawings
- Numbering, coding and identification procedures for engineering drawings, associated lists and documents   referenced on these associated lists
- Locations for marking on engineering drawings
- Methods for revision of engineering drawings and methods for recording such revisions
- Requirements for preparation of associated lists

6.3.3 Corporate Standards

US and International standards are comprehensive documents. However, they are created as general standards to cover the needs of many industries. The standards contain information that is used by all types of industries and is presented in a way that is useful to most of industry. However, many corporations have found the need to supplement or amend the standards to make it more useful for their particular industry.

Often corporate dimensioning standards are supplements based on an existing standard (e.g., ASME, ISO) with additions or exceptions described. Typically, corporate supplements include four types of information:
- Choose an option when the standard offers several ways to specify a tolerance.
- Discourage the use of certain tolerancing specifications that may be too costly for the types of products produced in a corporation.

- Include a special dimensioning specification that is unique to the corporation.
- Clarify a concept, which is new or needs further explanation from the standard.

Often the Standards default condition for tolerances is to a more restrictive condition regardless of product function. Corporate standards can be used to revise the standards defaults to reduce cost based on product function. An example of this is the simultaneous tolerancing requirement in ASME Y14.5M - 1994 (4.5.12). The rule creates simultaneous tolerancing as a default condition for geometric controls with identical datum references regardless of the product function. Simultaneous tolerancing reduces manufacturing tolerances which adds cost to produce the part. Although, in some cases it may be necessary to have this type of requirement, it is often not required by the function of the part. Some corporate dimensioning standards amend the ASME Y14.5M - 1994 standard so that the simultaneous tolerancing rule is not the default condition.

Another example of a corporate standard is the Auto Industry addendum to ASME Y14.5M - 1994. In 1994, representatives from General Motors, Ford and Chrysler formed a working group sanctioned by USCAR to create an Auto Industry addendum to Y14.5M - 1994. The Auto Industry addendum amends the Y14.5M - 1994 standard to create dimensioning conventions to be used by the auto industry.

Many corporations are moving from using corporate standards to using national or international standards. An addendum is often used to cover special needs of the corporation. The corporate dimensioning addendums are often only a few pages long, in place of several hundred pages the corporate standards used to be. (Reference 5)

6.3.4 Multiple Dimensioning Standards

Multiple dimensioning standards are problematic in industry for three reasons:
- Because there are several dimensioning standards used in industry, the drawing user must be cautious to understand which standards apply to each drawing. Drawing users need to be skilled in interpreting several dimensioning standards.
- The dimensioning standards appear to be similar, so differences are often subtle, but significant. Drawing users need to have the skills to recognize the differences among the various standards and how they affect the interpretation of the drawing.
- Not only are there different standards, but there are multiple revision dates for each standard. Drawing users need to be familiar with each version of a standard and how it affects the interpretation of a drawing.

There are four steps that can be taken to reduce confusion on dimensioning standards. (Reference 5)

1. Maintain or have immediate access to a library of the various dimensioning standards. This applies to both current and past versions of standards.
2. Ensure each drawing used is clearly identified for the dimensioning standards that apply.
3. Develop several employees to be fluent in the various dimensioning standards. These employees will be the company experts for drawing interpretation issues. They should also keep abreast of new developments in the standards field.
4. Train all employees who use drawings to recognize which standard applies to each drawing.

6.4 Future of Dimensioning Standards

As the world evolves toward a global marketplace, there is a greater need to create common dimensioning standards. The authors predict a single global dimensioning standard will evolve in the future.

Product development is becoming an international collaboration among engineers, manufacturers, and suppliers. Members of a product development team used to be located in close proximity to one another, working together to produce a product. In the global marketplace, collaborating parties geographically separated by thousands of miles, several time zones, and different languages, must effectively define and/or interpret product specifications. Therefore it is becoming important to create a common dimensioning and tolerancing standard to firmly anchor product specifications as drawings are shared and used throughout the product lifecycle.

6.5 Effects of Technology

Technology has infiltrated all aspects of product development, from product design and development to the inspection of manufactured parts. Computer Aided Design (CAD) helps engineers design products as well as document and check their specifications. Coordinate Measuring Machines (CMMs) help inspect geometric characteristics of parts with respect to their dimensions and tolerances while reducing the subjectivity of hand gaging.

A single dimensioning standard would effectively increase the use and accuracy of automated tools such as CAD and CMM. CAD software with automated GD&T checkers would require less maintenance by computer programmers to keep standards information current if they were able to concentrate on a single common standard.

To increase the use of automated inspection equipment such as a CMM, a more math-based dimensioning and tolerancing standard is required. Only math-based standards are defined to the degree necessary to eliminate ambiguity during the inspection process.

6.6 New Dimensioning Standards

One possible future for Geometric Dimensioning and Tolerancing is a new standard for defining product specifications without symbols, feature control frames, dimensions or tolerances that can be read from a blueprint. Instead, there may come a time when all current GD&T information can be incorporated into a 3-D computer model of the part. The computer model would be used directly to design, manufacture and inspect the product. An ASME subcommittee is currently working on standard Y14.41 that would define just such a standard.

6.7 References

1. DeRaad, Scott, and Alex Krulikowski. 1997. *Quick Comparison of Dimensioning Standards - 1997 Edition.* Wayne, Michigan: Effective Training Inc.
2. Henzold, G. 1995. *Handbook of Geometrical Tolerancing - Design, Manufacturing and Inspection.* Chichester, England: John Wiley & Sons Ltd.
3. International Standards Organization. 1981-1995. "Various GD&T Standards" International Standards Organization: Switzerland.
   ISO 1101-1983   ISO 8015-1985   ISO 10578-1992
   ISO 1660-1987   ISO 5458-1987   ISO 10579-1993
   ISO 2692-1988   ISO 5460-1985   ISO 129-1985
   ISO 2768-1989   ISO 5459-1981   ISO TR 14638-1995
4. Krulikowski, Alex. 1998. *Fundamentals of Geometric Dimensioning and Tolerancing, 2ed.* Detroit, Michigan: Delmar Publishers.

5. Krulikowski, Alex. 1998. *Advanced Concepts of GD&T*. Wayne, Michigan: Effective Training Inc.
6. Other Web Servers Providing Standards Information. June 17, 1998. In http://www.iso.ch/infoe/stbodies.html. Internet.
7. The American Society of Mechanical Engineers. 1995. *ASME Y14.5M-1994, Dimensioning and Tolerancing*. New York, New York: The American Society of Mechanical Engineers.

Chapter 7

Mathematical Definition of Dimensioning and Tolerancing Principles

Mark A. Nasson
Draper Laboratory
Cambridge, Massachusetts

Mr. Nasson is a principal staff engineer at Draper Laboratory and has twenty years of experience in precision metrology, dimensioning and tolerancing, and quality management. Since 1989, he has been a member of various ASME subcommittees pertaining to dimensioning, tolerancing, and metrology, and presently serves as chairman of the ASME Y14.5.1 subcommittee on mathematical definition of dimensioning and tolerancing principles. Mr. Nasson is also a member of the US Technical Advisory Group (TAG) to ISO Technical Committee (TC) 213 on Geometric Product Specification. Mr. Nasson is an ASQ certified quality manager.

7.1 Introduction

This chapter describes a relatively new item on the dimensioning and tolerancing standards scene: mathematically based definitions of geometric tolerances. You will learn how and why such definitions came to be, how to apply them, what they have accomplished for us, and where these definitions may take us in the not-too-distant future.

7.2 Why Mathematical Tolerance Definitions?

After reading this chapter, I hope and trust that you will be asking the reverse question: Why *not* mathematical definitions of tolerances? As you will see, a number of interesting events combined to open the door for their creation. In short, though, mechanical tolerancing is a much more complex discipline than most people realize, and it requires a similar level of treatment as has proven to be necessary for the nominal geometric design discipline (CAD/solid modeling).

Although the seeds for mathematical tolerance definitions were planted well before the early 1980s, a special event of that era indirectly helped trigger a realization of their need. The arrival of the personal computer quite suddenly and dramatically decreased the cost of computing power. As a result, vendors of metrology equipment, predominantly coordinate measuring machines (CMMs) began offering affordably priced measurement systems with integrated personal computers. Also, a number of individuals developed homegrown systems for their companies (as did this author) by pairing an older measuring system that they already owned with a newly purchased personal computer. Just as personal computers have affected us in countless other ways, they also contributed to the resurgence of the coordinate measuring machine.

Another device also contributed to the resurgence of coordinate measuring machines: the touch trigger probe, originally developed in the U.K. by Renishaw. Prior to this invention, conventional coordinate measuring machines used a "hard" probe (a steel sphere) for establishing contact with part features. Not only were hard probes slow to use, but they also were capable of disturbing the part, and even damaging it if the inspector failed to exercise sufficient care. Touch probes improved this state of affairs by enabling the coordinate measuring machine to significantly overtravel after the part feature was triggered upon initial contact. Productivity and accuracy were both improved with touch probes.

The advent of touch probe technology and the availability of relatively inexpensive computing power through new microprocessors enabled quick and sophisticated collection, processing, and display of measurement data. That was the good news of the early 1980s. The bad news? The many instances of software applications developed for metrology equipment did not interpret geometric dimensioning and tolerancing uniformly. Although the personal computer helped us recognize a number of underlying problems with tolerancing and metrology (and hence, for much of manufacturing), other key events helped us further diagnose problems and even chart out plans for resolving them. Writing and using mathematical tolerance definitions were among the suggested corrective actions.

7.2.1 Metrology Crisis (The GIDEP Alert)

In September of 1988, Mr. Richard Walker of Westinghouse Corp. issued a GIDEP Alert[1] against the data reduction software from five unnamed CMM vendors. Himself aware of inconsistency problems with CMM software for some time through painful experience, Mr. Walker sought to bring this serious state of affairs to public light by issuing the GIDEP Alert. Typically, GIDEP issues alerts against specific manufacturer's product lines or production lots with quality concerns. In this case, the problem was not attributable to just one CMM vendor; this was an industry-wide problem and was not confined to the metrology industry. It was a serious symptom of a larger problem. First, though, let's deal with the subject of the GIDEP Alert.

Ideally, and not unreasonably, we expect that a measurement process for a given part (say flatness as measured by a CMM) will yield repeatable results. The degree of repeatability depends on many factors such as the number of points sampled, point sampling strategy, stability of the part, and probing force. Each of these factors comes into play on measurements performed on a single, given CMM.

---

[1] GIDEP (Government-Industry Data Exchange Program, http://www.gidep.corona.navy.mil) is an organization of government and industry participants who share technical information with each other regarding product research, design, development, and production. One function of GIDEP is to issue alerts to its members that pertain to nonconforming parts, processes, etc. In this case, the subjects of the alert were nonconforming software algorithms.

But what about the repeatability of measurements of the same part as performed by CMMs from different manufacturers? Potential contributors to repeatability in this context are the differences in mechanical stability between the CMMs *and the software algorithms used to process the sampled point coordinate data*. It's the latter with which Mr. Walker's GIDEP Alert dealt. Suspicious of inconsistencies between measurement results obtained by different CMMs, Mr. Walker crafted ingeniously simple, but strategically chosen sets of point coordinate data to test the performance of CMM software algorithms for calculating measured values of flatness, parallelism, straightness, and perpendicularity. A data set that could be solved graphically without any algorithms was strategically selected. So not only did Mr. Walker check for consistency between the five CMMs tested, but he also checked for *correctness*.

The results were rather shocking. The worst offending algorithm in one case reported results that were 37% worse than the actual results; in other words, the algorithm indicated that the part feature was worse than it actually was. In another case, the worst offending algorithm reported results that were 50% better than the actual results, indicating that the part feature was better than it actually was. These results led to the realization that many CMM software algorithms were unreliable. Coupling this fact with an increasingly wide awareness that different measurement techniques applied to the same parameter yielded different results, a true metrology crisis was in effect.

In true Ralph Nader spirit, Mr. Walker acted on behalf of the customers of metrology equipment vendors. Rather than letting the potential impact on the CMM vendors determine how he handled this discovery, he publicized this information to educate and warn CMM users and the customers of their results. He resisted those that preferred him to keep silent while these problems were solved behind closed doors. Instead, the GIDEP Alert served as a beacon to those who experienced similar problems and had the motivation and technical ability to do something about it. Mr. Walker was criticized by many for his actions—a sure sign that he was on to something.

7.2.2 Specification Crisis

The GIDEP Alert convincingly illustrated the unstable situation with metrology software. However, it is crucial to recognize that the metrology crisis was actually a symptom of the true problem. The inherent ambiguity in the text-based definitions of mechanical tolerances enabled the writing of varied and incorrect computer algorithms for processing inspection data. Though text-based definitions seem to have served engineering well for many years, the robustness and rigor required by computerization has revealed a number of underlying problems. Without the ability to unambiguously specify and assign tolerance controls to mechanical parts, we cannot expect to be able to uniformly verify the adherence of actual parts to those specifications. Thus, one could accurately say that the specification crisis spawned the metrology crisis.

7.2.3 National Science Foundation Tolerancing Workshop

Under a grant from the National Science Foundation, the ASME Board on Research and Development conducted a workshop with invited guests of varied manufacturing backgrounds from a number of domestic and international companies. Held soon after release of the GIDEP Alert, this workshop sought to identify research opportunities in the field of tolerancing of mechanical parts. These research opportunities were determined on the basis of unsolved problems or technological gaps hampering the effectiveness of various engineering disciplines. Among the recommendations generated by the workshop was that mathematically based definitions of mechanical tolerances should be written in order to remove ambiguities and reduce misuse. This recommendation paved the way for the establishment of a body whose sole purpose was to meet that goal.

7.2.4 A New National Standard

In January of 1989 the Y14.5.1 "ad hoc" subcommittee on mathematization of geometric tolerances held its inaugural meeting in Longboat Key, Florida. In approximately fifteen meetings held over five years' time, Chairman Richard Walker led an inspired group of volunteers to the publication of a new national standard, ASME Y14.5.1M-1994 - Mathematical Definition of Dimensioning and Tolerancing Principles. The continually surprising degree of effort that was necessary to write this document provided constant confirmation that the document was truly needed. Some ambiguities were known before mathematization efforts began, but many other subtle problems were revealed as the subcommittee members took on the challenge of unequivocally specifying what was previously conveyed through written word and figures drawn from specific examples.

7.3 What Are Mathematical Tolerance Definitions?

7.3.1 Parallel, Equivalent, Unambiguous Expression

Mathematical tolerance definitions are a reiteration of the tolerance definitions that appear in textual form in the Y14.5 standard. In many cases, actual mathematical expressions describe geometric constraints on regions of points in space yielding a mathematical/geometrical description of the tolerance zone for each tolerance type. However, tolerance types are only part of the story. The Y14.5.1 standard handles the crucial subject of datum reference frame construction not with mathematical equations, but with mathematical formulations that are expressed textually with supporting tables and logical expressions. In any case, the contents of the Y14.5.1 standard have a direct tracing to an unambiguous mathematical basis. The unfortunate tradeoff is that they are not readily assimilated by human beings, but they are easily converted into programming code.

7.3.2 Metrology Independent

The developers of the Y14.5.1 mathematical standard diligently maintained at arm's length (or farther!) any influences from current measurement techniques and technology on the mathematical tolerance definitions. There was a frequent tendency to think in terms of inspection procedures when trying to mathematically describe some characteristic of a geometric tolerance, but it was resisted. Measurability was never a criterion that prevailed during the deliberations of the Y14.5.1 subcommittee. The reason was simple: tolerancing is a design function, and it must not be encumbered by metrology, a downstream activity in the product life cycle. Today's state-of-the-art in measurement technology eventually becomes yesterday's obsolescence. Desired features and capabilities for dimensioning and tolerancing that enable precise specification of part functionality and producibility should drive technology development in metrology. To have specified mathematical tolerance definitions in terms of industry-accepted measurement techniques would surely have made the definitions more recognizable, but generality would have been sacrificed.

7.4 Detailed Descriptions of Mathematical Tolerance Definitions

7.4.1 Introduction

This section contains introductory material necessary to read and understand mathematical tolerance definitions as they appear in the Y14.5.1 standard. Those readers with a physics and/or mathematics background may bypass the section on vectors that follows. Section 7.4.3 presents some key terms and concepts specific to the Y14.5.1 standard. The remaining sections cover a selection of actual mathematical tolerance definitions. Note that not all aspects of the Y14.5.1 standard are covered here, and that this chapter is designed to provide the reader with enough background to enable him/her to make effective use of the standard.

7.4.2 Vectors

This section contains a brief overview of vectors and the manner in which they are handled in mathematical expressions. Those readers with a physics and/or mathematics background will not find it necessary to read further. The material is included, however, because not all users of geometric dimensioning and tolerancing have had exposure to it, and it is the basis of the definitions that follow.

Vectors are abstract geometric entities that describe direction and magnitude (length). A position vector can describe every point in space, which is simply a line drawn from the origin to the point. Vectors also exist between points in space. The magnitude of a vector is its length as measured from its starting point to its end point. A vector of arbitrary length is typically designated by a letter with an arrow ($\vec{A}$) over it. Graphically, vectors are shown as a line with an arrow at one end; the length of the line represents the vector's magnitude, while the arrow represents its direction. See Fig. 7-1.

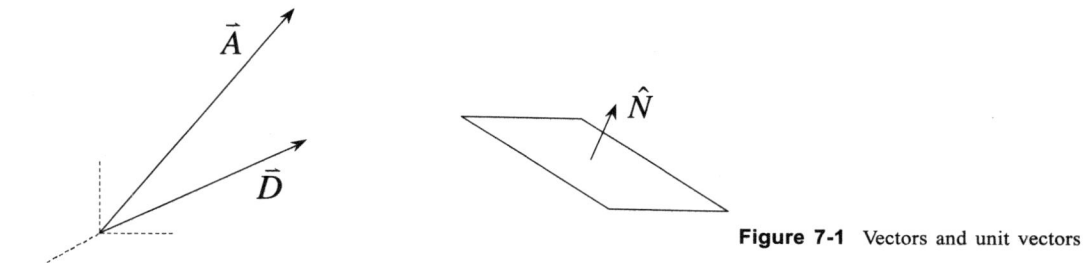

Figure 7-1 Vectors and unit vectors

A special type of vector is the *unit* vector which, not surprisingly, is of unit length. Unit vectors are often used to define or specify the direction of an axis or the direction of a plane's normal; a unit vector is appropriate for such purposes because it is the direction and not the magnitude that is important. A unit vector is typically designated by a letter with a hat, or carat, ($\hat{T}$) over it.

7.4.2.1 Vector Addition and Subtraction

Vectors may be added and subtracted to create other vectors. Two vectors are added by overlapping the starting point of one vector on the end point of the other vector. The resultant vector, or sum vector, is that vector that extends from the starting point of the first vector to the end point of the second vector. See Fig. 7-2.

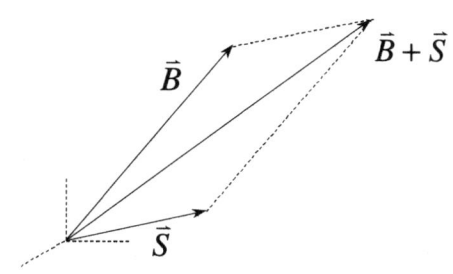

Figure 7-2 Vector addition

Vector subtraction is performed analogously. In Fig. 7-3, the vector $\vec{C} - \vec{R}$ is obtained by adding the negative of vector $\vec{R}$ (which simply points in the opposite direction as $\vec{R}$) to vector $\vec{C}$.

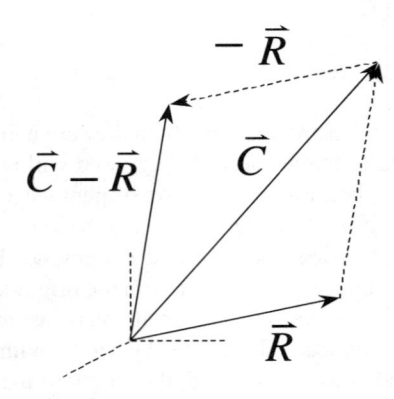

Figure 7-3 Vector subtraction

Vectors may be translated in space without affecting their behavior in mathematical expressions, so long as their length and direction are preserved. For instance, it is common to draw a difference vector as starting at the end point of the "subtrahend" vector ($\bar{R}$ in Fig. 7-3) and ending at the end point of the "minuend" vector ($\bar{C}$ in Fig. 7-3).

7.4.2.2 Vector Dot Products

Vectors may be multiplied in two different ways: by dot product and by cross product. Rules for vector products are different than for products between numbers. Dot products and cross products always involve two vectors. Cross products are discussed in the next section.

The result of a dot product is always a scalar, which is just a fancy term for a number. A dot product is equal to the product of the numerical magnitude of the vectors, which in turn is multiplied by the cosine of the angle between the vectors. The mathematical expression for the dot product between vectors $\bar{A}$ and $\bar{B}$ is $\bar{A} \bullet \bar{B}$. Naturally, for two unit vectors that are $45°$ apart, their dot product is $(1)(1)\cos(45) = 0.707$. Also, when two vectors have a dot product that equals 0, they must be perpendicular, regardless of their magnitude, because the cosine of $90°$ is 0. And when two unit vectors have a dot product equal to 1, they must be parallel because the cosine of $0°$ is 1. Two unit vectors that point in opposite directions yield a dot product of -1 because the cosine of $180°$ is -1.

When a vector is multiplied with a unit vector via a dot product, the result equals the length of the component of the original vector that is pointing in the direction of the unit vector. The mathematical definitions of geometric tolerances make use of these dot product characteristics.

7.4.2.3 Vector Cross Products

Unlike a vector dot product which yields a number, the result of a vector cross product is always another vector. The mathematical expression for the cross product between vectors $\bar{A}$ and $\bar{B}$ is $\bar{A} \times \bar{B}$, the result of which we will express as $\bar{C}$. By definition, vector $\bar{C}$ is perpendicular to the plane defined by the first two vectors. The magnitude of the vector $\bar{C}$ is equal to the product of the magnitudes of the vectors $\bar{A}$ and $\bar{B}$, which in turn is multiplied by the sine of the angle between $\bar{A}$ and $\bar{B}$. So when two unit vectors are perpendicular, their cross product is another unit vector that is perpendicular to the first two unit vectors;

this because the sine of 90° is 1. And when any two vectors are parallel (or antiparallel), their cross product is a vector of length 0 because the sine of 0° and 180° is 0. The mathematical definitions of geometric tolerances make use of these properties of vector cross products.

7.4.3 Actual Value/Measured Value

A subtle but important distinction exists between the actual value and the measured value of a quantity. Soon after beginning its work program, the Y14.5.1 subcommittee quickly recognized the need to clearly draw this distinction. An *actual value* of a measured quantity is the inherently true value. It is the value that would be obtained by a measurement process that is perfect in every way; that is, a measurement process that has no measurement error or uncertainty associated with it, and which makes use of all of the information that is contained in the item being measured (i.e., the infinite number of data points that a surface consists of). In less esoteric terms, it is the value that we always hope to obtain, but never really can. The actual value can never be obtained because every measurement process has some degree of error and uncertainty associated with it, however small. Moreover, discrete measurement techniques operate on a relatively small subset of the infinite number of points of which a surface is comprised. Even though we can never obtain the actual value, it is important to have a concrete definition of it as well as an understanding of the reasons for its elusiveness.

The *measured value* of a quantity is self-explanatory. Quite simply, it is the value generated by a measurement process. A measured value is an estimate of the actual value; it has an uncertainty associated with it. The goal of any measurement process is to obtain a measured value that approximates the actual value within some tolerable level of uncertainty. The uncertainty associated with a measurement process depends on many factors such as the quantity of data sampled, the data sampling strategy, environmental effects, and so on. This uncertainty is never zero, and the degree to which it is minimized amounts to an economic decision based on the time required to conduct the measurement and the expense of the personnel and equipment employed.

It is not uncommon for the distinction between the measured value and the actual value to become blurred, and this may occasionally contribute to miscommunications between design engineers and metrologists. Early on, the Y14.5.1 subcommittee wrestled with these notions and decided that the scope of its work concerned itself solely with actual values and not with measured values. (The issues surrounding measured values were to be taken up by another subcommittee.) That is not to say that mathematical definitions somehow enable us to obtain actual values. Rather, the mathematical definitions presented in the Y14.5.1 standard focus on the geometric controls that the various tolerance types exert on part features. Further, the tolerance types operate not only on actual, tangible part features, but also more importantly on conceptual models of those part features that exist only on drawings or CAD/solid model representations. The genesis of a manufactured product is a representation of the product that is repeatedly modified, typically involving tradeoffs, in response to various constraints upon it. Allowable geometric variation of the product is one constraint, and the intent of the Y14.5.1 subcommittee was to create mathematical definitions of tolerance types that would be applicable to this conceptual design stage of product development. Accordingly, the notion of an actual value is appropriate.

In fact, in writing mathematical definitions it was crucial to maintain this "separation of church and state" as it were. The potential difficulty in obtaining a reliable measured value of a tolerance was of little or no concern during the development of the Y14.5.1 standard. The philosophy is that it is more important to arm a design engineer with flexible tools to uniquely specify a tolerance design rather than to compromise that ability in favor of easing the eventual measurements required to prove conformance of an actual part to those tolerances. It is inappropriate to standardize tolerances around the state-of-the-art in metrology because it is continually changing.

7.4.4 Datums

7.4.4.1 Candidate Datums/Datum Reference Frames

Datums are geometric entities of perfect form that are derived from datum features specified on a drawing. The configuration of one or more datums as specified in a feature control frame results in a datum reference frame. A datum reference frame essentially amounts to a coordinate system that is located and oriented on the datum features of the part, and from which the location and orientation of other part features are controlled.

For two reasons, a given datum feature may yield more than one datum. Most easy to visualize is the situation whereby a primary datum feature of size is referenced at maximum material condition (MMC) and is manufactured at a size between its maximum material size and its least material size. By the rules of Y14.5, the datum may assume any size, location, and orientation between the datum feature and its MMC limit. These potentially numerous datums form a candidate datum set.

Another reason why a set of candidate datums may result from a given datum feature has to do with the fact that actual datum features, like all actual features, necessarily have form error. Form error often undermines the effectiveness of the rules that Y14.5 specifies in section 4.4.1 for associating perfect form datums to imperfect form datum features. These rules are ideally intended to isolate a single datum from a datum feature, but in practice they reduce the size of the candidate datum set, hopefully to a reasonable extent. For instance, consider a nominal flat surface specified as a primary datum, an actual instance of which has form error consisting of small raised areas scattered all over the surface in such a way that a conceptual, perfect form datum feature (a perfectly flat plane) does not engage the actual surface in just one, unique orientation. In fact, there are multiple sets of three raised areas that provide stable engagement. Each results in a potentially valid datum, and they collectively form the candidate datum set.

Thus, we say in general that a datum feature results in a set of candidate datums. Since each datum in a datum reference frame has (or may have) multiple candidate datums, there are potentially a multitude of candidate datum reference frames. What are we to do with all of these candidates? It is reasonable to conclude that one has the freedom to search among the candidate datum reference frame set for a datum reference frame that yields acceptable evaluations of all tolerances. One could also search for a datum reference frame that collectively minimizes (in some unspecified sense) the departure of all of the features controlled with respect to the datum reference frame. Regardless, if a datum reference frame can be found that yields acceptable evaluations of all tolerances, then the part is considered to be acceptable.

7.4.4.2 Degrees of Freedom

The balance of the discussion on datums will focus on degrees of freedom. A datum reference frame can be thought of as a coordinate system that is fixed to datum features on the part according to rules of association and precedence. If we think of a coordinate system as being represented by three mutually perpendicular axes, then the process of establishing a datum reference frame amounts to a series of positioning and orienting operations of these axes relative to datum features on the part. These positioning and orienting operations take place with respect to a fixed "world" coordinate system.

A datum reference frame has three positional degrees of freedom, and three orientational degrees of freedom within the world coordinate system. In other words, the origin of a datum reference frame may be independently located along three world coordinate system axes. Similarly, the three planes formed by the three pairs of datum reference frame axes have angular relationships to the three planes formed by pairs of world coordinate system axes. The establishment of a datum reference frame equates to a systematic reduction of its available degrees of freedom within the world coordinate system. A datum reference frame that has no available degrees of freedom is said to be fully constrained.

Note that it is not always necessary to fully constrain a datum reference frame. Consider a part that only has an orientation tolerance applied to a feature with respect to another datum feature. One can see that it is not necessary or productive to position the datum reference frame in any manner because the orientation of the feature with respect to the datum is not affected by location of the datum nor of the feature.

The rules of datum precedence embodied in Y14.5 can be expressed in terms of degrees of freedom. A primary datum may arrest one or more of the original six degrees of freedom. A secondary datum may arrest one or more additional *available* degrees of freedom; that is, a secondary datum may not arrest or modify any degrees of freedom that the primary datum arrested. A tertiary datum may also arrest any available degrees of freedom, though there may be none after the primary and secondary datums have done their job; in such a case, a tertiary datum is superfluous and can only add confusion.

The Y14.5.1 standard contains several tables that capture the finite number of ways that datum reference frames may be constructed using the geometric entities points, lines, and planes. Included are conditions between the primary, secondary, and tertiary datums for each case.

7.4.5 Form Tolerances

Form tolerances are characterized by the fact that the tolerance zones are not referenced to a datum reference frame. Form tolerances do not control the form of a feature with respect to another feature, nor with respect to a coordinate system established by other features. Form tolerances are often used to refine the inherent form control imparted by a size tolerance, but not always. Therefore, the mathematical definitions presented in this section reflect the *independent* application of form tolerances. The mathematical description of the net effect of simultaneously applied multiple tolerance types to a feature is not covered in this chapter.

Although form tolerances are conceptually simple, too many users of geometric dimensioning and tolerancing seem to attribute erroneous characteristics to them, most notably that the orientation and/or location of the tolerance zone are related to a part feature. As stated in the prior paragraph, form tolerances are independent of part features or datum reference frames. The mathematical definitions that appear below describe in vector form the geometric elements of the tolerance zones associated with form tolerances; these geometric elements are axes, planes, points, and curves in space. The description of these geometric elements must not be misconstrued to mean that they are specified up front as part of the *application* of a form tolerance to a nominal feature; they are not. The geometric elements of form tolerances are dependent only on the characteristics of the toleranced feature itself, and this is information that cannot be known until the feature actually exists and is measured.

7.4.5.1 Circularity

A circularity tolerance controls the form error of a sphere or any other feature that has nominally circular cross sections (there are some exceptions). The cross sections are taken in a plane that is perpendicular to some *spine*, which is a term for a curve in space that has continuous first derivative (or tangent). The circularity tolerance zone for a particular cross-section is an annular area on the cross-section plane, which is centered on the spine. Because circularity is a form tolerance, the tolerance zone is not related to a datum reference frame, nor is the spine specified as part of the tolerance application. Note that the circularity definition described here is consistent with the ANSI/ASME Y14.5M-1994 definition, but is not entirely consistent with the 1982 version of the standard. See the end of this section for a fuller explanation.

The mathematical definition of a circularity tolerance consists of equations that put constraints on a set of points denoted by $\bar{P}$ such that these points are in the circularity tolerance zone, and no others.

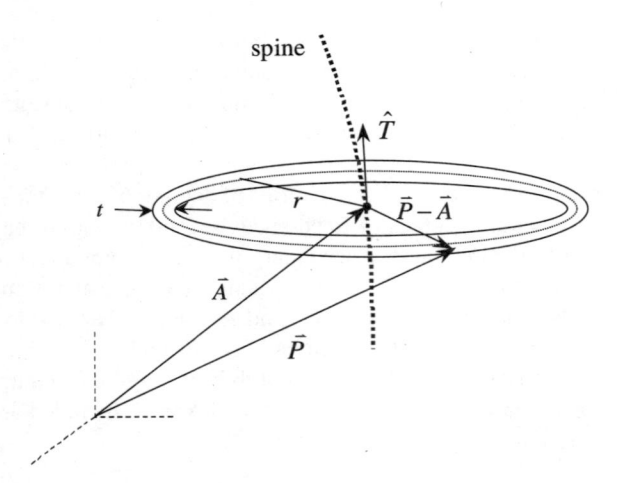

Figure 7-4 Circularity tolerance zone definition

Consider on Fig. 7-4 a point $\vec{A}$ on a spine, and a unit vector $\hat{T}$ which points in the direction of the tangent to the spine at $\vec{A}$.

The set of points $\vec{P}$ on the cross-section that passes through $\vec{A}$ is defined by Eq. (7.1) as follows.

$$\hat{T} \bullet (\vec{P} - \vec{A}) = 0 \qquad (7.1)$$

The zero dot product between the vectors $\hat{T}$ and $(\vec{P} - \vec{A})$ indicates that these vectors are perpendicular to one another. Since we know that $\hat{T}$ is perpendicular to the spine at $\vec{A}$, and $\vec{P} - \vec{A}$ is a vector that points from $\vec{A}$ to $\vec{P}$, then the points $\vec{P}$ must be on a plane that contains $\vec{A}$ and that is perpendicular to $\hat{T}$. Thus, we have defined all of the points that are on the cross section. Next, we need to restrict this set of points to be only those in the circularity tolerance zone.

As was stated above, the circularity tolerance zone consists of an annular area, or the area between two concentric circles that are centered on the spine. The difference in radius between these circles is the circularity tolerance t.

$$\left| |\vec{P} - \vec{A}| - r \right| \leq \frac{t}{2} \qquad (7.2)$$

Eq. (7.2) says that there is a reference circle at a distance r from the spine, and that the points $\vec{P}$ must be no farther than half of the circularity tolerance from it, either toward or away from the spine. This equation completes the mathematical description of the circularity tolerance zone for a particular cross section.

To verify that a measured feature conforms to a circularity tolerance, one must establish that the measured points meet the restrictions imposed by Eqs. (7.1) and (7.2). In geometric terms, one must find a spine that has the circularity tolerance zones that are created according to Eqs. (7.1) and (7.2), containing all of the measured points. The reader will likely find this definition of circularity foreign, so some explanation is in order.

As was stated earlier in this section, the details of circularity that are discussed here correspond to the ANSI/ASME Y14.5M-1994 standard, which contains some changes from the 1982 version. The 1982 version of the standard, as written, required that cross sections be taken perpendicular to a *straight* axis, and that the circularity tolerance zones be centered on that straight axis, thereby effectively limiting the application of circularity to surfaces of revolution. In order to expand the applicability of circularity tolerances to other features that have circular cross sections, such as tail pipes and waveguides, the definition of circularity was modified such that circularity controls form error with respect to a *curved* "axis" (a spine) rather than a straight axis. The 1994 standard preserves the centering of the circularity tolerance zone on the spine.

Unfortunately, the popular interpretation of circularity does not correspond to either the 1982 or the 1994 versions of Y14.5M. Rather, a metrology standard (B89.3.1-1972, Measurement of Out of Roundness) seems to have implicitly provided an alternative definition of circularity by virtue of the measurement techniques that it describes. The main difference between the B89 metrology standard and the Y14.5M tolerance definition standard is that the B89 standard does not require the circularity tolerance zone to be centered on the axis. Instead, various fitting criteria are provided for obtaining the "best" center of the tolerance zone for a given cross section. Without delving into the details of the B89.3.1-1972 standard, suffice it to say that the four criteria are least squares circle (LSC), minimum radial separation (MRS), maximum inscribed circle (MIC), and minimum circumscribed circle (MCC).

There is a rather serious geometrical ramification to allowing the circularity tolerance zone to "float." Consider in Fig.7-5 a three-dimensional figure known as an elliptical cylinder which is created by translating or extruding an ellipse perpendicular to the plane in which it lies. Obviously, such a figure has elliptical cross sections, but it also has perfectly *circular* cross sections if taken perpendicular to a properly titled axis.

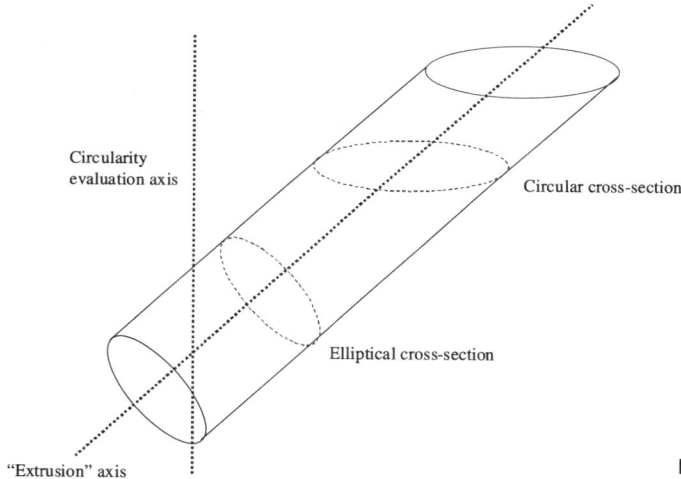

Figure 7-5 Illustration of an elliptical cylinder

Thus, a perfectly formed elliptical cylinder (even one with high eccentricity) would have no circularity error as measured according to the B89.3.1-1972 standard. Of course, any sensible, well-trained metrologist would intuitively select an axis for evaluating circularity that closely matches the axis of symmetry of the feature, and would thus find significant circularity error. However, as tolerancing and metrology progress toward computer-automated approaches (as the design and solid modeling disciplines already have), we must depend less and less on subjective judgment and intuition. It is for this reason that the relevant standards committees have recognized these issues with circularity tolerances and measurements, and they are working toward their resolution.

Creation of a mathematical definition of circularity revealed the inconsistency between the Y14.5M-1982 definition of circularity and common measurement practice as described in B89.3.1-1972, and also revealed subtle but potentially significant problems with the latter. This example illustrates the value that mathematical definitions have brought to the tolerancing and metrology disciplines.

7.4.5.2 Cylindricity

A cylindricity tolerance controls the form error of cylindrically shaped features. The cylindricity tolerance zone consists of a set of points between a pair of coaxial cylinders. The axis of the cylinders has no predefined orientation or location with respect to the toleranced feature, nor with respect to any datum reference frame. Also, the cylinders have no predefined size, although their difference in radii equals the cylindricity tolerance t.

We mathematically define a cylindricity tolerance zone as follows. A cylindricity axis is defined by a unit vector $\hat{T}$ and a position vector $\bar{A}$ as illustrated in Fig. 7-6.

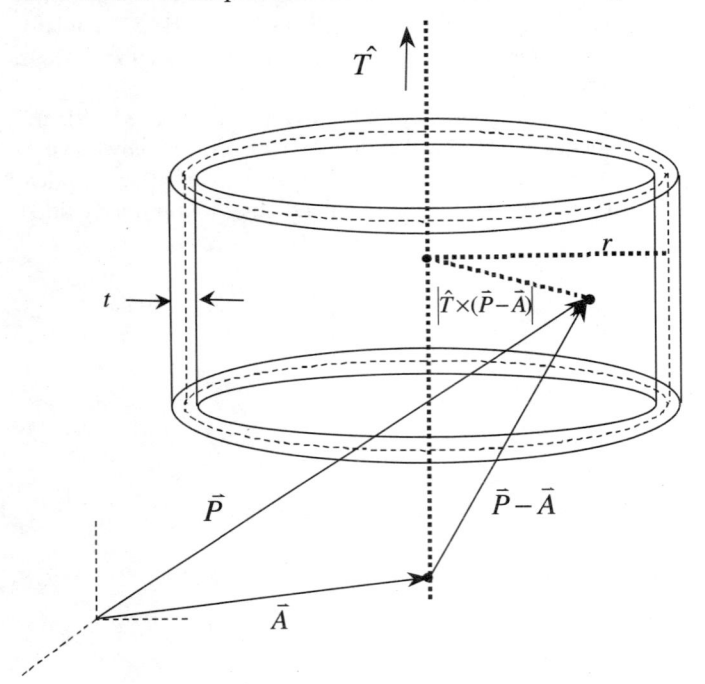

Figure 7-6 Cylindricity tolerance definition

If we consider the unit vector $\hat{T}$, which points parallel to the cylindricity axis, to be anchored at the end of the vector $\bar{A}$, one can see from Fig. 7-6 that the distance from the cylindricity axis to point $\bar{P}$ is obtained by multiplying the length of the unit vector $\hat{T}$ (equal to one by definition) by the length of the vector $\bar{P} - \bar{A}$, and by the sine of the angle between $\hat{T}$ and $\bar{P} - \bar{A}$. The mathematical operations just described are those of the vector cross product. Thus, the distance from the axis to a point $\bar{P}$ is expressed mathematically as $\left| \hat{T} \times (\bar{P} - \bar{A}) \right|$. To generate a cylindricity tolerance zone, the points $\bar{P}$ must be restricted to be between two coaxial cylinders whose radii differ by the cylindricity tolerance t.

Eq. (7.3) constrains the points $\bar{P}$ such that their distance from the surface of an imaginary cylinder of radius r is less than half of the cylindricity tolerance.

$$\left| \left| \hat{T} \times (\bar{P} - \bar{A}) \right| - r \right| \leq \frac{t}{2} \tag{7.3}$$

If, when assessing a feature for conformance to a cylindricity tolerance, we can find an axis whose direction and location in space are defined by $\hat{T}$ and $\vec{A}$, and a radius r such that all of the points of the actual feature consist of a subset of these points $\vec{P}$, then the feature meets the cylindricity tolerance.

7.4.5.3 Flatness

A flatness tolerance zone controls the form error of a nominally flat feature. Quite simply, the toleranced surface is required to be contained between two parallel planes that are separated by the flatness tolerance. See Fig. 7-7.

To express a flatness tolerance mathematically, we define a reference plane by an arbitrary locating point $\vec{A}$ on the plane and a unit direction $\hat{T}$ that points in a direction normal to the plane. The quantity

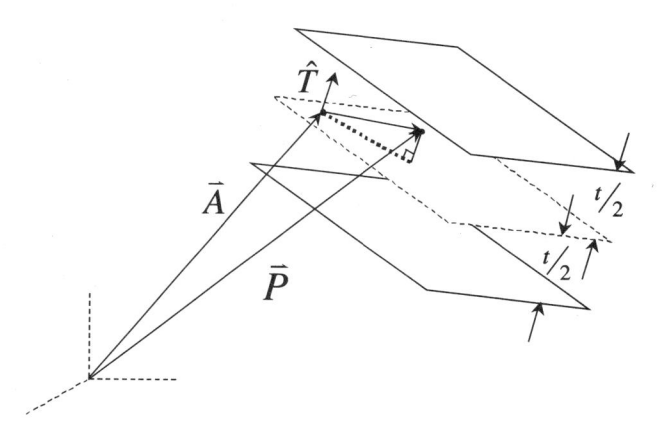

Figure 7-7 Flatness tolerance definition

$\vec{P} - \vec{A}$ is the vector distance from the reference plane's locating point to any other point $\vec{P}$. Of more interest though is the component of that distance in the direction normal to the reference plane. This is obtained by taking the dot product of $\vec{P} - \vec{A}$ and $\hat{T}$.

$$\left| \hat{T} \bullet (\vec{P} - \vec{A}) \right| \leq \frac{t}{2} \tag{7.4}$$

Eq. (7.4) requires that the points $\vec{P}$ be within a distance equal to half of the flatness tolerance from the reference plane.

In mathematical terms, to determine conformance of a measured feature to a flatness tolerance, we must find a reference plane from which the distances to the farthest measured point to each side of the reference plane are less than half of the flatness tolerance.

Note that Eq. (7.4) is not as general as it could be. The true requirement for flatness is that the *sum* of the normal distances of the most extreme points of the feature to each side of the reference plane be no more than the flatness tolerance. Stated differently, although Eq. (7.4) is not incorrect, there is no requirement that the reference plane equally straddle the most extreme points to either side. In fact, many coordinate measuring machine software algorithms for flatness will calculate a least squares plane through the measured data points and assess the distances to the most extreme points to each side of this plane. In general, the least squares plane will not equally straddle the extreme points, but it may serve as an adequate reference plane nevertheless.

7.5 Where Do We Go from Here?

Release of the Y14.5.1 standard in 1994 addressed one of the major recommendations that emanated from the NSF Tolerancing Workshop. However, the work of the Y14.5.1 subcommittee is not complete. The Y14.5.1 standard represents an important first step in increasing the formalism of geometric tolerancing, but many other things must happen before we can claim to have resolved the metrology crisis. The good news is that things are happening. Research efforts related to tolerancing and metrology have accelerated over the time frame since the GIDEP Alert, and we are moving forward.

7.5.1 ASME Standards Committees

Though five years have passed since the release of the Y14.5.1 standard, it is difficult to discern the impact that it has had on the practitioners of geometric tolerancing. However, the impact that it has had on the standards development scene is easier to measure. Advances in standards work are greatly facilitated when standards developers have a minimal dependence on subjective interpretations of the standardized materials. Indeed, it is the specific duty and responsibility of standards developers to define their subject matter in objectively interpretable terms; otherwise standardization is not achieved. The Y14.5.1 standard, and the philosophy that it embodies, provides a means for ensuring a lack of ambiguity in standardized definitions of tolerances.

Despite the alphanumeric subcommittee designation (Y14.5.1), which suggests that it sit below the Y14.5 subcommittee, the Y14.5.1 subcommittee has the same reporting relationship to the Y14 main committee, as does the Y14.5 subcommittee. The new Y14.5.1 effort was truly a parallel effort to that of Y14.5 (though certainly not entirely independent). Its value has been sufficiently demonstrated within the subcommittees to the extent that the leaders of each group are establishing a much closer degree of collaboration. The result will undoubtedly be better standards, better tools for specifying allowable part variation, less disagreement between suppliers and customers regarding acceptability of parts, and better and cheaper products.

7.5.2 International Standards Efforts

The impact of the Y14.5.1 standard extends to the international standards scene as well. Over the past few years, the International Organization for Standardization (ISO) has been engaged in a bold effort to integrate international standards development across the disciplines from design through inspection. As a participating member body to this effort, the United States has made its share of contributions. Among these contributions are mathematical definitions of form tolerances. These definitions are closely derived from the Y14.5.1 versions, but customized to reflect the particular detailed differences, where they exist, between the Y14.5 definitions and the ISO definitions. As other ISO standards are developed or revised, additional mathematical tolerance definitions will be part of the package.

7.5.3 CAE Software Developers

Aside from standards developers, computer aided engineering (CAE) software developers should be the key group of users of mathematical tolerance definitions. Recalling the lack of uniformity and correctness in CMM software as brought to light by the GIDEP Alert, it should not be difficult to see the need for programmers of CAE systems (including design, tolerancing, and metrology) to know the detailed aspects of the tolerance types and code their software accordingly. In some cases, this can be achieved by coding the mathematical expressions from the Y14.5.1 standard directly into their software.

We are not yet aware of the actual extent of usage of the mathematical tolerance definitions from the Y14.5.1 standard among CAE software developers. Where vendors of such software claim compliance to US dimensioning and tolerancing standards, customers should rightly expect that the vendor owns a copy of the Y14.5.1 standard and has ensured that its algorithms are consistent with its requirements. It might be reasonable to assume that this is not the case across the board, and it would be a worthy endeavor to determine the extent of any such lack of compliance. As of this writing, ten years have passed since the GIDEP Alert, and perhaps the time is right to see whether the situation has improved with metrology software.

7.6 Acknowledgments

The groundbreaking Y14.5.1 standard was the result of a collective effort by a team of talented and unique individuals with diverse but related backgrounds. This author was but one contributor to the effort, and I would like to sincerely thank the other contributors for their wit, wisdom, and camaraderie; I learned quite a lot from them through this process. Rather than list them here, I refer the reader to page v of the standard for their names and their sponsoring organizations. At the top of that list is Mr. Richard Walker who demonstrated notable dedication and leadership through several years of intense development.

Unlike many other countries, standards of these types in the United States are voluntarily specified and observed by customers and suppliers rather than mandated by government. Moreover, the standards are developed primarily with private funding by companies that have an interest in the field and have personnel with the proper expertise. These companies enable committee members to contribute to standards development by providing them with travel expenses for meetings and other tools and resources needed for such work.

7.7 References

1. British Standards Institution. 1989. *BS 7172, British Standard Guide to Assessment of Position, Size and Departure from Nominal Form of Geometric Features.* United Kingdom. British Standards Institution.
2. Hocken, R.J., J. Raja, and U. Babu. Sampling Issues in Coordinate Metrology. *Manufacturing Review* 6(4): 282-294.
3. James/James. 1976. *Mathematical Dictionary - 4th Edition.* New York, New York: Van Nostrand.
4. Srinivasan V., H.B. Voelcker, eds. 1993. *Proceedings of the 1993 International Forum on Dimensional Tolerancing and Metrology*, CRTD-27. New York, New York: The American Society of Mechanical Engineers.
5. The American Society of Mechanical Engineers. 1972. *ANSI B89.3.1 - Measurement of Out-of-Roundness.* New York, New York: The American Society of Mechanical Engineers.
6. The American Society of Mechanical Engineers. 1994. *ASME Y14.5 - Dimensioning and Tolerancing.* New York, New York: The American Society of Mechanical Engineers.
7. The American Society of Mechanical Engineers. 1994. *ASME Y14.5.1 - Mathematical Definition of Dimensioning and Tolerancing Principles.* New York, New York: The American Society of Mechanical Engineers.
8. Tipnis V. 1990. *Research Needs and Technological Opportunities in Mechanical Tolerancing,* CRTD-15. New York, New York: The American Society of Mechanical Engineers.
9. Walker, R.K. 1988. *CMM Form Tolerance Algorithm Testing, GIDEP Alert, #X1-A-88-01A.*
10. Walker R.K., V. Srinivasan. 1994. Creation and Evolution of the ASME Y14.5.1M Standard. *Manufacturing Review* 7(1): 16-23.

Chapter 8

Statistical Tolerancing

Vijay Srinivasan, Ph.D
IBM Research and Columbia University
New York

Dr. Vijay Srinivasan is a research staff member at the IBM Thomas J. Watson Research Center, Yorktown Heights, NY. He is also an adjunct professor in the Mechanical Engineering Department at Columbia University, New York, NY. He is a member of ASME Y14.5.1 and several of ISO/TC 213 Working Groups. He is the Convener of ISO/TC 213/WG 13 on Statistical Tolerancing of Mechanical Parts. He holds membership in ASME and SIAM.

8.1 Introduction

Statistical tolerancing is an alternative to worst-case tolerancing. In worst-case tolerancing, the designer aims for 100% interchangeability of parts in an assembly. In statistical tolerancing, the designer abandons this lofty goal and accepts at the outset some small percentage of failures of the assembly.

Statistical tolerancing is used to specify a population of parts as opposed to specifying a single part. Statistical tolerances are usually, but not always, specified on parts that are components of an assembly. By specifying part tolerances statistically the designer can take advantage of cancellation of geometrical errors in the component parts of an assembly — a luxury he does not enjoy in worst-case tolerancing. This results in economic production of parts, which then explains why statistical tolerancing is popular in industry that relies on mass production.

In addition to gain in economy, statistical tolerancing is important for an integrated approach to statistical quality control. It is the first of three major steps - specification, production, and inspection - in any quality control process. While national and international standards exist for the use of statistical methods in production and inspection, none exists for product specification. For example, ASME Y14.5M-1994 focuses mainly on the worst-case tolerancing. By using statistical tolerancing, an integrated statistical approach to specification, production, and inspection can be realized.

8-2　Chapter Eight

Since 1995, ISO (International Organization for Standardization) has been working on developing standards for statistical tolerancing of mechanical parts. Several leading industrial nations, including the US, Japan, and Germany are actively participating in this work which is still in progress. This chapter explains what ISO has accomplished thus far toward standardizing statistical tolerancing. The reader is cautioned that everything reported in this chapter is subject to modification, review, and voting by ISO, and should not be taken as the final standard on statistical tolerancing.

8.2　Specification of Statistical Tolerancing

Statistical tolerancing is a language that has syntax (a symbol structure with rules of usage) and semantics (explanation of what the symbol structure means). This section describes the syntax and semantics of statistical tolerancing.

Statistical tolerancing is specified as an extension to the current geometrical dimensioning and tolerancing (GD&T) language. This extension consists of a statistical tolerance symbol and a statistical tolerance frame, as described in the next two paragraphs. Any geometrical characteristic or condition (such as size, distance, radius, angle, form, location, orientation, or run-out, including MMC, LMC, and envelope requirement) of a feature may be statistically toleranced. This is accomplished by assigning an actual value to a chosen geometrical characteristic in each part of a population. Actual values are defined in ASME Y14.5.1M-1994. (See Chapter 7 for details about the Y14.5.1M-1994 standard that provides mathematical definitions of dimensioning and tolerancing principles.) Some experts think that statistically toleranced features should be produced by a manufacturing process that is in a state of statistical control for the statistically toleranced geometrical characteristic; this issue is still being debated.

The statistical tolerance symbol first appeared in ASME Y14.5M-1994. It consists of the letters ST enclosed within a hexagonal frame as shown, for example, in Fig. 8-1. For size, distance, radius, and angle characteristics the ST symbol is placed after the tolerances specified according to ASME Y14.5M-1994 or ISO 129. For geometrical tolerances (such as form, location, orientation, and run-out) the ST symbol is placed after the geometrical tolerance frame specified according to ASME Y14.5M-1994 or ISO 1101. See Figs. 8-2 and 8-3 for further examples.

The statistical tolerance frame is a rectangular frame, which is divided into one or more compartments. It is placed after the ST symbol as shown in Figs. 8-1, 8-2, and 8-3. Statistical tolerance requirements can be indicated in the ST frame in one of the three ways defined in sections 8.2.1, 8.2.2, and 8.2.3.

8.2.1　Using Process Capability Indices

Three sets of process capability indices are defined as follows.

- $Cp = \dfrac{U-L}{6\sigma}$,

- $Cpk = \min(Cpl, Cpu)$, where $Cpl = \dfrac{\mu - L}{3\sigma}$ and $Cpu = \dfrac{U - \mu}{3\sigma}$, and

- $Cc = \max(Ccl, Ccu)$ where $Ccl = \dfrac{\tau - \mu}{\tau - L}$ and $Ccu = \dfrac{\mu - \tau}{U - \tau}$.

In these definitions L is the lower specification limit, U is the upper specification limit, τ is the target value, μ is the population mean, and σ is the population standard deviation.

The process capability indices are nondimensional parameters involving the mean and the standard deviation of the population. The nondimensionality is achieved using the upper and lower specification limits. Cp is a measure of the spread of the population about the average. Cc is a measure of the location of the average of the population from the target value. Cpk is a measure of both the location and the spread of the population.

All of these five indices need not be used at the same time. Numerical lower limits for Cp, Cpk (or Cpu, Cpl) and numerical upper limit for Cc (or Ccu, Ccl) are indicated as shown in Fig. 8-1 using the $\geq$ and $\leq$ symbols. Cpu and Ccu are used instead of Cpk and Cc, respectively, for all geometrical tolerances (form, location, orientation, and run-out) specified at RFS (Regardless of Feature Size). The requirement here is that the mean and the standard deviation of the population of actual values should be such that all the specified indices are within the indicated limits.

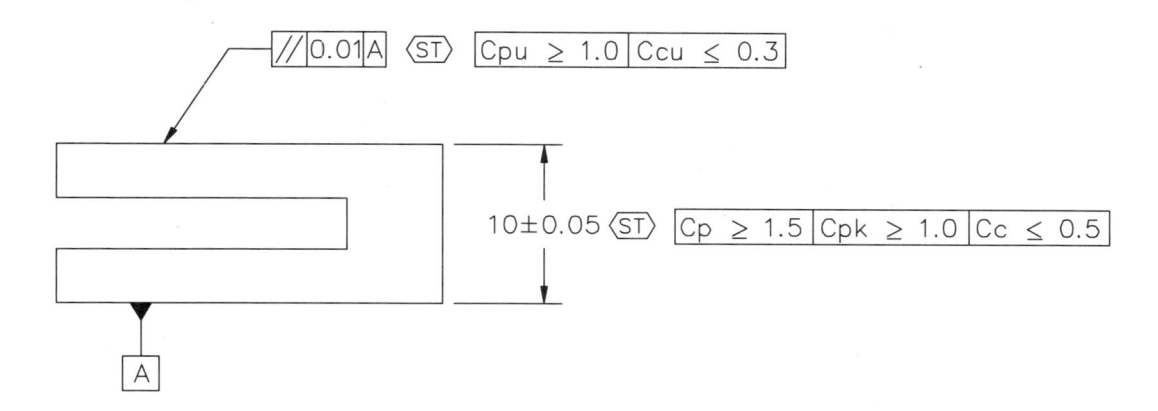

Figure 8-1 Statistical tolerancing using process capability indices

For the example illustrated in Fig. 8-1, the population of actual values for the specified size should have its Cp value at or above 1.5, Cpk value at or above 1.0, and Cc value at or below 0.5. For the indicated parallelism, the population of out-of-parallelism values (that is, the actual values for parallelism) should have its Cpu value at or above 1.0, and its Ccu value at or below 0.3.

Limits on the process capability indices also imply limits on the mean and the standard deviation of the population of actual values through the formulas shown at the beginning of this section. Such limits on μ and σ can be visualized as zones in the μ–σ plane, as described in section 8.3.1. To derive the limits on μ and σ, values of L, U, and τ should be obtained from the specification. For the example illustrated in Fig. 8-1, consider the size first. From the size specification, the lower specification limit $L = 9.95$, the upper specification limit $U = 10.05$, and the target value $\tau = 10.00$ because it is the midpoint of the allowable size variation. Next consider the specified parallelism, from which it can be inferred that $L = 0.00$, $U = 0.01$, and $\tau = 0.00$ because zero is the intended target value.

Using Cpl, Cpu, or Cpk in the ST tolerance frame implies only that these values should be within the limits indicated. Caution must be exercised in any further interpretation, such as the fraction of population lying outside the L and/or U limits, because it requires further assumption about the type of distribution, such as normality, of the population. Note that such additional assumptions are not part of the specification, and their invocation, if any, should be separately justified.

8-4   Chapter Eight

Process capability indices are used quite extensively in industrial production, both in the US and abroad, to quantify manufacturing process capability and process potential. Their use in product specification may seem to be in conflict with the time-honored "process independence" principle of the ASME Y14.5. This apparent conflict is false; the process capability indices do not dictate what manufacturing process should be used — they place demand only on some statistical characteristics of whatever process that is chosen.

Issues raised in the last two paragraphs have led to some rethinking of the use of the phrase "process capability indices" in statistical tolerancing. We will come back to this point in section 8.5, after the introduction of a powerful concept called population parameter zones in section 8.3.1.

8.2.2   Using RMS Deviation Index

RMS (root-mean-square) deviation index is defined as $Cpm = \dfrac{U-L}{6\sqrt{\sigma^2 + (\mu - \tau)^2}}$. A numerical lower limit for Cpm is indicated as shown in Fig. 8-2 using the $\geq$ symbol. The requirement here is that the mean and standard deviation of the population of actual values should be such that the Cpm index is within the specified limit.

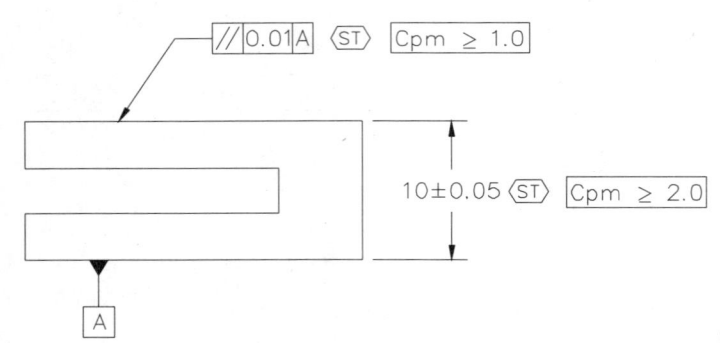

Figure 8-2   Statistical tolerancing using RMS deviation index

For the example illustrated in Fig. 8-2, the population of actual values for the size should have a Cpm value that is greater than or equal to 2.0. For the specified parallelism, the population of out-of-parallelism values (that is, the actual values for parallelism) should have a Cpm value that is greater than or equal to 1.0.

Cpm is called the RMS deviation index because $\sqrt{\sigma^2 + (\mu - \tau)^2}$ is the square root of the mean of the square of the deviation of actual values from the target value τ. Limiting Cpm also limits the mean and the standard deviation, and this can be visualized as a zone in the μ–σ plane. Section 8.3.1 describes such zones. To derive the limits on μ and σ, values for L, U, and τ should be obtained from the specification of Fig. 8-2 as explained in section 8.2.1.

Cpm is closely related to Taguchi's quadratic cost function, which states that the total cost to society of producing a part whose actual value deviates from a specified target value increases quadratically with the deviation. Specifying an upper limit for Cpm is equivalent to specifying an upper limit to the average cost of parts according to the quadratic cost function. This methodology is popular in some Japanese industries.

8.2.3 Using Percent Containment

A tolerance interval or upper limit followed by the P symbol and a numerical value of the percent ending with a % symbol is indicated as shown in Fig. 8-3. The tolerance range indicated inside the ST frame should be smaller than the tolerance range indicated outside the ST frame before the ST symbol. The requirement here is that the entire population of actual values should be contained within the limits indicated before the ST symbol; the percentage following the P symbol inside the ST frame indicates the minimum percentage of the population of actual values that should be contained within the limits indicated within the ST frame before the ST symbol; the remaining population should be contained in the remaining tolerance range proportionately.

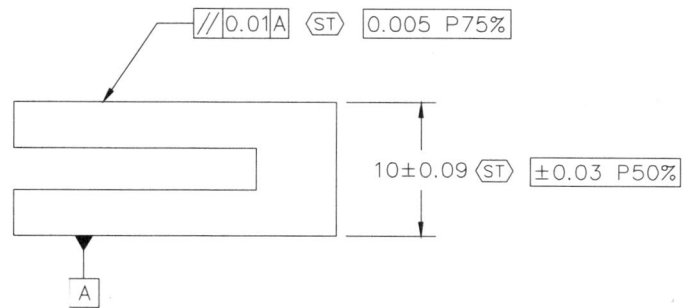

Figure 8-3 Statistical tolerancing using percent containment

In the example illustrated in Fig. 8-3 for the specified size, the entire population should be contained within 10 ± 0.09; at least 50% of the population should be contained within 10 ± 0.03; no more than 25% should be contained within $10^{-0.03}_{-0.09}$ and no more than 25% should be contained within $10^{+0.09}_{+0.03}$. For the specified parallelism, the entire population of out-of-parallelism values (that is, the actual values for the parallelism) should be less than 0.01 and at least 75% of this population of values should be less than 0.005.

Percent containment statements are best visualized using distribution functions. A distribution function, denoted $Pr[X \leq x]$, is the probability that the random variable X is less than or equal to a value x. Distribution functions are also known as cumulative distribution functions in some engineering literature. A distribution function is a nondecreasing function of x, and it varies between 0 and 1. It is possible to visually represent the percent containment requirements as zones that contain acceptable distribution functions, as shown in section 8.3.2.

Using percent containment is popular in some German industries. It is a simple but powerful way to indicate directly the percentage of populations that should lie within certain intervals.

8.3 Statistical Tolerance Zones

Statistical tolerance zone is a useful tool to visualize what is being specified and to compare different types of specifications. It is also a powerful concept that unifies several seemingly disparate practices of statistical tolerancing in industry today. A statistical tolerance zone can be either a population parameter

8-6   Chapter Eight

zone (PPZ) or distribution function zone (DFZ). PPZs are based on parametric statistics, and DFZs are based on nonparametric statistics.

8.3.1   Population Parameter Zones

A PPZ is a region in the mean - standard deviation plane, as shown in Fig. 8-4. In this example, the shaded PPZ on the left is the zone that corresponds to the statistical specification of size in Fig. 8-1, and the shaded PPZ on the right is the zone that corresponds to the statistical specification of parallelism in Fig. 8-1. Vertical lines that limit the PPZ arise from limits on Cc, Ccu or Ccl because they limit only the mean; the top horizontal line comes from limiting Cp because it limits only the standard deviation; the slanted lines are due to limits on Cpk, Cpu or Cpl because they limit both the mean and the standard deviation. If the (μ, σ) point for a given population of geometrical characteristics lies within the PPZ, then the population is acceptable; otherwise it is rejected.

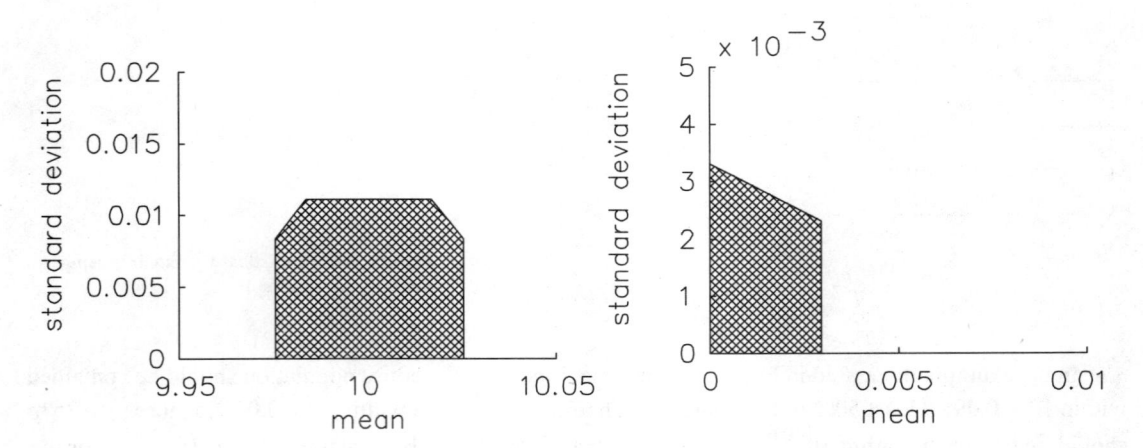

Figure 8-4 Population parameter zones for the specifications in Fig. 8.1

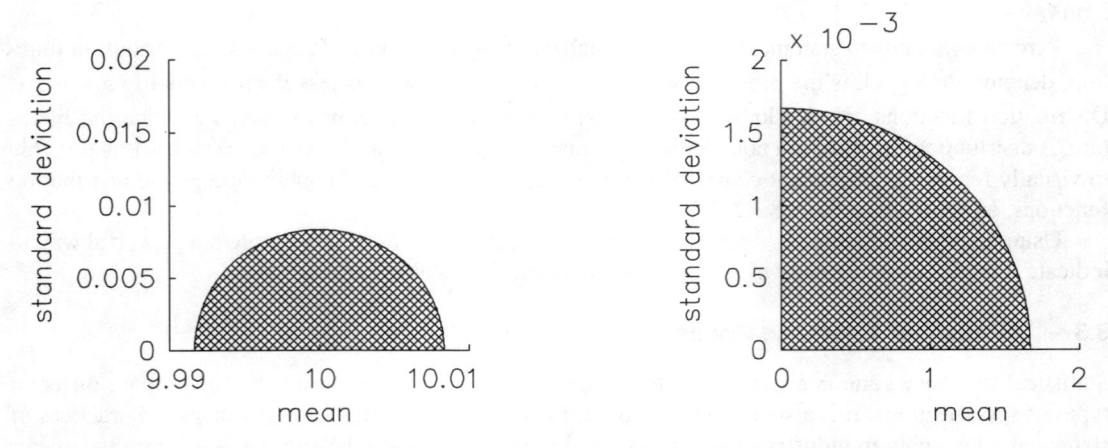

Figure 8-5 Population parameter zones for the specifications in Fig. 8.2

PPZs can be defined for specifications that use the RMS deviation index as well. Fig. 8-5 illustrates the PPZs for the specifications in Fig. 8-2. Here the zones are bounded by circular arcs. Again, the interpretation is that all (μ,σ) points that lie inside the zone correspond to acceptable populations, and points that lie outside the zone correspond to populations that are not acceptable per specification.

8.3.2 Distribution Function Zones

A DFZ is a region that lies between an upper and a lower distribution function, as shown in Fig. 8-6. Any population whose distribution function lies within the shaded zone is acceptable; if not, it is rejected.

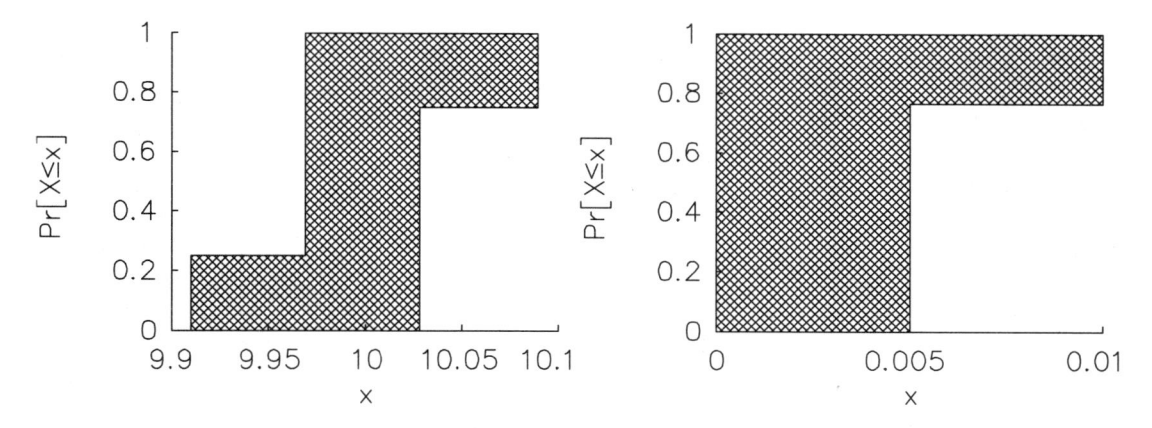

Figure 8-6 Population parameter zones for the specifications in Fig. 8.3

8.4 Additional Illustrations

Figs. 8-7 through 8-10 illustrate valid uses of statistical tolerancing in several examples. Though not exhaustive, these illustrations help in understanding valid specifications of statistical tolerancing.

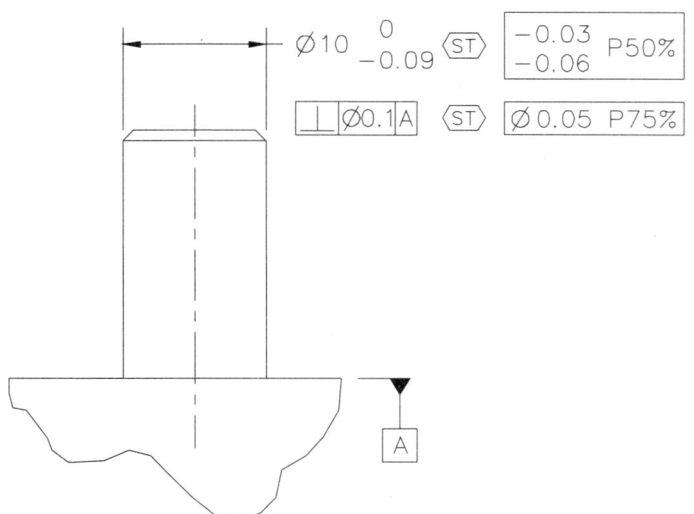

Figure 8-7 Additional illustration of specifying percent containment

8-8 Chapter Eight

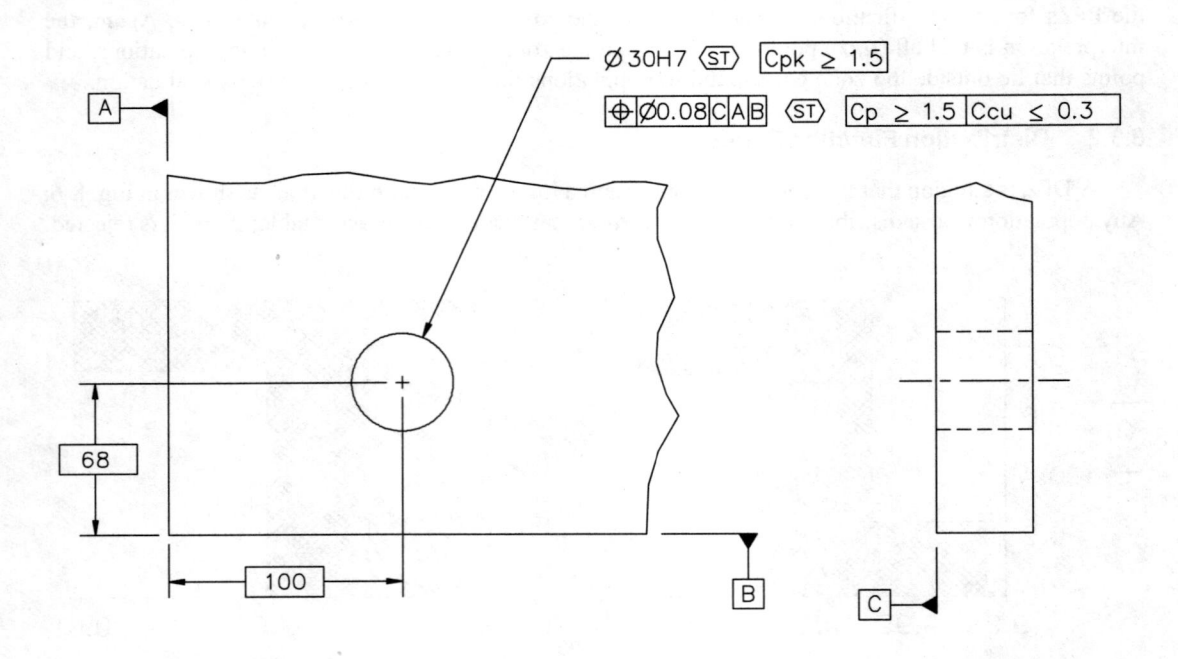

Figure 8-8 Illustration specifying process capability indices

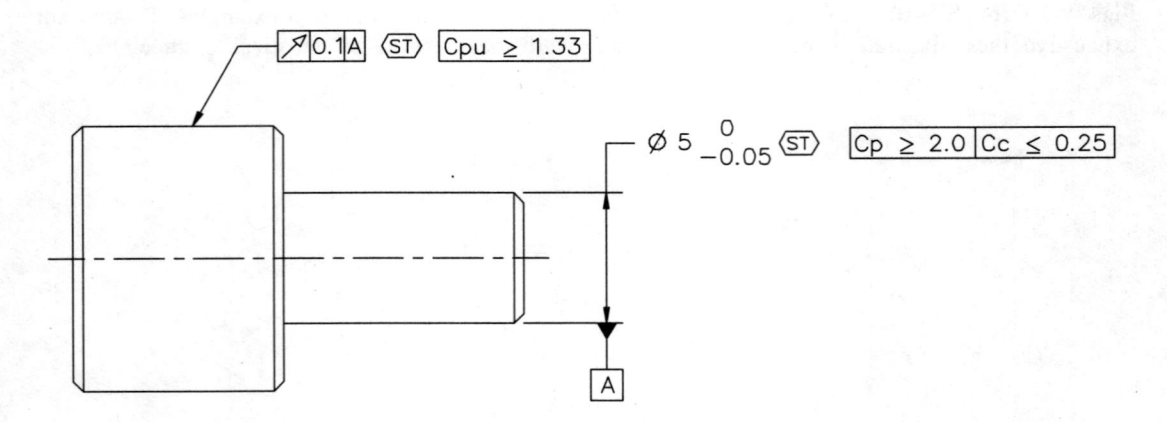

Figure 8-9 Additional illustration specifying process capability indices

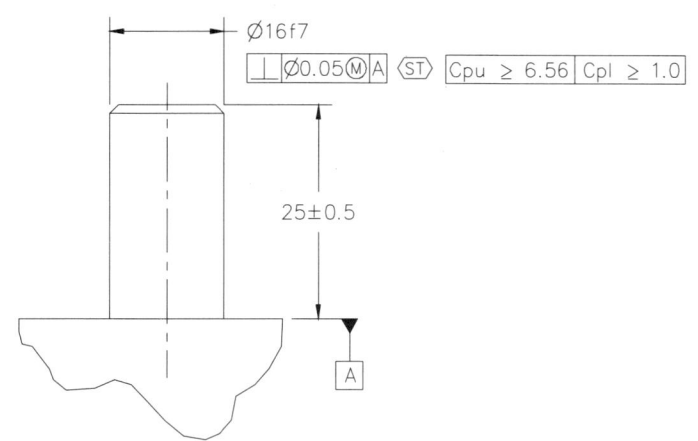

Figure 8-10 Illustration of statistical tolerancing under MMC

8.5 Summary and Concluding Remarks

This chapter dealt with the language of statistical tolerancing of mechanical parts. Statistical tolerancing is applicable when parts are produced in large quantities and assumptions about statistical composition of part deviations while assembling products can be justified. The economic case for statistical tolerancing can indeed be very compelling. In this chapter, three ways of indicating statistical tolerancing were described, and the associated statistical tolerance zones were illustrated. Population parameter zone (PPZ) and distribution function zone (DFZ) are the two most relevant new concepts that are driving the design of the ISO statistical tolerancing language.

Statistical tolerancing is deliberately designed as an extension to the current GD&T language. This has some disadvantages. It might be, for example, a better idea to indicate the statistical tolerance zones directly in the specifications. However, acceptance of statistical tolerancing by industry is greatly enhanced if it is designed as an extension to an existing popular language.

It was indicated earlier that some believe that statistically controlled parts should be produced by a manufacturing process that is in a state of statistical control. Strictly speaking, this is not a necessary condition for the success of statistical tolerancing. However, it is a good practice to insist on a state of statistical control, which can be achieved by the use of statistical process control methodologies for the manufacturing process. This is particularly true if a company has implemented just-in-time delivery, a practice in which one may not have the luxury of drawing a part at random from an existing bin full of parts. As mentioned in the body of this chapter, this issue is still being debated within ISO.

Similarly, there is a vigorous debate within ISO on the use of the phrase "process capability indices" indicated symbolically by Cp, Cpl, Cpu, Cpk, Ccl, Ccu, Cc, and Cpm. This debate is fueled by a current lack of ISO standardized interpretation of the meaning of these indices. To circumvent this controversy, these symbols may be replaced by Fp, Fpl, Fpu, Fpk, Fcl, Fcu, Fc, and Fpm, respectively, but without changing their functional relationship to L, U, μ, σ, and τ. The intent is to preserve the powerful notion of population parameter zones, which is an important concept for statistical tolerancing, while avoiding the use of the nonstandard phrase "process capability indices." This move may also open up the syntax to accept any user-defined function of population parameters.

A typical design problem is a tolerance allocation (also known as tolerance synthesis) problem. Here, given a tolerable variation in an assembly-level characteristic, the designer decides what are the tolerable variations in part-level geometrical characteristics. In general, this is a difficult problem. A more tractable problem is that of tolerance analysis, wherein given part-level geometrical variations the designer predicts what is the variation in an assembly-level characteristic. These are the types of problems that a designer faces in industry everyday. Both analytical and numerical (e.g., Monte-Carlo simulations) methods have been developed to solve the statistical tolerance analysis problem. Discussion of statistical tolerance analysis or synthesis is, however, beyond the scope of this chapter.

Acknowledgment and a Disclaimer

The author would like to express his deep gratitude to numerous colleagues who participated, and continue to participate, in the ASME and ISO standardization efforts. Standardization is a truly community affair, and he has merely reported their collective effort. Although the work described in this chapter draws heavily from the ongoing ISO efforts in standardization of statistical tolerancing, opinions expressed here are his own and not that of ISO or any of its member bodies.

8.6   References

1. Duncan, A.J. 1986. *Quality Control and Industrial Statistics*. Homewood, IL: Richard B.Irwin, Inc.
2. Kane, V.E. 1986. Process Capability Indices. *Journal of Quality Technology*, 18 (1), pp. 41-52.
3. Kotz, S. and N.L. Johnson. 1993. *Process Capability Indices*. London: Chapman & Hall.
4. Srinivasan, V. 1997. *ISO Deliberates Statistical Tolerancing*. Paper presented at 5th CIRP Seminar on Computer-Aided Tolerancing, April 1997, Toronto, Canada.

Chapter 9

Traditional Approaches to Analyzing Mechanical Tolerance Stacks

Paul Drake

9.1 Introduction

Tolerance analysis is the process of taking known tolerances and analyzing the combination of these tolerances at an assembly level. This chapter will define the process for analyzing tolerance stacks. It will show how to set up a loop diagram to determine a nominal performance/assembly value and four techniques to calculate variation from nominal.

The most important goal of this chapter is for the reader to understand the *assumptions* and *risks* that go along with each tolerance analysis method.

9.2 Analyzing Tolerance Stacks

Fig. 9-1 describes the tolerance analysis process.

9.2.1 Establishing Performance/Assembly Requirements

The first step in the process is to identify the requirements for the system. These are usually requirements that determine the "performance" and/or "assembly" of the system. The system requirements will, either directly, or indirectly, flow down requirements to the mechanical subassemblies. These requirements usually determine what needs to be analyzed. In general, a requirement that applies for most mechanical subassemblies is that parts must fit together. Fig. 9-2 shows a cross section of a motor assembly. In this example, there are several requirements.

- Requirement 1. The gap between the shaft and the inner bearing cap must always be greater than zero to ensure that the rotor is clamped and the bearings are preloaded.
- Requirement 2. The gap between the housing cap and the housing must always be greater than zero to ensure that the stator is clamped.

9-2 Chapter Nine

```
1. Establish the Performance Requirements
        ↓
2. Draw a Loop Diagram
        ↓
3. Convert All Dimensions to Mean Dimension with an Equal Bilateral Tolerance
        ↓
4. Calculate the Mean Value for the Performance Requirement
        ↓
5. Determine the Method of Analysis
        ↓
6. Calculate the Variation for the Performance Requirement
```

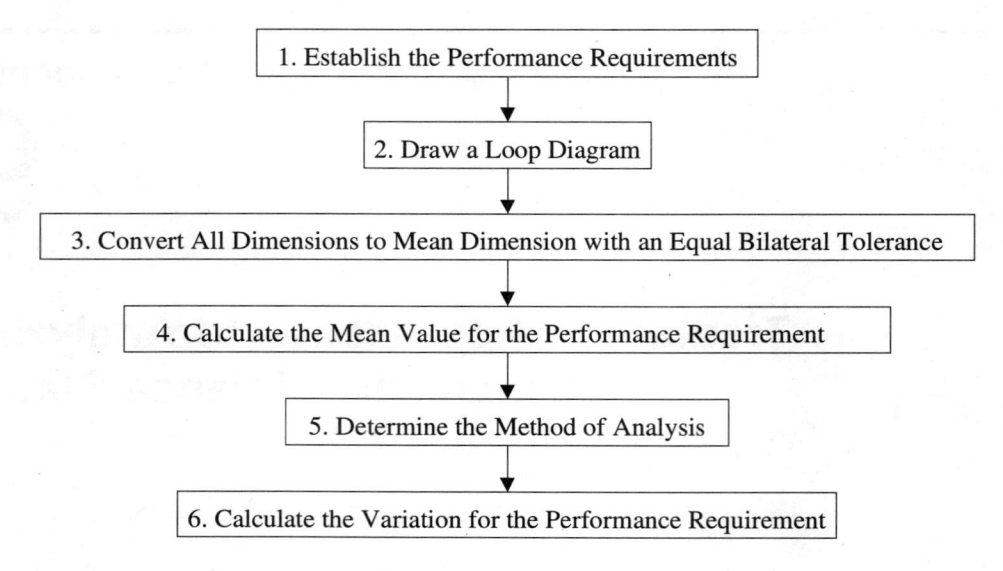

Figure 9-1 Tolerance analysis process

- Requirement 3. The mounting surfaces of the rotor and stator must be within ±.005 for the motor to operate.
- Requirement 4. The bearing outer race must always protrude beyond the main housing, so that the bearing stays clamped.
- Requirement 5. The thread of the bearing cap screw must have a minimum thread engagement of .200 inches.
- Requirement 6. The bottom of the bearing cap screw thread must never touch the bottom of the female thread on the shaft.
- Requirement 7. The rotor and stator must never touch. The maximum radial distance between the rotor and stator is .020.

Other examples of performance/assembly requirements are:
- Thermal requirements, such as contact between a thermal plane and a heat sink,
- Amount of "squeeze" on an o-ring
- Amount of "preload" on bearings
- Sufficient "material" for subsequent machining processes
- Aerodynamic requirements
- Interference requirements, such as when pressing pins into holes
- Structural requirements
- Optical requirements, such as alignment of optical elements

The second part of Step 1 is to convert each requirement into an assembly gap requirement. We would convert each of the previous requirements to the following.
- Requirement 1. Gap 1 ≥ 0
- Requirement 2. Gap 2 ≥ 0

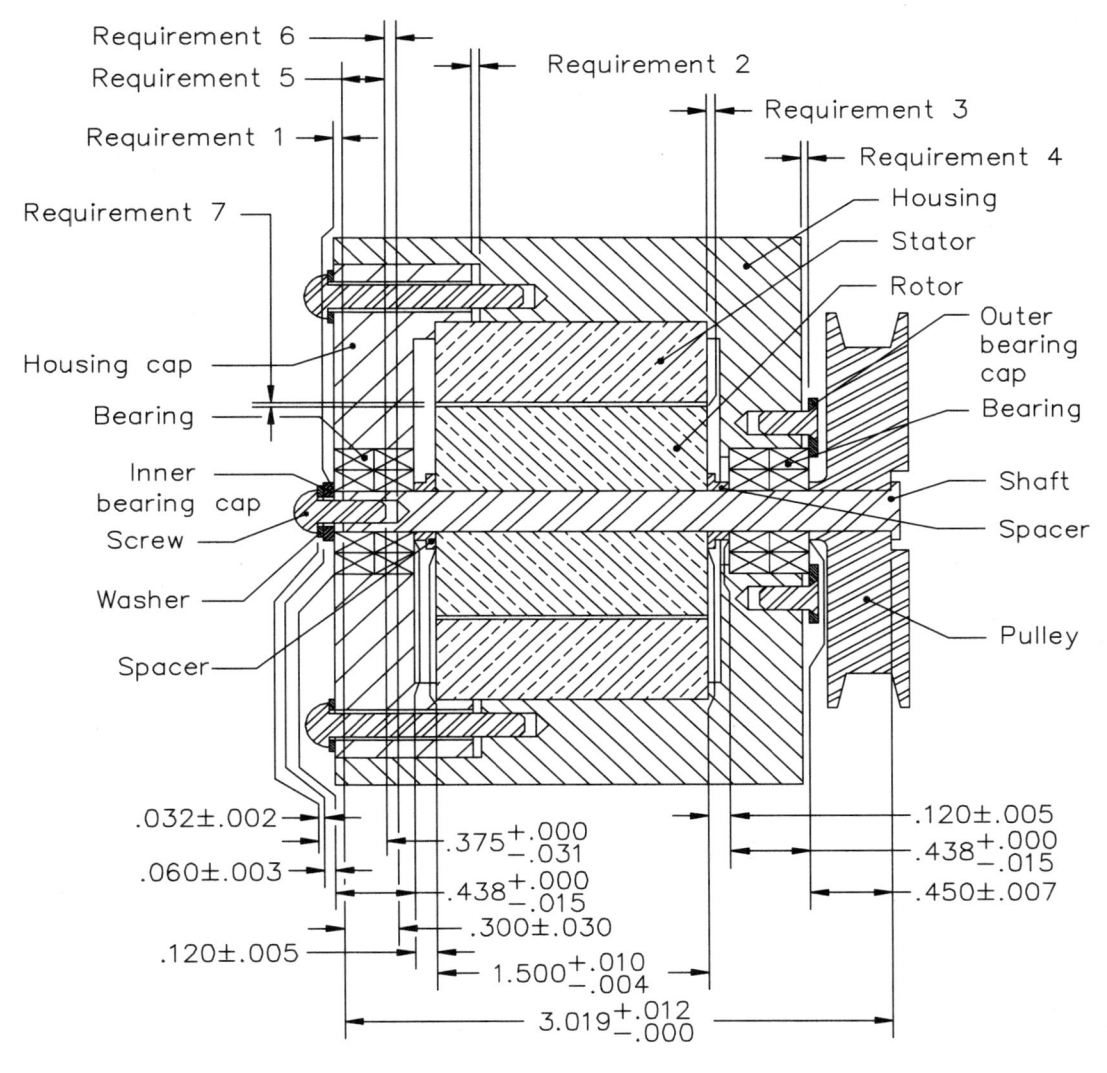

Figure 9-2 Motor assembly

- Requirement 3. Gap $3 \geq .005$
- Requirement 4. Gap $4 \geq 0$
- Requirement 5. Gap $5 \geq .200$
- Requirement 6. Gap $6 \geq 0$
- Requirement 7. Gap $7 \geq 0$ and $\leq .020$

9.2.2 Loop Diagram

The loop diagram is a graphical representation of each analysis. Each requirement requires a separate loop diagram. Simple loop diagrams are usually horizontal or vertical. For simple analyses, vertical loop diagrams will graphically represent the dimensional contributors for vertical "gaps." Likewise, horizontal loop diagrams graphically represent dimensional contributors for horizontal "gaps." The steps for drawing the loop diagram follow.

1. For horizontal dimension loops, start at the surface on the left of the gap. Follow a complete dimension loop, to the surface on the right. For vertical dimension loops, start at the surface on the bottom of the gap. Follow a complete dimension loop, to the surface on the top.
2. Using vectors, create a "closed" loop diagram from the starting surface to the ending surface. Do not include gaps when selecting the path for the dimension loop. Each vector in the loop diagram represents a dimension.
3. Use an arrow to show the direction of each "vector" in the dimension loop. Identify each vector as positive (+), or negative (–), using the following convention.
    For horizontal dimensions:
        Use a + sign for dimensions followed from left to right.
        Use a – sign for dimensions followed from right to left.
    For vertical dimensions:
        Use a + sign for dimensions followed from bottom to top.
        Use a – sign for dimensions followed from top to bottom.
4. Assign a variable name to each dimension in the loop. (For example, the first dimension is assigned the variable name A, the second, B.)

Fig. 9-3 shows a horizontal loop diagram for Requirement 6.

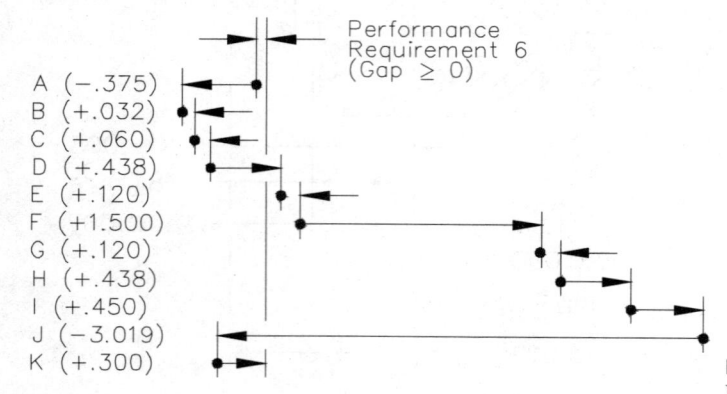

A (−.375)
B (+.032)
C (+.060)
D (+.438)
E (+.120)
F (+1.500)
G (+.120)
H (+.438)
I (+.450)
J (−3.019)
K (+.300)

Figure 9-3 Horizontal loop diagram for Requirement 6

5. Record sensitivities for each dimension. The magnitude of the sensitivity is the value that the gap changes, when the dimension changes 1 unit. For example, if the gap changes .001 when the dimension changes .001, then the magnitude of the sensitivity is 1 (.001/.001). On the other hand, if the gap changes .0005 for a .001 change in the dimension, then the sensitivity is .5 (.0005/.001).

If the dimension vector is positive (pointing to the right for horizontal loops, or up for vertical loops), enter a positive sensitivity. If a dimension with a positive sensitivity increases, the gap will also increase.

If the vector is negative (pointing to the left for horizontal loops, or down for vertical loops), enter a negative sensitivity. If a dimension with a negative sensitivity increases, the gap will decrease. Note, in Fig. 9-3, all of the sensitivities are equal to ±1.

6. Determine whether each dimension is "fixed" or "variable." A fixed dimension is one in which we have no control, such as a vendor part dimension. A variable dimension is one that we can change to influence the outcome of the tolerance stack. (This will become important later, because we will be able to "adjust" or "resize" the variable dimensions and tolerances to achieve a desired assembly performance. We are not able to resize fixed dimensions or tolerances.)

9.2.3 Converting Dimensions to Equal Bilateral Tolerances

In Fig. 9-2, there were several dimensions that were toleranced using unilateral tolerances (such as .375 +.000/-.031, 3.019 +.012/-.000 and .438 +.000/-.015) or unequal bilateral tolerances (such as +1.500 +.010/-.004). If we look at the length of the shaft, we see that there are several different ways we could have applied the tolerances. Fig. 9-4 shows several ways we can dimension and tolerance the length of the shaft to achieve the same upper and lower tolerance limits (3.031/3.019). From a design perspective, all of these methods perform the same function. They give a boundary within which the dimension is acceptable.

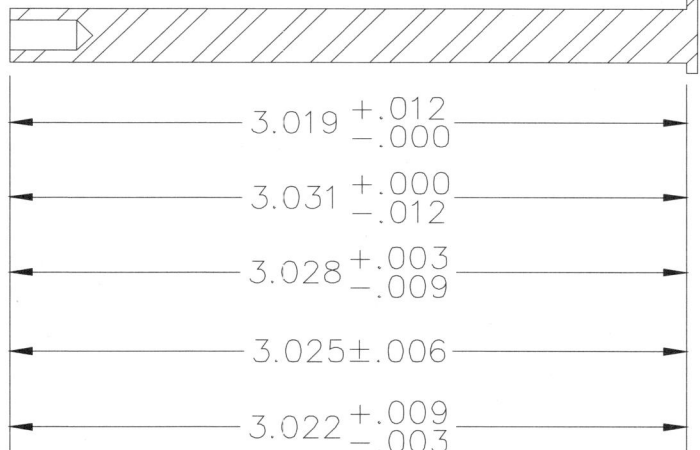

Figure 9-4 Methods to dimension the length of a shaft

The designer might think that changing the nominal dimension has an effect on the assembly. For example, a designer may dimension the part length as 3.019 +.012/-.000. In doing so, the designer may falsely think that this will help minimize the gap for Requirement 1. A drawing, however, doesn't give preference to any dimension within the tolerance range.

Fig. 9-5 shows what happens to the manufacturing yield if the manufacturer "aims" for the dimension stated on the drawing and the process follows the normal distribution. In this example, if the manufacturer aimed for 3.019, half of the parts would be outside of the tolerance zone. Since manufacturing shops want to maximize the yield of each dimension, they will aim for the nominal that yields the largest number of good parts. This helps them minimize their costs. In this example, the manufacturer would aim for 3.025. This allows them the highest probability of making good parts. If they aimed for 3.019 or 3.031, half of the manufactured parts would be outside the tolerance limits.

As in the previous example, many manufacturing processes are normally distributed. Therefore, if we put any unilateral, or unequal bilateral tolerances on dimensions, the manufacturer would convert them to a mean dimension with an equal bilateral tolerance. The steps for converting to an equal bilateral tolerance follow.

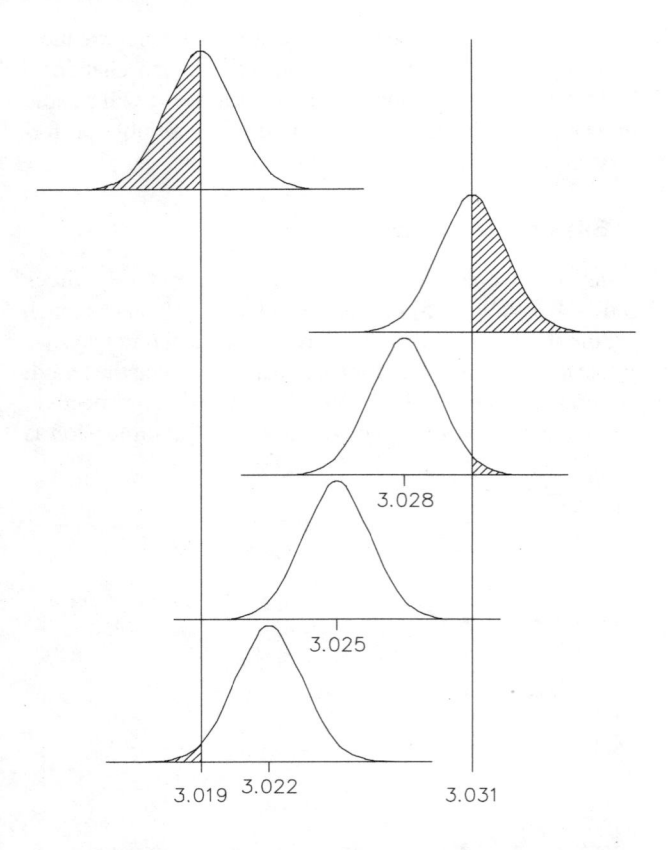

Figure 9-5 Methods of centering manufacturing processes

1. Convert the dimension with tolerances to an upper limit and a lower limit. (For example, 3.028 +.003/-.009 has an upper limit of 3.031 and a lower limit of 3.019.)
2. Subtract the lower limit from the upper limit to get the total tolerance band. (3.031-3.019=.012)
3. Divide the tolerance band by two to get an equal bilateral tolerance. (.012/2=.006)
4. Add the equal bilateral tolerance to the lower limit to get the mean dimension. (3.019 +.006=3.025). Alternately, you could subtract the equal bilateral tolerance from the upper limit. (3.031-.006=3.025)

As a rule, designers should use equal bilateral tolerances. Sometimes, using equal bilateral tolerances may force manufacturing to use nonstandard tools. In these cases, we should not use equal bilateral tolerances. For example, we would not want to convert a drilled hole diameter from ∅.125 +.005/.001 to ∅.127±.003. In this case, we want the manufacturer to use a standard ∅.125 drill. If the manufacturer sees ∅.127 on a drawing, he may think he needs to build a special tool. In the case of drilled holes, we would also want to use an unequal bilateral tolerance because the mean of the drilling process is usually larger than the standard drill size. These dimensions should have a larger plus tolerance than minus tolerance.

As we will see later, when we convert dimensions to equal bilateral tolerances, we don't need to keep track of which tolerances are "positive" and which tolerances are "negative" because the positive tolerances are equal to the negative tolerances. This makes the analysis easier. Table 9-1 converts the necessary dimensions and tolerances to mean dimensions with equal bilateral tolerances.

Table 9-1 Converting to mean dimensions with equal bilateral tolerances

Original Dimension/Tolerance	Mean Dimension with Equal Bilateral Tolerance
.375+.000/-.031	.3595 +/- .0155
.438+.000/-.015	.4305 +/- .0075
1.500+.010/-.004	1.503 +/- .007
3.019+.012/-.000	3.025 +/- .006

9.2.4 Calculating the Mean Value (Gap) for the Requirement

The first step in calculating the variation at the gap is to calculate the mean value of the requirement. The mean value at the gap is:

$$d_g = \sum_{i=1}^{n} a_i d_i \quad (9.1)$$

where

d_g = the mean value at the gap. If d_g is positive, the mean "gap" has clearance, and if d_g is negative, the mean "gap" has interference.

n = the number of independent variables (dimensions) in the stackup

a_i = sensitivity factor that defines the direction and magnitude for the ith dimension. In a one-dimensional stackup, this value is usually +1 or –1. Sometimes, in a one-dimensional stackup, this value may be +.5 or -.5 if a radius is the contributing factor for a diameter callout on a drawing.

d_i = the mean value of the ith dimension in the loop diagram.

Table 9-2 shows the dimensions that are important to determine the mean gap for Requirement 6. We have assigned Variable Name to each dimension so that we can write a loop equation. We have also added

Table 9-2 Dimensions and tolerances used in Requirement 6

Description	Variable Name	Mean Dimension	Sensitivity	Fixed/ Variable	+/- Equal Bilateral Tolerance
Screw thread length	A	.3595	-1	Fixed	.0155
Washer length	B	.0320	1	Fixed	.0020
Inner bearing cap turned length	C	.0600	1	Variable	.0030
Bearing length	D	.4305	1	Fixed	.0075
Spacer turned length	E	.1200	1	Variable	.0050
Rotor length	F	1.5030	1	Fixed	.0070
Spacer turned length	G	.1200	1	Variable	.0050
Bearing length	H	.4305	1	Fixed	.0075
Pulley casting length	I	.4500	1	Variable	.0070
Shaft turned length	J	3.0250	-1	Variable	.0060
Tapped hole depth	K	.3000	1	Variable	.0300

a column titled Fixed/Variable. This identifies which dimensions and tolerances are "fixed" in the analysis, and which ones are allowed to vary (variable). Typically, we have no control over vendor items, so we treat these dimensions as fixed. As we make adjustments to dimensions and tolerances, we will only change the "variable" dimensions and tolerances.

The mean for Gap 6 is:

$$\text{Gap } 6 = a_1d_1 + a_2d_2 + a_3d_3 + a_4d_4 + a_5d_5 + a_6d_6 + a_7d_7 + a_8d_8 + a_9d_9 + a_{10}d_{10} + a_{11}d_{11}$$

$$\text{Gap } 6 = (-1)A + (1)B + (1)C + (1)D + (1)E + (1)F + (1)G + (1)H + (1)I + (-1)J + (1)K$$

$$\text{Gap } 6 = (-1).3595 + (1).0320 + (1).0600 + (1).4305 + (1).1200 + (1)1.5030 + (1).1200 + (1).4305 + (1).4500 + (-1)3.0250 + (1).0300$$

$$\text{Gap } 6 = .0615$$

9.2.5 Determine the Method of Analysis

Eq. (9.1) only calculates the nominal value for the gap. The next step is to analyze the variation at the gap. Historically, mechanical engineers have used two types of tolerancing models to analyze these variations: 1) a "worst case" (WC) model, and 2) a "statistical" model. Each approach offers tradeoffs between piecepart tolerances and assembly "quality." In Chapters 11 and 14, we will see that there are other methods based on the optimization of piecepart and assembly quality and the optimization of total cost.

Fig. 9-6 shows how the assumptions about the pieceparts affect the requirements (gaps), using the worst case and statistical methods. In this figure, the horizontal axis represents the manufactured dimension. The vertical axis represents the number of parts that are manufactured at a particular dimension on the horizontal axis.

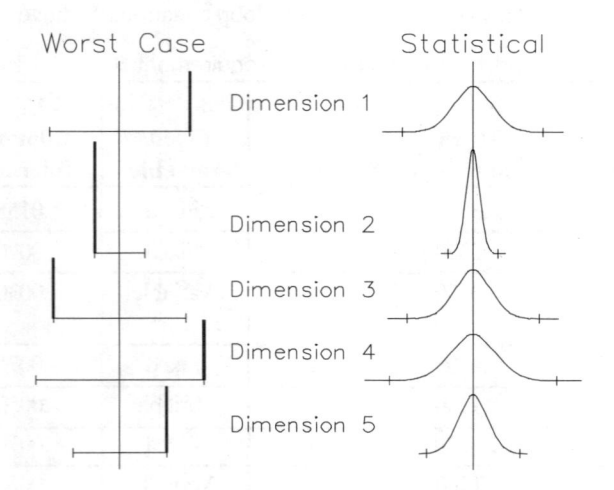

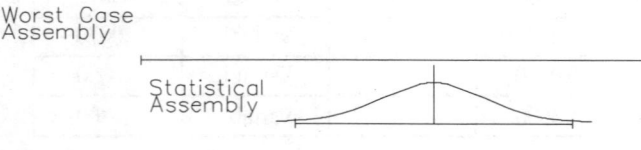

Figure 9-6 Combining piecepart variations using worst case and statistical methods

In the Worst Case Model, we verify that the parts will perform their intended function 100 percent of the time. This is oftentimes a conservative approach. In the statistical modeling approach, we assume that most of the manufactured parts are centered on the mean dimension. This is usually less conservative than a worst case approach, but it offers several benefits which we will discuss later. There are two traditional statistical methods; the Root Sum of the Squares (RSS) Model, and the Modified Root Sum of the Squares (MRSS) Model.

9.2.6 Calculating the Variation for the Requirement

During the design process, the design engineer makes tradeoffs using one of the three classic models. Typically, the designer analyzes the requirements using worst case tolerances. If the worst case tolerances met the required assembly performance, the designer would stop there. On the other hand, if this model did not meet the requirements, the designer increased the piecepart tolerances (to make the parts more manufacturable) at the risk of nonconformance at the assembly level. The designer would make trades, using the RSS and MRSS models.

The following sections discuss the traditional Worst Case, RSS, and MRSS models. Additionally, we discuss the Estimated Mean Shift Model that includes Worst Case and RSS models as extreme cases.

9.2.6.1 Worst Case Tolerancing Model

The Worst Case Model, sometimes referred to as the "Method of Extremes," is the simplest and most conservative of the traditional approaches. In this approach, the tolerance at the interface is simply the sum of the individual tolerances.

The following equation calculates the expected variation at the gap.

$$t_{wc} = \sum_{i=1}^{n} |a_i t_i| \qquad (9.2)$$

where

t_{wc} = maximum expected variation (equal bilateral) using the Worst Case Model.
t_i = equal bilateral tolerance of the ith component in the stackup.

The variation at the gap for Requirement 6 is:

$t_{wc} = |(-1).0155|+|(1).0030|+|(1).0050|+|(1).0075|+|(1).0050|+|(1).0070|+|(1).0050|$
$\qquad +|(1).0075|+|(1).0070|+|(-1).0060|+|(1).0300|$

$t_{wc} = .0955$

Using the Worst Case Model, the minimum gap is equal to the mean value minus the "worst case" variation at the gap. The maximum gap is equal to the mean value plus the "worst case" variation at the gap.

Minimum gap = $d_g - t_{wc}$
Maximum gap = $d_g + t_{wc}$

The maximum and minimum assembly gaps for Requirement 6 are:

Minimum Gap 6 = $d_g - t_{wc}$ = .0615 - .0955 = -.0340
Maximum Gap 6 = $d_g + t_{wc}$ = .0615 + .0955 = .1570

The requirement for Gap 6 is that the minimum gap must be greater than 0. Therefore, we must increase the minimum gap by .0340 to meet the minimum gap requirement. One way to increase the minimum gap is to modify the dimensions (d_i's) to increase the nominal gap. Doing this will also increase the maximum gap of the assembly by .0340. Sometimes, we can't do this because the maximum requirement may not allow it, or other requirements (such as Requirement 5) won't allow it. Another option is to reduce the tolerance values (t_i's) in the stackup.

Resizing Tolerances in the Worst Case Model

There are two ways to reduce the tolerances in the stackup.
1. The designer could randomly change the tolerances and analyze the new numbers, or
2. If the original numbers were "weighted" the same, then all variable tolerances (those under the control of the designer) could be multiplied by a "resize" factor to yield the minimum assembly gap. This is the correct approach if the designer assigned original tolerances that were equally producible.

Resizing is a method of allocating tolerances. (See Chapters 11 and 14 for further discussion on tolerance allocation.) In allocation, we start with a desired assembly performance and determine the piecepart tolerances that will meet this requirement. The resize factor, F_{wc}, scales the original worst case tolerances up or down to achieve the desired assembly performance. Since the designer has no control over tolerances on purchased parts (fixed tolerances), the scaling factor only applies to variable tolerances. Eq. (9.2) becomes:

$$t_{wc} = \sum_{j=1}^{p} a_j t_{jf} + \sum_{k=1}^{q} a_k t_{kf}$$

where,
    a_j = sensitivity factor for the jth, fixed component in the stackup
    a_k = sensitivity factor for the kth, variable component in the stackup
    t_{jf} = equal bilateral tolerance of the jth, fixed component in the stackup
    t_{kv} = equal bilateral tolerance of the kth, variable component in the stackup
    p = number of independent, fixed dimensions in the stackup
    q = number of independent, variable dimensions in the stackup

The resize factor for the Worst Case Model is:

$$F_{wc} = \frac{d_g - g_m - \sum_{j=1}^{p} |a_j t_{jf}|}{\sum_{k=1}^{q} a_k t_{kv}}$$

where
    g = minimum value at the (assembly) gap. This value is zero if no interference or clearance is allowed.

The new variable tolerances ($t_{kv,wc,\,resized}$) are the old tolerances multiplied by the factor F_{wc}.

$$t_{kv,wc,resized} = F_{wc}\, t_{kv}$$

$t_{kv,wc,resized}$ = equal bilateral tolerance of the kth, variable component in the stackup after resizing using the Worst Case Model.

Fig. 9-7 shows the relationship between the piecepart tolerances and the assembly tolerance before and after resizing.

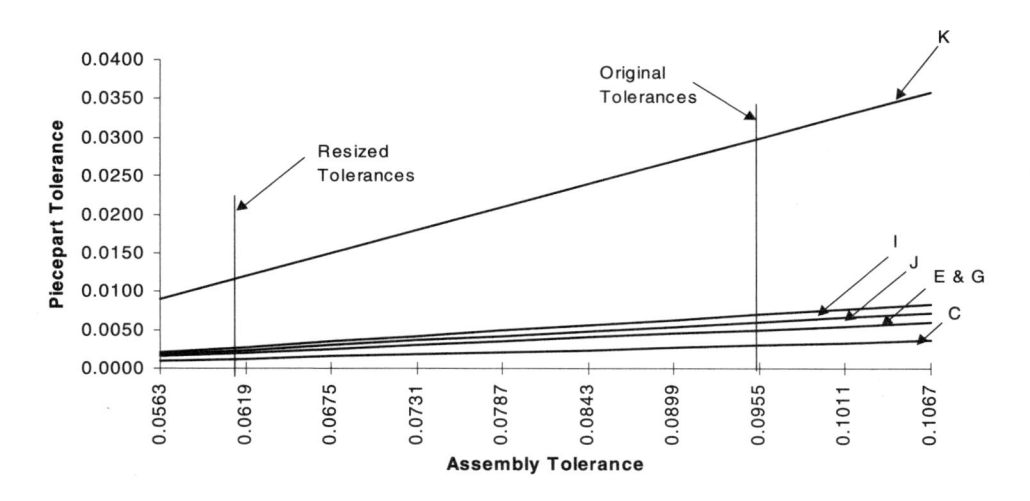

Figure 9-7 Graph of piecepart tolerances versus assembly tolerance before and after resizing using the Worst Case Model

The resize factor for Requirement 6 equals .3929. (For example, .0030 is resized to .3929*.0030 = .0012.) Table 9-3 shows the new (resized) tolerances that would give a minimum gap of zero.

Table 9-3 Resized tolerances using the Worst Case Model

Variable Name	Mean Dimension	Fixed/ Variable	+/- Equal Bilateral Tolerance	Resized +/- Equal Bilateral Tolerance ($t_{iv,wc,resized}$)
A	.3595	Fixed	.0155	
B	.0320	Fixed	.0020	
C	.0600	Variable	.0030	.0012
D	.4305	Fixed	.0075	
E	.1200	Variable	.0050	.0020
F	1.5030	Fixed	.0070	
G	.1200	Variable	.0050	.0020
H	.4305	Fixed	.0075	
I	.4500	Variable	.0070	.0027
J	3.0250	Variable	.0060	.0024
K	.3000	Variable	.0300	.0118

As a check, we can show that the new maximum expected assembly gap for Requirement 6, using the resized tolerances, is:

$t_{wc,resized}$ = .0155+.0020+.0012+.0075+.0020+.0070+.0020+.0075+.0027+.0024+.0118

$t_{wc,resized}$ = .0616

The variation at the gap is:

Minimum Gap 6 = $d_g - t_{wc,resized}$ = .0615 - .0616 = -.0001

Maximum Gap 6 = $d_g + t_{wc,resized}$ = .0615 + .0616 = .1231

Assumptions and Risks of Using the Worst Case Model

In the worst case approach, the designer does not make any assumptions about how the individual piecepart dimensions are distributed within the tolerance ranges. The only assumption is that all pieceparts are within the tolerance limits. While this may not always be true, the method is so conservative that parts will probably still fit. This is the method's major advantage.

The major disadvantage of the Worst Case Model is when there are a large number of components or a small "gap" (as in the previous example). In such applications, the Worst Case Model yields small tolerances, which will be costly.

9.2.6.2 RSS Model

If designers cannot achieve producible piecepart tolerances for a given requirement, they can take advantage of probability theory to increase them. This theory is known as the Root Sum of the Squares (RSS) Model.

The RSS Model is based on the premise that it is more likely for parts to be manufactured near the center of the tolerance range than at the ends. Experience in manufacturing indicates that small errors are usually more numerous than large errors. The deviations are bunched around the mean of the dimension and are fewer at points farther from the mean dimension. The number of manufactured pieces with large deviations from the mean, positive or negative, may approach zero as the deviations from the mean increase.

The RSS Model assumes that the manufactured dimensions fit a statistical distribution called a *normal curve*. This model also assumes that it is unlikely that parts in an assembly will be randomly chosen in such a way that the worst case conditions analyzed earlier will occur.

Derivation of the RSS Equation*

We'll derive the RSS equation based on statistical principles of combinations of standard deviations. To make our derivation as generic as possible, let's start with a function of independent variables such as $y=f(x_1,x_2,...,x_n)$. From this function, we need to be able to calculate the standard deviation of y, or σ_y. But how do we find σ_y if all we have is information about the components x_i? Let's start with the definition of σ_y.

$$\sigma_y^2 = \frac{\sum_{i=1}^{r}(y_i - \mu_y)^2}{r}$$

---

*Derived by Dale Van Wyk and reprinted by permission of Raytheon Systems Company

where,

μ_y = the mean of the random variable y
r = the total number of measurements in the population of interest

Let $\Delta_y = y_i - \mu_y$

If Δ_y is small, which is usually the case, $\Delta_y \approx dy = \dfrac{\partial f}{\partial x_1} dx_1 + \dfrac{\partial f}{\partial x_2} dx_2 + \ldots + \dfrac{\partial f}{\partial x_n} dx_n$ (9.3)

Therefore, $\sigma_y^2 = \dfrac{\sum_{i=1}^{r} dy_i^2}{r}$ (9.4)

From Eq. (9.3),

$$dy^2 = \left(\dfrac{\partial f}{\partial x_1} dx_1 + \dfrac{\partial f}{\partial x_2} dx_2 + \ldots + \dfrac{\partial f}{\partial x_n} dx_n \right)^2$$

$$= \left(\dfrac{\partial f}{\partial x_1} \right)^2 (dx_1)^2 + \left(\dfrac{\partial f}{\partial x_2} \right)^2 (dx_2)^2 + \ldots + \left(\dfrac{\partial f}{\partial x_n} \right)^2 (dx_n)^2$$

$$+ \sum_{j=1}^{n} \sum_{k=1}^{n} \left[\left(\dfrac{\partial f}{\partial x_j} \right) \left(\dfrac{\partial f}{\partial x_k} \right) (dx_j)(dx_k) \right]_{j \neq k}$$

If all the variables x_i are independent, $\sum_{j=1}^{n} \sum_{k=1}^{n} \left[\left(\dfrac{\partial f}{\partial x_j} \right) \left(\dfrac{\partial f}{\partial x_k} \right) (dx_j)(dx_k) \right]_{j \neq k} = 0$

The same would hold true for all similar terms. As a result,

$$\sum_{i=1}^{r} (dy_i)^2 = \sum_{i=1}^{r} \left[\left(\dfrac{\partial f}{\partial x_1} \right)^2 (dx_1)^2 + \left(\dfrac{\partial f}{\partial x_2} \right)^2 (dx_2)^2 + \ldots + \left(\dfrac{\partial f}{\partial x_n} \right)^2 (dx_n)^2 \right]_i$$

Each partial derivative is evaluated at its mean value, which is chosen as the nominal. Thus,

$\dfrac{\partial f}{\partial x_i} = C_i$

where C_i is a constant for each x_i,

$$\sum_{i=1}^{r} (dy_i)^2 = \left(\dfrac{\partial f}{\partial x_1} \right)^2 \sum_{i=1}^{r} (dx_1)_i^2 + \left(\dfrac{\partial f}{\partial x_2} \right)^2 \sum_{i=1}^{r} (dx_2)_i^2 + \ldots + \left(\dfrac{\partial f}{\partial x_n} \right)^2 \sum_{i=1}^{r} (dx_n)_i^2$$ (9.5)

Using the results of Eq. (9.5) and inserting into Eq. (9.4)

$$\sigma_y^2 = \frac{\left(\frac{\partial f}{\partial x_1}\right)^2 \sum_{i=1}^{r}(dx_1)_i^2 + \left(\frac{\partial f}{\partial x_2}\right)^2 \sum_{i=1}^{r}(dx_2)_i^2 + \ldots + \left(\frac{\partial f}{\partial x_n}\right)^2 \sum_{i=1}^{r}(dx_n)_i^2}{r}$$

$$\sigma_y^2 = \left(\frac{\partial f}{\partial x_1}\right)^2 \frac{\sum_{i=1}^{r}(dx_1)_i^2}{r} + \left(\frac{\partial f}{\partial x_2}\right)^2 \frac{\sum_{i=1}^{r}(dx_2)_i^2}{r} + \ldots + \left(\frac{\partial f}{\partial x_n}\right)^2 \frac{\sum_{i=1}^{r}(dx_n)_i^2}{r} \tag{9.6}$$

$$\sigma_y^2 = \left(\frac{\partial f}{\partial x_1}\right)^2 \sigma_{x_1}^2 + \left(\frac{\partial f}{\partial x_2}\right)^2 \sigma_{x_2}^2 + \ldots + \left(\frac{\partial f}{\partial x_n}\right)^2 \sigma_{x_n}^2$$

Now, let's apply this statistical principle to tolerance analysis. We'll consider each of the variables x_i to be a dimension, D_i, with a tolerance, T_i. If the nominal dimension, D_i, is the same as the mean of a normal distribution, we can use the definition of a standard normal variable, Z_i, as follows. (See Chapters 10 and 11 for further discussions on Z.)

$$Z_i = \frac{USL_i - D_i}{\sigma_i} = \frac{T_i}{\sigma_i}$$

$$\sigma_i = \frac{T_i}{Z_i} \tag{9.7}$$

If the piececparts are randomly selected, this relationship applies for the function y as well as for each T_i.

For one-dimensional tolerance stacks, $y = \sum_{i=1}^{n} a_i D_i$ where each a_i represents the sensitivity.

In this case, $\frac{\partial y}{\partial x_i} = a_i$ and Eq. (9.6) becomes

$$\sigma_y^2 = a_1^2 \sigma_{x_1}^2 + a_2^2 \sigma_{x_2}^2 + \ldots + a_n^2 \sigma_{x_n}^2 \tag{9.8}$$

When you combine Eq. (9.7) and Eq. (9.8), $\left(\frac{T_y}{Z_y}\right)^2 = \left(\frac{a_1 T_1}{Z_1}\right)^2 + \left(\frac{a_2 T_2}{Z_2}\right)^2 + \ldots + \left(\frac{a_n T_n}{Z_n}\right)^2 \tag{9.9}$

If all of the dimensions are equally producible, for example if all are exactly 3σ tolerances, or all are 6σ tolerances, $Z_y = Z_1 = Z_2 = \ldots = Z_n$. In addition, let $a_1 = a_2 = \ldots = a_n = +/-1$.

Eq. (9.9) will then reduce to $T_y^2 = T_1^2 + T_2^2 + \ldots + T_n^2$

or $T_y = \sqrt{T_1^2 + T_2^2 + \ldots + T_n^2} \tag{9.10}$

which is the classical RSS equation.

Let's review the assumptions that went into the derivation of this equation.

- All the dimensions D_i are statistically independent.
- The mean value of D_i is large compared to s_i. The recommendation is that D_i/σ_i should be greater than five.
- The nominal value is truly the mean of D_i.
- The distributions of the dimensions are Gaussian, or normal.
- The pieceparts are randomly assembled.
- Each of the dimensions is equally producible.
- Each of the sensitivities has a magnitude of 1.
- Z_i equations assume equal bilateral tolerances.

The validity of each of these assumptions will impact how well the RSS prediction matches the reality of production.

Note that while Eq. (9.10) is the classical RSS equation, we should generally write it as follows so that we don't lose sensitivities.

$$t_{rss} = \sqrt{a_1^2 t_1^2 + a_2^2 t_2^2 + \ldots + a_n^2 t_n^2} \tag{9.11}$$

Historically, Eq. (9.11) assumed that all of the component tolerances (t_i) represent a $3\sigma_i$ value for their manufacturing processes. Thus, if all the component distributions are assumed to be normal, then the probability that a dimension is between $\pm t_i$ is 99.73%. If this is true, then the assembly gap distribution is normal and the probability that it is $\pm t_{rss}$ between is 99.73%.

Although most people have assumed a value of $\pm 3\sigma$ for piecepart tolerances, the RSS equation works for "equal σ" values. If the designer assumed that the input tolerances were $\pm 4\sigma$ values for the piecepart manufacturing processes, then the probability that the assembly is between $\pm t_{rss}$ is 99.9937 (4σ).

The 3σ process limits using the RSS Model are similar to the Worst Case Model. The minimum gap is equal to the mean value minus the RSS variation at the gap. The maximum gap is equal to the mean value plus the RSS variation at the gap.

Minimum 3σ process limit $= d_g - t_{rss}$
Maximum 3σ process limit $= d_g + t_{rss}$

Using the original tolerances for Requirement 6, t_{rss} is:

$$t_{rss} = \begin{bmatrix} (-1)^2 .0155^2 + (1)^2 .0020^2 + (1)^2 .0030^2 + (1)^2 .0075^2 + (1)^2 .0050^2 + (1)^2 .0070^2 + \\ (1)^2 .0050^2 + (1)^2 .0075^2 + (1)^2 .0070^2 + (-1)^2 .0060^2 + (1)^2 .0300^2 \end{bmatrix}^{\frac{1}{2}}$$

$t_{rss} = .0381$

The three sigma variation at the gap is:
Minimum 3σ process variation for Gap 6 $= d_g - t_{rss} = .0615 - .0381 = .0234$
Maximum 3σ process variation for Gap 6 $= d_g + t_{rss} = .0615 + .0381 = .0996$

Resizing Tolerances in the RSS Model

Using the RSS Model, the minimum gap is greater than the requirement. As in the Worst Case Model, we can resize the variable tolerances to achieve the desired assembly performance. As before, the scaling factor only applies to variable tolerances.

The resize factor, F_{rss}, for the RSS Model is:

$$F_{rss} = \sqrt{\frac{(d_g - g_m)^2 - \sum_{j=1}^{p}(a_j t_{jf})^2}{\sum_{k=1}^{q}(a_k t_{kv})^2}}$$

The new variable tolerances ($t_{kv,rss,resized}$) are the old tolerances multiplied by the factor F_{rss}.

$$t_{kv,rss,resized} = F_{rss}\, t_{kv}$$

$t_{kv,rss,resized}$ = equal bilateral tolerance of the kth, variable component in the stackup after resizing using the RSS Model.

Fig. 9-8 shows the relationship between the piecepart tolerances and the assembly tolerance before and after resizing.

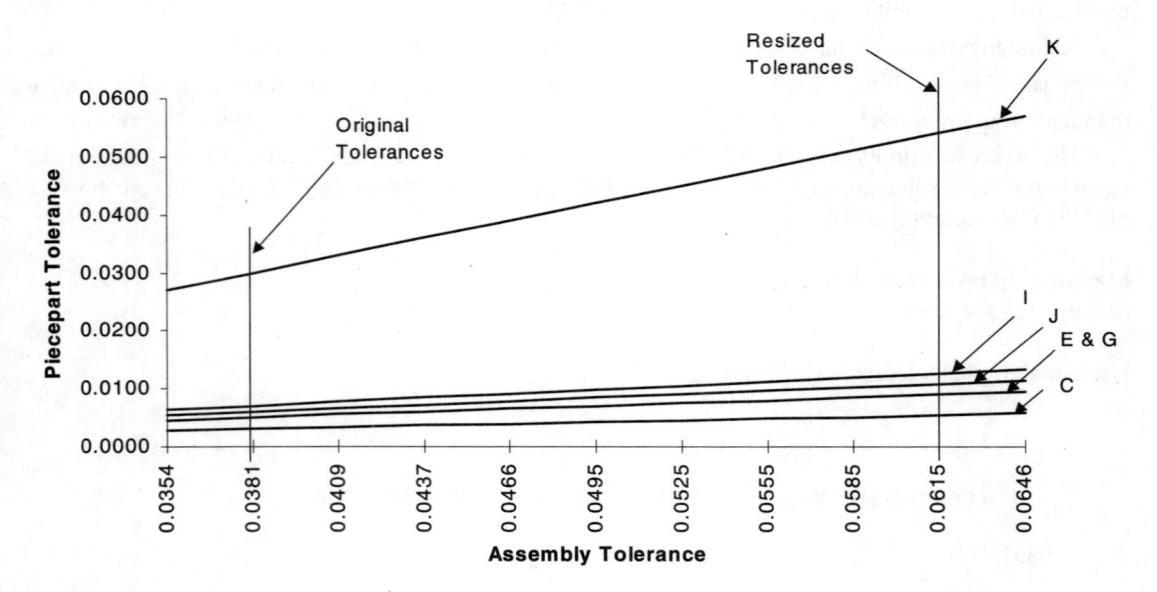

Figure 9-8 Graph of piecepart tolerances versus assembly tolerance before and after resizing using the RSS Model

The new variable tolerances are the old tolerances multiplied by the factor F_{rss}.
The resize factor for Requirement 6 is 1.7984. (For example, .0030 is resized to 1.7984*.0030 = .0054.)

Table 9-4 shows the new tolerances that would give a minimum gap of zero.

Table 9-4 Resized tolerances using the RSS Model

Variable Name	Mean Dimension	Fixed/Variable	Original +/- Equal Bilateral Tolerance	Resized +/- Equal Bilateral Tolerance ($t_{iv,rss,resized}$)
A	.3595	Fixed	.0155	
B	.0320	Fixed	.0020	
C	.0600	Variable	.0030	.0054
D	.4305	Fixed	.0075	
E	.1200	Variable	.0050	.0090
F	1.5030	Fixed	.0070	
G	.1200	Variable	.0050	.0090
H	.4305	Fixed	.0075	
I	.4500	Variable	.0070	.0126
J	3.0250	Variable	.0060	.0108
K	.3000	Variable	.0300	.0540

As a check, we can show that the new maximum expected assembly gap for Requirement 6, using the resized tolerances, is:

$$t_{rss,resized} = \begin{bmatrix} (-1)^2 .0155^2 + (1)^2 .0020^2 + (1)^2 .0054^2 + (1)^2 .0075^2 + (1)^2 .0090^2 + (1)^2 .0070^2 + \\ (1)^2 .0090^2 + (1)^2 .0075^2 + (1)^2 .0126^2 + (-1)^2 .0108^2 + (1)^2 .0540^2 \end{bmatrix}^{\frac{1}{2}}$$

$t_{rss,resized} = .0615$

The variation at the gap is:

Minimum 3σ process variation for Gap 6 = $d_g - t_{rss,resized}$ = .0615 - .0615 = 0
Maximum 3σ process variation for Gap 6 = $d_g + t_{rss,resized}$ = .0615 + .0615 = .1230

Assumptions and Risks of Using the RSS Model

The RSS Model yields larger piecepart tolerances for a given assembly gap, but the risk of defects at assembly is higher. The RSS Model assumes:

a) Piecepart tolerances are tied to process capabilities. This model assumes that when the designer changes a tolerance, the process capabilities will also change.

b) All process distributions are centered on the midpoint of the dimension. It does not allow for mean shifts (tool wear, etc.) or for purposeful decentering.

c) All piecepart dimensions are independent (covariance equals zero).

d) The bad parts are thrown in with the good in the assembly. The RSS Model does not take into account part screening (inspection).
e) The parts included in any assembly have been thoroughly mixed and the components included in any assembly have been selected at random.
f) The RSS derivation assumes equal bilateral tolerances.

Remember that by deriving the RSS equation, we made the assumption that all tolerances (t_i's) were equally producible. This is usually not the case. The only way to know if a tolerance is producible is by understanding the process capability for each dimension. The traditional assumption is that the tolerance (t_i) is equal to 3σ, and the probability of a defect at the gap will be about .27%. In reality, it is very unlikely to be a 3σ value, but rather some unknown number.

The RSS Model is better than the Worst Case Model because it accounts for the tendency of pieceparts to be centered on a mean dimension. In general, the RSS Model is not used if there are less than four dimensions in the stackup.

9.2.6.3 Modified Root Sum of the Squares Tolerancing Model

In reality, the probability of a worst case assembly is very low. At the other extreme, empirical studies have shown that the RSS Model does not accurately predict what is manufactured because some (or all) of the RSS assumptions are not valid. Therefore, an option designers can use is the RSS Model with a "correction" factor. This model is called the Modified Root Sum of the Squares Method.

$$t_{mrss} = C_f \sqrt{a_1^2 t_1^2 + a_2^2 t_2^2 + \ldots + a_n^2 t_n^2}$$

where
C_f = correction factor used in the MRSS equation.
t_{mrss} = expected variation (equal bilateral) using the MRSS model.

Several experts have suggested correction factors (C_f) in the range of 1.4 to 1.8 (References 1,4,5 and 6). Historically, the most common factor is 1.5.

The variation at the gap is:

Minimum gap = $d_g - t_{mrss}$
Maximum gap = $d_g + t_{mrss}$

In our example, we will use the correction factor suggested in Reference 2.

$$C_f = \frac{0.5(t_{wc} - t_{rss})}{t_{rss}(\sqrt{n} - 1)} + 1$$

This correction factor will always give a t_{mrss} value that is less than t_{wc}. In our example, C_f is:

$$C_f = \frac{0.5(.0955 - .0381)}{.0381(\sqrt{11} - 1)} + 1$$

$C_f = 1.3252$

Using the original tolerances for Requirement 6, t_{mrss} is:

$$t_{mrss} = 1.3252 \begin{bmatrix} (-1)^2 .0155^2 + (1)^2 .0020^2 + (1)^2 .0030^2 + (1)^2 .0075^2 + (1)^2 .0050^2 + (1)^2 .0070^2 + \\ (1)^2 .0050^2 + (1)^2 .0075^2 + (1)^2 .0070^2 + (-1)^2 .0060^2 + (1)^2 .0300^2 \end{bmatrix}^{\frac{1}{2}}$$

$t_{mrss} = .0505$

The variation at the gap is:

Minimum Gap 6 = $d_g - t_{mrss}$ = .0615 - .0505 = .0110
Maximum Gap 6 = $\bar{d}_g t_{mrss}$ = .0615 + .0505 = .1120

Resizing Tolerances in the RSS Model

Similar to the RSS Model, the minimum gap using the MRSS Model is greater than the requirement. Like the other models, we can resize the variable tolerances to achieve the desired assembly performance. The equation for the resize factor, F_{mrss}, is much more complex for this model. The value of F_{mrss} is a root of the following quadratic equation.

$$aF_{mrss}^2 + bF_{mrss} + c = 0$$

where

$$a = 0.25 \left(\sum_{k=1}^{q} a_k t_{kv} \right)^2 - 2.25 \sum_{k=1}^{q} (a_k t_{kv})^2 + 3\sqrt{n} \sum_{k=1}^{q} (a_k t_{kv})^2 - n \sum_{k=1}^{q} (a_k t_{kv})^2$$

$$b = 0.5 \sum_{k=1}^{q} (a_k t_{kv}) \sum_{j=1}^{p} (a_j t_{jf}) + \left(\sum_{k=1}^{q} a_k t_{kv} \right) (d_g - g_m) - \sqrt{n} \left(\sum_{k=1}^{q} a_k t_{kv} \right) (d_g - g_m)$$

$$c = 0.25 \left(\sum_{j=1}^{p} a_j t_{jf} \right)^2 + (d_g - g_m)^2 - 2\sqrt{n}(d_g - g_m)^2 + n(d_g - g_m)^2 + \left(\sum_{j=1}^{p} a_j t_{jf} \right) (d_g - g_m)$$

$$- \sqrt{n} \left(\sum_{j=1}^{p} a_j t_{jf} \right) (d_g - g_m) - 2.25 \sum_{j=1}^{p} (a_j t_{jf})^2 + 3\sqrt{n} \sum_{j=1}^{p} (a_j t_{jf})^2 - n \sum_{j=1}^{p} (a_j t_{jf})^2$$

Therefore,

$$F_{mrss} = \frac{-b - \sqrt{b^2 - 4ac}}{2a}$$

Fig. 9-9 shows the relationship between the piecepart tolerances and the assembly tolerance before and after resizing.

The new variable tolerances $(t_{kv, mrss, resized})$ are the old tolerances multiplied by the factor F_{mrss}.

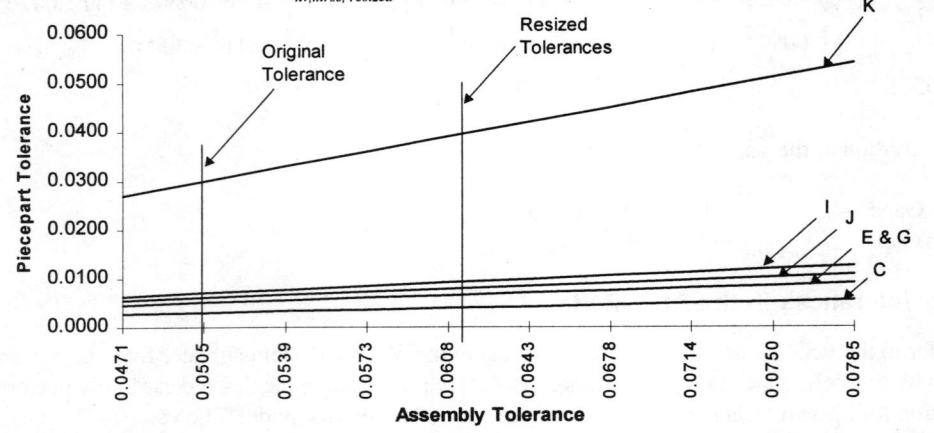

Figure 9-9 Graph of piecepart tolerances versus assembly tolerance before and after resizing using the MRSS Model

$$t_{kv, mrss, resized} = F_{mrss} \, t_{kv}$$

$t_{kv, mrss, resized}$ = equal bilateral tolerance of the kth, variable component in the stackup after resizing using the MRSS Model.

The resize factor for Requirement 6 is 1.3209. (For example, .0030 is resized to 1.3209*.0030 = .0040.) Table 9-5 shows the new tolerances that would give a minimum gap of zero.

Table 9-5 Resized tolerances using the MRSS Model

Variable Name	Mean Dimension	Fixed/ Variable	Original +/- Equal Bilateral Tolerance	Resized +/- Equal Bilateral Tolerance ($t_{iv, mrss, resized}$)
A	.3595	Fixed	.0155	
B	.0320	Fixed	.0020	
C	.0600	Variable	.0030	.0040
D	.4305	Fixed	.0075	
E	.1200	Variable	.0050	.0066
F	1.5030	Fixed	.0070	
G	.1200	Variable	.0050	.0066
H	.4305	Fixed	.0075	
I	.4500	Variable	.0070	.0092
J	3.0250	Variable	.0060	.0079
K	.3000	Variable	.0300	.0396

As a check, we show the following calculations for the resized tolerances.

$t_{wc,\,resized} = .0155 + .0020 + .0040 + .0075 + .0066 + .0070 + .0066 + .0075 + .0092 + .0079 + .0396$

$t_{wc,\,resized} = .1134$

$$t_{rss,\,resized} = \begin{bmatrix} (-1)^2.0155^2 + (1)^2.0020^2 + (1)^2.0040^2 + (1)^2.0075^2 + (1)^2.0066^2 + (1)^2.0070^2 + \\ (1)^2.0066^2 + (1)^2.0075^2 + (1)^2.0092^2 + (-1)^2.0079^2 + (1)^2.0396^2 \end{bmatrix}^{\frac{1}{2}}$$

$t_{rss,\,resized} = .0472$

$$C_{f,\,resized} = \frac{0.5(.1134 - .0472)}{.0472(\sqrt{11} - 1)} + 1$$

$C_{f,\,resized} = 1.3032$

$$t_{mrss,\,resized} = 1.3032 \begin{bmatrix} (-1)^2.0155^2 + (1)^2.0020^2 + (1)^2.0040^2 + (1)^2.0075^2 + (1)^2.0066^2 + (1)^2.0070^2 + \\ (1)^2.0066^2 + (1)^2.0075^2 + (1)^2.0092^2 + (-1)^2.0079^2 + (1)^2.0396^2 \end{bmatrix}^{\frac{1}{2}}$$

$t_{mrss,\,resized} = .0615$

As a check, we can show that the expected assembly gap for Requirement 6, using the resized tolerances, is:

Minimum Gap 6 = $d_g - t_{mrss,resized}$ = .0615 - .0615 = .0000
Maximum Gap 6 = $d_g + t_{mrss,resized}$ = .0615 + .0615 = .1230

Assumptions and Risks of Using the MRSS Model

The uncertainty associated with the MRSS Model is that there is no mathematical reason for the factor C_f. The correction factor can be thought of as a "safety" factor. The more the RSS assumptions depart from reality, the higher the safety factor should be.

The MRSS Model also has other problems.

a) It applies the same "safety" factor to all the tolerances, even though they don't deviate from the RSS assumptions equally.

b) If fixed correction factors proposed in the literature are used, the MRSS tolerance can be larger than the worst case stackup. This problem is eliminated with the use of the calculated C_f shown here.

c) If the tolerances are equal and there are only two of them, the MRSS assembly tolerance will always be larger than the worst case assembly tolerance when using the calculated correction factor.

The MRSS Model is generally considered better than the RSS and Worst Case models because it tries to model what has been measured in the real world.

9.2.6.4 Comparison of Variation Models

Table 9-6 summarizes the Worst Case, RSS, and MRSS models for Requirement 6. The "Resized" columns show the tolerances that will give a minimum expected gap value of zero, and a maximum expected gap value of .1230 inch. As expected, the worst case tolerance values are the smallest. In this example, the resized RSS tolerance values are approximately three times greater than the worst case tolerances. It is obvious that the RSS tolerances will yield more pieceparts. The MRSS resized tolerance values fall between the worst case (most conservative) and RSS (most risk of assembly defects) values.

Table 9-6 Comparison of results using the Worst Case, RSS, and MRSS models

			\multicolumn{6}{c}{Tolerance Analysis}
			Worst Case		RSS		MRSS	
Mean Dim.	Sens.	Dim. Type	Original	Resized	Original	Resized	Original	Resized
.3595	-1.0000	Variable	.0155		.0155		.0155	
.0320	1.0000	Fixed	.0020		.0020		.0020	
.0600	1.0000	Variable	.0030	.0012	.0030	.0054	.0030	.0040
.4305	1.0000	Fixed	.0075		.0075		.0075	
.1200	1.0000	Variable	.0050	.0020	.0050	.0090	.0050	.0066
1.5030	1.0000	Fixed	.0070		.0070		.0070	
.1200	1.0000	Variable	.0050	.0020	.0050	.0090	.0050	.0066
.4305	1.0000	Fixed	.0075		.0075		.0075	
.4500	1.0000	Variable	.0070	.0027	.0070	.0126	.0070	.0092
3.0250	-1.0000	Variable	.0060	.0024	.0060	.0108	.0060	.0079
.3000	1.0000	Variable	.0300	.0118	.0300	.0540	.0300	.0396
Nominal Gap			.0615	.0615	.0615	.0615	.0615	.0615
Minimum Gap			-.0340	.0001	.0234	.0000	.0110	.0000
Expected Variation			.0955	.0616	.0381	.0615	.0505	.0615

Table 9-7 summarizes the tradeoffs for the three models. All the models have different degrees of risk of defects. The worst case tolerances have the least amount of risk (i.e. largest number of assemblies within the expected assembly requirements). Because of the tight tolerances we will reject more pieceparts. Worst case also implies that we are doing 100% inspection. Since we have to tighten up the tolerances to meet the assembly specification, the number of rejected pieceparts increases. Therefore, this model has the highest costs associated with it. The RSS tolerances will yield the least piecepart cost at the expense of a lower probability of assembly conformance. The MRSS Model tries to take the best of both of these models. It gives a higher probability of assembly conformance than the RSS Model, and lower piecepart costs than the Worst Case Model.

Within their limitations, the traditional tolerancing models have worked in the past. The design engineer, however, could not quantify how well they worked. He also could not quantify how cost effective the tolerance values were. Obviously, these methods cannot consistently achieve quality goals. One way to achieve quality goals is to eliminate the assumptions that go along with the classical tolerancing models. By doing so, we can quantify (sigma level, defects per million opportunities (dpmo)) the tolerances and optimize tolerances for maximum producibility. These issues are discussed in Chapter 11, Predicting Assembly Quality.

Table 9-7 Comparison of analysis models

Consideration	Worst Case Model	RSS Model	MRSS Model
Risk of Defect	Lowest	Highest	Middle
Cost	Highest	Lowest	Middle
Assumptions about component processes	None	The process follows a normal distribution. The mean of the process is equal to the nominal dimension. Processes are independent.	The process follows a normal distribution. The mean of the process distribution is not necessarily equal to the nominal dimension.
Assumptions about drawing tolerances	Dimensions outside the tolerance range are screened out.	The tolerance is related to a manufacturing process capability. Usually the tolerance range is assumed to be the +/- 3 sigma limit of the process.	The tolerance is related to a manufacturing process capability. Usually the tolerance range is assumed to be the +/- 3 sigma limit of the process.
Assumptions about expected assembly variation	100% of the parts are within the maximum and minimum performance range.	The assembly distribution is normal. Depending on the piecepart assumptions, a percentage of the assemblies will be between the minimum and maximum gap. Historically, this has been 99.73%. Some out of specification parts reach assembly.	99.73% of the assemblies will be between the minimum and maximum gap. The correction factor (C_f) is a safety factor.

9.2.6.5 Estimated Mean Shift Model

Generally, if we don't have knowledge about the processes for manufacturing a part, such as a vendor part, we are more inclined to use the Worst Case Model. On the other hand, if we have knowledge about the processes that make the part, we are more inclined to use a statistical model. Chase and Greenwood proposed a tolerancing model that blends the Worst Case and RSS models. (Reference 6) This *Estimated Mean Shift Model* is:

$$t_{ems} = \sum_{i=1}^{n} |m_i a_i t_i| + \sum_{i=1}^{n} \sqrt{\left((1-m_i)^2 a_i^2 t_i^2\right)}$$

where
 m_i = the mean shift factor for the *i*th component

In this model, the mean shift factor is a number between 0 and 1.0 and represents the amount that the midpoint is estimated to shift as a fraction of the tolerance range. If a process were closely controlled, we would use a small mean shift, such as .2. If we know less about the process, we would use higher mean shift factors.

Using a mean shift factor of .2 for the variable components and .8 for the fixed components, the expected variation for Requirement 6 is:

$$t_{ems} = |.8(-1).0155| + |.8(1).0020| + |.2(1).0030| + |.8(1).0075| + |.2(1).0050| +$$
$$|.8(1).0070| + |.2(1).0050| + |.8(1).0075| + |.2(1).0070| + |.2(-1).0060| + |.2(1).0300| +$$

$$\left[\begin{array}{l} (.2(-1).0155) + (.2^2 (1)^2 .0020^2) + (.8^2 (1)^2 .0030^2) + (.2^2 (1)^2 .0075^2) + \\ (.8^2 (1)^2 .0050^2) + (.2^2 (1)^2 .0070^2) + (.8^2 (1)^2 .0050^2) + (.2^2 (1)^2 .0075^2) + \\ (.8^2 (1)^2 .0070^2) + (.8^2 (-1)^2 .0060^2) + (.8^2 (1)^2 .0300^2) \end{array} \right]^{\frac{1}{2}}$$

$$t_{ems} = .0690$$

The first part of the Estimated Mean Shift Model is the sum of the mean shifts and is similar to the Worst Case Model. Notice if we set the mean shift factor to 1.0 for all the components, t_{ems} is equal to .0955, which is the same as t_{wc}. The second part of the model is the sum of the statistical components. Notice if we used a mean shift factor of zero for all of the components, t_{ems} is equal to .0381, which is the same as t_{rss}.

The two major advantages of the Estimated Mean Shift Model are:
- It allows flexibility in the design. Some components may be modeled like worst case, and some may be modeled statistically.
- The model can be used to estimate designs (using conservative shift factors), or it can accept manufacturing data (if it is available).

9.3 Analyzing Geometric Tolerances

The previous discussions have only included tolerances associated with dimensions in the tolerance analysis. We have not yet addressed how to model geometric tolerances in the loop diagram.

Generally, geometric controls will restrain one or several of the following attributes:
- Location of the feature
- Orientation of the feature
- Form of the feature

The most difficult task when modeling geometric tolerances is determining which of the geometric controls contribute to the requirement and how these controls should be modeled in the loop diagram. Because the geometric controls are interrelated, there are no hard and fast rules that tell us how to include geometric controls in tolerance analyses. Since there are several modeling methods, sometimes we include GD&T in the model, and sometimes we do not.

Generally, however, if a feature is controlled with geometric tolerances, the following apply.
- If there is a location control on a feature in the loop diagram, we will usually include it in the analysis.
- If there is an orientation control on a feature in the loop diagram, we may include it in the analysis as long as the location of the feature is not a contributor to the requirement.

- If there is a form control on a feature in the loop diagram, we may include it in the analysis as long as the location, orientation, or size of the feature is not a contributor to the requirement. Any time parts come together, however, we have surface variations that introduce variations in the model.
- Geometric form and orientation controls on datum features are usually not included in loop diagrams. Since datums are the "starting points" for measurements, and are defined as the geometric counterparts (high points) of the datum feature, the variations in the datum features usually don't contribute to the variation analysis.

There is a difference between a GD&T control (such as a form control) and a feature variation (such as form variation). If we add a GD&T control to a stack, we add to the output. Therefore, we should only include the GD&T controls that add to the output.

GD&T controls are generally used only in worst case analyses. Previously we said that the Worst Case Model assumes 100% inspection. Since GD&T controls are the specification limits for inspection, it makes sense to use them in this type of analysis. In a statistical analysis, however, we either make assumptions about the manufacturing processes (as shown previously), or use real data from the manufacturing processes (as shown in Chapter 11). Since the manufacturing processes are sources of variation, they should be inputs to the statistical analyses. Since GD&T controls are not sources of variation, they should not be used in a statistical analysis.

The following sections show examples of how to model geometric tolerances. The examples are single part stacks, but the concepts can be applied to stacks with multiple components.

9.3.1 Form Controls

Form controls should seldom be included in a variation analysis. For nonsize features, the location, or orientation tolerance usually controls the extent of the variation of the feature. The form tolerance is typically a refinement of one of these controls. If a form control is applied to a size feature (and the Individual Feature of Size Rule applies from ASME Y14.5), the size tolerance is usually included in the variation analysis. In these cases, the form tolerance boundary is inside the size tolerance boundary, the location tolerance boundary, or the orientation tolerance boundary, so the form control is not modeled.

If form tolerances are used in the loop diagram, they are modeled with a nominal dimension equal to zero, and an equal bilateral tolerance equal to half the form tolerance:

Fig. 9-10 shows an assembly with four parts. In this example, the requirement is for the Gap to be greater than zero. For this requirement, the following applies to the form controls.

- Flatness of .002 on the substrate is not included in the loop diagram because it is a refinement of the size dimension (.040 ± .003).
- Flatness of .001 on the substrate is not included in the loop diagram because it is a datum.
- Flatness of .002 on the heatsink is included in the loop diagram.
- Flatness of .002 on the housing is not included in the loop diagram because it is a refinement of the location tolerance.
- Flatness of .004 on the housing is not included in the loop diagram because it is a datum.
- Flatness of .006 on the housing is not included in the loop diagram because it is a refinement of the location.

Figure 9-10 Substrate package

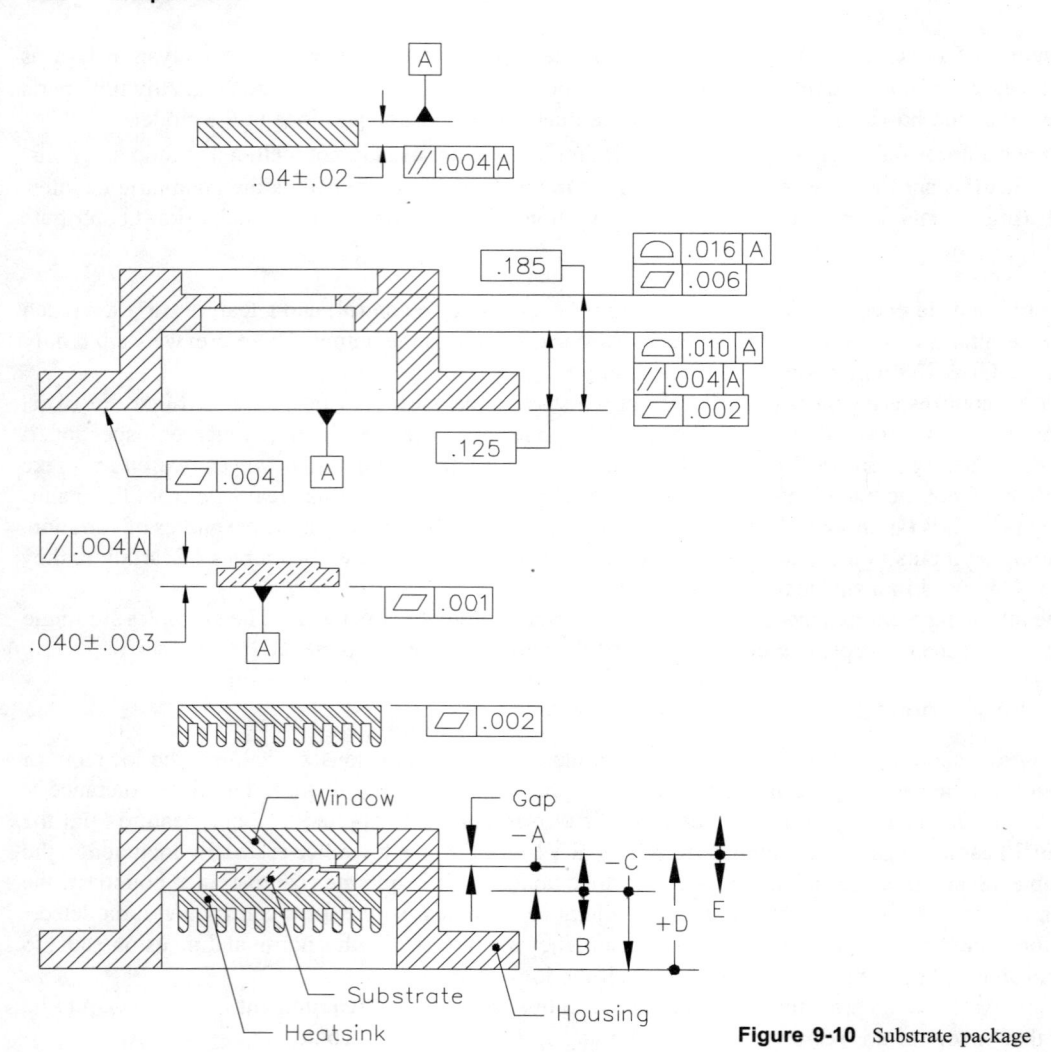

9.3.2 Orientation Controls

Like form controls, we do not often include orientation controls in a variation analysis. Typically we determine the feature's worst-case tolerance boundary using the location or size tolerance.

If orientation tolerances are used in the loop diagram, they are modeled like form tolerances. They have a nominal dimension equal to zero, and an equal bilateral tolerance equal to half the orientation tolerance.

In Fig. 9-10, the following describes the application of the orientation controls to the Gap analysis.

- Parallelism of .004 to datum A on the Substrate is not included in the loop diagram because it is a refinement of the size dimension (.040±.003).
- Parallelism of .004 to datum A on the Housing is not included in the loop diagram because it is a refinement of the location tolerance.
- Parallelism of .004 to datum A on the Window is included in the loop diagram.

Therefore, the equation for the Gap in Fig. 9-10 is: Gap = -A+B-C+D+E
where
    A = .040 ±.003
    B = 0 ±.001
    C = .125 ±.005
    D = .185 ±.008
    E = 0 ±.002

9.3.3 Position

There are several ways to model a position geometric constraint. When we use position at regardless of feature size (RFS), the size of the feature, and the location of the feature are treated independently. When we use position at maximum material condition (MMC) or at least material condition (LMC), the size and location dimensions cannot be treated independently. The following sections show how to analyze these situations.

9.3.3.1 Position at RFS

Fig. 9-11 shows a hole positioned at RFS.

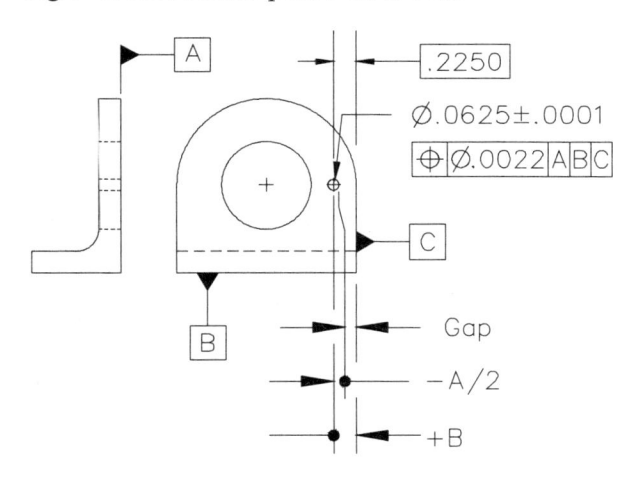

Figure 9-11 Position at RFS

The equation for the Gap in Fig. 9-11 is: Gap = -A/2+B
where
    A = .0625 ±.0001
    B = .2250 ±.0011

9.3.3.2 Position at MMC or LMC

As stated earlier, when we use position at MMC or LMC, the size and location dimensions should be combined into one component in the loop diagram. We can do this using the following method.

1) Calculate the largest "outer" boundary allowed by the dimensions and tolerances.
2) Calculate the smallest "inner" boundary allowed by the dimensions and tolerances.
3) Convert the inner and outer boundary into a nominal diameter with an equal bilateral tolerance.

9.3.3.3 Virtual and Resultant Conditions

When calculating the internal and external boundaries for features of size, it is helpful to understand the following definitions from ASME Y14.5M-1994.

Virtual Condition: A constant boundary generated by the collective effects of a size feature's specified MMC or LMC and the geometric tolerance for that material condition.

- The virtual condition (outer boundary) of an external feature, called out at MMC, is equal to its maximum material condition plus its tolerance at maximum material condition.
- The virtual condition (inner boundary) of an internal feature, called out at MMC, is equal to its maximum material condition minus its tolerance at maximum material condition.
- The virtual condition (inner boundary) of an external feature, called out at LMC, is equal to its least material condition minus its tolerance at least material condition.
- The virtual condition (outer boundary) of an internal feature, called out at LMC, is equal to its least material condition plus its tolerance at least material condition.

Resultant Condition: The variable boundary generated by the collective effects of a size feature's specified MMC or LMC, the geometric tolerance for that material condition, the size tolerance, and the additional geometric tolerance derived from its specified material condition.

- The smallest resultant condition (inner boundary) of an external feature, called out at MMC, is equal to its least material condition minus its tolerance at least material condition.
- The largest resultant condition (outer boundary) of an internal feature, called out at MMC, is equal to its least material condition plus its tolerance at least material condition.
- The largest resultant condition (outer boundary) of an external feature, called out at LMC, is equal to its maximum material condition plus its tolerance at maximum material condition.
- The smallest resultant condition (inner boundary) of an internal feature, called out at LMC, is equal to its maximum material condition minus its tolerance at maximum material condition.

9.3.3.4 Equations

We can use the following equations to calculate the inner and outer boundaries.

For an external feature at MMC
 outer boundary = VC = MMC + Geometric Tolerance at MMC
 inner boundary = (smallest) RC = LMC − Tolerance at LMC

For an internal feature at MMC
 inner boundary = VC = MMC − Geometric Tolerance at MMC
 outer boundary = (largest) RC = LMC + Tolerance at LMC

For an external feature at LMC
 inner boundary = VC = LMC − Geometric Tolerance at LMC
 outer boundary = (largest) RC = MMC + Tolerance at MMC

For an internal feature at LMC
 outer boundary = VC = LMC + Geometric Tolerance at LMC
 inner boundary = (smallest) RC = MMC − Tolerance at MMC

Converting an Internal Feature at MMC to a Nominal Value with an Equal Bilateral Tolerance

Fig. 9-12 shows a hole that is positioned at MMC.

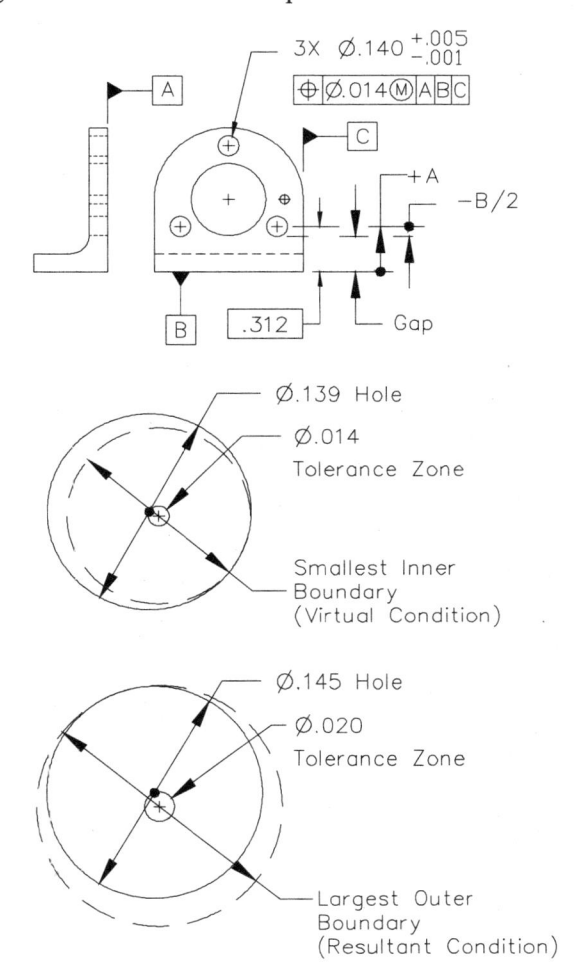

Figure 9-12 Position at MMC—internal feature

The value for B in the loop diagram is:
- Largest outer boundary = $\varnothing.145 + \varnothing.020 = \varnothing.165$
- Smallest inner boundary = $\varnothing.139 - \varnothing.014 = \varnothing.125$
- Nominal diameter = $(\varnothing.165 + \varnothing.125)/2 = \varnothing.145$
  Equal bilateral tolerance = $\varnothing.020$

For position at MMC, an easier way to convert this is:
  LMC ± (total size tolerance + tolerance in the feature control frame)
  = $\varnothing.145 \pm (.006 + .014) = .145 \pm .020$

The equation for the Gap in Fig. 9-12 is: Gap = A−B/2
where
  A = .312 ±0 and B = .145 ±.020

Converting an External Feature at MMC to a Nominal Value with an Equal Bilateral Tolerance

Fig. 9-13 shows a pin positioned at MMC.

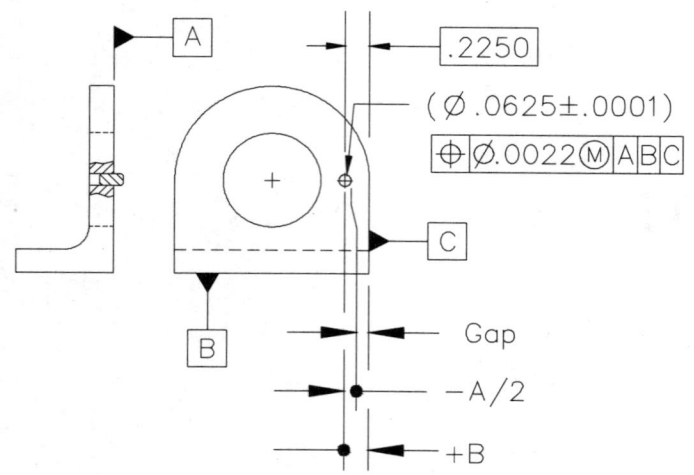

Figure 9-13 Position at MMC—external feature

The value for B in the loop diagram is:
- Largest outer boundary = ⌀.0626 + ⌀.0024 = ⌀.0650
- Smallest inner boundary = ⌀.0624 − ⌀.0022 = ⌀.0602
- Nominal diameter = (⌀.0650 + ⌀.0602)/2 = ⌀.0626
  Equal bilateral tolerance = ⌀.0024

As shown earlier, the easier conversion for position at MMC, is:
  LMC ±(total size tolerance + tolerance in the feature control frame)
  = ⌀.0626 ±(.0002+.0022) = .0626+/−.0024

The equation for the Gap in Fig. 9-13 is: Gap = −A/2+B
where
  A = .0626 ±.0024
  B = .2250 ±0

Converting an Internal Feature at LMC to a Nominal Value with an Equal Bilateral Tolerance

Fig. 9-14 shows a hole that is positioned at LMC.

The value for B in the loop diagram is:
- Largest outer boundary = ⌀.52 + ⌀.03 = ⌀.55
- Smallest inner boundary = ⌀.48 − ⌀.07 = ⌀.41
- Nominal diameter = (⌀.55 + ⌀.41)/2 = ⌀.48
  Equal bilateral tolerance = ⌀.07

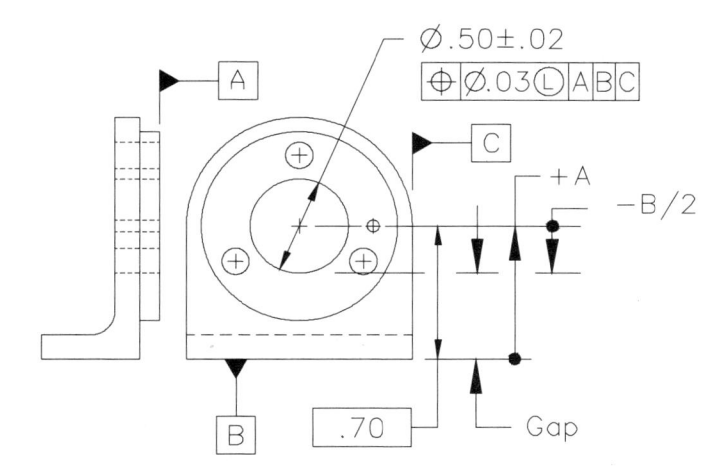

Figure 9-14 Position at LMC—internal feature

For position at LMC, an easier way to convert this is:
MMC ±(total size tolerance + tolerance in the feature control frame)
= ∅.48 ± (.04+.03) = .48 ± .07

The equation for the Gap in Fig. 9-14 is: Gap = A − B/2
where
  A = .70   ±0
  B = .48   ±.07

Converting an External Feature at LMC to a Nominal Value with an Equal Bilateral Tolerance

Fig. 9-15 shows a "boss" that is positioned at LMC.

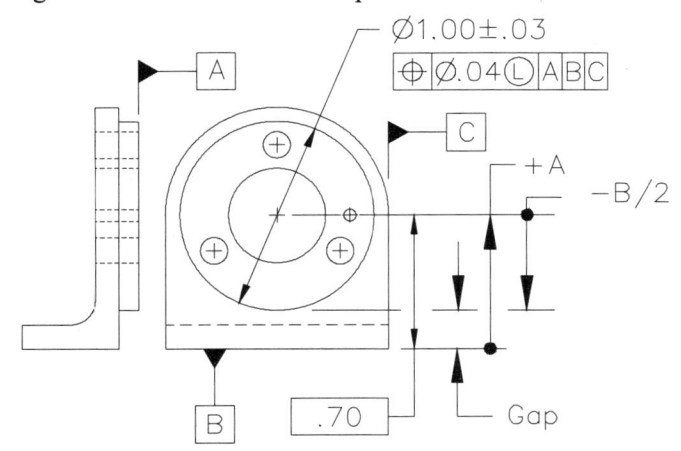

Figure 9-15 Position at LMC—external feature

The value for B in the loop diagram is:
- Largest outer boundary = ∅1.03 + ∅.10 = ∅1.13
- Smallest inner boundary = ∅.97 − ∅.04 = ∅.93
- Nominal diameter = (∅1.13 + ∅.93)/2 = ∅1.03
  Equal bilateral tolerance = ∅.10

As shown earlier, the easier conversion for position at LMC is:

MMC ±(total size tolerance + tolerance in the feature control frame)
= ∅1.03 ±(.06+.04) = 1.03 +/-.10

The equation for the Gap in Fig. 9-15 is: Gap = A-B/2
where
  A = .70   ±0
  B = 1.03   ±.10

9.3.3.5 Composite Position

Fig. 9-16 shows an example of composite positional tolerancing.

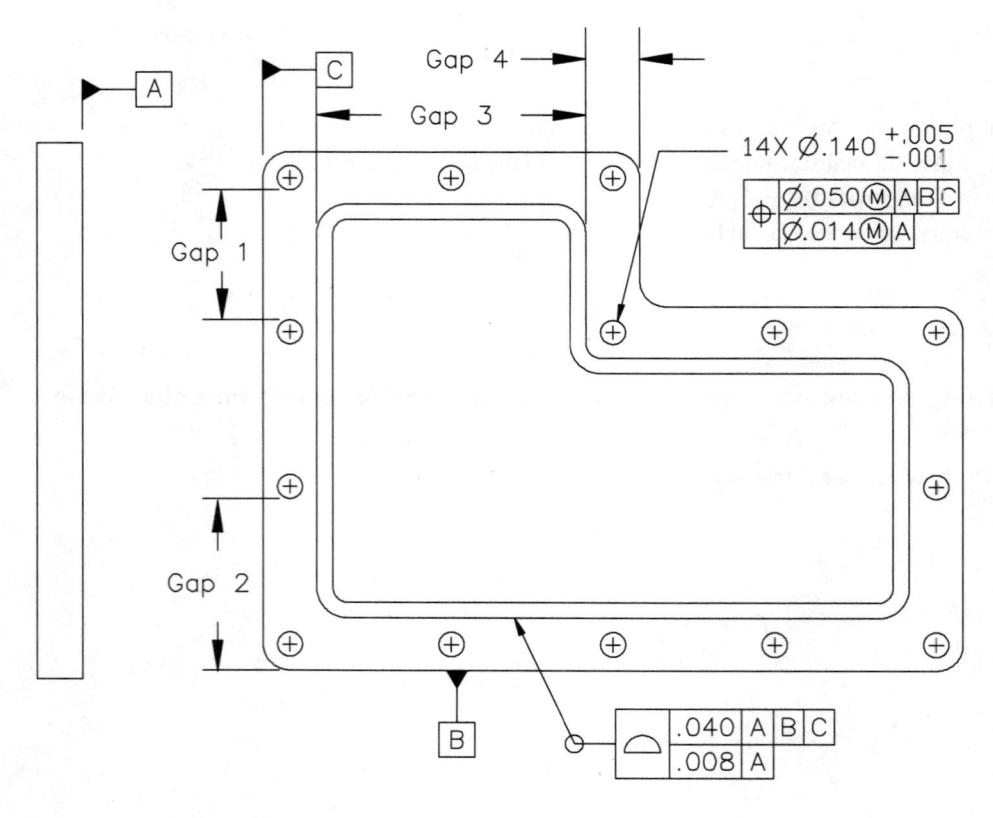

Figure 9-16 Composite position and composite profile

Composite positional tolerancing introduces a unique element to the variation analysis; an understanding of which tolerance to use. If a requirement only includes the pattern of features and nothing else on the part, we use the tolerance in the lower segment of the feature control frame. Since Gap 1 in Fig. 9-16 is controlled by two features within the pattern, we use the tolerance of ∅.014 to calculate the variation for Gap 1.

Gap 2, however, includes variations of the features back to the datum reference frame. In this situation, we use the tolerance in the upper segment of the feature control frame (∅.050) to calculate the variation for Gap 2.

9.3.4 Runout

Analyzing runout controls in tolerance stacks is similar to analyzing position at RFS. Since runout is always RFS, we can treat the size and location of the feature independently. We analyze total runout the same as circular runout, because the worst-case boundary is the same for both controls.

Fig. 9-17 shows a hole that is positioned using runout.

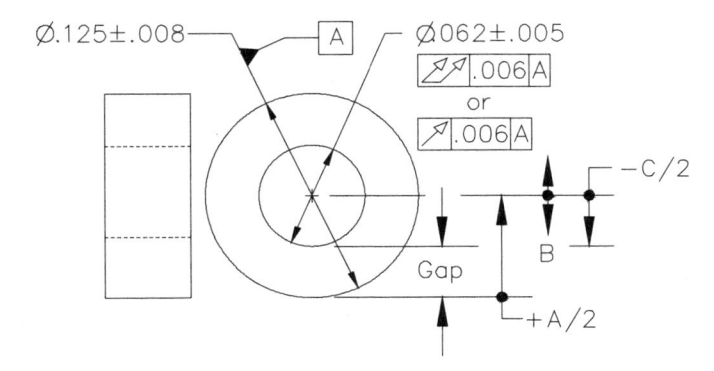

Figure 9-17 Circular and total runout

We model the runout tolerance with a nominal dimension equal to zero, and an equal bilateral tolerance equal to half the runout tolerance.

The equation for the Gap in Fig. 9-17 is: Gap = +A/2 + B − C/2
where

$A = .125 \quad \pm .008$
$B = 0 \quad \pm .003$
$C = .062 \quad \pm .005$

9.3.5 Concentricity/Symmetry

Analyzing concentricity and symmetry controls in tolerance stacks is similar to analyzing position at RFS and runout.

Fig. 9-18 is similar to Fig. 9-17, except that a concentricity tolerance is used to control the ∅.062 feature to datum A.

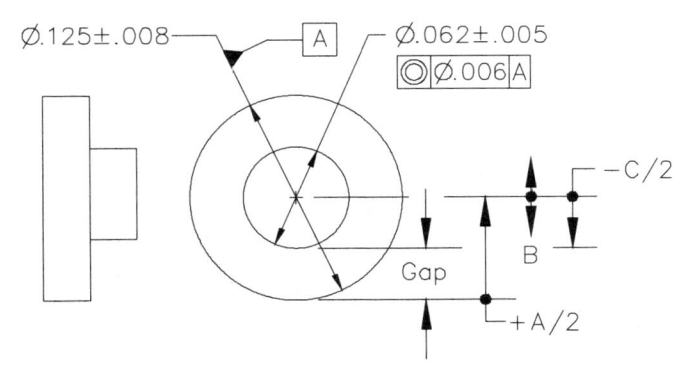

Figure 9-18 Concentricity

The loop diagram for this gap is the same as for runout. The equation for the Gap in Fig. 9-18 is:
Gap = +A/2 + B − C/2
where
$\quad$ A = .125 $\quad$ ±.008
$\quad$ B = 0 $\quad$ ±.003
$\quad$ C = .062 $\quad$ ±.005

Symmetry is analogous to concentricity, except that it is applied to planar features. A loop diagram for symmetry would be similar to concentricity.

9.3.6 Profile

Profile tolerances have a basic dimension locating the true profile. The tolerance is depicted either equal bilaterally, unilaterally, or unequal bilaterally. For equal bilateral tolerance zones, the profile component is entered as a nominal value. The component is equal to the basic dimension, with an equal bilateral tolerance that is half the tolerance in the feature control frame.

9.3.6.1 Profile Tolerancing with an Equal Bilateral Tolerance Zone

Fig. 9-19 shows an application of profile tolerancing with an equal bilateral tolerance zone.

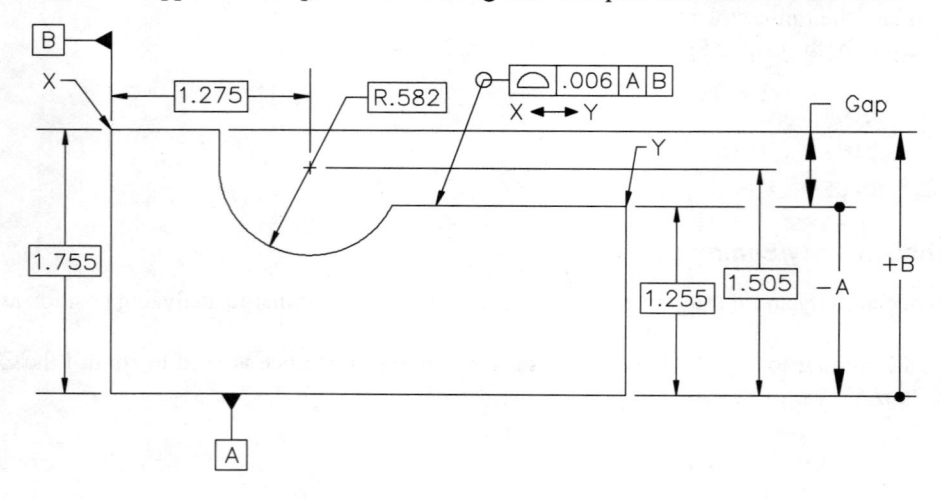

Figure 9-19 Equal bilateral tolerance profile

The equation for the Gap in Fig. 9-19 is: Gap = -A+B
where
$\quad$ A = 1.255 $\quad$ ±.003
$\quad$ B = 1.755 $\quad$ ±.003

9.3.6.2 Profile Tolerancing with a Unilateral Tolerance Zone

Fig. 9-20 shows a figure similar to Fig. 9-19 except the equal bilateral tolerance was changed to a unilateral tolerance zone.

The equation for the Gap is the same as Fig. 9-19: Gap = −A + B

Figure 9-20 Unilateral tolerance profile

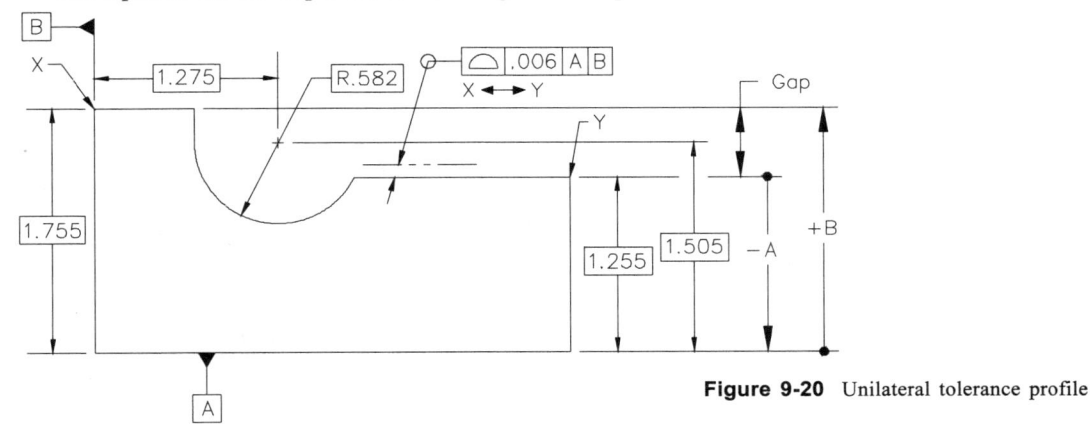

In this example, however, we need to change the basic dimensions and unilateral tolerances to mean dimensions and equal bilateral tolerances.
Therefore,
  A = 1.258 ±.003
  B = 1.758 ±.003

9.3.6.3 Profile Tolerancing with an Unequal Bilateral Tolerance Zone

Fig. 9-21 shows a figure similar to Fig. 9-19 except the equal bilateral tolerance was changed to an unequal bilateral tolerance zone.

The equation for the Gap is the same as Fig. 9-19: Gap = −A + B

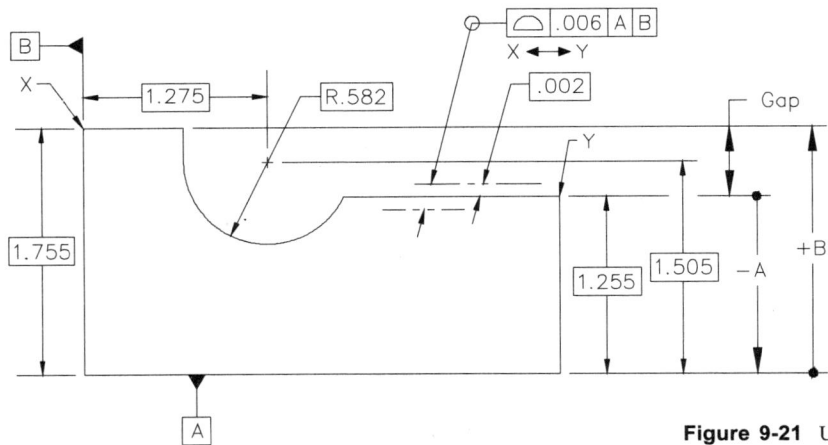

Figure 9-21 Unequal bilateral tolerance profile

9-36 Chapter Nine

As we did in Fig. 9-20, we need to change the basic dimensions and unequal bilateral tolerances to mean dimensions and equal bilateral tolerances.
Therefore,

A = 1.254 ±.003
B = 1.754 ±.003

9.3.6.4 Composite Profile

Composite profile is similar to composite position. If a requirement only includes features within the profile, we use the tolerance in the lower segment of the feature control frame. If the requirement includes variations of the profile back to the datum reference frame, we use the tolerance in the upper segment of the feature control frame.

Fig. 9-16 shows an example of composite profile tolerancing. Gap 3 is controlled by features within the profile, so we would use the tolerance in the lower segment of the profile feature control frame (∅.008) to calculate the variation for Gap 3.

Gap 4, however, includes variations of the profiled features back to the datum reference frame. In this situation, we would use the tolerance in the upper segment of the profile feature control frame (∅.040) to calculate the variation for Gap 4.

9.3.7 Size Datums

Fig. 9-22 shows an example of a pattern of features controlled to a secondary datum that is a feature of size.

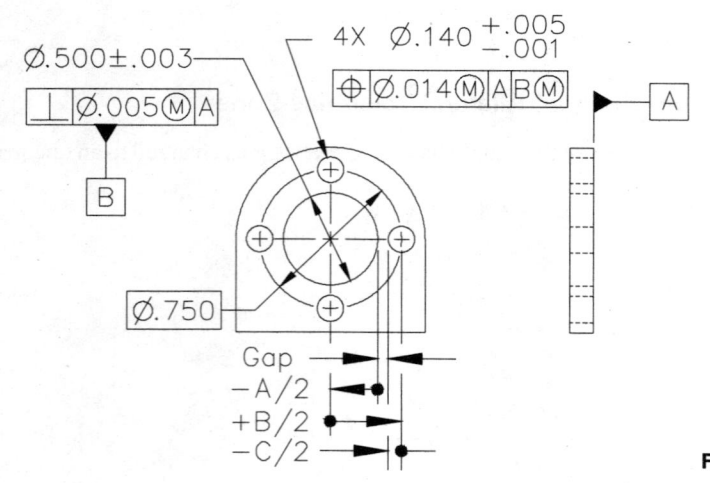

Figure 9-22 Size datum

In this example, ASME Y14.5 states that the datum feature applies at its virtual condition, even though it is referenced in its feature control frame at MMC. (Note, this argument also applies for secondary and tertiary datums invoked at LMC.) In the tolerance stack, this means that we will get an additional "shifting" of the datum that we need to include in the loop diagram.

The way we handle this in the loop diagram is the same way we handled features controlled with position at MMC or LMC. We calculate the virtual and resultant conditions, and convert these boundaries into a nominal value with an equal bilateral tolerance.

The value for A in the loop diagram is:
- Largest outer boundary = ⌀.503 + ⌀.011 = ⌀.514
- Smallest inner boundary = ⌀.497 − ⌀.005 = ⌀.492
- Nominal diameter = (⌀.514 + ⌀.492)/2 = ⌀.503
- Equal bilateral tolerance = ⌀.011

An easier way to convert to this radial value is:
LMC ±(total size tolerance + tolerance in the feature control frame)
= ⌀.503 ±(.006+.005) = .503±.011

The value for C in the loop diagram is:
- Largest outer boundary = ⌀.145 + ⌀.020 = ⌀.165
- Smallest inner boundary = ⌀139 − ⌀.014 = ⌀.125
- Nominal diameter = (⌀.165 + ⌀.125)/2 = ⌀.145
- Equal bilateral tolerance = ⌀.020

An easier way to convert to this radial value is:
LMC ±(total size tolerance + tolerance in the feature control frame)
= ⌀.145 ±(.006+.014) = .145±.020

The equation for the Gap in Fig. 9-22 is: Gap = − A/2 + B/2 − C/2

where
A = .503   ±.011
B = .750   ±0
C = .145   ±.020

9.4   Abbreviations

Variable   **Definition**

a_i — sensitivity factor that defines the direction and magnitude for the ith dimension. In a one-dimensional stackup, this value is usually +1 or -1. Sometimes, in a one-dimensional stackup, this value may be +.5 or -.5 if a radius is the contributing factor for a diameter callout on a drawing.

a_j — sensitivity factor for the jth, fixed component in the stackup

a_k — sensitivity factor for the kth, variable component in the stackup

C_f — correction factor used in the MRSS equation

$C_{f,resized}$ — correction factor used in the MRSS equation, using resized tolerances

$\dfrac{\partial f}{\partial x_i}$ — partial derivative of function y with respect to x_i

d_g — the mean value at the gap. If d_g is positive, the mean "gap" has clearance, and if d_g is negative, the mean "gap" has interference

d_i — the mean value of the ith dimension in the loop diagram

D_i	dimension associated with i^{th} random variable x_i
F_{wc}	resize factor that is multiplied by the original tolerances to achieve a desired assembly performance using the Worst Case Model
F_{mrss}	resize factor that is multiplied by the original tolerances to achieve a desired assembly performance using the MRSS Model
F_{rss}	resize factor that is multiplied by the original tolerances to achieve a desired assembly performance using the RSS Model
g_m	minimum value at the (assembly) gap. This value is zero if no interference or clearance is allowed.
μ_y	mean of random variable y
n	number of independent variables (dimensions) in the equation (stackup)
p	number of independent, fixed dimensions in the stackup
q	number of independent, variable dimensions in the stackup
r	the total number of measurements in the population of interest
σ_y	standard deviation of function y
t_i	equal bilateral tolerance of the ith component in the stackup
T_i	tolerance associated with ith random variable x_i
t_{jf}	equal bilateral tolerance of the jth, fixed component in the stackup
t_{kv}	equal bilateral tolerance of the kth, variable component in the stackup
$t_{kv,wc,resized}$	equal bilateral tolerance of the kth, variable component in the stackup after resizing, using the Worst Case Model
$t_{kv,rss,resized}$	equal bilateral tolerance of the kth, variable component in the stackup after resizing, using the RSS Model
$t_{kv,mrss,resized}$	equal bilateral tolerance of the kth, variable component in the stackup after resizing, using the MRSS Model
t_{mrss}	expected assembly gap variation (equal bilateral) using the MRSS Model
$t_{mrss,resized}$	the expected variation (equal bilateral) using the MRSS Model and resized tolerances
t_{rss}	the expected variation (equal bilateral) using the RSS Model
$t_{rss,resized}$	the expected variation (equal bilateral) using the RSS Model and resized tolerances
t_{wc}	maximum expected variation (equal bilateral) using the Worst Case Model
$t_{wc,resized}$	maximum expected variation (equal bilateral) using the Worst Case Model and resized tolerances
USL_i	upper specification limit of the ith dimension
x_i	ith independent variable
y	function consisting of n independent variables $(x_1,...,x_n)$
Z_i	standard normal transform of ith dimension
Z_y	standard normal transform of y

9.5 Terminology

MMC = Maximum Material Condition: The condition in which a feature of size contains the maximum amount of material within the stated limits of size.

LMC = Least Material Condition: The condition in which a feature of size contains the least amount of material within the stated limits of size.

VC = Virtual Condition: A constant boundary generated by the collective effects of a size feature's specified MMC or LMC material condition and the geometric tolerance for that material condition.

RC = Resultant Condition: The variable boundary generated by the collective effects of a size feature's specified MMC or LMC material condition, the geometric tolerance for that material condition, the size tolerance, and the additional geometric tolerance derived from the feature's departure from its specified material condition.

9.6 References

1. Bender, A. May 1968. Statistical Tolerancing as it Relates to Quality Control and the Designer *Society of Automotive Engineers, SAE paper No. 680490.*
2. Braun, Chuck, Chris Cuba, and Richard Johnson. 1992. Managing Tolerance Accumulation in Mechanical Assemblies. *Texas Instruments Technical Journal.* May-June: 79-86.
3. Drake, Paul and Dale Van Wyk. 1995. Classical Mechanical Tolerancing (Part I of II). *Texas Instruments Technical Journal.* Jan.-Feb: 39-46.
4. Gilson, J. 1951. *A New Approach to Engineering Tolerances.* New York, NY: Industrial Press.
5. Gladman, C.A. 1980. Applying Probability in Tolerance Technology: Trans. Inst. Eng. Australia. *Mechanical Engineering* ME5(2): 82.
6. Greenwood, W.H., and K. W. Chase. May 1987. A New Tolerance Analysis Method for Designers and Manufacturers. *Transactions of the ASME Journal of Engineering for Industry.* 109. 112-116.
7. Hines, William, and Douglas Montgomery. 1990. *Probability and Statistics in Engineering and Management Sciences.* New York, New York: John Wiley and Sons.
8. Kennedy, John B., and Adam M. Neville. 1976. *Basic Statistical Methods for Engineers and Scientists.* New York, NY: Harper and Row.
9. The American Society of Mechanical Engineers. 1995. *ASME Y14.5M-1994, Dimensioning and Tolerancing.* New York, NY: The American Society of Mechanical Engineers.
10. Van Wyk, Dale and Paul Drake. 1995. Mechanical Tolerancing for Six Sigma (Part II). *Texas Instruments Technical Journal.* Jan-Feb: 47-54.

Chapter 10

Statistical Background and Concepts

Ron Randall
Ron Randall & Associates, Inc.
Dallas, Texas

Ron Randall is an independent consultant specializing in applying the principles of Six Sigma quality. Since the 1980s, Ron has applied Statistical Process Control and Design of Experiments principles to engineering and manufacturing at Texas Instruments Defense Systems and Electronics Group. While at Texas Instruments, he served as chairman of the Statistical Process Control Council, a Six Sigma Champion, Six Sigma Master Black Belt, and a Senior Member of the Technical Staff. His graduate work has been in engineering and statistics with study at SMU, the University of Tennessee at Knoxville, and NYU's Stern School of Business under Dr. W. Edwards Deming. Ron is a Registered Professional Engineer in Texas, a senior member of the American Society for Quality, and a Certified Quality Engineer. Ron served two terms on the Board of Examiners for the Malcolm Baldrige National Quality Award.

10.1 Introduction

Statistics do a fine job of enumerating what has already occurred. Industry's most urgent needs are to estimate what will happen in the future. Will the product be profitable? How often will defects occur? The job of statistics is to help estimate the future based on the past.

When designing any part or system, it is necessary to estimate and account for the variation that is likely to occur in the parts, materials, and product features. Statistics can help estimate or model the most likely outcome, and how much variation there is likely to be in that outcome. From these models, estimates of manufacturability and product performance can be made long before production. Knowledge of the probabilities of defects prior to production is important to the financial success of the product. Changes to the design or manufacturing processes that are completed prior to production are far less costly than changes made during production or changes made after the product is fielded. Statistics can help estimate these probabilities.

10.2 Shape, Locations, and Spread

Historical data or data from a designed experiment when displayed in a histogram will:
- Have a shape
- Have a location relative to some important values such as the average or a specification limit
- Have a spread of values across a range.

For example, Fig. 10-1 contains full indicator movement (FIM) runout values of 1,000 steel shafts, measured in thousandths of an inch (mils). Ideally, these 1,000 shafts would all be the same, but the histogram begins to reveal some information about these shafts and the processes that made them. The thousand data points are displayed in a histogram in Fig. 10-1. A histogram displays the frequency (how often) a range of values is present. The histogram has a shape, its location is concentrated between the values 0.000 and 0.005, and is spread out between the values 0 and 0.030. The range that occurs most often is 0.000 to 0.002, but there are many shafts that are larger than this. Statistics can help quantify the histogram. With knowledge of the type of distribution (shape), the mean of the sample (location), and the standard deviation of the sample (spread), one can estimate the chance that a shaft will exceed a certain value like a specification. We will come back to this example later.

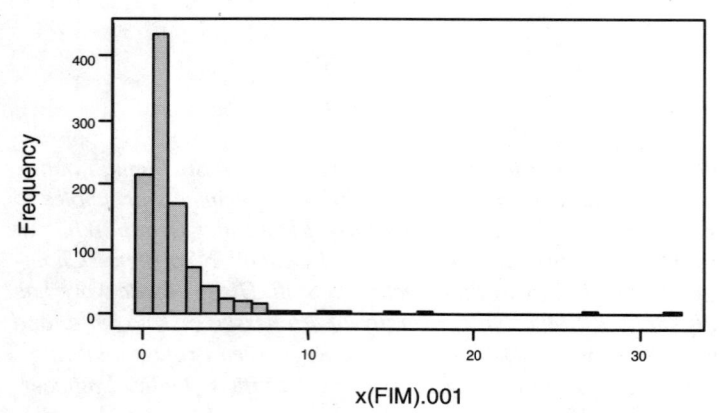

Figure 10-1 Histogram of runout (FIM) data

10.3 Some Important Distributions

Data that is measured on a continuous scale like inches, ohms, pounds, volts, etc. is referred to as variables data. Data that is classified by pass or fail, heads or tails, is called attributes data. Variables data may be more expensive to gather than attributes data, but is much more powerful in its ability to make estimates about the future.

10.3.1 The Normal Distribution

The normal distribution is a mathematical model. All mathematical models are wrong, in that there is always some error. Some models are useful. This is one of them.

Karl Frederick Gauss described this distribution in the eighteenth century. Gauss found that repeated measurements of the same astronomical quantity produced a pattern like the curve in Fig. 10-2. This pattern has since been found to occur almost everywhere in life. Heights, weights, IQs, shoe sizes,

various standardized test scores, economic indicators, and a host of measurements in service and manufacturing are all examples of where the normal distribution applies. (Reference 4) A normal distribution:
- Has one central value (the average).
- Is symmetrical about the average.
- Tails off asymptotically in each direction.

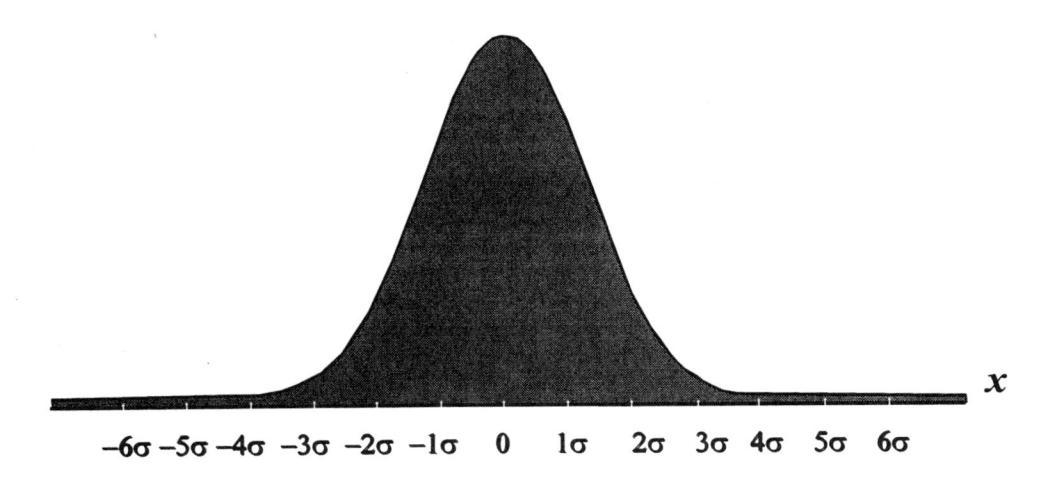

Figure 10-2 The normal distribution

The normal distribution is defined by: $$f(x) = \frac{1}{\sigma\sqrt{2\pi}} e^{-(1/2)[(x-\mu)/\sigma]^2}$$

The mean (μ) is: $$\mu = \sum_{i=1}^{N} \frac{x_i}{N}$$

The standard deviation (σ) is: $$\sigma = \sqrt{\sum_{i=1}^{N}\left(\frac{x_i - \mu}{N}\right)^2}$$

where
N is the size of the population
x_i is value of the ith component in the population

It is important to note that the definitions for the mean (μ) and the standard deviation (σ) are not dependent on the distribution $f(x)$. We will see other functions later, but the definitions for the mean and the standard deviation are the same.

Data that appear to be normally distributed occur often in science and engineering. In my many years of practice and study, I have never seen a perfectly normal distribution. To illustrate, the following histograms (Figs. 10-3 to 10-6) were generated by picking random numbers from a true normal distribution with a mean of 10 and a standard deviation of 1.

Five samples from a true normal distribution yield a histogram with very little information (Fig. 10-3). The curve is a normal distribution with an average and a standard deviation calculated from the five samples. It is used to compare the data with a normal curve produced from that data.

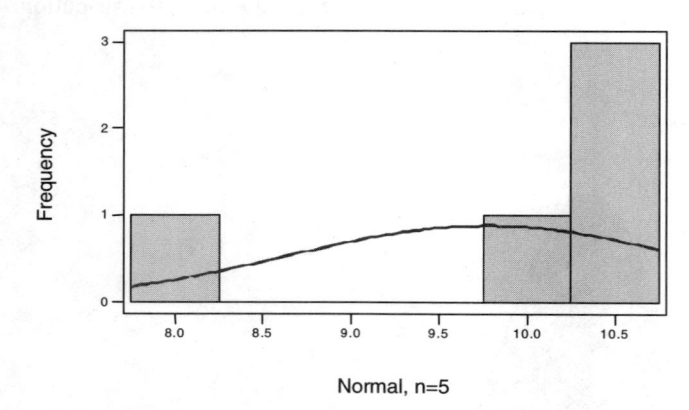

Figure 10-3 Histogram of normal, n=5, with normal curve

When 50 samples are taken from a normal distribution we see the following histogram and a normal curve generated from the 50 samples (Fig. 10-4). Here we begin to see a central tendency between 10.0 and 10.5 and a gradual decline in frequency as we move away from the center.

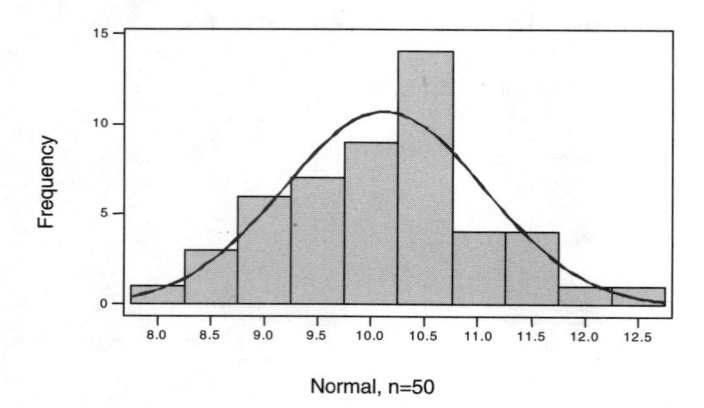

Figure 10-4 Histogram of normal, n=50, with normal curve

The histogram for 500 samples (Fig. 10-5) was taken from a truly normal distribution. Even with 500 samples the histogram does not quite fit the normal model. In this example, the mode (highest peak) is around 9.75.

The histogram for 5000 samples (Fig. 10-6) taken from a normal distribution is still not a perfect fit. Be aware of this behavior when you examine data and distributions. There are statistical tests for judging whether or not a distribution could be from a normal distribution. In these examples, all of the histograms passed the Anderson-Darling test for normality. (Reference 1)

How do I calculate the percent of the population that will be beyond a certain value?

The mathematical answer is to integrate the function $f(x)$. The practical answer is to use a Z table found in statistics books (see Appendix at the end of this chapter), or a statistical software package like *Minitab 12*. (Reference 6) Statisticians long ago prepared a table called a Z table to make this easier.

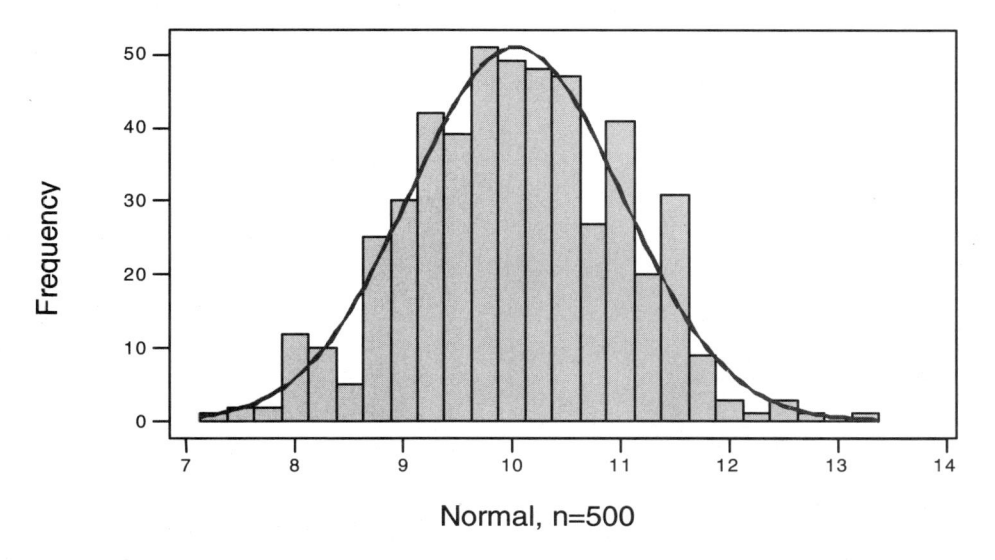

Figure 10-5 Histogram of normal, n=500, with normal curve

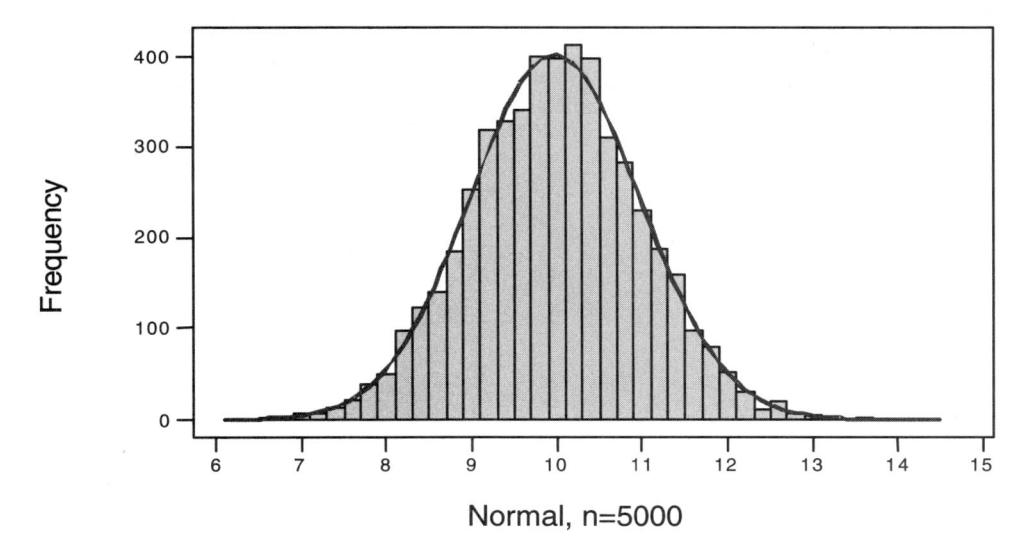

Figure 10-6 Histogram of normal, n=5000, with normal curve

There are different types of Z tables. The Appendix shows a Z table for the unilateral tail area under a normal curve beyond a given Z value. To use the table, we need a Z value. Z is a statistic that is defined as:

$Z = (x-\mu)/\sigma$, where:

    x is a value we are interested in, a specification limit, for example
    μ is the mean (average)
    σ is the standard deviation

Continuing with Fig. 10-7 as an example, suppose we are interested in knowing the probability of x being greater than 2.5σ. (Remember that σ is a value that has a unit of measure like inches.) Using the Z table in the Appendix for $Z = 2.5$, we find the value 0.00621, which is the probability that x will be greater than 2.5σ.

Z Statistic

$$Z = \frac{x - \mu}{\sigma}$$

$$= \frac{2.5\sigma - 0}{\sigma}$$

$$= 2.5$$

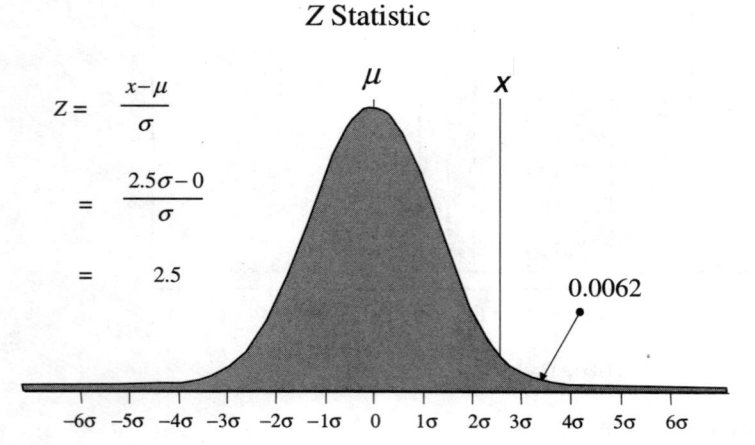

0.0062

Figure 10-7 Z Statistic

What if the histogram does not look like a normal distribution?

There are many continuous distributions that occur in science and engineering that are not normal. Some of the most common continuous distributions are:
1. Beta
2. Cauchy
3. Exponential
4. Gamma
5. Laplace
6. Logistic
7. Lognormal
8. Weibull

We will look at the lognormal briefly here for illustration, although I think it is best to refer to texts on statistics and reliability for more detail. (References 3 and 4)

10.3.2 Lognormal Distribution

Recall the above example of the FIM of the shafts. (Fig. 10-1) Certainly this is not normally distributed. Fig. 10-8 is a test for normality. The plot points do not follow the expected line for a normal distribution and the p value is 0.000. The chance that this data came from a normal distribution is almost zero.

This has the shape of a lognormal distribution, which occurs often in mechanical and electrical measurements. The measurements tend to stack up near zero because that is the natural limit. For example, shafts cannot be better than zero FIM and electrical resistance cannot be less than zero.

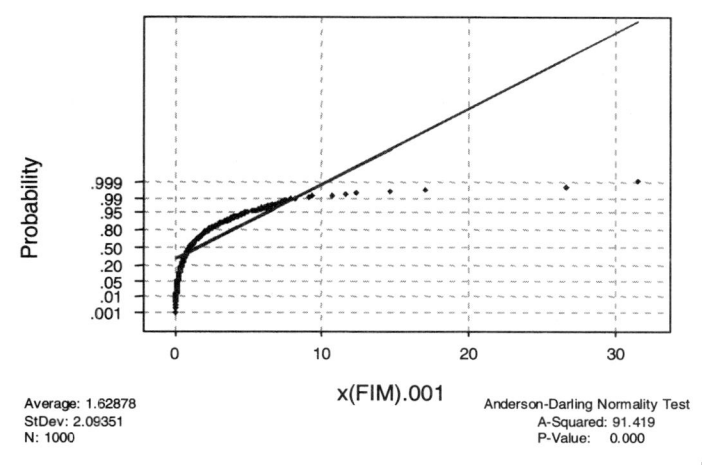

Figure 10-8 Normality test FIM

There are two ways to handle the lognormal distribution. One is to transform the value of the x's by using the relationship:

$y = ln(x)$,

And plot a new histogram (Fig. 10-9).

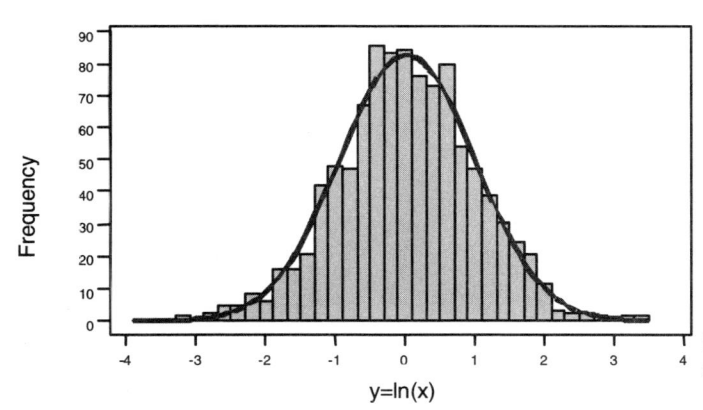

Figure 10-9 Histogram of transformed FIM measurements

This new histogram looks like a good approximation to a normal curve. It passes the Anderson-Darling test for normality (Fig. 10-10), and we can now apply the usual statistics to this transformed set of data.

The second way to work with lognormal distributions is to perform the calculations directly on the lognormal data using a statistical software package like *Minitab 12*. This software can calculate and plot all the relevant statistics from most distributions.

In either case, we can determine the probability of exceeding a value like a specification limit.

The probabilities are *additive* for each dimension or feature of a part or system. This additive property allows a design team to estimate the probability of a defect at any level in the system.

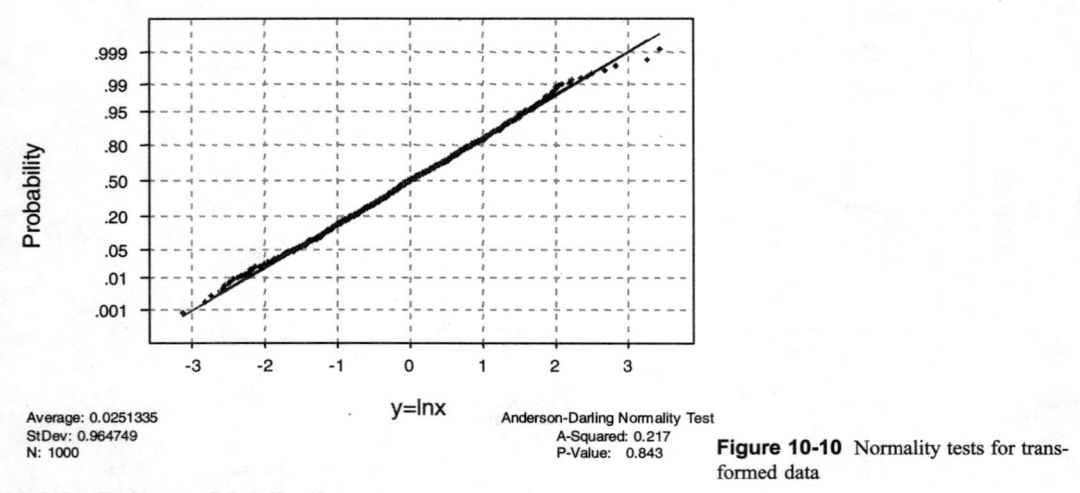

[Figure 10-10: Normality plot with axes showing Probability (.001 to .999) vs y=lnx (-3 to 3)]

Average: 0.0251335
StDev: 0.964749
N: 1000

Anderson-Darling Normality Test
A-Squared: 0.217
P-Value: 0.843

Figure 10-10 Normality tests for transformed data

10.3.3 Poisson Distribution

Discrete data that is classified by pass or fail, heads or tails, is called attributes data. Attributes data can be distributed according to:

- A uniform distribution of probability
- The hypergeometric distribution
- The binomial distribution or
- The Poisson distribution

Figure 10-11 shows an example of attributes data.

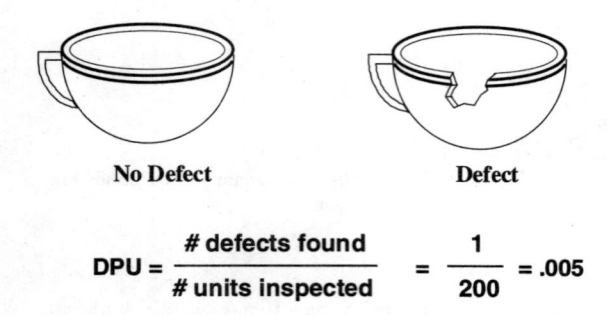

[No Defect cup illustration] [Defect cup illustration]

$$\text{DPU} = \frac{\text{\# defects found}}{\text{\# units inspected}} = \frac{1}{200} = .005$$

Figure 10-11 Attributes data

The Poisson can be applied to many randomly occurring phenomena over time or space. Consider the following scenarios:

- The number of disk drive failures per month for a particular type of disk drive
- The number of dental cavities per 12-year-old child
- The number of particles per square centimeter on a silicon wafer
- The number of calls arriving at an emergency dispatch station per hour
- The number of defects occurring in a day's production of radar units
- The number of chocolate chips per cookie

The Poisson can model each of these scenarios. The Poisson random variable is characterized by the form "the number of occurrences per unit interval," where an occurrence could be a defect, a mechanical or electrical failure, an arrival, a departure, or a chocolate chip. The unit could be a unit of time, or a unit of space, or a physical unit like a radar or a cookie, or a person.

The probability distribution function for the Poisson is: $P(X = x) = (\lambda^x e^{-\lambda})/x!$
where
P is the probability that a single unit has x occurrences
λ is a positive constant representing "the average number of occurrences per unit interval"
x is a nonnegative integer and is the specified number of occurrences per unit interval
e is the number whose natural logarithm is 1, and is equal to approximately 2.71828.

For example, suppose we had the following information about a product:
- 1,000 units were inspected and 519 defects were observed.

We want to:
- calculate the number of defects per unit (DPU), and
- estimate the number of units that have exactly three defects (X=3).

The overall rate (λ) that defects occur is: 519/1000 = 0.519 defects per unit (DPU). For X = 3 defects (exactly 3 defects on a unit), the probability is:

$$P(X = 3) = [(\lambda^3)(e^{-\lambda})]/3!$$
$$\lambda = 519/1000 = 0.519$$
$$P(X = 3) = 0.01387$$

The probability that a unit has exactly 3 defects is 0.01387. So, for 1,000 units we would expect 14 units to have exactly 3 defects each. Table 10-1 enumerates the distribution of the 519 defects.

Table 10-1 Distribution of defects

X (number of defects)	P(X)	Number of Units	Defects
0	0.5951	595	0
1	0.3088	309	309
2	0.0802	80	160
3	0.0139	14	42
4	0.0018	2	8
5	0.0002	0	0
6	0.0000	0	0
7	0.0000	0	0
Total	1.0000	1,000	519

The distribution appears graphically in Fig. 10-12.

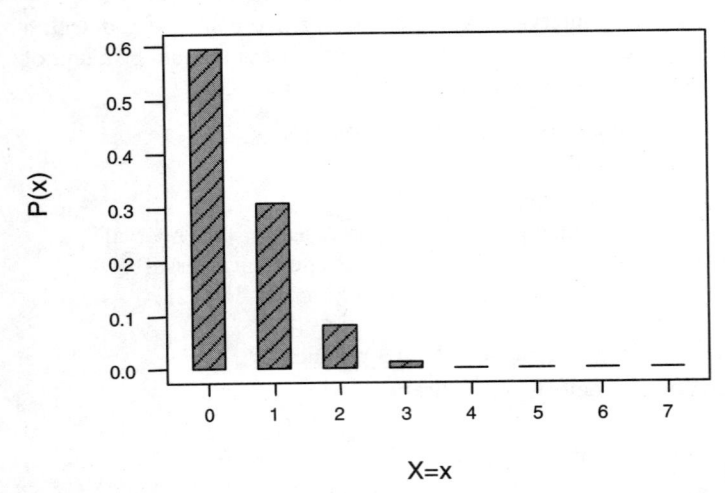

Figure 10-12 Plot of Poisson probabilities

How do I estimate yield from DPU?

To produce a unit of product with zero defects, we need to know the probability of zero defects. Recalling the Poisson equation above,

$$P(X = x) = (\lambda^x e^{-\lambda})/x!$$

Substituting DPU for λ, and solving for x = 0, we have $P(0) = e^{-DPU}$

To yield good product, there must be no defects. Therefore, the first time yield is : $FTY = e^{-DPU}$. First time yield is a function of how many defects there are. Zero *DPU* means that *FTY*=100%. This agrees with our intuition that if there are no defects, the yield must be 100%.

How do I estimate parts per million (PPM) from yield?

PPM is a measure of the estimated number of defects that are expected from a process if a million units were made. Parts per million defective is: $PPM = (1-FTY)(1,000,000)$.

10.4 Measures of Quality and Capability

10.4.1 Process Capability Index

Historically, process capability has been defined by industry as + or - 3σ (Fig. 10-13). For any one feature or process output, plus or minus 3 sigma gives good results 99.73% of the time with a normal

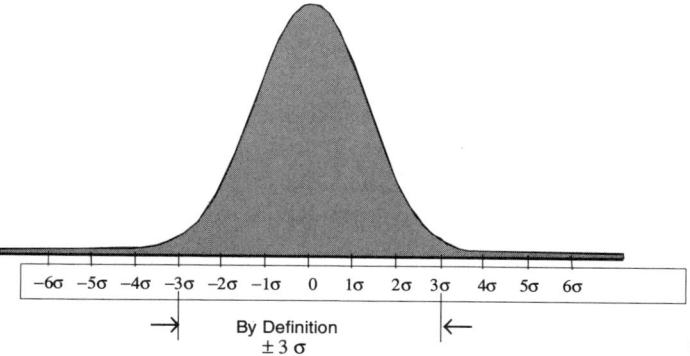

Figure 10-13 Process capability

distribution. This is certainly adequate, especially when dealing with a few features. From this concept came the Process Capability Index (Cp), defined in Fig. 10-14.

$$Cp = \frac{\text{Spec Width}}{\text{Mfg Capability}} = \frac{USL - LSL}{\pm 3\sigma}$$

"Concurrent Engineering Index"
Design / Manufacturing

Figure 10-14 Capability index

The automotive industry, with leadership from Ford Motor Company, set the design standard of Cp=1.33 in the early 1980s, which corresponds to a process capability of ±4 sigma (Fig. 10-15). This standard has been upgraded since that time, but it is important to note that the product designers had a standard to meet, and that *implied knowing the capability of the process.*

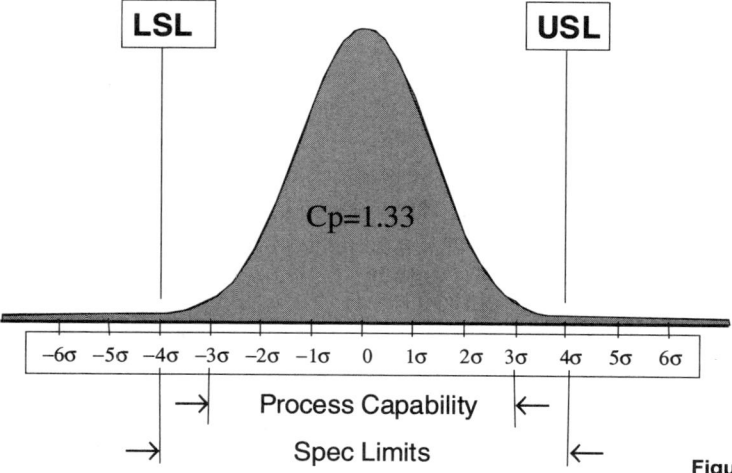

Figure 10-15 Capability index at ± 4 sigma

The Cp index can be thought of as the *concurrent engineering index*. The design engineers have responsibility for the specifications (the numerator), and the process engineers have responsibility for the capability (the denominator). Today's integrated product teams should know the Cp index for each critical-to-quality characteristic.

10.4.2 Process Capability Index Relative to Process Centering (Cpk)

The Cp index has a shortcoming. It does not account for shifts and drifts that occur during the long-term course of manufacturing. Another index is needed to account for shifts in the centering. See Fig. 10-16.

With Six Sigma, the process mean can shift 1.5 standard deviations (see Chapter 1) even when the process is monitored using modern statistical process control (SPC). Certainly, once the shift is detected, corrective action is taken, but the ability to detect a shift in the process on the next sample is small. (It can be shown that for the common x-bar and range chart method with sample size of 5, the probability of detecting a 1.5 sigma shift on the next sample is about 0.50.)

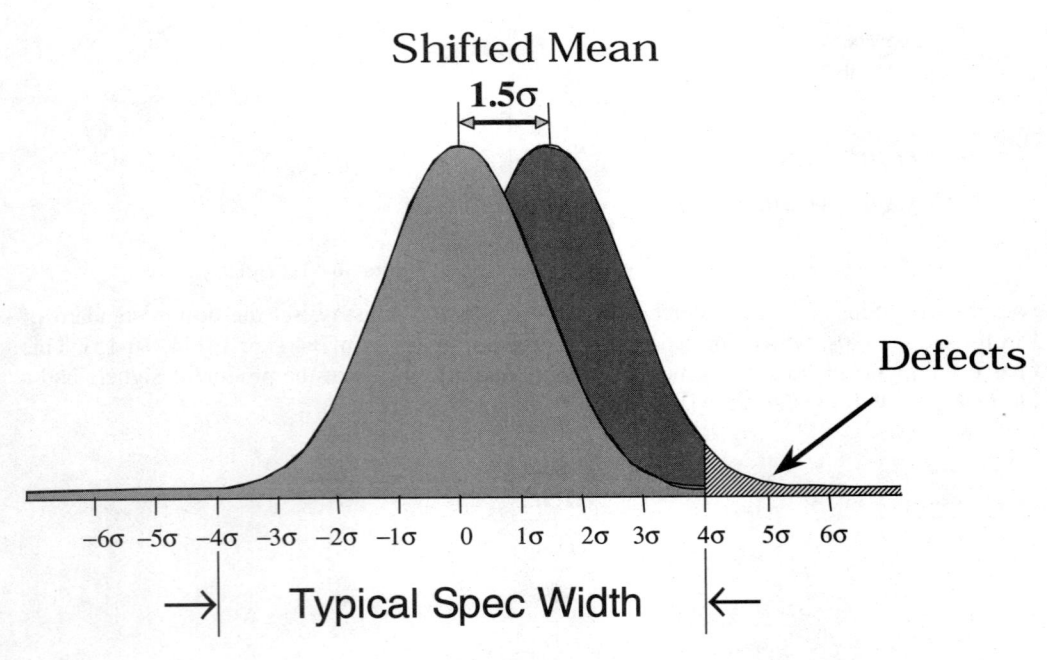

Figure 10-16 The reality

Another index is needed to indicate process centering. Cpk is the process capability index adjusted for centering. It is defined as:

Cpk = Cp(1-k)

where k is the ratio of the amount the center has moved off target divided by the amount from the center to the nearest specification limit. See Fig. 10-17.

If the design target is ±6 sigma, then Cp = 2, and Cpk = 1.5. If every critical-to-quality (CTQ) characteristic is at ±6 sigma, then the probability of all the CTQs being good simultaneously is very high. There would be only 3.4 defects for every 1 million CTQs. See Figs. 10-17 and 10-18.

$Cp = 2$
$k = {}^a/_b$
$a = 1.5\sigma$
$b = 6\sigma$
$Cpk = Cp(1-k)$
$= 2(1-.25) = 1.5$

Shifted Mean

3.4 ppm

$-6\sigma\ -5\sigma\ -4\sigma\ -3\sigma\ -2\sigma\ -1\sigma\ 0\ 1\sigma\ 2\sigma\ 3\sigma\ 4\sigma\ 5\sigma\ 6\sigma$

Spec Limits

Process Capability

Figure 10-17 Cp and Cpk at Six Sigma

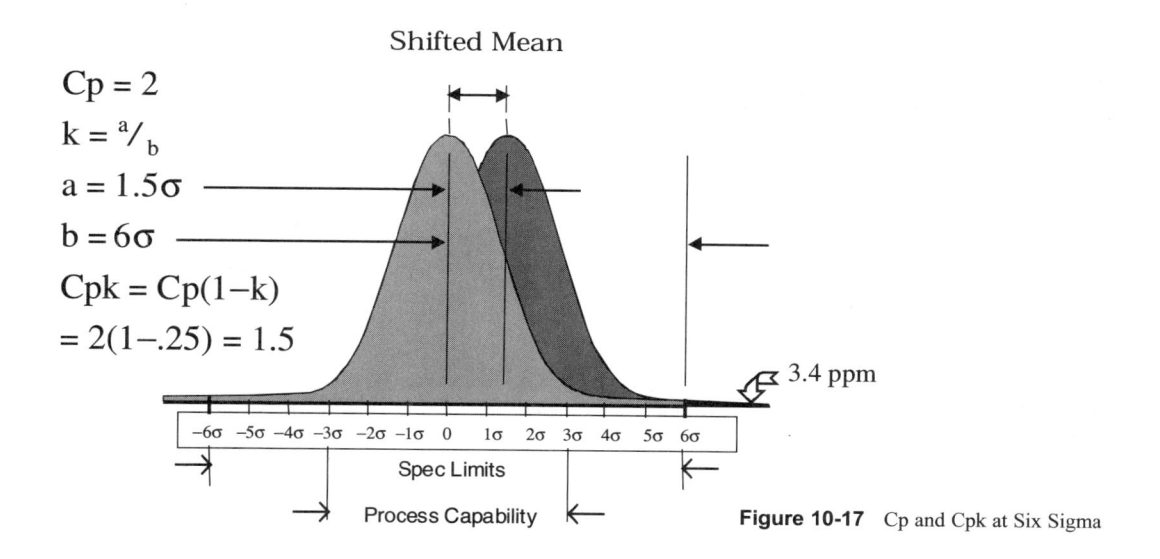

Distribution Shifted 1.5σ

CTQs	±3σ	±4σ	±5σ	±6σ
1	93.32%	99.379%	99.9767%	99.99966%
10	50.08	93.96	99.768	99.9966
30	12.57	82.95	99.30	99.99
50	---	73.24	98.84	99.98
100	---	53.64	97.70	99.966
150	---	39.28	96.57	99.948
200	---	28.77	95.45	99.931
300	---	15.43	93.26	99.897
400	---	8.28	91.11	99.862
500	---	4.44	89.02	99.828
800	---	00.69	83.02	99.724
1200	---	00.06	75.63	99.587

Figure 10-18 Yields through multiple CTQs

10.5 Summary

"We should design products in light of that variation which we know is inevitable rather than in the darkness of chance." –Mikel J. Harry

Estimating the variation that will occur in the parts, materials, processes, and product features is the responsibility of the design team. Estimates of product performance and manufacturability can be made long before production. Statistics can help estimate the most likely outcome, and how much variation there is likely to be in that outcome. Changes made early in the design process are easier and less costly than changes made after production has started. Six Sigma design is the application of statistical techniques to analyze and optimize the inherent system design margins. The objective is a design that can be built error free.

10.6 References

1. D'Augostino and M.A. Stevens, Eds. 1986. *Goodness-of-Fit Techniques.* New York, NY: Marcel Dekker.
2. Harry, Mikel, and J.R. Lawson. 1990. *Six Sigma Producibility Analysis and Process Characterization.* Schaumburg, Illinois: Motorola University Press.
3. Juran, J.M. and Frank M. Gryna. 1988. *Juran's Quality Control Handbook.* 4th ed. New York, NY: McGraw-Hill.
4. Kiemele, Mark J., Stephen R. Schmidt, and Ronald J. Berdine. 1997. *Basic Statistics: Tools for Continuous Improvement.* 4th ed. Colorado Springs, Colorado: Air Academy Press.
5. Microsoft Corporation, 1997, Microsoft® Excel 97 SR-1. Redmond, Washington: Microsoft Corporation.
6. Minitab, Inc. 1997. Minitab Release 12 for Windows. State College, PA: Minitab, Inc.

10.7 Appendix

Table of Unilateral Tail Under the Normal Curve Beyond Selected Z Values

	0	0.01	0.02	0.03	0.04	0.05	0.06	0.07	0.08	0.09
0	5.0000E-01	4.9601E-01	4.9202E-01	4.8803E-01	4.8405E-01	4.8006E-01	4.7608E-01	4.7210E-01	4.6812E-01	4.6414E-01
0.1	4.6017E-01	4.5620E-01	4.5224E-01	4.4828E-01	4.4433E-01	4.4038E-01	4.3644E-01	4.3251E-01	4.2858E-01	4.2465E-01
0.2	4.2074E-01	4.1683E-01	4.1294E-01	4.0905E-01	4.0517E-01	4.0129E-01	3.9743E-01	3.9358E-01	3.8974E-01	3.8591E-01
0.3	3.8209E-01	3.7828E-01	3.7448E-01	3.7070E-01	3.6693E-01	3.6317E-01	3.5942E-01	3.5569E-01	3.5197E-01	3.4827E-01
0.4	3.4458E-01	3.4090E-01	3.3724E-01	3.3360E-01	3.2997E-01	3.2636E-01	3.2276E-01	3.1918E-01	3.1561E-01	3.1207E-01
0.5	3.0854E-01	3.0503E-01	3.0153E-01	2.9806E-01	2.9460E-01	2.9116E-01	2.8774E-01	2.8434E-01	2.8096E-01	2.7760E-01
0.6	2.7425E-01	2.7093E-01	2.6763E-01	2.6435E-01	2.6109E-01	2.5785E-01	2.5463E-01	2.5143E-01	2.4825E-01	2.4510E-01
0.7	2.4196E-01	2.3885E-01	2.3576E-01	2.3270E-01	2.2965E-01	2.2663E-01	2.2363E-01	2.2065E-01	2.1770E-01	2.1476E-01
0.8	2.1186E-01	2.0897E-01	2.0611E-01	2.0327E-01	2.0045E-01	1.9766E-01	1.9489E-01	1.9215E-01	1.8943E-01	1.8673E-01
0.9	1.8406E-01	1.8141E-01	1.7879E-01	1.7619E-01	1.7361E-01	1.7106E-01	1.6853E-01	1.6602E-01	1.6354E-01	1.6109E-01
1	1.5866E-01	1.5625E-01	1.5386E-01	1.5151E-01	1.4917E-01	1.4686E-01	1.4457E-01	1.4231E-01	1.4007E-01	1.3786E-01
1.1	1.3567E-01	1.3350E-01	1.3136E-01	1.2924E-01	1.2714E-01	1.2507E-01	1.2302E-01	1.2100E-01	1.1900E-01	1.1702E-01
1.2	1.1507E-01	1.1314E-01	1.1123E-01	1.0935E-01	1.0749E-01	1.0565E-01	1.0383E-01	1.0204E-01	1.0027E-01	9.8525E-02
1.3	9.6800E-02	9.5098E-02	9.3417E-02	9.1759E-02	9.0123E-02	8.8508E-02	8.6915E-02	8.5343E-02	8.3793E-02	8.2264E-02
1.4	8.0757E-02	7.9270E-02	7.7804E-02	7.6358E-02	7.4934E-02	7.3529E-02	7.2145E-02	7.0781E-02	6.9437E-02	6.8112E-02
1.5	6.6807E-02	6.5522E-02	6.4255E-02	6.3008E-02	6.1780E-02	6.0571E-02	5.9380E-02	5.8207E-02	5.7053E-02	5.5917E-02
1.6	5.4799E-02	5.3699E-02	5.2616E-02	5.1551E-02	5.0503E-02	4.9471E-02	4.8457E-02	4.7460E-02	4.6479E-02	4.5514E-02
1.7	4.4565E-02	4.3633E-02	4.2716E-02	4.1815E-02	4.0930E-02	4.0059E-02	3.9204E-02	3.8364E-02	3.7538E-02	3.6727E-02
1.8	3.5930E-02	3.5148E-02	3.4380E-02	3.3625E-02	3.2884E-02	3.2157E-02	3.1443E-02	3.0742E-02	3.0054E-02	2.9379E-02
1.9	2.8717E-02	2.8067E-02	2.7429E-02	2.6804E-02	2.6190E-02	2.5588E-02	2.4998E-02	2.4419E-02	2.3852E-02	2.3296E-02
2	2.2750E-02	2.2216E-02	2.1692E-02	2.1178E-02	2.0675E-02	2.0182E-02	1.9699E-02	1.9226E-02	1.8763E-02	1.8309E-02
2.1	1.7865E-02	1.7429E-02	1.7003E-02	1.6586E-02	1.6177E-02	1.5778E-02	1.5386E-02	1.5004E-02	1.4629E-02	1.4262E-02
2.2	1.3904E-02	1.3553E-02	1.3209E-02	1.2874E-02	1.2546E-02	1.2225E-02	1.1911E-02	1.1604E-02	1.1304E-02	1.1011E-02
2.3	1.0724E-02	1.0444E-02	1.0170E-02	9.9031E-03	9.6419E-03	9.3867E-03	9.1375E-03	8.8940E-03	8.6563E-03	8.4242E-03
2.4	8.1975E-03	7.9762E-03	7.7602E-03	7.5494E-03	7.3436E-03	7.1428E-03	6.9468E-03	6.7556E-03	6.5691E-03	6.3871E-03
2.5	6.2096E-03	6.0365E-03	5.8677E-03	5.7030E-03	5.5425E-03	5.3861E-03	5.2335E-03	5.0848E-03	4.9399E-03	4.7987E-03

10-16　Chapter Ten

	0	0.01	0.02	0.03	0.04	0.05	0.06	0.07	0.08	0.09
2.6	4.6611E-03	4.5270E-03	4.3964E-03	4.2691E-03	4.1452E-03	4.0245E-03	3.9069E-03	3.7924E-03	3.6810E-03	3.5725E-03
2.7	3.4668E-03	3.3640E-03	3.2640E-03	3.1666E-03	3.0718E-03	2.9796E-03	2.8899E-03	2.8027E-03	2.7178E-03	2.6353E-03
2.8	2.5550E-03	2.4769E-03	2.4011E-03	2.3273E-03	2.2556E-03	2.1858E-03	2.1181E-03	2.0522E-03	1.9883E-03	1.9261E-03
2.9	1.8657E-03	1.8070E-03	1.7500E-03	1.6947E-03	1.6410E-03	1.5888E-03	1.5381E-03	1.4889E-03	1.4411E-03	1.3948E-03
3	1.3498E-03	1.3062E-03	1.2638E-03	1.2227E-03	1.1828E-03	1.1441E-03	1.1066E-03	1.0702E-03	1.0349E-03	1.0007E-03
3.1	9.6755E-04	9.3539E-04	9.0421E-04	8.7400E-04	8.4471E-04	8.1632E-04	7.8882E-04	7.6217E-04	7.3636E-04	7.1135E-04
3.2	6.8713E-04	6.6367E-04	6.4095E-04	6.1896E-04	5.9766E-04	5.7704E-04	5.5708E-04	5.3776E-04	5.1906E-04	5.0097E-04
3.3	4.8346E-04	4.6652E-04	4.5013E-04	4.3427E-04	4.1894E-04	4.0411E-04	3.8977E-04	3.7590E-04	3.6249E-04	3.4953E-04
3.4	3.3700E-04	3.2489E-04	3.1318E-04	3.0187E-04	2.9094E-04	2.8038E-04	2.7017E-04	2.6032E-04	2.5080E-04	2.4160E-04
3.5	2.3272E-04	2.2415E-04	2.1587E-04	2.0788E-04	2.0017E-04	1.9272E-04	1.8554E-04	1.7860E-04	1.7191E-04	1.6545E-04
3.6	1.5922E-04	1.5322E-04	1.4742E-04	1.4183E-04	1.3644E-04	1.3124E-04	1.2623E-04	1.2140E-04	1.1674E-04	1.1225E-04
3.7	1.0793E-04	1.0376E-04	9.9739E-05	9.5868E-05	9.2138E-05	8.8546E-05	8.5086E-05	8.1753E-05	7.8543E-05	7.5453E-05
3.8	7.2477E-05	6.9613E-05	6.6855E-05	6.4201E-05	6.1646E-05	5.9187E-05	5.6822E-05	5.4545E-05	5.2355E-05	5.0249E-05
3.9	4.8222E-05	4.6273E-05	4.4399E-05	4.2597E-05	4.0864E-05	3.9198E-05	3.7596E-05	3.6057E-05	3.4577E-05	3.3155E-05
4	3.1789E-05	3.0476E-05	2.9215E-05	2.8003E-05	2.6839E-05	2.5721E-05	2.4648E-05	2.3617E-05	2.2627E-05	2.1676E-05
4.1	2.0764E-05	1.9888E-05	1.9047E-05	1.8241E-05	1.7466E-05	1.6723E-05	1.6011E-05	1.5327E-05	1.4671E-05	1.4042E-05
4.2	1.3439E-05	1.2860E-05	1.2305E-05	1.1773E-05	1.1263E-05	1.0774E-05	1.0306E-05	9.8568E-06	9.4264E-06	9.0140E-06
4.3	8.6189E-06	8.2403E-06	7.8777E-06	7.5303E-06	7.1976E-06	6.8790E-06	6.5739E-06	6.2817E-06	6.0020E-06	5.7343E-06
4.4	5.4780E-06	5.2327E-06	4.9979E-06	4.7732E-06	4.5582E-06	4.3525E-06	4.1558E-06	3.9675E-06	3.7875E-06	3.6153E-06
4.5	3.4506E-06	3.2932E-06	3.1426E-06	2.9987E-06	2.8611E-06	2.7295E-06	2.6038E-06	2.4837E-06	2.3689E-06	2.2592E-06
4.6	2.1544E-06	2.0543E-06	1.9586E-06	1.8673E-06	1.7800E-06	1.6967E-06	1.6171E-06	1.5412E-06	1.4686E-06	1.3994E-06
4.7	1.3333E-06	1.2702E-06	1.2101E-06	1.1526E-06	1.0978E-06	1.0455E-06	9.9562E-07	9.4803E-07	9.0263E-07	8.5934E-07
4.8	8.1805E-07	7.7868E-07	7.4115E-07	7.0536E-07	6.7124E-07	6.3872E-07	6.0772E-07	5.7818E-07	5.5003E-07	5.2320E-07
4.9	4.9764E-07	4.7329E-07	4.5009E-07	4.2800E-07	4.0695E-07	3.8691E-07	3.6782E-07	3.4965E-07	3.3234E-07	3.1587E-07
5	3.0019E-07	2.8526E-07	2.7105E-07	2.5753E-07	2.4466E-07	2.3242E-07	2.2077E-07	2.0969E-07	1.9915E-07	1.8912E-07

Statistical Background and Concepts   10-17

	0	0.01	0.02	0.03	0.04	0.05	0.06	0.07	0.08	0.09
5.1	1.7958E-07	1.7051E-07	1.6189E-07	1.5369E-07	1.4589E-07	1.3848E-07	1.3143E-07	1.2473E-07	1.1837E-07	1.1231E-07
5.2	1.0656E-07	1.0110E-07	9.5910E-08	9.0978E-08	8.6293E-08	8.1843E-08	7.7616E-08	7.3602E-08	6.9790E-08	6.6170E-08
5.3	6.2733E-08	5.9469E-08	5.6371E-08	5.3431E-08	5.0640E-08	4.7991E-08	4.5477E-08	4.3091E-08	4.0827E-08	3.8680E-08
5.4	3.6642E-08	3.4709E-08	3.2876E-08	3.1137E-08	2.9488E-08	2.7924E-08	2.6441E-08	2.5035E-08	2.3702E-08	2.2438E-08
5.5	2.1240E-08	2.0104E-08	1.9028E-08	1.8008E-08	1.7042E-08	1.6126E-08	1.5258E-08	1.4436E-08	1.3657E-08	1.2919E-08
5.6	1.2221E-08	1.1559E-08	1.0932E-08	1.0338E-08	9.7764E-09	9.2443E-09	8.7405E-09	8.2636E-09	7.8121E-09	7.3848E-09
5.7	6.9804E-09	6.5976E-09	6.2354E-09	5.8927E-09	5.5684E-09	5.2616E-09	4.9714E-09	4.6968E-09	4.4371E-09	4.1915E-09
5.8	3.9592E-09	3.7395E-09	3.5318E-09	3.3353E-09	3.1496E-09	2.9740E-09	2.8081E-09	2.6512E-09	2.5029E-09	2.3627E-09
5.9	2.2303E-09	2.1051E-09	1.9868E-09	1.8751E-09	1.7695E-09	1.6698E-09	1.5755E-09	1.4865E-09	1.4024E-09	1.3230E-09
6	1.2481E-09	1.1773E-09	1.1104E-09	1.0473E-09	9.8765E-10	9.3138E-10	8.7825E-10	8.2811E-10	7.8078E-10	7.3611E-10
6.1	6.9395E-10	6.5417E-10	6.1663E-10	5.8121E-10	5.4779E-10	5.1626E-10	4.8651E-10	4.5845E-10	4.3199E-10	4.0702E-10
6.2	3.8348E-10	3.6128E-10	3.4034E-10	3.2060E-10	3.0198E-10	2.8443E-10	2.6788E-10	2.5228E-10	2.3758E-10	2.2372E-10
6.3	2.1065E-10	1.9834E-10	1.8674E-10	1.7580E-10	1.6550E-10	1.5579E-10	1.4665E-10	1.3803E-10	1.2991E-10	1.2226E-10
6.4	1.1506E-10	1.0827E-10	1.0188E-10	9.5864E-11	9.0196E-11	8.4858E-11	7.9833E-11	7.5100E-11	7.0645E-11	6.6450E-11
6.5	6.2502E-11	5.8784E-11	5.5285E-11	5.1992E-11	4.8892E-11	4.5975E-11	4.3229E-11	4.0646E-11	3.8214E-11	3.5927E-11
6.6	3.3775E-11	3.1750E-11	2.9845E-11	2.8053E-11	2.6367E-11	2.4781E-11	2.3290E-11	2.1887E-11	2.0568E-11	1.9327E-11
6.7	1.8160E-11	1.7063E-11	1.6032E-11	1.5062E-11	1.4150E-11	1.3293E-11	1.2487E-11	1.1729E-11	1.1017E-11	1.0348E-11
6.8	9.7185E-12	9.1272E-12	8.5715E-12	8.0493E-12	7.5585E-12	7.0974E-12	6.6641E-12	6.2570E-12	5.8745E-12	5.5151E-12
6.9	5.1775E-12	4.8604E-12	4.5625E-12	4.2827E-12	4.0198E-12	3.7730E-12	3.5411E-12	3.3234E-12	3.1189E-12	2.9269E-12
7	2.7466E-12	2.5773E-12	2.4183E-12	2.2691E-12	2.1290E-12	1.9974E-12	1.8740E-12	1.7580E-12	1.6492E-12	1.5471E-12
7.1	1.4512E-12	1.3612E-12	1.2768E-12	1.1975E-12	1.1232E-12	1.0534E-12	9.8787E-13	9.2642E-13	8.6875E-13	8.1465E-13
7.2	7.6389E-13	7.1627E-13	6.7159E-13	6.2968E-13	5.9036E-13	5.5348E-13	5.1888E-13	4.8643E-13	4.5600E-13	4.2745E-13
7.3	4.0068E-13	3.7558E-13	3.5203E-13	3.2995E-13	3.0925E-13	2.8983E-13	2.7163E-13	2.5456E-13	2.3855E-13	2.2355E-13
7.4	2.0948E-13	1.9629E-13	1.8393E-13	1.7234E-13	1.6148E-13	1.5129E-13	1.4175E-13	1.3280E-13	1.2441E-13	1.1655E-13
7.5	1.0919E-13	1.0228E-13	9.5813E-14	8.9749E-14	8.4068E-14	7.8743E-14	7.3754E-14	6.9080E-14	6.4700E-14	6.0596E-14

	0	0.01	0.02	0.03	0.04	0.05	0.06	0.07	0.08	0.09
7.6	5.6750E-14	5.3148E-14	4.9773E-14	4.6611E-14	4.3648E-14	4.0873E-14	3.8274E-14	3.5839E-14	3.3558E-14	3.1421E-14
7.7	2.9420E-14	2.7546E-14	2.5790E-14	2.4146E-14	2.2606E-14	2.1164E-14	1.9813E-14	1.8548E-14	1.7364E-14	1.6255E-14
7.8	1.5216E-14	1.4243E-14	1.3333E-14	1.2480E-14	1.1682E-14	1.0934E-14	1.0234E-14	9.5786E-15	8.9651E-15	8.3906E-15
7.9	7.8529E-15	7.3494E-15	6.8781E-15	6.4370E-15	6.0239E-15	5.6373E-15	5.2754E-15	4.9367E-15	4.6196E-15	4.3228E-15
8	4.0450E-15	3.7850E-15	3.5417E-15	3.3139E-15	3.1008E-15	2.9013E-15	2.7145E-15	2.5398E-15	2.3763E-15	2.2233E-15
8.1	2.0801E-15	1.9460E-15	1.8206E-15	1.7033E-15	1.5935E-15	1.4907E-15	1.3946E-15	1.3046E-15	1.2205E-15	1.1417E-15
8.2	1.0680E-15	9.9906E-16	9.3455E-16	8.7420E-16	8.1773E-16	7.6491E-16	7.1548E-16	6.6924E-16	6.2599E-16	5.8552E-16
8.3	5.4766E-16	5.1224E-16	4.7911E-16	4.4812E-16	4.1913E-16	3.9201E-16	3.6664E-16	3.4291E-16	3.2071E-16	2.9994E-16
8.4	2.8052E-16	2.6236E-16	2.4536E-16	2.2947E-16	2.1460E-16	2.0070E-16	1.8769E-16	1.7553E-16	1.6415E-16	1.5351E-16
8.5	1.4356E-16	1.3425E-16	1.2554E-16	1.1740E-16	1.0979E-16	1.0267E-16	9.6007E-17	8.9779E-17	8.3954E-17	7.8507E-17
8.6	7.3412E-17	6.8648E-17	6.4193E-17	6.0026E-17	5.6130E-17	5.2486E-17	4.9079E-17	4.5892E-17	4.2913E-17	4.0126E-17
8.7	3.7521E-17	3.5084E-17	3.2806E-17	3.0675E-17	2.8683E-17	2.6820E-17	2.5078E-17	2.3449E-17	2.1926E-17	2.0501E-17
8.8	1.9169E-17	1.7924E-17	1.6760E-17	1.5671E-17	1.4653E-17	1.3701E-17	1.2810E-17	1.1978E-17	1.1200E-17	1.0472E-17
8.9	9.7916E-18	9.1553E-18	8.5604E-18	8.0042E-18	7.4841E-18	6.9978E-18	6.5431E-18	6.1180E-18	5.7204E-18	5.3487E-18
9	5.0012E-18	4.6762E-18	4.3724E-18	4.0883E-18	3.8227E-18	3.5744E-18	3.3421E-18	3.1250E-18	2.9220E-18	2.7322E-18
9.1	2.5547E-18	2.3888E-18	2.2336E-18	2.0885E-18	1.9529E-18	1.8260E-18	1.7074E-18	1.5966E-18	1.4929E-18	1.3959E-18
9.2	1.3053E-18	1.2206E-18	1.1413E-18	1.0672E-18	9.9795E-19	9.3317E-19	8.7260E-19	8.1597E-19	7.6301E-19	7.1350E-19
9.3	6.6720E-19	6.2391E-19	5.8343E-19	5.4559E-19	5.1020E-19	4.7710E-19	4.4616E-19	4.1723E-19	3.9017E-19	3.6487E-19
9.4	3.4122E-19	3.1910E-19	2.9841E-19	2.7907E-19	2.6099E-19	2.4407E-19	2.2826E-19	2.1347E-19	1.9964E-19	1.8671E-19
9.5	1.7462E-19	1.6331E-19	1.5274E-19	1.4285E-19	1.3360E-19	1.2495E-19	1.1687E-19	1.0930E-19	1.0223E-19	9.5617E-20
9.6	8.9432E-20	8.3648E-20	7.8238E-20	7.3179E-20	6.8448E-20	6.4023E-20	5.9885E-20	5.6015E-20	5.2395E-20	4.9010E-20
9.7	4.5844E-20	4.2883E-20	4.0114E-20	3.7524E-20	3.5101E-20	3.2836E-20	3.0716E-20	2.8734E-20	2.6880E-20	2.5146E-20
9.8	2.3525E-20	2.2008E-20	2.0589E-20	1.9261E-20	1.8020E-20	1.6859E-20	1.5772E-20	1.4756E-20	1.3806E-20	1.2917E-20
9.9	1.2085E-20	1.1307E-20	1.0579E-20	9.8985E-21	9.2616E-21	8.6658E-21	8.1084E-21	7.5870E-21	7.0992E-21	6.6429E-21

This table was generated using Microsoft ® Excel (Reference 4) and the Z equation from Reference 2.

Chapter 11

Predicting Assembly Quality (Six Sigma Methodologies to Optimize Tolerances)

Dale Van Wyk
Raytheon Systems Company
McKinney, Texas

Mr. Van Wyk has more than 14 years of experience with mechanical tolerance analysis and mechanical design at Texas Instruments' Defense Group, which became part of Raytheon Systems Company. In addition to direct design work, he has developed courses for mechanical tolerancing and application of statistical principles to systems design. He has also participated in development of a U.S. Air Force training class, teaching techniques to use statistics in creating affordable products. He has written several papers and delivered numerous presentations about the use of statistical techniques for mechanical tolerancing. Mr. Van Wyk has a BSME from Iowa State University and a MSME from Southern Methodist University.

11.1 Introduction

We introduced the traditional approaches to tolerance analysis in Chapter 9. At that time, we noted several assumptions and limitations that (perhaps not obvious to you) are particularly important in the root sum of squares and modified root sum of squares techniques. These assumptions and limitations introduce some risk that defects will occur during the assembly process. The problem: There is no way to understand the magnitude of this risk or to estimate the number of defects that will occur. For example, if you change a tolerance from .010 to .005, the RSS Model would assume that a different process with a higher precision would be used to manufacture it. This is not necessarily true.

11.2 What Is Tolerance Allocation?

In this chapter, we will introduce and demonstrate methods of tolerance allocation. Fig. 11-1 shows how tolerance allocation differs from tolerance analysis. Tolerance analysis is a process where we assign tolerances to each component and determine how well we meet a goal or requirement. If we don't meet the goal, we reassign or resize the tolerances until the goal is met. It is by nature an iterative process.

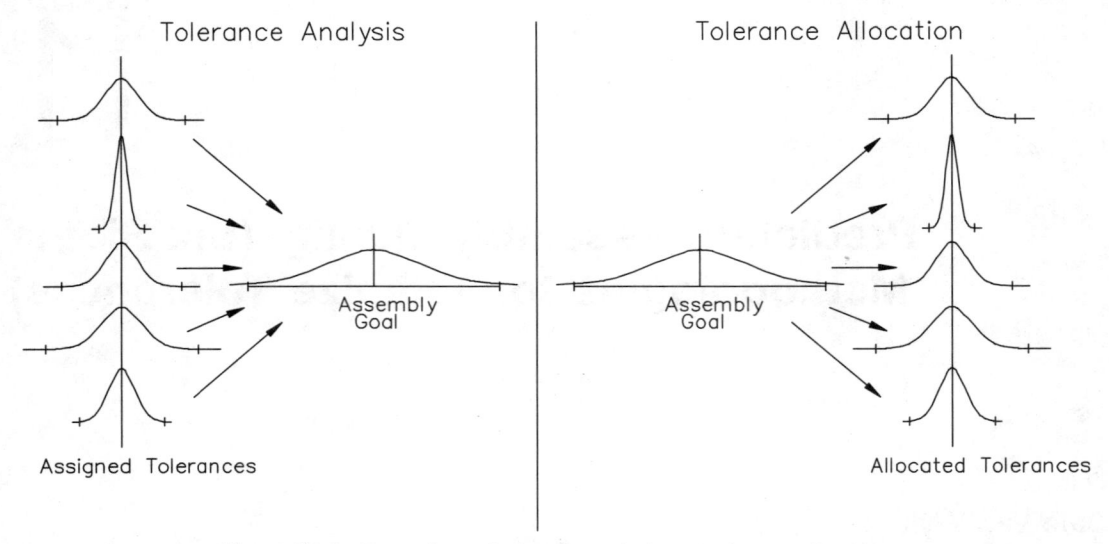

Figure 11-1 Comparison of tolerance analysis and tolerance allocation

With tolerance allocation, we will present methods that will allow us to determine the tolerance to assign to each of the components with the minimum number of iterations. We will start with the defined goal for the assembly, decide how each component part will be manufactured, and allocate tolerances so that the components can be economically produced and the assembly will meet its requirements.

11.3 Process Standard Deviations

Prior to performing a tolerance allocation, we need to know how we're going to manufacture each component part. We'll use this information, along with historical knowledge about how the process has performed in the past, to select an expected value for the standard deviation of the process. We will use this in a similar manner to what was introduced in Chapter 10 and make estimates of both assembly and component defect rates. In addition we will use data such as this to assign tolerances to each of the components that contribute to satisfying an assembly requirement.

In recent years, many companies have introduced statistical process control as a means to minimize defects that occur during the manufacturing process. This not only works very well to detect processes that are in danger of producing defective parts prior to the time defects arise, but also provides data that can be used to predict how well parts can be manufactured even before the design is complete. Of interest to us is the data collected on individual features. For example, suppose a part is being designed and is expected to be produced using a milling operation. A review of data for similar parts manufactured using a milling process shows a typical standard deviation of .0003 inch. We can use this data as a basis for allocating tolerances to future designs that will use a similar process. It is extremely important to understand how the parts are going to be manufactured prior to assigning standard deviations. Failure to do so will yield unreliable results, and potentially unreliable designs. For example, if you conduct an analysis assuming a feature will be machined on a jig bore, and it is actually manufactured on a mill, the latter is less precise, and has a larger standard deviation. This will lead to a higher defect rate in production than predicted during design.

If data for your manufacturing operations is not available, you can estimate a standard deviation from tables of recommended tolerances for various machine tools. Historically, most companies have considered a process with a Cp of 1 as desirable. (See Chapters 2 and 10 for more discussion of Cp.) Using that as a criterion, you can estimate a standard deviation for many manufacturing processes by finding a recommended tolerance in a handbook such as Reference 1 and dividing the tolerance by three to get a standard deviation. Table 11-1 shows some estimated standard deviations for various machining processes that we'll use for the examples in this book.

This chapter will introduce four techniques that use process standard deviations to allocate tolerances. These techniques will allow us to meet specific goals for defect rates that occur during assembly and fabrication. All four techniques should be used as design tools to assign tolerances to a drawing that will meet targeted quality goals. The choice of a particular technique will depend on the assumptions (and associated risks) with which you are comfortable. To compare the results of these analyses with the more traditional approaches, we will analyze the same problem that was used in Chapter 9. See Fig. 11-2.

Even with a statistical analysis, some assumptions need to be made. They are as follows:
- The distributions that characterize the expected ranges of each variable dimension are normal. This assumption is more important when estimating the defect rates for the components than for the assembly. If

Table 11-1 Process standard deviations that will be used in this chapter

Process	Standard Deviation (in.)	Process	Standard Deviation (in.)
N/C end milling	.00026	JB end milling	.000105
N/C side milling	.00069	JB side milling	.000254
N/C side milling, > 6.0 in.	.00093	JB bore holes < .13 diameter	.000048
N/C drilling holes (location)	.00076	JB bore holes < .13 diameter	.000056
N/C drilling holes (diameter)	.00056	JB bore holes (location)	.000054
N/C tapped holes (depth)	.0025	JB drilling holes (location)	.000769
N/C bore/ream holes (diameter)	.00006	JB countersink (diameter)	.001821
N/C bore/ream holes (location)	.00022	JB reaming (diameter)	.000159
N/C countersink (location)	.00211	JB reaming (location)	.000433
N/C end mill parallel < 16 sq. in	.00020	JB end mill parallel < 16 sq. in.	.000090
N/C end mill parallel > 16 sq. in	.00047	JB end mill parallel > 16 sq. in.	.000232
N/C end mill flat < 16 sq. in	.00019	JB end mill flat < 16 sq. in.	.000046
N/C end mill flat > 16 sq. in	.00027	JB end mill flat > 16 sq. in.	.000132
N/C bore perpendicular < .6 deep	.00020	JB bore perpendicular < .6 deep	.000107
N/C bore perpendicular > .6 deep	.00031	JB bore perpendicular > .6 deep	.000161
Turning ID	.000127		
Turning OD	.000132	Treypan ID	.000127
Bore/ream ID	.000111	Turning lengths	.000357
Grinding, surface	.000029	Grinding, lap	.000027
Grinding, ID	.000104	Grinding, tub	.000031
Grinding, OD	.000029		

Process	Standard Deviation (in.)	Process	Standard Deviation (in.)
Aluminum Casting		Steel Casting	
Cast up to .250	.000830	Cast up to .250	.000593
Cast up to .500	.001035	Cast up to .500	.001060
Cast up to .1.00	.001597	Cast up to 1.00	.001346
Cast up to 2.00	.002102	Cast up to 2.00	.002099
Cast up to 3.00	.002662	Cast up to 3.25	.003064
Cast up to 4.00	.003391	Cast up to 4.25	.003921
Cast up to 5.00	.003997	Cast up to 5.25	.005118
Cast up to 6.00	.004389	Cast up to 6.25	.005784
Cast up to 7.00	.005418	Cast up to 7.25	.007427
Cast up to 8.00	.006464	Cast up to 8.25	.007699
Cast up to 9.00	.006879	Cast up to 9.25	.008317
Cast up to 10.00	.008085	Cast up to 10.00	.009596
Cast up to 11.00	.008126	Cast up to 11.00	.011711
Cast over 11.00	.008725	Cast over 11.00	.011743
Cast flat < 2 sq. in.	.001543	Cast flat < 2 sq. in.	.001520
Cast flat < 4 sq. in.	.002003	Cast flat < 4 sq. in.	.002059
Cast flat < 6 sq. in.	.002860	Cast flat < 6 sq. in.	.003108
Cast flat < 8 sq. in.	.003828	Cast flat < 8 sq. in.	.004131
Cast flat < 10 sq. in.	.004534	Cast flat < 10 sq. in.	.004691
Cast flat 10+ sq. in.	.005564	Cast flat 10+ sq. in.	.005635
Cast straight < 2 in.	.001965	Cast straight < 2 in.	.002197
Cast straight < 4 in.	.004032	Cast straight < 4 in.	.004167
Cast straight < 6 in.	.004864	Cast straight < 6 in.	.005240
Cast straight < 8 in.	.007087	Cast straight < 8 in.	.006695
Cast straight < 10 in.	.007597	Cast straight < 10 in.	.007559
Cast straight over 10 in.	.009040	Cast straight over 10 in.	.009289

the distribution for the components is significantly different than a normal distribution, the estimated defect rate may be incorrect by an order of magnitude or more. Assembly distributions tend to be closer to normal as the number of components in the stack increase because of the central limit theorem (Reference 9). Therefore, the error will tend to decrease as the number of dimensions in the stack increase. How important are these errors? Usually, they don't really matter. If our estimated defect rate is high, we have a problem that we need to correct before finishing our design. If our design has a low estimated defect rate, an error of an order of magnitude is still a small number. In either case, the error is of little relevance.

- The mean of the distribution for each dimension is equal to the nominal value (the center of the tolerance range). If specific information about the mean of any dimension is known, that value should be substituted

in place of the nominal number in the dimension loop. An example where this might apply is the tendency to machine toward maximum material condition for very tightly toleranced parts.
- Each of the dimensions in the stack is statistically independent of all others. This means that the value (or change in value) of one has no effect on the value of the others. (Reference 7)

Tolerances on some dimensions, such as purchased parts, are not usually subject to change. In the following methods, their impact will be considered to act in a worst case manner. For example, if a dimension is $3.00 \pm .01$ in., it will affect the gap as if it is really fixed at 2.09 or 3.01 with no tolerance. We choose the minimum or maximum value based on which one minimizes the gap.

11.4 Worst Case Allocation

In many cases, a product needs to be designed so that assembly is assured, regardless of the particular combination of dimensions within their respective tolerance ranges. It is also desirable to assign the individual tolerances in such a way that all are equally producible. The technique to accomplish this using known process standard deviations is called worst case allocation. Fig. 11-2 shows a motor assembly similar to Fig. 9-2 that we will use as an example problem to demonstrate the technique.

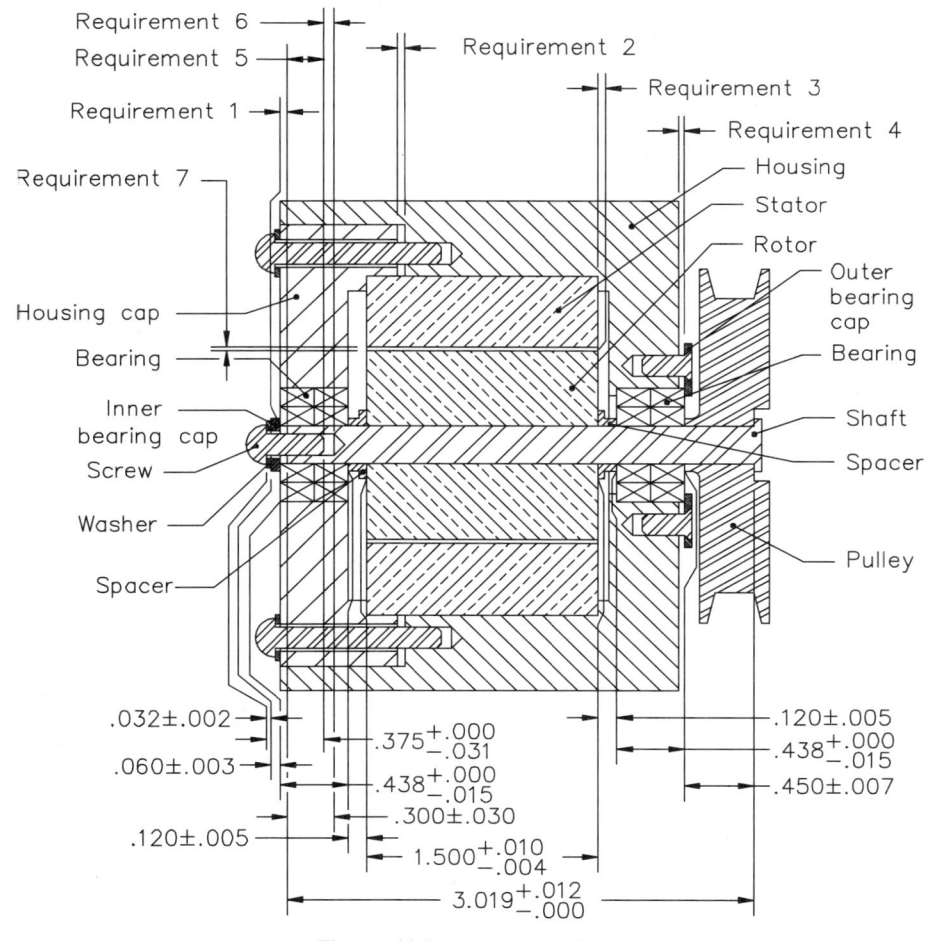

Figure 11-2 Motor assembly

11.4.1 Assign Component Dimensions

The process follows the flow chart shown in Fig. 11-3, the worst case allocation flow chart. The first step is to determine which of the dimensions in the model contribute to meeting the requirement. We identify these dimensions by using a loop diagram identical to the one shown in Fig. 9-3, which we've repeated in Fig. 11-4 for your convenience. In this case, there are 11 dimensions contributing to the result. We'll allocate tolerances to all except the ones that are considered fixed. Thus, there are five dimensions that have tolerances and six that need to be allocated. The details are shown in Table 11-2.

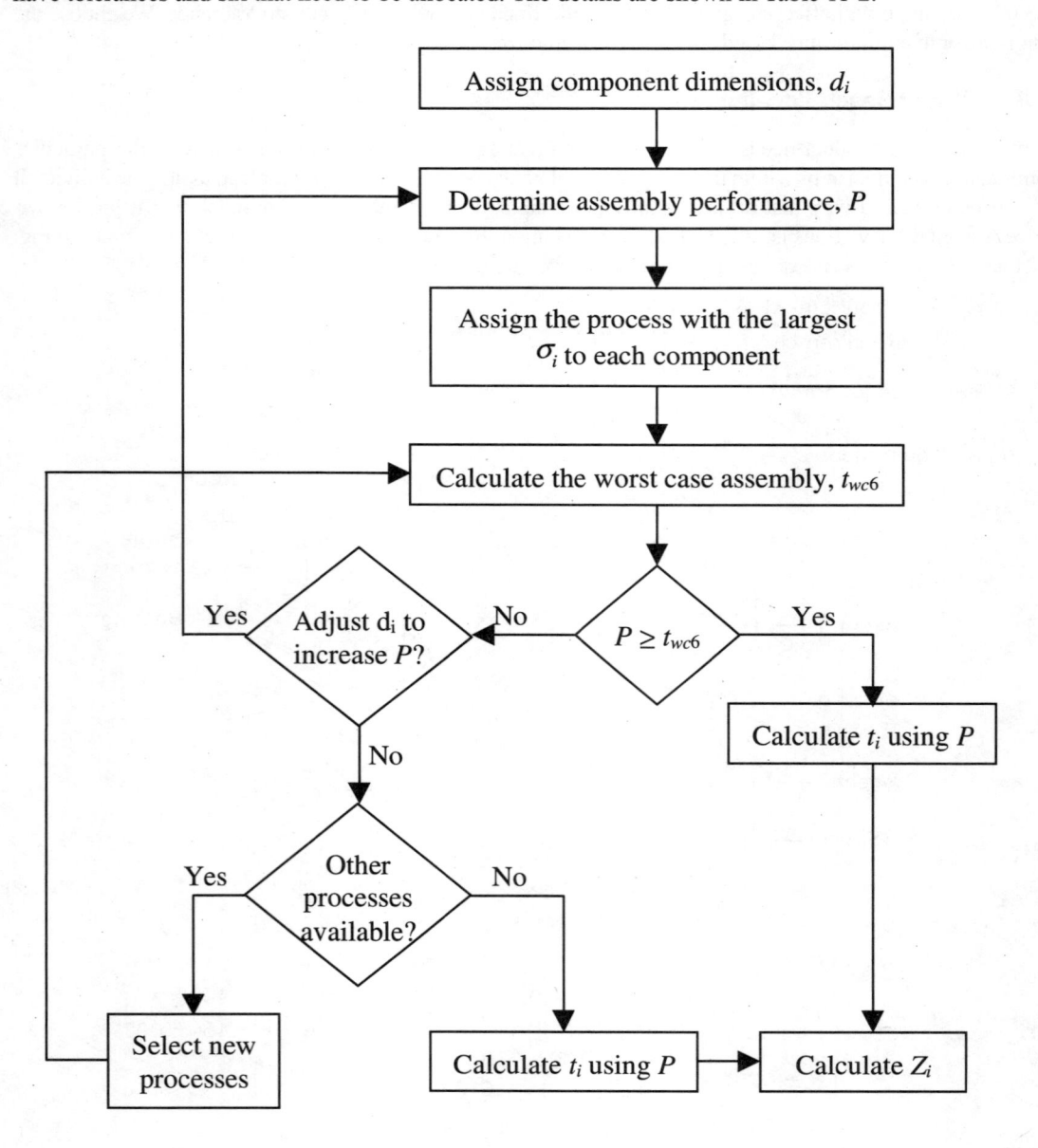

Figure 11-3 Worst case allocation flow chart

Predicting Assembly Quality (Six Sigma Methodologies to Optimize Tolerances)   11-7

```
A (−.375)
B (+.032)
C (+.060)
D (+.438)
E (+.120)
F (+1.500)
G (+.120)
H (+.438)
I (+.450)
J (−3.019)
K (+.300)
```

Performance Requirement 6 (Gap ≥ 0)

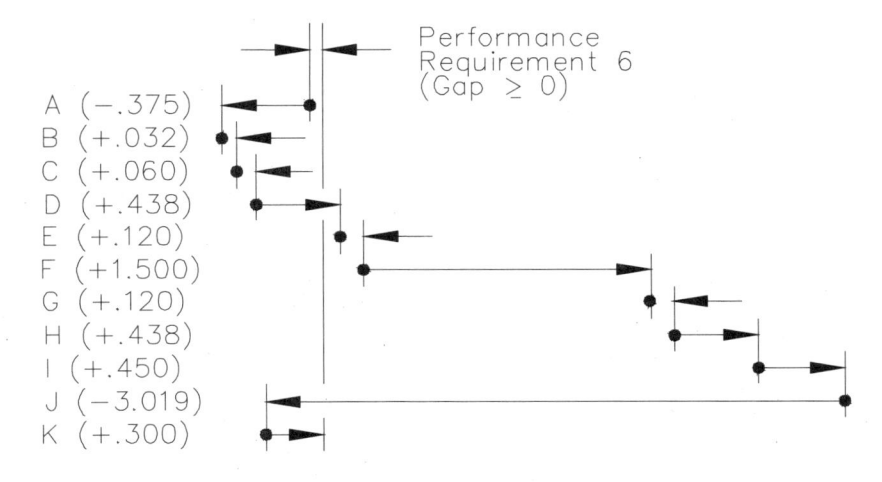

Figure 11-4 Dimension loop for Requirement 6

11.4.2 Determine Assembly Performance, *P*

The second step is to calculate the assembly performance, *P*. This is found using Eq. (11.1). While it is similar to Eq. (9.1) that was used to calculate the mean gap in Chapter 9, there are some additional terms here. The first term represents the mean gap and the result is identical to Eq. (9.1). This value is adjusted by two added terms. The first added term, $\Sigma |a_j t_{if}|$, accounts for the effect of the fixed tolerances. In this case, we calculate the sum of the tolerances and subtract them from the mean gap. The effect is that we treat fixed tolerances as worst case. The second added term is an adjustment on the gap to account for instances where you need to keep the minimum gap greater than zero. For example, suppose we want to

Table 11-2 Data used to allocate tolerances for Requirement 6

Variable Name	Mean Dimension (in.)	Sensitivity	Fixed/Variable	±Tolerance (in.)	Standard Deviation (in.)	Process
A	.3595	-1	Fixed	.0155		
B	.0320	1	Fixed	.0020		
C	.0600	1	Variable		.000357	Turning length
D	.4305	1	Fixed	.0075		
E	.1200	1	Variable		.000357	Turning length
F	1.5030	1	Fixed	.0070		
G	.1200	1	Variable		.000357	Turning length
H	.4305	1	Fixed	.0075		
I	.4500	1	Variable		.00106	Steel casting up to .500
J	3.0250	-1	Variable		.000357	Turning length
K	.3000	1	Variable		.0025	N/C tapped hole depth

ensure a certain ease of assembly for two parts. We may establish a minimum gap of .001 in. so they don't bind when using a manual assembly operation. Then we would set g_m to .001 in. The sum, P, is the amount that we have to allocate to the rest of the dimensions in the stack. For Requirement 6, assembly ease is not a concern, so we'll set g_m to .000 in.

$$P = \sum_{i=1}^{n} a_i d_i - \sum_{j=1}^{p} |a_j t_{jf}| - g_m \tag{11.1}$$

where
     $n=$ number of independent variables (dimensions) in the stackup
     $p=$ number of fixed independent dimensions in the stackup

For Requirement 6,

$$\sum_{j=1}^{p} |a_j t_{jf}| = |(-1).0155| + |(1).0020| + |(1).0075| + |(1).0070| + |(1).0075| = .0395 \text{ in.}$$

$g_m = .000$ in.

$P = (-1).3595 + (1).0320 + (1).0600 + (1).4305 + (1).1200 + (1)1.5030 + (1).1200 + (1).4305$
    $+ (1).4500 + (-1)3.0250 + (1).3000 - .0395 - .000$
   $= .022$ in.

Thus, we have .022 in. to allocate to the six dimensions that do not have fixed tolerances.

11.4.3 Assign the Process With the Largest σ_i to Each Component

The next step on the flow chart in Fig. 11-3 is to choose the manufacturing process with the largest standard deviation for each component. For the allocation we are completing here, we will use the processes and data in Table 11-1. If you have data from your manufacturing facility, you should use it for the calculations. Table 11-2 shows the standard deviations selected for the components in the motor assembly that contribute to Requirement 6.

11.4.4 Calculate the Worst Case Assembly, t_{wc6}

The term t_{wc6} that is calculated in Eq. (11.2) can be thought of as the gap that would be required to meet 6σ or another design goal.

$$t_{wc6} = 6.0 \sum_{i=1}^{n-p} |a_i \sigma_i| \tag{11.2}$$

In the examples that follow, we'll assume the design goal is 6σ, which is a very high-quality design. If we use the equations as written, our design will have quality levels near 6σ. If our design goal is something less than or greater than 6σ, we can modify Eqs. (11.2) and (11.3) by changing the 6.0 to the appropriate value that represents our goal. For example, if our goal is 4.5σ, Eq. (11.2) becomes:

$$t_{wc6} = 4.5 \sum_{i=1}^{n-p} |a_i \sigma_i|$$

Using the process standard deviations shown in Table 11-2, t_{wc6} for Requirement 6 is calculated below.

$$t_{wc6} = 6.0\big(|(1).000357| + |(1).000357| + |(1).000357| + |(1).00106| + |(-1).000357| + |(1).0025|\big) = .0299$$

11.4.5 Is $P \geq t_{wc6}$?

If P is smaller than t_{wc6}, the amount we have to allocate is less than what is required for a 6σ design. If P is greater than or equal to t_{wc6}, the tolerances we can allocate will be greater than or equal to 6σ. In our case, the former is true, so we have some decisions to make.

The first choice would be to evaluate all the dimensions and decide if any can be changed that will increase P. The amount to change any component depends on the sensitivity and design characteristics. The sensitivity tells us whether to increase or decrease the size of the dimension. (Dimensions with arrows to the right and up in the loop diagram are positive; left and down are negative.) If the dimension has a positive sensitivity, making the nominal dimension larger will make P larger. Conversely, if you increase the nominal value of a dimension with a negative sensitivity, the gap will get smaller. The amount of change in the size of the gap depends on the magnitude. Sensitivities with a magnitude of +1 or –1 will change the gap .001 in. if a dimension is changed by .001 in. Suppose we change the depth of the tapped hole from .300 in. to .310 in. Following the flow chart in Fig. 11-3, we need to recalculate P, which is now .032 in. Thus, we will exceed our design goal.

If we evaluate the design and find that we can't change any of the dimensions, a second option is to select processes that have smaller standard deviations. If some are available, we would have to recalculate t_{wc6} and compare it to P. In general, it takes relatively large changes in standard deviations to make a significant impact on t_{wc6}. This option, then, can have a considerable effect on product cost.

If we follow the flow chart in Fig.11-3 and neither of these options are acceptable, we will have a design that does not meet our quality goal. However, it may be close enough that we can live with it. The key is the producibility of the component tolerances. If they can be economically produced, then the design is acceptable. If not, we may have to reconsider the entire design concept and devise an alternative approach. For the purposes of this example, we'll assume that design or process changes are not possible, so we have to assign the best tolerances possible. After that we can evaluate whether or not they are economical.

We'll use Eq. (11.3) to calculate the component tolerances. Looking at the terms in Eq. (11.3), we see that P and t_{wc6} will be the same for all the components. Thus, components manufactured with similar processes (equal standard deviations) will have equal tolerances. We'll have three different tolerances because we have three different standard deviations: .000357 in. for turned length, .0025 in. for tapped hole depth, and .00106 in. for the cast pulley.

$$t_i = 6.0 \left(\frac{P}{t_{wc6}}\right) \sigma_i \tag{11.3}$$

First, for the dimensions made on a Numerical Controlled (N/C) lathe:

$$t = 6.0 \left(\frac{.022}{.0299}\right).000357$$
$$= .0016 \text{ in.}$$

11-10 Chapter Eleven

For the dimensions made by casting (pulley):

$$t = 6.0 \left(\frac{.022}{.0299} \right) .00106$$

$$= .0046 \text{ in.}$$

Finally, for the tapped hole depth:

$$t = 6.0 \left(\frac{.022}{.0299} \right) .0025$$

$$= .011 \text{ in.}$$

Table 11-3 contains the final allocated tolerances.

Table 11-3 Final allocated and fixed tolerances to meet Requirement 6

Variable Name	Mean Dimension (in.)	Fixed/ Variable	± Tolerance (in.)	Allocated ± Tolerance (in.)
A	.3595	Fixed	.0155	
B	.0320	Fixed	.0020	
C	.0600	Variable		.0016
D	.4305	Fixed	.0075	
E	.1200	Variable		.0016
F	1.5030	Fixed	.0070	
G	.1200	Variable		.0016
H	.4305	Fixed	.0075	
I	.4500	Variable		.0046
J	3.0250	Variable		.0016
K	.3000	Variable		.011

11.4.6 Estimating Defect Rates

We have to complete two more tasks to finish the analysis. The first will be to verify that all the dimensions with allocated tolerances are equally producible. Our definition of producibility in this case will be the estimated defect rate. Eq. (11.4) defines a term Z_i that represents the number of standard deviations (sigmas) that are between the nominal value of a dimension and the tolerance limits. If we assume that the components are produced with a process that approximates a normal distribution, then we can use some standard tables to estimate the defect rate.

$$Z_i = \frac{t_i}{\sigma_i} \tag{11.4}$$

The method to calculate the defect rate depends on the nature of the standard deviation used and the way the data was collected. For example, suppose the standard deviation represented a sample rather than the total population. Since we're usually interested in long-term versus short-term yields, the sample may not represent what will happen over a long period of time. We have a couple of techniques to use to adjust the calculation to account for long-term effects. The first one involves a shift in the mean; the second an inflation of the value of the standard deviation. In both cases, we'll use Eq. (11.4) and assume the component dimensions will be normally distributed.

For the dimensions that are manufactured on the N/C lathe, the tolerance is .0016 in. and the standard deviation is .000357 in. If we use the mean shift model, we'll calculate Z directly from Eq. (11.4).

$$Z_1 = \frac{.0016}{.000357} = 4.48$$

We now reduce the value of Z_1 by 1.5, which is equivalent to shifting the mean by 1.5 standard deviations (Reference 5). Thus, we will look in a table of values from a standard normal distribution (see Chapter 10 Appendix) with $Z = 4.48 - 1.5 = 2.98$. The defect rate is equal to the area to the right of the T_U line in Fig. 11-5 that represents the component dimension tolerance limit (far right). From the Z value we just calculated, the estimated defect rate will be .0014, or the yield on this dimension will be 99.86%. Since the mean has been shifted, it is only necessary to get the value from one tail of the distribution. The other tail is very small in comparison and its effect is negligible.

When doing this calculation, we take a shortcut to simplify the technique. When we assume a mean shift of 1.5 standard deviations, we make no mention of the direction that the mean shifts. Our example (Fig. 11-5) showed the mean shifting $+1.5\sigma$. We could have shown it shifting 1.5σ in the negative direction just as easily. We are actually assuming that the shift happens in both directions with an equal probability. Therefore, the complete equation could more properly be written as $.5*.0014 + .5*.0014 = .0014$, which is the same number as before.

The second way to adjust the defect rate estimate is to inflate the value of the standard deviation. Usually, the factor chosen is based on data from statistical process control and is between 33% and 50%. We'll use 33% here. The new value for the standard deviation is:

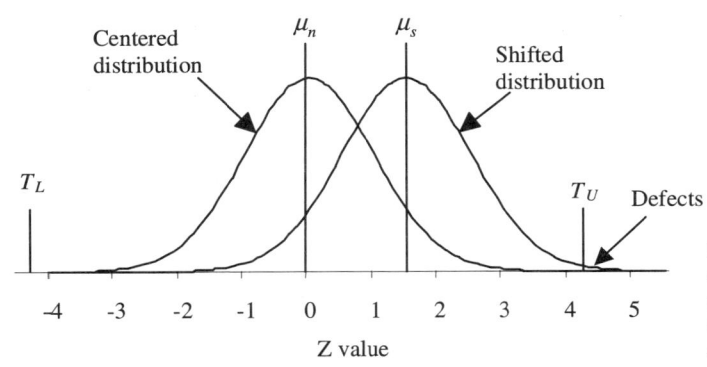

Figure 11-5 Effect of shifting the mean of a normal distribution to the right. T_L is the lower tolerance limit, T_U the upper tolerance limit, μ_n is the unshifted mean, and μ_s is the shifted mean

$.000357(1.33) = .000475$ in.

and

$$Z_1 = \frac{.0016}{.000475} = 3.37$$

We can look up Z from a table of tail area of a normal distribution (see Appendix of Chapter 10). The estimated defect rate is .00075 or the yield is 99.92%. Note that in this case, we double the value from the table so that both tails of the distribution are included. This is necessary because, as shown in Fig. 11-6, the area in both tails is the same and one is not negligible compared to the other.

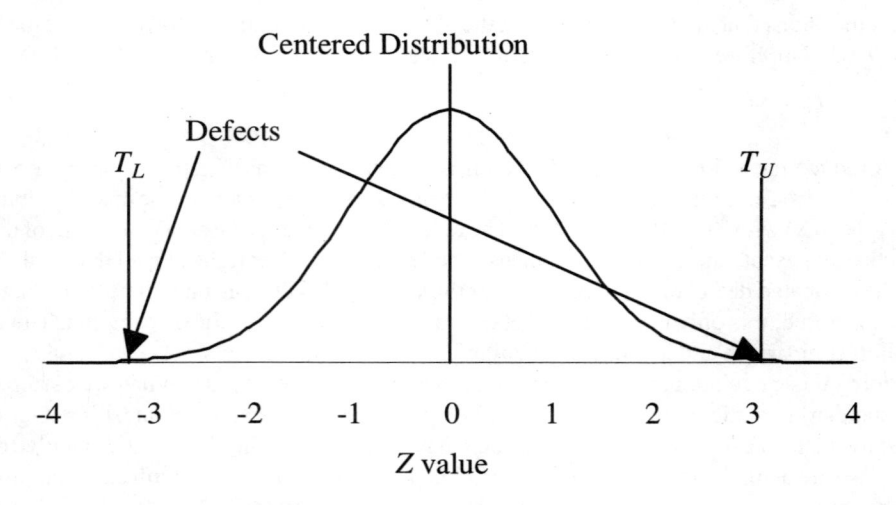

Figure 11-6 Centered normal distribution. Both tails are significant.

Normally, we don't expect the answer to be the same for both methods. The one you choose should be based on your knowledge about the manufacturing process and the data collected.

The tolerances for the pulley and the tapped hole depth are determined in similar manner and are .0046 in. and .011 in. respectively. If we follow the same process as above, we can verify that the estimated defect rates for these two dimensions are identical to the lathe parts and they are equally producible.

11.4.7 Verification

Finally, we should verify that the tolerances will meet Requirement 6. We'll use Eq. (9.2) to ensure that we can assemble the components as desired.

$$t_{wc} = \sum_{i=1}^{n} |a_i t_i|$$

$$\sum_{i=1}^{n} |a_i t_i| = |(-1).0155| + |(1).0020| + |(1).0016| + |(1).0075| + |(1).0016| + |(1).0070| + |(1).0016| + |(1).0075|$$

$$+ |(1).0046| + |(-1).0016| + |(1).011|$$

$$= .0615 \text{ in.}$$

Recall that Requirement 6 is a minimum gap of zero. Using the worst case allocation technique, we were able to quickly assign tolerances so that the minimum gap is .0615 in. - .0615 in. = .0000 in. This meets our performance requirement with a single pass through the process. While the tolerances added up exactly to the worst case requirement in this case, they often do not because of rounding errors.

11.4.8 Adjustments to Meet Quality Goals

In the previous sections, we quickly allocated tolerances that met Requirement 6, but without meeting our quality goal of 6σ producibility. We briefly discussed the other options presented by the flow chart in Fig. 11-3. The first and most desirable choice is to modify the nominal component dimensions so that P is greater than or equal to t_{wc6}. It is clear that changing any combination of the dimensions so that P is increased by $t_{wc6} - P = .0299$ in. $- .022$ in. $= .0079$ in. will accomplish the task. We can look at Table 11-2 to give us guidance about how to change component dimensions. The sensitivity for each dimension is the key factor. Increasing a dimension with a positive sensitivity will increase P, while increasing a dimension with a negative sensitivity will make P smaller. Also, it is generally not practical to change any of the dimensions with fixed tolerances, since the dimension is usually fixed as well. Therefore, we can increase P by changing the thickness of the inner bearing cap (component dimension C) from .060 in. to .068 in. We can easily calculate a new value of P using Eq. (11.1) and find it is now .030 in. Since P is now greater than t_{wc6}, we can allocate tolerances that meet our quality and assembly goal simultaneously.

It would be a less desirable choice if we decided to try to change our processes to try to make t_{wc6} smaller. Even though the mathematics of the problem don't seem to steer us away from this option, reality does. The first problem is that our unit costs would rise as we move to more precise processes. Second, it usually takes many process changes to make a significant change in t_{wc6}, compounding the cost penalty. If we end up in a situation where we can't alter P, it is often better to either review the entire design concept and consider other approaches to achieving the design's objective or accept the lower assembly producibility from our original allocation.

A third option we could consider is a statistical allocation technique that we will discuss in later sections of this chapter.

11.4.9 Worst Case Allocation Summary

Let's recap the important points about worst case allocation.

- Tolerances will combine to meet assembly requirements at worst case.
- Tolerances are allocated with a minimum of iteration.
- Worst case allocation will lead to tolerances that are equally producible, based on estimated defect rates.
- Tolerances that are manufactured using similar manufacturing processes will be assigned the same values.
- Choosing the most economical processes (largest standard deviation) first can help lead to the lowest cost design.
- Data from the manufacturing floor will lead to predictable quality levels.
- Since we are performing a worst case analysis, the predicted assembly yield is 100%.

11.5 Statistical Allocation

Although worst case allocation will lead to a design with each dimension equally producible, it can cause tighter tolerances than are necessary. In a manner similar to what is used for traditional RSS analysis, we will statistically combine standard deviations to determine an expected variation of the assembly, which will allow a prediction of the number of defects that may occur. Then we will allocate tolerances to each of the component dimensions so that each of them is equally producible and will be larger than we achieved with the worst case allocation model.

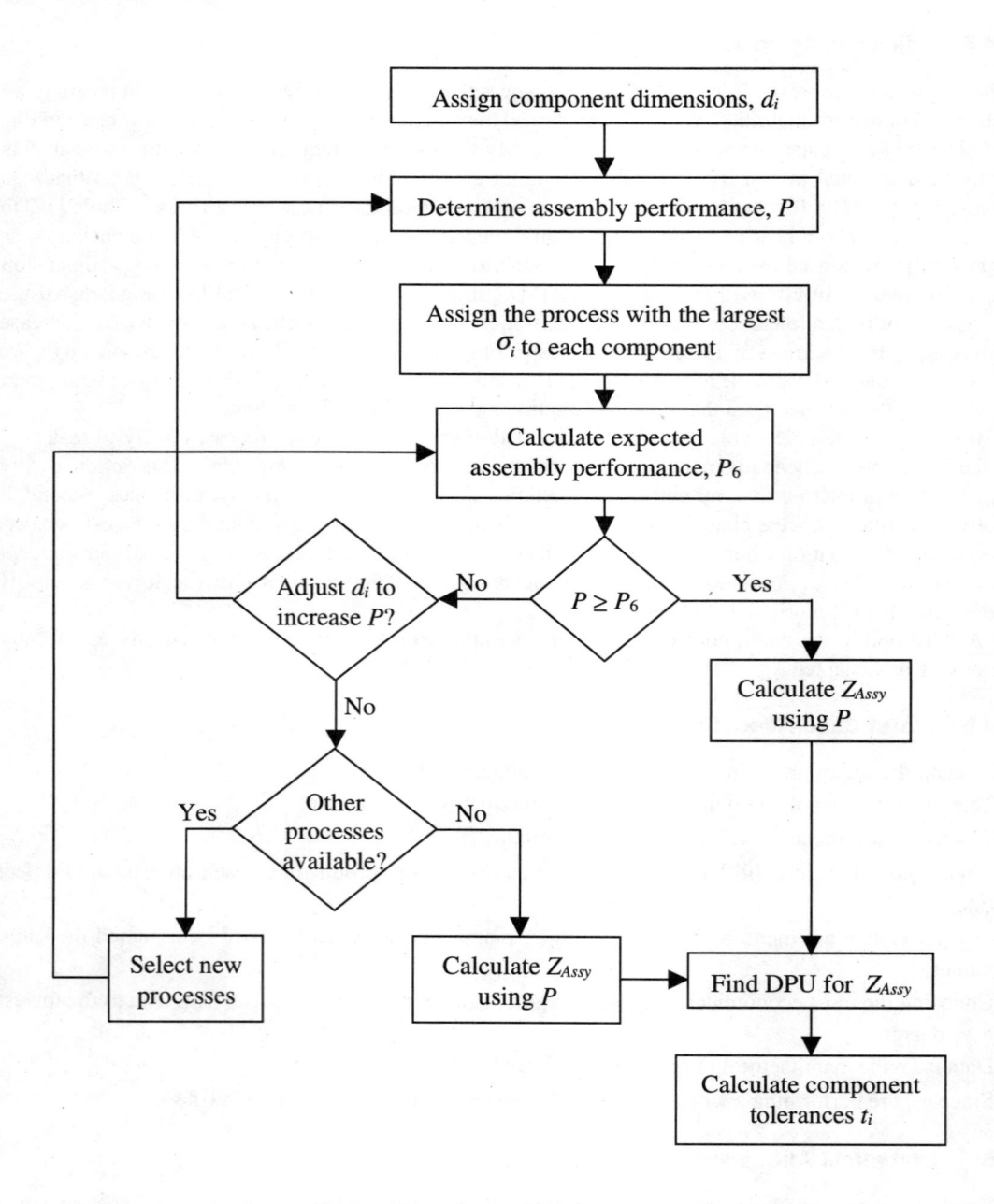

Figure 11-7 Statistical allocation flow chart

Looking at the statistical allocation flow chart shown in Fig. 11-7, there is an obvious similarity to the one used for worst case allocation. The differences are primarily in the equations used to calculate the terms.

11.5.1 Calculating Assembly Variation and Defect Rate

In Chapter 9, Eq. (9.8) was developed during derivation of the RSS technique. It shows how standard deviations of each of the dimensions in a tolerance analysis can be combined to yield a standard deviation of the gap.

$$\sigma_{Assy} = \sqrt{\sum_{i=1}^{n} (a_i \sigma_i)^2} \tag{11.5}$$

The use of Eq. (11.5) requires that all the variables (dimensions) be statistically independent. Two (or more) variables are considered statistically independent if the value (or change in value) of one has no effect on the value of the other(s). (Reference 8)

Eq. (11.5) gives us the ability to estimate the defect rate at the assembly level in the same manner that we calculated it for the component dimensions with worst case allocation. The standard deviations (σ_is) used in the equation are the same ones from Table 11-1 that we used during worst case allocation. Thus,

$$Z_{Assy} = \frac{P}{\sigma_{Assy}} \tag{11.6}$$

From Z_{Assy} we can find the estimated assembly defect rate using the same techniques introduced in section 11.4.6.

11.5.2 First Steps in Statistical Allocation

Referring to the process flow chart in Fig. 11-7, the first three steps are identical to the ones for worst case allocation. For Requirement 6, the component dimensions, P, and standard deviations are the same ones we used in sections 11.4 through 11.4.7 and shown in Table 11-2. Recall that P is the clearance between the end of the screw and the bottom of the tapped hole and that it has a value of .022 in. We determined the value for P using Eq. (11.1) and it consists of the nominal gap that is reduced by the effect of fixed tolerances and the minimum clearance requirement.

11.5.3 Calculate Expected Assembly Performance, P_6

The next step is slightly different than for worst case allocation, but the meaning is similar. Like t_{wc6}, P_6 can be thought of as the goal to meet a particular assembly defect objective. When using Eq. (11.7) below, the goal would be 6σ.

$$P_6 = 6.0 \sigma_{Assy} \tag{11.7}$$

Inserting the values from Table 11-2 into Eqs. (11.5) and (11.7) for Requirement 6,

$$\sigma_{Assy} = \sqrt{(1(.000357))^2 + (1(.000357))^2 + (1(.000357))^2 + (1(.00106))^2 + (-1(.000357))^2 + (1(.0025))^2}$$
$$= .00281 \text{ in.}$$

and

$$P_6 = 6.0(.00281)$$
$$= .01685 \text{ in}$$

11.5.4 Is $P \geq P_6$?

If P is smaller than P_6, the amount we have to allocate is less than what is required for both the assembly and components to be a 6σ design. Conversely, if P is greater than or equal to P_6, we can allocate tolerances so that the assembly and all the component dimensions that contribute to Requirement 6 will be greater than or equal to 6σ. In our case, the former is true, so we can allocate the tolerances to each of the component dimensions.

Before we allocate the tolerances, though, let's evaluate the expected assembly defect rate. Once again, the standard deviations we are using are considered short-term values, so the calculated standard deviation for the assembly is a short-term value. Thus, we'll have to adjust it so we can estimate the assembly defect rate we will see over an extended period of time. We'll use the same two techniques as in section 11.4.6 along with Eq. (11.6).

Using the mean shift model, as shown in Fig. 11-5,

$$Z_{Assy} = \frac{.022}{.00281}$$
$$= 7.83$$

From a table of the standard normal distribution with $Z = Z_{Assy} - 1.5 = 6.33$, the tail area in the normal distribution is $1.8(10^{-10})$. Before we can estimate the assembly defect rate, we need to think about the condition where acceptable assembly occurs. When we calculated defect rates for the component dimensions using the worst case allocation technique, we needed to be concerned about parts that were manufactured both above and below the tolerance limits. For the assembly we are evaluating, we are concerned if the gap becomes too small, but larger gaps are not expected to cause any problems. Thus, we won't consider large gaps to be defects and the estimated defect rate will be half the area of the tail area, or $9.0(10^{-11})$.

If we choose to inflate the standard deviation, the same factor of 33% that we used earlier is appropriate. The adjusted standard deviation is:

$.00281(1.33) = .00374$ in.

and

$$Z_{Assy} = \frac{.022}{.00374}$$
$$= 5.88$$

Again looking in a table of areas from a standard normal distribution, we find that the area beyond the value of 5.88 is $2.5(10^{-9})$. Since this value is for a unilateral tail area and we are only concerned with one side of the distribution, there is no need to double the value. Therefore, the estimated assembly defect rate using the inflation technique is $2.5(10^{-9})$.

Regardless of the method we use to transform our values from short term to long term, there is very little chance of a defect occurring with this assembly.

When we use the normal distribution to estimate assembly defect rates, there are a couple of assumptions we're making that are worth noting. First, we are assuming the assembly distribution is indeed normal. If each of the component distributions is normal, then the assembly distribution will be normal for these kinds of problems (linear combinations). If some of the component distributions are non-normal, then the assembly distribution is also non-normal. The error that results may or may not be significant, and is relatively difficult to determine through direct analytical means. (Reference 4) A commonsense approach will help us decide if it is important or not. If we have a situation like the one that we've just evaluated, our estimation errors could be incorrect by two or three orders of magnitude and we would still have very low defect rates. In cases similar to this, it makes little difference whether the distribution is normal or not; we still have a very slight chance that an assembly will be defective. If the defect rate is much higher, the error caused by the shape of the distribution may become significant. In these cases, a Monte Carlo simulation (Reference 2) or a second-order technique (Reference 4) can be used to find a better estimate of the shape of the assembly distribution and the defect rate.

A second assumption we make is that there is no inspection of component parts. When we inspect parts, we rework or discard the defects, and the final distribution might look like Fig. 11-8 instead of a full normal distribution. While this looks pretty significant, it is not usually so. The distribution shown in Fig. 11-8 is truncated at about $\pm 2\sigma$. Parts with such a high defect rate are not desirable in production. If we suspect that this will occur, a Monte Carlo technique is a good alternative to use to estimate defect rates. We could also consider a worst case allocation approach. In most cases, the effect of the truncation on the assembly defect rate is negligible and ignoring it immensely simplifies the calculations.

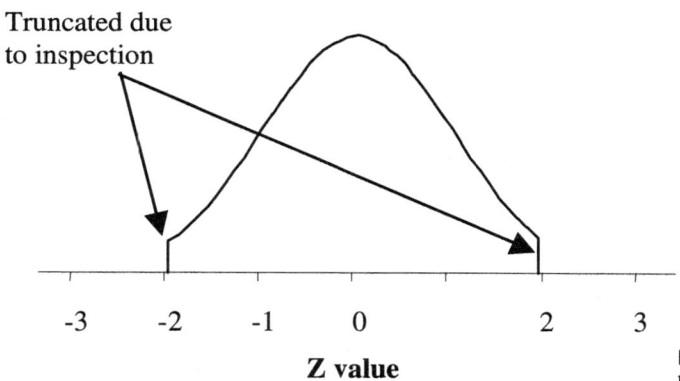

Figure 11-8 Normal distribution that has been truncated due to inspection

11.5.5 Allocating Tolerances

There are two different approaches we can use to allocate the tolerances. The first, statistical allocation, is to allocate tolerances to each of the component dimensions to meet a specific quality goal. For example, if our goal is 6σ, we would use Eq. (11.8), which allocates tolerances to each dimension that are 6 times the standard deviation.

$$t_i = 6.0 \sigma_i \qquad (11.8)$$

With this technique, the tolerance for the dimensions created by turning on an N/C lathe is

$t = 6.0\,(.000357)$
$= .0021$ in.

For the dimensions made by casting (pulley):

$t = 6.0\,(.00106)$
$= .0064$ in.

Finally, for the tapped hole depth:

$t = 6.0\,(.0025)$
$= .015$ in.

The results for all the dimensions are shown in Table 11-4.

Table 11-4 Fixed and statistically allocated tolerances for Requirement 6

Variable Name	Mean Dimension (in.)	Fixed/ Variable	± Tolerance (in.)	Statistically Allocated ± Tolerance (in.)
A	.3595	Fixed	.0155	
B	.0320	Fixed	.0020	
C	.0600	Variable		.0021
D	.4305	Fixed	.0075	
E	.1200	Variable		.0021
F	.5030	Fixed	.0070	
G	.1200	Variable		.0021
H	.4305	Fixed	.0075	
I	.4500	Variable		.0064
J	3.0250	Variable		.0021
K	.3000	Variable		.015

A second method for statistically allocating tolerances, RSS allocation, would give us component tolerances that have the same estimated defect rate as the assembly.

$$t_i = Z_{Assy}\sigma_i \qquad (11.9)$$

We can also express the same relationship as

$$t_i = \frac{P}{\sigma_{Assy}} \sigma_i \qquad (11.10)$$

or

$$t_i = \left(\frac{P}{\sqrt{\sum_{j=1}^{n}(a_j\sigma_j)^2}}\right)\sigma_i$$

Since we've already calculated Z_{Assy}, we'll use the simplest of these equations, Eq. (11.9), to calculate tolerances.

First, for the dimensions made on an N/C lathe:

$t = 7.83(.000357)$
$= .0028$ in.

For the dimensions made by casting (pulley):

$t = 7.83(.00106)$
$= .0083$ in.

Finally, for the tapped hole depth:

$t = 7.83(.0025)$

$= .0196$ in.

The tabulated results for the RSS allocation method are shown in Table 11-5. When we compare the results in Table 11-4 that were calculated with the first method, we see the tolerances are larger. This is a consequence of magnitude of the performance requirement, represented here by P, compared to a specific goal for defect rate. In this case, P is larger than required to meet a specific defect goal (e.g., 6σ that is represented by P_6). Therefore, restricting the allocated tolerance to the 6σ goal makes it smaller than if it is calculated based on the assembly defect rate. On the other hand, when P is smaller than P_6 the allocated tolerance will be greater for the first method than the second. The assembly defect rate is the same for both cases because we are assuming there is no parts screening or inspection at the component level.

Table 11-5 Fixed and RSS allocated tolerances for Requirement 6

Variable Name	Mean Dimension (in.)	Fixed/ Variable	± Tolerance (in.)	RSS Allocated ± Tolerance (in.)
A	.3595	F	.0155	
B	.0320	F	.0020	
C	.0600	V		.0028
D	.4305	F	.0075	
E	.1200	V		.0028
F	1.5030	F	.0070	
G	.1200	V		.0028
H	.4305	F	.0075	
I	.4500	V		.0083
J	3.0250	V		.0028
K	.3000	V		.0197

If we use RSS allocation, the calculated component tolerances will equal P when combined using the RSS analysis from Chapter 9, Eq. (9.11).

$t_{Assy} = \sqrt{.0028^2 + .0028^2 + .0028^2 + .0083^2 + .0028^2 + .0197^2}$

$= .022$ in.

We didn't fully discuss the options on the flow chart in Fig. 11-7 that we would explore if P was less than P_6. They are the same as with worst case allocation. The first choice would be to modify one or more of the component dimensions so that P is greater than or equal to P_6. If this is not an option, a more costly alternative is to select different processes with smaller standard deviations. Finally, if both of these are impractical or prohibitively expensive, the design concept can be re-evaluated.

11.5.6 Statistical Allocation Summary

Let's recap the important points about these two statistical allocation techniques.
- Tolerances allocated using the statistical techniques are larger than the ones allocated with the worst case technique.
- Predicting assembly quality quantifies the risk that is being taken with a statistical allocation.
- Tolerances are allocated to take advantage of the statistical nature of manufacturing processes.
- Tolerances are allocated with a minimum of iteration.
- Statistical allocation will lead to tolerances that will meet specific goals for defect rate.
- RSS allocation will lead to tolerances that will combine, using the RSS analysis technique, to meet the assembly requirement,
- Tolerances that are manufactured using similar manufacturing processes will be assigned the same values.
- Choosing the most economical processes (largest standard deviation) first can help lead to the lowest cost design.
- Data from the manufacturing floor will lead to predictable quality levels.

11.6 Dynamic RSS Allocation

The next two techniques we'll investigate are modifications of Motorola's dynamic RSS and static RSS methods from Reference 7. Both follow the flow chart of Fig. 11-7, so we'll highlight the differences instead of rigorously following the chart. The primary difference is the way that P_6 is calculated. We will allocate tolerances in a manner similar to the RSS allocation technique.

Motorola's equation for dynamic RSS is repeated below:

$$Z_F = \frac{\sum_{i=1}^{n} N_i V_i B_i - F}{\sqrt{\sum_{i=1}^{n} \left(\frac{T_i B_i}{3 \text{Cpk}_i} \right)^2}} \qquad (11.11)$$

Let's relate these terms to the same ones we've been using. First, Z_F is the same as Z_{Assy}. V_i is +1 or −1 depending on the direction of the arrow in the loop diagram and B_i is the magnitude of the sensitivity. Combined, $V_i B_i$ is equal to a_i, N_i is the same as d_i, and F is g_m.

Now let's look at the denominator. Harry and Stewart derive this in Reference 6 by defining a term

$$\sigma_{adj} = \frac{T}{3\text{Cpk}} \qquad (11.12)$$

where Cpk is a capability index commonly used in statistical process control. We'll use the definition of Cpk and a second index, Cp, to define a convenient way to use σ_{adj}. (See Chapters 2 and 10 for more explanations about Cp and Cpk.) The equations defining Cp and Cpk are:

$$\text{Cp} = \frac{USL - LSL}{6\sigma} \qquad (11.13)$$

where USL is the maximum allowable size of a feature and LSL is the minimum allowable size. Therefore, USL - LSL = 2T.

$$Cpk = Cp(1-k) \qquad (11.14)$$

Combining equations (11.12), (11.13), and (11.14),

$$\frac{USL - LSL}{6\sigma}(1-k) = \frac{T}{3\sigma_{adj}} \qquad (11.15)$$

Whenever we do a statistical analysis or allocation, the tolerance must be equal bilateral as explained in Chapter 9. Thus, $USL - LSL = 2T$. Substituting into Eq. (11.15) and simplifying gives us

$$\sigma_{adj} = \frac{\sigma}{(1-k)} \qquad (11.16)$$

The adjusted value of the standard deviation in Eq. (11.16) includes the transformation from a short-term value to a long-term one. Thus, it is similar to the adjustments we made to the standard deviation in section 11.4.6. The way we inflated the standard deviation in section 11.4.6 was by multiplying it by a factor that was between 1.33 and 1.50.

Substituting all these terms into Eq. (11.11) and recalling that V_i is either $+1$ or -1 gives us

$$Z_{Assy} = \frac{\sum_{i=1}^{n} a_i d_i - g_m}{\sqrt{\sum_{i=1}^{n}\left[a_i\left(\frac{1}{1-k_i}\right)\sigma_i\right]^2}} \qquad (11.17)$$

This equation is beginning to look very similar to the statistical allocation model from section 11.5 through 11.5.6. The primary difference is that the standard deviations from Table 11-1 are adjusted by an inflation factor, $\frac{1}{(1-k)}$, prior to calculating the assembly standard deviation. Eq. (11.17) also does not account for the effect of fixed tolerances, which can be easily incorporated by subtracting them from the numerator. The equation is now

$$Z_{Assy} = \frac{\sum_{i=1}^{n} a_i d_i - \sum_{j=1}^{p} |a_j t_{jf}| - g_m}{\sqrt{\sum_{i=1}^{n}\left[a_i\left(\frac{1}{1-k_i}\right)\sigma_i\right]^2}} \qquad (11.18)$$

Comparing the numerator of Eq. (11.18) to Eq. (11.1), we find that it is identical to P. Simplifying,

$$Z_{Assy} = \frac{P}{\sqrt{\sum_{i=1}^{n}\left[a_i\left(\frac{1}{1-k_i}\right)\sigma_i\right]^2}}$$

For Requirement 6, P is .022 in. We'll use the values of $\frac{1}{(1-k)}$ from Table 11-6 for each dimension. We'll also use the same values for the standard deviations for the component dimensions as before. From Eq. (11.14) we see that the values to use for $(1 - k)$ are available from SPC data or we can make estimates based on process knowledge.

Table 11-6 Standard deviation inflation factors and DRSS allocated tolerances for Requirement 6

Variable Name	Mean Dimension (in.)	$\frac{1}{(1-k)}$	DRSS Allocated ± Tolerance (in.)
A	.3595		
B	.0320		
C	.0600	1.05	.0025
D	.4305		
E	.1200	1.22	.0029
F	1.5030		
G	.1200	1.13	.0027
H	.4305		
I	.4500	1.27	.0088
J	3.0250	1.33	.0031
K	.3000	1.18	.0195

The denominator is the standard deviation of the assembly. Since it is calculated using different assumptions than previously, we'll call it σ_{DAssy}.

$$\sigma_{DAssy} = \sqrt{\begin{array}{l}(1((1.05).000357))^2 + (1((1.22).000357))^2 + (1((1.13).000357))^2 \\ + (1((1.27).00106))^2 + (-1((1.33).000357))^2 + (1((1.18).0025))^2\end{array}}$$ (11.19)

$= .00335$ in.

We'll find P_6 by modifying Eq. (11.7), renaming the term P_{D6}.

$P_{D6} = 4.5\sigma_{DAssy}$

$= 4.5(.00335)$

$= .0151$

We changed the 6.0 to 4.5 because the former value is based on short-term standard deviations. Since the value of σ_{DAssy} calculated in Eq. (11.19) is based on long-term effects, it would be inappropriate to include them again when calculating P_{D6}. Since $P \geq P_{D6}$, we can follow the flow chart of Fig. 11-7 and calculate Z_{Assy}.

$Z_{Assy} = \dfrac{0.022}{0.00335}$

$= 6.57$

Remember, we adjusted the standard deviations for the components before calculating σ_{Assy}, so there is no need to account for long-term effects by reducing the value of Z_{Assy} to simulate a 1.5σ shift or to multiply σ_{Assy} by an adjustment factor. Therefore, we estimate the assembly defect rate from Z_{Assy} by finding 6.57 in the table for tail areas of a standard normal distribution. Thus, the estimated defect rate is $4.1(10^{11})$. Next we'll allocate tolerances by modifying Eq. (11.10).

$$t_i = \frac{P}{\sigma_{DAssy}} \left(\frac{1}{1-k_i}\right) \sigma_i$$

For dimension C, which is made on an N/C lathe:

$$t_C = 6.57(1.05)(.000357)$$
$$= .0025 \text{ in.}$$

The tolerances for the remaining dimensions are calculated similarly and shown in Table 11-6. Comparing the tolerances calculated by the DRSS allocation method and RSS allocation shows that some are larger with one method and some with the other. This is because we chose different values of k for each dimension. Had we chosen identical values of k for each dimension, use of the DRSS method would have given the same tolerances that we calculated using RSS allocation.

Once again, we can easily confirm that the tolerances will equal P if we combine them using the RSS analysis from Chapter 9, Eq. (9.11).

$$t_{Assy} = \sqrt{.0025^2 + .0029^2 + .0027^2 + .0088^2 + .0031^2 + .0195^2}$$
$$= .022 \text{ in.}$$

11.7 Static RSS Analysis

A second technique from Reference 6 is called static RSS analysis. We can't use this technique to directly allocate tolerances, but we can use it to make another estimate of assembly defect rates. The concept behind Motorola's static RSS technique is to assume a mean shift on each component dimension that is equal to 1.5 standard deviations. Further, the shift will occur in the direction that will be most likely to cause an interference or a failure to meet the requirement. For example, the 1.5σ shift for .450 dimension has the effect of reducing its mean value to .4484 ($.450 - 1.5(.00106)$), which makes the gap smaller. The easiest way to implement this approach is to define a new parameter, P_{SRSS}, as follows:

$$P_{SRSS} = \sum_{i=1}^{n} a_i d_i - \sum_{j=1}^{p} |a_j t_{jf}| - g_m - 1.5 \sum_{q=1}^{n-p} \sigma_q$$

P_{SRSS} will be used to calculate Z_{Assy} and estimate the assembly defect rate.

Let's calculate P_{SRSS}. Comparing the first three terms to Eq. (11.1), we see they are equal to P, or .022 in. The fourth term is

$$1.5 \sum_{q=1}^{n-p} \sigma_q = (1.5)(.000357 + .000357 + .000357 + .00106 + .000357 + .0025)$$
$$= .0075$$

Now it is easy to calculate P_{SRSS}.

$$P_{SRSS} = .022 - .0075$$
$$= .0145$$

11-24   Chapter Eleven

Now we calculate Z_{Assy} using P_{SRSS}, using Eq. (11.6) with P_{SRSS} in place of P.

$$Z_{Assy} = \frac{P_{SRSS}}{\sigma_{Assy}}$$
$$= \frac{.0145}{.00281}$$
$$= 5.16$$

We can estimate the assembly defect rate by looking in a table of areas for the tail of a normal distribution in the same manner as before. For 5.16, the area in one tail, and thus the estimated assembly defect rate is $1.31(10^{-7})$.

11.8   Comparison of the Techniques

For educational purposes, we need to compare the results of the four allocation techniques (Table 11-7). The smallest tolerances result when we use worst case allocation. When we use worst case allocation, we eliminate the risk of assembly defects occurring. Sometimes this may be worthwhile, but in this case it's probably not. Each of the other three defect estimation techniques shows a very low probability of a defect occurring. The difference in the assembly defect rates is the benefit of worst case allocation. The penalty is component parts that are more difficult to produce. In our example, the tolerances for the RSS allocation technique are almost twice as large as for the worst case allocation. The benefit for worst case is that we eliminate a $6.0(10^{-11})$ probability of a defect occurring. As you can see, it's not a very large benefit in this case.

Table 11-7   Comparison of the allocated tolerances for Requirement 6

Variable Name	Mean Dimension (in.)	Worst Case Allocated ±Tolerance (in.)	Statistically Allocated ±Tolerance (in.)	RSS Allocated ±Tolerance (in.)	DRSS Allocated ±Tolerance (in.)
C	.0600	.0016	.0021	.0028	.0025
E	.1200	.0016	.0021	.0028	.0029
G	.1200	.0016	.0021	.0028	.0027
I	.4500	.0046	.0064	.0083	.0088
J	3.0250	.0016	.0021	.0028	.0031
K	.3000	.011	.015	.0197	.0195
Assembly defect rate		.00	$9.0(10^{-11})$	$9.0(10^{-11})$	$4.1(10^{-11})$

Are there times when it makes sense to use worst case allocation? Absolutely! If there are less than four dimensions that contribute to a tolerance stack, it is often better. First, the difference between tolerances allocated by worst case and statistical techniques is smaller with fewer dimensions. Also, the effect of some of the assumptions is greater with fewer dimensions. For example, suppose that some of the mean values are not located at nominal. If there are a large number of dimensions in the stack, they will tend to balance out. If there are only a few, they might not, and there can be a significant effect on assembly producibility.

Another case where worst case might be justified is when safety is involved. Depending on the consequences of an assembly failure, we may not be able to afford even a small probability of a defect.

In most cases, the benefits (larger tolerances) of either statistical, RSS or DRSS allocation will outweigh the risk of an assembly defect. In fact, by estimating the assembly defect rate, we can make a decision with each of the three about whether the risk of a defect is acceptable. If it is not, we can evaluate the design at worst case, or make some change in the design concept to alleviate the problem.

11.9 Communication of Requirements

Ideally, if we assign a tolerance using a technique such as statistical allocation, we can notify the fabrication shop and the manufacturing process could be appropriately controlled. In the past, there has been no mechanism to use on an engineering drawing to communicate the assumptions made when assigning a tolerance to a dimension. This can lead to unexpected defects if the manufacturing shop does not treat a statistical tolerance appropriately.

A way to communicate statistical design intent is with the ⟨ST⟩ symbol that is available within ASME Y14.5M-1994 (Reference 10). Examples of statistical tolerances on drawings are shown in Fig. 11-9.

In Fig. 11-9 (a) and (c), the ⟨ST⟩ symbol designates the dimension has a tolerance that was statistically allocated. In addition to the symbol, a note is required. Although the exact wording of the note is not specified in the standard, one possibility suggested in ASME 14.5M-1994 is: "Features identified as statistically toleranced ⟨ST⟩ shall be produced with statistical process controls."

If there is a possibility that the parts will not be produced with SPC, the designer may choose to tolerance the dimensions as shown in Fig. 11-9 (b). This method gives the manufacturing shop an option to inspect at smaller limits if SPC is not used. In this case, the standard suggests the note might read: "Features identified as statistically toleranced ⟨ST⟩ shall be produced with statistical process controls or to the more restrictive arithmetic limits." The actual wording of the note is at the user's discretion.

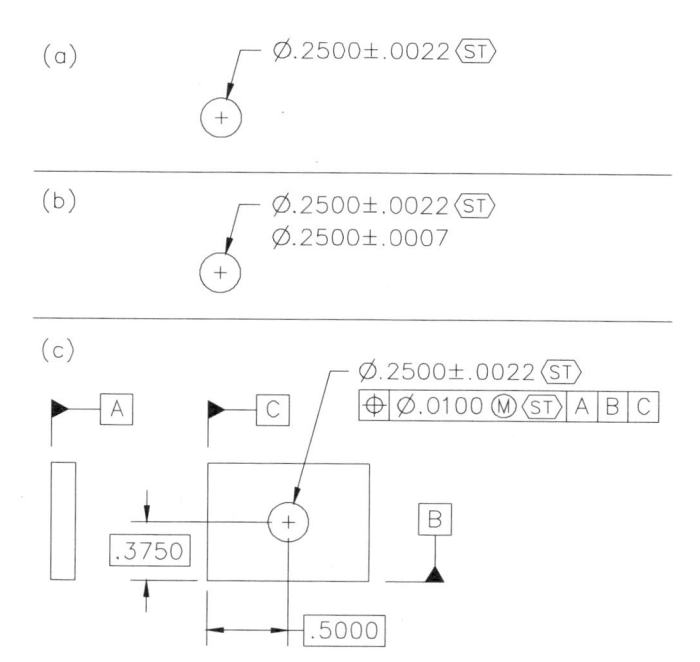

Figure 11-9 Three options for designating a statistically derived tolerance on an engineering drawing

11.10 Summary

Table 11-7 shows a comparison between worst case, statistical RSS, and DRSS allocation. As with the classical models, the worst case allocation method yields the smallest tolerances, and is the more conservative design. With worst case allocation, we don't make any prediction about defect rate, because it is assumed that parts screening will eliminate any possibility of a defect (not always the case).

We need detailed information about the expected manufacturing process for all of the allocation models. The best data is from our own operations. If none is available, then we can make estimates from recommended tolerance tables or use Table 11-1 in this chapter. The use of any of these techniques will have equal validity within the limitations of the applicable assumptions.

When comparing traditional techniques with the ones presented in this chapter, the primary difference between them is the amount of knowledge used to establish tolerances. In traditional worst case analyses, for example, we make decisions based on opinions about producibility. However, worst case allocation assigns tolerances that are equally producible based on process standard deviations. Clearly, the second method is more likely to produce products that will meet predictable quality levels.

Similarly, a comparison between traditional RSS and statistical, RSS or DRSS allocation reveals little difference in the basic principles. However, the allocation models overcome many of the assumptions that are inherent in RSS. In addition, they provide an estimate of assembly defect rates.

One requirement of the statistical, RSS or DRSS allocation techniques is that the manufacturing operations understand the assumptions that were made during design. This will ensure that the choice of process standard deviations used during design will be consistent with the method chosen to fabricate the parts. Perhaps the best way to accomplish this will be the ⟨ST⟩ symbol that is referenced in ASME Y14.5 M - 1994.

The question could be asked about whether it is ever desirable to use the traditional methods. There might be an occasional situation where all the tolerances being analyzed are purchased parts, or otherwise not under the design engineer's control. This situation is very rare. The techniques presented in this chapter are much better approaches because they take advantage of process standard deviations that have not been previously available, and eliminate the most dangerous of the assumptions inherent in the traditional methods.

11.11 Abbreviations

Variable	Definition
a_i, a_j, VB_i	sensitivity factor that defines the direction and magnitude for the ith, jth and nth dimension. In a one-dimensional stack, this is usually +1 or -1. Sometimes, it may be +.5 or -.5 if a radius is the contributing factor for a diameter called out on a drawing.
d_i, N_i	mean dimension of the ith component in the stack.
g_m, F	minimum gap required for acceptable performance
n	number of independent dimensions in the stackup
p	number of independent fixed dimensions in the stackup
P	nominal gap that is available for allocating tolerances
P_6	gap required to meet assembly quality goal
P_{D6}	gap required to meet assembly quality goal when using DRSS allocation
P_{SRSS}	expected gap when performing a static RSS analysis

σ_i — process standard deviation for the ith component in the stack

σ_{Assy}, σ_{DAssy} — standard deviation of a tolerance stack

σ_{adj} — adjusted standard deviation used in the DRSS allocation method

t_i, T_i — allocated equal bilateral tolerance for the ith component in the stack

t_{jf} — tolerance value of the jth fixed (purchased parts) component in the stack

t_{wc6} — assembly performance criterion (parameter) for the worst case allocation method

t_{wc} — worst case tolerance of an assembly stack

Z_i — a measure of the width of the process distribution as compared to the spec limits of the ith component dimension (standard normal transform)

Z_{Assy}, Z_F — a measure of the width of the assembly distribution as compared to the assembly requirement (standard normal transform)

T_U, USL — upper limit of a tolerance range

T_L, LSL — lower limit of a tolerance range

Cpk, Cp — capability indices

11.12   References

1. Bralla, James G. 1986. *Handbook of Product Design for Manufacturing*. New York, NY: McGraw-Hill Book Company.
2. Creveling, C.M. 1997. *Tolerance Design*. Reading, MA: Addison-Wesley Longman.
3. Drake, Paul and Dale Van Wyk. 1995. Classical Mechanical Tolerancing (Part I of II). *Texas Instruments Technical Journal*. Jan-Feb:39-46.
4. Glancy, Charles. 1994. A Second-Order Method for Assembly Tolerance Analysis. Master's thesis. Brigham Young University.
5. Harry, Mikel, and J.R. Lawson. 1990. *Six Sigma Producibility Analysis and Process Characterization*. Schaumburg, Illinois: Motorola University Press.
6. Harry, Mikel, and R. Stewart. 1988. *Six Sigma Mechanical Design Tolerancing*. Schaumburg, Illinois: Motorola University Press.
7. Hines, William, and Douglas Montgomery. 1990. *Probability and Statistics in Engineering and Management Sciences*. New York, NY: John Wiley and Sons.
8. Kennedy, John B., and Adam M. Neville. 1976. *Basic Statistical Methods for Engineers and Scientists*. New York, NY: Harper and Row.
9. Kiemele, Mark J. and Stephen R. Schmidt. 1991. *Basic Statistics. Tools for Continuous Improvement*. Colorado Springs, Colorado: Air Academy Press.
10. The American Society of Mechanical Engineers. 1995. *ASME Y14.5M-1994, Dimensioning and Tolerancing*. New York, NY: The American Society of Mechanical Engineers.
11. Van Wyk, Dale. 1993. Use of Tolerance Analysis to Predict Defects. *Six Sigma—Reaching Our Goal* white paper. Dallas, Texas: Texas Instruments.
12. Van Wyk, Dale and Paul Drake. 1995. Mechanical Tolerancing for Six Sigma (Part II). *Texas Instruments Technical Journal*. Jan-Feb: 47-54.

Chapter

12

Multi-Dimensional Tolerance Analysis (Manual Method)

Dale Van Wyk

Dale Van Wyk
Raytheon Systems Company
McKinney, Texas

Mr. Van Wyk has more than 14 years of experience with mechanical tolerance analysis and mechanical design at Texas Instruments' Defense Group, which became part of Raytheon Systems Company. In addition to direct design work, he has developed courses for mechanical tolerancing and application of statistical principles to systems design. He has also participated in development of a U.S. Air Force training class, teaching techniques to use statistics in creating affordable products. He has written several papers and delivered numerous presentations about the use of statistical techniques for mechanical tolerancing. Mr. Van Wyk has a BSME from Iowa State University and a MSME from Southern Methodist University.

12.1 Introduction

The techniques for analyzing tolerance stacks that were introduced in Chapter 9 were demonstrated using a one-dimensional example. By one-dimensional, we mean that all the vectors representing the component dimensions can be laid out along a single coordinate axis. In many analyses, the contributing dimensions are not all along a single coordinate axis. One example is the Geneva mechanism shown in Fig. 12-1. The tolerances on the C, R, S, and L will all affect the proper function of the mechanism. Analyses like we showed in Chapters 9 and 11 are insufficient to determine the effects of each of these tolerances. In this chapter, we'll demonstrate two methods that can be used to evaluate these kinds of problems.

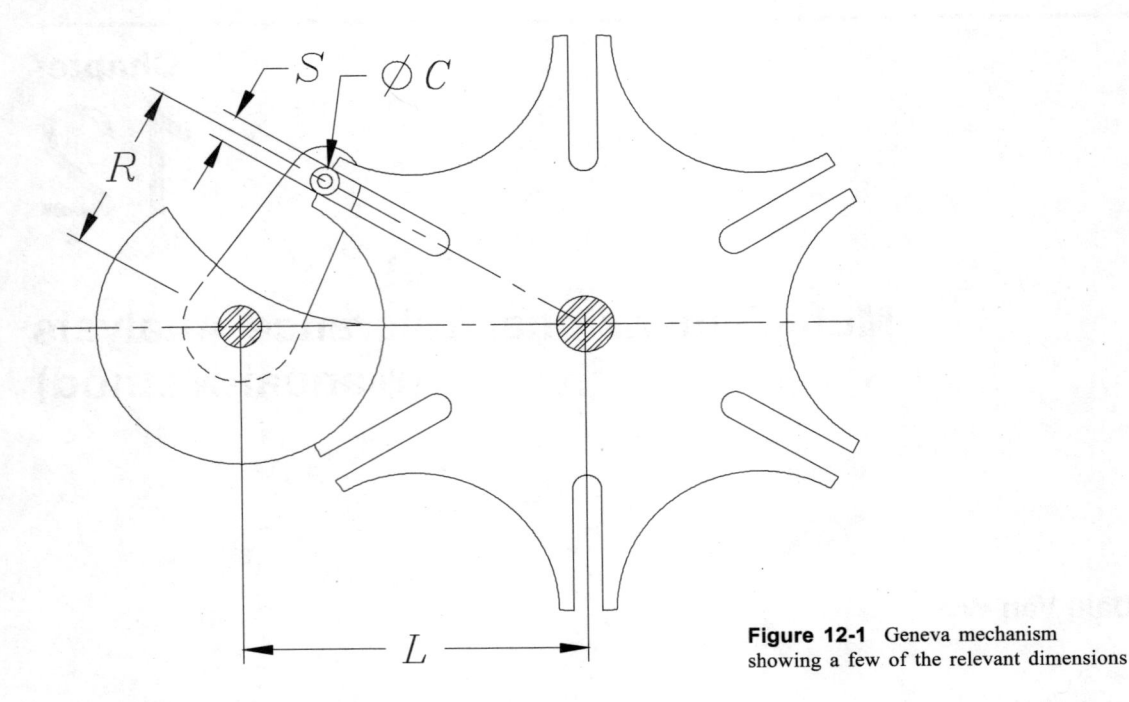

Figure 12-1 Geneva mechanism showing a few of the relevant dimensions

The following sections describe a systematic procedure for modeling and analyzing manufacturing variation within 2-D and 3-D assemblies. The key features of this system are:
1. A critical assembly dimension is represented by a vector loop, which is analogous to the loop diagram in 1-D analysis.
2. An explicit expression is derived for the critical assembly feature in terms of the contributing component dimensions.
3. The resulting expression is used to calculate tolerance sensitivities, either by partial differentiation or numerical methods.

A key benefit is that, once the expression is derived, this method easily solves for new nominal values directly as the design changes.

12.2 Determining Sensitivity

Recall the equations for worst case and RSS tolerance analysis equation from Chapter 9 (Eqs. 9.2 and 9.11).

$$t_{wc} = \sum_{i=1}^{n} |a_i t_i| \tag{12.1}$$

$$t_{rss} = \sqrt{\sum_{i=1}^{n} (a_i t_i)^2} \tag{12.2}$$

The technique we'll demonstrate for multidimensional tolerance analysis uses these same equations but we'll need to develop another way to determine the value of the sensitivity, a_i, in Eqs. (12.1) and (12.2) above. We noted in Chapter 9 that sensitivity is an indicator of the effect of a dimension on the stack. In one-dimensional stacks, the sensitivity is almost always either +1 or -1 so it is often left out of the one-dimensional tolerance equations. For the Geneva mechanism in Fig. 12-1, an increase in the distance L between the centers of rotation of the crank and the wheel require a change in the diameter, C, of the bearing, the width of the slot, S, and the length, R, of the crank. However, it won't be a one-to-one relationship like we usually have with a one-dimensional problem, so we need a different way to find sensitivity.

To see how we're going to determine sensitivity, let's start by looking at Fig. 12-2. If we know the derivative (slope) of the curve at point A, we can estimate the value of the function at points B and C as follows:

$$F(B) \approx F(A) + \Delta x \frac{dy}{dx}$$

and

$$F(C) \approx F(A) - \Delta x \frac{dy}{dx}$$

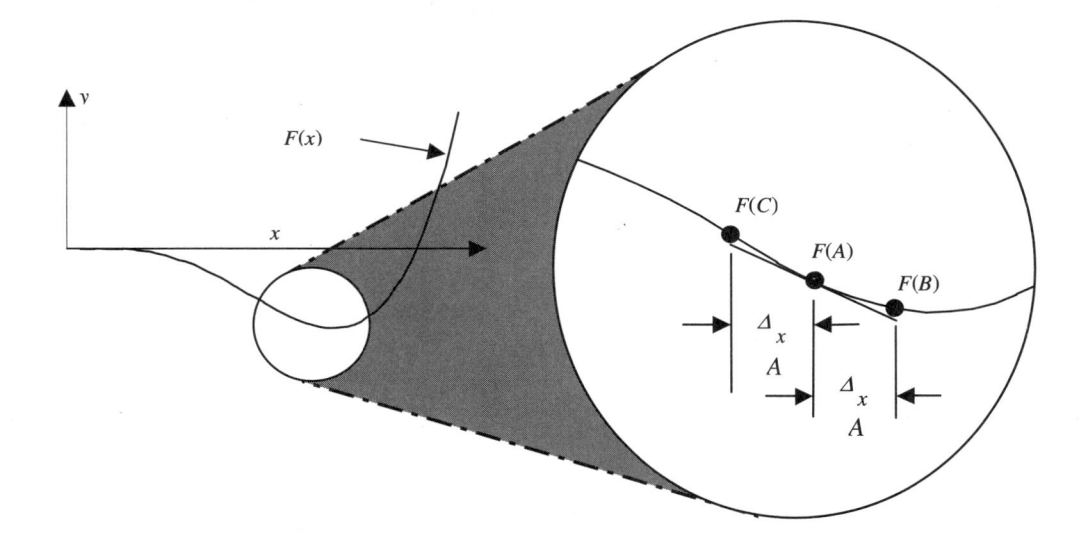

Figure 12-2 Linearized approximation to a curve

We'll use the same concept for multidimensional tolerance analysis. We can think of the tolerance as Δx, and use the sensitivity to estimate the value of the function at the tolerance extremes. As long as the tolerance is small compared to the slope of the curve, this provides a very good estimate of the effects of tolerances on the gap.

With multidimensional tolerance analysis, we usually have several variables that will affect the gap. Our function is an n-space surface instead of a curve, and the sensitivities are found by taking partial derivatives with respect to each variable. For example, if we have a function $\Theta(y_1, y_2, \ldots y_n)$, the sensitivity of Θ with respect to y_1 is

$$a_1 = \frac{\partial \Theta}{\partial y_1}\bigg|_{Nominal\ Values}$$

Therefore we evaluate the partial derivative at the nominal values of each of the variables. Remember that the nominal value for each variable is the center of the tolerance range, or the value of the dimension when the tolerances are equal bilateral. Once we find the values of all the sensitivities, we can use any of the tolerance analysis or allocation techniques in Chapters 9 and 11.

12.3 A Technique for Developing Gap Equations

Developing a gap equation is the key to performing a multidimensional tolerance analysis. We'll show one method to demonstrate the technique. While we're using this method as an example, any technique that will lead to an accurate gap equation is acceptable. Once we develop the gap equation, we'll calculate the sensitivities using differential calculus and complete the problem using any tolerance analysis or allocation technique desired. A flow chart listing the steps is shown in Fig. 12-3.

We'll solve the problem shown in Fig. 12-4. While this problem is unlikely to occur during the design process, its use demonstrates techniques that are helpful when developing gap equations.

Step 1. Define requirement of interest

The first thing we need to do with any tolerance analysis or allocation is to define the requirement that we are trying to satisfy. In this case, we want to be able to install the two blocks into the frame. We conducted a study of the expected assembly process, and decided that we need to have a minimum clearance of .005 in. between the top left corner of Block 2 and the Frame. We will perform a worst case analysis using the dimensions and tolerances in Table 12-1. The variable names in the table correspond to the variables shown in Fig. 12-4.

Step 2. Establish gap coordinate system

Our second step is establishing a coordinate system at the gap. We know that the shortest distance that will define the gap is a straight line, so we want to locate the coordinate sys-

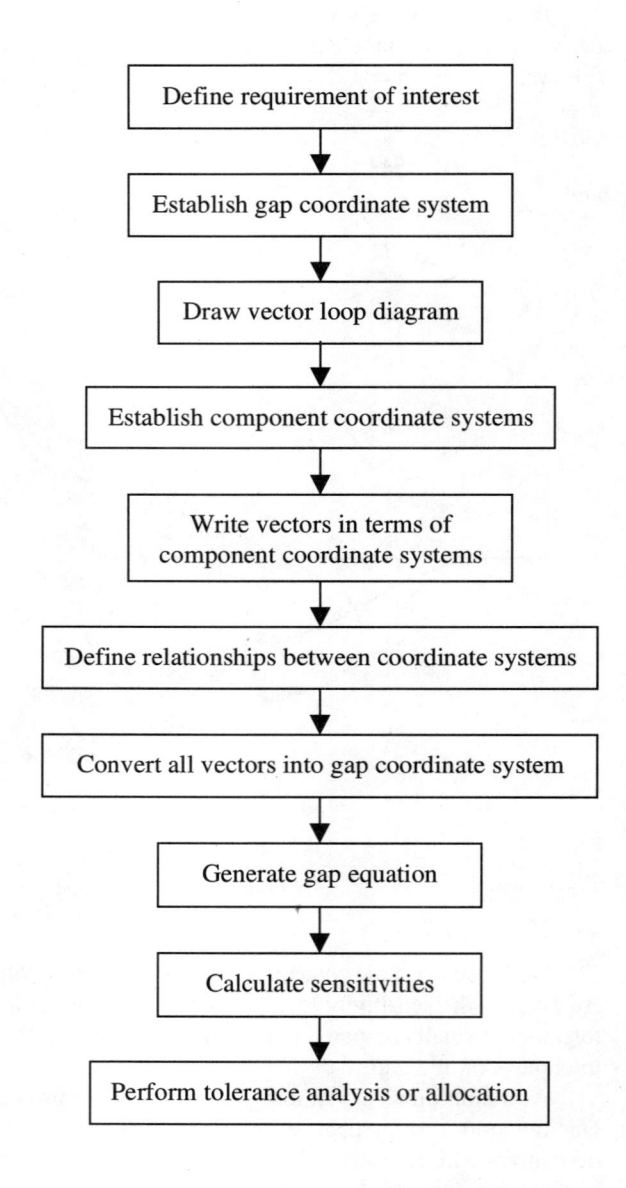

Figure 12-3 Multidimensional tolerancing flow chart

Multi-Dimensional Tolerance Analysis (Manual Method)   12-5

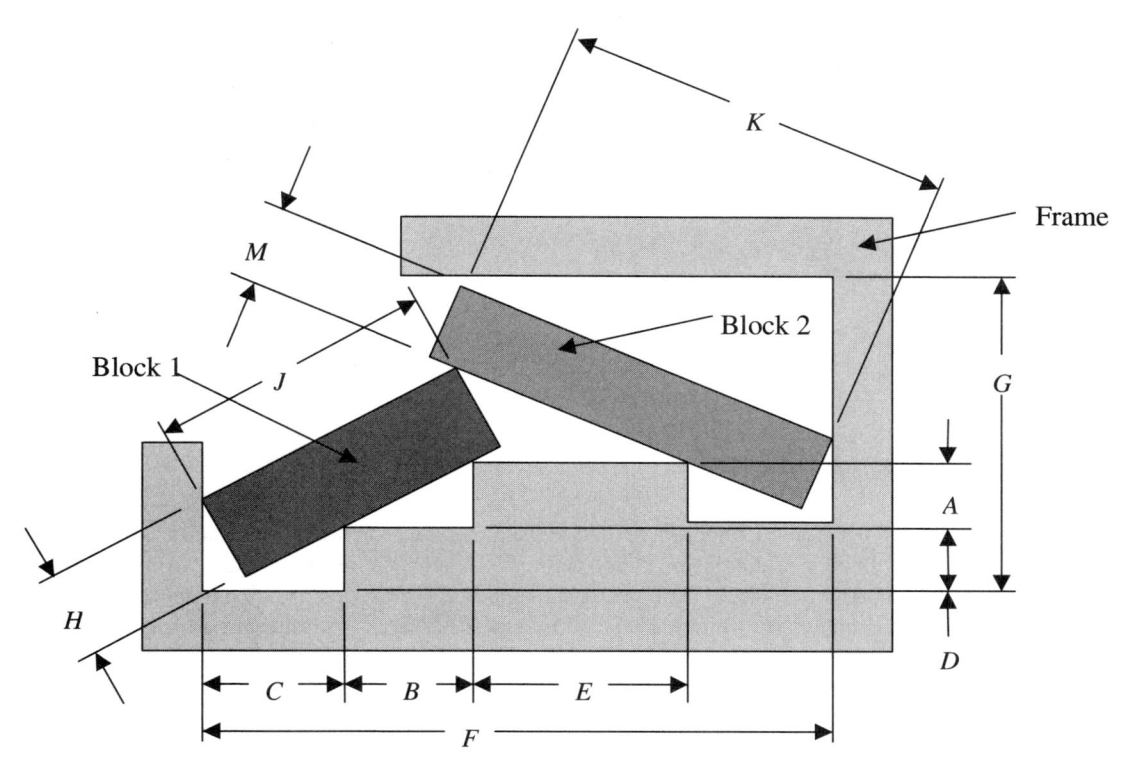

Figure 12-4 Stacked blocks we will use for an example problem

Table 12-1   Dimensions and tolerances corresponding to the variable names in Fig. 12-4

Variable Name	Mean Dimension (in.)	Tolerance (in.)
A	.875	.010
B	1.625	.020
C	1.700	.012
D	.875	.010
E	2.625	.020
F	7.875	.030
G	4.125	.010
H	1.125	.020
J	3.625	.015
K	5.125	.020
M	1.000	.010

tem along that line. We set the origin at one side of the gap and one of the axes will point to the other side, along the shortest direction. It's not important which side of the gap we choose for the origin. Coordinate system $\{u_1, u_2\}$ is shown in Fig. 12-5 and represents a set of unit vectors.

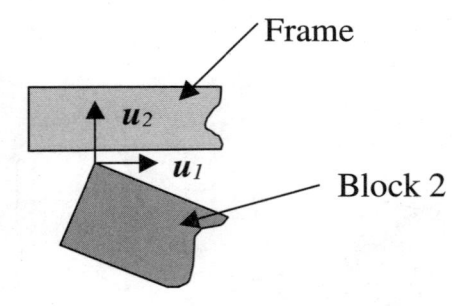

Figure 12-5 Gap coordinate system $\{u_1, u_2\}$

Step 3. Draw vector loop diagram

Now we'll have to draw a vector loop diagram similar to the dimension loop diagram constructed in section 9.2.2. Just like we did with the one-dimensional loop diagram, we'll start at one side of the gap and work our way around to the other. Anytime we go from one part to another, it must be through a point or surface of contact. When we've completed our analysis, we want a positive result to represent a clearance and a negative result to represent an interference. If we start our vector loop at the origin of the gap coordinate system, we'll finish at a more positive location on the axis, and we'll achieve the desired result.

For our example problem, there are several different vector loops we can chose. Two possibilities are shown in Fig. 12-6. The solution to the problem will be the same regardless of which vector loop we choose, but some may be more difficult to analyze than others. It's generally best to choose a loop that has a minimum number of vectors that need the length calculated. In Loop T, vectors T_2 and T_3 need the length calculated while Loop S has five vectors with undefined lengths. We can find lengths of the vectors S_5 and S_6 through simple one-dimension analysis, but S_2, S_4, and S_6 will require more work. So it appears that Loop T may provide easier calculations.

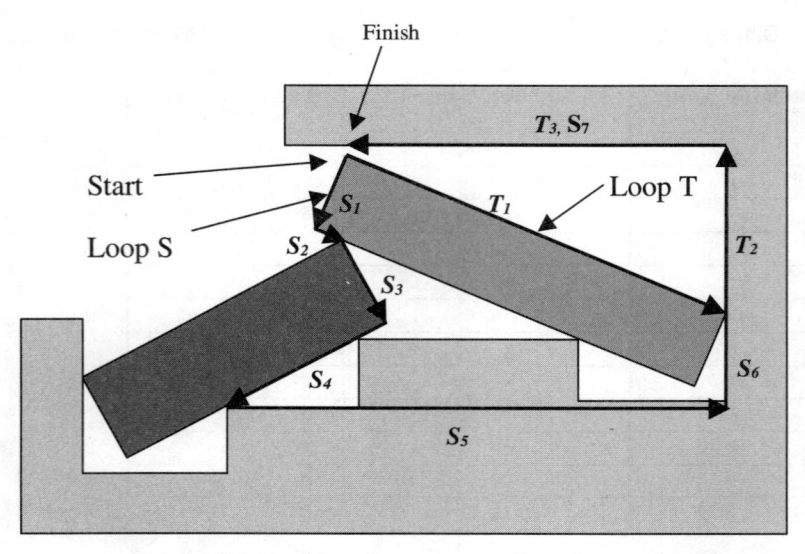

Figure 12-6 Possible vector loops to evaluate the gap of interest

Multi-Dimensional Tolerance Analysis (Manual Method)   12-7

As an alternative, look at the vector loop in Fig. 12-7. It has only three vectors with unknown length, one of which (x_9) is a linear combination of other dimensions. For vectors x_2 and x_{10}, we can calculate the length relatively easily. This is the loop we will use to analyze the problem.

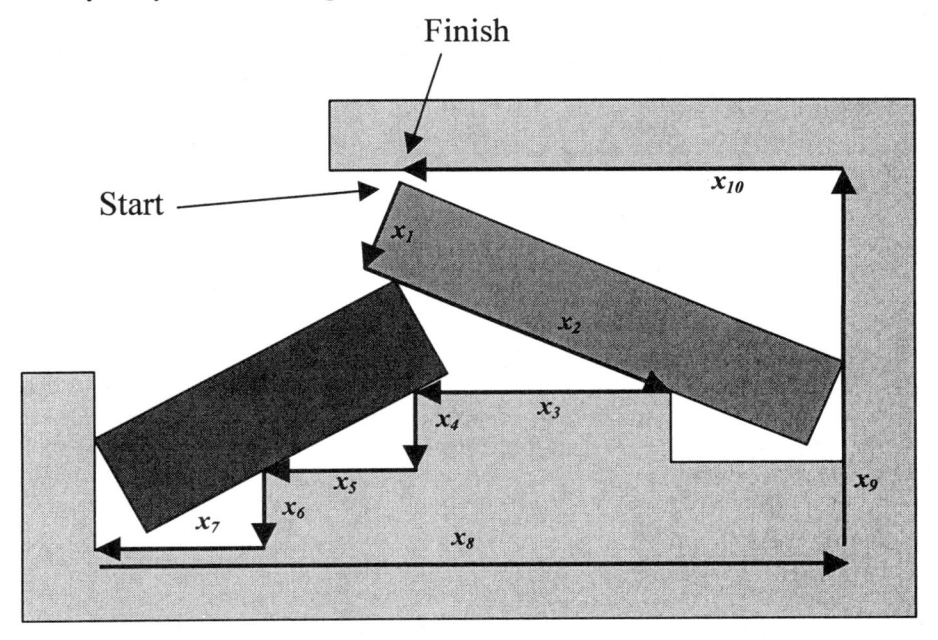

Figure 12-7 Vector loop we will use to analyze the gap. It presents easier calculations of unknown vector lengths.

Step 4. Establish component coordinate systems

The next step is establishing component coordinate systems. The number needed will depend on the configuration of the assembly. The idea is to have a coordinate system that will align with every component dimension and vector that will contribute to the stack. One additional coordinate system is needed and is shown in Fig. 12-8.

Coordinate system $\{v_1, v_2\}$ is needed for the vectors on Block 2. The dimensions on the frame align with $\{u_1, u_2\}$ so an additional coordinate system is not needed for them. Dimensions J and H on Block 1 do not contribute directly to a vector length so they do not need a coordinate system.

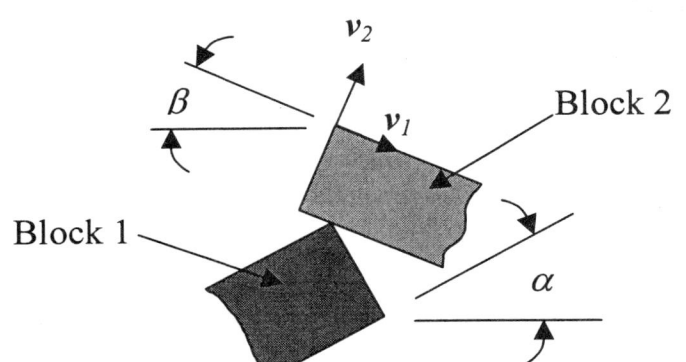

Figure 12-8 Additional coordinate system needed for the vectors on Block 2

Step 5. Write vectors in terms of component coordinate systems

The vectors in Fig. 12-7 are listed below in terms of their coordinate systems, angle β, and the dimensional variables in Table 12-1.

$x_1 = -Mv_2$

$x_2 = \left(K - \dfrac{F - C - B - E - M \sin \beta}{\cos \beta} \right) v_1$

$x_3 = -Eu_1$
$x_4 = -Au_2$
$x_5 = -Bu_1$
$x_6 = -Du_2$
$x_7 = -Cu_1$
$x_8 = Fu_1$
$x_9 = Gu_2$
$x_{10} = K\cos\beta \, u_1$

Angle β is not known yet, so we'll have to calculate it. Angle α contributes to the value of β, and is also needed. The equations for angles α and β are shown below.

$\alpha = \arctan\left(\dfrac{A}{B}\right)$

$= \arctan\left(\dfrac{.875}{1.625}\right)$

$= 28.30°$

$\beta = \arctan\left[\dfrac{\left(J - \dfrac{C - H\sin\alpha}{\cos\alpha} - \sqrt{A^2 + B^2} \right)\sin\alpha + H\cos\alpha}{E - \left(J - \dfrac{C - H\sin\alpha}{\cos\alpha} - \sqrt{A^2 + B^2} \right)\cos\alpha + H\sin\alpha} \right]$

$= \arctan\left[\dfrac{\left(3.625 - \dfrac{1.700 - 1.125(.4741)}{.8805} - \sqrt{.875^2 + 1.625^2} \right)(.4741) + 1.125(.8805)}{2.625 - \left(3.625 - \dfrac{1.700 - 1.125(.4741)}{.8805} - \sqrt{.875^2 + 1.625^2} \right)(.8805) + 1.125(.4741)} \right]$

$= 23.62°$

Step 6. Define relationships between coordinate systems

In order to relate the vectors in Step 5 to the gap, we will have to transform them into the same coordinate system as the gap. Thus, we'll have to convert vectors x_1 and x_2 into coordinate system $\{u_1, u_2\}$. One method follows.

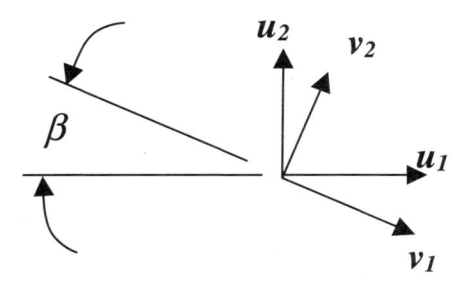

Figure 12-9 Relationship between coordinate systems $\{u_1,u_2\}$ and $\{v_1,v_2\}$

	u_1	u_2
v_1	$\cos\beta$	$-\sin\beta$
v_2	$\sin\beta$	$\cos\beta$

Fig. 12-9 shows the $\{u_1,u_2\}$ and $\{v_1,v_2\}$ coordinate systems and the angle β between them. To build a transformation between the two coordinate systems, we'll find the components of v_1 and v_2 in the directions of the unit vectors u_1 and u_2. For example, the component of v_1 in the u_1 direction is $\cos\beta$. The component of v_1 in the u_2 direction is $-\sin\beta$. The sign of the sine is negative because the component is pointing in the opposite direction as the positive u_2 axis. The table is completed by performing a similar analysis with vector v_2.

A matrix, Z, can be defined as follows:

$$Z = \begin{bmatrix} \cos\beta & -\sin\beta \\ \sin\beta & \cos\beta \end{bmatrix}$$

Multiplying Z by and $\{u_1,u_2\}^T$ will give us a transformation matrix that we can use to convert any vector in the $\{v_1,v_2\}$ coordinate system to the $\{u_1,u_2\}$ coordinate system.
Let $Q = Z\{u_1,u_2\}^T$

$$Q = \begin{bmatrix} \cos\beta & -\sin\beta \\ \sin\beta & \cos\beta \end{bmatrix} \begin{bmatrix} u_1 \\ u_2 \end{bmatrix}$$

$$Q = \begin{bmatrix} \cos\beta\, u_1 - \sin\beta\, u_2 \\ \sin\beta\, u_1 + \cos\beta\, u_2 \end{bmatrix}$$

Now we can transform any vector in the $\{v_1,v_2\}$ coordinate system to the $\{u_1,u_2\}$ coordinate system by multiplying it by Q.

Let's see how this works by transforming the vector $2v_1 + v_2$ to the $\{u_1,u_2\}$ coordinate system. We start by representing the vector as a matrix $[2\ 1]$.

$$2v_1 + v_2 = \begin{bmatrix} 2 & 1 \end{bmatrix} \begin{bmatrix} \cos\beta\, u_1 - \sin\beta\, u_2 \\ \sin\beta\, u_1 + \cos\beta\, u_2 \end{bmatrix}$$
$$= 2(\cos\beta\, u_1 - \sin\beta\, u_2) + \sin\beta\, u_1 + \cos\beta\, u_2$$
$$= (2\cos\beta + \sin\beta)u_1 + (\cos\beta - 2\sin\beta)u_2$$

Step 7. Convert all vectors into gap coordinate system

For our problem, we need all the vectors x_i that we found in Step 5 to be represented in the $\{u_1,u_2\}$ coordinate system. The only ones that need converting are x_1 and x_2.

$$x_1 = -Mv_2$$
$$= -M\begin{bmatrix}0 & 1\end{bmatrix}\begin{bmatrix}\cos\beta u_1 - \sin\beta u_2 \\ \sin\beta u_1 + \cos\beta u_2\end{bmatrix}$$
$$= -M(\sin\beta u_1 + \cos\beta u_2)$$

Similarly,

$$x_2 = \left(K - \frac{F-C-B-E-M\sin\beta}{\cos\beta}\right)v_1$$
$$= \left(K - \frac{F-C-B-E-M\sin\beta}{\cos\beta}\right)\begin{bmatrix}1 & 0\end{bmatrix}\begin{bmatrix}\cos\beta u_1 - \sin\beta u_2 \\ \sin\beta u_1 + \cos\beta u_2\end{bmatrix}$$
$$= \left(K - \frac{F-C-B-E-M\sin\beta}{\cos\beta}\right)(\cos\beta u_1 - \sin\beta u_2)$$

Step 8. Generate gap equation

To generate the gap equation now is very easy. We only need to observe that no components in the u_1 direction affect the gap. Thus, all we need to do is take the components in the u_2 direction and add them together.

$$Gap = -M\cos\beta + \left(K - \frac{F-C-B-E-M\sin\beta}{\cos\beta}\right)(-\sin\beta) - A - D + G \tag{12.3}$$

Now we have to insert the nominal values of each of the dimensions along with the values of the $\sin\beta$ and $\cos\beta$ into Eq. (12.3).

$$Gap = -1.000(.9162) + \left(5.125 - \frac{7.875 - 1.700 - 1.625 - 2.625 - 1.00(.4007)}{.9162}\right)(-.4007)$$
$$\quad -.875 - .875 + 4.125$$
$$= .0719$$

This is the nominal value of the gap.

Step 9. Calculate sensitivities

Next we need to calculate the sensitivities, which we'll find by evaluating the partial derivatives at the nominal value for each of the dimensions. As an example to the approach, we'll find the sensitivity for variable E, and provide tabulated results for the other variables.

Since β is a function of E, we'll have to apply the chain rule for partial derivatives. Let's start by redefining the gap as a function of β and E, say $Gap = \Psi(\beta, E)$. All the other terms will be treated as constants. Then,

$$\frac{\partial Gap}{\partial E} = \frac{\partial \Psi}{\partial E}\frac{dE}{dE} + \frac{\partial \Psi}{\partial \beta}\frac{\partial \beta}{\partial E}$$

Solving for each of the terms,

$$\frac{\partial \Psi}{\partial E} = -\tan \beta$$
$$= -.4373$$

$$\frac{dE}{dE} = 1$$

$$\frac{\partial \Psi}{\partial \beta} = \frac{F - C - B - E - M \sin \beta - K (\cos \beta)^3}{(\cos \beta)^2}$$
$$= \frac{7.875 - 1.700 - 1.625 - 2.625 - 1.000(.4007) - 5.125(.9162)^3}{(.9162)^2}$$
$$= -2.8796$$

$$\frac{\partial \beta}{\partial E} = \frac{-\left(J - \dfrac{C - H \sin \alpha}{\cos \alpha} - \sqrt{A^2 + B^2}\right) \sin \alpha - H \cos \alpha}{\left[\left(E - \left(J - \dfrac{C - H \sin \alpha}{\cos \alpha} - \sqrt{A^2 + B^2}\right) \cos \alpha + H \sin \alpha\right)^2 + \left(\left(J - \dfrac{C - H \sin \alpha}{\cos \alpha} - \sqrt{A^2 + B^2}\right) \sin \alpha + H \cos \alpha\right)^2\right]}$$

$$= \frac{-\left(3.625 - \dfrac{1.700 - 1.125(.4741)}{.8805} - \sqrt{.875^2 + 1.625^2}\right)(.4741) - 1.125(.8805)}{\left[\left(2.625 - \left(3.625 - \dfrac{1.700 - 1.125(.4741)}{.8805} - \sqrt{.875^2 + 1.625^2}\right)(.8805) + 1.125(.4741)\right)^2 + \left(\left(3.625 - \dfrac{1.700 - 1.125(.4741)}{.8805} - \sqrt{.875^2 + 1.625^2}\right)(.4741) + 1.125(.8805)\right)^2\right]}$$

$$= -.1331$$

$$\frac{\partial Gap}{\partial E} = -.4373(1) + (-2.8796)(-.1331)$$
$$= -.0540$$

Table 12-2 contains the sensitivities of the remaining variables. While calculating sensitivities manually is difficult for many gap equations, there are many software tools that can calculate them for us, simplifying the task considerably.

Table 12-2 Dimensions, tolerances, and sensitivities for the stacked block assembly

Variable Name	Mean Dimension (in.)	Tolerance (in.)	Sensitivity
A	.875	.010	-.5146
B	1.625	.020	.1567
C	1.700	.012	.4180
D	.875	.010	-1.0000
E	2.625	.020	-.0540
F	7.875	.030	.4372
G	4.125	.010	1.0000
H	1.125	.020	-.9956
J	3.625	.015	-.7530
K	5.125	.020	-.4006
M	1.000	.010	-1.0914

Step 10. Perform tolerance analysis or allocation

Now that we have calculated a nominal gap (.0719 in.) and all the sensitivities, we can use any of the analysis or allocation methods in Chapters 9 and 11. In Step 1, we decided to perform a worst case analysis. Using Eq. (12.1),

$$t_{wc} = |(-.5146)(.010)| + |(.1567)(.020)| + |(.4180)(.012)| + |(-1)(.010)| +$$
$$|(-.0540)(.020)| + |(.4372)(.030)| + |(1)(.010)| + |(-.9956)(.020)| +$$
$$|(-.7530)(.015)| + |(-.4006)(.020)| + |(1)(.010)|$$
$$= .0967$$

The minimum gap expected at worst case will be .0719 - .0967 = -.0248 in.

The negative number indicates that we can have an interference at worst case, and we do not satisfy our assembly requirement of a minimum clearance of .005 in.

12.4 Utilizing Sensitivity Information to Optimize Tolerances

Since we don't meet our assembly requirement, we need to consider some alterations to the design. We can use the sensitivities to help us make decisions about what we should target for change. For example, dimensions B and E have small sensitivities, so changing the tolerance on them will have little effect on the gap. To reduce the magnitude of the worst case tolerance stack, we would target the dimensions with the largest sensitivity first.

Also, the sensitivities help us decide which dimension we should consider changing to increase the gap. It takes a large change in a dimension with a small sensitivity to make a significant change in the gap. For example, making Dimension E .018 in. smaller will make the gap only about .001 in. larger. Conversely, making Dimension M .001 in. smaller will make the gap slightly more than .001 in. larger. If our goal is to correct the problem of assembly fit without changing the design any more than necessary, working with the dimensions with the largest sensitivities will be advantageous.

The simplest solution would be to increase the opening in the frame, Dimension G, from 4.125 in. to 4.160 in. which will provide the clearance we need. However, if we assume the thickness of the top of the frame can't change, that will cause us to increase the size of the frame. That could be a problem. So instead, we'll change one of the internal dimensions on the frame, making Dimension A equal to .815 in. With this change, the nominal gap will be .1044 in., worst case tolerance stack is .0980 in. and the minimum clearance is .0064 in.

The worst case tolerance stack increased because many of the sensitivities changed when A was changed. This is because we evaluate the partial derivatives at the nominal value of the dimensions, so when the nominal value of A was changed, we changed the calculated result. Another way to think of it is that we moved to a different point in our design space when we changed Dimension A, so the slope changed in several different directions.

The final dimensions, tolerances and sensitivities are shown in Table 12-3.

Table 12-3 Final dimensions, tolerances and sensitivities of the stacked block assembly

Variable Name	Mean Dimension (in.)	Tolerance (in.)	Sensitivity
A	.815	.010	-.5605
B	1.625	.020	.1642
C	1.700	.012	.3846
D	.875	.010	-1.0000
E	2.625	.020	-.0552
F	7.875	.030	.4488
G	4.125	.010	1.0000
H	1.125	.020	-.9811
J	3.625	.015	-.7450
K	5.125	.020	-.4094
M	1.000	.010	-1.0961

12.5 Summary

In this section, we've demonstrated a technique for analyzing tolerances for multi-dimensional problems. While this is an approximate method, the results are very good as long as tolerances are not too large compared to the curvature of the n-space surface represented by the gap equation. It's good to remember that once we have found the gap equation and calculated the sensitivities, we can use any of the analysis or allocation techniques discussed in Chapters 9 and 11.

An important point to reiterate is that we show one method for developing a gap equation. While this will give accurate results, it may be more cumbersome at times than deriving the equation directly from the geometry of the problem. In general, the more complicated problems will be easier to solve using the technique shown here because it helps break the problem into smaller pieces that are more convenient to evaluate.

In this section, we evaluated an assembly that is not similar to ones found during the design process, but the technique works equally well on typical design problems. In fact, one thing very powerful about this technique is that it is not limited to traditional tolerance stacks. For example, we can use it to evaluate the effect of tolerances on the magnitude of the maximum stress in a loaded, cantilevered beam. Once we have developed the stress equation, we can calculate the sensitivities and determine the effect of things like the length, width and thickness of the beam, location of the load, and material properties such as the modulus of elasticity and yield strength. It even works well for electrical problems, such as evaluating the range of current we'll see in a circuit due to tolerances on the electrical components.

Chapter 13

Multi-Dimensional Tolerance Analysis (Automated Method)

**Kenneth W. Chase, Ph.D.
Brigham Young University
Provo, Utah**

Dr. Chase has taught mechanical engineering at the Brigham Young University since 1968. An advocate of computer technology, he has served as a consultant to industry on numerous projects involving engineering software applications. He served as a reviewer of the Motorola Six Sigma Program at its inception. He also served on an NSF select panel for evaluating tolerance analysis research needs. In 1984, he founded the ADCATS consortium for the development of CAD-based tools for tolerance analysis of mechanical assemblies. More than 30 sponsored graduate theses have been devoted to the development of the tolerance technology contained in the CATS software. Several faculty and students are currently involved in a broad spectrum of research projects and industry case studies on statistical variation analysis. Past and current sponsors include Allied Signal, Boeing, Cummins, FMC, Ford, GE, HP, Hughes, IBM, Motorola, Sandia Labs, Texas Instruments, and the US Navy.

13.1 Introduction

In this chapter, an alternative method to the one described in Chapter 12 is presented. This method is based on vector loop assembly models, but with the following distinct differences:

1. A set of rules is provided to assure a valid set of vector loops is obtained. The loops include only those controlled dimensions that contribute to assembly variation. All dimensions are datum referenced.
2. A set of kinematic modeling elements is introduced to assist in identifying the adjustable dimensions within the assembly that change to accommodate dimensional variations.

3. In addition to describing variation in assembly gaps, a comprehensive set of assembly tolerance requirements is introduced, which are useful to designers as performance requirements.
4. Algebraic manipulation to derive an explicit expression for the assembly feature is eliminated. This method operates equally well on implicit assembly equations. The loop equations are solved the same way every time, so it is well suited for computer automation.

This chapter distinguishes itself from Chapter 12 by replacing differentiation of a complicated assembly expression with a single matrix operation, which determines all necessary tolerance sensitivities simultaneously. Since the matrix only contains sines and cosines, derivations are simple. As with the method shown in Chapter 12, this method may also include other sources of variation, such as position tolerance, parallelism error, or profile variations.

13.2 Three Sources of Variation in Assemblies

There are three main sources of variation, which must be accounted for in mechanical assemblies:
1. Dimensional variations (lengths and angles)
2. Geometric form and feature variations (position, roundness, angularity, etc.)
3. Kinematic variations (small adjustments between mating parts)

Dimensional and form variations are the result of variations in the manufacturing processes or raw materials used in production. Kinematic variations occur at assembly time, whenever small adjustments between mating parts are required to accommodate dimensional or form variations.

The two-component assembly shown in Figs. 13-1 and 13-2 demonstrates the relationship between dimensional and form variations in an assembly and the small kinematic adjustments that occur at assembly time. The parts are assembled by inserting the cylinder into the groove until it makes contact on the two sides of the groove. For each set of parts, the distance U will adjust to accommodate the current value of dimensions A, R, and θ. The assembly resultant U represents the nominal position of the cylinder, while $U + \Delta U$ represents the position of the cylinder when the variations ΔA, ΔR, and $\Delta \theta$ are present. This adjustability of the assembly describes a kinematic constraint, or a closure constraint on the assembly.

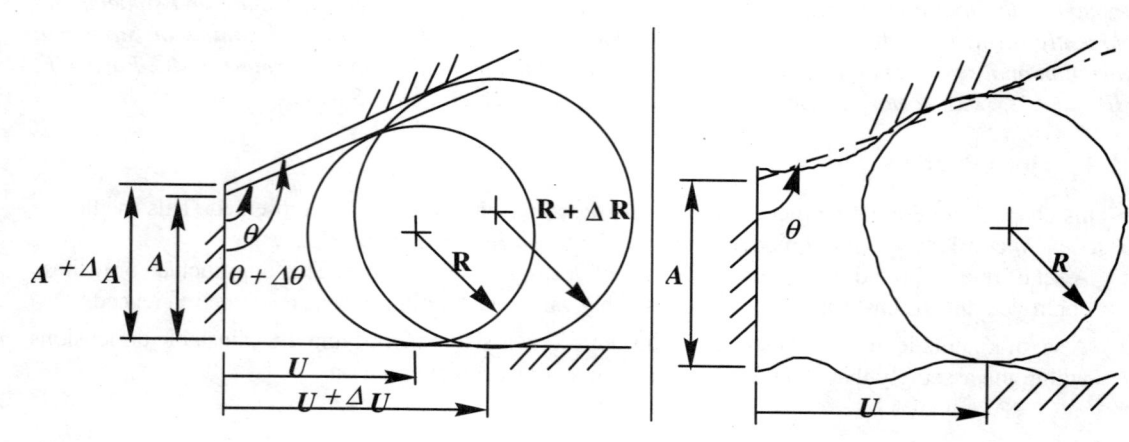

Figure 13-1 Kinematic adjustment due to component dimension variations

Figure 13-2 Adjustment due to geometric shape variations

It is important to distinguish between component and assembly dimensions in Fig. 13-1. Whereas A, R, and θ are component dimensions, subject to random process variations, distance U is not a component dimension. It is a resultant assembly dimension. U is not a manufacturing process variable, it is a kinematic assembly variable. Variations in U can only be measured after the parts are assembled. A, R, and θ are the independent random sources of variation in this assembly. They are the inputs. U is a dependent assembly variable. It is the output.

Fig. 13-2 illustrates the same assembly with exaggerated geometric feature variations. For production parts, the contact surfaces are not really flat and the cylinder is not perfectly round. The pattern of surface waviness will differ from one part to the next. In this assembly, the cylinder makes contact on a peak of the lower contact surface, while the next assembly may make contact in a valley. Similarly, the lower surface is in contact with a lobe of the cylinder, while the next assembly may make contact between lobes.

Local surface variations such as these can propagate through an assembly and accumulate just as size variations do. Thus, in a complete assembly model all three sources of variation must be accounted for to assure realistic and accurate results.

13.3 Example 2-D Assembly – Stacked Blocks

The assembly in Fig. 13-3 illustrates the tolerance modeling process. It consists of three parts: a Block, resting on a Frame, is used to position a Cylinder, as shown. There are four different mating surface conditions that must be modeled. The gap G, between the top of the Cylinder and the Frame, is the critical assembly feature we wish to control. Dimensions a through f, r, R, and θ are dimensions of component features that contribute to assembly variation. Tolerances are estimates of the manufacturing process variations. Dimension g is a utility dimension used in locating gap G.

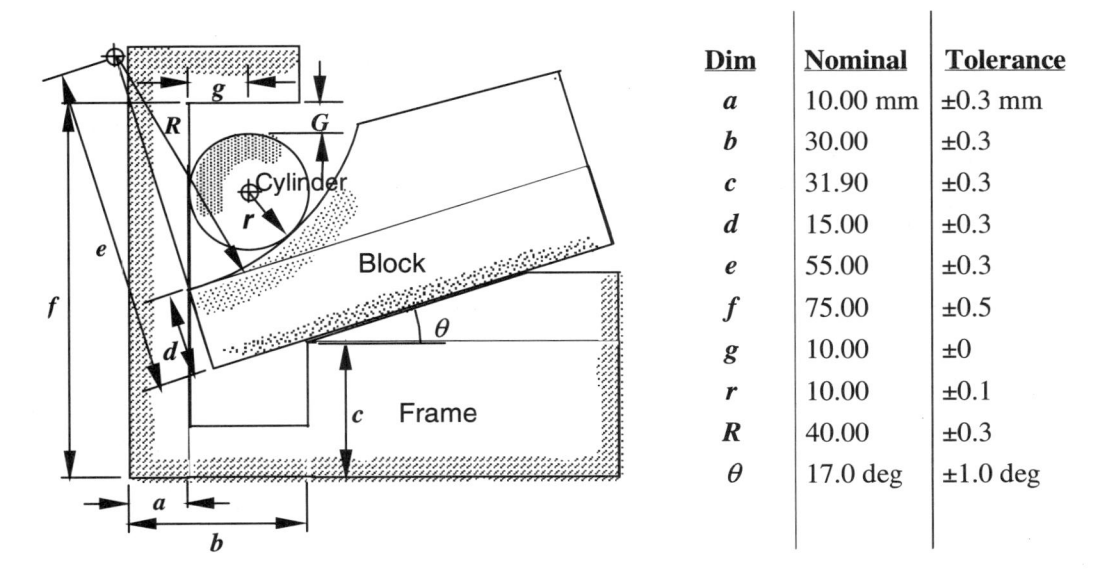

Dim	Nominal	Tolerance
a	10.00 mm	±0.3 mm
b	30.00	±0.3
c	31.90	±0.3
d	15.00	±0.3
e	55.00	±0.3
f	75.00	±0.5
g	10.00	±0
r	10.00	±0.1
R	40.00	±0.3
θ	17.0 deg	±1.0 deg

Figure 13-3 Stacked blocks assembly

13.4 Steps in Creating an Assembly Tolerance Model

Step 1. Create an assembly graph

An assembly graph is a simplified diagram representing an assembly. All geometry and dimensions are removed. Only the mating conditions between the parts are shown. Each part is shown as a balloon. The

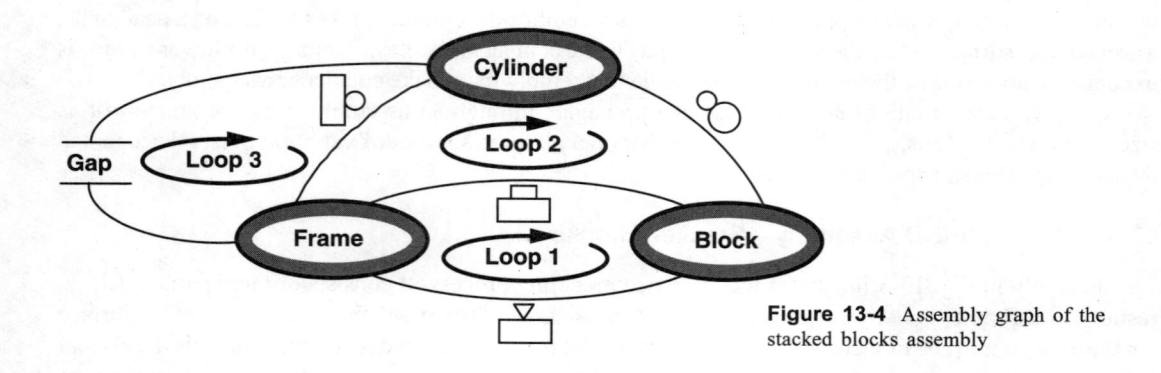

Figure 13-4 Assembly graph of the stacked blocks assembly

contacts or joints between mating parts are shown as arcs or edges joining the corresponding parts. Fig. 13-4 shows the assembly graph for the sample problem.

The assembly graph lets you see the relationship between the parts in the assembly. It also reveals by inspection how many loops (dimension chains) will be required to build the tolerance model. Loops 1 and 2 are closed loop assembly constraints, which locate the Block and Cylinder relative to the Frame. Loop 3 is an open loop describing the assembly performance requirement. A systematic procedure for defining the loops is illustrated in the steps that follow.

Symbols have been added to each edge identifying the type of contact between the mating surfaces. Between the Block and Frame there are two contacts: plane-to-plane and edge-to-plane. These are called Planar and Edge Slider joints, respectively, after their kinematic counterparts.

Only six kinematic joint types are required to describe the mating part contacts occurring in most 2-D assemblies, as shown in Fig. 13-5. Arrows indicate the degrees of freedom for each joint, which permit relative motion between the mating surfaces. Also shown are two datum systems described in the next section.

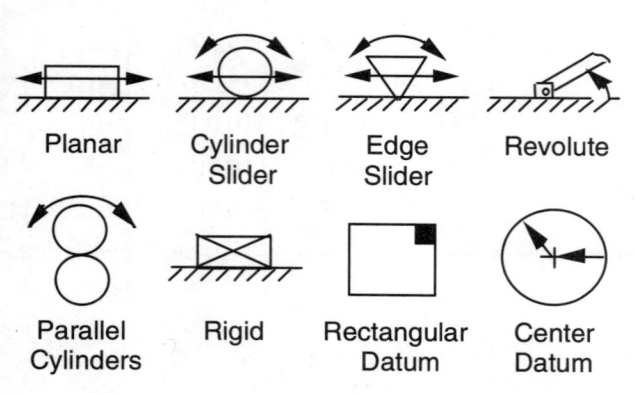

Figure 13-5 2-D kinematic joint and datum types

Step 2. Locate the datum reference frame for each part

Creating the tolerance model begins with an assembly drawing, preferably drawn to scale. Elements of the tolerance model are added to the assembly drawing as an overlay. The first elements added are a set of local coordinate systems, called *Datum Reference Frames*, or DRFs. Each part must have its own DRF. The DRF is used to locate features on a part. You probably will choose the datum planes used to define the parts. But, feel free to experiment. As you perform the tolerance analysis, you may find a different dimensioning scheme that reduces the number of variation sources or is less sensitive to variation. Identifying such effects and recommending appropriate design changes is one of the goals of tolerance analysis.

In Fig. 13-6, the Frame and Block both have rectangular DRFs located at their lower left corners, with axes oriented along orthogonal surfaces. The Cylinder has a cylindrical DRF system at its center. A second center datum has been used to locate the center of the large arc on the Block. This is called a *feature datum* and it is used to locate a single feature on a part. It represents a virtual point on the Block and must be located relative to the Block DRF.

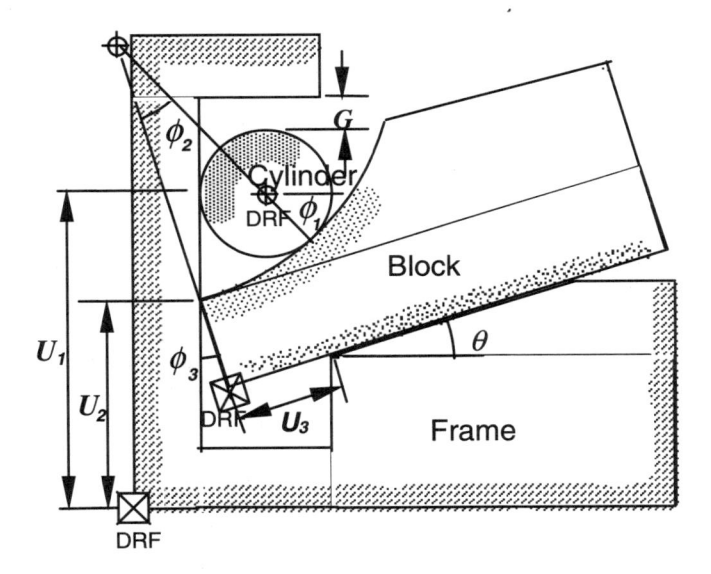

Figure 13-6 Part datums and assembly variables

Also shown in Fig. 13-6 are the assembly variables occurring within this assembly. U_1, U_2, and U_3 are adjustable dimensions determined by the sliding contacts between the parts. ϕ_1, ϕ_2, and ϕ_3 define the adjustable rotations that occur in response to dimensional variations. Each of the adjustable dimensions is associated with a kinematic joint. Dimension G is the gap whose variation must be controlled by setting appropriate tolerances on the component dimensions.

Step 3. Locate kinematic joints and create datum paths

In Fig. 13-7, the four kinematic joints in the assembly are located at points of contact and oriented such that the joint axes align with the adjustable assembly dimensions (called the *joint degrees of freedom*). This is done by inspection of the contact surfaces. There are simple modeling rules for each joint type. Joint 1 is an edge slider. It represents an edge contacting a planar surface. It has two degrees of freedom: it can slide along the contact plane (U_2) and rotate relative to the contact point (ϕ_3). Of course, it is constrained not to slide or rotate by contact with mating parts, but a change in dimensions a, b, c, d, or θ will cause U_2 and ϕ_3 to adjust accordingly.

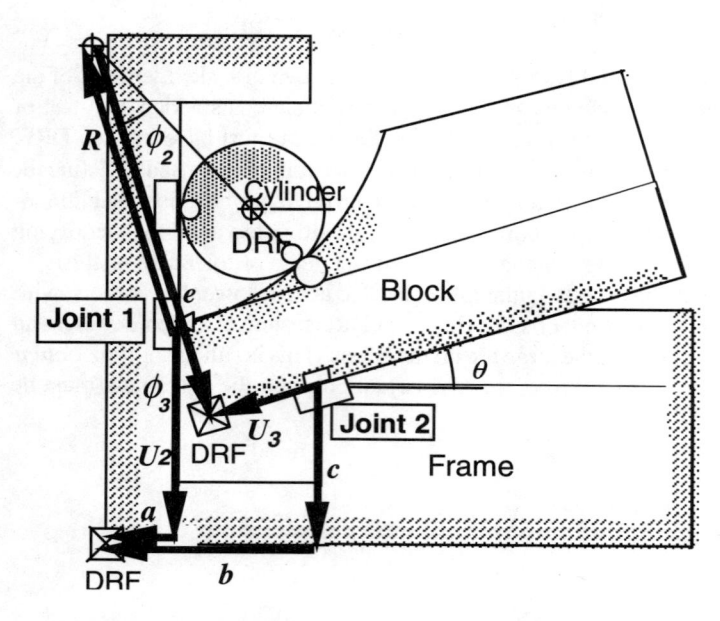

Figure 13-7 Datum paths for Joints 1 and 2

Joint 2 is a planar joint describing sliding contact between two planes. U_3 locates a reference point on the contacting surface relative to the Block DRF. U_3 is constrained by the corner of the Block resting against the vertical wall of the Frame.

In Fig. 13-8, Joint 3 locates the contact point between the Cylinder and the Frame. A cylinder slider has two degrees of freedom: U_1 is in the sliding plane and ϕ_1 is measured at the center datum of the Cylinder. Joint 4 represents contact between two parallel cylinders. The point of contact on the Cylinder is located by ϕ_1; on the Block, by ϕ_2. Joints 3 and 4 are similarly constrained. However, changes in component dimensions cause adjustments in the points of contact from one assembly to the next.

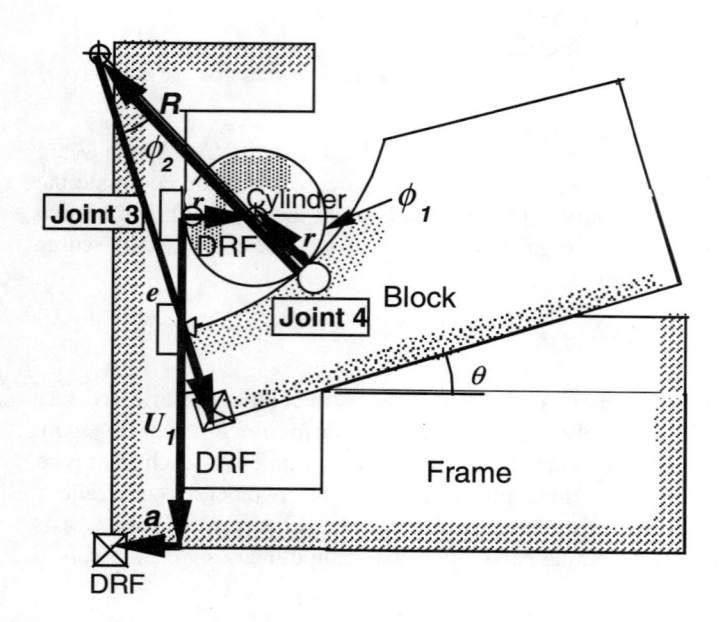

Figure 13-8 Datum paths for Joints 3 and 4

The vectors overlaid on Figs. 13-7 and 13-8 are called the *datum paths*. A datum path is a chain of dimensions that locates the point of contact at a joint with respect to a part DRF. For example, Joint 2 in Fig. 13-7 joins the Block to the Frame. The point of contact must be defined from both the Frame and Block DRFs. There are two vector paths that leave Joint 2. U_3 lies on the sliding plane and points to the Block DRF. Vectors c and b point to the Frame DRF. The two datum paths for Joint 1 are: vectors U_2 and a leading to the Frame DRF, and arc radius R and vector e, leading to the Block DRF. In Fig. 13-8, Joint 3 is located by radius r pointing to the Cylinder DRF, and U_1 and a defining the path to the Frame DRF. The contact point for Joint 4 is located by a second radius r pointing to the Cylinder DRF and arc radius R and e leading to the Block DRF.

Modeling rules define the path a vector loop must follow to cross a joint. Fig. 13-9 shows the correct vector paths for crossing four 2-D joints. The rule states that the loop must enter and exit a joint through the local joint datums. For the Planar and Edge Slider joints, a vector U (either incoming or outgoing) must lie in the sliding plane. Local Datum 2 represents a reference point on the sliding plane, from which the contact point is located. For the Cylindrical Slider joint, the incoming vector passes through center datum of the cylinder, follows a radius vector to the contact point and leaves through a vector in the sliding plane. The path through the parallel cylinder joint passes from the center datum of one cylinder to the center datum of the other, passing through the contact point and two colinear radii in between.

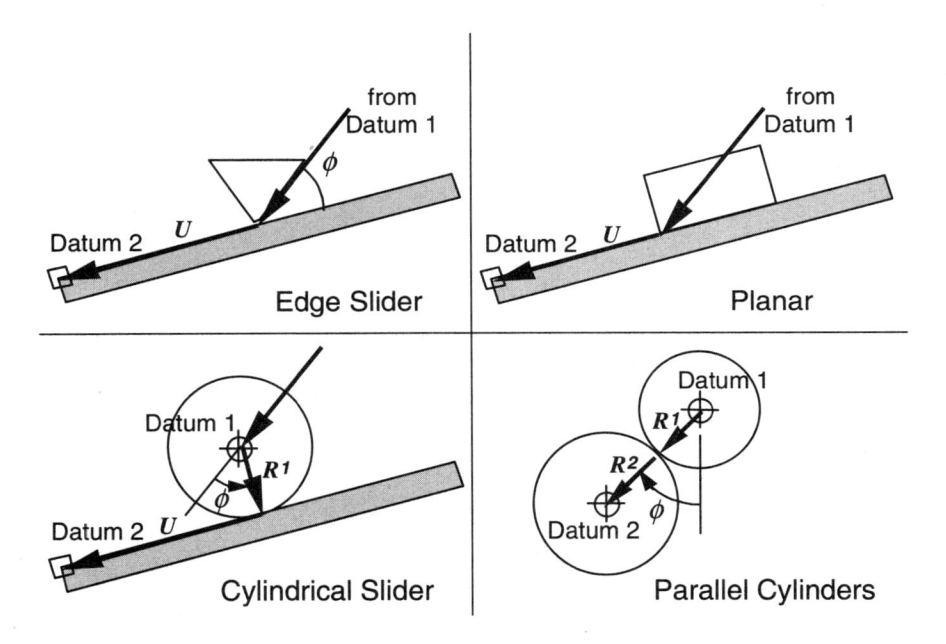

Figure 13-9 2-D vector path through the joint contact point

As we created the two datum paths from each joint, we were in fact creating the incoming and outgoing vectors for each joint. Although they were both drawn as outgoing vector paths, when we combine them to form the vector loops, one of the datum paths will be reversed in direction to correspond to the vector loop direction.

Each joint introduces kinematic variables into the assembly, which must be included in the vector model. The rules assure that the kinematic variables introduced by each joint are included in the loop, namely, the vector U in each sliding plane, and the relative angle ϕ.

13-8   Chapter Thirteen

Each datum path must follow controlled engineering dimensions or adjustable assembly dimensions. This is a critical task, as it determines which dimensions will be included in the tolerance analysis. All joint degrees of freedom must also be included in the datum paths. They are the unknown variations in the assembly tolerance analysis.

Step 4. Create vector loops

Vector loops define the assembly constraints that locate the parts of the assembly relative to each other. The vectors represent the dimensions that contribute to tolerance stackup in the assembly. The vectors are joined tip-to-tail, forming a chain, passing through each part in the assembly in succession.

A vector loop must obey certain modeling rules as it passes through a part. It must:
- Enter through a joint
- Follow the datum path to the DRF
- Follow a second datum path leading to another joint, and
- Exit to the next adjacent part in the assembly

This is illustrated schematically in Fig. 13-10. Thus, vector loops are created by simply linking together the datum paths. By so doing, all the dimensions will be datum referenced.

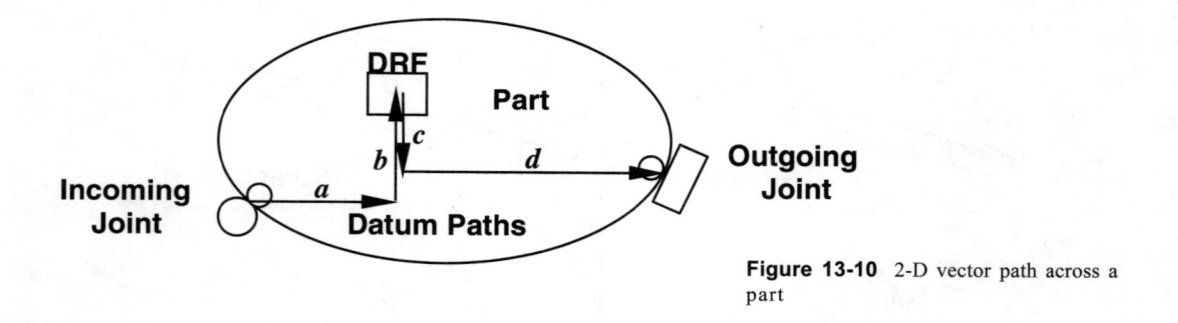

Figure 13-10 2-D vector path across a part

Additional modeling rules for vector loops include:
- Loops must pass through every part and every joint in the assembly.
- A single vector loop may not pass through the same part or the same joint twice, but it may start and end in the same part.
- If a vector loop includes the exact same dimension twice, in opposite directions, the dimension is redundant and must be omitted.
- There must be enough loops to solve for all of the kinematic variables (joint degrees of freedom). You will need one loop for each of the three variables.

Two closed loops are required for the example assembly, as we saw in the assembly graph of Fig. 13-4. The resulting loops are shown in Figs. 13-11 and 13-12. Notice how similar the loops are to the datum paths of Figs. 13-7 and 13-8. Also, notice that some of the vectors in the datum paths were reversed to keep all the vectors in each loop going in the same direction.

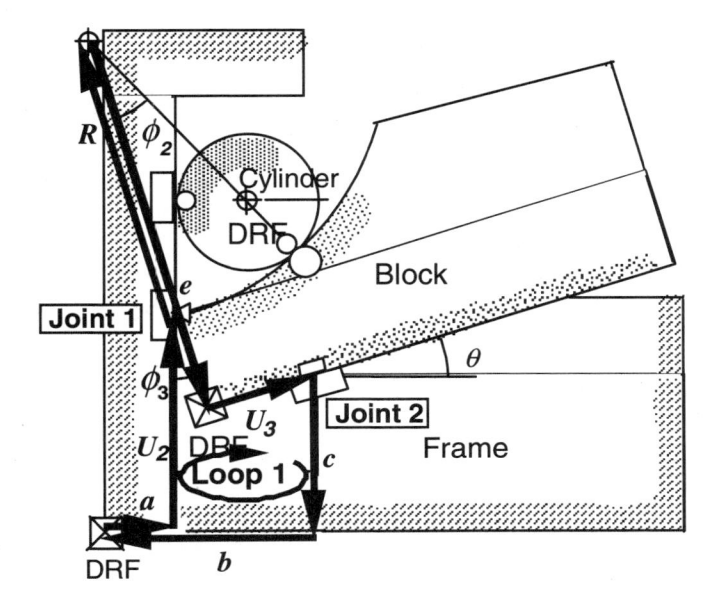

Figure 13-11 Assembly Loop 1

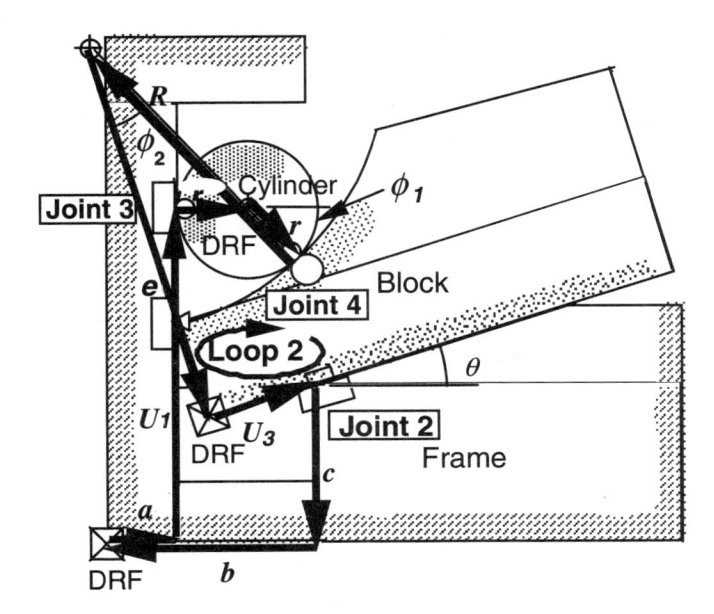

Figure 13-12 Assembly Loop 2

Step 5. Add geometric variations

Geometric variations of form, orientation, and location can introduce variation into an assembly. Such variations can accumulate statistically and propagate kinematically the same as size variations. The manner in which geometric variation propagates across mating surfaces depends on the nature of the contact. Fig. 13-13 illustrates this concept.

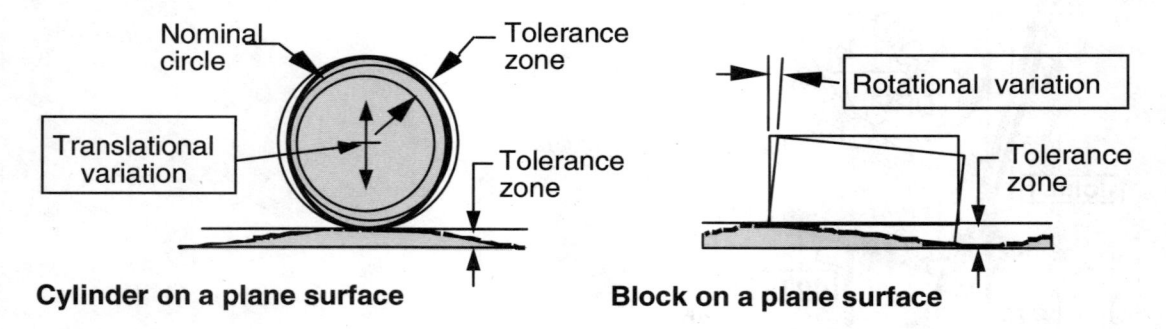

Figure 13-13 Propagation of 2-D translational and rotational variation due to surface waviness

Consider a cylinder on a plane, both of which are subject to surface waviness, represented by a tolerance zone. As the two parts are brought together to be assembled, the cylinder could rest on the top of a hill or down in a valley of a surface wave. Thus, for this case, the center of the cylinder will exhibit translational variation from assembly-to-assembly in a direction normal to the surface. Similarly, the cylinder could be lobed, as shown in the figure, resulting in an additional vertical translation, depending on whether the part rests on a lobe or in between.

In contrast to the cylinder/plane joint, the block on a plane shown in Fig. 13-13 exhibits rotational variation. In the extreme case, one corner of the block could rest on a waviness peak, while the opposite corner could be at the bottom of the valley. The magnitude of rotation would vary from assembly-to-assembly. Waviness on the surface of the block would have a similar effect.

In general, for two mating surfaces, we would have two independent surface variations that introduce variation into the assembly. How it propagates depends on the nature of the contact, that is, the type of kinematic joint. While there is little or no published data on typical surface variations for manufacturing processes, it is still instructive to insert estimates of variations and calculate the magnitude of their possible contribution. Fig. 13-14 illustrates several estimated geometric variations added to the sample assembly model. Only one variation is defined at each joint, since both mating surfaces have the same sensitivity. Examining the percent contribution to the gap variation will enable us to determine which surfaces should have a GD&T tolerance control.

Step 6. Define performance requirements

Performance requirements are engineering design requirements. They apply to assemblies of parts. In tolerance analysis, they are the specified limits of variation of the assembly features that are critical to product performance, sometimes called the *key characteristics* or *critical feature tolerances*. Several examples were illustrated in Chapter 9 for an electric motor assembly. Simple fits between a bearing and shaft, or a bearing and housing, would only involve two parts, while the radial and axial clearance between the armature and housing would involve a tolerance stackup of several parts and dimensions.

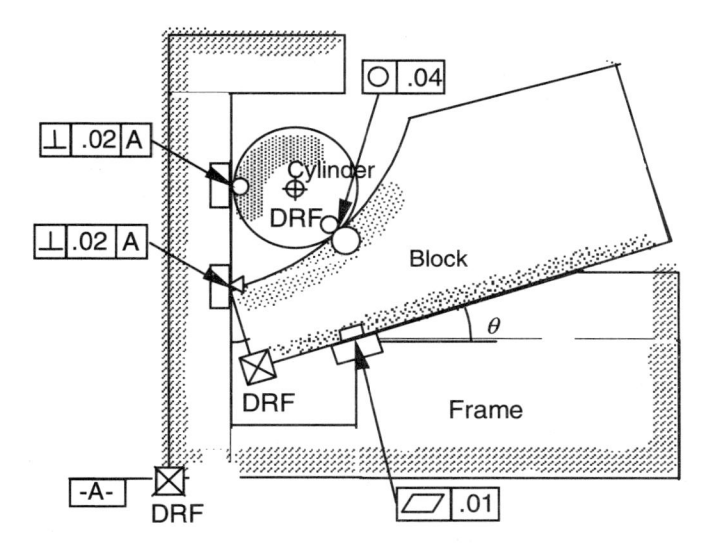

Figure 13-14 Applied geometric variations at contact points

Component tolerances are set as a result of analyzing tolerance stackup in an assembly and determining how each component dimension contributes to assembly variation. Processes and tooling are selected to meet the required component tolerances. Inspection and gaging equipment and procedures are also determined by the resulting component tolerances. Thus, we see that the performance requirements have a pervasive influence on the entire manufacturing enterprise. It is the designer's task to transform each performance requirement into assembly tolerances and corresponding component tolerances.

There are several assembly features that commonly arise in product design. A fairly comprehensive set can be developed by examining geometric dimensioning and tolerancing feature controls and forming a corresponding set for assemblies. Fig. 13-15 shows a basic set that can apply to a wide range of assemblies.

Note that when applied to an assembly feature, parallelism applies to two surfaces on two different parts, while GD&T standards only control parallelism between two surfaces on the same part. The same can be said about the other assembly controls, with the exception of position. Position tolerance in GD&T relates assemblies of two parts, while the position tolerance in Fig. 13-15 could involve a whole chain of intermediate parts contributing variation to the position of mating features on the two end parts. An example of the application of assembly tolerance controls is the alignment requirements in a car door assembly. The gap between the edge of the door and the door frame must be uniform and flush (parallel in two planes). The door striker must line up with the door lock mechanism (position).

Each assembly feature, such as a gap or parallelism, requires an open loop to describe the variation. You can have any number of open loops in an assembly tolerance model, one per critical feature. Closed loops, on the other hand, are limited to the number of loops required to locate all of the parts in the assembly. It is a unique number determined by the number of parts and joints in the assembly.

$$L = J - P + 1$$

where L is the required number of loops, J is the number of joints, and P is the number of parts. For the example problem:

$$L = 4 - 3 + 1 = 2$$

which is the number we determined by inspection of the assembly graph.

Figure 13-15 Assembly tolerance controls

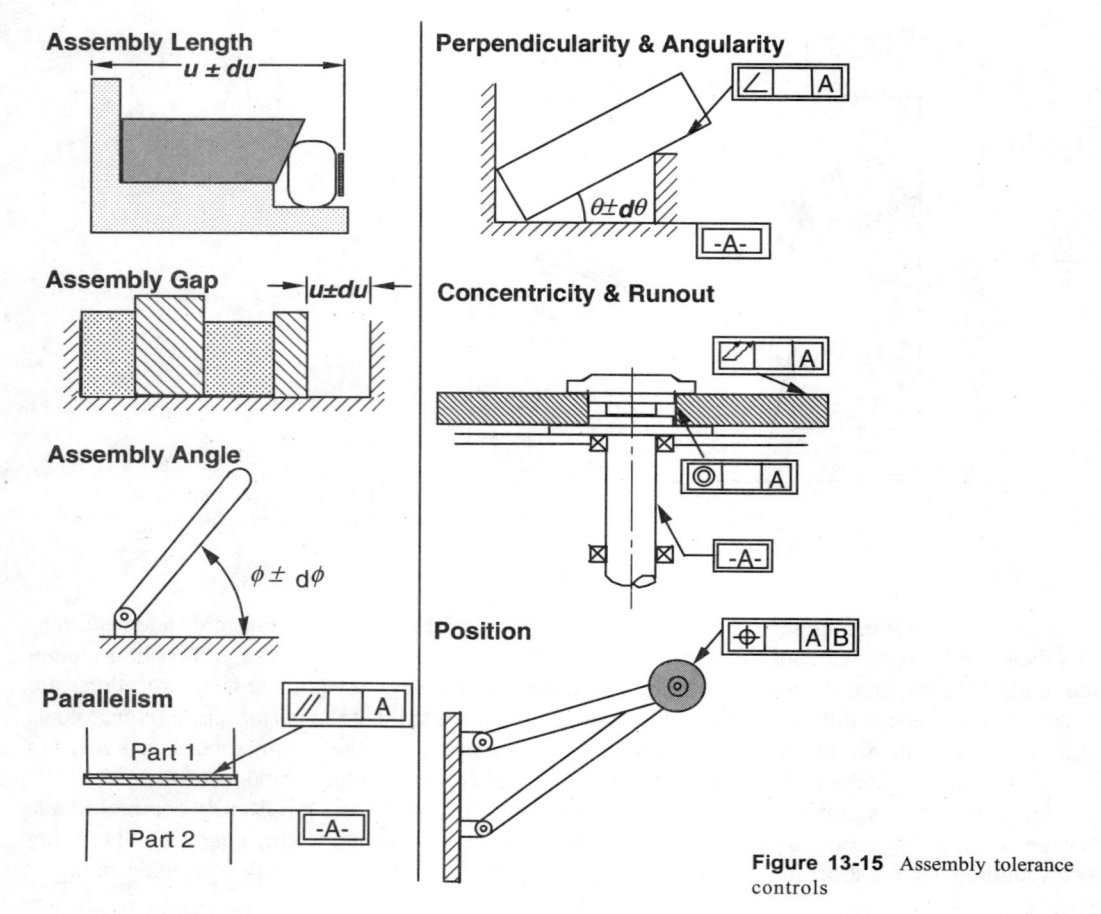

The example assembly has a specified gap tolerance between a cylindrical surface and a plane, as shown in Fig. 13-6. The vector loop describing the gap is shown in Fig. 13-16. It begins with vector *g*, on one side of the gap, proceeds from part-to-part, and ends at the top of the cylinder, on the opposite side of the gap. Note that vector *a,* at the DRF of the Frame, appears twice in the same loop in opposite directions. It is therefore redundant and both vectors must be eliminated. Vector *r* also appears twice in the cylinder; however, the two vectors are not in opposite directions, so they must both be included in the loop.

Vector *g*, incidentally, is not a manufactured dimension. It is really a kinematic variable, which adjusts to locate the point on the gap opposite the highest point on the cylinder. It was given zero tolerance, because it does not contribute to the variation of the gap.

The steps illustrated above describe a comprehensive system for creating assembly models for tolerance analysis. With just a few basic elements, a wide variety of assemblies may be represented. Next, we will illustrate the steps in performing a variational analysis of an assembly model.

13.5   Steps in Analyzing an Assembly Tolerance Model

In a 2-D or 3-D assembly, component dimensions can contribute to assembly variation in more than one direction. The magnitude of the component contributions to the variation in a critical assembly feature is determined by the product of the process variation and the tolerance sensitivity, summed by worst case

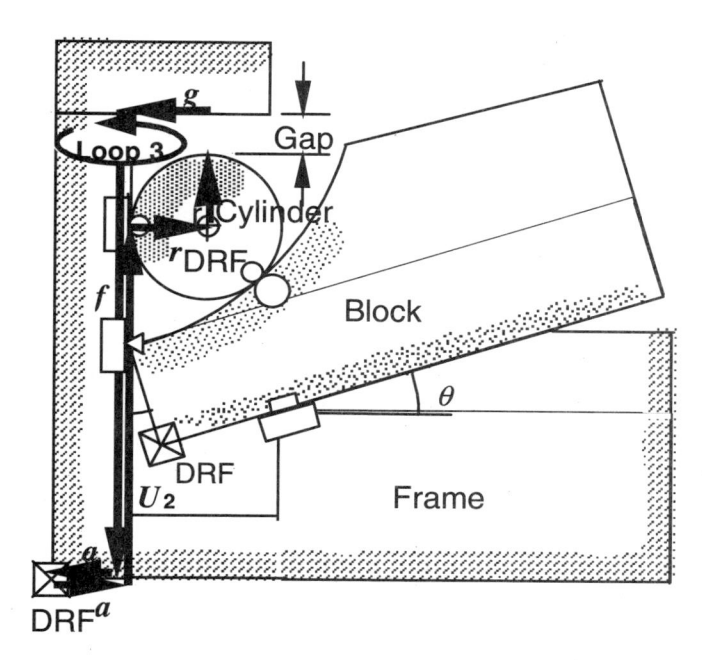

Figure 13-16 Open loop describing critical assembly gap

or Root Sum Squared (RSS). If the assembly is in production, actual process capability data may be used to predict assembly variation. If production has not yet begun, the process variation is approximated by substituting the specified tolerances for the dimensions, as described earlier.

The tolerance sensitivities may be obtained numerically from an explicit assembly function, as illustrated in Chapter 12. An alternative procedure will be demonstrated, which does not require the derivation of an explicit assembly function. It is a systematic method, which may be applied to any vector loop assembly model.

Step 1. Generate assembly equations from vector loops

The first step in an analysis is to generate the assembly equations from the vector loops. Three scalar equations describe each closed vector loop. They are derived by summing the vector components in the x and y directions, and summing the vector rotations as you trace the loop. For closed loops, the components sum to zero. For open, they sum to a nonzero gap or angle.

The equations describing the stacked block assembly are shown below. For Closed Loops 1 and 2, h_x, h_y, and h_θ are the sums of the x, y, and rotation components, respectively. See Eqs. (13.1) and (13.2). Both loops start at the lower left corner, with vector a. For Open Loop 3, only one scalar equation (Eq. (13.6)) is needed, since the gap has only a vertical component. Open loops start at one side of the gap and end at the opposite side.

Closed Loop 1

$$\begin{aligned}
h_x &= a\cos(0) + U_2\cos(90) + R\cos(90+\phi_3) + e\cos(90+\phi_3-180) + U_3\cos(\theta) \\
&\quad + c\cos(-90) + b\cos(-180) = 0 \\
h_y &= a\sin(0) + U_2\sin(90) + R\sin(90+\phi_3) + e\sin(90+\phi_3-180) + U_3\sin(\theta) \\
&\quad + c\sin(-90) + b\sin(-180) = 0 \\
h_\theta &= 0 + 90 + \phi_3 - 180 + 90 - \theta - 90 - 90 + 180 = 0
\end{aligned}$$

(13.1)

Closed Loop 2

$$h_x = a\cos(0) + U_1\cos(90) + r\cos(0) + r\cos(-\phi_1) + R\cos(-\phi_1+180) + e\cos(-\phi_1-\phi_2)$$
$$+ U_3\cos(\theta) + c\cos(-90) + b\cos(-180) = 0$$
$$h_y = a\sin(0) + U_1\sin(90) + r\sin(0) + r\sin(-\phi_1) + R\sin(-\phi_1+180) + e\sin(-\phi_1-\phi_2)$$
$$+ U_3\sin(\theta) + c\sin(-90) + b\sin(-180) = 0 \quad (13.2)$$
$$h_\theta = 0 + 90 - 90 - \phi_1 + 180 - \phi_2 - 180 + 90 - \theta - 90 - 90 + 180 = 0$$

Open Loop 3

$$Gap = r\sin(-90) + r\sin(180) + U_1\sin(-90) + f\sin(90) + g\sin(0) \quad (13.3)$$

The loop equations relate the assembly variables: U_1, U_2, U_3, ϕ_1, ϕ_2, ϕ_3, and Gap to the component dimensions: a, b, c, e, f, g, r, R, and θ. We are concerned with the effect of small changes in the component variables on the variation in the assembly variables.

Note the uniformity of the equations. All h_x components are in terms of the cosine of the angle the vector makes with the x-axis. All h_y are in terms of the sine. In fact, just replace the cosines in the h_x equation with sines to get the h_y equation. The loop equations always have this form. This makes the equations very easy to derive. In a CAD implementation, equation generation may be automated.

The h_θ equations are the sum of relative rotations from one vector to the next as you proceed around the loop. Counterclockwise rotations are positive. Fig. 13-17 traces the relative rotations for Loop 1. A final rotation of 180 is added to bring the rotations to closure.

While the arguments of the sines and cosines in the h_x and h_y equations represent the absolute angle from the x-axis, the angles are expressed as the sum of relative rotations up to that point in the loop. Using relative rotations is critical to the correct assembly model behavior. It allows rotational variations to propagate correctly through the assembly.

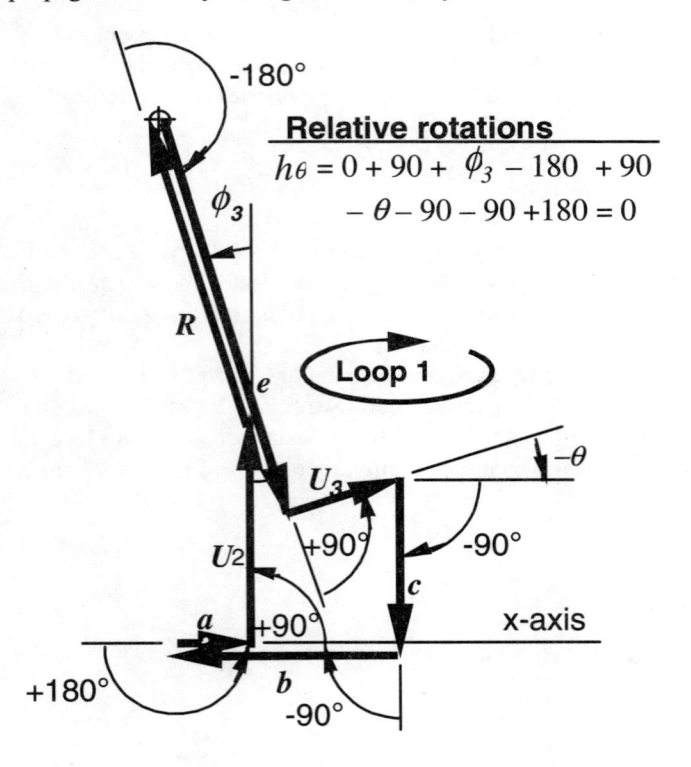

Figure 13-17 Relative rotations for Loop 1

A shortcut was used for the arguments for vectors U_2, c, and b. The sum of relative rotations was replaced with their known absolute directions. The sum of relative angles for U_2 is $(-\theta_1 - \theta_2 + 90)$, but it must align with the angled plane of the frame (θ). Similarly, vectors b and c will always be vertical and horizontal, respectively, regardless of the preceding rotational variations in the loop. Replacing the angles for U, C, and b is equivalent to solving the h_θ equation for θ and substituting in the arguments to eliminate some of the angle variables. If you try it both ways, you will see that you get the same results for the predicted variations. The results are also independent of the starting point of the loop. We could have started with any vector in the loop.

Step 2. Calculate derivatives and form matrix equations

The loop equations are nonlinear and implicit. They contain products and trigonometric functions of the variables. To solve for the assembly variables in this system of equations would require a nonlinear equation solver. Fortunately, we are only interested in the change in assembly variables for small changes in the components. This is readily accomplished by linearizing the equations by a first-order Taylor's series expansion.

Eq. (13.4) shows the linearized equations for Loop 1.

$$\begin{aligned}
\delta h_x &= \frac{\partial h_x}{\partial a}\delta a + \frac{\partial h_x}{\partial b}\delta b + \frac{\partial h_x}{\partial c}\delta c + \frac{\partial h_x}{\partial e}\delta e + \frac{\partial h_x}{\partial r}\delta r + \frac{\partial h_x}{\partial R}\delta R + \frac{\partial h_x}{\partial \theta}\delta\theta \\
&+ \frac{\partial h_x}{\partial \phi_1}\delta\phi_1 + \frac{\partial h_x}{\partial \phi_2}\delta\phi_2 + \frac{\partial h_x}{\partial \phi_3}\delta\phi_3 + \frac{\partial h_x}{\partial U_1}\delta U_1 + \frac{\partial h_x}{\partial U_2}\delta U_2 + \frac{\partial h_x}{\partial U_3}\delta U_3 \\
\delta h_y &= \frac{\partial h_y}{\partial a}\delta a + \frac{\partial h_y}{\partial b}\delta b + \frac{\partial h_y}{\partial c}\delta c + \frac{\partial h_y}{\partial e}\delta e + \frac{\partial h_y}{\partial r}\delta r + \frac{\partial h_y}{\partial R}\delta R + \frac{\partial h_y}{\partial \theta}\delta\theta \\
&+ \frac{\partial h_y}{\partial \phi_1}\delta\phi_1 + \frac{\partial h_y}{\partial \phi_2}\delta\phi_2 + \frac{\partial h_y}{\partial \phi_3}\delta\phi_3 + \frac{\partial h_y}{\partial U_1}\delta U_1 + \frac{\partial h_y}{\partial U_2}\delta U_2 + \frac{\partial h_y}{\partial U_3}\delta U_3 \quad (13.4)\\
\delta h_z &= \frac{\partial h_z}{\partial a}\delta a + \frac{\partial h_z}{\partial b}\delta b + \frac{\partial h_z}{\partial c}\delta c + \frac{\partial h_z}{\partial e}\delta e + \frac{\partial h_z}{\partial r}\delta r + \frac{\partial h_z}{\partial R}\delta R + \frac{\partial h_z}{\partial \theta}\delta\theta \\
&+ \frac{\partial h_z}{\partial \phi_1}\delta\phi_1 + \frac{\partial h_z}{\partial \phi_2}\delta\phi_2 + \frac{\partial h_z}{\partial \phi_3}\delta\phi_3 + \frac{\partial h_z}{\partial U_1}\delta U_1 + \frac{\partial h_z}{\partial U_2}\delta U_2 + \frac{\partial h_z}{\partial U_3}\delta U_3
\end{aligned}$$

where

δa represents a small change in dimension a, and so on.

Note that the terms have been rearranged, grouping the component variables a, b, c, e, r, R, and θ together and assembly variables U_1, U_2, U_3, ϕ_1, ϕ_2, and ϕ_3 together. The Loop 2 and Loop 3 equations may be expressed similarly.

Performing the partial differentiation of the respective h_x, h_y, and h_θ equations yields the coefficients of the linear system of equations. The partials are easy to perform because there are only sines and cosines to deal with. Eq. (13.5) shows the partials of the Loop 1 h_x equation.

Component Variables

$$\frac{\partial h_x}{\partial a} = \cos(0)$$

$$\frac{\partial h_x}{\partial b} = \cos(-180)$$

$$\frac{\partial h_x}{\partial c} = \cos(-90)$$

$$\frac{\partial h_x}{\partial e} = \cos(270+\varphi_3)$$

$$\frac{\partial h_x}{\partial r} = 0$$

$$\frac{\partial h_x}{\partial R} = \cos(90+\varphi_3)$$

$$\frac{\partial h_x}{\partial \theta} = -U_3 \sin(\theta\theta)$$

Assembly Variables

$$\frac{\partial h_x}{\partial \varphi_1} = 0$$

$$\frac{\partial h_x}{\partial \varphi_2} = 0$$

$$\frac{\partial h_x}{\partial \varphi_3} = -R\sin(90+\varphi_3) - e\sin(270+\varphi_3)$$

$$\frac{\partial h_x}{\partial U_1} = 0 \qquad (13.5)$$

$$\frac{\partial h_x}{\partial U_2} = \cos(90)$$

$$\frac{\partial h_x}{\partial U_3} = \cos(\theta)$$

Each partial is evaluated at the nominal value of all dimensions. The nominal component dimensions are known from the engineering drawings or CAD model. The nominal assembly values may be obtained by querying the CAD model.

The partial derivatives above are not the tolerance sensitivities we seek, but they can be used to obtain them.

Step 3. Solve for assembly tolerance sensitivities

The linearized loop equations may be written in matrix form and solved for the tolerance sensitivities by matrix algebra. The six closed loop scalar equations can be expressed in matrix form as follows:

$$[A]\{\delta X\} + [B]\{\delta U\} = \{0\}$$

where:

[A] is the matrix of partial derivatives with respect to the component variables,
[B] is the matrix of partial derivatives with respect to the assembly variables,
$\{\delta X\}$ is the vector of small variations in the component dimensions, and
$\{\delta U\}$ is the vector of corresponding closed loop assembly variations.

We can solve for the closed loop assembly variations in terms of the component variations by matrix algebra:

$$\{\delta U\} = -[B^{-1}A]\{\delta X\} \qquad (13.6)$$

The matrix $[B^{-1}A]$ is the matrix of tolerance sensitivities for the closed loop assembly variables. Performing the inverse of the matrix $[B]$ and multiplying $[B^{-1}A]$ may be carried out using a spreadsheet or other math utility program on a desktop computer or programmable calculator.

Multi-Dimensional Tolerance Analysis (Automated Method) 13-17

For the example assembly, the resulting matrices and vectors for the closed loop solution are:

$$\{\delta X\} = \begin{Bmatrix} \delta a \\ \delta b \\ \delta c \\ \delta e \\ \delta r \\ \delta R \\ \delta \theta \end{Bmatrix} \qquad \{\delta U\} = \begin{Bmatrix} \delta U_1 \\ \delta U_2 \\ \delta U_3 \\ \delta \phi_1 \\ \delta \phi_2 \\ \delta \phi_3 \end{Bmatrix}$$

$$[A] = \begin{bmatrix} \frac{\partial h_x}{\partial a} & \frac{\partial h_x}{\partial b} & \frac{\partial h_x}{\partial c} & \frac{\partial h_x}{\partial e} & \frac{\partial h_x}{\partial r} & \frac{\partial h_x}{\partial R} & \frac{\partial h_x}{\partial \theta} \\ \frac{\partial h_y}{\partial a} & \frac{\partial h_y}{\partial b} & \frac{\partial h_y}{\partial c} & \frac{\partial h_y}{\partial e} & \frac{\partial h_y}{\partial r} & \frac{\partial h_y}{\partial R} & \frac{\partial h_y}{\partial \theta} \\ \frac{\partial h_\theta}{\partial a} & \frac{\partial h_\theta}{\partial b} & \frac{\partial h_\theta}{\partial c} & \frac{\partial h_\theta}{\partial e} & \frac{\partial h_\theta}{\partial r} & \frac{\partial h_\theta}{\partial R} & \frac{\partial h_\theta}{\partial \theta} \\ \frac{\partial h_x}{\partial a} & \frac{\partial h_x}{\partial b} & \frac{\partial h_x}{\partial c} & \frac{\partial h_x}{\partial e} & \frac{\partial h_x}{\partial r} & \frac{\partial h_x}{\partial R} & \frac{\partial h_x}{\partial \theta} \\ \frac{\partial h_y}{\partial a} & \frac{\partial h_y}{\partial b} & \frac{\partial h_y}{\partial c} & \frac{\partial h_y}{\partial e} & \frac{\partial h_y}{\partial r} & \frac{\partial h_y}{\partial R} & \frac{\partial h_y}{\partial \theta} \\ \frac{\partial h_\theta}{\partial a} & \frac{\partial h_\theta}{\partial b} & \frac{\partial h_\theta}{\partial c} & \frac{\partial h_\theta}{\partial e} & \frac{\partial h_\theta}{\partial r} & \frac{\partial h_\theta}{\partial R} & \frac{\partial h_\theta}{\partial \theta} \end{bmatrix}$$

$$= \begin{bmatrix} 1 & -1 & 0 & \cos(270+\phi_3) & 0 & \cos(90+\phi_3) & -U_3 \sin(\theta) \\ 0 & 0 & -1 & \sin(270+\phi_3) & 0 & \sin(90+\phi_3) & U_3 \cos(\theta) \\ 0 & 0 & 0 & 0 & 0 & 0 & -1 \\ 1 & -1 & 0 & \cos(-\phi_1-\phi_2) & 1+\cos(-\phi_1) & \cos(-\phi_1+180) & -U_3 \sin(\theta) \\ 0 & 0 & -1 & \sin(-\phi_1-\phi_2) & \sin(-\phi_1) & \sin(-\phi_1+180) & U_3 \cos(\theta) \\ 0 & 0 & 0 & 0 & 0 & 0 & -1 \end{bmatrix}$$

$$= \begin{bmatrix} 1 & -1 & 0 & .2924 & 0 & -.2924 & -4.7738 \\ 0 & 0 & -1 & -.9563 & 0 & .9563 & 15.6144 \\ 0 & 0 & 0 & 0 & 0 & 0 & -1 \\ 1 & -1 & 0 & .2924 & 1.7232 & -.7232 & -4.7738 \\ 0 & 0 & -1 & -.9563 & -.6907 & .6907 & 15.6144 \\ 0 & 0 & 0 & 0 & 0 & 0 & -1 \end{bmatrix}$$

$$[B] = \begin{bmatrix} \frac{\partial h_x}{\partial U_1} & \frac{\partial h_x}{\partial U_2} & \frac{\partial h_x}{\partial U_3} & \frac{\partial h_x}{\partial \varphi_1} & \frac{\partial h_x}{\partial \varphi_2} & \frac{\partial h_x}{\partial \varphi_3} \\ \frac{\partial h_y}{\partial U_1} & \frac{\partial h_y}{\partial U_2} & \frac{\partial h_y}{\partial U_3} & \frac{\partial h_y}{\partial \varphi_1} & \frac{\partial h_y}{\partial \varphi_2} & \frac{\partial h_y}{\partial \varphi_3} \\ \frac{\partial h_\theta}{\partial U_1} & \frac{\partial h_\theta}{\partial U_2} & \frac{\partial h_\theta}{\partial U_3} & \frac{\partial h_\theta}{\partial \varphi_1} & \frac{\partial h_\theta}{\partial \varphi_2} & \frac{\partial h_\theta}{\partial \varphi_3} \\ \frac{\partial h_x}{\partial U_1} & \frac{\partial h_x}{\partial U_2} & \frac{\partial h_x}{\partial U_3} & \frac{\partial h_x}{\partial \varphi_1} & \frac{\partial h_x}{\partial \varphi_2} & \frac{\partial h_x}{\partial \varphi_3} \\ \frac{\partial h_y}{\partial U_1} & \frac{\partial h_y}{\partial U_2} & \frac{\partial h_y}{\partial U_3} & \frac{\partial h_y}{\partial \varphi_1} & \frac{\partial h_y}{\partial \varphi_2} & \frac{\partial h_y}{\partial \varphi_3} \\ \frac{\partial h_\theta}{\partial U_1} & \frac{\partial h_\theta}{\partial U_2} & \frac{\partial h_\theta}{\partial U_3} & \frac{\partial h_\theta}{\partial \varphi_1} & \frac{\partial h_\theta}{\partial \varphi_2} & \frac{\partial h_\theta}{\partial \varphi_3} \end{bmatrix}$$

$$= \begin{bmatrix} 0 & \cos(90) & \cos(\theta) & 0 & 0 & \begin{bmatrix} -R\sin(90+\phi_3)-e\sin(270+\phi_3) \\ R\cos(90+\phi_3)+e\cos(270+\phi_3) \end{bmatrix} \\ 0 & \sin(90) & \sin(\theta) & 0 & 0 & \\ 0 & 0 & 0 & 0 & 0 & 1 \\ \cos(90) & 0 & \cos(\theta) & \begin{bmatrix} r\sin(-\phi_1) \\ R\sin(180-\phi_1) \\ e\sin(-\phi_1-\phi_2) \end{bmatrix} & e\sin(-\phi_1-\phi_2) & 0 \\ \sin(90) & 0 & \sin(\theta) & \begin{bmatrix} -r\cos(-\phi_1) \\ -R\cos(180-\phi_1) \\ -e\cos(-\phi_1-\phi_2) \end{bmatrix} & -e\cos(-\phi_1-\phi_2) & 0 \\ 0 & 0 & 0 & -1 & -1 & 0 \end{bmatrix}$$

$$= \begin{bmatrix} 0 & 0 & .95631 & 0 & 0 & 14.3446 \\ 0 & 1 & .29237 & 0 & 0 & 4.3856 \\ 0 & 0 & 0 & 0 & 0 & 1 \\ 0 & 0 & .95631 & -31.8764 & -52.5968 & 0 \\ 1 & 0 & .29237 & 5.6144 & -16.0804 & 0 \\ 0 & 0 & 0 & -1 & -1 & 0 \end{bmatrix}$$

$$[B^{-1}] = \begin{bmatrix} .7413 & 0 & -10.6337 & -1.0470 & 1 & 38.9901 \\ -.3057 & 1 & 0 & 0 & 0 & 0 \\ 1.0457 & 0 & -15 & 0 & 0 & 0 \\ -.0483 & 0 & .6923 & .0483 & 0 & -2.5384 \\ .0483 & 0 & -.6923 & -.0483 & 0 & 1.5384 \\ 0 & 0 & 1 & 0 & 0 & 0 \end{bmatrix}$$

$$\{\delta U\} = -[B^{-1}A]\{\delta X\} \tag{13.7}$$

$$\begin{Bmatrix} \delta U_1 \\ \delta U_2 \\ \delta U_3 \\ \delta\phi_1 \\ \delta\phi_2 \\ \delta\phi_3 \end{Bmatrix} = \begin{bmatrix} .3057 & -.3057 & 1 & 1.0457 & 2.494885 & -1.2311 & 11.2825 \\ .3057 & -.3057 & 1 & 1.0457 & -1 & -1.0457 & -17.0739 \\ -1.0457 & 1.0457 & 0 & -.3057 & 0 & .3057 & -10.0080 \\ 0 & 0 & 0 & 0 & -.0832 & .0208 & -1.8461 \\ 0 & 0 & 0 & 0 & .0832 & -.0208 & .8461 \\ 0 & 0 & 0 & 0 & 0 & 0 & 1 \end{bmatrix} \begin{Bmatrix} \delta a \\ \delta b \\ \delta c \\ \delta e \\ \delta r \\ \delta R \\ \delta\theta \end{Bmatrix}$$

Estimates for variation of the assembly performance requirements are obtained by linearizing the open loop equations by a procedure similar to the closed loop equations. In general, there will be a system of nonlinear scalar equations which may be linearized by Taylor's series expansion. Grouping terms as before, we can express the linearized equations in matrix form:

$$\{\delta V\} = [C]\{\delta X\} + [E]\{\delta U\} \tag{13.8}$$

where

$\{\delta V\}$ is the vector of variations in the assembly performance requirements,
$[C]$ is the matrix of partial derivatives with respect to the component variables,
$[E]$ is the matrix of partial derivatives with respect to the assembly variables,
$\{\delta X\}$ is the vector of small variations in the component dimensions, and
$\{\delta U\}$ is the vector of corresponding closed loop assembly variations.

We can solve for the open loop assembly variations in terms of the component variations by matrix algebra, by substituting the results of the closed loop solution. Substituting for $\{\delta U\}$:

$$\{\delta V\} = [C]\{\delta X\} - [E][B^{-1}A]\{\delta X\}$$
$$= [C - EB^{-1}A]\{\delta X\}$$

The matrix $[C - E B^{-1}A]$ is the matrix of tolerance sensitivities for the open loop assembly variables. The $B^{-1}A$ terms come from the closed loop constraints on the assembly. The $B^{-1}A$ terms represent the effect of small internal kinematic adjustments occurring at assembly time in response to dimensional variations. The internal adjustments affect the $\{\delta V\}$ as well as the $\{\delta U\}$.

It is important to note that you cannot simply solve for the values of $\{\delta U\}$ in Eq. (13.6) and substitute them directly into Eq. (13.8), as though $\{\delta U\}$ were just another component variation. If you do, you are treating $\{\delta U\}$ as though it is independent of $\{\delta X\}$. But $\{\delta U\}$ depends on $\{\delta X\}$ through the closed loop constraints. You must evaluate the full matrix $[C - E B^{-1}A]$ to obtain the tolerance sensitivities. Allowing the $B^{-1}A$ terms to interact with C and E is necessary to determine the effect of the kinematic adjustments on $\{\delta V\}$. Treating them separately is similar to taking the absolute value of each term, then summing for Worst Case, rather than summing like terms before taking the absolute value. The same is true for RSS analysis. It is similar to squaring each term, then summing, rather than summing like terms before squaring.

For the example assembly, the equation for $\{\delta V\}$ reduces to a single scalar equation for the Gap variable.

13-20   Chapter Thirteen

$$\delta Gap = \frac{\partial Gap}{\partial a}\delta a + \frac{\partial Gap}{\partial b}\delta b + \frac{\partial Gap}{\partial c}\delta c + \frac{\partial Gap}{\partial e}\delta e + \frac{\partial Gap}{\partial f}\delta f + \frac{\partial Gap}{\partial g}\delta g$$
$$+ \frac{\partial Gap}{\partial r}\delta r + \frac{\partial Gap}{\partial R}\delta R + \frac{\partial Gap}{\partial \theta}\delta\theta + \frac{\partial Gap}{\partial U_1}\delta U_1 + \frac{\partial Gap}{\partial U_2}\delta U_2 + \frac{\partial Gap}{\partial U_3}\delta U_3$$
$$+ \frac{\partial Gap}{\partial \phi_1}\delta\phi_1 + \frac{\partial Gap}{\partial \phi_2}\delta\phi_2 + \frac{\partial Gap}{\partial \phi_3}\delta\phi_3$$

$$\delta Gap = [\sin(-90) + \sin(180)]\,\delta r + \sin(90)\,\delta f + \sin(0)\,\delta g + \sin(-90)\delta U_1$$
$$= -\delta r + \delta f - \delta U_1$$

Substituting for δU_1 from the closed loop results (Eq. (13.7)) and grouping terms:

$$\delta Gap = -\delta r + \delta f - (.3057\delta a - .3057\delta b + \delta c + 1.0457\delta e + 2.4949\delta r - 1.2311\delta R + 11.2825\delta\theta) \quad (13.9)$$
$$= -.3057\delta a + .3057\delta b - \delta c - 1.0457\delta e - 3.4949\delta r + 1.2311\delta R - 11.2825\delta\theta$$

While Eq. (13.9) expresses the assembly variation δGap in terms of the component variations δX, it is not an estimate of the tolerance accumulation. To estimate accumulation, you must use a model, such as Worst Case or Root Sum Squares.

Step 4.   Form Worst Case and RSS expressions

As has been shown earlier, estimates of tolerance accumulation for δU or δV may be calculated by summing the products of the tolerance sensitivities and component variations:

Worst Case   **RSS**

$$\delta U \text{ or } \delta V = \sum |S_{ij}|\,\delta x_j \qquad \delta U \text{ or } \delta V = \sqrt{\sum (S_{ij}\delta_{xj})^2}$$

S_{ij} is the tolerance sensitivities of assembly features to component variations. If the assembly variable of interest is a closed loop variable δU, S_{ij} is obtained from the appropriate row of the $B^{-1}A$ matrix. If δV_i is wanted, S_{ij} comes from the $[C\text{-}E\,B^{-1}A]$ matrix. If measured variation data are available, δx_j is the $\pm 3\sigma$ process variation. If production of parts has not begun, δx_j is usually taken to be equal to the $\pm 3\sigma$ design tolerances on the components.

In the example assembly, length U_1 is a closed loop assembly variable. U_1 determines the location of the contact point between the Cylinder and the Frame. To estimate the variation in U_1, we would multiply the first row of $[B^{-1}A]$ with $\{\delta X\}$ and sum by Worst Case or RSS.

Worst Case:
$$\delta U_1 = |S_{11}|\delta a + |S_{12}|\delta b + |S_{13}|\delta c + |S_{14}|\delta e + |S_{15}|\delta r + |S_{16}|\delta R + |S_{17}|\delta\theta$$
$$= |.3057|\,0.3 + |-.3057|\,0.3 + |1|\,0.3 + |1.0457|\,0.3 + |2.4949|\,0.1 + |-1.2311|\,0.3 + |11.2825|\,0.01745$$
$$= \pm 1.6129 \text{ mm}$$

RSS:

$$\delta U_1 = [(S_{11}\delta a)^2 + (S_{12}\delta b)^2 + (S_{13}\delta c)^2 + (S_{14}\delta e)^2 + (S_{15}\delta r)^2 + (S_{16}\delta R)^2 + (S_{17}\delta\theta)^2]^{.5}$$
$$= [(.3057 \cdot 0.3)^2 + (-.3057 \cdot 0.3)^2 + (1 \cdot 0.3)^2 + (1.0457 \cdot 0.3)^2 + (2.4949 \cdot 0.1)^2 + (-1.2311 \cdot 0.3)^2 +$$
$$(11.2825 \cdot 0.01745)^2]^{.5}$$
$$= \pm 0.6653 \text{ mm}$$

Note that the tolerance on θ has been converted to ± 0.01745 radians since the sensitivity is calculated per radian.

For the variation in the Gap, we would multiply the first row of $[C\text{-}EB^{-1}A]$ with $\{\delta X\}$ and sum by Worst Case or RSS. Vector $\{\delta X\}$ is extended to include δf and δg.

Worst Case:

$$\delta Gap = |S_{11}|\delta a + |S_{12}|\delta b + |S_{13}|\delta c + |S_{14}|\delta e + |S_{15}|\delta r + |S_{16}|\delta R + |S_{17}|\delta\theta + |S_{18}|\delta f + |S_{19}|\delta g$$
$$= |-.30573|\,0.3 + |.30573|\,0.3 + |-1|\,0.3 + |-1.04569|\,0.3 + |-3.4949|\,0.1 + |1.2311|\,0.3$$
$$+ |-11.2825|\,0.01745 + |1|\,0.5 + |0|\,0$$
$$= \pm 2.2129 \text{ mm}$$

RSS:

$$\delta Gap = [(S_{11}\delta a)^2 + (S_{12}\delta b)^2 + (S_{13}\delta c)^2 + (S_{14}\delta e)^2 + (S_{15}\delta r)^2 + (S_{16}\delta R)^2 + (S_{17}\delta\theta) + (S_{18}\delta f)^2 + (S_{19}\delta g)^2]^{.5}$$
$$= [(-.30573 \cdot 0.3)^2 + (.30573 \cdot 0.3)^2 + (-1 \cdot 0.3)^2 + (-1.04569 \cdot 0.3)^2 + (-3.4949 \cdot 0.1)^2 + (1.2311 \cdot 0.3)^2$$
$$+ (-11.2825 \cdot 0.01745)^2 + (1 \cdot 0.5)^2 + (0 \cdot 0)^2]^{.5}$$
$$= \pm 0.8675 \text{ mm}$$

By forming similar expressions, we may obtain estimates for all the assembly variables (Table 13-1).

Table 13-1 Estimated variation in open and closed loop assembly features

Assembly Variable	Mean or Nominal	WC ±δ_U	RSS ±δ_U
U_1	59.0026 mm	1.6129 mm	0.6653 mm
U_2	41.4708 mm	1.5089 mm	0.6344 mm
U_3	16.3279 mm	0.9855 mm	0.4941 mm
ϕ_1	43.6838°	2.68°	1.94°
ϕ_2	29.3162°	1.68°	1.04°
ϕ_3	17.0000°	1.00°	1.00°
Gap	5.9974 mm	2.2129 mm	0.8675 mm

Step 5. Evaluation and design iteration

The results of the variation analysis are evaluated by comparing the predicted variation with the specified design requirement. If the variation is greater or less than the specified assembly tolerance, the expressions can be used to help decide which tolerances to tighten or loosen.

13.5.5.1 Percent Rejects

The percent rejects may be estimated from Standard Normal tables by calculating the number of standard deviations from the mean to the upper and lower limits (UL and LL).

The only assembly feature with a performance requirement is the Gap. The acceptable range for proper performance is: Gap = 6.00 ±1.00 mm. Calculating the distance from the mean Gap to UL and LL in units equal to the standard deviation of the Gap:

$$Z_{UL} = \frac{UL - \mu_{Gap}}{\sigma_{Gap}} = \frac{7.000 - 5.9974}{0.2892} = 3.467\sigma \qquad R_{UL} = 263 \text{ ppm}$$

$$Z_{LL} = \frac{LL - \mu_{Gap}}{\sigma_{Gap}} = \frac{5.000 - 5.9974}{0.2892} = -3.449\sigma \qquad R_{LL} = 281 \text{ ppm}$$

The total predicted rejects are 544 ppm.

13.5.5.2 Percent Contribution Charts

The percent contribution chart tells the designer how each dimension contributes to the total Gap variation. The contribution includes the effect of both the sensitivity and the tolerance. The calculation is different for Worst Case or RSS variation estimates.

Worst Case

$$\% Cont = \frac{\left| \frac{\partial Gap}{\partial x_j} \cdot \delta x_j \right|}{\sum \left| \frac{\partial Gap}{\partial x_i} \cdot \delta x_i \right|}$$

RSS

$$\% Cont = \frac{\left(\frac{\partial Gap}{\partial x_j} \cdot \delta x_j \right)^2}{\sum \left(\frac{\partial Gap}{\partial x_i} \cdot \delta x_i \right)^2}$$

It is common practice to present the results as a bar chart, sorted according to magnitude. The results for the sample assembly are shown in Fig. 13-18.

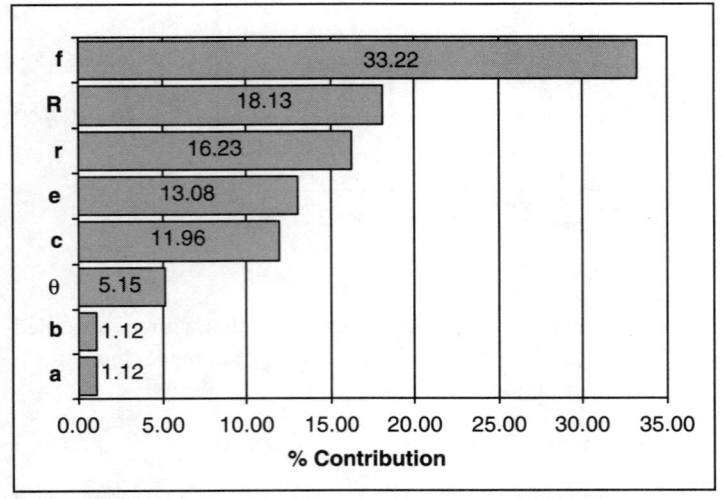

Figure 13-18 Percent contribution chart for the sample assembly

Variable	% Contribution
f	33.22
R	18.13
r	16.23
e	13.08
c	11.96
θ	5.15
b	1.12
a	1.12

It is clear that the outside dimension of the Gap, f, is the principal contributor, followed by the radius R. This plot shows the designer where to focus design modification efforts.

Simply changing the tolerances on a few dimensions can change the chart dramatically. Suppose we tighten the tolerance on f, since it is relatively easy to control, and loosen the tolerances on R and e, since they are more difficult to locate and machine with precision. We will say the Cylinder is vendor-supplied, so it cannot be modified. Table 13-2 shows the new tolerances.

Table 13-2 Modified dimensional tolerance specifications

Dimension	±Tolerance Original	±Tolerance Modified
a	0.3 mm	0.3 mm
b	0.3 mm	0.3 mm
c	0.3 mm	0.3 mm
e	0.3 mm	0.4 mm
r	0.1 mm	0.1 mm
R	0.3 mm	0.4 mm
θ	1.0°	1.0°
f	0.5 mm	0.4 mm

Now, R and e are the leading contributors, while f has dropped to third. Of course, changing the tolerances requires modification of the processes. See Fig. 13-19. Tightening the tolerance on f, for example, might require changing the feed or speed or number of finish passes on a mill.

Since it is the product of the sensitivity times the tolerance that determines the percent contribution, the sensitivity is also an important variation evaluation aid.

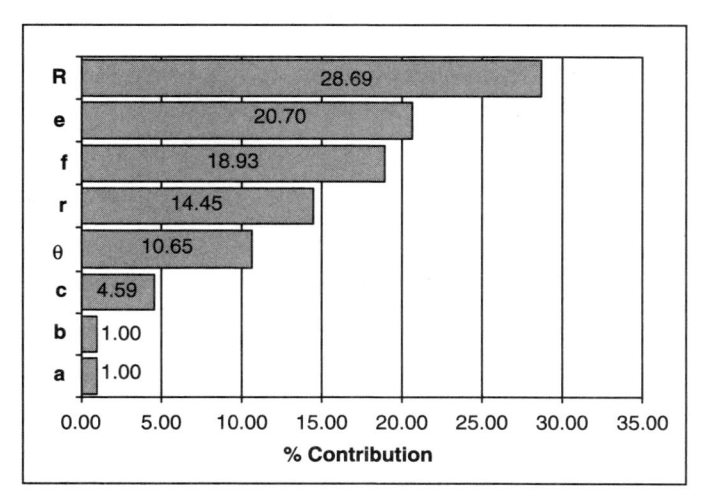

Figure 13-19 Percent contribution chart for the sample assembly with modified tolerances

Variable	% Contribution
R	28.69
e	20.70
f	18.93
r	14.45
θ	10.65
c	4.59
b	1.00
a	1.00

13.5.5.3 Sensitivity Analysis

The tolerance sensitivities tell how the arrangement of the parts and the geometry contribute to assembly variation. We can learn a great deal about the role played by each dimension by examining the sensitivities. For the sample assembly, Table 13-3 shows the calculated Gap sensitivities.

Table 13-3 Calculated sensitivities for the Gap

Dimension	Sensitivity
a	-0.3057
b	0.3057
c	-1.0
e	-1.0457
r	-3.4949
R	1.2311
θ	-11.2825
f	1.0

Note that the sensitivity of θ is calculated per radian.

For a 1.0 mm change in a or b, the Gap will change by 0.3057 mm. The negative sign for a means the Gap will decrease as a increases. For each mm increase in c, the Gap decreases an equal amount. This behavior becomes clear on examining Fig. 13-12. As a increases 1.0 mm, the Block is pushed up the inclined plane, raising the Block and Cylinder by the tan(17°) or 0.3057 and decreasing the Gap. As b increases 1.0 mm, the plane is pushed out from under the Block, causing it to lower the same amount. Increasing c 1.0 mm, causes everything to slide straight up, decreasing the Gap.

Dimensions e, r, R, and θ are more complex because several adjustments occur simultaneously. As r increases, the Cylinder grows, causing it to slide up the wall, while maintaining contact with the concave surface of the Block. As the Cylinder rises, the Gap decreases. As R increases, the concave surface moves deeper into the block, causing the Cylinder to drop, which increases the Gap. Increasing e causes the Block to thicken, forcing the front corner up the wall and pushing the Block up the plane. The net effect is to raise the concave surface, decreasing the Gap. Increasing θ causes the Block to rotate about the front edge of the inclined plane, while the front corner slides down the wall. The wedge angle between the concave surface and the wall decreases, squeezing the Cylinder upward and decreasing the Gap. The large sensitivities for r and θ are offset by their small corresponding tolerances.

13.5.5.4 Modifying Geometry

The most common geometry modification is to change the nominal values of one or more dimensions to center the nominal value of a gap between its UL and LL. For example, if we wanted to change the Gap specifications to be 5.00 ±1.000 mm, we could simply increase the nominal value of c by 1.00 mm. Since the sensitivity of the Gap to c is –1.0, the Gap will decrease by 1.0 mm.

Similarly, the sensitivities may be modified by changing the geometry. Since the sensitivities are partial derivatives, which are evaluated at the nominal values of the component dimensions, they can only be changed by changing the nominal values. An interesting exercise is to modify the geometry of the example assembly to make the Gap insensitive to variation in θ; that is, to make the sensitivity of θ go to zero. You will need nonlinear equation solver software to solve the original loop equations (Eqs. (13-4), (13-5), and (13-6)), for a new set of nominal assembly values. Solve for the kinematic assembly variables: U_1, U_2, U_3, ϕ_1, ϕ_2, and ϕ_3, corresponding to your new nominal dimensions: a, b, c, e, r, R, θ, f, and Gap.

The sensitivity of θ will decrease to nearly zero if we increase b to a value of 40 mm. We must also increase c to 35 mm to reduce the nominal Gap back to 6.00 mm. The [A], [B], [C], and [E] matrices will all need to be re-evaluated and solved for the variations. The modified results are shown in Table 13-4.

Table 13-4 Calculated sensitivities for the Gap after modifying geometry

Dimension	Nominal	±Tolerance	Sensitivity
a	10 mm	0.3	-0.3057
b	40 mm	0.3	0.3057
c	35 mm	0.3	-1.0
e	55 mm	0.4	-1.0457
r	10 mm	0.1	-3.4949
R	40 mm	0.4	1.2311
θ	17°	1.0°	-0.3478
f	75 mm	0.4	1.0

Notice that the only sensitivity to change was θ (per radian). This is due to the lack of coupling of b and c with the other variables. The calculated variations are shown in Table 13-5.

Table 13-5 Variation results for modified nominal geometry

Assembly Variable	Mean or Nominal	WC $+\delta_U$	RSS $+\delta_U$
U_1	59.0453 mm	1.6497 mm	0.7659 mm
U_2	41.5135 mm	1.9088 mm	0.8401 mm
U_3	26.7848 mm	0.9909 mm	0.4908 mm
ϕ_1	43.6838°	2.80°	1.97°
ϕ_2	29.3162°	1.80°	1.08°
ϕ_3	17°	1.00°	1.00°
Gap	5.9547 mm	2.1497 mm	0.8980 mm

The new percent contribution chart is shown in Fig. 13-20. Based on the low sensitivity, you could now increase the tolerance on θ without affecting the Gap variation.

Step 6. Report results and document changes

The final step in the assembly tolerance analysis procedure is to prepare the final report. Figures, graphs, and tables are preferred. Comparison tables and graphs will help to justify design decisions. If you have several iterations, it is wise to adopt a case numbering scheme to identify each table and graph with its corresponding case. A list of case numbers with a concise summary of the distinguishing feature for each would be appreciated by the reader.

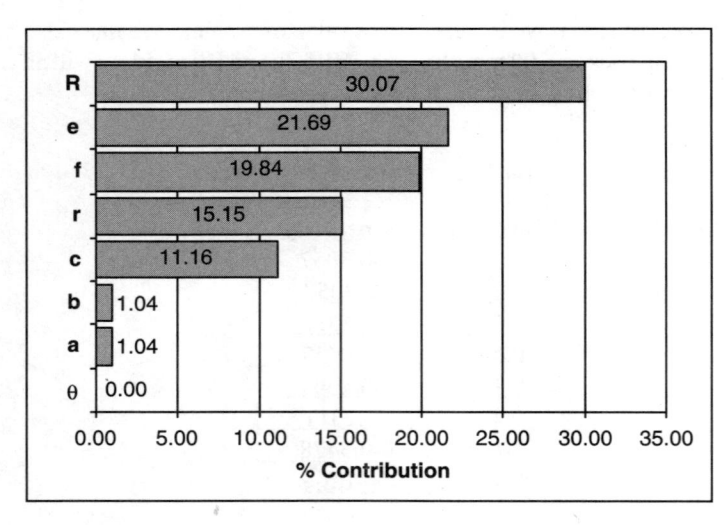

[Bar chart showing % Contribution: R 30.07, e 21.69, f 19.84, r 15.15, c 11.16, b 1.04, a 1.04, θ 0.00]

Figure 13-20 Modified geometry yields zero θ contribution

13.6 Summary

The preceding sections have presented a systematic procedure for modeling and analyzing assembly variation. Some of the advantages of the modeling system include:

- The three main sources of variation may be included: dimensions; geometric form, location, and orientation; and kinematic adjustments.
- Assembly models are constructed of vectors and kinematic joints, elements with which most designers are familiar.
- A variety of assembly configurations may be represented with a few basic elements.
- Modeling rules guide the designer and assist in the creation of valid models.
- It can be automated and integrated with a CAD system to achieve fully graphical model creation.

Advantages of the analysis system include:

- The assembly functions are readily derived from the graphical model.
- Nonlinear, implicit systems of equations are readily converted to a linear system. Tolerance sensitivities are determined by a single, standard, matrix algebra operation.
- Statistical algorithms estimate tolerance stackup accurately and efficiently without requiring repeated simulations.
- Once expressions for the variation in assembly features have been derived, they may be used for tolerance allocation or "what-if?" studies without repeating the assembly analysis.
- Variation parameters useful for evaluation and design are easily obtained, such as: the mean and standard deviation of critical assembly features, sensitivity and percent contribution of each component dimension and geometric form variation, percent rejects, and quality level.
- Tolerance analysis models combine design requirements with process capabilities to foster open communication between design and manufacturing and reasoned, quantitative decisions.
- It can be automated to totally eliminate manual derivation of equations or equation typing.

A CAD-based tolerance analysis system based on the procedures demonstrated previously has been developed. The basic organization of the Computer-Aided Tolerancing System (CATS) is shown schematically in Fig. 13-21. The system has been integrated with a commercial 3-D CAD system, so it looks and feels like the designer's own system. Many of the manual tasks of modeling and analysis described above have been converted to graphical functions or automated.

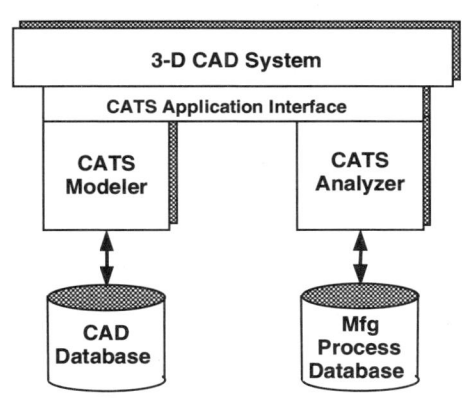

Figure 13-21 The CATS System

Tolerance analysis has become a mature engineering design tool. It is a quantitative tool for concurrent engineering. Powerful statistical algorithms have been combined with graphical modeling and evaluation aids to assist designers by bringing manufacturing considerations into their design decisions. Process selection, tooling, and inspection requirements may be determined early in the product development cycle. Performing tolerance analysis on the CAD model creates a virtual prototype for identifying variation problems before parts are produced. Designers can be much more effective by designing assemblies that work in spite of manufacturing process variations. Costly design changes to accommodate manufacturing can be reduced. Product quality and customer satisfaction can be increased. Tolerance analysis could become a key factor in maintaining competitiveness in today's international markets.

13.7　References

1. Carr, Charles D. 1993. "A Comprehensive Method for Specifying Tolerance Requirements for Assemblies." Master's thesis. Brigham Young University.
2. Chase, K. W. and A. R. Parkinson. 1991. A Survey of Research in the Application of Tolerance Analysis to the Design of Mechanical Assemblies. *Research in Engineering Design.* 3(1): 23-37.
3. Chase, K. W. and Angela Trego. 1994. *AutoCATS Computer-Aided Tolerancing System - Modeler User Guide.* ADCATS Report, Brigham Young University.
4. Chase, K. W., J. Gao and S. P. Magleby. 1995. General 2-D Tolerance Analysis of Mechanical Assemblies with Small Kinematic Adjustments. *Journal of Design and Manufacturing.* 5(4):263-274.
5. Chase, K. W., J. Gao and S. P. Magleby. 1998. Tolerance Analysis of 2-D and 3-D Mechanical Assemblies with Small Kinematic Adjustments. In *Advanced Tolerancing Techniques.* pp. 103-137. New York: John Wiley.
6. Chase, K. W., J. Gao, S. P. Magleby and C. D. Sorenson. 1996. Including Geometric Feature Variations in Tolerance Analysis of Mechanical Assemblies. *IIE Transactions.* 28(10): 795-807.
7. Fortini, E.T. 1967. *Dimensioning for Interchangeable Manufacture.* New York, New York: Industrial Press.
8. The American Society of Mechanical Engineers. 1995. *ASME Y14.5M-1994, Dimensioning and Tolerancing.* New York, New York: The American Society of Mechanical Engineers.

Chapter 14

Minimum-Cost Tolerance Allocation

Kenneth W. Chase, Ph.D.
Brigham Young University
Provo, Utah

Dr. Chase has taught mechanical engineering at the Brigham Young University since 1968. An advocate of computer technology, he has served as a consultant to industry on numerous projects involving engineering software applications. He served as a reviewer of the Motorola Six Sigma Program at its inception. He also served on an NSF select panel for evaluating tolerance analysis research needs. In 1984, he founded the ADCATS consortium for the development of CAD-based tools for tolerance analysis of mechanical assemblies. More than 30 sponsored graduate theses have been devoted to the development of the tolerance technology contained in the CATS software. Several faculty and students are currently involved in a broad spectrum of research projects and industry case studies on statistical variation analysis. Past and current sponsors include Allied Signal, Boeing, Cummins, FMC, Ford, GE, HP, Hughes, IBM, Motorola, Sandia Labs, Texas Instruments, and the US Navy.

14.1   Tolerance Allocation Using Least Cost Optimization

A promising method of tolerance allocation uses optimization techniques to assign component tolerances that minimize the cost of production of an assembly. This is accomplished by defining a cost-versus-tolerance curve for each component part in the assembly. An optimization algorithm varies the tolerance for each component and searches systematically for the combination of tolerances that minimize the cost.

14.2   1-D Tolerance Allocation

Fig. 14-1 illustrates the concept simply for a three component assembly. Three cost-versus-tolerance curves are shown. Three tolerances (T_1, T_2, T_3) are initially selected. The corresponding cost of production is $C_1 + C_2 + C_3$. The optimization algorithm tries to increase the tolerances to reduce cost; however, the specified assembly tolerance limits the tolerance size. If tolerance T_1 is increased, then tolerance T_2 or T_3 must decrease to keep from violating the assembly tolerance constraint. It is difficult to tell by inspection which combination will be optimum, but you can see from the figure that a decrease in T_2 results in a significant increase in cost, while a corresponding decrease in T_3 results in a smaller increase in cost. In this manner, one could manually adjust tolerances until no further cost reduction is achieved. The optimization algorithm is designed to find the minimum cost automatically. Note that the values of the set of optimum tolerances will be different when the tolerances are summed statistically than when they are summed by worst case.

Total Cost:

$$C_{tot} = C_1 + C_2 + C_3$$

Constraint:

$$T_{tot} = T_1 + T_2 + T_3 \quad \text{[Worst Case]}$$

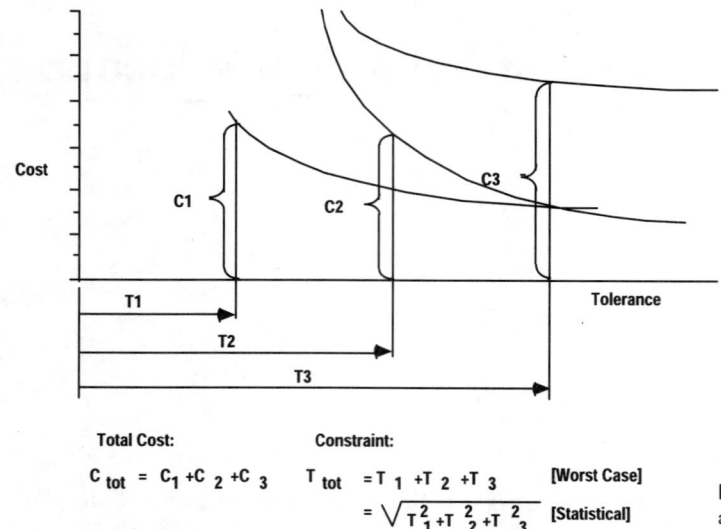

$$= \sqrt{T_1^2 + T_2^2 + T_3^2} \quad \text{[Statistical]}$$

Figure 14-1 Optimal tolerance allocation for minimum cost

A necessary factor in optimum tolerance allocation is the specification of cost-versus-tolerance functions. Several algebraic functions have been proposed, as summarized in Table 14-1. The Reciprocal Power function: $C = A + B/tol^k$ includes the Reciprocal and Reciprocal Squared rules for integer powers of k. The constant coefficient A represents fixed costs. It may include setup cost, tooling, material, and prior operations. The B term determines the cost of producing a single component dimension to a specified tolerance and includes the charge rate of the machine. Costs are calculated on a per-part basis. When tighter tolerances are called for, speeds and feeds may be reduced and the number of passes increased, requiring more time and higher costs. The exponent k describes how sensitive the process cost is to changes in tolerance specifications.

Table 14-1 Proposed cost-of-tolerance models

Cost Model	Function	Author	Ref
Reciprocal Squared	$A + B/tol^2$	Spotts	Spotts 1973 (Reference 11)
Reciprocal	$A + B/tol$	Chase & Greenwood	Chase 1988 (Reference 3)
Reciprocal Power	$A + B/tol^k$	Chase et al.	Chase 1989 (Reference 4)
Exponential	$A\, e^{-B(tol)}$	Speckhart	Speckhart 1972 (Reference 10)

Little has been done to verify the form of these curves. Manufacturing cost data are not published since they are so site-dependent. Even companies using the same machines would have different costs for labor, materials, tooling, and overhead.

A study of cost versus tolerance was made for the metal removal processes over the full range of nominal dimensions. This data has been curve fit to obtain empirical functions. The form was found to follow the reciprocal power law. The results are presented in the Appendix to this chapter. The original cost study is decades old and may not apply to modern numerical controlled (N/C) machines.

A closed-form solution for the least-cost component tolerances was developed by Spotts. (Reference 11) He used the method of Lagrange Multipliers, assuming a cost function of the form $C=A+B/tol^2$. Chase extended this to cost functions of the form $C=A+B/tol^k$ as follows: (Reference 4)

$$\frac{\partial}{\partial T_i}(Cost\_function) + \lambda \frac{\partial}{\partial T_i}(Constraint) = 0 \quad (i=1,\ldots n)$$

$$\frac{\partial}{\partial T_i}\left(\sum (A_j + B_j/T_j^{k_j})\right) + \lambda \frac{\partial}{\partial T_i}\left(\sum T_j^2 - T_{asm}^2\right) = 0 \quad (i=1,\ldots n)$$

$$\lambda = \frac{k_i B_i}{2T_i^{(k_i+2)}} \quad (i=1,\ldots n)$$

Eliminating λ by expressing it in terms of T_1 (arbitrarily selected):

$$T_i = \left(\frac{k_i B_i}{k_1 B_1}\right)^{1/(k_i+2)} T_1^{(k_1+2)/(k_i+2)} \tag{14.1}$$

Substituting for each of the T_i in the assembly tolerance sum:

$$T_{ASM}^2 = T_1^2 + \sum \left(\frac{k_i B_i}{k_1 B_1}\right)^{2/(k_i+2)} T_1^{2(k_1+2)/(k_i+2)} \tag{14.2}$$

The only unknown in Eq. (14.2) is T_1. One only needs to iterate the value of T_1 until both sides of Eq. (14.2) are equal to obtain the minimum cost tolerances. A similar derivation based on a worst case assembly tolerance sum yields:

$$T_{ASM} = T_1 + \sum \left(\frac{k_i B_i}{k_1 B_1}\right)^{1/(k_i+1)} T_1^{(k_1+1)/(k_i+1)} \tag{14.3}$$

A graphical interpretation of this method is shown in Fig. 14-2 for a two-part assembly. Various combinations of the two tolerances may be selected and summed statistically or by worst case. By summing the cost corresponding to any T_1 and T_2, contours of constant cost may be plotted. You can see that cost decreases as T_1 and T_2 are increased. The limiting condition occurs when the tolerance sum equals the assembly requirement T_{ASM}. The worst case limit describes a straight line. The statistical limit is an ellipse. T_1 and T_2 values must not be outside the limit line. Note that as the method of Lagrange Multipliers assumes, the minimum cost tolerance value is located where the constant cost curve is tangent to the tolerance limit curve.

14.3 1-D Example: Shaft and Housing Assembly

The following example is based on the shaft and housing assembly shown in Fig. 14-3. Two bearing sleeves maintain the spacing of the bearings to match that of the shaft. Accumulation of variation in the assembly results in variation in the end clearance. Positive clearance is required.

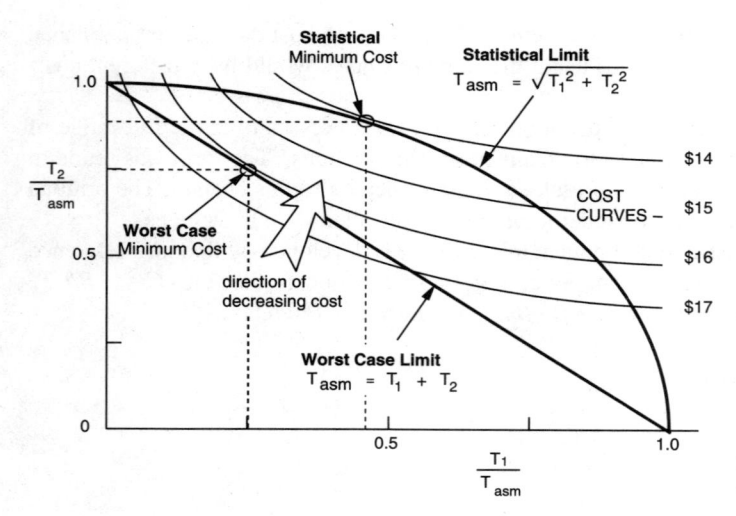

Figure 14-2 Graphical interpretation of minimum cost tolerance allocation

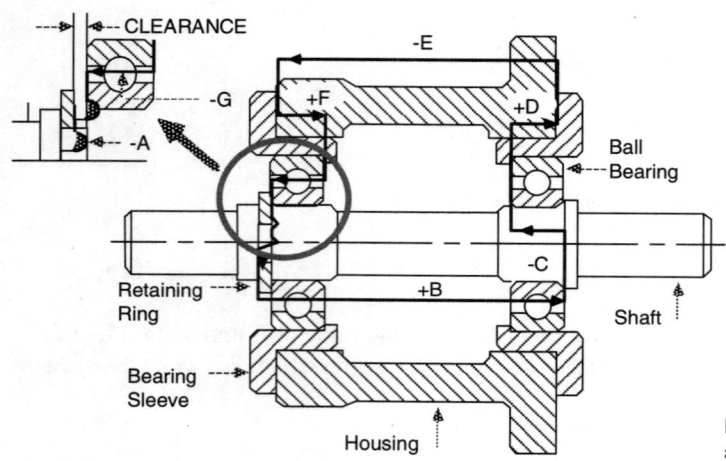

Figure 14-3 Shaft and housing assembly

Initial tolerances for parts B, D, E, and F are selected from tolerance guidelines such as those illustrated in Fig. 14-4. The bar chart shows the typical range of tolerance for several common processes. The numerical values appear in the table above the bar chart. Each row of the numerical table corresponds to a different nominal size range. For example, a turned part having a nominal dimension of .750 inch can be produced to a tolerance ranging from ±.001 to ±.006 inch, depending on the number of passes, rigidity of the machine, and fixtures. Tolerances are chosen initially from the middle of the range for each dimension and process, then adjusted to match the design limits and reduce production costs.

Table 14-2 shows the problem data. The retaining ring (A) and the two bearings (C and G) supporting the shaft are vendor-supplied, hence their tolerances are fixed and must not be altered by the allocation process. The remaining dimensions are all turned in-house. Initial tolerance values for B, D, E, and F were selected from Fig. 14-4, assuming a midrange tolerance. The critical clearance is the shaft end-play, which is determined by tolerance accumulation in the assembly. The vector diagram overlaid on the figure is the assembly loop that models the end-play.

Minimum-Cost Tolerance Allocation 14-5

RANGE OF SIZES		TOLERANCES ±3σ								
FROM	THROUGH									
0.000	0.599	0.00015	0.0002	0.0003	0.0005	0.0008	0.0012	0.002	0.003	0.005
0.600	0.999	0.00015	0.00025	0.0004	0.0006	0.001	0.0015	0.0025	0.004	0.006
1.000	1.499	0.0002	0.0003	0.0005	0.0008	0.0012	0.002	0.003	0.005	0.008
1.500	2.799	0.00025	0.0004	0.0006	0.001	0.0015	0.0025	0.004	0.006	0.010
2.800	4.499	0.0003	0.0005	0.0008	0.0012	0.002	0.003	0.005	0.008	0.012
4.500	7.799	0.0004	0.0006	0.001	0.0015	0.0025	0.004	0.006	0.010	0.015
7.800	13.599	0.0005	0.0008	0.0012	0.002	0.003	0.005	0.008	0.012	0.020
13.600	20.999	0.0006	0.001	0.0015	0.0025	0.004	0.006	0.010	0.015	0.025

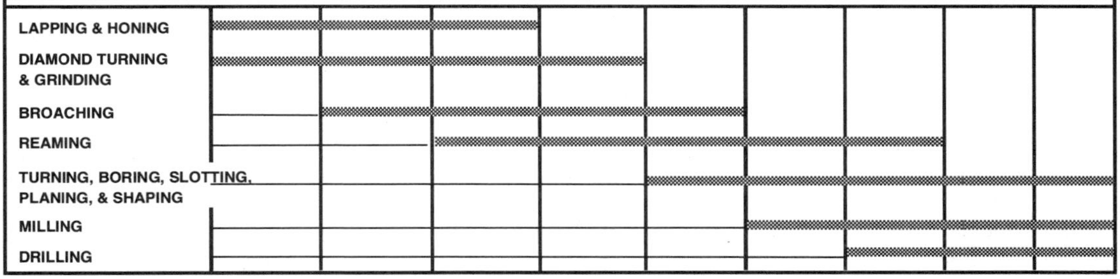

LAPPING & HONING
DIAMOND TURNING & GRINDING
BROACHING
REAMING
TURNING, BORING, SLOTTING, PLANING, & SHAPING
MILLING
DRILLING

Figure 14-4 Tolerance range of machining processes (Reference 12)

Table 14-2 Initial Tolerance Specifications

Dimension	Nominal	Initial Tolerance	Process Tolerance Limits	
			Min Tol	Max Tol
A	.0505	.0015*	*	*
B	8.000	.008	.003	.012
C	.5093	.0025*	*	*
D	.400	.002	.0005	.0012
E	7.711	.006	.0025	.010
F	.400	.002	.0005	.0012
G	.5093	.0025*	*	*

* Fixed tolerances

The average clearance is the vector sum of the average part dimensions in the loop:

Required Clearance $= .020 \pm .015$
Average Clearance $= -A + B - C + D - E + F - G$
$= -.0505 + 8.000 - .5093 + .400 - 7.711 + .400 - .5093$
$= .020$

The worst case clearance tolerance is obtained by summing the component tolerances:

$$T_{SUM} = T_A + T_B + T_C + T_D + T_E + T_F + T_G$$
$$= +.0015 + .008 + .0025 + .002 + .006 + .002 + .0025$$
$$= .0245 \text{ (too large)}$$

To apply the minimum cost algorithm, we must set $T_{SUM} = (T_{ASM}$ - fixed tolerances) and substitute for T_D, T_E, and T_F in terms of T_B, as in Eq. (14.3).

$$T_{ASM} - T_A - T_C - T_G = T_B + \left(\frac{k_D B_D}{k_B B_B}\right)^{1/(k_D+1)} T_B^{(k_B+1)/(k_D+1)} +$$

$$\left(\frac{k_E B_E}{k_B B_B}\right)^{1/(k_E+1)} T_B^{(k_B+1)/(k_E+1)} + \left(\frac{k_F B_F}{k_B B_B}\right)^{1/(k_F+1)} T_B^{(k_B+1)/(k_F+1)}$$

Inserting values into the equation yields:

$$.015 - .0015 - .0025 - .0025 = T_B + \left(\frac{(.46823)(.07202)}{(.43899)(.15997)}\right)^{1/(1.46823)} T_B^{(1.43899)/(1.46823)} +$$

$$\left(\frac{(.46537)(.12576)}{(.43899)(.15997)}\right)^{1/(1.46537)} T_B^{(1.43899)/(1.46537)} + \left(\frac{(.46823)(.07202)}{(.43899)(.15997)}\right)^{1/(1.46823)} T_B^{(1.43899)/(1.46823)}$$

The values of k and B for each nominal dimension were obtained from the fitted cost-tolerance functions for the turning process listed in the Appendix of this chapter. Using a spreadsheet program, calculator with a "Solve" function, or other math utility, the value of T_B satisfying the above expression can be found. T_B can then be substituted into the individual expressions to obtain the corresponding values of T_D, T_E, and T_F, and the predicted cost.

$T_B = .0025$

$$T_D = T_F = \left(\frac{(.46823)(.07202)}{(.43899)(.15997)}\right)^{1/(1.46823)} T_B^{(1.43899)/(1.46823)} = .0017$$

$$T_E = \left(\frac{(.46537)(.12576)}{(.43899)(.15997)}\right)^{1/(1.46537)} T_B^{(1.43899)/(1.46537)} = .0025$$

$$C = A_B + B_B(T_B)^{k_B} + A_D + B_D(T_D)^{k_D} + A_E + B_E(T_E)^{k_E} + A_F + B_F(T_F)^{k_F} = \$11.07$$

Numerical results for the example assembly are shown in Table 14-3.

The setup cost is coefficient A in the cost function. Setup cost does not affect the optimization. For this example, the setup costs were all chosen as equal, so they would not mask the effect of the tolerance allocation. In this case, they merely added $4.00 to the assembly cost for each case.

Parts A, C, and G are vendor-supplied. Since their tolerances are fixed, their cost cannot be changed by reallocation, so no cost data is included in the table.

The statistical tolerance allocation results were obtained by a similar procedure, using Eq. (14.2).

Note that in this example the assembly cost increased when worst case allocation was performed. The original tolerances, when summed by worst case, give an assembly variation of .0245 inch. This exceeds the specified assembly tolerance limit of .015 inch. Thus, the component tolerances had to be tightened, driving up the cost. When summed statistically, however, the assembly variation was only .0011 inch. This was less than the spec limit. The allocation algorithm increased the component tolerances, decreasing the cost. A graphical comparison is shown in Fig. 14-5. It is clear from the graph that tolerances for B and E were tightened in the Worst Case Model, while D and F were loosened in the Statistical Model.

Table 14-3 Minimum cost tolerance allocation

Dimension	Tolerance Cost Data			Original Tolerance	Allocated Tolerances	
	Setup A	Coefficient B	Exponent k		Worst Case	Stat. ±3σ
A		*	*	.0015*	.0015*	.0015*
B	$1.00	.15997	.43899	.008	.00254	.0081
C		*	*	.0025*	.0025*	.0025*
D	1.00	.07202	.46823	.002	.001736	.00637
E	1.00	.12576	.46537	.006	.002498	.00792
F	1.00	.07202	.46823	.002	.001736	.00637
G		*	*	.0025*	.0025*	.0025*
Assembly Variation				.0245(WC) .0111(RSS)	.0150(WC)	.0150(RSS)
Assembly Cost				$9.34	$11.07	$8.06
Acceptance Fraction					1.000	.9973
"True Cost"					$11.07	$8.08

*Fixed tolerances

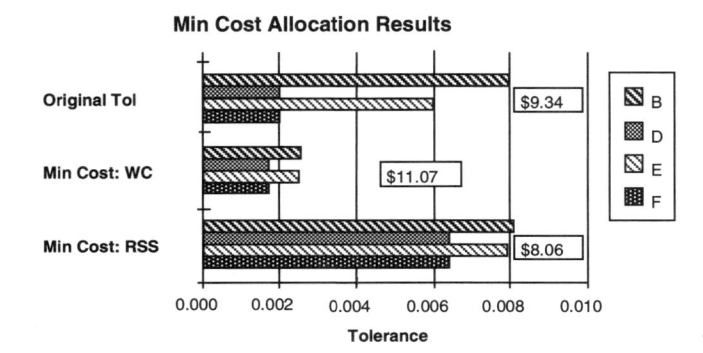

Figure 14-5 Comparison of minimum cost allocation results

14.4 Advantages/Disadvantages of the Lagrange Multiplier Method

The advantages are:
- It eliminates the need for multiple-parameter iterative solutions.
- It can handle either worst case or statistical assembly models.
- It allows alternative cost-tolerance models.

The limitations are:

- Tolerance limits cannot be imposed on the processes. Most processes are only capable of a specified range of tolerance. The designer must check the resulting component tolerances to make sure they are within the range of the process.
- It cannot readily treat the problem of simultaneously optimizing interdependent design specifications. That is, when an assembly has more than one design specification, with common component dimensions contributing to each spec, some iteration is required to find a set of shared tolerances satisfying each of the engineering requirements.

Problems exhibiting multiple assembly requirements may be optimized using nonlinear programming techniques. Manual optimization may be performed by optimizing tolerances for one assembly spec at a time, then choosing the lowest set of shared component tolerance values required to satisfy all assembly specs simultaneously.

14.5 True Cost and Optimum Acceptance Fraction

The "True Cost" in Table 14-4 is defined as the total cost of an assembly divided by the acceptance fraction or yield. Thus, the total cost is adjusted to include a share of the cost of the rejected assemblies. It does not include, however, any parts that might be saved by rework or the cost of rejecting individual component parts.

An interesting exercise is to calculate the optimum acceptance fraction; that is, the rejection rate that would result in the minimum True Cost. This requires an iterative solution. For the example problem, the results are shown in Table 14-4:

Table 14-4 Minimum True Cost

Cost Model	ΣA	$Z_{assembly}$	Optimum Acceptance Fraction	True Cost
$A + B/tol^k$	$4.00	2.03	.9576	$7.67
$A + B/tol^k$	$8.00	2.25	.9756	$11.82

The results indicate that loosening up the tolerances will save money on production costs, but will increase the cost of rejects. By iterating on the acceptance fraction, it is possible to find the value that minimizes the combined cost of production and rejects. Note, however, that the setup costs were set very low. If setup costs were doubled, as shown in the second row of the table, the cost of rejects would be higher, requiring a higher acceptance level.

In the very probable case where individual process cost-versus-tolerance curves are not available, an optimum acceptance fraction for the assembly could be based instead on more available cost-per-reject data. The optimum acceptance fraction could then be used in conjunction with allocation by proportional scaling or weight factors to provide a meaningful cost-related alternative to allocation by least cost optimization.

14.6 2-D and 3-D Tolerance Allocation

Tolerance allocation may be applied to 2-D and 3-D assemblies as readily as 1-D. The only difference is that each component tolerance must be multiplied by its tolerance sensitivity, derived from the geometry as described in Chapters 9, 11, and 12. The proportionality factors, weight factors, and cost factors are still obtained as described above, with sensitivities inserted appropriately.

14.7 2-D Example: One-way Clutch Assembly

The application of tolerance allocation to a 2-D assembly will be demonstrated on the one-way clutch assembly shown in Fig. 14-6. The clutch consists of four different parts: a hub, a ring, four rollers, and four springs. Only a quarter section is shown because of symmetry. During operation, the springs push the rollers into the wedge-shaped space between the ring and the hub. If the hub is turned counterclockwise, the rollers bind, causing the ring to turn with the hub. When the hub is turned clockwise, the rollers slip, so torque is not transmitted to the ring. A common application for the clutch is a lawn mower starter. (Reference 5)

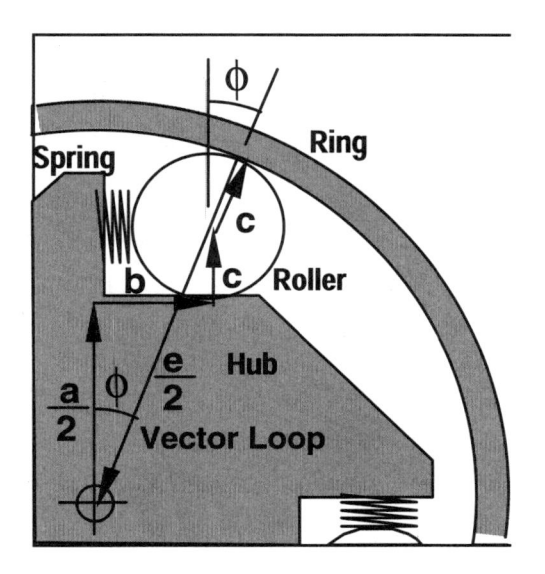

Figure 14-6 Clutch assembly with vector loop

The contact angle ϕ between the roller and the ring is critical to the performance of the clutch. Variable b, is the location of contact between the roller and the hub. Both the angle ϕ and length b are dependent assembly variables. The magnitude of ϕ and b will vary from one assembly to the next due to the variations of the component dimensions a, c, and e. Dimension a is the width of the hub; c and $e/2$ are the radii of the roller and ring, respectively. A complex assembly function determines how much each dimension contributes to the variation of angle ϕ. The nominal contact angle, when all of the independent variables are at their mean values, is 7.0 degrees. For proper performance, the angle must not vary more than ±1.0 degree from nominal. These are the engineering design limits.

The objective of variation analysis for the clutch assembly is to determine the variation of the contact angle relative to the design limits. Table 14-5 below shows the nominal value and tolerance for the three independent dimensions that contribute to tolerance stackup in the assembly. Each of the independent variables is assumed to be statistically independent (not correlated with each other) and a normally distributed random variable. The tolerances are assumed to be ±3σ.

Table 14-5 Independent dimensions for the clutch assembly

Dimension	Nominal	Tolerance
Hub width - a	2.1768 in.	.004 in.
Roller radius - c	.450 in.	.0004 in.
Ring diameter - e	4.000 in.	.0008 in.

14.7.1 Vector Loop Model and Assembly Function for the Clutch

The vector loop method (Reference 2) uses the assembly drawing as the starting point. Vectors are drawn from part-to-part in the assembly, passing through the points of contact. The vectors represent the independent and dependent dimensions that contribute to tolerance stackup in the assembly. Fig. 14-6 shows the resulting vector loop for a quarter section of the clutch assembly.

The vectors pass through the points of contact between the three parts in the assembly. Since the roller is tangent to the ring, both the roller radius c and the ring radius e are collinear. Once the vector loop is defined, the implicit equations for the assembly can easily be extracted. Eqs. (14.4) and (14.5) shows the set of scalar equations for the clutch assembly derived from the vector loop. h_x and h_y are the sum of vector components in the x and y directions. A third equation, h_θ, is the sum of relative angles between consecutive vectors, but it vanishes identically.

$$h_x = 0 = b + c \sin(\phi) - e \sin(\phi) \tag{14.4}$$

$$h_y = 0 = a + c + c \cos(\phi) - e \cos(\phi) \tag{14.5}$$

Eqs. (14.4) and (14.5) may be solved for ϕ explicitly:

$$\phi = \cos^{-1}\left(\frac{a+c}{e-c}\right) \tag{14.6}$$

The sensitivity matrix [S] can be calculated from Eq. (14.6) by differentiation or by finite difference:

$$[S] = \begin{bmatrix} \frac{\partial \phi}{\partial a} & \frac{\partial \phi}{\partial c} & \frac{\partial \phi}{\partial e} \\ \frac{\partial b}{\partial a} & \frac{\partial b}{\partial c} & \frac{\partial b}{\partial e} \\ \frac{\partial a}{\partial a} & \frac{\partial a}{\partial c} & \frac{\partial a}{\partial e} \end{bmatrix} = \begin{bmatrix} -2.6469 & -10.5483 & 2.6272 \\ -103.43 & -440.69 & 104.21 \end{bmatrix}$$

The tolerance sensitivities for $\delta\phi$ are in the top row of [S]. Assembly variations accumulate or stackup statistically by root-sum-squares:

$$\delta\phi = \sqrt{\sum\left(\left(S_{ij}\delta x_j\right)\right)^2}$$

$$= \sqrt{(S_{11}\delta a)^2 + (S_{12}\delta c)^2 + (S_{13}\delta e)^2}$$

$$= \sqrt{((-2.6469)(.004))^2 + ((-10.5483)(.0004))^2 + ((2.6272)(.0008))^2}$$

$$= .01159 \text{ radians} = .664 \text{ degrees}$$

where $\delta\phi$ is the predicted 3σ variation, δx_j is the set of 3σ component variations.

By worst case:

$$\delta\phi = \sum |S_{ij}||\delta x_j|$$

$$= |S_{11}||\delta a| + |S_{12}||\delta c| + |S_{13}||\delta e|$$

$$= (2.6469)(.004) + (10.5483)(.0004) + (2.6272)(.0008)$$

$$= .01691 \text{ radians} = .9688 \text{ degrees}$$

where $\delta\phi$ is the predicted extreme variation.

14.8 Allocation by Scaling, Weight Factors

Once you have RSS and worst case expressions for the predicted variation $\delta\phi$, you may begin applying various allocation algorithms to search for a better set of design tolerances. As we try various combinations, we must be careful not to exceed the tolerance range of the selected processes. Table 14-6 shows the selected processes for dimensions *a*, *c*, and *e* and the maximum and minimum tolerances obtainable by each, as extracted from the Appendix for the corresponding nominal size.

Table 14-6 Process tolerance limits for the clutch assembly

Part	Dimension	Process	Nominal (inch)	Sensitivity	Minimum Tolerance	Maximum Tolerance
Hub	a	Mill	2.1768	-2.6469	.0025	.006
Roller	c	Lap	.9000	-10.548	.00025	.00045
Ring	e	Grind	4.0000	2.62721	.0005	.0012

14.8.1 Proportional Scaling by Worst Case

Since the rollers are vendor-supplied, only tolerances on dimensions *a* and *e* may be altered. The proportionality factor P is applied to δa and δe, while $\delta \phi$ is set to the maximum tolerance of ±.017453 radians (±1°).

$$\delta \phi = \Sigma |S_{ij}| \delta x_j$$
$$.017453 = |S_{11}| P \delta a + |S_{12}| \delta c + |S_{13}| P \delta e$$
$$.017453 = (2.6469) P(.004) + (10.5483)(.0004) + (2.6272) P(.0008)$$

Solving for P:
$$P = 1.0429$$
$$\delta a = (1.0429)(.004) = .00417 \text{ in.}$$
$$\delta e = (1.0429)(.0008) = .00083 \text{ in.}$$

14.8.2 Proportional Scaling by Root-Sum-Squares

$$\delta \phi = \sqrt{\Sigma \left(\left(S_{ij} \delta x_j \right) \right)^2}$$
$$.017453 = \sqrt{(S_{11} P \delta a)^2 + (S_{12} \delta c)^2 + (S_{13} P \delta e)^2}$$
$$.017453 = \sqrt{((-2.6469) P(.004))^2 + ((-10.5483)(.0004))^2 + ((2.6272) P(.0008))^2}$$

Solving for P:
$$P = 1.56893$$
$$\delta a = (1.56893)(.004) = .00628 \text{ in.}$$
$$\delta e = (1.56893)(.0008) = .00126 \text{ in.}$$

Both of these new tolerances exceed the process limits for their respective processes, but by less than .001in each. You could round them off to .006 and .0012. The process limits are not that precise.

14.8.3 Allocation by Weight Factors

Grinding the ring is the more costly process of the two. We would like to loosen the tolerance on dimension *e*. As a first try, let the weight factors be $w_a = 10$, $w_e = 20$. This will change the ratio of the two tolerances and scale them to match the 1.0 degree limit. The original tolerances had a ratio of 5:1. The final ratio will be the product of 1:2 and 5:1, or 2.5:1. The sensitivities do not affect the ratio.

$$\delta\phi = \sqrt{\sum\left((S_{ij}\delta x_j)^2\right)}$$

$$.017453 = \sqrt{(S_{11}P(10/30)\delta a)^2 + (S_{12}\delta c)^2 + (S_{13}P(20/30)\delta e)^2}$$

$$.017453 = \sqrt{((-2.6469)P(10/30)(.004))^2 + ((-10.5483)(.0004))^2 + ((2.6272)P(20/30)(.0008))^2}$$

Solving for P:
$P = 4.460$
$\delta a = (4.460)(10/30)(.004) = .00595$ in.
$\delta e = (4.460)(20/30)(.0008) = .00238$ in.

Evaluating the results, we see that δa is within the .006in limit, but δe is well beyond the .0012 inch process limit. Since δa is so close to its limit, we cannot change the weight factors much without causing δa to go out of bounds. After several trials, the best design seemed to be equal weight factors, which is the same as proportional scaling. We will present a plot later that will make it clear why it turned out this way.

From the preceding examples, we see that the allocation algorithms work the same for 2-D and 3-D assemblies as for 1-D. We simply insert the tolerance sensitivities into the accumulation formulas and carry them through the calculations as constant factors.

14.9 Allocation by Cost Minimization

The minimum cost allocation applies equally well to 2-D and 3-D assemblies. If sensitivities are included in the derivation presented in Section 14.1, Eqs. (14.1) through (14.3) become:

Table 14-7 Expressions for minimum cost tolerances in 2-D and 3-D assemblies

Worst Case	RSS
$T_i = \left(\dfrac{k_i B_i S_1}{k_1 B_1 S_i}\right)^{1/(k_i+1)} T_1^{(k_1+1)/(k_i+1)}$	$T_i = \left(\dfrac{k_i B_i S^2_1}{k_1 B_1 S^2_i}\right)^{1/(k_i+2)} T_1^{(k_1+2)/(k_i+2)}$
$T_{ASM} = S_1 T_1 + \sum S_i \left(\dfrac{k_i B_i S_1}{k_1 B_1 S_i}\right)^{1/(k_i+1)} T_1^{(k_1+1)/(k_i+1)}$	$T^2_{ASM} = S_1^2 T_1^2 + \sum S_i^2 \left(\dfrac{k_i B_i S^2_1}{k_1 B_1 S^2_i}\right)^{2/(k_i+2)} T_1^{2(k_1+2)/(k_i+2)}$

The cost data for computing process cost is shown in Table 14-8:

Table 14-8 Process tolerance cost data for the clutch assembly

Part	Dimension	Process	Nominal (inch)	Sensitivity	B	k	Minimum Tolerance	Maximum Tolerance
Hub	a	Mill	2.1768	-2.6469	.1018696	.45008	.0025	.006
Roller	c	Lap	.9000	-10.548	.000528	1.130204	.00025	.00045
Ring	e	Grind	4.0000	2.62721	.0149227	.79093	.0005	.0012

14.9.1 Minimum Cost Tolerances by Worst Case

To perform tolerance allocation using a Worst Case Stackup Model, let $T_1 = \delta a$, and $T_i = \delta e$, then $S_1 = S_{11}$, $k_1 = k_a$, and $B_1 = B_a$, etc.

$$T_{ASM} = |S_{11}|\delta a + |S_{12}|\delta c + |S_{13}|\delta e$$

$$= |S_{11}|\delta a + |S_{12}|\delta c + |S_{13}|\left(\frac{k_e B_e S_{11}}{k_a B_a S_{13}}\right)^{1/(k_e+1)} \delta a^{(k_a+1)/(k_e+1)}$$

$$.017453 = 2.6469\,\delta a + 10.5483\,(.0004) + 2.6272\left(\frac{(.79093)(.0149227)(2.6469)}{(.45008)(0.1018696)(2.6272)}\right)^{1/(1.79093)} \delta a^{(1.45008)/(1.79093)}$$

The only unknown is δa, which may be found by iteration. δe may then be found once δa is known. Solving for δa and δe:

$\delta a = .00198$ in.

$$\delta e = \left(\frac{(.79093)(.0149227)(2.6469)}{(.45008)(0.1018696)(2.6272)}\right)^{1/(1.79093)} .00198^{(1.45008)/(1.79093)} = .00304 \text{ in.}$$

The cost corresponding to holding these tolerances would be reduced from C= $5.42 to C= $3.14.

Comparing these values to the process limits in Table 14-6, we see that δa is below its lower process limit (.0025 < δa < .006), while δe is much larger than the upper process limit (.0005 < δe < .0012). If we decrease δe to the upper process limit, δa can be increased until T_{ASM} equals the spec limit. The resulting values and cost are then:

$\delta a = .0038$ in. $\delta e = .0012$ in.   C = $4.30

The relationship between the resulting three pairs of tolerances is very clear when they are plotted as shown in Fig. 14-7. Tol e and Tol a are plotted as points in 2-D tolerance space. The feasible region is bounded by a box formed by the upper and lower process limits, which is cut off by the Worst Case limit curve. The original tolerances of (.004, .0008) lie within the feasible region, nearly touching the WC Limit. Extending a line through the original tolerances to the WC Limit yields the proportional scaling results found in section 14.2 (.00417, .00083), which is not much improvement over the original tolerances. The minimum cost tolerances (OptWC) were a significant change, but moved outside the feasible region. The feasible point of lowest cost (Mod WC) resulted at the intersection of the upper limit for Tol e and the WC Limit (.0038, .0012).

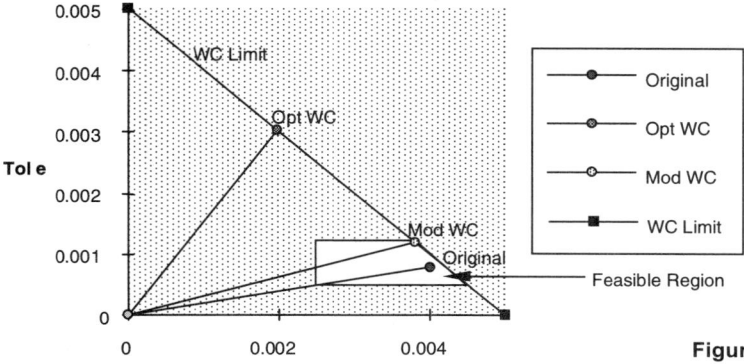

Figure 14-7 Tolerance allocation results for a Worst Case Model

14-14   Chapter Fourteen

This type of plot really clarifies the relationship between the three results. Unfortunately, it is limited to a 2-D graph, so it is only applicable to an assembly with two design tolerances.

14.9.2 Minimum Cost Tolerances by RSS

Repeating the minimum cost tolerance allocation using the RSS Stackup Model:

$$T_{ASM}^2 = (S_{11}\delta a)^2 + (S_{12}\delta c)^2 + (S_{13}\delta e)^2$$

$$= (S_{11}\delta a)^2 + (S_{12}\delta c)^2 + (S_{13})^2 \left(\frac{k_e B_e S_{11}}{k_a B_a S_{13}}\right)^{2/(k_e+2)} \delta a^{2(k_a+2)/(k_e+2)}$$

$$(.017453)^2 = (2.6469\,\delta a)^2 + ((10.5483)(.0004))^2$$

$$+\ 2.6272^2 \left(\frac{(.79093)(.0149227)(2.6469)}{(.45008)(.1018696)(2.6272)}\right)^{2/(2.79093)} \delta a^{2(2.45008)/(2.79093)}$$

Solving for δa by iteration and δe as before:
$\delta a = .00409$ in.

$$\delta e = \left(\frac{(.79093)(.0149227)(2.6469)}{(.45008)(.1018696)(2.6272)}\right)^{1/(2.79093)} (.00409)^{(2.45008)/(2.79093)}$$

$$= .00495 \text{ in.}$$

The cost corresponding to holding these tolerances would be reduced from C= $5.42 to C= $2.20.

Comparing these values to the process limits in Table 14-6, we see that δa is now safely within its process limits ($.0025 < \delta a < .006$), while δe is still much larger than the upper process limit ($.0005 < \delta e < .0012$). If we again decrease δe to the upper process limit as before, δa can be increased until it equals the upper process limit. The resulting values and cost are then:

$\delta a = .006$ in. $\delta e = .0012$ in.   C = $4.07

The plot in Fig. 14-8 shows the three pairs of tolerances. The box containing the feasible region is entirely within the RSS Limit curve. The original tolerances of (.004, .0008) lie near the center of the feasible region. Extending a line through the original tolerances to the RSS Limit yields the proportional scaling results found in section 14.2 (.00628, .00126), both of which lie just outside the feasible region. The

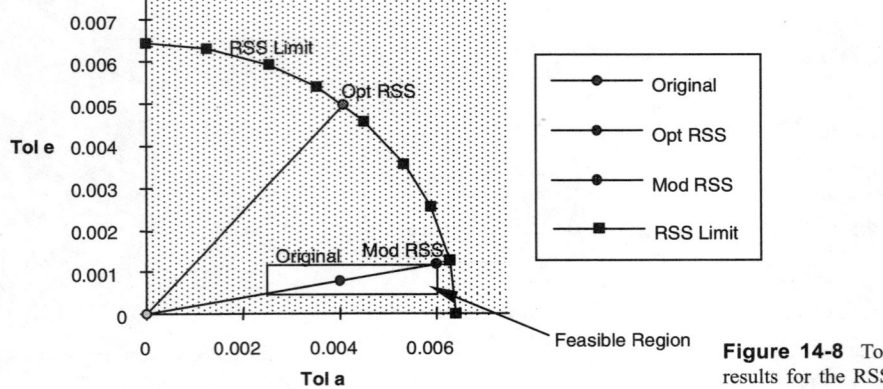

Figure 14-8 Tolerance allocation results for the RSS Model

minimum cost tolerances (OptRSS) were a significant change, but moved far outside the feasible region. The feasible point of lowest cost (ModRSS) resulted at the upper limit corner of the feasible region (.006, .0012).

Comparing Figs. 14-7 and 14-8, we see that the RSS Limit curve intersects the horizontal and vertical axes at values greater than .006 inch, while the WC Limit curve intersects near .005 inch tolerance. The intersections are found by letting Tol a or Tol e go to zero in the equation for T_{ASM} and solving for the remaining tolerance. The RSS and WC Limit curves do not converge to the same point because the fixed tolerance δc is subtracted from T_{ASM} differently for WC than RSS.

14.10 Tolerance Allocation with Process Selection

Examining Fig. 14-7 further, the feasible region appears very small. There is not much room for tolerance design. The optimization preferred to drive Tol e to a much larger value. One way to enlarge the feasible region is to select an alternate process for dimension e. Instead of grinding, suppose we consider turning. The process limits change to (.002< δe <.008), with $B_e = .118048$ $k_e = -.45747$. Table 14-9 shows the revised data.

Table 14-9 Revised process tolerance cost data for the clutch assembly

Part	Dimension	Process	Nominal (inch)	Sensitivity	B	k	Minimum Tolerance	Maximum Tolerance
Hub	a	Mill	2.1768	-2.6469	.1018696	.45008	.0025	.006
Roller	c	Lap	.9000	-10.548	.000528	1.130204	.00025	.00045
Ring	e	Turn	4.0000	2.62721	.118048	.45747	.002	.008

Milling and turning are processes with nearly the same precision. Thus, B_e and B_a are nearly equal as are k_e and k_a. The resulting RSS allocated tolerances and cost are:

$\delta a = .00434$ in. $\delta e = .00474$ in.   C = $2.54

The new optimization results are shown in Fig. 14-9. The feasible region is clearly much larger and the minimum cost point (Mod Proc) is on the RSS Limit curve on the region boundary. The new optimum point has also changed from the previous result (Opt RSS) because of the change in B_e and k_e for the new process.

The resulting WC allocated tolerances and cost are:

$\delta a = .00240$ in. $\delta e = .00262$ in.   C = $3.33

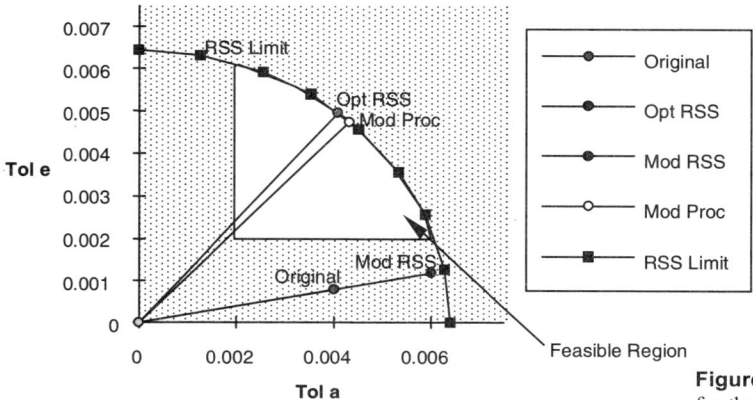

Figure 14-9 Tolerance allocation results for the modified RSS Model

The modified optimization results are shown in Fig. 14-10. The feasible region is the smallest yet due to the tight Worst Case (WC) Limit. The minimum cost point (Mod Proc) is on the WC Limit curve on the region boundary.

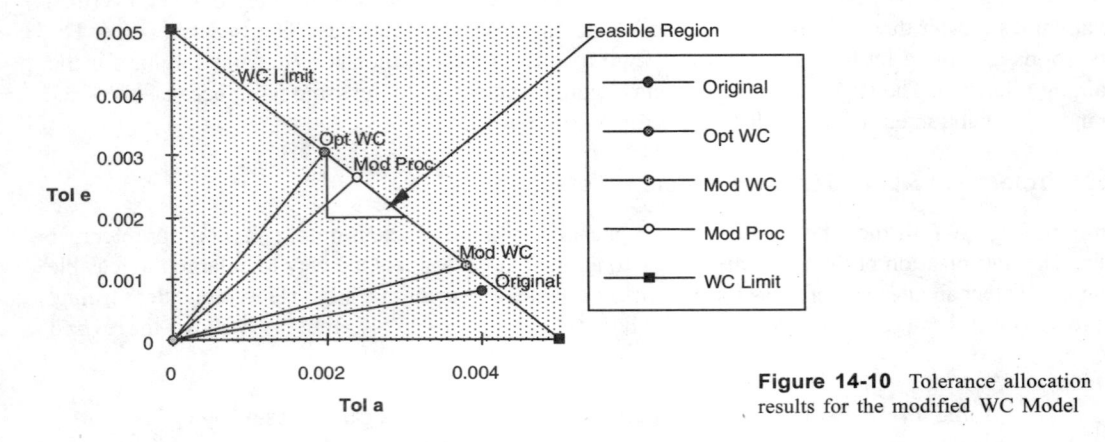

Figure 14-10 Tolerance allocation results for the modified WC Model

Cost reductions can be achieved by comparing cost functions for alternate processes. If cost-versus-tolerance data are available for a full range of processes, process selection can even be automated. A very systematic and efficient search technique, which automates this task, has been published. (Reference 4) It compares several methods for including process selection in tolerance allocation and gives a detailed description of the one found to be most efficient.

14.11 Summary

The results of WC and RSS cost allocation of tolerances are summarized in the two bar charts, Figs. 14-11 and 14-12. The changes in magnitude of the tolerances are readily apparent. Costs have been added for comparison.

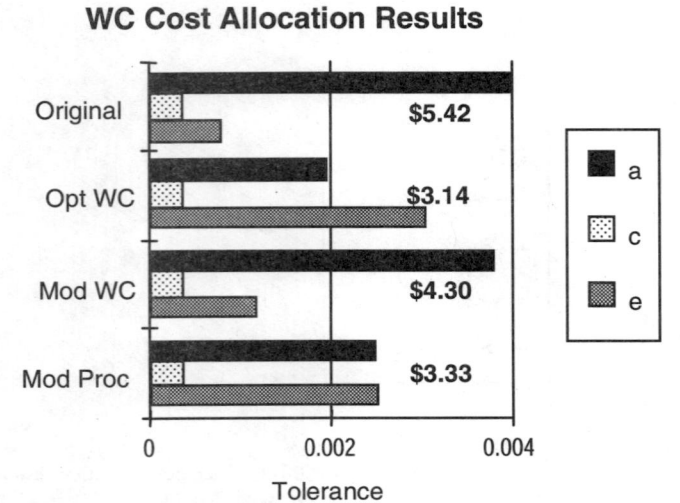

Figure 14-11 Tolerance allocation results for the WC Model

RSS Cost Allocation Results

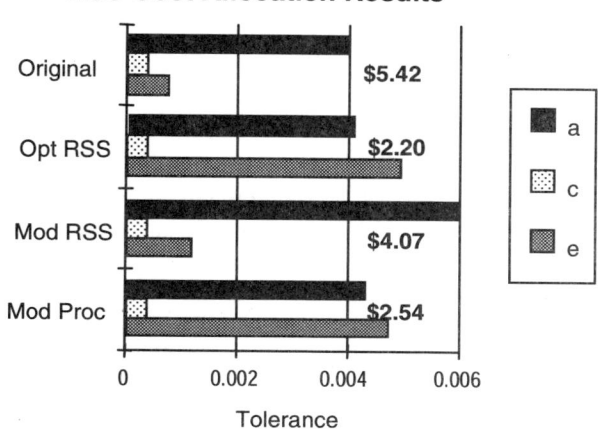

Figure 14-12 Tolerance allocation results for the RSS Model

Summarizing, the original tolerances for both WC and RSS were safely within tolerance constraints, but the costs were high. Optimization reduced the cost dramatically; however, the resulting tolerances exceeded the recommended process limits. The modified WC and RSS tolerances were adjusted to conform to the process limits, resulting in a moderate decrease in cost, about 20%. Finally, the effect of changing processes was illustrated, which resulted in a cost reduction near the first optimization. Only the allocated tolerances remained in the new feasible region.

A designer would probably not attempt all of these cases in a real design problem. He would be wise to rely on the RSS solution, possibly trying WC analysis for a case or two for comparison. Note that the clutch assembly only had three dimensions contributing to the tolerance stack. If there had been six or eight, the difference between WC and RSS would have been much more significant.

It should be noted that tolerances specified at the process limit may not be desirable. If the process is not well controlled, it may be difficult to hold it at the limit. In such cases, the designer may want to back off from the limits to allow for process uncertainties.

14.12 References

1. Chase, K. W. and A. R. Parkinson. 1991. *A Survey of Research in the Application of Tolerance Analysis to the Design of Mechanical Assemblies: Research in Engineering Design.* 3(1):23-37.
2. Chase, K. W., J. Gao and S. P. Magleby. 1995. General 2-D Tolerance Analysis of Mechanical Assemblies with Small Kinematic Adjustments. *Journal of Design and Manufacturing.* 5(4): 263-274.
3. Chase, K.W. and W.H. Greenwood. 1988. Design Issues in Mechanical Tolerance Analysis. *Manufacturing Review.* March, 50-59.
4. Chase, K. W., W. H. Greenwood, B. G. Loosli and L. F. Hauglund. 1989. Least Cost Tolerance Allocation for Mechanical Assemblies with Automated Process Selection. *Manufacturing Review.* December, 49-59.
5. Fortini, E.T. 1967. *Dimensioning for Interchangeable Manufacture.* New York, New York: Industrial Press.
6. Greenwood, W.H. and K.W. Chase. 1987. A New Tolerance Analysis Method for Designers and Manufacturers *Journal of Engineering for Industry, Transactions of ASME.* 109(2):112-116.
7. Hansen, Bertrand L. 1963. *Quality Control: Theory and Applications.* Paramus, New Jersey: Prentice-Hall.
8. Jamieson, Archibald. 1982. *Introduction to Quality Control.* Paramus, New Jersey: Reston Publishing.
9. Pennington, Ralph H. 1970. *Introductory Computer Methods and Numerical Analysis.* 2nd ed. Old Tappan, New Jersey: MacMillan.
10. Speckhart, F.H. 1972. Calculation of Tolerance Based on a Minimum Cost Approach. *Journal of Engineering for Industry, Transactions of ASME.* 94(2):447-453.
11. Spotts, M.F. 1973. Allocation of Tolerances to Minimize Cost of Assembly. *Journal of Engineering for Industry, Transactions of the ASME.* 95(3):762-764.

12. Trucks, H.E. 1987. *Designing for Economic Production*. 2nd ed., Dearborn, MI: Society of Manufacturing Engineers.
13. U.S. Army Management Engineering Training Activity, Rock Island Arsenal, IL. (Original report is out of print)

14.13 Appendix

Cost-Tolerance Functions for Metal Removal Processes

Although it is well known that tightening tolerances increases cost, adjusting the tolerances on several components in an assembly and observing its effect on cost is an impossible task. Until you have a mathematical model, you cannot effectively optimize the allocation of tolerance in an assembly. Elegant tools for minimum cost tolerance allocation have been developed over several decades. However, they require empirical functions describing the relationship between tolerance and cost.

Cost-versus-tolerance data is very scarce. Very few companies or agencies have attempted to gather such data. Companies who do, consider it proprietary, so it is not published. The data is site and machine-specific and subject to obsolescence due to inflation. In addition, not all processes are capable of continuously adjustable precision.

Metal removal processes have the capability to tighten or loosen tolerances by changing feeds, speeds, and depth of cut or by modifying tooling fixtures, cutting tools and coolants. The workpiece may also be modified, switching to a more machinable alloy or modifying geometry to achieve greater rigidity.

A noteworthy study by the US Army in the 1940s experimentally determined the natural tolerance range for the most common metal removal processes. (Reference 13) They also compared the cost of the various processes and the relative cost of tightening tolerances. Relative costs were used to eliminate the effects of inflation. The resulting chart, Table 14A-1, appears in References 7 and 8. Least squares curve fits were performed at Brigham Young University and are presented here for the first time. The Reciprocal Power equation, $C = A + B/T^k$, presented in Chapter 14, was used as the empirical function. Fig. 14A-1 shows a typical plot of the original data and the fitted data. The curve fit procedure was a standard nonlinear method described in Reference 9, which uses weighted logarithms of the data to convert to a linear regression problem. Results are tabulated in Table 14A-2 and plotted in Figs. 14A-2 and 14A-3.

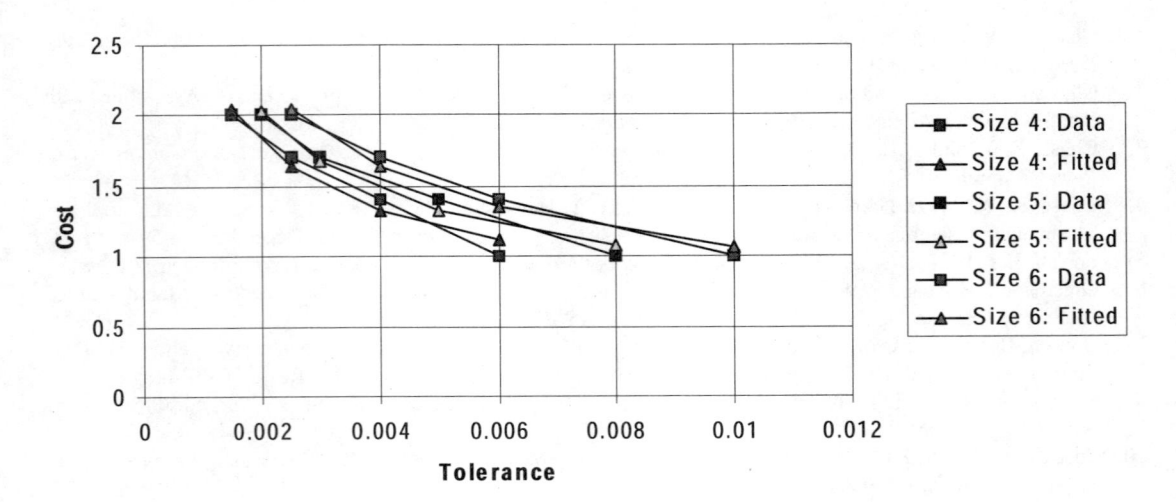

Figure 14A-1 Plot of cost-versus-tolerance for fitted and raw data for the turning process

Minimum-Cost Tolerance Allocation   14-19

Table 14A-1 Relative cost of obtaining various tolerance levels

Range of Sizes (in.)		Tolerances (in.)									
From	To										
0.000	0.599	0.0002	0.00025	0.0004	0.0005	0.0008	0.0012	0.0020	0.0030	0.0050	
0.600	0.999	0.00025	0.0003	0.00045	0.0006	0.0010	0.0015	0.0025	0.0040	0.0060	
1.000	1.499	0.0003	0.0004	0.0005	0.0008	0.0012	0.0020	0.0030	0.0050	0.0080	
1.500	2.799	0.0004	0.0005	0.0006	0.0010	0.0015	0.0025	0.0040	0.0060	0.0100	
2.800	4.499	0.0005	0.0006	0.0008	0.0012	0.0020	0.0030	0.0050	0.0080		
4.500	7.799	0.0006	0.0007	0.0010	0.0015	0.0025	0.0040	0.0060	0.0100		
7.800	13.599	0.0007	0.0008	0.0012	0.0020	0.0030	0.0050	0.0080	0.0120		
13.600	20.999	0.0008	0.0010	0.0015	0.0025	0.0040	0.0060	0.0100	0.0150		

21.00 and over follow same tolerancing trends

Process	Relative Cost of Tightening Tolerance*									Process Cost
Lap and Hone	200%	180%	100%							300%
Grind, Diamond Turn and Bore	200%	180%	140%	100%						300%
Broach		200%	175%	140%	100%					200%
Ream				175%	140%	100%				175%
Turn, Bore, Slot, Plane, and Shape					200%	170%	140%	100%		100%
Mill						150%	125%	100%		100%
Drill								175%	100%	100%

*Total relative cost for a given process is the percentage product of the tolerance tightening cost and the process cost (200%*300%=600%) Reproduced from Reference 2.

Table 14A-2 Cost-tolerance functions for metal removal processes

Size Range	A	B	k	Min Tol	Max Tol
Lap / Hone					
0.000-0.599		0.00189378	0.9508781	0.0002	0.0004
0.600-0.999		0.00052816	1.1302036	0.00025	0.00045
1.000-1.499		0.00220173	0.9808618	0.0003	0.0005
1.500-2.799		0.00033129	1.2590875	0.0004	0.0006
2.800-4.499		0.00026156	1.3269297	0.0005	0.0008
4.500-7.799		0.00038119	1.3073528	0.0006	0.001
7.800-13.599		0.00059824	1.2716314	0.0007	0.0012
13.600-20.999		0.00427422	1.0221757	0.0008	0.0015
Grind / Diamond turn					
0.000-0.599		0.02484363	0.6465727	0.0002	0.0005
0.600-0.999		0.01525616	0.7221989	0.00025	0.0006
1.000-1.499		0.0205072	0.7039047	0.0003	0.0008
1.500-2.799		0.0133561	0.7827624	0.0004	0.001
2.800-4.499		0.01492268	0.790932	0.0005	0.0012
4.500-7.799		0.02467047	0.7413291	0.0006	0.0015
7.800-13.599		0.05119944	0.6548091	0.0007	0.002
13.600-20.999		0.08317908	0.6017646	0.0008	0.0025
Broach					
0.000-0.599		0.0438552	0.548619	0.00025	0.0008
0.600-0.999		0.04670538	0.55230115	0.0003	0.001
1.000-1.499		0.04071362	0.58686634	0.0004	0.0012
1.500-2.799		0.048524	0.579761	0.0005	0.0015
2.800-4.499		0.0637591	0.559608	0.0006	0.002
4.500-7.799		0.0922923	0.521758	0.0007	0.0025
7.800-13.599		0.144046	0.46957	0.0008	0.003
13.600-20.999		0.171785	0.45907	0.001	0.004
Ream					
0.000-0.599		0.03245261	0.6000163	0.0005	0.0012
0.600-0.999		0.04682158	0.565492	0.0006	0.0015
1.000-1.499		0.04204992	0.6021191	0.0008	0.002
1.500-2.799		0.04809684	0.6021191	0.001	0.0025
2.800-4.499		0.06929088	0.565492	0.0012	0.003
4.500-7.799		0.09203907	0.5409254	0.0015	0.004
Turn / bore / shape					
0.000-0.599		0.07201641	0.46822793	0.0008	0.003
0.600-0.999		0.085969502	0.45747142	0.001	0.004
1.000-1.499		0.101233386	0.44723008	0.0012	0.005
1.500-2.799		0.11800302	0.4389869	0.0015	0.006
2.800-4.499		0.11804756	0.45747142	0.002	0.008
4.500-7.799		0.12576137	0.46536684	0.0025	0.01
7.800-13.599		0.15997103	0.4389869	0.003	0.012
13.600-20.999		0.15300611	0.46822793	0.004	0.015
Mill					
0.000-0.599		0.0862308	0.4259173	0.0012	0.003
0.600-0.999		0.10878812	0.4044547	0.0015	0.004
1.000-1.499		0.09544417	0.4431399	0.002	0.005
1.500-2.799		0.10186958	0.4500798	0.0025	0.006
2.800-4.499		0.14399071	0.4044547	0.003	0.008
4.500-7.799		0.12976209	0.4431399	0.004	0.01
7.800-13.599		0.13916564	0.4500798	0.005	0.012
13.600-20.999		0.17114563	0.4259173	0.006	0.015
Drill					
0.000-0.599		0.00301435	1.0955124	0.003	0.005
0.600-0.999		0.00085791	1.3801824	0.004	0.006
1.000-1.499		0.00318631	1.1906627	0.005	0.008
1.500-2.799		0.00644133	1.0955124	0.006	0.01
2.800-4.499		0.00223316	1.3801824	0.008	0.012

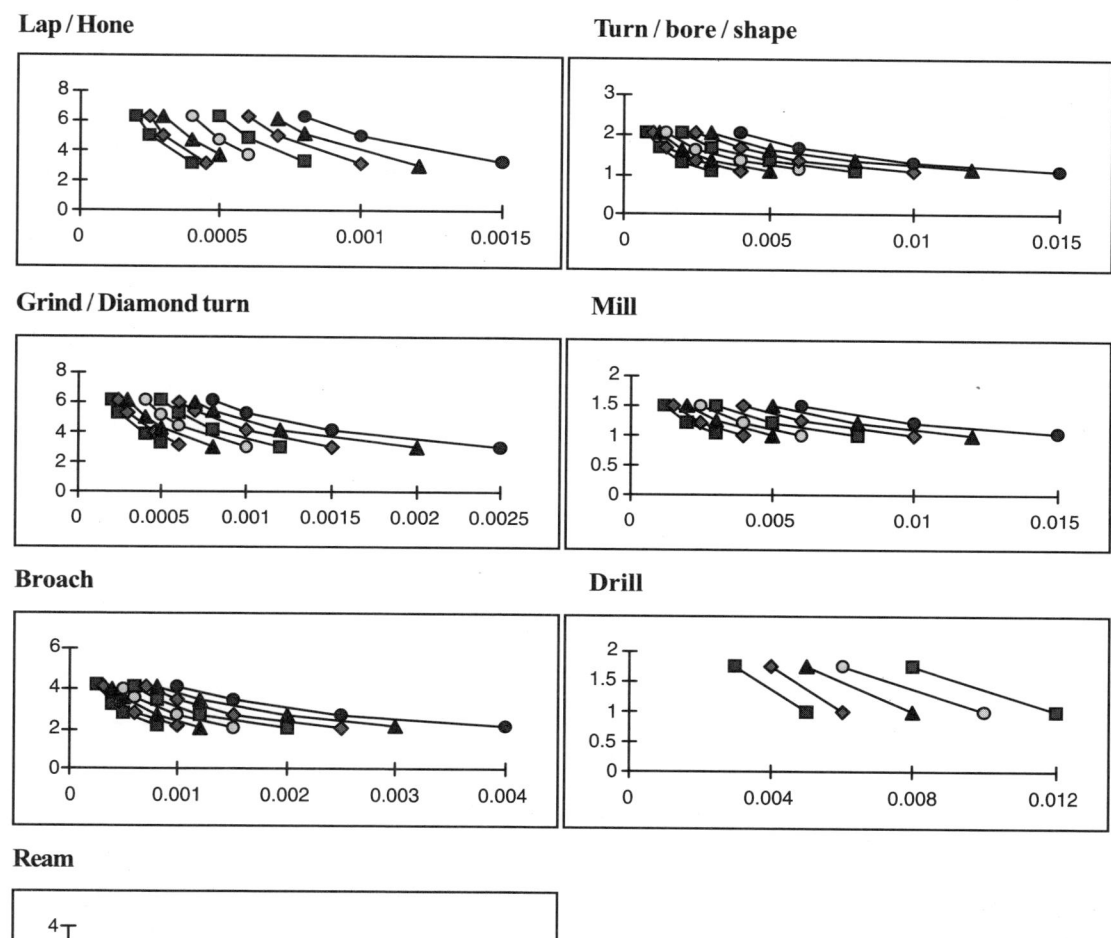

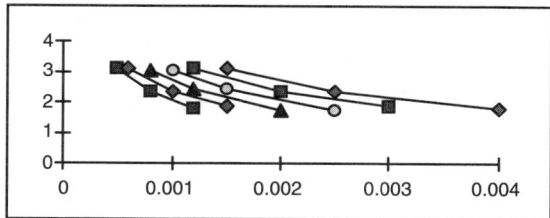

Figure 14A-2 Plot of fitted cost versus tolerance functions

14-22   Chapter Fourteen

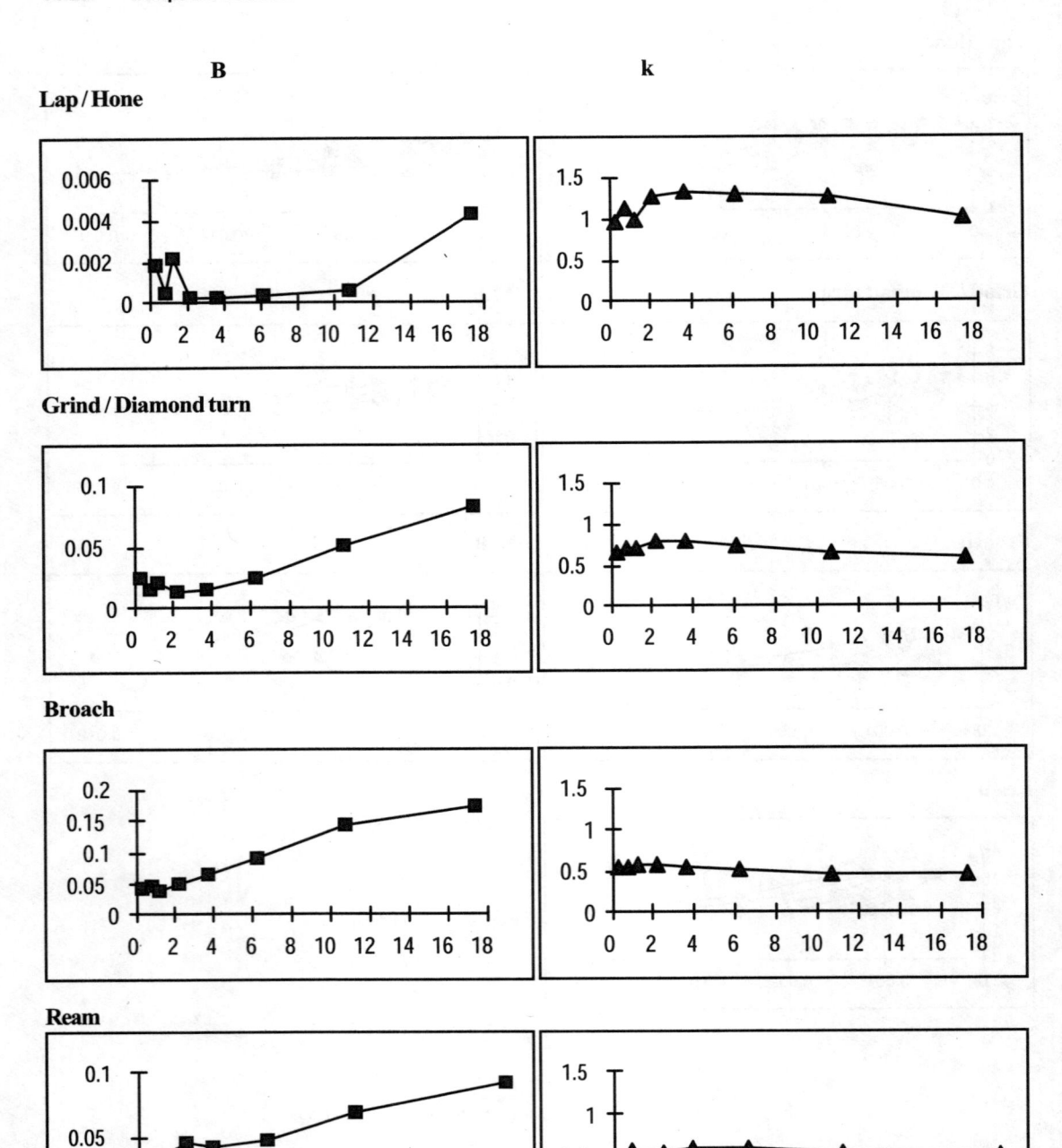

Figure 14A-3 Plot of coefficients versus size for cost-tolerance functions

B

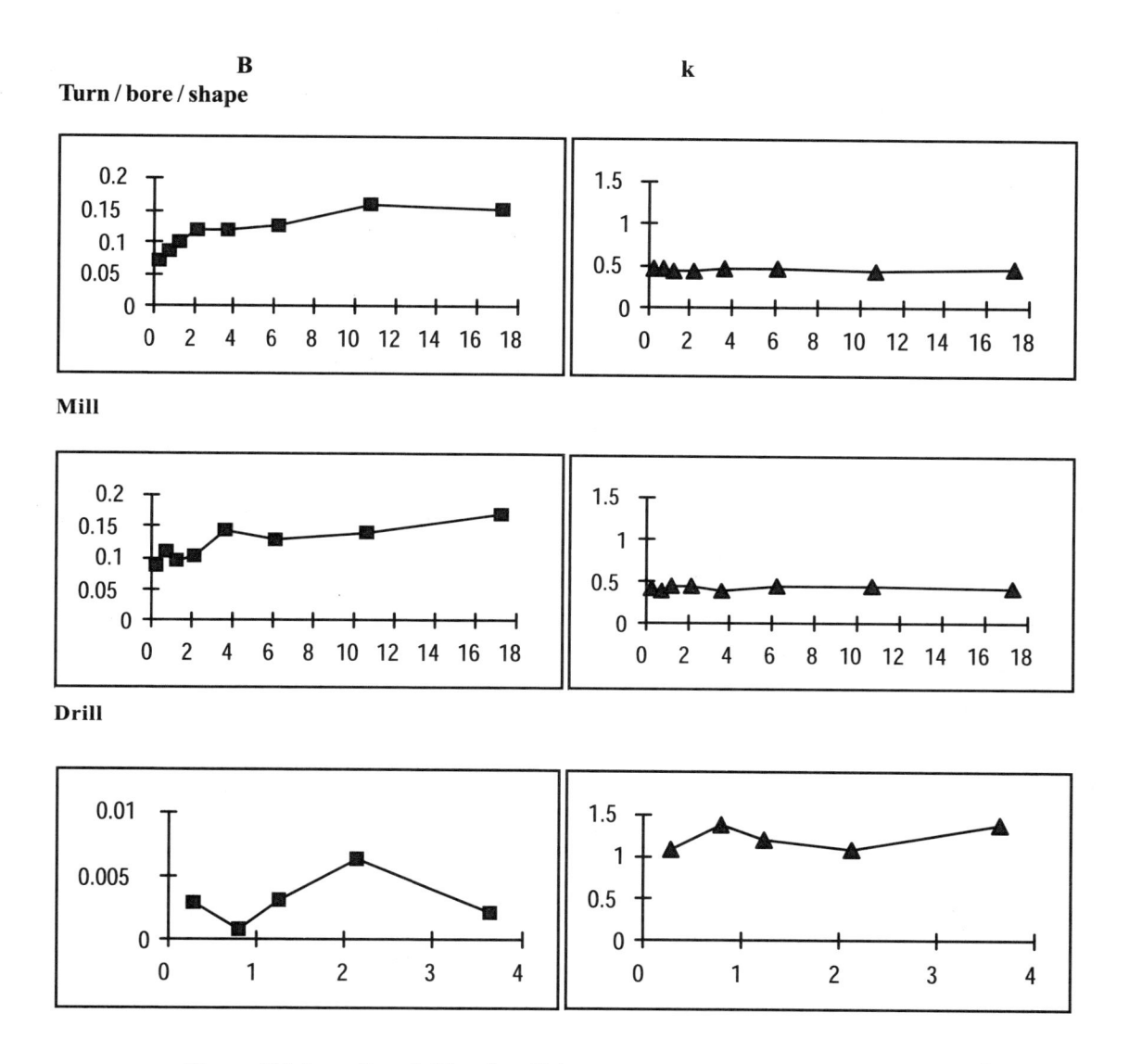

Turn / bore / shape

k

Mill

Drill

Figure 14A-3 continued Plot of coefficients versus size for cost-tolerance functions

Chapter 15

Automating the Tolerancing Process

Charles Glancy
James Stoddard
Marvin Law

Charles Glancy
Raytheon Systems Company
Dallas, Texas

Mr. Glancy is a senior software developer for the CE/TOL SixSigma Tolerance Optimization System at Raytheon Systems Company. Charles received his master's degree in mechanical engineering from Brigham Young University in 1994. At BYU, Mr. Glancy was a research assistant for Dr. Kenneth Chase, founder of the Association for the Development for Computer-Aided Tolerancing Systems (ADCATS). His research included three-dimensional tolerance analysis algorithm development and a system for second-order approximations for nonlinear tolerance analysis. He has written a thesis "A Second-Order Method for Assembly Tolerance Analysis" and co-authored a paper, "A Comprehensive System for Computer-Aided Tolerance Analysis of 2-D and 3-D Mechanical Assemblies."

James Stoddard
Raytheon Systems Company
Dallas, Texas

Mr. Stoddard is a senior software developer of CE/TOL SixSigma Tolerance Optimization System, a tolerance analysis application developed by Raytheon Systems Company. He received his master's degree in mechanical engineering from Brigham Young University. As a graduate student, Mr. Stoddard worked with Dr. Kenneth Chase, founder of ADCATS, on research related to the automation of the tolerance modeling process. In his thesis, "Characterizing Kinematic Variation in Assemblies from Geometric Constraints," he developed an approach to automatic kinematic joint recognition.

Marvin Law
Raytheon Systems Company
Dallas, Texas

Mr. Law is a senior software developer at Raytheon Systems Company. He is involved in researching, designing, and implementing the CE/TOL SixSigma Tolerance Optimization System. Marvin received his master's degree in mechanical engineering from Brigham Young University in 1996. At BYU, Mr. Law was a research assistant for Dr. Kenneth Chase, founder of the Association for the Development for Computer-Aided Tolerancing Systems (ADCATS). For his graduate thesis, "Multivariate Statistical Analysis of Assembly Tolerance Specifications," he developed methods for mathematically characterizing and performing simultaneous statistical analysis of multiple design requirements.

15.1   Background Information

The steady increase of computing capability over the past several years has made powerful engineering analysis tools, such as Computer-Aided Design (CAD) and Finite Element Analysis, available to every engineer. Computer-Aided Tolerancing (CAT) systems that use the CAD geometry to derive mathematical tolerance models are now becoming available. These CAT systems hold great promise in automating tolerancing tasks that used to be performed by hand or with computer spreadsheets, outside of the CAD environment.

This chapter will introduce an automated tolerance analysis process and discuss the different component technologies available that can be used to automate the steps in the tolerancing process.

15.1.1   Benefits of Automation

In general, computer automation can provide great benefits. For tolerance analysis, automation can simplify the tolerance modeling and analysis process, increase the analysis accuracy, reduce analysis time, and reduce calculation errors. An automated tolerance analysis method can also be augmented to include tolerance optimization. Automation can be used to improve communication between design and manufacturing personnel. Furthermore, a CAT system that is integrated with a CAD system can keep the tolerance data synchronized with the CAD model.

15.1.2   Overview of the Tolerancing Process

The tolerancing process begins with two competing pieces of information: the design requirements that must be met to ensure performance and quality, and the manufacturing process capability that can be achieved with the tools available. As shown in Fig. 15-1, the tolerancing process is the means by which these competing requirements are balanced.

A tolerance model is constructed by first deriving design measurements from design requirements. A model function must then be defined to serve as a mathematical relationship between input variables and design measurements. Finally, the input variables must be derived from the manufacturing process capabilities.

Once constructed the tolerance model can be used to perform tolerance analysis or allocation. The terms "analysis" and "allocation" refer to moving through the tolerance model in opposite directions.

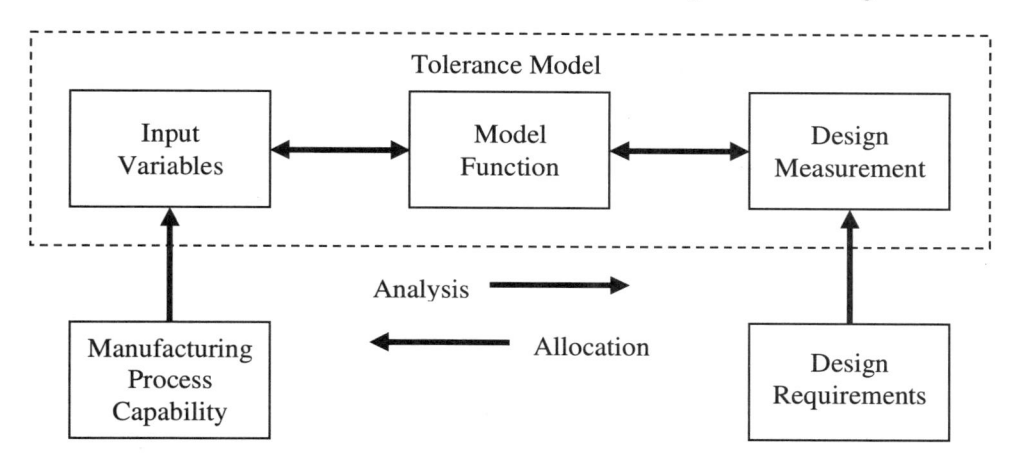

Figure 15-1 Tolerancing process

Tolerance analysis is the process of finding the output quality of a design measurement from the supplied input variables. Tolerance allocation, on the other hand, is the process of finding a set of values for the input variables that will give a desired quality for each design measurement. See Chapter 11 and Fig. 11-1.

The following three sections will discuss aspects of this tolerancing process including model creation, analysis, and optimization in more detail. They will focus on what considerations need to be made in deciding how to automate the various steps of the tolerancing process.

15.2 Automating the Creation of the Tolerance Model

15.2.1 Characterizing Critical Design Measurements

The first step in building a tolerance model is to define the critical design requirements that will be analyzed. Many design requirements are initially posed in qualitative form rather than quantitative form. For example, a design requirement that a circuit card must easily slide into a slot must be translated into insertion force and ultimately to clearance measurements. It is therefore a necessary step of any tolerance modeling process to characterize all qualitative design requirements as quantitative design measurements.

Automation of the characterization process requires the definition of a finite set of design measurements. This set must be general enough to mathematically characterize all the classes of design requirements that may exist. Typical types of design measurements include:

- Gap - Measurable distance between two features along a specified direction
- Angle - Measurable angle between two specified surfaces about a specified axis
- Position - Measurable deviation from a specified location within a specified plane

This set is general enough that most design requirements can be described with one or more of these design measurements.

With an automation tool the process by which a design measurement is defined is also important. This process must be intuitive and easy to use. In cases where a tolerance analysis tool is integrated with a CAD system, the process can be simplified by mapping the definition of the design measurement to physical features within the geometry. This gives associativity and context to the definition of the critical design measurement.

15.2.2 Characterizing the Model Function

The second step in the model creation process is to define the model function. The model function characterizes, in a mathematical form, all the behaviors and interactions that exist in real-world parts and assemblies. In order to properly define this function, all sources of variation and how they propagate must be understood. Understanding the form of the model function and the simplifying assumptions used to limit the scope of the tolerance model are also important.

15.2.2.1 Model Definition

Two significant classifications of variation are manufacturing process variation and assembly process variation. Manufacturing process variation describes all the variation that is introduced in the steps of the manufacturing process plan. These variations may be the result of machining error, setup error, tooling error, or tool wear.

Assembly process variation describes the variations that are introduced as parts are brought together to form assemblies. Assembly fixture error and fastening process error are two examples of assembly process variation.

The model function must take into account how these sources of variation will combine to affect the variation of critical features in the assembly. The features referenced during manufacturing setup determine how variation will accumulate within a part. Dimension chains or dimension paths are the terms typically used to refer to this accumulation. Automation of dimension path creation can greatly simplify the tolerance modeling process. The difficulty lies in trying to include the effects of the manufacturing and assembly process plan before the plan exists. When this plan does not exist the dimensioning scheme used for design may be used with some simple assumptions about tolerances and process capability.

At the assembly level, variation propagates either through small kinematic adjustments or through small part deformations. Small kinematic adjustments in the relative position of components occur as a

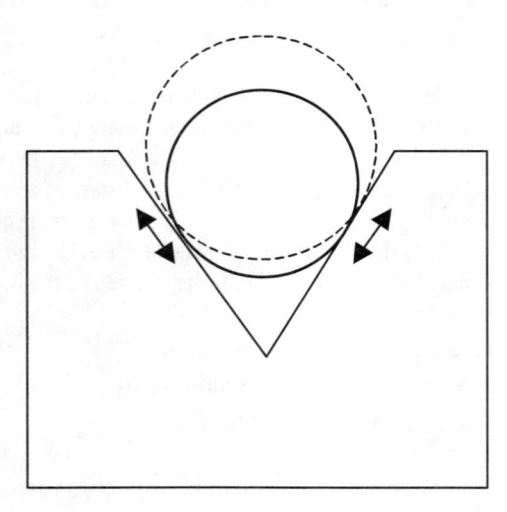

Figure 15-2 Small kinematic adjustments

result of variation in the assembled components, which are exactly constrained. For example, as the diameter of a cylinder in Fig. 15-2 increases, it will rest at a different location within an angled groove.

A complete model function must be able to account for these small kinematic adjustments. One way of characterizing these adjustments is to overlay the mating contacts within the assembly with a kinematic model. The kinematic model describes all mating contacts with kinematic joints and all parts as linkages.

The degrees of freedom are appropriately defined to correctly describe the nature of each contact. The kinematic model can then be solved to find the resulting position of the assembled components.

If the assembly is overconstrained so that parts cannot adjust their relative positions to account for variation, deformation of the components will occur. This is typically the case when sheetmetal parts are used. Sheetmetal parts are brought together by fixtures and rigidly fastened together. Once the fixtures are removed, the resulting assembly deforms to minimize its internal stress state. These deformation adjustments can be described by overlaying a finite element model of the components. This finite element model can then be solved to find the stresses and strains that will result from variation in the component parts and predict how the assembly will deform.

A comprehensive model function will include the effects of all these sources of variation and their corresponding methods of propagation.

15.2.2.2 Model Form

The model function must be captured in mathematical form for computer automation. It must be determined whether an exact or an approximation model will be used. Explicit equations $(y = f(x_1 ... x_n))$ rather than implicit equations $(y = f(y, x_1 ... x_n))$ are desired to perform tolerance analysis because analytical rather than brute force methods can be used. Exact models, however, can often only be expressed in implicit form for complex assembly models that include all sources of variation.

An alternative to an exact mathematical model is an approximation model. This approximation model can be of any order, but typically a first- or second-order approximation is used. The approximation model is defined by finding sensitivities of critical features to each input variable of interest. These sensitivities can be reasoned geometrically or calculated numerically. Once the sensitivity model is produced, it can be used as the basis for analytical algorithms of tolerance analysis and optimization.

One useful mathematical model of the assembly is the CAD model. A CAD model has a full mathematical definition of the assembly that can be interrogated through the CAD system's native or programmatic interface to extract valuable information. Critical features and dimensioning schemes can be identified from the CAD model. CAD systems that are parametric or variational geometry based can be perturbed to find sensitivities directly. Assembly based CAD systems that have meaningful assembly constraints can also provide definition for the assembly process variation. The CAD model is therefore a good starting point in defining the mathematical tolerance model.

15.2.2.3 Model Scope

The definition of an absolutely complete and correct model is often inefficient and unnecessary. By making simplifying assumptions, the complexity of the model can be reduced without losing significant accuracy. It is important, however, to understand the implications of these simplifying assumptions because making the wrong assumptions can lead to invalid results.

One of the most common assumptions is the simplification of 3-D problems to 1-D or 2-D stackups. The world is 3-D and the variations in an assembly interact three-dimensionally. Therefore, a truly accurate model will describe all the 3-D relationships that exist in an assembly. Historically, tolerance analyses have been simplified to 1-D stackups because many were performed by hand. One-D models ignore the effects of most assembly processes on a design measurement and include only the effects of linear variations along a single direction. This may be sufficient for assemblies that have only planar interfaces that are all at right angles to one another and do not involve complex assembly processes. Two-D models start to include the interdependencies that are introduced at the assembly level, but the variation is still restricted to a single plane. Reducing models to 1-D or 2-D may simplify a model function, but is not appropriate in all cases.

Another simplifying assumption is to reduce the number of parts and/or features included in the study. Not all features of all parts affect every design requirement. Ignoring irrelevant parts and features limits the complexity of the assembly function without losing accuracy. In addition to features that have no effect, there may be some features that have only minor effects on the variation in the assembly measurements. Cosmetic and manufacturabilty features such as fillets and rounds often fall into this category. Again, it is important to understand the effects of such simplifying assumptions on the accuracy of the model.

15.2.3   Characterizing Input Variables

The final step in the building of a tolerance model is the characterization of the input variables. The model function is the means of transforming how a change in the inputs will change the outputs. The input variables to the model function are assumed to vary based on variation in the different manufacturing and assembly processes. Tolerance ranges are also supplied for each variable as a limit of acceptable variation. The discussion of the analysis process in the next section will show that the type of analysis performed drives the type and form of the input data. Worst case analysis only requires tolerance limits while statistical analysis requires a defined distribution on the variation of each variable.

Input variable data can come from several sources. The variable definitions, along with some or all of the tolerance data, can be extracted from a CAD system. The statistical distribution information must come from manufacturing data, as will be discussed in section 15.5.

A complete tolerance model is therefore composed of quantitative design measurements, a comprehensive model function and characterized input variables. This comprehensive tolerance model becomes the basis from which tolerance analysis algorithms can be performed.

15.3   Automating Tolerance Analysis

While many tolerance analysis algorithms are simple enough to be applied without automation, there are great benefits in automating tolerance analysis calculations. Automating the analysis calculations can reduce effort and errors. Also, with automation, more advanced analysis methods can be implemented to provide greater accuracy than simple analysis methods.

The Worst Case and RSS methods discussed in Chapter 9, and the DRSS and SRSS methods discussed in Chapter 11 are all simple enough to be used without automation. For example, the RSS method is frequently used to solve simple 1-D tolerance stacks by hand. Very little data is required to use these four methods. The formulas for each of these methods only require tolerances, derivatives and, in some cases, Cpk values as inputs. Of course, these four methods are also easily automated by programming a computer spreadsheet or programming software code.

There are two advanced tolerance analysis methods that are not easily applied without some form of automation: the Method of System Moments and Monte Carlo Simulation. While both these methods are more complicated to implement and require more input data, both offer better accuracy and more capability than Worst Case, RSS, DRSS, or SRSS. Commercial CAT systems are generally based on one of these two methods. The next two sections will describe these advanced methods in detail.

15.3.1   Method of System Moments

The RSS, DRSS, and SRSS methods are all derived from a more general method, the Method of System Moments (MSM). MSM is a statistical method that estimates the first four statistical moments of a function of random variables. These four statistical moments are mean, variance, skewness, and kurtosis. MSM consists of four equations that relate to each of the four statistical moments. With the model function expressed in this form,

$$y = f(x_i), \quad i = 1, 2, 3 \ldots n$$

the four equations for MSM are:

$$\mu_1 = 0 \tag{15.1}$$

$$\mu_2 = \sum_{i=1}^{n} \left(\frac{\partial y}{\partial x_i} \right)^2 \mu_2(x_i) \tag{15.2}$$

$$\mu_3 = \sum_{i=1}^{n} \left(\frac{\partial y}{\partial x_i} \right)^3 \mu_3(x_i) \tag{15.3}$$

$$\mu_4 = \sum_{i=1}^{n} \left(\frac{\partial y}{\partial x_i} \right)^4 \mu_4(x_i) + \sum_{i=1}^{n-1} \sum_{j=i+1}^{n} 6 \left(\frac{\partial y}{\partial x_i} \right)^2 \left(\frac{\partial y}{\partial x_j} \right)^2 \mu_2(x_i) \mu_2(x_j) \tag{15.4}$$

where:

$\frac{\partial y}{\partial x_i}$ is the partial derivative of the function with respect to the ith variable,

$\mu_i(x_j)$ is the ith statistical moment of the jth variable, and

μ_i is the ith raw statistical moment of the function.

Eqs. (15.1 through 15.4) are the four raw moments of the model function. These four raw moments can be easily converted to mean, variance, skewness and kurtosis. The first equation is the mean shift, the second equation is the variance, and the third and fourth equations are related to the skewness and kurtosis, respectively.

Eq. (15.1), the mean shift, is included because the mean shift is not zero for the second-order version of MSM. The four equations given above are based on a linear, or first-order, Taylor's Series approximation of the model function. The four MSM equations can also be developed using a second-order Taylor's Series approximation. A second-order approximation improves the accuracy of the approximation for nonlinear functions. The trade-off with the second-order formulation is that the four MSM equations become much more complex. The four second-order MSM equations can be found in Cox. (Reference 3)

The RSS, DRSS, and SRSS are first-order MSM methods derived from Eq. (15.2), the variance equation. Taking the square root of Eq. (15.2) yields the RSS formula, a formula for the standard deviation of the model function. (See Chapter 9 for another derivation of the RSS formula.) Unlike the RSS, DRSS, and SRSS methods, however, MSM allows the input variable to be characterized by any statistical distribution, including nonnormal distributions. Note that the four MSM equations include the first four statistical moments of the input variables. These four moments are calculated from the probability distributions of the input variables.

In summary, MSM is an advanced tolerance analysis method similar to RSS, but more general. MSM adds the capability of nonnormal input variables and a nonnormal estimate of the model function. Also, if a second-order approximation is used, MSM can provide a more accurate approximation for nonlinear model functions. The computation time for MSM is very small. In addition, once sensitivities are calculated, only the four MSM equations need to be re-evaluated whenever the distribution characteristics of the input variables change. This quality makes MSM very attractive for rapid design iteration.

15.3.2 Monte Carlo Simulation

Monte Carlo Simulation (MCS) is another advanced tolerance analysis method. MCS is a statistical technique based on random number generation. For the MCS method, each input variable is characterized by a statistical distribution. A random value is selected from each input variable distribution and then plugged into the model function. The resulting function value is then stored. To simulate manufacturing, the process of randomly selecting the input values and then storing the resultant function value is repeated many times. The stored function values can be plotted in a histogram, used to calculate the standard deviation of the model function or used to calculate other metrics. The sample size, the number of times the simulation is run, determines the accuracy of the analysis. The larger the sample size, the more accurate the analysis. A typical sample size is 5000 assemblies. Obviously, this type of method must be automated.

In contrast to MSM, MCS does not use an approximation of the model function. No derivatives are required for MCS. This can be useful if the model function happens to be discontinuous. However, since MCS evaluates the model function many times, the computation time of MCS can be significant, especially if high levels of accuracy are needed. Also, if any input variable's distribution is modified, the entire simulation must be re-run.

Tolerance analysis benchmarks have been performed which show the first-order MSM method to have about the same accuracy as MCS with a sample size of 30,000 assemblies. (Reference 5) These same benchmarks showed the second-order MSM to have about the same accuracy as MCS with a sample size of 100,000 assemblies. (Reference 6) The accuracy and speed of MSM makes it a good candidate for CAT systems.

Table 15-1 compares the features of the two advanced tolerance analysis methods. Selecting which analysis method to implement between MSM and MCS is mostly a matter of determining whether the function to be analyzed is continuous. If derivatives can be calculated, MSM provides a solution that is more suited to design iteration because of its fast analysis. Furthermore, the derivatives used by MSM can also be used to automate tolerance optimization.

Table 15-1 Advanced tolerance analysis methods: MSM versus MCS

	Method of System Moments	Monte Carlo Simulation
Fast Analysis	√	
Nonlinear Analysis	√*	√
Nonnormal Inputs	√	√
Nonnormal Output	√	√
Discontinuous Functions		√

*Using a second-order approximation

15.3.3 Distribution Fitting

Distribution fitting is an important automation issue for the MSM and MCS tolerance analysis methods. A distribution must be fit to the output of both MSM and MCS in order for quality metrics such as sigma, PPM, DPU, etc., to be calculated. For the MSM method, the four statistical moments of the model function are fit with a distribution. For MCS, a distribution is fit to the histogram of the simulations. Distribution fitting is automated by using tabular data or numerical methods for known distribution types. The distribution types that are most commonly automated are the normal distribution, Lambda distribution, and the Pearson and Johnson families of distributions. (References 8 and 9)

In addition to fitting a distribution to the output of the MSM and MCS methods, the distribution types of the input variables must also be defined. Ideally, for the input variables, the designer can define specific distributions based on actual manufacturing data. If this data is not available, however, a distribution can be assumed from the tolerance value. For example, frequently it is assumed that variables are normally distributed, the mean is equal to the nominal, and the standard deviation is equal to one-third the tolerance value.

15.4 Automating Tolerance Optimization

One of the biggest benefits of automating the tolerance analysis algorithm is the opportunity to combine the automated analysis method with a tolerance optimization method. Tolerance optimization is the process of finding the optimal set of tolerances to meet certain design objectives. These design objectives might be assembly cost, assembly quality, and/or part quality. Tolerance optimization and allocation methods are presented in Chapter 11 and Chapter 14.

The analysis methods based on derivatives such as the Method of System Moments (MSM) have an advantage over Monte Carlo Simulation (MCS) with respect to optimization. These derivatives provide valuable information to optimization methods so that an optimal solution may be found quickly and efficiently. The MCS method has been successfully used with optimization methods, but in order to have reasonable computation time, sample sizes are usually set at 500 assemblies. Accuracy is sacrificed at sample sizes this small.

15.5 Automating Communication Between Design and Manufacturing

Automating the creation, analysis, and optimization of the tolerance model is the first part of the tolerance automation process. Automating the communication between design and manufacturing is the second part.

One of the main purposes of automating the tolerancing process is to reduce problems in the transition of a product from design to manufacturing. A major cause of transition problems is a lack of communication. Designers often don't understand manufacturing processes and capabilities. Manufacturing personnel may be unsure of the design intent and what is important to performance. These are the same issues addressed by concurrent engineering. Automating the communication between design and manufacturing is analogous to automating the application of concurrent engineering principles (Fig. 15-3).

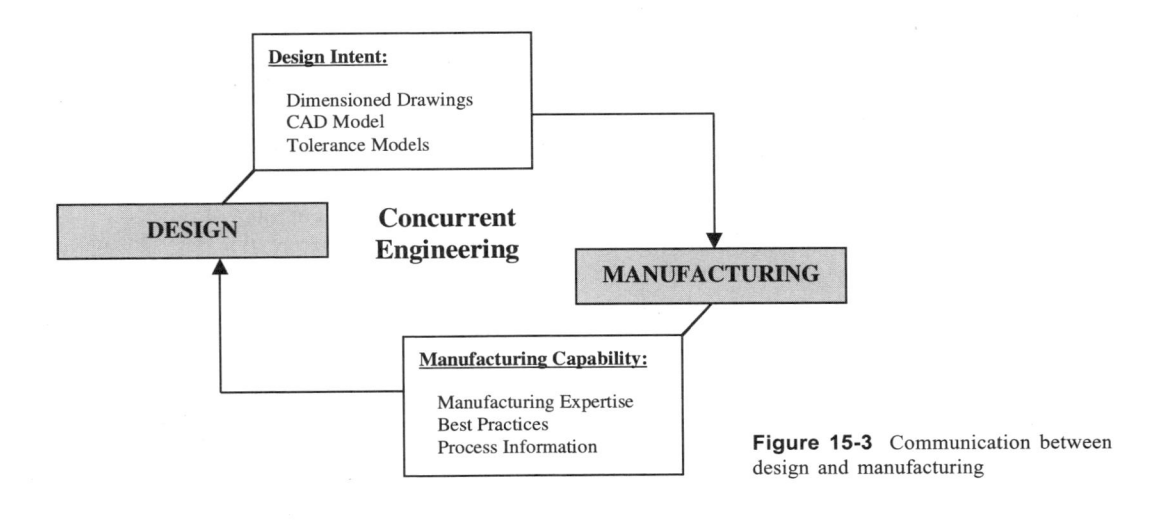

Figure 15-3 Communication between design and manufacturing

Improved communication between designers and manufacturing personnel can be defined in terms of deliverables from one group to the other. The deliverable from manufacturing to design is manufacturing process information. The deliverable from design to the manufacturing personnel is the product definition. The purpose of tolerance automation at this level is to simplify the delivery and use of these "deliverables."

15.5.1 Manufacturing Process Capabilities

A central tenet of concurrent engineering is that accounting for manufacturing capabilities early in the design cycle produces designs that are easier to build, less costly, and more robust. To accomplish concurrent engineering, designers need to understand what manufacturing processes will be used to produce the parts, along with the associated process capabilities. Giving the designers accurate process capability information allows them to predict approximate yields before production begins and to tailor their design to the available manufacturing processes.

Including manufacturing personnel in design teams is a common way to communicate process capabilities. (See Chapter 2, section 2.2.2.1.) Though effective, this is resource-intensive, often inconvenient to schedule, and may be overkill for some of the information needed by designers. Automation can simplify the transfer of some of the more common pieces of manufacturing information. One effective way to accomplish this is to provide the designers with a database of manufacturing process capabilities. (Reference 4)

15.5.1.1 Manufacturing Process Capability Database

Ideally, a database of manufacturing process capabilities should represent all the information necessary to make intelligent decisions about how to manufacture a design. It would include the types and capabilities of manufacturing processes used in-house. It would include the types and capabilities of manufacturing processes used by the vendors that supply the company with components. It would also include real-world application information, such as machine setup issues, fixturing, production cells, what machines can be used for various feature types, and rules of thumb related to manufacturing process planning.

As discussed in sections 15.2 and 15.3, performing statistical tolerance analysis requires characterizing the variation of the input variables of the tolerance model function as statistical distributions. By definition, a manufacturing process capability database would automate the characterization of the tolerance model input variable distributions.

Most companies do not have the resources to create a database of this caliber for their designers. However, it is realistic for most companies to characterize and catalogue, at a minimum, their manufacturing process capabilities and store them in a database. The knowledge of how to use that information to select manufacturing processes will still need to come from the manufacturing personnel. Once the manufacturing processes are selected, the designers will be able to use the manufacturing process capability information from the database to refine their design and check performance and producibility requirements.

To build a useful manufacturing process capability database, a company needs to look at its historical manufacturing process performance. Many companies have accumulated large amounts of process capability data through using SPC (Statistical Process Control) methods. Unfortunately, this data is usually not used effectively beyond the manufacturing floor. If process data is collected correctly, it can be used to form the basis of a process capability library. Proper gathering of data involves issues beyond the scope of this chapter. See Reference 7 and Chapter 17 for further details on collecting and developing process capability models.

15.5.1.2   Database Administration

The database form, organization, and location must be well planned to successfully automate the exchange of manufacturing process capabilities.

There are several formats that can be used to store the distribution information for each manufacturing process. The most direct is fitting a specific distribution to the process data and storing the distribution type and parameters. A second approach is to extract the first four moments from the process data and storing those values directly. This approach is especially appropriate if MSM analysis is performed. A third approach is to assume a distribution type and store a tolerance value and process capability index (Cp/Cpk). The distribution parameters are then derived from the tolerance and capability index values. Normal and uniform distributions are commonly used in this manner. Various combinations and modifications of these formats can also be used. The format selected may depend in part on what standard quality metrics the company uses. See Chapter 8 for methods of specifying statistical tolerances.

Manufacturing process capability data must be organized so that both designers and manufacturing can readily find the applicable manufacturing process information. For example, the data could be organized according to machine type, material type, feature type, feature size, and variation type (i.e., length or angular variation) for each manufacturing process. Additional organization factors might include vendor name, lead-time required, cost data, and surface finish capability.

Finally, the data must be placed in a location that is accessible to the designers. The most desirable setup would allow the designers to access the data from directly inside their tolerance analysis tool. This requires either that the tool itself provide an internal mechanism for storing a library of process information, or both the manufacturing process database and the tolerance analysis tool support a common database format. At the same time, the content of the data must be controlled so that it can only be updated by following a defined procedure.

15.5.2   Design Requirements and Assumptions

A second way to automate communication is for the designers to deliver a more complete definition of the design to manufacturing. Information frequently missing from the design definition is a tolerance model describing what design requirements are most important, and how those design requirements are affected by manufacturing variation. One of the products of the tolerancing process on a design should be a set of reusable tolerance models. The tolerance models and their results can then be delivered along with the rest of the design definition to manufacturing.

Providing tolerance models to manufacturing can help automate several critical production tasks. First, it helps automate troubleshooting manufacturing problems. The tolerance analysis model should identify both the design requirements and the driving dimensions (input variables). Each design requirement is driven by some critical subset of part dimensions. Not all part dimensions are relevant to a particular design requirement. When issues arise in meeting a design requirement, the tolerance model will provide visibility into what the primary variation contributors to the requirement are. This visibility helps automate finding the source of manufacturing problems.

Second, it helps automate predicting the impact of manufacturing process changes. The manufacturing processes used to produce a part may need to be changed in order to reduce costs, free up a specific machine tool for other production runs, or act as a substitute when the original machine breaks down. If manufacturing has access to the original tolerance models, they can pull up the relevant studies and change the assumptions to reflect the new process, and check conformance to the design requirements.

Third, it simplifies communicating design and manufacturing problems back to the designers. By using the same tolerance models, both design and manufacturing have a common frame of reference and can speak a common language when problems arise. The process of identifying the problem and finding a solution can be much quicker.

Fourth, it helps evaluate the usability of parts that are out of specification. For example, batches of parts may come in with mean shifts or excessive dimensional variations. With both manufacturing process capability data and a tolerance model accessible, the tolerance model can be updated to test the effect on the design requirements and see if the parts can be accepted.

15.6 CAT Automation Tools

Sections 15.2 through 15.5 discussed principles of automating the tolerancing process in terms of the creation, analysis, and optimization of tolerance analysis models, as well as methods of automating the transfer of information between design and manufacturing. The practical way these principles can be realized is by implementing them in a tolerance analysis tool.

There are a growing number of tolerance analysis tools marketed commercially, and even more that have been developed internally by various companies. Whether or not a specific tolerance analysis tool is suitable for a company's efforts to automate their tolerancing process is determined by the capability and usability of the tool.

15.6.1 Tool Capability

When selecting CAT tools, it's important to distinguish between specialized tools and general-purpose tools. Specialized tools are optimized for a specific type of tolerance analysis, such as optical lenses or electrical connector interfaces. General-purpose tools are generic enough to adapt to many common analysis situations — mechanisms, fixturing, assembly process variations, and others.

Defining the capability requirements of a tool requires understanding the common tolerance analysis situations seen in the company. Answering this requires conscientiously collecting information from a variety of designers and manufacturing personnel, and not simply relying on the judgment of one or two "experts" in the company. Individuals tend to develop tunnel vision about what types of tolerance analysis are important. It is important that a CAT tool comprehends the majority of the analysis situations and simplifies the current analysis methods.

While tool capability is very important, it is not the only criteria to consider when shopping for CAT tools. Several usability issues must be considered. In many ways, the usability issues eclipse the importance of tool capability. Sections 15.6.2 through 15.6.8 will discuss issues related to the usability of CAT tools.

15.6.2 Ease of Use

Ease of use is the single most important factor in determining the success of a CAT tool's deployment. If the tool is not easy to use, acceptance among designers and manufacturing personnel is unlikely. Defining what is easy to use is highly subjective, but several general characteristics should be considered.
- The user interface should have an intuitive layout. The information should be well organized with the most important data readily accessible.
- Model creation should follow a logical process that uses a clearly defined set of operations. The model creation process should be designed around a systematic approach that can be generically applied to a wide range of problem types.
- Model creation should be quick. Time is a scarce resource to designers. Few industries have the luxury of long tolerance analysis cycles. If the designers cannot quickly create a model, run the analysis, and get on to their next task, they are likely to use another means to analyze the tolerances or skip it altogether.

- The tool should have useful documentation. The tool's documentation is often the last place searched for answers to questions. However, when it is finally referred to, the user should find that the documentation is well organized and contains useful examples. The documentation should be available both on-line and as hard copy.

The importance of a CAT tool's ease of use cannot be overemphasized.

15.6.3 Training

The nature of tolerance analysis requires training. Tolerance analysis covers a wide range of specialized concepts: dimensioning, tolerancing, GD&T standards, optimization, statistics, mechanisms, kinematics, manufacturing, inspection, SPC, and others. The amount of training required is determined by the background of the trainee, the difficulty of the tool, the quality of the training program, and the complexity of the analyses to be performed. Purchased tools should provide training classes and materials. Companies that develop CAT tools in-house bear the burden of developing classes and materials to train its users.

15.6.4 Technical Support

The complexity of tolerance analysis guarantees that questions will arise about the use or behavior of a CAT tool. Extra assistance may be needed to understand problems in specific application situations. Software bugs will also occur. There must be resources available to answer the users' questions and assist in workarounds until fixes are available.

Commercially purchased tools should have a help line and a mechanism for distributing technical information (such as known bugs and workarounds). Help-line access usually requires a company to purchase a software maintenance package in addition to the tolerance analysis tool itself.

If tools are developed in-house, help-line resources must be budgeted yearly and skilled help-line personnel developed internally to support the users.

15.6.5 Data Management and CAD Integration

Computer-based tolerance analysis tools generate data files that must be maintained. Tolerance model files developed for a specific CAD model need to be stored with that CAD model. This may also be true of the analysis output files. To this end, the tolerance analysis files should integrate smoothly with the company's CM/PDM (Configuration Management/Product Data Management) system.

To help the designers achieve concurrent engineering, the CAT tool should work natively with the CAD system. The easier it is to keep the CAD model and the tolerance model in sync, the better. Having the CAT tool integrated with the CAD system also helps the manufacturing and quality control personnel find and use the tolerance models when they need them.

15.6.6 Reports and Records

Documenting a tolerance study and distributing the results should be quick and easy. The reports themselves should have a format that covers the important information. At a minimum, the reports should include:
- Output statistical/worst case variation plots
- Sensitivity/Percent contribution pareto of each performance or fit requirement to the part dimensions
- Part dimensions, manufacturing variations, and process capability metrics.

Reports need to be modifiable by the user. They should be output as straight text or another common format that can be easily read and edited by a word processor. Any graphic should also be output in a standard format that can be easily imported into a word processor.

15.6.7 Tool Enhancement and Development

It is unlikely that any existing tool on the market will meet all the requirements of a company. The CAT tool industry is still relatively immature and is changing rapidly. Therefore it's important to understand a CAT tool's future development path. Issues to understand include:

- What future enhancements are planned for the tool?
- Do future enhancements address all the outstanding issues (e.g., missing functionality) that the company has with the tool?
- Is there an effective mechanism for entering enhancement requests and bug reports?
- How rapidly is the tool being improved?
- If it is a commercial product, is the tool provider stable? If it is a tool developed in-house, does it have a stable funding source?

It is vital that the selected CAT tool is growing and the tool provider is reliable. If it is, the investment in a CAT tool has a far greater chance of delivering real returns to the company in terms of improved quality and reduced cost.

15.6.8 Deployment

The issue of deploying a CAT tool in a company is too large to address within the scope of this chapter. However, some questions that must be answered relative to deployment include:

- Who has responsibility for implementing the tool in the company?
- How much effort will be required internally to install and maintain the tool?
- Does the tool work on company-supported hardware and operating system versions?

In short, a deployment plan must comprehend all the infrastructure required to install and maintain the CAT tool.

15.7 Summary

Automation can provide great benefits to the tolerancing process. Through automation, tolerance model creation and analysis can be simplified and accuracy improved. The time it takes to develop an optimal dimension scheme for a design can be greatly reduced. Automation can also improve the communication between design and manufacturing and help develop a more concurrent engineering environment. Finally, careful consideration of the important capability and usability issues will enable the successful selection and deployment of tolerance automation tools.

15.8 References

1. Bralla, James G. 1996. *Design For Excellence*. New York: McGraw-Hill, Inc.
2. Bralla, James G. 1986. *Handbook of Product Design for Manufacturing: A Practical Guide to Low-Cost Production*. New York: McGraw-Hill, Inc.
3. Cox, N.D. 1979. Tolerance Analysis by Computer. *Journal of Quality Technology*. 11(2):80-87.
4. Creveling, C.M. 1997. *Tolerance Design*. Reading, Massachusetts: Addison Wesley Longman, Inc.

5. Gao, Jinsong. 1993. "Nonlinear Tolerance Analysis of Mechanical Assemblies." Dissertation, Mechanical Engineering Department, Brigham Young University.
6. Glancy, Charles. 1994. A Second-Order Method for Assembly Tolerance Analysis. Master's thesis. Mechanical Engineering Department, Brigham Young University.
7. Harry, Mikel, and J.R. Lawson. 1992. *Six Sigma Producibility Analysis and Process Characterization.* Reading, Massachusetts: Addison Wesley Longman, Inc.
8. Johnson, N.L. 1965. Tables to facilitate fitting S_U frequency curves. *Biometrika* 52(3 and 4):547-558.
9. Ramberg, J.S., P.R. Tadikamalla, E.J. Dudewicz, E.F. Mykytha. 1979. A Probability Distribution and Its Uses in Fitting Data. *Technometrics.* 21(2):201-214.
10. Stoddard, James. 1995. Characterizing Kinematic Variation in Assemblies from Geometric Constraints. Master's thesis. Mechanical Engineering Department. Brigham Young University.

Chapter 16

Working in an Electronic Environment

Paul Matthews
Ultrak
Lewisville, TX

Paul Matthews has been practicing mechanical design engineering for the past 12 years. In his 10 years of experience with Texas Instruments, he was part of the design team for the F-117 Stealth Fighter infrared night sight and a major author of the Mechanical Product Development Process for the Defense System and Electronics Group. At TI, he gained a high proficiency at 3-D solid modeling using ProENGINEER and developed several standard best practices for modeling and data management. For the past two years he has been employed as a design mechanical engineer and division director at Ultrak, specializing in the design of larger volume commercial and professional security-related CCTV products.

16.1   Introduction

One question I've dealt with as a mechanical engineer is: "Why generate so many paper drawings and documents to get a product built?" A simple answer to this question is to provide a manufacturer information on how to make the product parts and assemblies. However, a more important and often forgotten reason is to *make a profit for the company that pays me.*

I get paid to design and build a product to sell. In today's environment, if I can't accomplish this faster than my competition, I might as well not do it at all. If I'm really paid to produce a product faster and better than my competition, will I have the time to generate 2-dimensional (2-D) paper documentation to capture the 3-dimensional (3-D) design information and notes referred to in the previous chapters? Will I ever consistently generate a drawing that everyone in the product life cycle interprets the same way? And will this drawing provide the information necessary to build the component? Even if I did, does a manufacturer use this information in a way that helps an improved product move faster to market?

The main reason for writing this chapter is to give you ideas for *capturing and sharing design information to manufacture products with minimal paper movement.* The ideas presented here are not limited to drawing dimensions and tolerances, but include all information associated with the product development process and the data formats used to better support today's rapid product development and production.

16.2 Paperless/Electronic Environment

16.2.1 Definition

I've been in several situations where design programs advertise hours saved by going to a paperless design and manufacturing environment. When asked how they do it, the responses usually indicate that drawings are transferred to the manufacturing facility by modem, e-mail, or LAN-based communications. After the drawings are downloaded, the manufacturing engineers print the files and pass the paper to the next person in the process. This saves numerous hours compared with the hand delivery of the same paper drawing. Yet this does not reflect the true meaning of "Electronic/Paperless Environment" that I want to discuss here. There's more to this environment than the speed in which electronic data can be transferred from point to point.

An electronic environment process has two distinct functions:
- To capture the design and manufacture information in a data format best suited to the person making the decisions for the particular process step.
- To share and reuse the captured information in concurrent engineering for later steps in the process.

For many of the designs done in industry today, this data format is a computer-aided engineering (CAE) database; a 3-D computer aided design (CAD) database, and various other formats for supporting notes. By putting less emphasis on paper documentation and more emphasis on a well-documented concurrent design/manufacture data capture and share process, the cycle time, cost, and quality of new designs is improved.

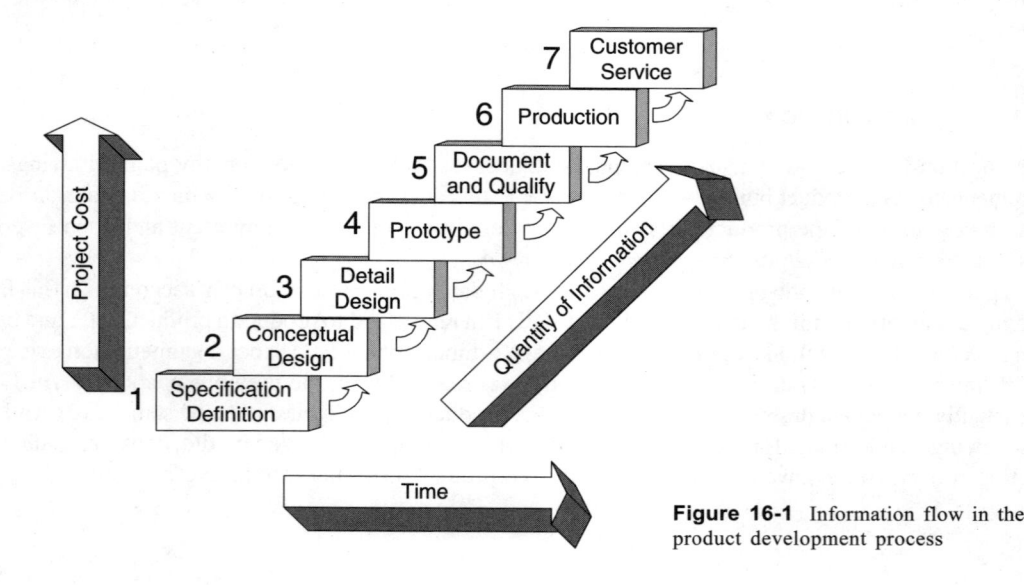

Figure 16-1 Information flow in the product development process

A typical product development process is shown in Fig. 16-1. During the product development process, the quantity of information increases rapidly and each prior process block's information supports the process block above it. The majority of this information is in several types of computer formats and each separate block in the process represents not only a process step, but possibly a different person, department and even company completing the task. It is critical to the process that this information is captured and seamlessly shared from block to block. As seen in the figure, the bigger the information overlap on the blocks, the shorter the time and inherently the increased strength of the product design process.

16.3 Development Information Tools

What we all want to do is make the product development process better. To make the process better, we need to capture and share design and manufacturing information in the most efficient way possible. The most efficient way, for some companies, is to use paper and pencil and many manila folders to navigate information through the development process. For the majority of the competing companies in the marketplace, the computer is used to help guide the information flow.

This section describes several techniques to help the product team with design and manufacturing information in electronic forms.

16.3.1 Product Development Automation Strategy

Electronic automation is a simple concept for most companies today. The best automation is generated from a simple idea put together with other ideas to form a completed tool. It starts with something known and builds on solutions until the requirements are met.

What generates a good automation solution?

- **Product Process Requirements Knowledge**

The product process must be defined. Often companies build automation and then figure out how the process needs to flow to use the automation that was constructed. Inherently, this forces the automation and process to iterate until a common compromise on both automation and process is met. Clearly, successful companies know what information is needed during the product life cycle and what the process needs to be to support the capture and flow of the information. The automation of the information flow becomes very well defined and simple to implement.

- **Automation Experience**

Solid experience is critical. To know when something worked before (or didn't work!) enables automation designers to think ahead and not waste time pursuing paths that will dead end later. A new technology is always alluring to automation designers, but may not be the best solution to the problem. Experience, with not only the latest and greatest technologies, but also the tried and true technologies, will usually generate the best solutions.

- **Process Tool Proficiency**

Tools are meant to help someone complete a task. When a person who generates automation is proficient in the process tool that the automation is designed for, the automation is stronger. The proficient tool user enhances the features in the process tool and does not construct the automation to force the desired outcome. A simple example is a person writing a Visual Basic script to add up a column of numbers in a spreadsheet program. Obviously, the spreadsheet program has built-in functions to do this task and a script would be foolish.

- **Imagination**

Without the ability to solve a problem in many different ways, automation designers can get easily stuck. There is always a way to complete the desired task. If you don't think of the best way to do it, your competitor will. Don't underestimate the importance of this point. Most often, the simple obvious choice is the right choice. In those situations, when the obvious choice does not produce the desired outcome, the automation designer needs to think outside the confines of previous solutions. Here is an example of a problem and a solution.

Process step: During this particular product development process step, a design team member is responsible for providing a marketing team member with a photorender of the new product for marketing literature, such as an advertisement for new company products.

Problem: The new product's 3-D solid model is so complex and has so many features, the photorender software used to automate this process step will not run to completion on the current computer system.

Solution: The automation designer develops the parameters associated with this size of the solid model and flags solid models this size or larger as candidates for Stereolithography and paint. After the scaled model is built and painted, a real picture can be taken.

In this example, the automation designer has the ability to think outside his expertise for a solution to the problem. A more powerful computer helps (by the way, you can never have enough!), but for this particular company, it was not cost justified for the number of products that fell into this category.

- **Automation Flexibility**

No product development process will remain fixed long enough to develop a full set of automation support. Automation that is built to endure modification in the process is very costly and almost impossible. The process must be able to change with the company's growth and expectations. When the process changes, the automation must be updated to support the change without major rework.

- **Support**

Like any tool, automation requires maintenance and repair. Support personnel are required to keep the tool current with the process and also with changing technologies. Automation that is left alone will slowly wilt like a plant without water. The difference is that the plant will show signs of fatigue, where the tool will just stop growing with the process. The first sign of trouble is when the product competitors beat you to market with better designs.

- **Luck**

Luck is a relative word. Anyone who claims they can control product development team expectations, keep key employees from leaving the company, and prevent lightning strikes to the main computer, has had incredible luck in their career. I prefer to anticipate bad luck (even expect it) and always be ready to regroup and attack.

The above concepts together create good process automation. Keep in mind, automation is not the most important point here. The main effort with any automation is to support the process that needs the automation. A tool never dictates what a process should be.

16.3.2 Master Model Theory

As computer software becomes more advanced, it enables the design team to capture more information into a single database. This single database is referred to as the master model. The information captured in this database appears in many forms. Some are listed in Table 16-1.

The master model is the controlling design database, capturing all relevant design data in one central location. The key to the master model concept is to generate the design and manufacturing process based around a focused design data set and use this master set to generate all supporting documents. Once captured, other engineering and manufacturing disciplines reference this information in formats best

Table 16-1 Information captured in a database

Information Type	Description
Graphical Data	The nominal geometrical representation of the design.
Graphical Data Attributes	Geometry attributes such as line colors, widths, and visibility.
Dimensional Attributes	Dimension and tolerance attributes associated with the geometry. Dimensional attributes provide the scale of the geometry.
Design Notes	Notes and design calculations used in the product process that may be needed for future revisions of the product.
Parameter Data	Information such as cost, part name, designer name, part number, material, and design revision are a few examples. The number of fields of parameter data can be quite large and provide excellent process automation opportunities.
Software-Generated Parameters	Calculations done by the software using designer parameters and attributes as inputs: mass properties, number of parts in an assembly, and measurement calculations are several possibilities.
Manufacturing Process Data	Manufacturing specifications needed to complete the fabrication of the design. Material finish, packaging/shipping requirements, surface roughness, special tool requirements, and regulatory conformance requirements are examples.

suited for what they need during any particular process step. When the master model is updated, supporting information is updated concurrently, with little interpretation. This update process can be very efficient if automated.

Fig. 16-2 shows a simple example of when the engineer decides to add a screw to an assembly. The most logical place for this to take place is in the CAD model, where he parametrically adds the screw model into the CAD database. The database is considered the master model in this case. Other documents are linked to this master model, and because of this, are directly updated with the new information. The principal point here is that all the other product design disciplines know to look at the master model for

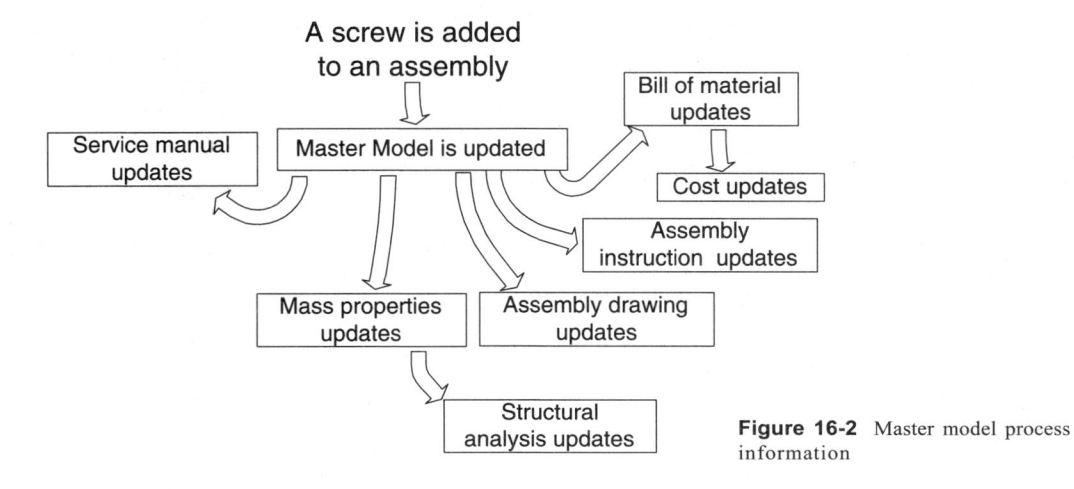

Figure 16-2 Master model process information

changing information. Once again, if this process is automated, very little effort is needed for this change to be cleanly incorporated across the product design group.

There are many examples of how the master model can be used in the product design process.

- Computer Aided Process Planning (CAPP) software for the manufacturing process uses the master model as the seed for generating detailed work-flow estimates and numerical-controlled (NC) code for machining.
- Purchasing may use the master model source as a guide for ordering purchased hardware for the assembly.
- The structural analysis of a part may automatically be recalculated for updated geometry. A document may be autogenerated showing inspection dimensions that fall below a certain process capability of a machining center.
- The tolerance analysis may be directly linked to the solid model CAD database, so that when the tolerance is changed in the model, the analysis is automatically updated.

Theoretically, information is captured one time in a single database file by one software program used by all disciplines of the product development process. In reality, this is unfortunately not the case. A printed circuit board assembly (PCBA) design is a good example. A PCBA will have a mechanical database to specify packaging constraints constructed in one CAD software, electrical schematic data to define the circuit in another CAD software, a circuit board layout for the etch runs, bill of materials in a third software, and possibly simulation data in a fourth. There are also numerous soldering specifications, material specifications, component data sheets and any other referenced document. All of these together capture the design intent for the product. One of the most important pieces to the success of the product process is to know the master model or master data set, and let this single data set control the design automation and reference.

The following is an example of a very common occurrence that illustrates the importance of the master model:

I used ProENGINEER™ solid modeling software to create the design database. It was common practice to take the 3-D solid ProENGINEER™ files and convert them (using a DXF conversion standard) to 2-D AutoCAD® files to generate the drawings. These drawings were taken to the shop where 3-D Computer Vision (CADDS4X) databases were generated to create the NC program. Remember the design database (master model) was ProENGINEER™.

Here are the problems:

- The design was interpreted five times, with each conversion moving farther away from the designer's thought.

  Designer thought → 3-D CAD→2-D Drawing→3-D CAM→ NC Program→Inspection
- When making changes, the change was updated and interpreted in at least four different databases. If the parts were measured with a coordinate measuring machine (CMM), this adds another interpretation.
- Each step in the process may have a different owner, department, or in some cases company involved to complete the process step.

This simple idea can provide a powerful tool for automation and a strong product process information set. Concentrate on the fundamental purpose behind the master model: *Focus all product team members to a common data set.* When the product team can quickly and easily find the needed information in a convenient format, the development process will flow smoothly.

16.3.3 Template Design

The most powerful technique for product development is the ability to quickly reuse information from past experience. In my opinion, 80% of all product design work has been done before, and when a company can capture this history and standardize it to boost new products, the company is successful.

Templates can be generated for everything. A template consists of known information that is formatted in such a way to enable the person using it to supply only minimal bits of new information. The template is complete when all the missing variables are supplied. This concept is critical in the product design process. It not only aids in the capture and format of information, but it tells the user when they are done and can go on to the next task. In the electronic environment, templates are linked to provide easy access and update to the master model.

Template strategy is important. As with any product development tool, the tool or template must directly support specific tasks in the process. Not only does the template need to support the process, it needs to properly link and reuse the information with other templates or tools in the process. Common variable attribute names should be generated and used to ensure the compatibility and consistency between the tools. The following list shows a basic procedure for generation of templates.

1. Define and document the complete product development process.
2. Determine the flow objects needed to complete the process. Flow objects are considered the bits of information passed from one process step to the next, the inputs or deliverables of a particular process step. Think of flow objects as the baton passed to the next runner in a relay race.
3. Generate the list of variable names or parameters needed to efficiently define the flow objects' information.
4. Group the parameters using timing requirements or functional disciplines. As an example, cost, size, and weight goals need to be known at the beginning of product design. Usually, marketing determines these constraints based on customer demands or expectations. The designer uses these goals as requirements during the design of the product and, during the design process, updates the parameters. This group of parameters (cost, size, and weight) begins with a marketing function and flows to the designer for ownership and update.
5. Capture the parameters or attributes in a template format best suited for the person making the decision. Once the parameters are captured, reformatting for reuse into other templates later in the process should not be a problem. The goal is to have the person who makes the decision enter the information only once.
6. Test the process templates. Remember my comment about luck earlier in the chapter. The product development process will change as fast as you generate these templates. Don't focus on designing the perfect process or the perfect set of parameters. Design the process, templates, and all other tools to be flexible to change. The idea is to improve the design process using a consistent means of capturing and communicating information, not to overly constrict or require data that has no positive effect on the design process.

Define→ Determine→ Generate→ Group→ Capture→ Test

16.3.3.1 Template Part and Assembly Databases

There are many feature-based CAD tools on the market today. A feature-based tool allows the user to build the geometry and design requirements by parametrically adding up small mathematical features into the final, sometimes complex database. When using these types of tools, the user does not have to start modeling the design from the first feature. This is not always obvious. However, many times parts and assembly databases have common information based on the classification of the model. By capturing these common elements and putting them in data models, you define templates.

A template part or assembly can be used to capture common information or modeling technique into a *starting database* to jump-start the model. These model databases are declared standard and are used as the base elements of a design. Since these elements are predefined, automation can be easily written to retrieve information needed.

Templates should not be confused with library components. The templates are starting points of a new design, where a library component is a complete configured data set that is not changed during the product development.

Common elements for a template database were shown in Table 16-1. Table 16-2 adds more detailed descriptions and suggestions for these elements.

Table 16-2 Examples of templates

Information Type	*Template Examples*
Graphical Data	Common starting geometry such as a cylinder for a lathe part or a rectangular chunk for a hog-out
Graphical Data Attributes	Defined entity colors and feature or drawing layers. Standard views such as front, back, right, left, top, bottom, and isometric
Dimensional Attributes	Standard dimensional scheme or modeling practice. Defined datum planes for the associated geometry. Standard units such as inch or millimeter. Material values such as density.
Engineering Design Notes	Engineer's name, employee number, computer name, and design location. References to other designs with similar characteristics.
Variable Attribute Data	Part cost, part name, part number, material description, design revision, drawing number, part title/description, revision level, current mass properties, vendor number, and customer number are a few examples. File attributes such as size of database, database location, and last modified date.
Software-Generated Parameters	Mathematical relationships in the database. Formatted mass property reports. Equations that may calculate estimated cost based on parameter information supplied during the design process.
Manufacturing Process Data	Standard material finishes and specifications. Reference to a standard tool list or feature list used for geometry generation. Tolerance limits for process capability calculation. Common raw material or stock parts.

16.3.3.2 Template Features

Similar to template parts and assemblies, common features can be generated and put into libraries to be shared by all. Often there are common feature groups that can be inserted into the model as a set. A common example would be two pinholes for location of a part to a mating part. The holes can have the correct tolerancing and dimension and also reference the correct pins to use in the assembly. Library features can have built-in knowledge parameters to pass on information such as cost of machining operations, process capabilities, NC machine code, tooling list, and design guidelines for using the particular feature. With this information available to the designer, the designer has the immediate ability to know the impact of using the feature before the feature is designed into the product. The designer also does not have to spend any extra time locating information that could easily be supplied as a parameter or attribute.

16.3.3.3 Templates for Analyses

It is very unlikely a designer will do an analysis new to the industry. I must have 30 spreadsheets that I've generated or acquired that perform specific design-related activities ranging from tolerance analysis to trade-off analysis of cost and scheduling of a new product. Once again, a company's success is dependent on the ability to use its resources to generate these common templates and build them into standards. Once standardized, electronic information can be shared between product team members for efficient design and manufacture of products.

16.3.3.4 Templates for Documentation

One of the most common uses for a template is a drawing. As seen in Chapter 4, drawings are made up of various elements put together to define a particular product. For commercial products, there is a limited number of manufacturing processes, materials, and drafting rules to generate product documentation. It is very possible to generate complete documentation directly from a master model with little or no user input. Current Internet and Intranet technologies can generate these pieces of documentation in the background without any designer effort.

Other common document templates used by other product development team members are shown in Table 16-3.

Table 16-3 Common document templates

Engineering Change Notices (Requests, Proposals, etc.)	Assembly Work Instructions
Material Requests	NC Machine Programming
Purchase Requisitions	Service Manuals
Marketing Information	Quality Control
Manufacturing Instructions	Budgets, Schedules

16.3.4 Component Libraries

Component libraries are very powerful resources for the product design team. Not only can the library provide a CAD model; it can include all necessary data associated with the respective library component. All parameters and attributes should be set to reflect all needed information about the component. With this data captured in the component, it is available throughout the development of the product.

When capturing components for libraries, keep in mind the following:
- Geometry must reflect the component as accurately as possible, but not provide so much detail that the application software is overloaded. As an example, an actual helical thread on a solid model of a screw is most likely too detailed.

- Geometry should be modeled at the mean of the manufacturing process. This is usually the center of the tolerance zone. To illustrate: A bearing which may be specified at .437 +.000/-.014 should be modeled at a process mean dimension of .430 ± .007.
- The attribute data must be correct and under configuration control so as not to be inadvertently changed.
- Library components should be controlled from a central distribution area for ease of update and configuration.
- Library components should be verified with any application software revision.

16.3.5 Information Verification

Information is easily entered *incorrectly*. Companies are increasing their dependence on the information captured in complex Master Models to support concurrent product development and manufacturing. The current problem with this dependence is the possible lack of control and verification of this information. Questionable user proficiency in the tools, growing product development processes, and constant change in personnel complicate the standardization, completeness, and integrity of the design data. In turn, the cost and quality of the developed products suffer.

Mechanical solid modeling tools are very powerful. Along with the strength and capability of the tool comes the complexity of the tool use. In my 10+ years of mechanical design using ProENGINEER™, I have seen many models that have grown into complex webs of features. One of the main issues is that the person modeling the design may not recognize the problem. Often, these designs were released for production without any verification to corporate modeling standards. After several weeks, when the design needed to be updated, the complex model was virtually destroyed in the process of update.

All product development data should go through an automated verification process prior to process step acceptance. This information can be used to determine schedule milestones, resource requirements, and verification of clean information flow to the next product development team member.

The following shows a few common examples of corporate standards to verify and document in a solid model to keep consistency in the quality of the databases:

- Adherence to corporate modeling standards
  - ✓ Model was started with a common template.
  - ✓ Corporate standard-defined features are used.
  - ✓ External references to other geometry are controlled.
  - ✓ Tolerances are correctly attached to features.
  - ✓ Parameter information follows corporate standards.
  - ✓ Model name convention follows data management standards.
- Model Completeness
  - ✓ Number and type of features reflect completeness of design.
  - ✓ Material has been defined.
  - ✓ Complexity of model.
  - ✓ Proportion of sketch dimensions per feature measures model complexity.
  - ✓ Number of parent/children features measures model dependence complexity.
  - ✓ Number of mathematical relations in the model shows design-captured information.

- ✓ Family tables or grouping information displays family parts.
- ✓ Regeneration or rebuild time helps determine computer hardware requirements.
- ✓ References to other data forms show relationships to other information.
- ✓ Total database file size helps determine archival requirements.
- ✓ Proportion of physical size of model versus physical volume gives insight into fabrication costs.
- Integrity of model database
  - ✓ A regeneration error list helps determine problems in the model.
  - ✓ Dimension values less than .01% of the model size can help determine questionable design.
  - ✓ Suppressed or hidden features list can determine modeling mistakes.

16.4   Product Information Management

The management and control of the product data is the key to a successful electronic environment.

A paper document is a fairly easy item to keep in revision control and requires very little knowledge to handle. A database, on the other hand, requires knowledge of the database format, knowledge of the software used to extract the required data, and hardware to support the electronic media. Many lawsuits have forced society into legal document frenzy. Okay, maybe I exaggerate a little. But no doubt, having a fully dimensioned, fully toleranced, printed drawing, makes any fabrication shop a little happier. The manufacturer wants to point to a piece of paper and say, "That's what I built." The drawing, then, acts as the common interface, the legal binding document, between the designer and the fabricator. There are several main elements to consider about product information management:

- The product team will NOT use an information management tool that inhibits the development process.
- The developing product must be defined well enough to fabricate and verify.
- Product data must be in a format that is supported throughout the life of the product.

There are several ways to manage the configuration of the product documentation. Each of these methods should be used to ensure the electronic data is under configuration control. The Master Model Theory really comes into play in this task. To have only one place to update and control information is much safer than several different places.

16.4.1   Configuration Management Techniques

Configuration and control of information is big business for many companies. There are hundreds of software developers selling their information management products. Each of these tools is designed to support a data management process, as suggested in section 16.3.1. To select the correct tool for the development team, choose the tool that supports the team's process.

Remember that the best tool for a job is the easiest and simplest to use to get the job done. This may result in no automation tool at all. If the product team understands the importance of data management, less formal control is needed and the data is instinctively controlled. On the other hand, if the team does not understand the importance, the process and associated tools need to be strict and authoritative to assure data is not inadvertently damaged.

16.4.2 Data Management Components

There are a few simple components to a data management philosophy. Fig. 16-3 shows the hierarchy and descriptions of these components.

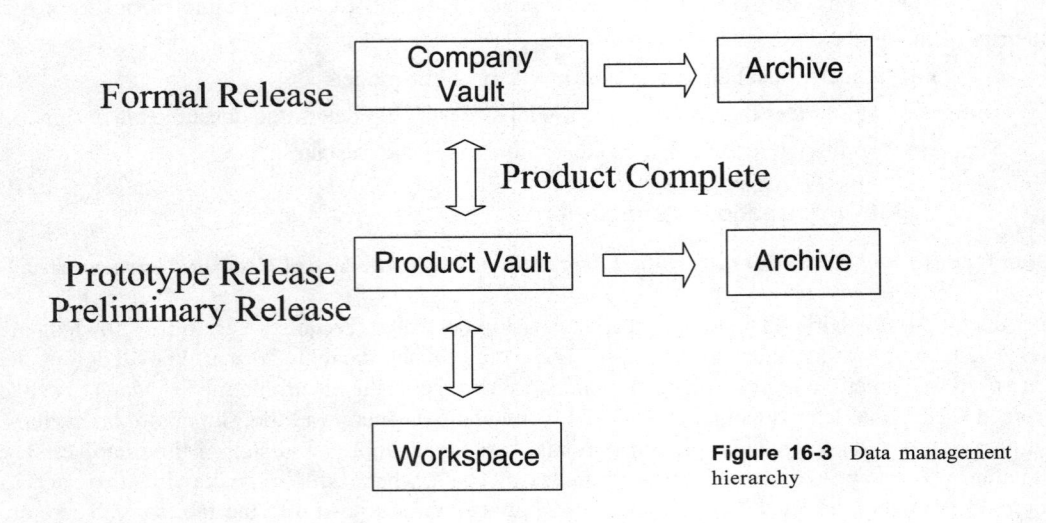

Figure 16-3 Data management hierarchy

16.4.2.1 Workspace

The workspace is the area where the daily development efforts take place. This is where the work is moved for update and the addition of information. Think of this area as the desk where the work is being done. The contributing development team member has full control over the data. They are responsible for all changes to the data and are also responsible for putting the data back at certain levels of completion. There is only one version of the data kept at this level and it is normally not archived or controlled.

16.4.2.2 Product Vault

A product vault is a place where the data is kept and controlled for the product. Multiple revisions may be captured and managed to ensure the product data is current and available to the complete product team. At this level, the data is archived for safety. Release levels may be set to ensure particular revisions, such as the release for a prototype part, are kept, When a particular part of the data is considered complete, it can be put into a preliminary release status to make sure it does not change while it waits for promotion into the company vault. This level may be thought of as a special locked office in the product area where everyone puts their information at the end of the day.

16.4.2.3 Company Vault

Formal release procedures are in place to submit data to the company vault. This level gives the entire company access to the information. Strict change management is in place. This level is archived at the company level to ensure the product data set is not lost or corrupted. The company vault is a crucial component because product development teams may not remain intact after the product is released.

16.4.3 Document Administrator

In any orchestra, there is a conductor. For data management this conductor is the document administrator. The focused effort of this product team member is to manage the data. This is not just a policing effort, or a sign-off block on a print, but a detailed understanding of data that emphasizes wrapping up the data in a consistent package. Verifying the file formats, modeling and documentation standards, release levels and where the data is stored are all responsibilities of the document administrator. This is a perfect application for the information mentioned in section 16.3.5.

16.4.4 File Cabinet Control

One of the simplest, lowest cost and most effective approaches to data management is the concept of file cabinet control. In a paper world, this would equate to (as the title suggests) a file cabinet. Each drawer on the file cabinet can be locked and unlocked by different people on the development team. Each paper folder in the cabinet drawer may represent a different revision of the product. In the computer world, this translates to folder permissions, computer access, and database filenames. Directory levels are set up to match with appropriate permission levels. This method may become cumbersome with larger product teams and higher administration efforts, but is very effective for small and medium product development efforts.

16.4.5 Software Automation

Product Data Management (PDM) software is available in many different levels to support the processes mentioned. The cost and level of detail on these packages range from low, such as a simple program used to copy the data to a different area, to very high, such as a total data management system that supports an entire company worldwide. Remember that no automation at all may be the best solution for the development team. Rely on the product development process to help pick the appropriate automation.

16.5 Information Storage and Transfer

The capture of product information is important, but without the storage and distribution of the information, the process comes to a halt. This section describes some of the most common information storage and distribution methods available. The world is changing fast in this area, and new methods and techniques appear every day. Don't limit the product team by what method has been used in the past and don't forget to support the development process with the methods you choose.

16.5.1 Internet

The World Wide Web (WWW) has grown enormously in the last few years. Many companies have both an Internet (outside the company security) and an Intranet (inside the company security).

The company Internet (outside) usually supports information distribution for the customer of the products. This allows very easy access and distribution of product specification information, troubleshooting tips, costing and sales-related information, software upgrades and patches, and many other customer-related service elements.

The company Intranet (inside) supports information and distribution of information for internal company use. Phone lists, human resources procedures and policies, technical data, and product specific development efforts are just a few examples. The Intranet is internal to the security of the company. Usually a firewall device inhibits outside hacking and provides the necessary security.

Both the Internet and Intranet are powerful with today's electronic information. When generating these systems keep several points in mind.

- Keep the focus of the system on the support of the process.
- Make sure there are support resources after the initial posting of information.
- Advertise where the information is located.
- Allow the structure and organization of the system to change with the process.
- Don't be scared to try new system technology.

16.5.2 Electronic Mail

E-mail has become one of the most used (and abused) forms of information transfer and distribution. Unlike an Intranet, information is pushed to the recipient but not able to be pulled when needed. This electronic communication is incredibly fast and convenient by allowing files to be attached with text and sent around the world in a matter of minutes.

There are several points about the use of e-mail.

- The e-mail you send can be intercepted and read by someone who really wants to get the data.
- E-mail is convenient, quick, and powerful. I sometimes find myself reading 10 to 20 e-mails daily addressed to "GROUP EVERYONE" sharing how someone in a different group may be leaving an hour early from work. Be aware of the groups you are sending the mail to and make sure the data is relevant to that group.
- The data you are sending may not necessarily be archived or kept. e-mail is like a paper letter that may get filed or thrown away.

16.5.3 File Transfer Protocol

Most transfers of files on the Intranet are transferred via FTP. Once connected to the Internet, this protocol allows not only getting data (use the command GET), but also putting data (use the command PUT). There are many software applications that support FTP and make it look and feel like a standard Windows-type program. If an application of this type is not available, a generic FTP program comes with Windows 95 and Windows NT; you guessed it, it's called FTP.

To GET or PUT a file using FTP follow these steps:

1. Logon to Internet
2. At a command prompt type: FTP HOST COMPUTER. The HOST COMPUTER is the FTP server with which you want to communicate.
3. Provide the appropriate login and password. For many servers you can use ANONYMOUS for the user and your e-mail address for the password.
4. Type BINARY. This sets the transfer mode to a binary protocol which will correctly transfer most files.
5. Type STATUS. This gives you status of the transfer.
6. Use GET to get a file from the server, PUT to put something onto the server.
7. EXIT logs off the server. QUIT leaves the FTP program.

16.5.4 Media Transfer

Transferring over the Internet is the fastest way to transfer data around the world. There are many times when a vendor or supplier does not have access to the Internet and another media needs to be used to transfer the information. Here are several media types commonly supported.

- CDROM. Writable CDROMs (WORM - write once read many) are very convenient for media data transfer up to about 650 megabytes. Almost every computer has a CDROM and can read the data. CDROMs are excellent because the data sent won't be accidentally erased or changed. There is a permanent record of what information was sent. Although CDROMs are common, there are different formats for the data. It is necessary to know which CD format is most versatile.
- Tape. There are many tape archive formats available ranging from 400 megabytes to more than 4 gigabytes. Although the tape can hold a lot of data, the data retrieval is cumbersome and slow.
- Floppy Disk. The 3.5 inch floppy is supported everywhere. It will hold up to 1.4 megabytes, is small, and very cost effective.

These different media are all useful, but the most powerful tool used during transfer, both electronic and by shipping media, is the ability to compress the data. There are different data compression algorithms and tools, but the most common are Zip utilities by PKWARE®. It is not uncommon to compress ASCII data formats by 80% as well as adding security encryption at the same time.

16.6 Manufacturing Guidelines

This book is titled as a dimensioning and tolerancing handbook. The chapter so far has delivered suggestions associated with electronic data; how to use it, control it, –and automate it. This section is devoted to providing some guidelines and best practices associated with the mechanical engineering development process, specifically the transfer of information to manufacturing for fabrication.

16.6.1 Manufacturing Trust

The most important aspect of working with a manufacturer and electronic data is *trust*. The customer must trust that the vendor will do their best and the vendor must trust that when they do their best, the customer will be satisfied. More often than not, a manufacturer will require a detailed drawing for inspection of the finished part. They do not necessarily need the drawing, but need the legal document to cover themselves if things do not go as planned. In the following sections, trust is a major element. Some of the new prototyping and manufacturing processes are higher risk to get a better delivery schedule or cost. The higher risk processes are more likely to have problems, and when the problems come up, the manufacturer needs to know he is part of the product team.

Another point to make in this section concerns the inspection methods used by the manufacturer. Although there will be some inspection to stabilize a production process, the movement of manufacturing is to verify processes. What this means is that the tolerances are not inspected if they fall within the manufacturing process capability. Only the tolerances outside the manufacturing process capability are verified and therefore only those tolerances and dimensions need to be relayed to the inspector.

16.6.2 Dimensionless Prints

A common compromise to no printed documentation is a dimensionless print. Basically, views are put onto a drawing format with dimensions and tolerances outside the process capability shown. Specific notes and processes are also captured on the print to allow easy access on the shop floor. This lets the database control the programming and majority of the features, yet allows paper control of inspection, notes, and processes. This also provides a printed document that can be used for better communication between the shop and change control.

CAD/CAM feature-based modeling software is able to capture tolerances associated with feature dimensions. Prior to passing a manufacturing database to NC programming, all dimension tolerances should be set to the mean of the manufacturing process, which is usually the center of the tolerance zone. This will force the geometry to regenerate at its nominal size and therefore the NC program will be written at the mean of the manufacturing process.

There are several standard pieces of information needed on a dimensionless print. These are usually called out in notes or in the title block of the drawing.

- Material. Specify the manufacturing material.
- Finish Processes. Specify processes such as heat treatment and surface finish.
- Manufacturing Process. Specify either the actual manufacturing process (possibly the machining center) or the general tolerance that drives the manufacturing process. A sample note may read, "All features in true profile of .030 relative to datums A (primary), B (secondary), and C (tertiary)."
- Marking Requirement. Specify any particular marking done on the part after finish.
- Design Model. Specify the 3-D model to be used for the geometry. Make sure to include enough information to clearly specify the exact model.

16.6.2.1 Sheetmetal

Many of today's commercial parts are designed and fabricated using sheetmetal or sheetmetal techniques to deliver the product in a fast, cost-effective manner. One reason sheetmetal has such success is the relatively limited number of machine operations that can be done on it in a production environment.

Sheetmetal comes to the manufacturer as a sheet, as the name suggests, and from there it is cut, punched, formed, and bent. Cutting, punching, and forming are all operations thought of as 2-D operations. The sheet is horizontal and some type of tool strikes the metal, usually at 90 degrees. After the 2-D operations are complete, the flat pattern is bent to the desired shape. More bending processes add more complexity, and make the parts more difficult to manufacture. After bending the material, the process is complete after the finish process and hardware is added.

Table 16-4 Information provided for sheetmetal process

Information Type	Description
Provided Documentation	Dimensionless print showing installed hardware
Provided Database	3-D wireframe IGES/DXF format 2-D views of all features IGES/DXF format Unfolded flat pattern with bend lines and bend allowances are shown in IGES/DXF format. Be aware that each manufacture will probably use a different bend allowance, so make sure the one you used is defined for reference.
Prototype Methods	Laser-cut metal flat patterns, cardboard, paper, and scissors
Tooling Needed	Nonstandard punches or forms
Automation Methods	Standard library templates of known punches and process capabilities

16.6.2.2 Injection Molded Plastic

Plastic parts are the most prevalent parts in today's commercial products. After initial tool production and design, plastic injection molded parts are very cost effective and part tolerances can be controlled consistently. In the past, injection-mold tools limited this manufacturing technique to parts with very high production numbers. Techniques are available to use the injection molding process on lower quantity part counts, with drastically reduced tooling costs.

Table 16-5   Information provided for injection molding process

Information Type	Description
Provided Documentation	Dimensionless print
Provided Database	3-D solid model native format (preferred) 3-D STL format 3-D IGES surfaced file
Prototype Methods	Stereolithography parts RTV silicone molds generated from SLA patterns Foam and glue
Tooling Needed	High cost production steel or aluminum tooling
Automation Methods	Mold flow-analysis programs

16.6.2.3 Hog-Out Parts

Parts manufactured from chunks of raw material that are cut away into the desired shape are often called hog-outs. Mills, lathes, saws, drills, and many other machines have been designed to cut away material from a piece of raw stock. This type of manufacturing is sometimes time-consuming and often inefficient if the final part does not closely resemble the raw material. The major benefit is that the end item product may not require any tooling or up-front expenditure. This not only saves in up-front cost, but also in lead-time to produce the first samples or prototypes. The process capability of a hog-out can be very good.

Table 16-6   Information provided for hog-out process

Information Type	Description
Provided Documentation	Dimensionless print
Provided Database	3-D solid model native format (preferred) 3-D STL format 3-D IGES surfaced file
Prototype Methods	Stereolithography parts RTV silicone molds generated from SLA patterns Foam and glue Fast turnaround time of Investment Cast prototypes is possible using a Stereolithography QUICKCAST part as the casting pattern. Limited quantity prototypes from steel, aluminum, and assorted other metals can be fabricated at relatively low cost
Tooling Needed	Tooling required dependent on casting process
Automation Methods	Standard library templates of known process capabilities

16.6.2.4 Castings

Castings are an excellent way to produce metallic parts with minimal secondary machining. By casting the near net shape with machine stock on secondary machined surfaces, the time for machining is greatly reduced. The cutting machine needs to only clean up the features whose tolerance is greater than the casting process.

Table 16-7   Information provided for casting process

Information Type	Description
Provided Documentation	Dimensionless print
Provided Database	3-D solid model native format (preferred) 3-D IGES surfaced file
Prototype Methods	Stereolithography parts RTV silicone molds generated from SLA patterns Foam and glue
Tooling Needed	Very little special tooling needed
Automation Methods	Standard library templates of known process capabilities

16.6.2.5 Rapid Prototypes

There are many different prototyping processes for mechanical parts. The most versatile and affordable is the Stereolithography (SLA) process. This process can generate an epoxy resin pattern directly off the solid model usually in a matter of days and can also be used to generate molds for rapid tooling for multiple parts.

The methodology for creating a SLA is simple and the hardware for the growing of the prototypes is becoming more affordable. A simple description of the process follows.

Step 1. A solid computer database is sliced up into cross sections.

Step 2. Starting at the base of a model on a platform, a laser sweeps out the cross section on a pool of resin. When the laser strikes the resin it solidifies.

Step 3. The platform is lowered very little and another cross section is swept.

Step 4. The process continues until the part has been grown.

Step 5. The part is removed from the vat of resin and chemically cleaned.

Step 6. The prototype is sanded to remove any ridges.

There are a few things to keep in mind when using the SLA process for models, patterns, and tooling.
- The process capability of the machines is fairly good, (+/- .005) but the parts may dimensionally move over time. Keeping the parts cool will help. Transporting the prototypes in your trunk in the middle of summer is not a good idea. I know this lesson first hand.
- There is usually handwork needed to clean up the model. The quality of this personal touch will vary with manufacturer.
- Some epoxy resin prototype material becomes brittle with age. Care must be taken not to crack the models during handling.
- For rapid tooling, account for any shrink in the molding material in the solid model of the pattern.

Table 16-8 Information provided for prototyping process

Information Type	Description
Provided Documentation	Dimensionless print
Provided Database	3-D STL format
Prototype Methods	N/A
Tooling Needed	N/A
Automation Methods	N/A

16.7 Database Format Standards

The information generated about a product during its design, manufacture, use, maintenance, and disposal is used for many purposes during its life cycle. The use may involve many computer systems, including some that may be located in different organizations. To support such uses, organizations need to represent their product information in a common computer-readable form that is required to remain complete and consistent when exchanged among different computer systems.

There are many different types of electronic databases used in today's product development process. This sometimes causes a barrier to sharing information efficiently. When configuring templates, CAD data sharing and any other product development tool, be aware of the data formats used.

16.7.1 Native Database

A native database is considered the database generated by the computer program used by the person inputting the information. For Microsoft Word, the file has an extension .DOC and it is the default format in which the software saves the file. When a Master model uses its native database type, it is most powerful due to absence of anything lost during a conversion to another format. That is why it is critical to pick product development tools that support common database file types.

One of the problems with native database formats is the lack of control from software revision to revision. The data format will usually change with the revision of the software, making backward database compatibility an issue. A native format is also generally saved in a proprietary binary file, making it difficult to extract data file information from outside the native software. Most all common formats (IGES, DXF, STEP) save the data in a clearly documented ASCII file, allowing the data in the file to be used by any third-party software.

16.7.2 2-D Formats

These formats are supported by most popular software when needing to import or export 2-D wireframe graphics.

16.7.2.1 Data eXchange Format (DXF)

Data eXchange Format (DXF) is the external format for AutoCAD®. It is a text-based representation of a 2-D drawing database. A DXF file can contain 2-D geometry, dimensions, drawing cosmetics, and entity layers. The DXF format is usually stable between different releases of AutoCAD®, although items are added to the specification as new entities are added to AutoCAD®. Most all vector software, both CAD software and Microsoft Office products, strongly support the DXF format. Whenever a drawing or line drawings need to be converted to a vector format for another application, a DXF file is most likely to satisfy everyone involved.

16.7.2.2 Hewlett-Packard Graphics Language (HPGL)

The Hewlett-Packard Graphics Language (HPGL) was developed over a number of years by the Hewlett-Packard Corporation for use in their line of plotters. HPGL has become a standard for plotter formats and is supported by almost all plotter manufacturers as a standard emulation. Most CAD systems have the capability of outputting the 2-D format, but very few have the ability to input the format. The HPGL format can be read by some Microsoft office products and seems to be a clean way to import 2-D geometry into programs such as Word.

16.8 3-D Formats

3-D wireframe/surfacing and solid modeling software support these 3-D conversion formats.

16.8.1 Initial Graphics Exchange Specification (IGES)

Initial Graphics Exchange Specification (IGES) data format is considered a neutral file scheme for CAD data. Most vector graphic programs can convert to and from this neutral file. The IGES standard supports not only vector information, but also 3-D b-spline surfaces. Using a neutral file format decreases the total number of translators needed and provides a file format transferable to virtually any 2-D or 3-D CAD platform. Appendix A shows a listing of popular IGES entities.

16.8.2 STandard for the Exchange of Product (STEP)

STEP is the ISO STandard for the Exchange of Product data. The STEP format is evolving to cover the whole Product Life Cycle for data sharing, storage, and exchange. This format supports wireframe, surfaced and solid geometry. Current CAD/CAE exchange standards like IGES, DXF, SET, and VDAFS will be replaced by STEP, as well as allow for complete descriptions in electronic form of all data related to product manufacture. STEP is open and extensible and will meet design and manufacturing needs well into the next century.

The STEP format is defined in publications produced by the US Product Data Association (US PRO) IGES/PDES Organization. The complete set of specifications for STEP is referred to as ISO 10303. This is an international standard.

The STEP format is organized as a series of documents with each part published separately. Application Protocols (APs) that reference generic parts of ISO 10303 are produced to meet specific data exchange needs required for a particular application. AP203, Configuration Controlled Three-Dimensional Designs of Mechanical Parts and Assemblies, is an International Standard (IS) version.

Products supporting STEP can implement this interface using different levels of data transfer. Each level provides various mechanisms to store, accept, and pass product definition data between heterogeneous systems in a consistent and standardized way.

16.8.3 Virtual Reality Modeling Language (VRML)

Similar to HTML, Virtual Reality Modeling Language (VRML) has emerged as a standard database structure for viewing solid shaded geometry on the Internet. The user can see the shaded geometry, and navigate around and through the shaded geometry. As with HTML, current releases of solid modeling CAD software support this standard.

16.8.4 STereoLithography (STL)

STereoLithography interface format (STL) was generated by 3-D Systems, the designers of Stereolithography Apparatus (SLA), to provide an unambiguous description of a solid part that could be interpreted by the SLA's software. The STL file is a "tessellated surface file" in which geometry is described by triangle shapes laid onto the geometry's surface. Associated with each triangle is a surface normal that is pointed away from the body of the part. This format could be described as being similar to a finite analysis model. When creating an STL file, care must be taken to generate the file with sufficient density so that the facets do not affect the quality of the part built by the SLA. The SLA file holds geometry information only and is used only in the interpretation of the part.

STL files represent the surfaces of a solid model as groups of small polygons. The system writes these polygons to an ASCII text or binary file. Fig. 16-4 shows the file format for an STL file.

```
solid Part1
    facet normal 0.000000e+000 0.000000e+000 1.000000e+000
        outer loop
            vertex 1.875540e-001 2.619040e-001 4.146040e-001
            vertex 1.875540e-001 2.319040e-001 4.146040e-001
            vertex 2.175540e-001 2.619040e-001 4.146040e-001
        endloop
    endfacet
endsolid
```

Figure 16-4 File format for one triangle in an STL file

16.9 General Information Formats

The formats in this section are not specifically designed to support CAD information. These formats are best suited for document templates, product database interrogations, and general distribution of text and pictures.

16.9.1 Hypertext Markup Language (HTML)

HyperText Markup Language (HTML) operates as a database designed for the World Wide Web. HTML code is a basic text file with formatting codes imbedded into the text. These formatting codes are read by specific client software and acted upon to format the text. Most everyone has had experience with HTML and its capabilities. What makes HTML very useful is the power of not being machine specific. Many documents and pictures can be linked on different machines, in different offices, even in different countries, and still appear as if they are all in one place. This virtual Master Model follows the general rules of the Master Model Theory, yet allows multiple areas for the data to be stored.

Current releases of several CAD programs are supporting the product development process as follows:

- Showing the product design on the web as it matures
- Allowing the simple capture of design information
- Having other support groups "look in" without interrupting the design flow

16.9.2 Portable Document Format (PDF)

Portable Document Format (PDF) is an electronic distribution format for documents. The PDF format is good because it keeps the document you are distributing in a format that looks almost exactly like the original. For distributing corporate standards, this format is nice because it can be configured to allow or disallow modifications and printing, as well as other security features. PDF files are compact, cross platform and can be viewed by anyone with a free Adobe® Acrobat Reader. This format and accompanying browser supports zooming in on text as well as page-specific indexing and printing.

16.10 Graphics Formats

These formats are used to support color graphics needed for silkscreen artwork, labels, and other graphic-intensive design activities. The formats may also be used to capture photographic information.

16.10.1 Encapsulated PostScript (EPS)

EPS stands for Encapsulated PostScript. PostScript was originally designed only for sending to a printer, but PostScript's ability to scale and translate makes it possible to embed pieces of PostScript and place them where you want on the page. These pieces of the file are usually EPS files. The file format is ASCII-text based, and can be edited with knowledge of the format.

Encapsulated PostScript files are supported by many graphics programs and also supported across different computing platforms. This format keeps the font references associated with the graphics. When transferring this file format to other programs, it is important to make sure they support the necessary fonts. The format also keeps the references to text and line objects. This allows editing of the objects by other supporting graphics programs.

This is a common file format when transferring graphic artwork for decals and labels to a vendor.

16.10.2 Joint Photographic Experts Group (JPEG)

The Joint Photographic Experts Group (JPEG) format is a standardized image compression mechanism used for digital photographic compression. The Joint Photographic Experts Group was the original committee that wrote the standard.

JPEG is designed for compressing either full-color or gray-scale images of natural, real-world scenes. It works well on photographs, naturalistic artwork, and similar material, but not so well on lettering, simple cartoons, or line drawings. When saving the JPEG file, the compression parameters can be adjusted to achieve the desired finished quality.

This is a common binary format for World Wide Web distribution and most web browsers support the viewing of the file. I use this format very often when I e-mail digital photographs of components to show my overseas vendors.

16.10.3 Tagged Image File Format (TIFF)

TIFF is a tag-based binary image file format that is designed to promote the interchange of digital image data. It is a standard for desktop images and is supported by all major imaging hardware and software developers. This nonproprietary industry standard for data communication has been implemented by most desktop publishing applications.

The format does not save any object information such as fonts or lines. It is strictly graphics data. This allows transfer to any other software with minimal risk of graphic data compatibility. This is a very common format for sending graphic data to vendors for the generation of labels and decals.

16.11 Conclusion

Some of the many techniques for electronic automation, information management, and manufacturing guidelines are presented in this chapter. This small sample has given you more tools to use in successful product development. The chapter also provides two main points to keep in mind in future projects:

Engineering and manufacturing data are critical components in the development process and need to be strategically planned. Computers and electronic data can offer huge possibilities for rapid development, but process success relies on understanding not only *what* can be done but also *why* it is done.

The age of the paper document is not gone yet, but successful corporations in the coming years will rely completely on *capturing and sharing design information to manufacture products with minimal paper movement.*

16.12 Appendix A IGES Entities

IGES Color Codes

IGES Code	Color
8	White
5	Yellow
2,6	Red
4,7	Blue

IGES Entity

Type	Name	Form
100	Circular Arc	
106	Copius Data	11-Polylines 31-Section 40-Witness Line 63-Simple Closed Planar Curve
108	Clipping Planes	
110	Line	
116	Point	
124	Transformation Matrix	
202	Angular Dimension	
206	Diameter Dimension	
210	General Label	
212	General Note	
214	Leader (Arrow)	
216	Linear Dimension	
218	Ordinate Dimension	
222	Radius Dimension	
228	General Symbol	
230	Sectioned Area	
304	Line Font Definition	
314	Color Definition	
404	Drawing	
406	Property Entity	15-Name 16-Drawing Size 17-Drawing Units
410	View Entities	

Chapter 17

Collecting and Developing Manufacturing Process Capability Models

Michael D. King
Raytheon Systems Company
Plano, Texas

Mr. King has more than 23 years of experience in engineering and manufacturing processes. He is a certified Six Sigma Black Belt and currently holds a European patent for quality improvement tools and techniques. He has one US patent pending, numerous copyrights for his work as a quality champion, and has been a speaker at several national quality seminars and symposiums. Mr. King conceptualized, invented, and developed new statistical tools and techniques, which led the way for significant breakthrough improvements at Texas Instruments and Raytheon Systems Company. He was awarded the "DSEG Technical Award For Excellence" from Texas Instruments in 1994, which is given to less than half of 1% of the technical population for innovative technical results. He completed his masters degree from Southern Methodist University in 1986.

17.1 Why Collect and Develop Process Capability Models?

In the recent past, good design engineers have focused on form, fit, and function of new designs as the criteria for success. As international and industrial competition increases, design criteria will need to include real considerations for manufacturing cost, quality, and cycle time to be most successful. To include these considerations, the designer must first understand the relationships between design features and manufacturing processes. This understanding can be quantified through prediction models that are based on process capability models. This chapter covers the concepts of how cost, quality, and cycle time criteria can be designed into new products with significant results!

In answer to the need for improved product quality, the concepts of Six Sigma and quality improvement programs emerged. The programs' initial efforts focused on improving manufacturing processes and using SPC (Statistical Process Control) techniques to improve the overall quality in our factories. We quickly realized that we would not achieve Six Sigma quality levels by only improving our manufacturing processes. Not only did we need to improve our manufacturing process, but we also needed to improve the quality of our new designs. The next generation of Six Sigma deployment involved using process capability data collected on the factory floor to influence new product designs prior to releasing them for production.

Next, quality prediction tools based on process capability data were introduced. These prediction tools allowed engineers and support organizations to compare new designs against historical process capability data to predict where problems might occur. By understanding where problems might occur, designs can easily be altered and tolerances reallocated to meet high-quality standards and avoid problem areas before they occur. It is critical that the analysis is completed and acted upon during the *initial design stage* of a new design because new designs are very flexible and adaptable to changes with the least cost impact. The concept and application of using historical quality process capability data to influence a design has made a significant impact on the resulting quality of new parts, assemblies, and systems.

While the concepts and application of Six Sigma techniques have made giant strides in quality, there are still areas of cost and cycle time that Six Sigma techniques do not take into account. In fact, if all designs were designed around only the highest quality processes, many products would be too expensive and too late for companies to be competitive in the international and industrial market place. This leads us to the following question: If we can be very successful at improving the quality of our designs by using historical process capability data, then can we use some of the same concepts using three-dimensional models to predict cost, quality, and cycle time? Yes. By understanding the effect of all three during the initial design cycle, our design engineers and engineering support groups can effectively design products having the best of all three worlds.

17.2 Developing Process Capability Models

By using the same type of techniques for collecting data and developing quality prediction models, we can successfully include manufacturing cost, quality, and cycle time prediction models. This is a significant step-function improvement over focusing only on quality! An interactive software tool set should include predictive models based on process capability history, cost history, cycle time history, expert opinion, and various algorithms. Example technology areas that could be modeled in the interactive prediction software tool include:

- Metal fabrication
- Circuit card assembly
- Circuit card fabrication
- Interconnect technology
- Microwave circuit card assembly
- Antenna / nonmetallic fabrication
- Optical assembly, optics fabrication
- RF/MW module technology
- Systems assembly

We now have a significant opportunity to design parts, assemblies, and systems while understanding the impact of design features on manufacturing cost, quality, and cycle time before the design is completed and sent to the factory floor. Clearly, process capability information is at the heart of the prediction tools and models that allow engineers to design products with accurate information and considerations for manufacturing cost, quality, and cycle time! In the following paragraphs, I will focus only on the quality prediction models and then later integrate the variations for cost and cycle time predictions.

17.3 Quality Prediction Models - Variable versus Attribute Information

Process capability data is generally collected or developed for prediction models using either variable or attribute type information. The process itself and the type of information that can be collected will determine if the information will be in the form of variable, attribute, or some combination of the two. In general, if the process is described using a standard deviation, this is considered variable data. Information that is collected from a percent good versus percent bad is considered attribute information. Some processes can be described through algorithms that include both a standard deviation and a percent good versus percent bad description.

17.3.1 Collecting and Modeling Variable Process Capability Models

The examples and techniques of developing variable models in this chapter are based on the premise of determining an average short-term standard deviation for processes to predict long-term results. Average short-term standard deviation is used because it better represents what the process is really capable of, without external influences placed upon it.

One example of a process where process capability data was collected from variable information is that of side milling on a numerically controlled machining center. Data was collected on a single dimension over several parts that were produced using the process of side milling on a numerically controlled machine. The variation from the nominal dimension was collected and the standard deviation was calculated. This is one of several methods that can be used to determine the capability of a variable process. The *capability* of the process is described mathematically with the standard deviation. Therefore, I recommend using SPC data to derive the standard deviation and develop process capability models.

Standard formulas based on Six Sigma techniques are used to compare the standard deviation to the tolerance requirements of the design. Various equations are used to calculate the defects per unit (dpu), standard normal transformation (Z), defects per opportunity (dpo), defects per million opportunities (dpmo), and first time yield (fty). The standard formulas are as follows (Reference 3):

dpu = dpo * number of opportunities for defects per unit
dpu = total opportunities * dpmo / 1000000
fty = e^{-dpu}
Z   = ((upper tolerance + lower tolerance)/2) / standard deviation of process
sigma = (SQRT(LN(1/dpo)^2)))-(2.515517 + 0.802853 * (SQRT(LN(1/dpo)^2))) + 0.010328 * (SQRT(LN(1/dpo)^2)))^2)/(1 + 1.432788 * (SQRT(LN(1/ (dpo)^2))) + 0.189269 * (SQRT(LN(1 / (dpo)^2)))^2 + 0.001308 * (SQRT(LN(1 / dpo)^2)))^3) +1.5
dpo = [(((((((1 + 0.049867347 * (z –1.5)) + 0.0211410061 * (z –1.5) ^2) + 0.0032776263 *(z -1.5)^3) + 0.0000380036 * (z –1.5)^4) + 0.0000488906 * (z –1.5)^5) + 0.000005383 * (z –1.5)^6)^ – 16)/2]
dpmo = dpo * 1000000

where
    dpmo = defects per million opportunities
    dpo = defects per opportunity
    dpu = defects per unit
    fty = first time yield percent (this only includes perfect units and does not include any scrap or rework conditions)

Let's look at an example. You have a tolerance requirement of ±.005 in 50 places for a given unit and you would like to predict the part or assembly's sigma level (Z value) and expected first time yield. (See Chapters 10 and 11 for more discussion on Z values.) You would first need to know the short-term standard deviation of the process that was used to manufacture the ±.005 feature tolerance. For this example, we will use .001305 as the standard deviation of the process. The following steps would be used for the calculation:

1. Divide the ±tolerance of .005 by the standard deviation of the process of .001305. This results in a predicted sigma of 3.83.
2. Convert the sigma of 3.83 to defects per opportunity (dpo) using the dpo formula. This formula predicts a dpo of .00995.
3. Multiply the dpo of .00995 times the opportunity count of 50, which was the number of places that the unit repeated the ±.005 tolerance. This results in a defect per unit (dpu) of .4975.
4. Use the (e^{-dpu}) first time yield formula to calculate the predicted yield based on the dpu. The result is 60.8% predicted first time yield.
5. The answer to the initial question is that the process is a 3.83 sigma process, and the part or assembly has a predicted first time yield of 60.8% based on a 3.83 sigma process being repeated 50 times on a given unit.

Typically a manufactured part or assembly will include several different processes. Each process will have a different process capability and different number of times that the processes will be applied. To calculate the overall predicted sigma and yield of a manufactured part or assembly, the following steps are required:

1. Calculate the overall dpu and opportunity count of each separate process as shown in the previous example.
2. Add all of the total dpu numbers of each process together to give you a cumulative dpu number.
3. Add the opportunity counts of each process together to give you a cumulative opportunity count number.
4. To calculate the cumulative first time yield of the part or assembly use the (e^{-dpu}) first time yield formula and the cumulative dpu number in the formula.
5. To calculate the sigma rollup of the part or assembly divide the cumulative dpu by the cumulative opportunity count to give you an overall (dpo) defect per opportunity. Now use the sigma formula to convert the overall dpo to the sigma rollup value.

When using an SPC data collection system to develop process capability models, you must have a very clear understanding of the process and how to set up the system for optimum results. For best results, I recommend the following:

- Select features and design tolerances to measure that are close to what the process experts consider to be just within the capability of the process.
- Calculate the standard deviations from the actual target value instead of the nominal dimension if they are different from each other.
- If possible, use data collected over a long period of time, but extract the short-term data in groups and average it to determine the standard deviation of a process.
- Use several different features on various types of processes to develop a composite view of a short-term standard deviation of a specific process.

Collecting and Developing Manufacturing Process Capability Models 17-5

Selecting features and design tolerances that are very close to the actual tolerance capability of the process is very important. If the design tolerances are very easily attained, the process will generally be allowed to vary far beyond its natural variation and the data will not give a true picture of the processes capability. For example, you may wish to determine the ability of a car to stay within a certain road width. See Fig. 17-1. To do this, you would measure how far a car varies from a target and record points along the road. Over a distance of 100 miles, you would collect all the points and calculate the standard deviation from the center of the road. The standard deviation would then be in with the previous formulas to predict how well the car might stay within a certain width tolerance of a given road. If the driver was instructed to do his or her best to keep the car in the center of a very narrow road, the variance would probably be kept at a minimum and the standard deviation would be kept to a minimum. However, if the road were three lanes wide, and the driver was allowed to drive in any of the three lanes during the 100-mile trip, the variation and standard deviation would be significantly larger than the same car and driver with the previous instructions.

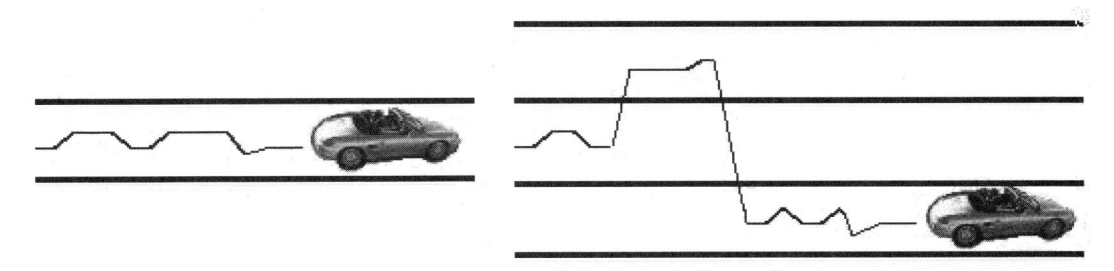

Figure 17-1 Narrow road versus three-lane road

This same type of activity happens with other processes when the specifications are very wide compared to the process capability. One way to overcome this problem is to collect data from processes that have close requirements compared to the processes' actual capability.

Standard deviations should be calculated from the actual target value instead of the nominal dimension if they are different from each other. This is very important because it improves the quality of your answer. Some processes are targeted at something other than the nominal for very good reasons. The actual process capability is the variation from a targeted position and that is the true process capability. For example, on a numerically controlled machining center side milling process that machines a nominal dimension of .500 with a tolerance of +. 005/–. 000, the target dimension would be .5025 and the nominal dimension would be .500. If the process were centered on the .500 dimension, the process would result in defective features. In addition to one-sided tolerance dimensions, individual preferences play an important role in determining where a target point is determined. See Fig. 17-2 for a graphical example of how data collected from a manufacturing process may have a shifting target.

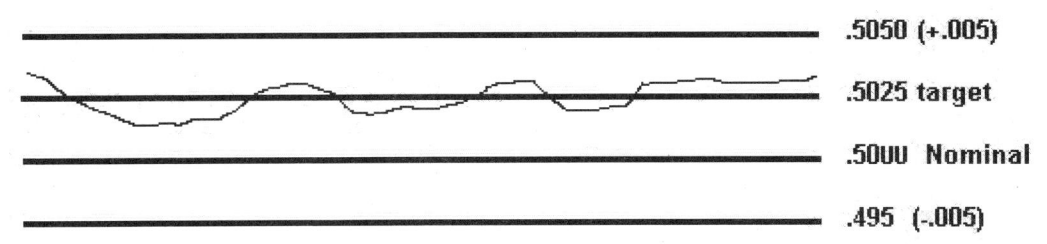

.5050 (+.005)
.5025 target
.50UU Nominal
.495 (-.005)

Figure 17-2 Data collected from a process with a shifted target

It is best to collect data from variable information over a long period of time using several different feature types and conditions. Once collected, organize the information into short-term data subgroups within a target value. Now calculate the standard deviation of the different subgroups. Then average the short-term subgroup information after discarding any information that swings abnormally too high or too low compared to the other information collected. See Fig. 17-3 for an example of how you may wish to group the short-term data and calculate the standard deviation from the new targets.

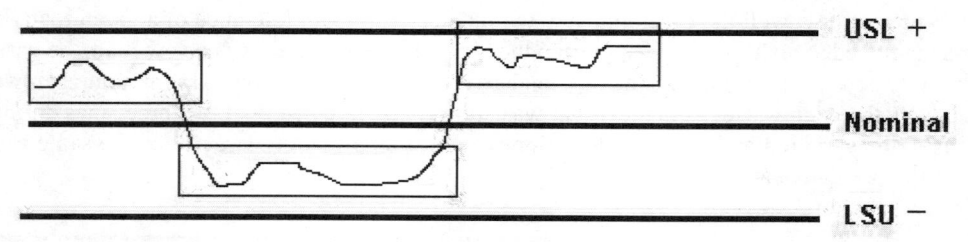

Figure 17-3 Averaging and grouping short-term data

A second method for developing process capability models and determining the standard deviation of a process might include controlled experiments. Controlled experiments are very similar to the SPC data collection process described above. The difference is in the selection of parts to sample and in the collection of data. You may wish to design a specific test part with various features and process requirements. The test parts could be run over various times or machines using the same processes under controlled conditions. Data collected would determine the standard deviation of the processes. Other controlled experiments might include collecting data on a few features of targeted parts over a certain period of time to result in a composite perspective of the given process or processes. Several different types of controlled experiments may be used to determine the process capability of a specific process.

A third method of determining the standard deviation of a given process is based on a process expert's knowledge. This process might be called the "five sigma rule of thumb" estimation technique for determining the process capability. To determine a five sigma tolerance of a specific process, talk to someone who is very knowledgeable about a given process or a process expert to estimate a tolerance that can be achieved 98%-99% of the time on a generally close tolerance dimension using a specific process. That feature should be a normal-type feature under normal conditions for manufacturing and would not include either the best case or worst case scenario for manufacturing. Once determined, divide that number by 5 and consider it the standard deviation. This estimation process gets you very close to the actual standard deviation of the process because a five sigma process when used multiple times on a given part or unit will result in a first time yield of approximately 98% - 99%.

Process experts on the factory floor generally have a very good understanding of process capability from the perspective of yield percents. This is typically a process that has a good yield with some loss, but is performing well enough not to change processes. This tolerance is generally one that requires close attention to the process, but is not so easily obtained that outside influences skew the natural variations and distort the data. Even though this method uses expert opinion to determine the short-term standard deviation and not actual statistical data, it is a quick method for obtaining valuable information when none is available. Historically, this method has been a very accurate and successful tool in estimating information (from process experts) for predicting process capability. In addition to using process experts, tolerances may be obtained from reference books and brochures. These tolerances should result in good quality (98%-100% yield expectations).

Models that are variable-based usually provide the most accurate predictors of quality. There are several different methods of determining the standard deviation of a process. However, the best method is to use all three of these techniques with a regressive method to adjust the models until they accurately predict the process capability. The five sigma rule of thumb will help you closely estimate the correct answer. Use it when other data is not available or as a check-and-balance against SPC data.

17.3.2 Collecting and Modeling Attribute Process Capability Models

Models that are variable models are attribute models. Defect information for attribute models is usually collected as percent good versus bad or yield. An example of an attribute process capability model would be the painting process. An attribute model can be developed for the painting process in several different ways based on the type of information that you have.

- At the simplest level, you could just assign an average defect rate for the process of painting.
- At higher levels of complexity, you could assign different defect rates for the various features of the painting process that affect quality.
- At an even higher level of complexity, you could add interrelationships among different features that affect the painting process.

17.3.3 Feature Factoring Method

The factoring method assigns a given dpmo to a process as a basis. In the model, all other major quality drivers are listed. Each quality driver is assigned a defect factor, which may be multiplied times the dpmo basis to predict a new dpmo if that feature is used on a given design. Factors may have either a positive or negative effect on the dpmo basis of an attribute model. Each quality driver may be either independent or dependent upon other quality drivers. If several features with defect factors are concurrently chosen, they will have a cumulative effect on the dpmo basis for the process. The factoring method gives significant flexibility and allows predictions at the extremes of both ends of the quality spectrum. See Fig. 17-4 for an example of the feature factoring methods flexibility with regards to predictions and dpmo basis.

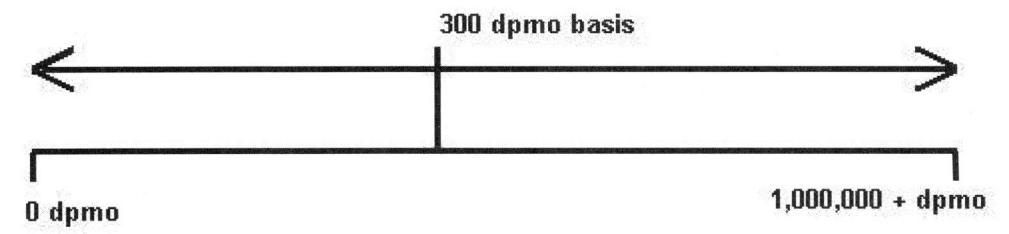

Figure 17-4 Feature factoring methodology flexibility

17.3.4 Defect-Weighting Methodology

This defect-weighting method assigns a best case dpmo and a worst case dpmo for the process similar to a guard-banding technique. Defect driver features are listed and different weights assigned to each. As different features are selected from the model, the defect weighting of each feature or selection reduces the process dpmo accordingly. Generally, when all the best features are selected, the process dpmo remains at its guard-banded best dpmo rating. And when most or all of the worst features with regards to quality are selected, the dpmo rating changes to the worst dpmo rating allowed under the guard-banding scenario.

The following steps describe the defect-weighting model.

1. Using either data collected or expert knowledge, determine the dpmo range of the process you are modeling.
2. Determine the various feature selections that affect the process quality.
3. Assign a number to each of the features that will represent its defect weight with regard to all of the other feature selections. The total of all selectable features must equal 1.0 and the higher the weight number, the higher the effect on the defect rating it will be. The features may be categorized so that you can choose one feature from each category with the totals of each category equal to 1.0.
4. Calculate the new dpmo prediction number by subtracting the highest dpmo number from the lowest dpmo number and multiplying that number times the total weight number. Then add that number to the lowest dpmo number to get the new dpmo number.

The formula is: The new process defect per million opportunity (dpmo) rating
= (highest dpmo number − lowest dpmo number)
× the cumulative weight numbers

For example, you may assign the highest dpmo potential to be 2,000 with the lowest dpmo at 100. If the cumulative weights of the features with defect ratings equal .5, then the new process dpmo rating would be a dpmo of 1,050 (2000 − 100 = 1,900; 1900 × .5 = 950; 950 + 100 = 1,050).

See Fig. 17-5 for a graphic of the defect-weighting methodology with regard to guard-banding and dpmo predictions. This defect-weighting method allows you to set the upper and lower limits of a given process dpmo rating. The method also includes design features that drive the number of defects. The design dpmo rating will vary between the dpmo minimum number and the dpmo maximum number. If the designer chooses features with the higher "weights," the design dpmo approaches the dpmo maximum. If the designer chooses features with lower "weights," the design dpmo approaches the dpmo minimum.

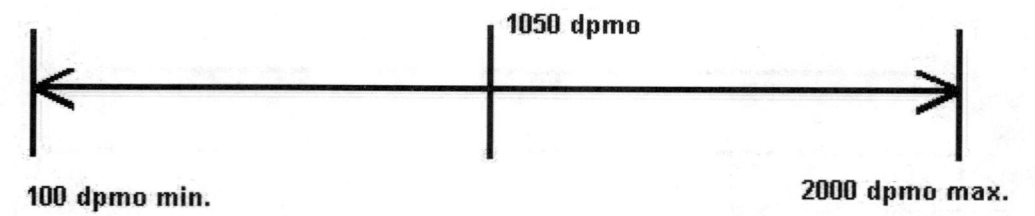

Figure 17-5 Dpmo-weighting and guard-banding technique

17.4 Cost and Cycle Time Prediction Modeling Variations

You might wish to use a combination of both or either of the two previously discussed modeling techniques for your cost and cycle time prediction models. Cost and cycle time may have several different definitions depending upon your needs and familiar terminology. For the purpose of this example, cost is defined as the cost of manufacturing labor and overhead. Cycle time is defined as the total hours required producing a product from order placement to final delivery. Cost and cycle time will generally have a very close relationship.

One method for predicting cost of a given product might be to associate a given time to each process feature of a given design. Multiply the associated process time by the hourly process rate and overhead.

Depending upon the material type and part size, you may wish to also assign a factor to different material types and part envelope sizes from some common material type and material size as a basis. Variations from that basis will either factor the manufacturing time and cost up or down. Additional factors may be applied such as learning curve factors and formulas for lot size considerations. Cost and cycle time models should also include factors related to the quality predictions to account for scrap and rework costs. The cycle time prediction portion of the model would be based upon the manufacturing hours required plus normal queue and wait time between processes. An almost unlimited number of factors can be applied to cost and cycle time prediction models. *Most important is to develop a methodology that gives you a basis from which to start.* Use various factors that will be applied to that basis to model cost and cycle time predictions.

Cost and cycle time predictions can be very valuable tools when making important design decisions. Using an interactive predictive model including relative cost predictions would easily allow real-time what-if scenarios. For example, a design engineer may decide to machine and produce a given part design from material A. Other options could have been material B, C or D, which have similar properties to material A. There may not be any difference in material A, B, C or D as far as fit, form or function of the design is concerned. However, material A could take 50% more process time to complete and thus be 50% more costly to produce.

Here is an example of how cycle time models might be influential. Take two different chemical corrosion resistance processes that yield the same results with similar costs. The difference might only be in the cycle time prediction model that highlights significant cycle time requirements of different processes due to where the corrosion resistance process is performed. Process A might be performed in-house or locally with a short cycle time. Process B might be performed in a different state or country only, which typically requires a significant cycle time. Overall, cost and cycle time prediction models are very powerful complements to quality prediction models. They can be very similar in concept or very different from either the attribute or variable models used in quality predictions.

17.5 Validating and Checking the Results of Your Predictive Models

Making sure your predictive models are accurate is a very important part of the model development process. The validation and checking process of process capability models is a very iterative process and may be done using various techniques. Model predictions should be compared to actual results with modifications made to the predictive model, data collection system, or interpretation of the data as needed. Models should be compared at the individual model level and at the part or assembly rollup level, which may include several processes. Validating the prediction model at the model level involves comparing actual process history to the answer predicted by the interactive model.

With variable models, the model level validation involves comparing both the standard deviation number and the actual part yields through the process versus the first time yield (fty) prediction of the process. The second step of the validation process for variable models requires talking with process experts or individuals that have a very good understanding of the process and its real-world process capabilities. One method of comparing variable prediction models, standard deviations, and expert opinion involves using the five sigma rule of thumb technique.

A 5.0 sigma rating at a specific tolerance will mathematically relate to a first time yield of 98%-99% when several opportunities are applied against it. The process experts selected should be individuals on the factory floor that have hands-on experience with the process rather than statisticians. A process expert can determine a specific standard deviation number. Ask them to estimate the tolerance that the process can produce consistently 98%-99% of the time on a close tolerance dimension. The answer given can be considered the estimated 5.0 sigma process. Using the five sigma rule of thumb technique, divide the tolerance given by the process experts by 5 to determine the standard deviation for the process. You would probably want to take a sampling of process experts to determine the number that you will be dividing by 5. Note that the way you phrase the question to the process experts is very critical. It is very important to ask the process experts the question with regard to the following criteria:

1. The process needs to be under normal process conditions.
2. The estimate is not based on either best or worst case tolerance capabilities.
3. The tolerance that will yield 98%-99% of the product on a consistent basis is based on a generally close tolerance and if the tolerance were any smaller, they would expect inconsistent yields from the process.

After receiving the answer from the process experts, repeat back to them the answer that they gave you and ask them if that is what they understood their answer to be. If they gave you an answer of ±.005, you might ask the following back to them: Under normal conditions, and a close tolerance dimension for that process, you would expect ±.005 to yield approximately 98%-99% product that would not require rework or scrap of the product? Would you expect the same process with ±.004 (four sigma) to yield approximately 75%-80% yields under normal conditions? If they answer "yes" to both of these answers, they probably have a good understanding of your previous questions and have given you a good answer to your question. If you question several process experts and generally receive the same answer, you can consider it a good estimation of a five sigma process under that tolerance.

Compare the estimated standard deviation from that of your SPC data collection system. If there is more than a 20% difference between the two, something is significantly wrong and you must revisit both sources of information to determine the right ones. The two standard deviation numbers should be within 5%-10% of each other for prediction models to be reasonable.

Overall, the best approach to validating variable models is to use a combination of all three techniques to determine the best standard deviation number to use for the process. To do this, compare:

1. The standard deviation derived from the average short-term SPC data.
2. The standard deviation derived from expert opinion and the five sigma rule of thumb method.
3. Using the standard deviations derived from the two methods listed above, enter them one at a time into the interactive prediction tool or equations. Then compare actual process yield results to predict yield predictions based on the two standard deviations and design requirements.

Attribute models are also validated at the model level by comparing actual results to predictive results of the individual model. Similarly, expert opinions are very valuable in validating the models when actual data at the model level cannot be extracted. The validation of attribute models can be achieved by reviewing a series of predictions under different combinations of selections with factory process experts. The process experts should be asked to agree or disagree with different model selection combinations and results. The models should be modified several times until the process experts agree with the model's resulting predictions. Actual historical data should be shared with the process experts during this process to better understand the process and information collected.

In addition to model validation at the individual model level, many processes and combinations of processes need to be validated at the part or assembly rollup level. Validation at the rollup level requires that all processes be rolled up together at either the part or subassembly level and actual results compared to predictions. For a cost rollup validation on a specific part, the cost predictions associated with all processes should be added together and compared to the total cost of the part for validation. For a quality rollup validation on a specific part, all dpu predictions should be added up and converted to yield for comparison to the actual yield of manufacturing that specific part.

17.6 Summary

Both international and industrial competition motivate us to stay on the cutting edge of technology with our designs and manufacturing processes. New technologies and innovative processes like those described in this chapter give design engineers significant competitive advantage and opportunity to design for success. Today's design engineers can work analytical considerations for manufacturing cost, quality, and cycle time into new designs *before* they are completed and sent to the factory floor.

The new techniques and technology described in this chapter have been recently implemented at a few technically aggressive companies in the United States with significant cost-saving results. The impact of this technology includes more than $50 million of documented cost savings during the first year of deployment at just one of the companies using the technology! With this kind of success, we need to continue to focus on adopting and using new technologies such as those described in this chapter.

17.7 References

1. Bralla, James G. 1986. *Handbook of Product Design for Manufacturing.* New York, New York: McGraw-Hill Book Co.
2. Dodge, Nathon. 1996. Michael King: Interview and Discussion about PCAT. *Texas Instruments Technical Journal.* 31(5):109-111.
3. Harry, Mikel J. and J. Ronald Lawson. 1992. *Six Sigma Producibility Analysis and Process Characterization.* Reading, Massachusetts: Addison-Wesley Publishing Company.
4. King, Michael. 1997. Designing for Success. Paper presented at Applied Statistical Tools and Techniques Conference, 15 October, 1997, at Raytheon TI Systems, Dallas, TX.
5. King, Michael. 1996. Improving Mechanical / Metal Fabrication Designs. *Process Capability Analysis Toolset Newsletter.* Dallas, Texas: Raytheon TI Systems
6. King, Michael. 1994. Integration and Results of Six Sigma on the DNTSS Program. Paper presented at Texas Instruments 1st Annual Process Capability Conference. 27 October, 1994, Dallas, TX.
7. King, Michael. 1994. Integrating Six Sigma Tools with the Mechanical Design Process. Paper presented at Six Sigma Black Belt Symposium. Chicago, Illinois.
8. King, Michael. 1992. Six Sigma Design Review Software. *TQ News Newsletter.* Dallas, Texas: Texas Instruments, Inc.
9. King, Michael. 1993. Six Sigma Software Tools. Paper presented at Six Sigma Black Belt Symposium. Rochester, New York.
10. King, Michael. 1994. Using Process Capability Data to Improve Casting Designs. Paper presented at International Casting Institute Conference. Washington, DC.

Chapter 18

Paper Gage Techniques

Martin P. Wright
Behr Climate Systems, Inc.
Fort Worth, Texas

Martin P. Wright is supervisor of Configuration Management for Behr Climate Systems, Inc. in Fort Worth, Texas, where he directs activities related to dimensional management consulting and company training programs. He has more than 20 years of experience utilizing the American National Standard on Dimensioning and Tolerancing and serves as a full-time, on-site consultant assisting employees with geometric tolerancing applications and related issues. Mr. Wright has developed several multilevel geometric tolerancing training programs for several major companies, authoring workbooks, study guides, and related class materials. He has instructed more than 4,500 individuals in geometric tolerancing since 1988.

Mr. Wright is currently an active member and Working Group leader for ASME Y14.5, which develops the content for the American National Standard on dimensioning and tolerancing. He also serves as a member of the US Technical Advisory Group (TAG) to ISO TC213 devoted to dimensioning, tolerancing, and mathematization practices for international standards (ISO). In addition to these standards development activities, Mr. Wright serves as a member and/or officer on six other technical standard subcommittees sponsored by the American Society of Mechanical Engineers (ASME).

18.1   What Is Paper Gaging?

Geometric Dimensioning and Tolerancing (GD&T) as defined by ASME Y14.5M-1994 provides many unique and beneficial concepts in defining part tolerances. The GD&T System allows the designer to specify round, three-dimensional (3-D) tolerance zones for locating round, 3-D features (such as with a pattern of holes). The system also offers expanded concepts, such as the maximum material condition (MMC) principle, that allows additional location tolerance based on the produced size of the feature.

(See Chapter 5.) These concepts work well in assuring that part features will function as required by the needs of the design, while maximizing all available production tolerances for the individual workpiece. Although these tolerancing concepts are beneficial for both design and manufacturing, their use can pose some unique problems for the inspector who must verify the requirements.

It is widely recognized that, in terms of inspection, the optimum means for verifying part conformance to geometric tolerancing requirements is through the use of a fixed-limit gage. (See Chapter 19.) This gage is essentially the physical embodiment of a 3-D, worst case condition of the mating part. If the part fits into the functional gage, the inspector may also be assured that it will assemble and interchange with its mating part. Since the gaging elements are fixed in size, the additional location tolerance allowed for a larger produced hole (or the dynamic "shift" of a datum feature subject to size variation) is readily captured by the functional gage. Additionally, functional gages are easily used by personnel with minimal inspection skills and they can significantly reduce overall inspection time. However, there are drawbacks to using functional gages. They are expensive to design, build, and maintain, and they require that a portion of the product tolerance be sacrificed (usually about 10%) to provide tolerance for producing the gage itself. For these reasons, use of functional gages is generally limited to cases where a large quantity of parts are to be verified and the reduced inspection time will offset the cost of producing the gage.

Verification of geometric tolerances for the vast majority of produced parts is accomplished through the use of data collected either manually in a layout inspection, or electronically using a Coordinate Measuring Machine (CMM). Either method requires the inspector to lock the workpiece into a frame of reference as prescribed by the engineering drawing and take actual measurements of the produced features. The inspector must then determine "X" and "Y" coordinate deviations for the produced features by comparing the actual measured values to the basic values as indicated on the drawing. Typically, these coordinate deviations are used in determining positional tolerance error for the produced feature through one of two methods: mathematical conversion of the coordinate deviations or by use of a paper gage.

Paper gaging is one of several common inspection verification techniques that may be used to ensure produced feature conformance to an engineering drawing requirement. This technique, also referred to as Soft Gaging, Layout Gaging, or Graphical Inspection Analysis, provides geometric verification through a graphical representation and manipulation of the collected inspection data. Cartesian coordinate deviations derived from the measurement process are plotted on to a coordinate grid, providing a graphical "picture" of the produced feature locations in relation to their theoretically "true" location.

Modern tolerancing methods as defined throughout ASME Y14.5M-1994 prescribe that round features, such as holes, be located within round tolerance zones. However, most dimensional inspection techniques measure parts in relation to a square, Cartesian coordinate system. Paper gaging provides a convenient and accurate method for converting these measured values into the round, polar coordinate values required in a positional tolerance verification. This is accomplished graphically by superimposing a series of rings over the coordinate grid that represents the positional tolerance zones.

18.2 Advantages and Disadvantages to Paper Gaging

Since the optimum means for a geometric tolerancing requirement is through the use of a fixed-limit gage, the primary advantage provided by paper gaging lies in its ability to verify tolerance limits similar to those of a hard gage. Paper gaging techniques graphically represent the functional acceptance boundaries for the feature, without the high costs of design, manufacture, maintenance, and storage required for a fixed-limit gage. Additionally, paper gaging does not require that any portion of the product tolerance be sacrificed for gage tolerance or wear allowance.

Paper gaging is also extremely useful in capturing dynamic tolerances found in datum features subject to size variation or feature-to-feature relationships within a pattern of holes. Neither of these can be effectively captured in a typical layout inspection. The ability to manipulate the polar coordinate overlay used in the paper gage technique gives the inspector a way to duplicate these unique tolerance effects.

Since it provides a visual record of the actual produced features, paper gaging can be an extremely effective tool for evaluating process trends and identifying problems. Unlike a hard gage, which simply verifies GO/NO-GO attributes of the workpiece, the paper gage can provide the operator with a clear illustration of production problems and the precise adjustment necessary to bring the process back into control. Factors such as tooling wear and misalignment can readily be detected during production through periodic paper gaging of verified parts. Additionally, paper gages can be easily stored using minimal, low-cost space.

The primary drawback to paper gage method of verification is that it is much more labor-intensive than use of a fixed-limit gage. Paper gaging requires a skilled inspector to extract actual measurements from the workpiece, then translate this data to the paper gage. For this reason, paper gaging is usually considered only when the quantity of parts to be verified is small, or when parts are to be verified only as a random sampling.

18.3 Discrimination Provided By a Paper Gage

With paper gaging, the coordinate grid and polar overlay are developed proportionately relative to one another and do not necessarily represent a specific measured value. Because they are generic in nature, the technique may be used with virtually any measurement discrimination. The spacing between the lines of the coordinate grid may represent .1 inch for verification of one part, and .0001 inch for another.

A typical inspection shop may only need to develop and maintain three or four paper gage masters. Each master set would represent a maximum tolerance range capability for that particular paper gage. The difference between them would be the number of grid lines per inch used for the coordinate grid. More grid lines per inch on the coordinate grid allow a wider range of tolerance to be effectively verified by the paper gage. However, an increase in the range of the paper gage lowers the overall accuracy of the plotted data. The inspector should always select an appropriate grid spacing that best represents the range of tolerance being verified.

18.4 Paper Gage Accuracy

A certain amount of error is inherent in any measurement method, and paper gages are no exception. The overall accuracy of a paper gage may be affected by factors such as error in the layout of the lines that make up the graphs, coefficient of expansion of the material used for the graphs or overlays, and the reliability of the inspection data. Most papers tend to expand with an increase in the humidity levels and, therefore, make a poor selection for grid layouts where fine precision is required. Where improved accuracy is required, Mylar is usually the material of choice since it remains relatively stable under normal changes in temperature and humidity.

By amplifying (enlarging) the grid scale, we can reduce the effects of layout error in the paper gage. Most grid layout methods will provide approximately a .010 inch error in the positioning of grid lines. From this, the apparent error provided by the grid as a result of the line positioning error of the layout may be calculated as follows:

$$\frac{\text{Line Position Error}}{\text{Scale Factor}} = \text{Apparent Layout Error}$$

For example, if a 10×10 to-the-inch grid is selected, with each line of the grid representing .001, a scale factor of 100-to-1 is provided, resulting in an apparent layout error for the grid of .0001 inch. However, if a

5 × 5 to-the-inch grid is selected, with each line of the grid representing .001, a scale factor of 200-to-1 is provided, resulting in an apparent layout error for the grid of only .00005 inch.

18.5 Plotting Paper Gage Data Points

It is extremely important for all users to plot data points on the coordinate grid of a paper gage in the same manner. This is a mandatory requirement in order to maintain consistency and to provide an accurate representation of the produced part. Inadvertently switching the X and Y values, or plotting the points in the wrong direction (plus or minus) will result in an inaccurate picture of the produced part features. This renders the paper gage useless as an effective process analysis tool.

On the engineering drawing, each hole or feature has a basic or "true" location specified. If the hole or feature were located perfectly, the measured value and the basic value would be the same. It could therefore be stated that the theoretical address of the hole or feature at true position is X=0, Y=0. Since geometric location tolerances are only concerned with the deviation from true position, the center of the coordinate grid may be used to represent the theoretical address for each feature being verified.

The data points represent deviations from true position and should always be plotted on the coordinate grid based on the relationship to its theoretical address and in a manner consistent with the view in which the holes are specified. For example, when plotting the X deviation for a hole, the data point is considered to have a plus X value where the feature falls to the right of its theoretical address, and a minus X value where it falls to the left of its theoretical address. When plotting the Y deviation, the data point is considered to have a plus Y value where the feature falls above its theoretical address, and a minus Y value where it falls below the theoretical address. See Fig. 18-1. Consistently following this methodology for plotting the data points will assure the reliability of the paper gage for both tolerance evaluation and process analysis.

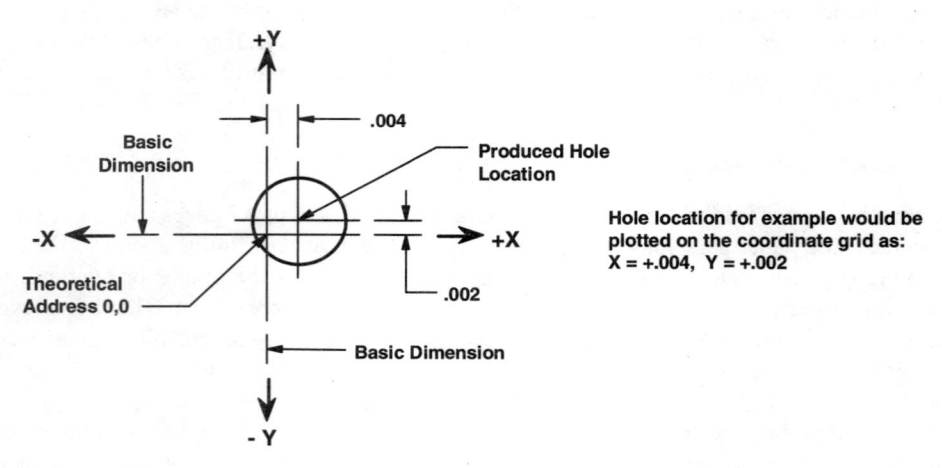

Figure 18-1 Directional indicators for data point plotting

18.6 Paper Gage Applications

The following examples illustrate some of the common applications for paper gages in evaluating part tolerances and analyzing process capabilities. Although these examples illustrate just a few of the many uses for a paper gage, they provide the reader with an excellent overview as to the effectiveness and versatility of this valuable manufacturing and inspection tool.

18.6.1 Locational Verification

Development of a functional gage to verify feature locations may not be practical or cost effective for many parts. For example, parts that will be produced in relatively small quantities, or parts that will fall under some type of process control where part verification will only be done on a random, periodic basis may not require production of a functional gage. For these parts, it may be more cost effective to verify the tolerances manually using data collected from a layout inspection. This data may then be transferred to a paper gage to verify the locational attributes of the features (similar to a fixed-limit gage) for only a fraction of the cost.

18.6.1.1 Simple Hole Pattern Verification

The following example illustrates how the paper gage may be used to verify the locational requirement of the hole pattern for the part shown in Fig. 18-2. The drawing states that the axis of each hole must lie within a Ø.010 tolerance zone when produced at their maximum material condition size limit of Ø.309. Since an MMC modifier has been specified, additional locational tolerance is allowed for the holes as they depart their MMC size limit (get larger) by an amount equal to the departure.

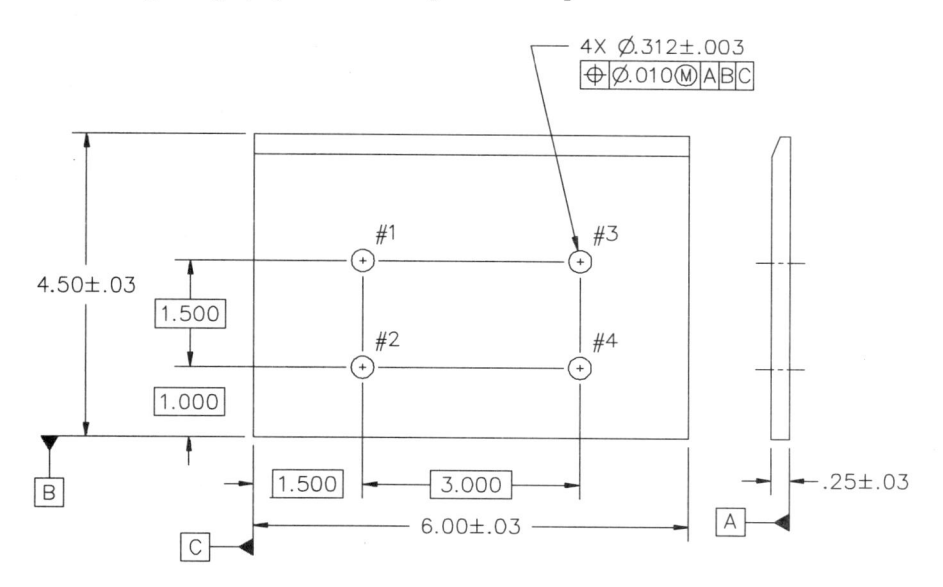

Figure 18-2 Example four-hole part

A layout inspection requires that the inspector collect actual measurements from the produced part and compare these with the tolerances indicated by the engineering drawing. The actual measurement data may be obtained electronically using a CMM or manually using a surface table and angle plate setup. The data collected from a layout inspection provides actual "X" and "Y" values for the location of features in relation to the measurement origin. That is, the measurement provided is always in relation to a Cartesian Coordinate frame of reference.

In evaluating the locational requirements for the hole pattern, the inspector must first verify that all holes fall within their acceptable limits of size. The inspector must also know the produced size of each hole in order to determine the amount of positional tolerance allowed for each hole. To determine the produced hole size, the inspector inserts the largest gage pin possible into each of the holes. This effectively defines the actual mating size of the hole, allowing the inspector to calculate the amount of

18-6 Chapter Eighteen

additional positional tolerance (bonus tolerance) allowed for location. The difference between the actual mating size and the specified MMC size is the allowed bonus tolerance. This tolerance may be added to the tolerance value specified in the feature control frame.

Once it has been determined that the hole sizes are within acceptable limits, the inspector must set up the part to measure the hole locations. He accomplishes this by relating the datum features specified by the feature control frame to the measurement planes of the inspector's equipment (i.e., surface table, angle plate). The inspector MUST use the datum features in the same sequence as indicated by the feature control frame. The final setup for the sample part shown above may resemble the part illustrated in Fig. 18-3.

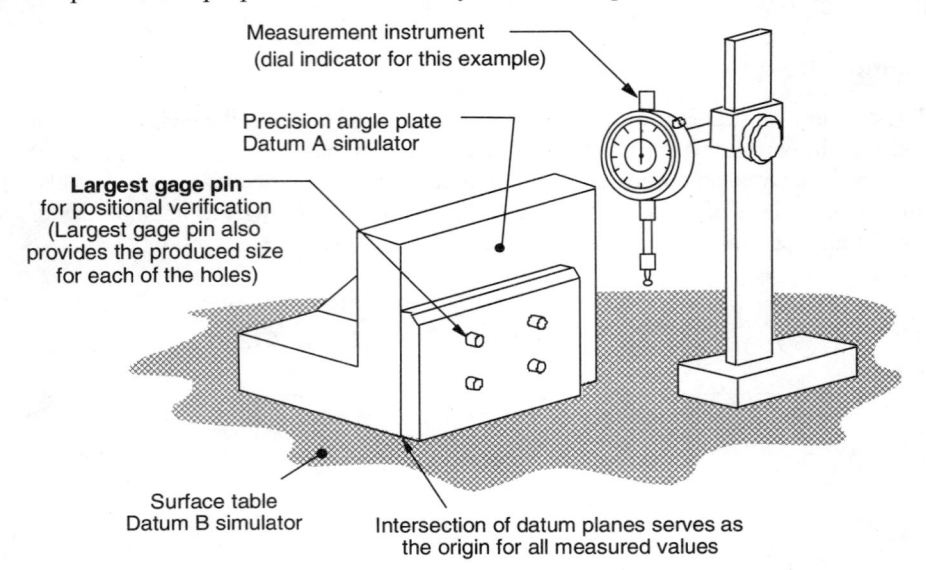

Figure 18-3 Layout inspection of four-hole part

The pins placed in the holes aid the inspector when measuring the hole location. Actual "X" and "Y" measurements are made to the surface of the pin and as near to the part face as practicable. With the size of each pin known, adding 1/2 of the pin's diameter to the measured value will provide the total actual measurement to the center of each hole.

Once the part is locked into the datum reference frame, measurements are made in an "X" and a "Y" direction and the data is recorded on the Inspection Report for final evaluation. This evaluation involves taking the coordinate data from the actual measurements and converting it into a round positional tolerance. Table 18-1 illustrates a sample Inspection Report that provides the data for paper gage evaluation of the hole pattern.

Table 18-1 Layout Inspection Report of four-hole part

NO.	FEATURE	MMC	ACTUAL	DEV.	ALLOW TOL.	X BASIC	X ACTUAL	X DEV	Y BASIC	Y ACTUAL	Y DEV	ACCEPT	REJECT
1	.312±.00	.309	.311	.002	Ø.012	1.500	1.503	+.003	2.500	2.501	+.001	X	
2	.312±.00	.309	.313	.004	Ø.014	1.500	1.505	+.005	1.000	.998	-.002	X	
3	.312±.00	.309	.312	.003	Ø.013	4.500	.496	-.004	2.500	2.497	-.003	X	
4	.312±.00	.309	.310	.001	Ø.011	4.500	.494	-.006	1.000	1.002	+.002		X

Using the data from the Inspection Report, the information is then transferred to the paper gage by plotting each of the holes on a coordinate grid as shown in Fig. 18-4. The center of the grid represents the basic or true position (theoretical address 0,0) for each of the holes. Their actual location in relation to their theoretical address is plotted on the grid using the X and Y deviations from the Inspection Report.

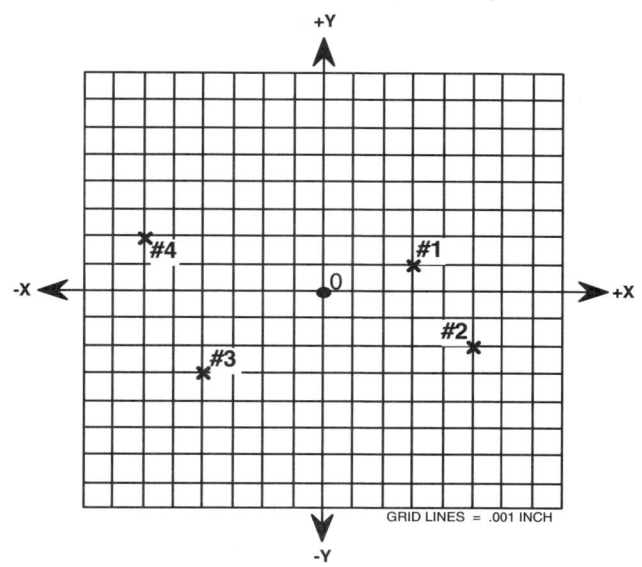

Figure 18-4 Plotting the holes on the coordinate grid

Once the holes have been plotted onto the coordinate grid, a polar coordinate system (representing the round positional tolerance zones) is laid over the coordinate grid. See Fig. 18-5. The rings of the polar coordinate system represent the range of positional tolerance zones as allowed by the drawing specifica-

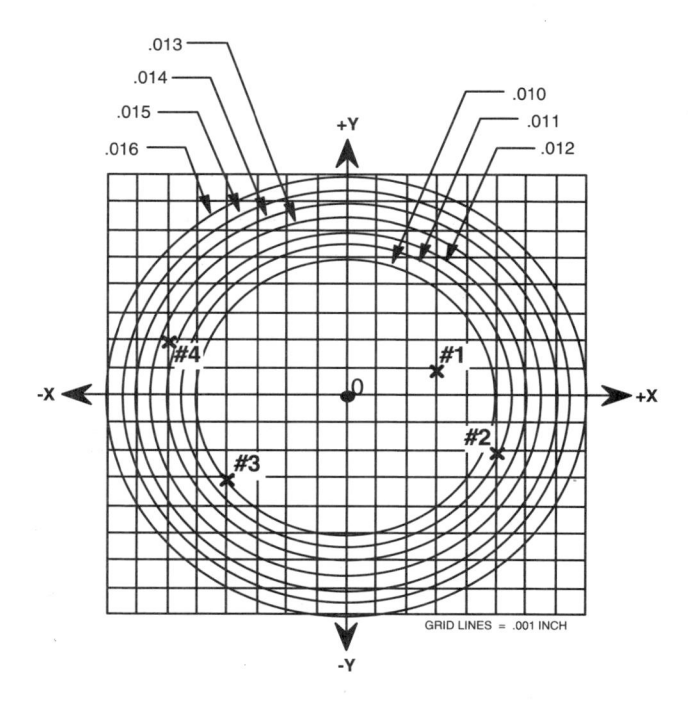

Figure 18-5 Overlaying the polar coordinate system

18-8 Chapter Eighteen

tion; Ø.010 positional tolerance allowed for a Ø.309 hole, up to Ø.016 allowed for a Ø.315 hole. With the center of the polar coordinate system aligned with the center of the coordinate grid, the inspector then visually verifies that each plotted hole falls inside its allowable position tolerance. If all the holes fall inside their zones, the part is good and the inspector is done.

For the example, all of the holes fall inside their respective tolerance zones, with the exception of hole #4 which is required to be inside a Ø.011 tolerance zone. However, the paper gage shows that the hole *does* fall inside a Ø.013 ring. With the MMC concept, the hole may be enlarged by Ø.002 to a size of Ø.312, which in turn increases the allowable positional tolerance to Ø.013. This brings the hole into compliance with the drawing specification.

18.6.1.2 Three-Dimensional Hole Pattern Verification

In the previous example, the holes were verified using a two-dimensional (2-D) analysis of the hole pattern using only measurements taken along the X and Y axes. This is a common practice used in reducing overall inspection time. By using only a 2-D analysis of the hole pattern, the inspector takes a calculated risk that the holes will remain relatively perpendicular based on known capabilities of the processes. Longer holes (usually 1/2-inch in length or longer) should be verified through a 3-D analysis of the hole pattern.

Fig. 18-6 illustrates the part used in the previous example except that the part thickness is greatly increased, making the length of the holes approximately 1-1/2 inches long. The part must be verified three-dimensionally to ensure that the entire length of the hole resides within the specified positional tolerance.

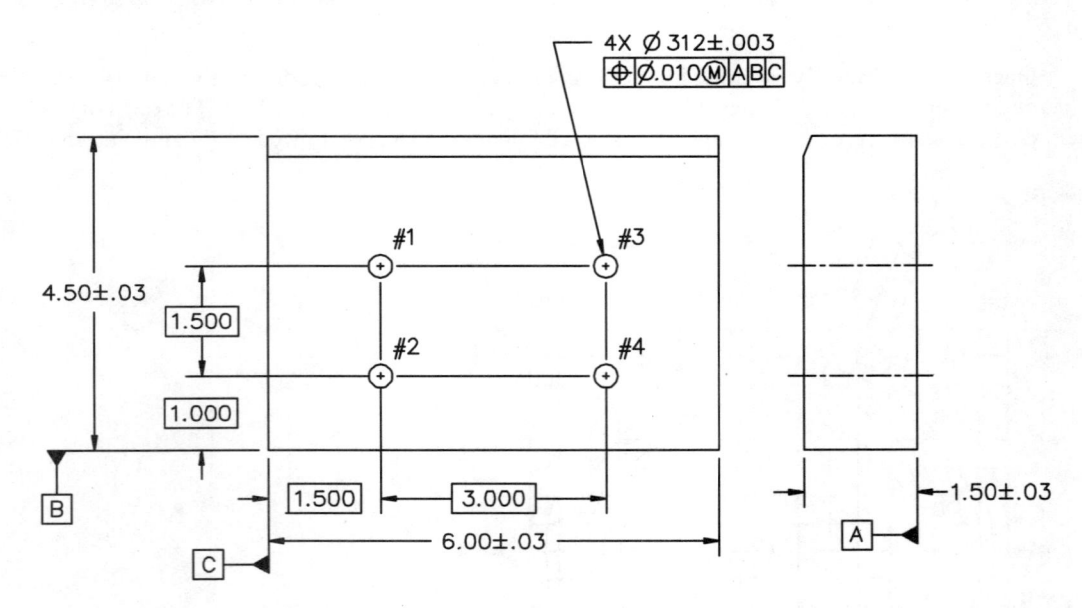

Figure 18-6 Example four-hole part with long holes

Setup and measurement of the workpiece is done in a manner similar to that used for the 2-D analysis except that the inspector must now collect two sets of measurements— one set for each end of the hole. Collecting data from each end of the hole allows the inspector to plot both ends of the hole axis on the coordinate grid of the paper gage: providing a 3-D rendering of the hole axis. Table 18-2 illustrates a sample Inspection Report used for a 3-D analysis of the hole pattern.

Paper Gage Techniques   18-9

Table 18-2 Inspection Report for part with long holes

NO.	FEATURE	FEATURE SIZE MMC	FEATURE SIZE ACTUAL	FEATURE SIZE DEV.	ALLOW TOL.	X LOCATION BASIC	X LOCATION ACTUAL	X LOCATION DEV	Y LOCATION BASIC	Y LOCATION ACTUAL	Y LOCATION DEV	ACCEPT	REJECT
1	.312±.003	.309	.312	.003	Ø.013	1.500	1.503	+.003	2.500	2.501	+.001	X	
						1.500	1.505	+.005	2.500	2.498	-.002	X	
2	.312±.003	.309	.311	.002	Ø.012	1.500	1.496	-.004	1.000	.997	-.003	X	
						1.500	1.494	-.006	1.000	1.002	+.002		X
3	.312±.003	.309	.313	.004	Ø.014	4.500	4.501	+.001	2.500	2.502	+.002	X	
						4.500	4.499	-.001	2.500	2.506	+.006	X	
4	.312±.003	.309	.312	.003	Ø.013	4.500	4.504	+.004	1.000	1.001	+.001	X	
						4.500	4.507	+.007	1.000	1.002	+.002		X

The Inspection Report reflects two sets of X and Y deviations for each hole, with each set representing the measured location of the hole axis. Both points are plotted on the coordinate grid and joined by a line to indicate that they represent the axis of a single hole. Fig. 18-7 illustrates the hole axes as they would appear after plotting on the coordinate grid.

As with the previous example, a polar coordinate system (representing the round positional tolerance zones) is laid over the coordinate grid as illustrated in Fig. 18-7 (right). With the center of the polar coordinate system aligned with the center of the coordinate grid, the inspector visually verifies that both ends of the hole axes reside inside its allowable position tolerance. This procedure creates the effect of a 3-D gage for the holes. For the example, both holes 2 and 4 would be rejected since one end of their axes lies outside the allowable tolerance zone.

When required, this technique also allows the individual perpendicularity for each hole to be easily measured. By circumscribing the smallest circle about the two points representing each hole axis, the actual perpendicularity for each individual hole can be derived. The actual perpendicularity must be less than, or equal to, the specified perpendicularity defined by the engineering drawing.

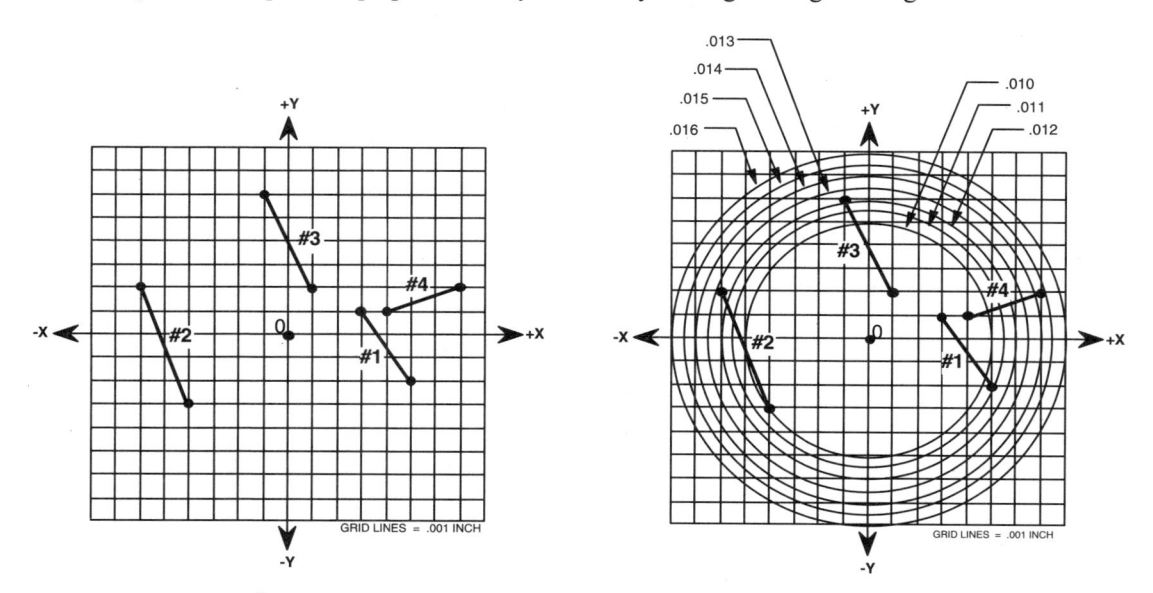

Figure 18-7 Plotting 3-dimensional hole data on the coordinate grid

18-10   Chapter Eighteen

18.6.1.3   Composite Positional Tolerance Verification

Composite positional tolerancing is a unique tolerance used in controlling patterns of two or more features. In this tolerancing method, the location of the entire pattern is less important than the relationship of features within the pattern. Verifying a composite positional tolerance using a fixed-limit gage would require the development of two separate gages, one for each requirement. However, with the paper gage, both requirements may be easily verified from a single set of measurements. Fig. 18-8 illustrates a composite position specification for the four-hole part used in previous examples.

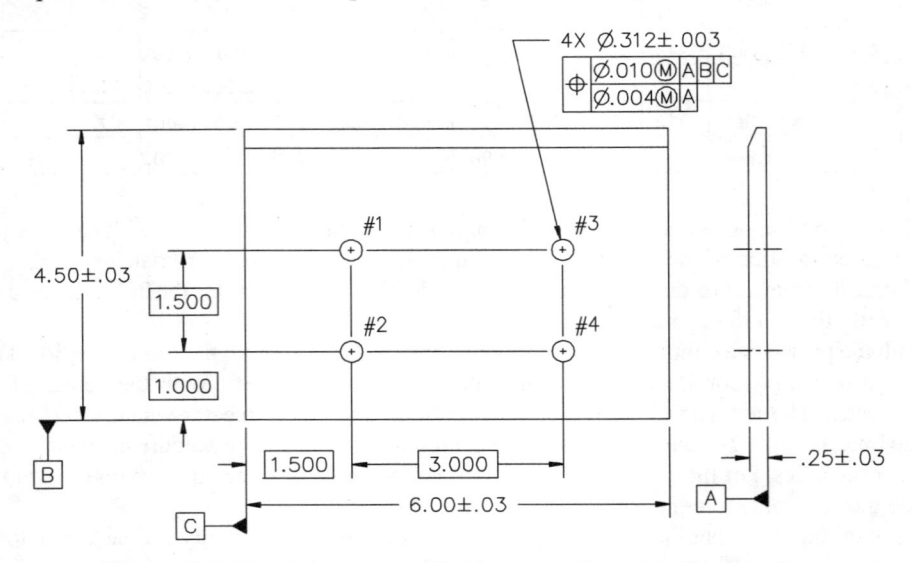

Figure 18-8   Four-hole part controlled by composite positional tolerancing

As in the previous examples, the inspector would set up the part, extract the measurements, and record the data on the Inspection Report as shown in Table 18-3. Note that the report reflects two allowable tolerances for each hole. The larger tolerance represents tolerance allowed by the upper segment of the feature control frame, with the smaller tolerance representing the tolerance allowed by the lower segment of the feature control frame.

Table 18-3   Inspection Report for composite position verification

		LAYOUT INSPECTION REPORT											
NO.	FEATURE	**FEATURE SIZE**			ALLOW TOL.	**X LOCATION**			**Y LOCATION**			ACCEPT	REJECT
		MMC	ACTUAL	DEV.		BASIC	ACTUAL	DEV	BASIC	ACTUAL	DEV		
1	.312 ±.003	.309	.310	.001	Ø.011	1.500	1.506	+.006	2.500	2.503	+.003		X
					Ø.005							X	
2	.312 ±.003	.309	.315	.006	Ø.016	1.500	1.505	+.005	1.000	1.006	+.006	X	
					Ø.010							X	
3	.312 ±.003	.309	.313	.004	Ø.014	4.500	4.506	+.006	2.500	2.499	−.001	X	
					Ø.008							X	
4	.312 ±.003	.309	.312	.003	Ø.013	4.500	4.501	+.001	1.000	1.005	+.005	X	
					Ø.007								X

Paper Gage Techniques   18-11

Verification of the upper segment is accomplished as in previous examples. A polar coordinate system (representing the round positional tolerance zones) is laid over the coordinate grid with the centers of both aligned as shown in Fig. 18-9. The inspector then visually verifies that each plotted hole falls inside its allowable position tolerance. If all the holes fall inside their zones, the part has passed the first requirement.

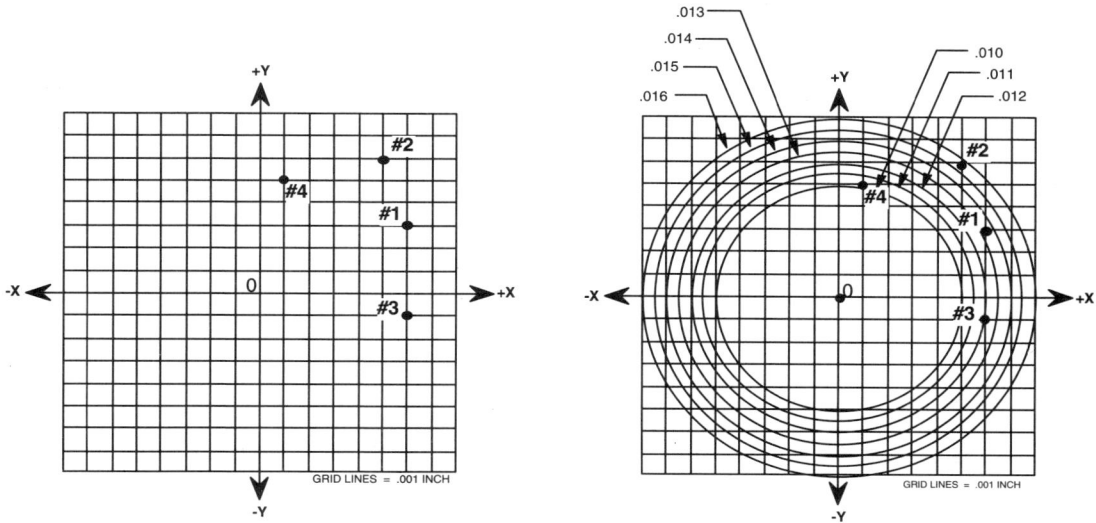

Figure 18-9 Paper gage verification of hole pattern location

Verification of the lower segment requires that a second set of smaller rings be laid over the same coordinate grid verifying the feature-to-feature relationship. Since the holes are not being measured back to the datums, the center of these smaller rings need not be aligned with the center of the coordinate grid. The overlay may be adjusted to an optimum position where all the holes fall inside their respective allowable tolerance zones, verifying that the holes are properly located one to the other. Fig. 18-10 illustrates the feature-to-feature verification for the example part.

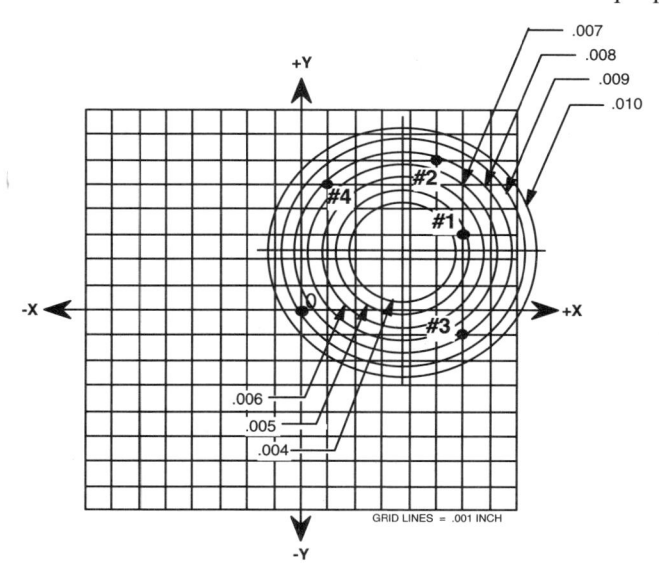

Figure 18-10 Paper gage verification of feature-to-feature location

18.6.2 Capturing Tolerance from Datum Features Subject to Size Variation

In one common assembly application, a pilot hole or diameter is used as a datum feature in locating a pattern of holes. Paper gaging is extremely useful in capturing dynamic tolerances that cannot be effectively captured in a typical layout inspection.

18.6.2.1 Datum Feature Applied on an RFS Basis

Verification in relation to a datum feature of size applied on a regardless of feature size (RFS) basis is done in a similar manner to datum features without size discussed earlier. For the part shown in Fig. 18-11, locational verification of the hole pattern requires that the inspector establish a datum reference frame from the high points of datum feature A (primary) and center on the pilot diameter B (secondary) regardless of its produced size. Establishing the secondary datum axis requires use of an actual mating envelope (smallest circumscribed cylinder perpendicular to datum plane A) as the true geometric counterpart for secondary datum B.

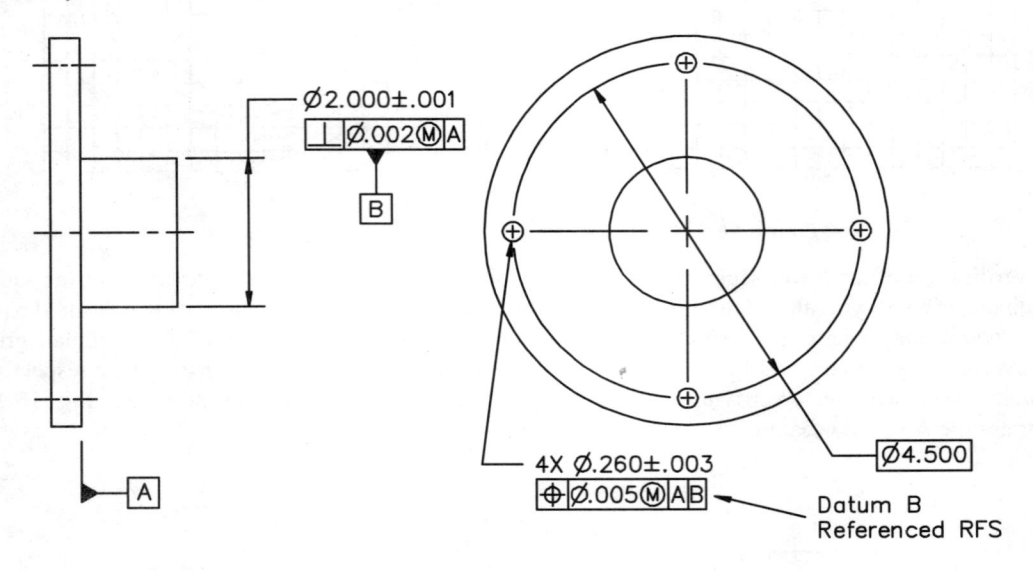

Figure 18-11 Datum feature subject to size variation—RFS applied

With the part locked into the datum reference frame, measurements are made in an "X" and "Y" direction and the data is recorded on the Inspection Report. The data is then transferred to the coordinate paper gage grid and converted into a round positional tolerance using the polar overlay. Since the datum feature has been referenced on an RFS basis, the polar overlay must remain centered on the coordinate grid to reflect the hole pattern centered on the datum feature, regardless of its produced size.

18.6.2.2 Datum Feature Applied on an MMC Basis

A fixed-limit boundary is used to represent the datum feature, where a datum feature of size is referenced on an MMC basis. For a primary datum feature of size, the boundary is the MMC size of the datum feature. For a secondary or tertiary datum feature of size, the boundary is the virtual condition of the datum feature. These boundaries are easily represented in a functional gage, allowing the datum feature to "rattle" around inside the boundary if the actual produced feature has departed its MMC or virtual condition size.

This rattle is commonly referred to as "datum shift" and is allowed to occur every time a datum feature of size is referenced on an MMC basis. However, unlike "bonus" tolerance, this shift allowance is not additive to the location tolerance indicated by the feature control frame for the holes. Rather, datum shift allows the pattern tolerance zone framework to shift off the datum axis (all the holes as a group) to get the controlled features in the tolerance zones.

This concept of allowing the actual datum feature to shift off the center of the datum simulator cannot be readily captured when verifying parts in a dimensional layout inspection. This is because conventional dimensional metrology equipment usually requires that the inspector "center-up" on features in order to take measurements. For a layout inspection, paper gaging may be the only way the inspector can capture these dynamic datum shift allowances.

Fig. 18-12 illustrates an example where a datum shift tolerance has been allowed for a geometric tolerance. The three holes and the outside shape are located in relation to the face (primary datum A) and the large diameter hole in the center (secondary datum B at MMC). Let's see how the datum shift tolerance might be captured by the inspector in this setup.

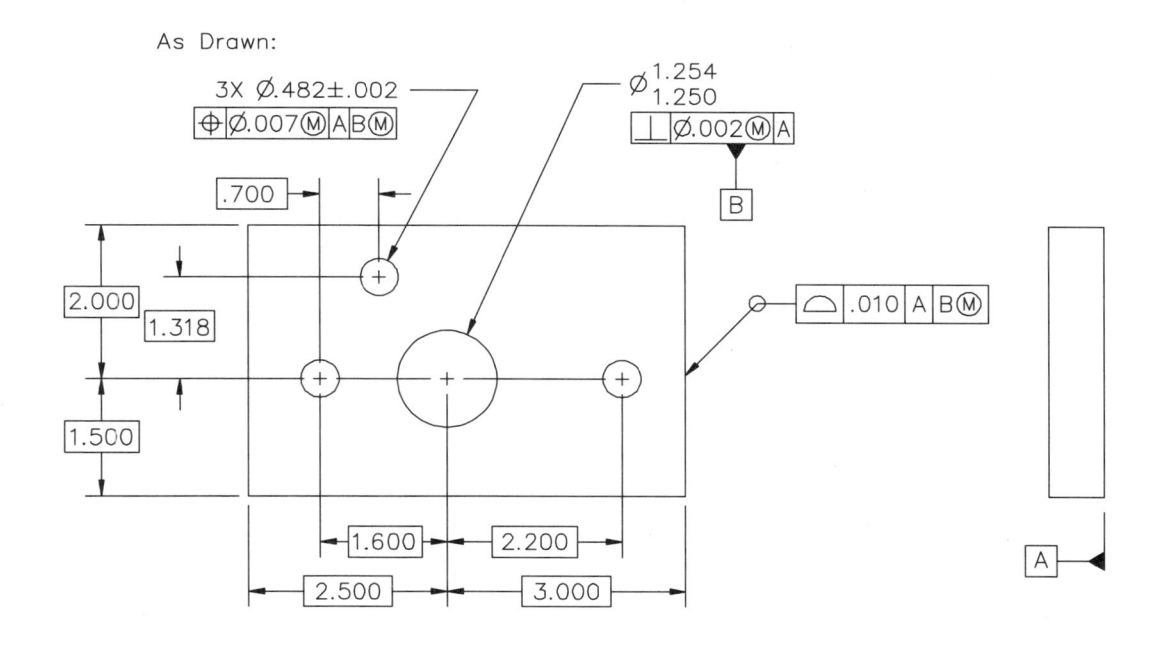

Figure 18-12 Paper gage verification for datum applied at MMC

A layout inspection of this part would begin with the inspector inserting the largest pins that could be placed inside the holes as a means of verifying their size. The part must then be locked into the datum reference frame by setting up to the face first (primary datum plane A) and centering on the large hole (secondary datum axis B). To provide direction for the measurements, one of the three smaller holes is arbitrarily selected to antirotate the part. The final measurement layout might resemble the setup illustrated in Fig. 18-13.

The inspector extracts actual measurements in an "X" and "Y" direction from the established frame of reference, as well as produced sizes and calculations for the allowable positional tolerances on each hole.

18-14 Chapter Eighteen

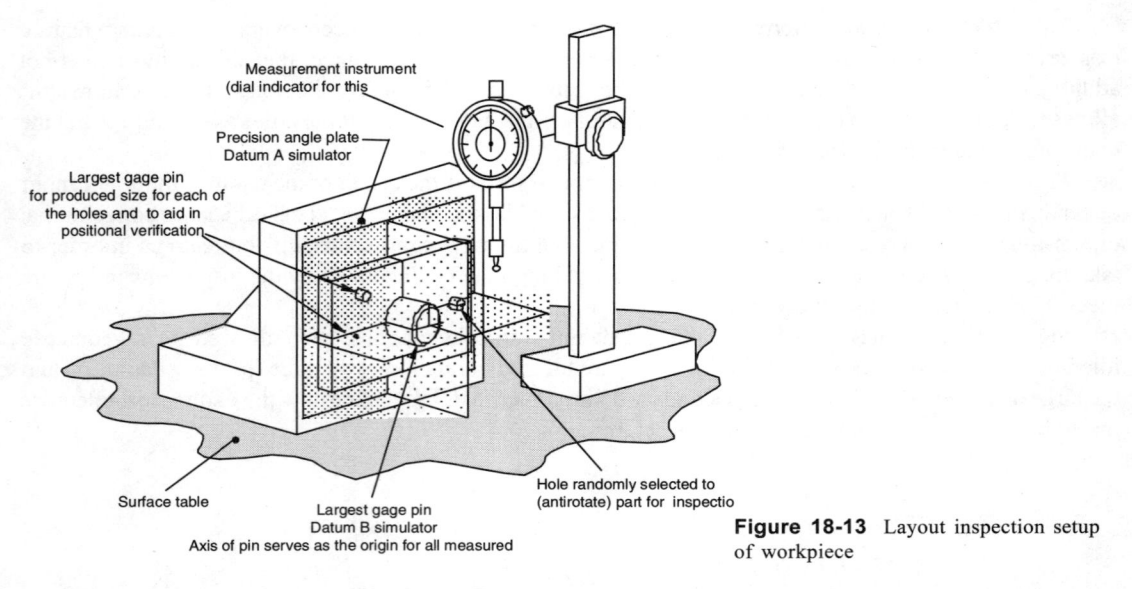

Figure 18-13 Layout inspection setup of workpiece

The amounts each hole deviated from the basic dimensions as defined by the engineering drawing are entered in the Inspection Report as "X" and "Y" deviations as shown in Fig. 18-14.

Produced Part:

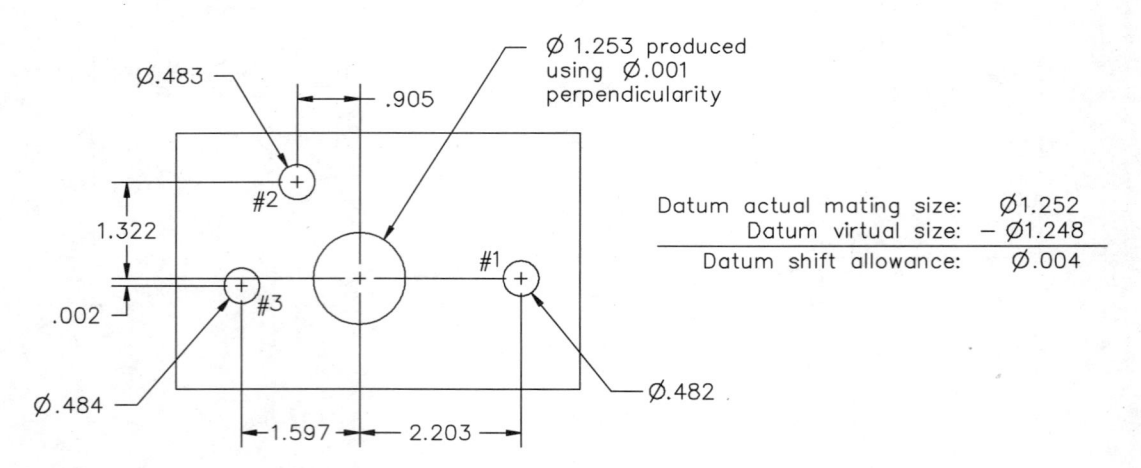

Datum actual mating size: Ø1.252
Datum virtual size: − Ø1.248
Datum shift allowance: Ø.004

LAYOUT INSPECTION REPORT

NO.	FEATURE	FEATURE SIZE			ALLOW TOL.	X LOCATION			Y LOCATION			ACCEPT	REJECT
		MMC	ACTUAL	DEV.		BASIC	ACTUAL	DEV	BASIC	ACTUAL	DEV		
1	.482±.002	.480	.482	.002	Ø.009	2.200	2.203	+.003	0	0	0	X	
2	.482±.002	.480	.483	.003	Ø.010	-.900	-.905	-.005	1.318	1.322	+.004	X	
3	.482±.002	.480	.484	.004	Ø.011	-1.600	-1.597	+.003	0	-.002	-.002	X	

Figure 18-14 Inspection Report — part allowing datum shift

Using the data from the Inspection Report, the information is transferred to the paper gage by plotting each of the holes on a coordinate grid (which represents the inspector's measurements) as shown in Fig. 18-15. The center of this grid represents the basic or true position for each of the holes, as well as the center of the datum reference frame. The actual hole locations relative to their true position is plotted on the grid using the X and Y deviations from the inspector's measurements.

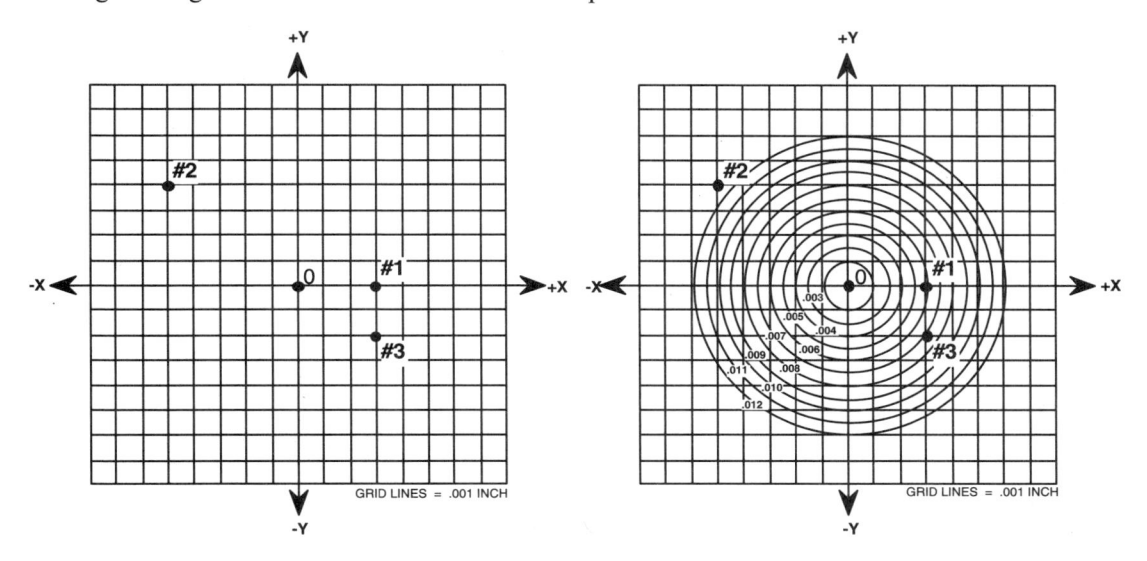

Figure 18-15 Verifying hole pattern prior to datum shift

Once the holes have been plotted onto the coordinate grid, a polar grid (representing the round positional tolerance zones) is laid over the coordinate grid as shown in Fig. 18-15 (right), with the centers of the two grids aligned. The inspector then looks to see that each plotted hole falls inside its total allowable position tolerance. If all the holes fall inside their zones, the part is good and the inspector is done.

But, for the example shown, hole #2 falls well outside the Ø.010 positional tolerance allowed for a Ø.483 hole when the polar grid is centered on the coordinate grid. Even enlarging the hole to its largest size of Ø.484 would not add enough bonus tolerance to make the part good. But, is the part really bad?

Remember that when the holes were inside their tolerance "rings," the two grids were aligned, with one on the center of the other (RFS). But the drawing references datum B on an MMC basis requiring that a fixed-limit, virtual condition cylinder represent the datum. Comparing the actual mating size of datum feature B to its calculated virtual condition size shows that there is a Ø.004 difference between the two. This difference reflects the shift tolerance allowed for the datum feature. This allowable shift may be translated to the hole verification by moving the polar grid such that the center of the coordinate grid remains inside a Ø.004 zone when measuring the holes as shown in Fig. 18-16.

This movement between the two grids represents the allowable shift derived from the datum feature's departure from virtual condition. When shifting the polar grid in this manner, care must be taken to assure that all of the holes fall within their respective tolerance zones. If the polar grid can be moved to an optimum position that accepts all of the holes in their tolerance zones without violating the datum shift tolerance zone, then the hole pattern is accepted as being within tolerance.

18-16   Chapter Eighteen

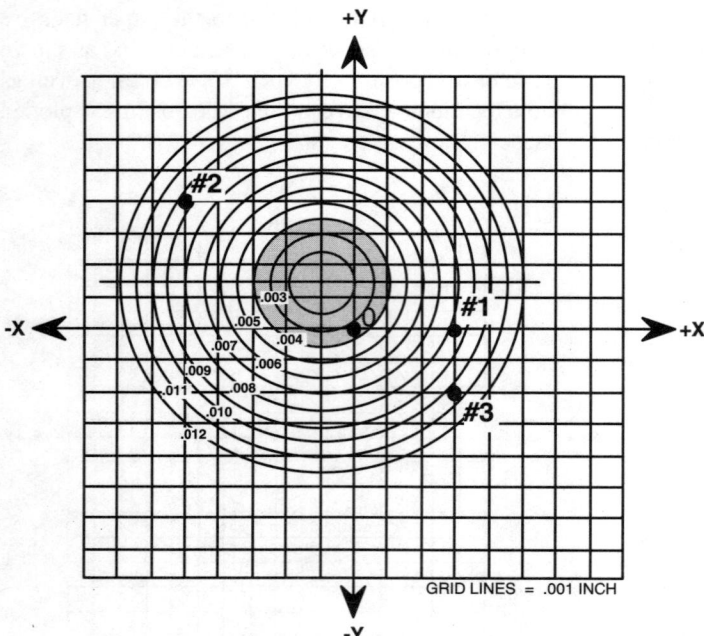

Figure 18-16 Verifying the hole pattern after datum shift

18.6.2.3 Capturing Rotational Shift Tolerance from a Datum Feature Applied on an MMC Basis

For the cylindrical part in Fig. 18-17, the hole pattern must be oriented in relation to the tertiary datum slot, referenced on an MMC basis. If the slot were to be simulated in a functional gage, a virtual condition width would be used as the true geometric counterpart for datum feature C. As the produced slot departed virtual condition (it is produced at a larger size and/or uses less of its allowed positional tolerance) the

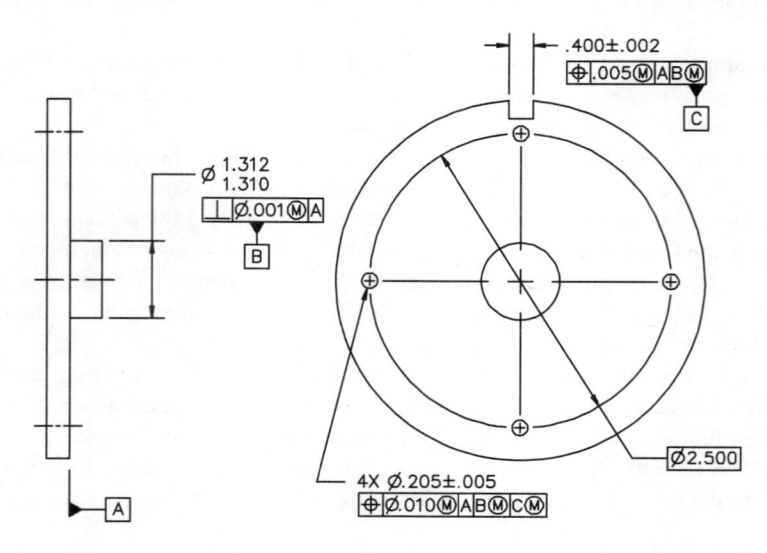

Figure 18-17 Part allowing rotational datum shift

Paper Gage Techniques 18-17

entire hole pattern, as a group, would be allowed to rotate in relation to the true geometric counterpart of datum feature C when verifying the position for the hole pattern.

As with previous examples, the inspector would lock the part into the datum reference frame as prescribed by the drawing and collect the measurement data for the hole locations. The extracted measurements would then be delineated on the Inspection Report as shown in Fig. 18-18.

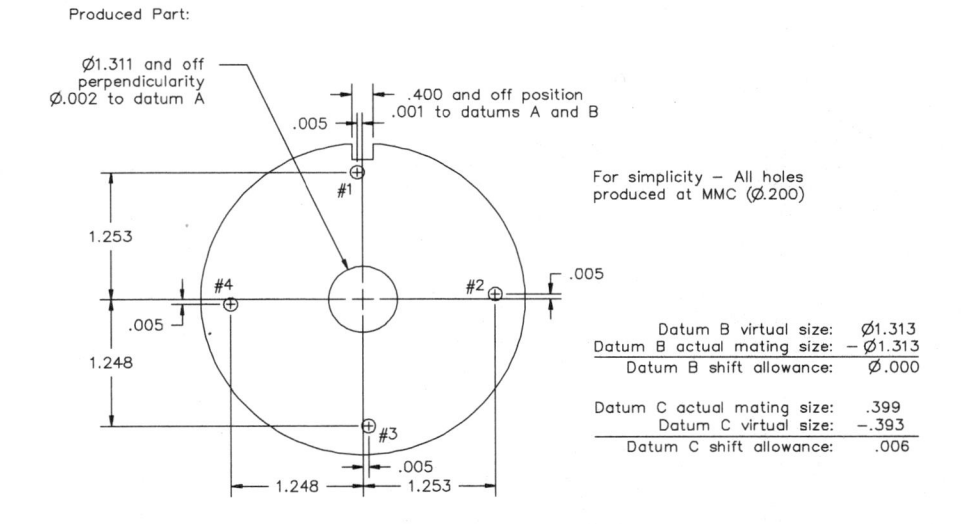

Produced Part:

Ø1.311 and off perpendicularity Ø.002 to datum A

.400 and off position .001 to datums A and B

For simplicity — All holes produced at MMC (Ø.200)

Datum B virtual size:	Ø1.313
Datum B actual mating size:	−Ø1.313
Datum B shift allowance:	Ø.000

Datum C actual mating size:	.399
Datum C virtual size:	−.393
Datum C shift allowance:	.006

LAYOUT INSPECTION REPORT

NO.	FEATURE	FEATURE SIZE			ALLOW TOL.	X LOCATION			Y LOCATION			ACCEPT	REJECT
		MMC	ACTUAL	DEV.		BASIC	ACTUAL	DEV	BASIC	ACTUAL	DEV		
1	.205±.005	.200	.200	0	Ø.010	0	−.005	−.005	1.250	1.253	+.003	X	
2	.205±.005	.200	.200	0	Ø.010	1.250	1.253	+.003	0	+.005	+.005	X	
3	.205±.005	.200	.200	0	Ø.010	0	+.005	+.005	−1.250	−1.248	+.002	X	
4	.205±.005	.200	.200	0	Ø.010	−1.250	−1.248	+.002	0	−.005	−.005	X	

Figure 18-18 Inspection Report—part allowing rotational datum shift

To focus on the datum shift derived from the slot, assume that all the holes are produced at MMC of Ø.200 and that the secondary datum pilot B is produced at its virtual condition, providing no datum shift itself. When the holes are plotted onto the grid as shown in Fig. 18-19, they all fall outside the Ø.010 positional tolerance allowed for a Ø.200 hole.

Since datum feature B was produced at its virtual condition (thereby allowing no datum shift), the polar grid must remain on the center of the coordinate grid. However, datum feature C (the slot) did depart from its virtual condition, allowing datum shift for the hole pattern in the form of rotation of the pattern.

Calculations show that the slot departed its virtual condition by .006 total. However, since the holes are closer to the center of rotation than is the slot, we may only realize a portion of the available .006 shift provided by the slot at the holes themselves. Since the holes lie roughly 80% of the distance from the rotational center to the center of the slot, it can be assumed that only about 80% of the .006 rotational shift tolerance will occur at the axis of the holes, or an estimated .005. This means that the hole pattern may be rotated by ±.0025 from its current position in an attempt to get all the holes inside the Ø.010 positional tolerance zone.

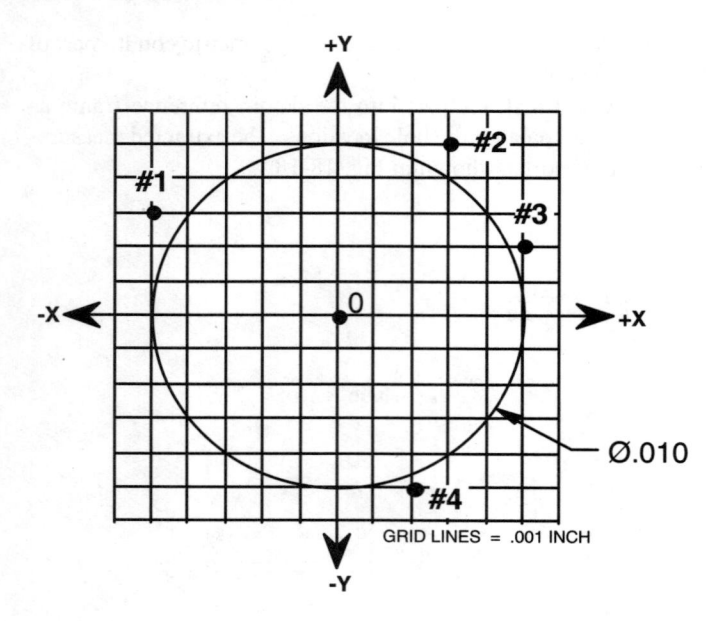

Figure 18-19 Verifying hole pattern prior to rotational shift

When the part is rotated, the holes will move (as a group) to a new location on the coordinate grid. If the part is rotated clockwise by .0025, hole #1 will shift to the right, hole #2 will shift down, hole #3 will shift to the left, and hole #4 will shift up. Fig. 18-20 illustrates how, after rotation, the pattern moves closer to the center, resulting in all of the hole axes falling well inside the allowable Ø.010 positional tolerance zone.

Use of the paper gage illustrated provides an approximate evaluation for the hole pattern. To prove the results, the inspector could reset the part for a second inspection using the new alignment for datum feature C.

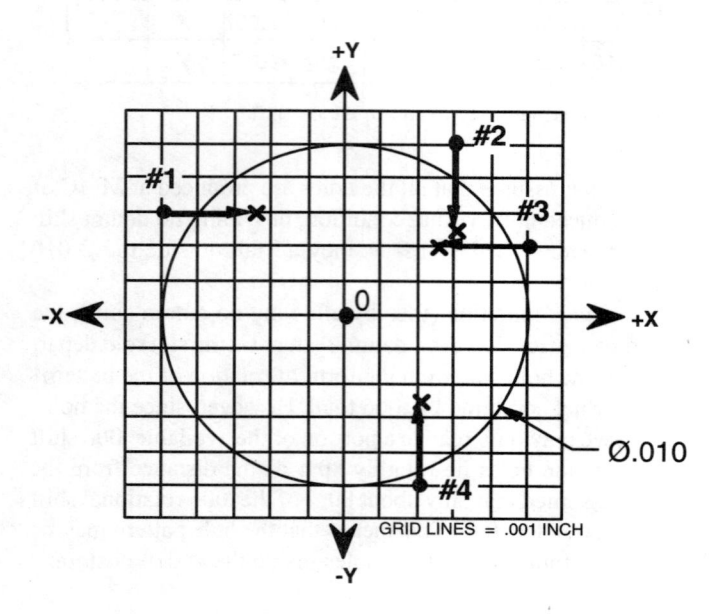

Figure 18-20 Verifying hole pattern after rotational datum shift

18.6.2.4 Determining the Datum from a Pattern of Features

Where a pattern of features, such as a hole pattern, are used as a datum feature at MMC, the true geometric counterpart of all holes in the pattern are used in establishing the datum. For the example shown in Fig. 18-21, the true geometric counterpart for the pattern of three round holes consists of three true cylinders representing the virtual condition of each hole in the pattern. (Using virtual condition cylinders compensates for any locational error between the holes.) When referenced on an MMC basis, the axis of the pattern may shift and/or rotate within the bounds of these cylinders as the holes in the pattern depart from virtual condition (i.e., they grow larger in size and/or use less positional tolerance).

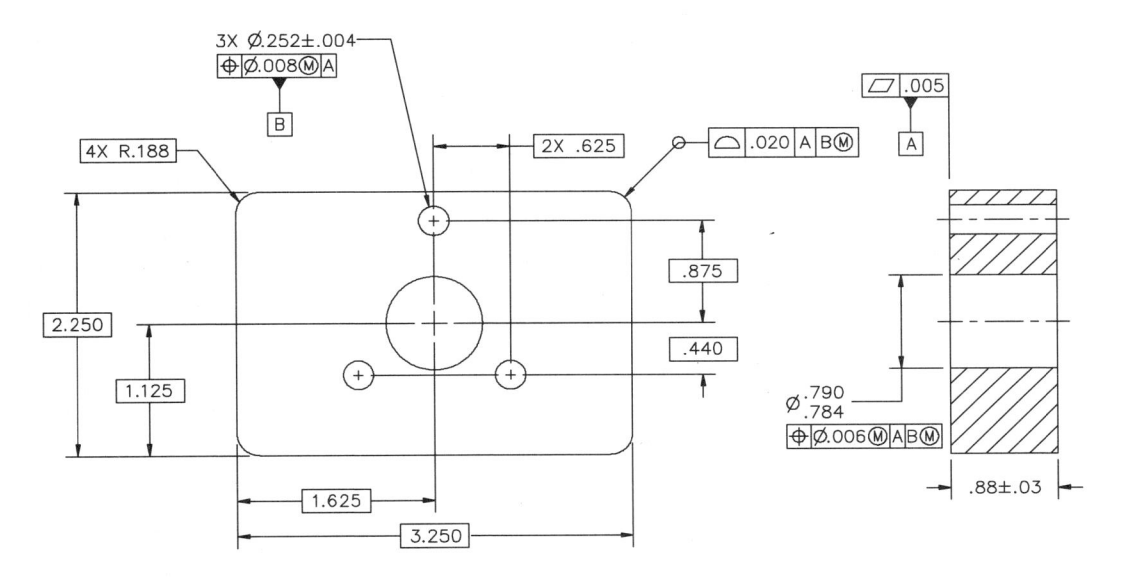

Figure 18-21 Example of datum established from a hole pattern

These virtual condition "cylinders" may be represented by pins in a functional gage. By simply dropping the part over the gage pins, the produced hole pattern will average over the pins, relating the part to datum axis B. But, development of a hard gage is not required to simulate the averaging of the feature pattern to establish the datum. The drawing in Fig. 18-21 shows a part where the three-hole pattern will serve as secondary datum feature B at MMC. Since this part will be made in a very small quantity, it would not be practical or cost effective to build a gage to simulate the datum. Verification of the geometric tolerances will be done using a conventional layout inspection and paper gaging.

To establish the datum reference frame from a pattern of holes in an open setup or CMM, the hole pattern must be "averaged" to find a "best fit" center for the pattern. This might be accomplished by randomly selecting any hole of the pattern from which to start measuring. The remaining holes may be checked to this "frame of reference" as well as other geometric tolerances related to the datum hole pattern. Fig. 18-22 illustrates the measurements extracted for the three-hole datum pattern where the inspector used the top hole as the starting point.

If all tolerances check within their respective zones, then the part is accepted. If the part checks to be bad, then the inspector may need to paper gage the actual measurements taken for the holes to find the pattern center. This would be done by plotting the holes on the grid and then graphically "squaring up" the pattern by rotating the holes about the datum setup hole until they are equally dispersed in relation to

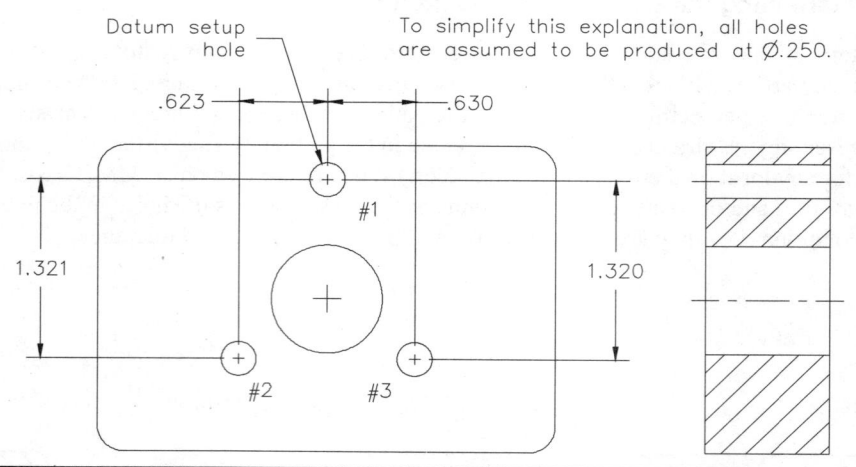

18-20 Chapter Eighteen

Figure 18-22 Inspection Report—hole pattern as a datum

NO.	FEATURE	FEATURE SIZE			ALLOW TOL.	X LOCATION			Y LOCATION			ACCEPT	REJECT
		MMC	ACTUAL	DEV.		BASIC	ACTUAL	DEV	BASIC	ACTUAL	DEV		
1	.252±.004	.248	.250	.002	Ø.010	0	0	0	0	0	0	X	
2	.252±.004	.248	.250	.002	Ø.010	-.625	-.623	+.002	-1.315	-1.321	-.006	X	
3	.252±.004	.248	.250	.002	Ø.010	.625	.630	+.005	-1.315	-1.320	-.005	X	

the coordinate grid centerlines as illustrated in Fig. 18-23 (left). To square up the pattern for this example, the part is rotated clockwise by .0035".

By circumscribing the smallest diameter about the plotted holes, the "axis of the feature pattern" (best-fit center) for the pattern of holes may be approximated. For the example in Fig. 18-23 (right), the inspector would need to reset the origin for measurement by -.00075 in the "X" direction and -.003 in the "Y" direction to get the actual measurements from the pattern center.

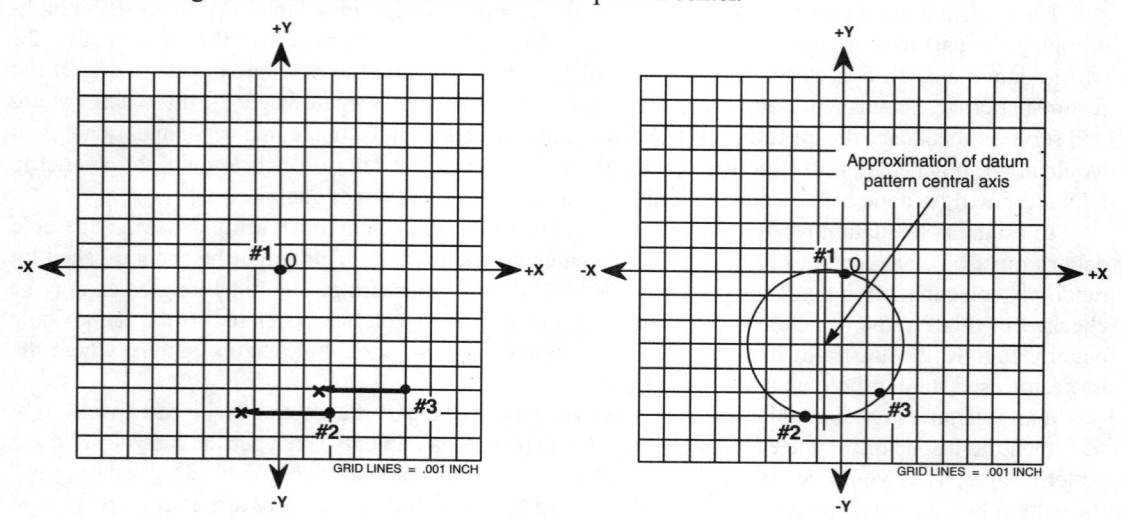

Figure 18-23 Determining the central datum axis from a hole pattern

Since the hole pattern is referenced on an MMC basis, the part would be allowed to shift and/or rotate in relation to the datum reference frame as the holes of the datum feature pattern depart from virtual condition. The amount of shift for the hole pattern may be determined on the paper gage by striking an arc representing the allowed positional tolerance for each of the plotted holes as shown in Fig. 18-24. The resulting area where the tolerance zones overlap approximates the pattern's departure from virtual condition (available datum shift tolerance).

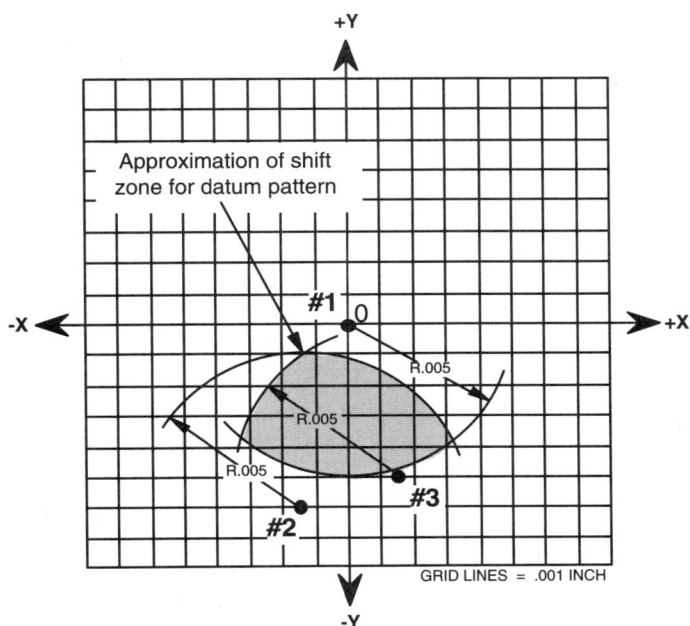

Figure 18-24 Approximating datum shift from a hole pattern

18.6.3 Paper Gage Used as a Process Analysis Tool

As stated earlier in the text, paper gaging techniques are excellent tools used in identifying problems during the manufacturing process. When the holes are plotted on the coordinate grid, they provide a graphical "picture" of the process that can help identify production problems and isolate their root cause. Periodic paper gage evaluations, combined with accepted statistical methods, can assist the operator in keeping the process in control *before* bad parts are produced. This can significantly reduce production costs by raising the usable output, lowering scrap rates, and eliminating wasted man-hours attempting to salvage defective parts. Fig. 18-25 illustrates several production problems that may be identified using paper gage techniques.

In Fig. 18-25 (a and b), it appears that the process is quite capable of producing the parts since the holes on both grids fall together in a relatively close grouping. The problem for these parts seems to be that the pattern has drifted off center; one pattern along the X axis (Fig. 18-25a) and the other along the Y axis (Fig. 18-25b). This may have resulted from movement of the stops used to locate the part in the machinery. It may have resulted from something preventing the part from coming down fully to the stops, such as excessive chips on the machine bed. The amount of correction required can be determined by circumscribing the smallest possible circle about the hole grouping. This roughly approximates the center of the pattern. By simply counting the grid lines between the center of this circle and the center of the coordinate grid, the operator may determine the amount of adjustment required to get the pattern back on center.

(a) Pattern Shift in X Axis

(b) Pattern Shift in Y Axis

(c) Process out of control

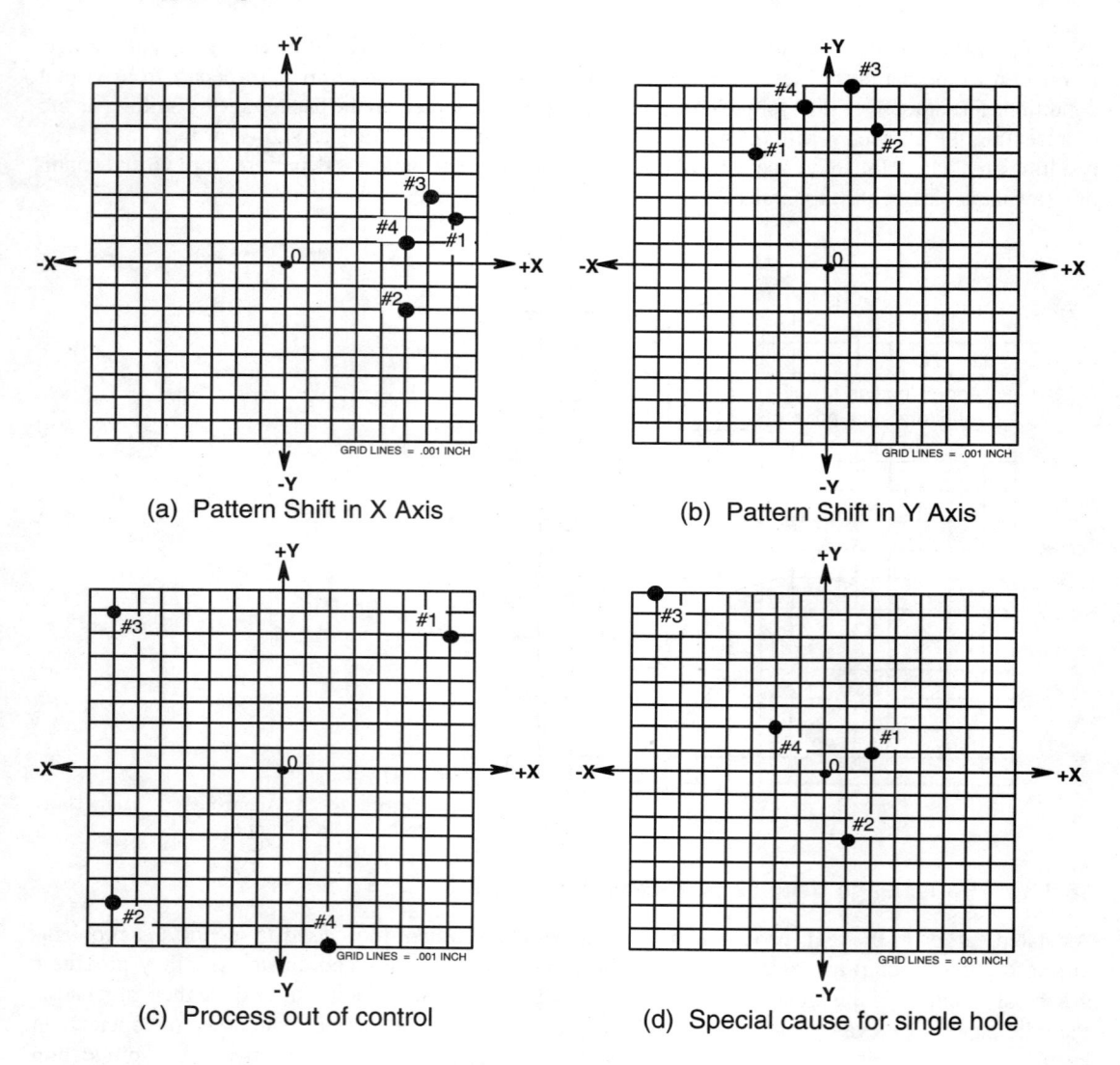

(d) Special cause for single hole

Figure 18-25 Process evaluation using a paper gage

The coordinate grid shown in Fig. 18-25(c) illustrates a hole pattern that is widely scattered over the coordinate grid and falls toward the extremes of the tolerance limits. The accuracy of the hole pattern is poor, and the reliability is questionable since a minor change in the process could result in one or more of the holes dropping outside their allowable tolerance. This could indicate an unstable or out-of-control process.

Fig. 18-25(d) illustrates a hole pattern where one of the holes (hole #3) has deviated to an extreme from the others. The remaining three holes fall as a group relatively close to the grid center, indicating a generally accurate and reliable process for the majority of the holes. This is a clear indicator that hole #3 deviated due to some special cause. Paper gaging additional parts would help to determine if this were a single occurrence or an ongoing problem requiring additional corrective action.

18.7 Summary

Paper gaging is an extremely valuable dimensional analysis tool used in verifying a wide range of geometric tolerance applications. As illustrated in this chapter, the technique allows for the easy translation of 2-D coordinate measurements extracted from traditional layout inspections into round 3-D tolerance zones for verifying part conformance. The technique also provides an effective means for capturing dynamic tolerances, such as datum shift allowance, which cannot be realized in a traditional layout inspection.

Simplicity of preparation and use, combined with the pictorial form of data presentation, makes a paper gage extremely easy for the average person to read and understand. When used appropriately, a paper gage can also save time and money in part inspection through its ability to represent part functional boundaries without the high cost of designing, building, and maintaining a traditional hard gage.

This chapter has also demonstrated how a paper gage may be used as a manufacturing problem-solving tool to quickly identify and correct problems during production. Periodic paper gage evaluations, combined with accepted statistical methods, can greatly aid the operator in keeping the process in control before bad parts are produced. This can help to lower production costs by raising usable part yield, lowering scrap rates, and eliminating wasted man-hours attempting to salvage defective product.

18.8 References

1. Foster, Lowell W. 1986. *Geometrics II, The Application of Geometric Tolerancing Techniques.* Minneapolis, MN: Addison-Wesley Publishing Company, Inc.
2. Neuman, Alvin G. 1995. *Geometric Dimensioning and Tolerancing Workbook.* Longboat Key, FL: Technical Consultants, Inc.
3. Pruitt, George O. 1983. *Graphical Inspection Analysis.* Doc No. NWC TM 5154. China Lake, CA: U.S. Naval Weapons Center.
4. The American Society of Mechanical Engineers. 1995. *ASME Y14.5M-1994, Dimensioning and Tolerancing.* New York, New York: The American Society of Mechanical Engineers.

Chapter 19

Receiver Gages — Go Gages and Functional Gages

James D. Meadows
Institute for Engineering & Design, Inc.
Hendersonville, Tennessee

James D. Meadows, president of the Institute for Engineering and Design, Inc., has instructed more than 20,000 professionals in Geometric Dimensioning and Tolerancing and related topics over the last 30 years. He is the author of two current hardcover textbooks, a workbook and a 14-hour, 12-tape applications-based video training program on GD&T per the ASME Y14.5M-1994 standard on dimensioning and tolerancing. He is an ASME Y14.5.2 Certified Senior Level Geometric Dimensioning and Tolerancing Professional. Mr. Meadows is a member of eight ANSI/ASME and ISO standards committees and serves as the chairman of the committee on Functional Gaging and Fixturing of Geometric Tolerances. He is a journeyman tool and die maker and a graduate of Wayne State University.

19.1 Introduction

Receiver gaging is one of the most effective ways to determine the functionality of workpiece features.

There are two members of the receiver or attribute gage family: Functional Gages and GO gages. Functional and GO gages both determine feature compliance with a fixed size boundary; hence they are considered attribute gages.

Functional Gages inspect compliance with a constant functional boundary commonly associated with a worst mating condition. This boundary is known as a maximum material condition (MMC) concept virtual condition boundary. Functional Gages are made to the MMC concept virtual condition boundary of the features they inspect, then toleranced so they represent a situation worse than the features will face in assembly conditions.

GO gages are used to determine compliance with the maximum material condition boundary of perfect form required by several American National Standards (ANSI B4.4, ASME Y14.5, and ASME Y14.5.1).

This type of measurement is a physical representation of the theoretical principles of geometric tolerancing of workpieces. It shows the datum feature simulation and virtual condition boundaries as pins and holes that are cylindrical, diamond-shaped, widths, and even oddly configured. It demonstrates that planar features are represented by planar rails and that datum features and controlled features are represented in gages and fixtures by the shape of the MMC or virtual condition they generate. It allows a theoretical boundary to take on a physical form that a person can actually hold in their hands, and, by doing so, is capable of making a difficult geometric concept easy to understand.

Functional and GO gaging are time-tested tools of 3-dimensional (3-D) measurement that determine whether or not workpiece features will actually fit into assemblies. They do this without the use of computers or software. They are reliable and low tech. If used in a well-balanced measurement plan in conjunction with other measurement tools, they can provide the confidence needed to accept produced parts on the basis that they will perform their intended function.

Gaging of this variety is sometimes viewed as inappropriate because it produces no variables data (specifically how a feature has departed from perfect geometric size, form, orientation or location) and is therefore incapable of assisting in the statistical process control of manufacturing methods. However, many measurement techniques that do produce variables data are not representative of worst case assembly conditions and collect very little 3-D data concerning worst case feature high point interference possibilities. The type of data collected by functional and GO gaging is considered attribute (good vs. bad) information.

Both variables data and attribute data have their place in a well-balanced measurement procedure. Unfortunately, many measurement professionals are led to believe that only one of the two types of measurement information is to be used. Therefore, they lose the benefits of the type they do not choose.

19.2 Gaging Fundamentals

In a perfect circumstance, fixed limit gages accept all features that conform to their tolerance specification and reject all features that do not conform to their tolerance specification. The GO gage and the Functional Gage should each completely receive the feature it is inspecting.

GO plug gages should enter holes over the full length of the hole when applied by hand without using extreme force. A GO cylindrical ring gage should pass over the entire length of a shaft when applied by hand. This inspects not only a violation of the maximum material condition size limit, but also the envelope of perfect form at maximum material condition that American National Standards require. The rule in ANSI is that size limits control the surface form of rigid features.

The international rule is not the same. In the International Organization for Standardization (ISO), size is independent of form. Therefore, according to the ISO policy, unless otherwise specified, size inspection does not require a full form GO gage. Simple cross-sectional inspection procedures are all that are necessary to verify size requirements.

In ANSI-approved documents, NOGO gages are designed to inspect violations of the least material condition (LMC) limit of size. The LMC limit of feature size is inspected with a NOGO gage (or a simulation of this gage). The NOGO gage is a cross-sectional checking device, treating a cylinder as though it was a stack of coins. Each coin in the stack represents a circular cross-section of the surface. Each cross-section must not measure less than the least material condition. Since the requirement is that the gage "not go" over the workpiece, the NOGO gage should not be able to pass into or over the workpiece feature being inspected at any orientation or location.

A Functional Gage pin must be able to fully engage the hole it is inspecting over the entire depth of the hole without extreme force being applied. A Functional Gage hole, which is a full form ring gage, should be able to receive the shaft being gaged over the full length of the shaft without extreme force being applied. If planar datum features are being simulated by the gage, the datum features on the workpiece must contact the datum feature simulators on the gage with the required contact specified by ASME Y14.5M-1994 and ASME Y14.5.1M-1994. If restraint is to be used to inspect the workpiece features while on the datum features, it must be specified in notes or other documents relating to the feature measurement requirements. If no restraint is to be used, or restraint insufficient to alter the measurement readings, no note is required. However, a Free State Inspection symbol may be used inside feature control frames to clarify that the part is not to be distorted by restraining forces during the inspection procedure.

19.3 Gage Tolerancing Policies

Gages must be toleranced. There are three gage tolerancing policies commonly practiced throughout the world. These policies are known as: Optimistic Tolerancing, Tolerant Tolerancing, and Absolute Tolerancing (also called the Pessimistic Tolerancing approach).

Optimistic Tolerancing is not an ANSI-recommended practice for gages. It assures that all parts within specifications will be accepted by the gage. Most of the technically out-of-tolerance parts being inspected by the gage will be rejected, but a small percentage of technically out-of-tolerance parts will be accepted. This policy is accomplished by tolerancing the gages from their appropriate MMC or MMC concept virtual condition boundary so that gage pins can only shrink and gage holes can only grow from these boundaries. This method subtracts material from the gage so that gagemaker's tolerances, wear allowances, form tolerances and measurement uncertainties all reside outside the workpiece limits of size and geometric control.

Tolerant Tolerancing is also not an ANSI-recommended practice for gages. It assures that most parts within specification will be accepted by the gage. Most of the parts outside the specification will be rejected by the gage. A small percentage of parts outside the specifications may be accepted by the gages or a small percentage of parts that are within the specifications may be rejected by the gages. This policy may either add or subtract material from the gage MMC boundary or MMC concept virtual condition boundary since the tolerance is both plus and minus around these boundaries. This means that some of the gagemaker's tolerances, the wear allowances, the form tolerances and the measurement uncertainties reside both within and outside of the workpiece limits of size and geometric control.

Absolute Tolerancing is recommended. This type of gage tolerancing means that gage pins are toleranced only on the plus side of their MMC concept virtual condition boundary (only allowing them to grow) and that gage holes are toleranced only on the minus side of their MMC concept virtual condition boundary (only allowing them to shrink). This has the effect of rejecting all parts not within tolerance and accepting all parts that are within tolerance except those borderline parts that fall within the range of the gage tolerance. Part features that are produced within the range of the gage tolerance are rejected as though they were not in compliance with their geometric tolerance, even though technically they are within the design specification limits. This is the price we must pay if we choose to accept no parts that have violated their tolerance.

Absolute Tolerancing is the ANSI-recommended practice of applying gage tolerances so that the gages will reject all workpiece features that reside outside of their specifications. This is to assure complete random interchangeability of mating parts in an assembly inspected by these gages. Gagemaker's tolerances, wear allowances, form tolerances and measurement uncertainties of the gage are all within the workpiece limits of size and geometric control. These gage tolerances add material to the gage. The gages are dimensioned at the MMC limit or MMC concept virtual condition limit of the feature being gaged, then toleranced so that gage pins can only get larger and gage holes can only get smaller. This policy is based on the gaging premise that all parts not within tolerance will be rejected, most parts that are within tolerance will be accepted, and a small percentage of in-tolerance parts that are considered near the borderline between good and bad will be rejected as though they had violated their tolerance requirements.

The ANSI-recommended amount of tolerance is 5% of the tolerance on the feature being gaged plus an optional 5% of the tolerance allowed for wear allowance. This recommendation is only a place from which to begin the decision as to what tolerance will be assigned to the gage. Using the Absolute Tolerancing method, the actual amount of tolerance chosen will depend on the number of parts the gage will accept and the number of parts one is willing to reject with the gage. It is a balance between the cost of the gage and the cost of the rejection of good parts by the gage. The smaller the gage tolerance, the more expensive the gage and the quicker the gage will wear beyond acceptable limits and begin to accept bad parts. On the other hand, the larger the gage tolerance, the less expensive the gage. However, the gage will run the risk of being produced at a size that will reject a larger number of produced parts that are within tolerance but near the borderline.

19.4 Examples of Gages

The following examples show a variety of workpieces and the gages to verify their conformance with common geometric tolerances. The gages may be toleranced using maximum material condition, least material condition, or regardless of feature size concepts. Each has advantages and disadvantages of cost and part acceptance.

19.4.1 Position Using Partial and Planar Datum Features

In Fig. 19-1 the workpiece is a simple rectangular part with two holes. The datum reference frame is constructed from three planar surfaces, two of which are partial datum features of limited specified length.

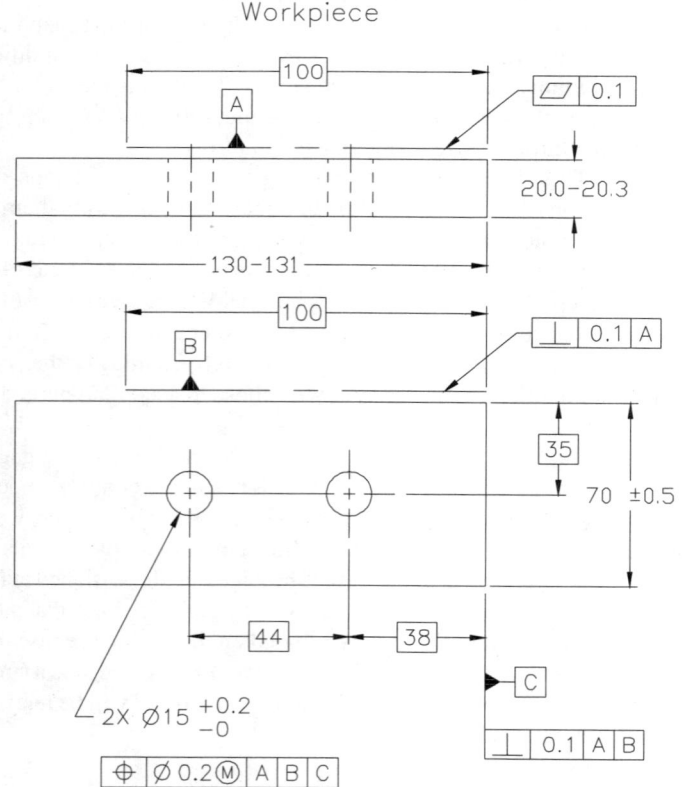

Figure 19-1 Position using partial and planar datum features

This is similar to using two datum target areas. The two partial datum features and the tertiary datum feature are first controlled and interrelated. The primary datum feature is given a flatness control. The secondary datum feature is given a perpendicularity control to only the primary datum plane formed by the three highest points within the primary datum feature. This controls both the orientation of the secondary datum feature and also its flatness. The tertiary datum feature is given a perpendicularity control to both the primary and secondary datums. Again, the perpendicularity control both forms and orients the tertiary datum feature. These three geometric characteristics of flatness, perpendicularity to one datum and then perpendicularity to two datums are used to give progressively more powerful geometric controls to the datum features. This not only gives them a needed interrelationship, but also implies a sequence of events for the reader of the drawing. These controls will also make the tolerancing of the gage easier, since the controls given to the gage elements will simply mimic the controls given to the part and use 5%-10% of the tolerance of the feature it represents.

The fourth and last geometric control shown is to position the two holes in the pattern to one another and to the three datum planes given by the three highest points of the primary datum feature, the two highest points of the secondary datum feature with respect to the primary datum plane, and the one highest point of the tertiary datum feature with respect to the primary datum plane and the secondary datum plane. Fig. 19-2 shows the gage for Fig. 19-1. The gage has, in order of consideration:

- A primary datum feature that is flat to within 10% of the flatness tolerance given to the primary datum feature on the workpiece,
- A secondary datum feature that is perpendicular to the primary datum plane to within 10% of the tolerance given to the secondary datum feature on the workpiece and,
- A tertiary datum feature that is perpendicular to the primary datum plane and the secondary datum plane to within 10% of the tolerance given to the tertiary datum feature on the workpiece.

Each datum feature simulator on the gage has enough surface area to entirely cover the datum feature from the workpiece it represents. It must try to hit the highest points of contact on the datum feature to properly construct the datum plane and unless it has enough surface area, it runs the risk of missing the appropriate high points and improperly establishing the datums. Too much surface area and the gage runs a similar risk of establishing nonfunctional and therefore inappropriate datums.

The gage also has two gage pins. Ideally, these gage pins will be at least as long as the holes they are gaging are deep. If these were simply GO gages meant to gage the maximum material condition of the holes, they would not be mounted on a plate, would have no relationship to the datum reference frame, and would be made at the maximum material condition of the holes. But these are Functional Gage pins meant to gage the positional requirement of the holes, so they are mounted and related to the datums and dimensioned to be at the virtual condition of the holes they are to inspect.

The size of the gage pins are dimensioned to begin at the virtual condition of the holes being gaged and go up in size tolerance by 10% of the size tolerance given to those holes. The gage pins also have a positional control based on 10% of the tolerance given to the holes they are gaging. If the workpiece is capable of being applied to the gage (as shown in the illustration), while maintaining its appropriate contact on the datum feature simulators, it is judged to be in compliance with the positional requirement. The size limits of the holes must be inspected separately.

One of the important requirements of workpieces to be gaged is that they are sufficiently defined to allow the gage designer/gagemaker to simply follow from control to control using 5%-10% of the tolerances that the workpiece shows. Unless the workpiece is complete in its definition, the gage designer cannot use it as a guide in the complete geometric definition of the gage. If necessary, the gage designer may add notes or even a procedural sheet to explain the proper use of the gage. As with all inspection, unless otherwise specified, the gage is to be used at 20 degrees Centigrade or 68 degrees Fahrenheit.

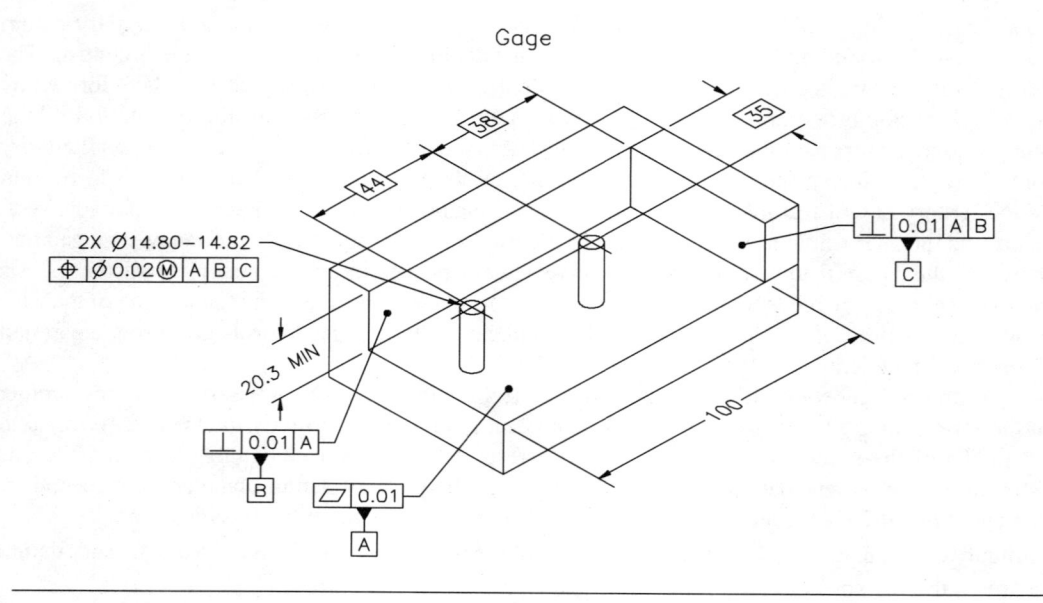

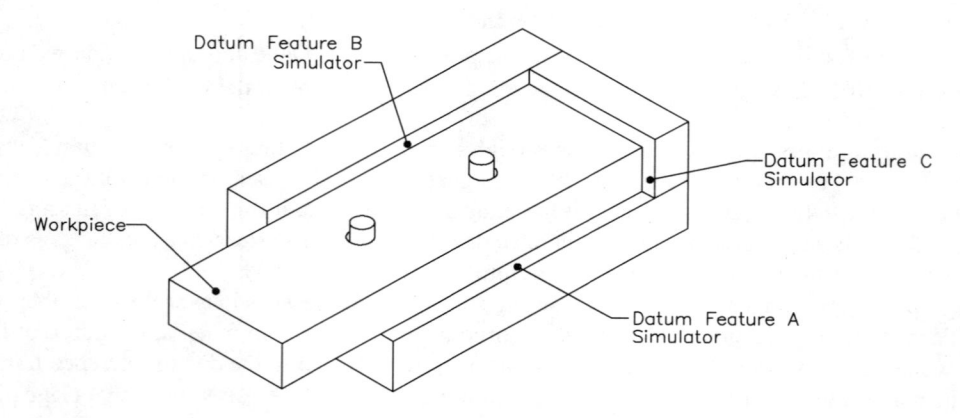

Figure 19-2 Gage for verifying two-hole pattern in Fig. 19-1

19.4.2 Position Using Datum Features of Size at MMC

Fig. 19-3 shows a workpiece that uses a planar primary datum feature, a secondary datum feature of size and a tertiary datum feature of size. By the time one gets to the tertiary datum feature of size, all spatial degrees of workpiece freedom have been eliminated by the primary and secondary datum features except angular orientation (what is commonly referred to as pattern rotation). The workpiece has been sufficiently defined to discuss the construction of the gage to inspect the position of the four-hole pattern. As is the case with many such workpieces, if the workpiece fits the gage used for the four-hole pattern's positional control, that gage will also inspect the position of the slot and the center hole's perpendicularity since they are represented on the gage as datum features for the four holes and they are represented at their virtual condition.

Receiver Gages — Go Gages and Functional Gages 19-7

Workpiece

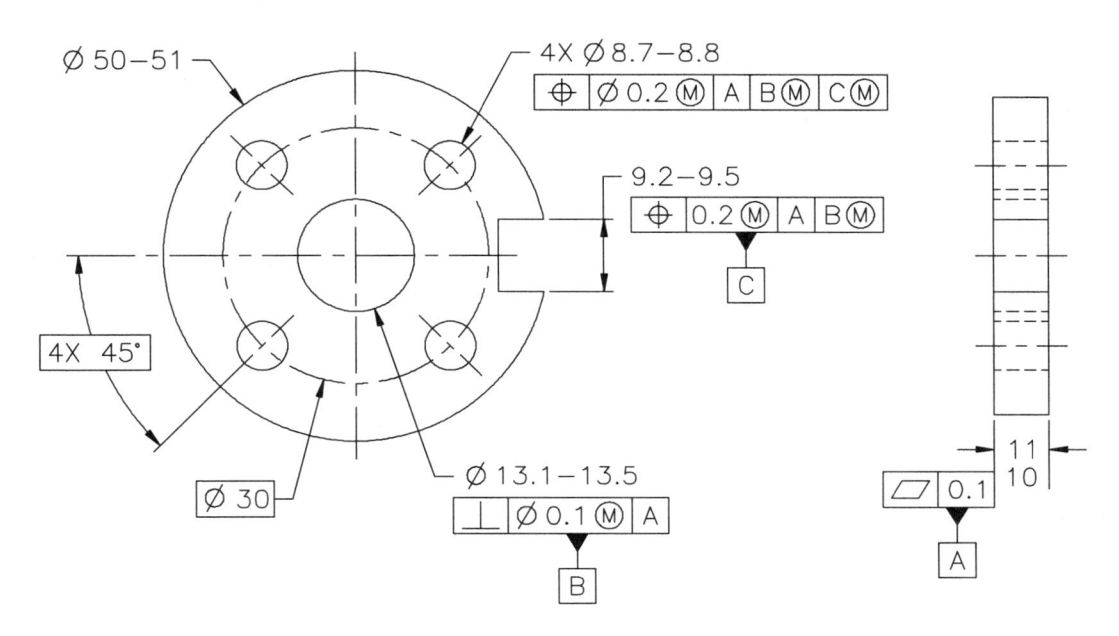

Figure 19-3 Position using datum features of size at MMC

A separate gage to inspect them individually would be considered redundant by most inspectors, since they would be represented at exactly the same size, orientation, and alignment as they are on the gage for the four-hole pattern. Again, as with Fig. 19-1, Fig. 19-3 has used a progressive geometric definition to make the workpiece complete enough to be both produced and inspected (at least for most of the purposes of this discussion).

1. The primary datum feature is controlled for 3-D form (flatness).
2. The secondary datum feature of size is controlled perpendicular to the primary datum plane.
3. The tertiary datum feature of size is controlled for position to the primary datum plane and the secondary datum axis.
4. The hole pattern is then controlled to the primary datum plane (for perpendicularity), the secondary datum axis (for location), and the tertiary datum centerplane (for angular orientation).

The maximum material condition concept has been used everywhere it is allowed for ease of manufacture and increased geometric tolerance while preserving functionality. The use of the MMC symbol after the geometric tolerances and also after the datum features of size will make it easy to represent them with gage pins at their appropriate constant boundary size (their virtual condition size). As in Fig. 19-1, each size tolerance and geometric tolerance has been mimicked by the gage that uses the same geometric characteristics and 10% of the tolerance on the workpiece. This geometric tolerance allows the gage pins to be only larger than the virtual condition boundary of the hole being represented so as to not accept a workpiece that exceeds its allowed tolerances.

This tolerancing of the gage pins to only get larger than the worst case boundary (and in the case of gage holes to only get smaller than the worst case boundary) being inspected will make the gages reject a small percentage of technically good parts that are near the borderline between good and bad. This way the gage doesn't accept a bad part. One must remember that this absolute tolerancing method is preferred by ANSI-approved documents, but is not the preferred practice in the ISO-approved documents on gaging.

The gage in Fig. 19-4 does not show the use of the maximum material condition symbol after the datum features of size. This will reduce the allowed inaccuracies in the gage, increase the chance of producing a more accurate gage and will accept more of the produced workpieces. Use of the regardless of feature size (RFS) concept after datum features of size on the gage design may increase the cost of the gage, but should more than make up for this additional cost by the gage's acceptance of a greater number of per-

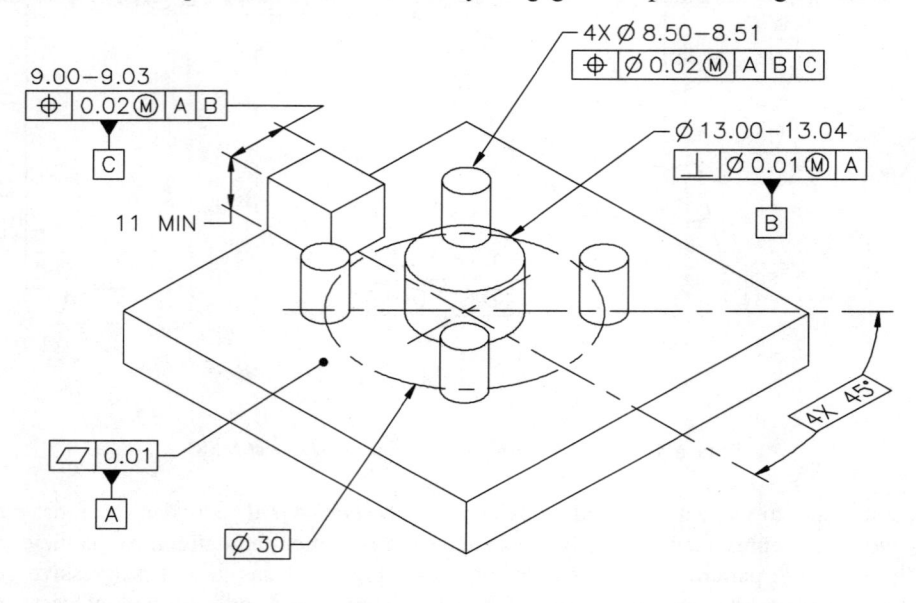

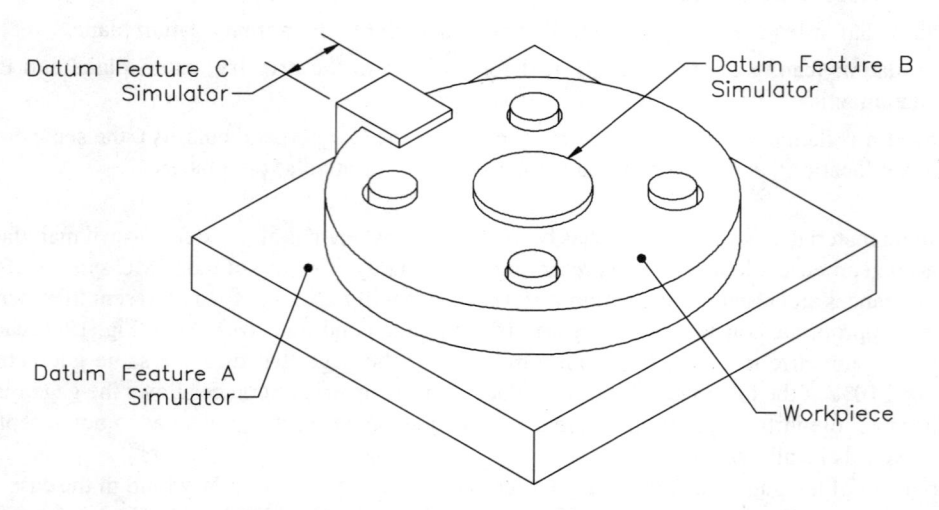

Figure 19-4 Gage for verifying four-hole pattern in Fig. 19-3

drawing technically good parts that are inspected by the gage. Even though the gage may use the regardless of feature size concept, it is commonly understood that receiver type gages, as discussed here, are most often used to inspect workpiece features and represent workpiece datum features that use the maximum material condition concept.

19.4.3 Position and Profile Using a Simultaneous Gaging Requirement

In Fig. 19-5, a simultaneous gaging requirement exists between a four-hole pattern and a profile control because both use exactly the same datum reference frame in exactly the same way. Both use a primary planar datum feature (A) and a secondary datum feature pattern of size (B) at maximum material condition. This creates a situation wherein, unless specified as a SEPARATE REQUIREMENT, the two geometric controls (position of the four-hole pattern and profile of the outside of the workpiece in the front view) must be inspected by the same gage. This is a more restrictive requirement than if both controls were allowed to use their own separate gage.

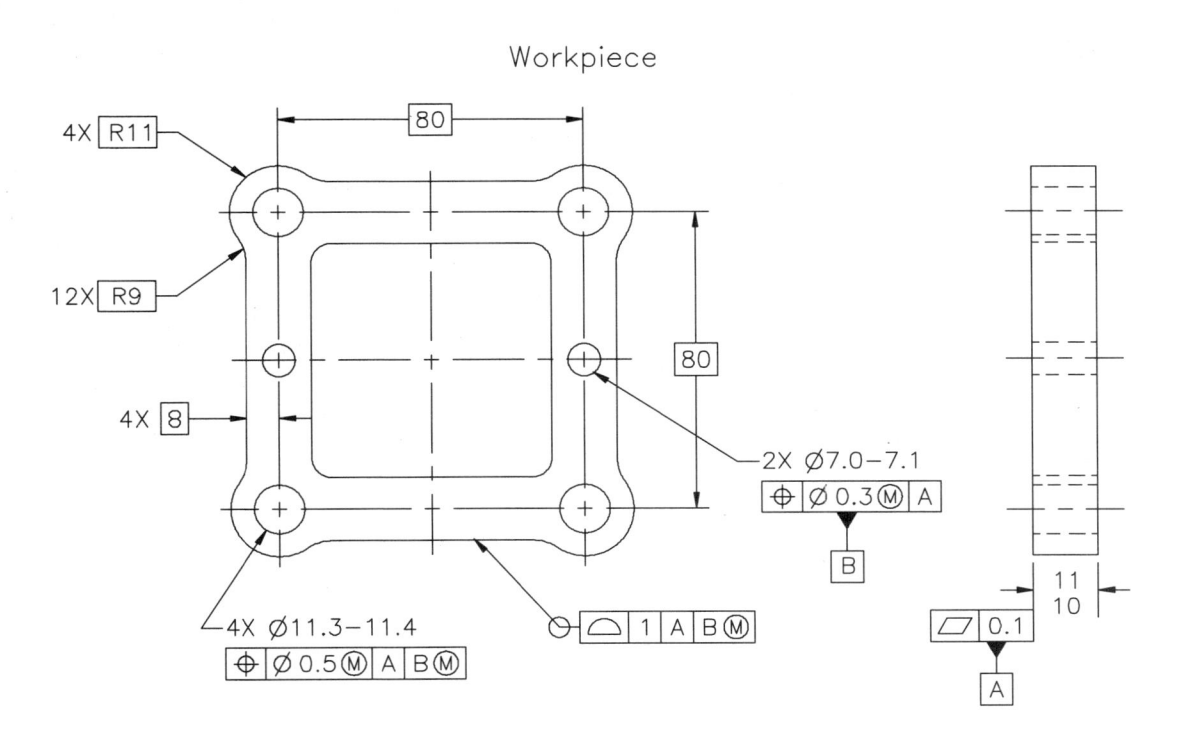

Figure 19-5 Position and profile using a simultaneous gaging requirement

For example, in a separate gaging requirement, the four-hole pattern could rock on datum A. This creates a different angle to be accepted than the rocked orientation on datum A used to accept the profile. Or as the datum pattern B holes grew from their virtual condition boundary toward their least material condition, the four-hole pattern as a group could shift to the left and the profile could shift to the right and be accepted. But in a simultaneous gaging requirement this would not be acceptable. Both the four holes and the profile would have to be accepted by one gage in one rocked orientation, with the four holes and the profile shifted in the same direction (if rock and shift were to occur).

Since Fig. 19-5 contains profile that is a geometric tolerance that cannot be referenced at maximum material condition, one may want to use a fixture to simulate only the datum features. See Fig. 19-6. If this is done, the gage/fixture will be capable of gaging the hole-to-hole requirement between the two holes in datum pattern B as well as their relationship to the primary datum plane A. It is also capable of stabilizing the workpiece to use a variables data collector such as a computerized coordinate measurement machine to measure the position of the four holes and the profile of the workpiece. The workpiece is stabilized in one orientation to measure the four holes and the profile controls. If the four holes and the profile meet their geometric tolerances when measured in that orientation, they may be considered as having met the SIMULTANEOUS REQUIREMENT condition of their inspection.

It is possible to create a complete gage that will not only represent the datum features, but also the four holes at their virtual condition (MMC concept) boundary and the worst case mating condition of the profile's outer boundary. Although the gage as shown in Fig. 19-7 for Fig. 19-5 will not gage the profile's inner boundary (which, if important, can be represented or inspected in other ways), the gage is

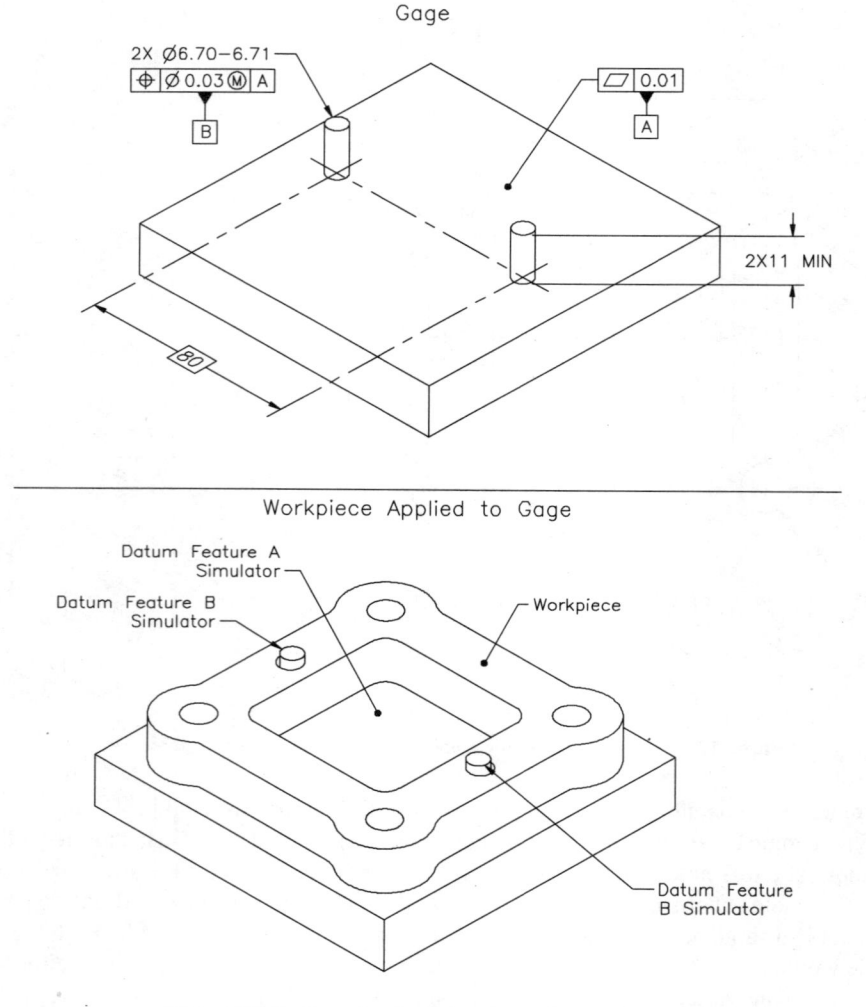

Figure 19-6 Gage for simulating datum features in Fig. 19-5

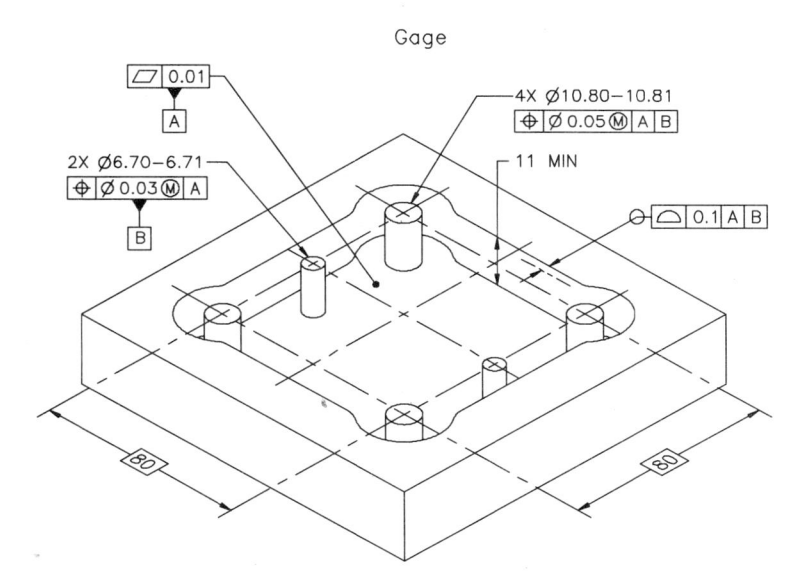

Receiver Gages — Go Gages and Functional Gages 19-11

Gage

Note: The nominal profile for the gage is the maximum part profile tolerance boundary. The profile tolerance on the gage is unilaterally in. The gage simultaneously verifies the hole locations and profile outer boundary. It does not verify the profile inner boundary.

Workpiece Applied to Gage

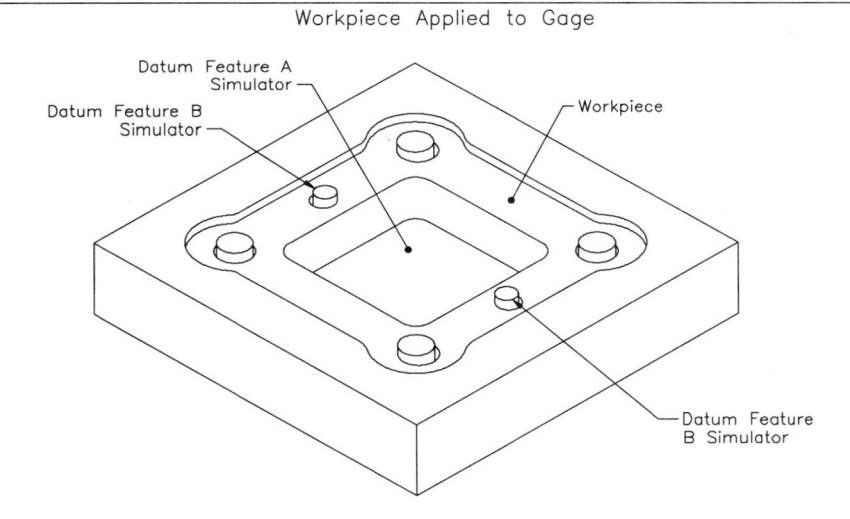

Figure 19-7 Gage for verifying four-hole pattern and profile outer boundary in Fig. 19-5

capable of inspecting the positional tolerance of the four holes, the outer boundary of the profile control, and the interrelationship between the four-hole pattern and the profile under the simultaneous requirement rule. The gage simulation for the profile has a nominal size that is the maximum part profile tolerance boundary. The profile tolerance on the gage (shown as 10% of the profile tolerance on the workpiece) is unilateral inside (as with all gage holes), allowing the gage tolerance to accept no profile that exceeds the outer boundary of the workpiece's profile tolerance.

19.4.4 Position Using Centerplane Datums

Fig. 19-8 shows a simultaneous gaging requirement for a four-hole pattern and a larger center hole. Each uses exactly the same datums in the same order of precedence with the same material condition symbols after the datum features. This creates the simultaneous gaging requirement. This is a very sequential geometric product definition.

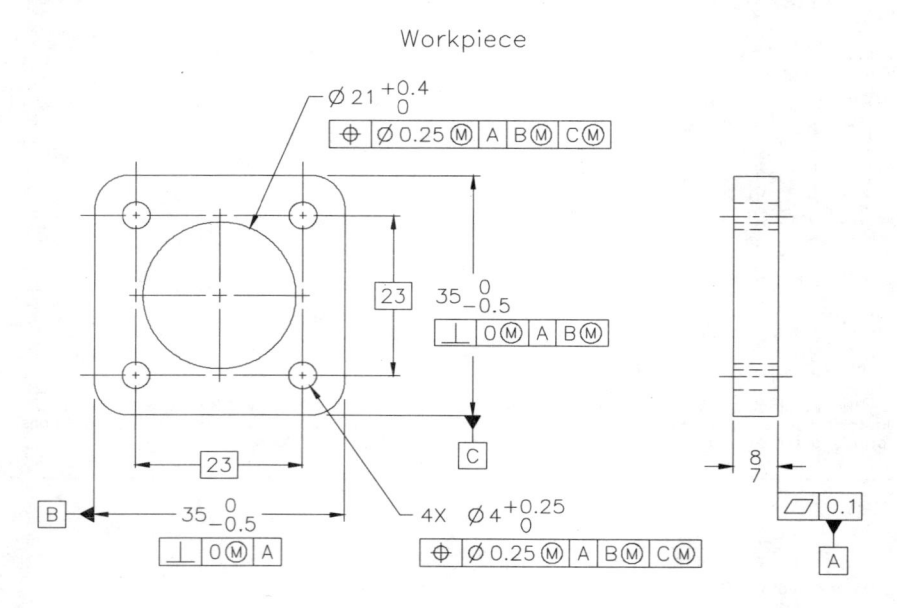

Figure 19-8 Position using centerplane datums

To understand the requirements, one might first look at the configurations and ignore the feature control frames. All four holes are shown centered to the hole in the middle and to the outside of the workpiece. The four holes are dimensioned 23 mm from each other, but since they are depicted centered to the center hole, we must assume each of the four holes is desired to be 11.5 mm from the center hole and from the middle of the workpiece. The hole in the center is exactly that; a hole we desire to be in the middle of the workpiece. The part is then geometrically toleranced in four steps. Step 1, the primary datum feature is identified and given a flatness tolerance. Step 2, the secondary datum feature is identified as one of the 35-mm widths creating a centerplane datum, and the datum feature that generates that centerplane is given a perpendicularity control back to the primary datum plane. Step 3, the tertiary datum feature is identified as the other 35-mm width creating a third datum plane which is also a centerplane datum. The datum feature that generates that centerplane is given a perpendicularity control back to the primary datum plane and the secondary datum centerplane. Step 4 is the simultaneous positional requirement of all five holes to each other and to the primary, secondary, and tertiary datum features. All geometric tolerances of perpendicularity and position are referenced at maximum material condition and use their datum features of size at maximum material condition. This makes it easy to represent each at a constant gage element size, either their MMC or their virtual condition, as applicable. Since in the case of the datum features of size a zero tolerance at MMC has been used, the MMC and the virtual condition are the same. Any gage that simulates these datum features will be able to gage their compliance with their given geometric tolerances and the geometric tolerances of the holes measured from them. The same Functional Gage will also be able to verify compliance with the 35-mm MMC size.

As shown in Fig. 19-9, step 1 on the gage shown represents datum feature A and gives it a flatness tolerance of 10% of the flatness tolerance on the workpiece. Step 2 on the gage represents datum feature B at a size of 35 mm plus zero and minus 10% of its size tolerance. It is then given exactly the same feature control frame the workpiece has on its datum feature B (10% of zero is still zero). Step 3 on the gage represents datum feature C at a size of 35 mm plus zero and minus 10% of its size tolerance. It is then given the same feature control frame the workpiece has on its datum feature C except it references its datum feature of size B at regardless of feature size. As explained in previous examples, this has the effect of increasing the cost of the gage by decreasing the allowed gage tolerance. However, it has a better chance of producing a gage that will accept more of the produced parts that are within their geometric tolerances. Step 4 on the gage represents all five controlled holes with gage pins. The gage pins begin at the virtual condition of the hole they represent and are toleranced for size with minus zero and plus 10% of the size tolerance of the hole. Then the gage pins are given a position tolerance of 10% of the position tolerance of the hole it represents to the datums simulated in steps 1-3.

Again, the datum features of size on the gage are referenced at regardless of feature size, even though the features they simulate are referenced at MMC. Keep in mind this is a personal choice. Gage datum

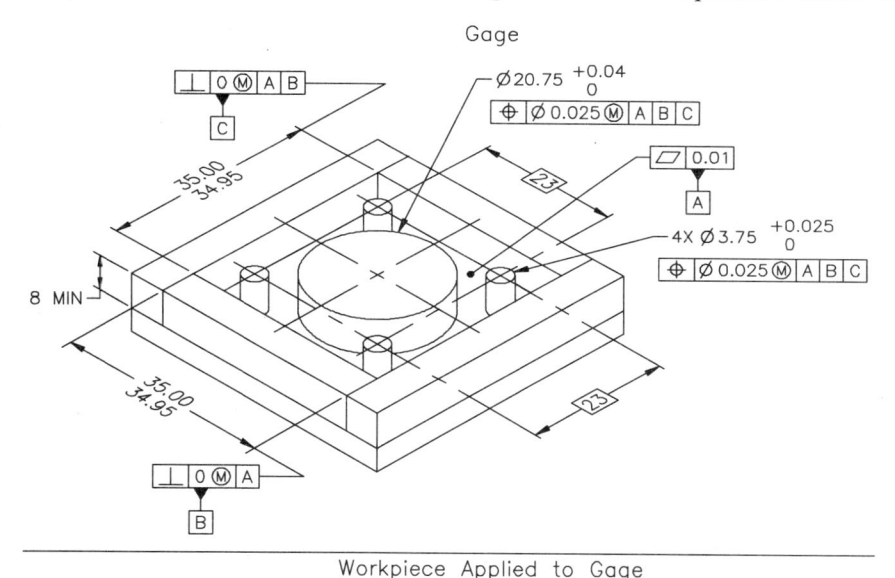

Figure 19-9 Gage for verifying four-hole pattern in Fig. 19-8

feature of size simulations may be referenced at MMC. This will make the gage tolerance larger, and potentially decreases the cost of the gage. It also runs the risk of the gage being made at a size, orientation, and location that rejects more of the technically in-tolerance workpieces it gages.

In these examples, a zero tolerance at MMC was used on the controlled datum features of size and therefore a zero tolerance at MMC was used on the gage simulation of the controlled datum features of size. For the purposes of gage tolerancing, one may consider that a workpiece using a geometric tolerance at MMC has a total tolerance that includes the size tolerance and the geometric tolerance. If one adds the size tolerance and the tolerance from the feature control frame on the feature being considered, a true sense of the total tolerance on the feature can be understood. When distributing tolerance on the gage, the tolerance distribution may be that 5%-10% of the total tolerance on the feature being gaged can be used in the size limits of its gaging element, and a zero tolerance at MMC used in its feature control frame. The effect on the gage of this method of tolerance distribution is usually a more cost-effective gage without the possibility that the gage will accept more or less of the parts that it inspects.

19.4.5 Multiple Datum Structures

In Fig. 19-10, the positional controls shown use zero at MMC for their geometric tolerances. This makes it easy to illustrate that the only tolerance available for the gage designer to take 5%-10% of is the difference between the MMC and the LMC of the controlled features. In each case, both for the center hole that becomes datum feature D and for the four holes that eventually are positioned to A, D at MMC, and B, a total of 2 mm is used as the size tolerance. This means that when the gage is produced, the gaging elements (pins) that are used to simulate these holes will use a percentage of the 2 mm as the total tolerance on the gage pin sizes and their orientation and location geometric tolerances. This tolerance can be split between the gage pin size and its geometric tolerance or simply used as size tolerance while the geometric tolerance uses zero at MMC, or zero at LMC.

Fig. 19-10 is sequentially toleranced, with a flatness control given to the primary planar datum feature, a perpendicularity tolerance given to the secondary planar datum feature back to the primary datum, and a perpendicularity tolerance given to the tertiary datum feature back to the primary and secondary datums.

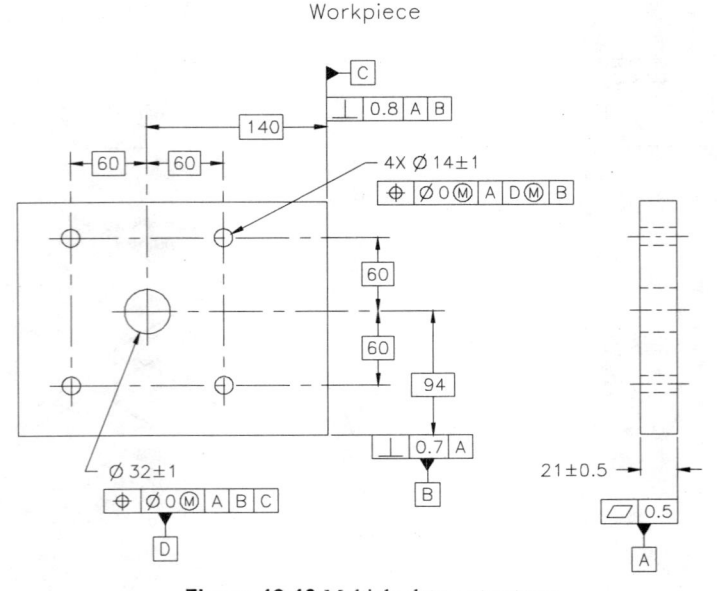

Figure 19-10 Multiple datum structures

This completes the first datum reference frame from which the center hole is positioned. The center hole is then made a datum feature (D) from which the outer four holes may be positioned for location on the X and Y axes while using datum A for perpendicularity and datum B for angular orientation.

Each geometric control is considered separately verifiable. If gaged, each positional control will be considered a different gage. Since each positional control uses a zero at MMC positional tolerance, the gages that inspect position will also be able to verify compliance with the MMC size envelope. The first gage verifies the position of the center hole. It consists of three planar datum feature simulators, each using exactly the same geometric control as the feature it represents. The only difference is that (as illustrated) a geometric tolerance of 10% of the feature it simulates has been used. The center hole being gaged is represented by a gage pin at the desired basic angle and distance from the datums (as depicted in Fig. 19-11). The gage pin is dimensioned at the virtual condition size of the hole it is gaging and is allowed to grow by 10% (0.2) of the tolerance on the hole. The gage pin is then given a positional tolerance of zero at MMC to the datum features used on the gage.

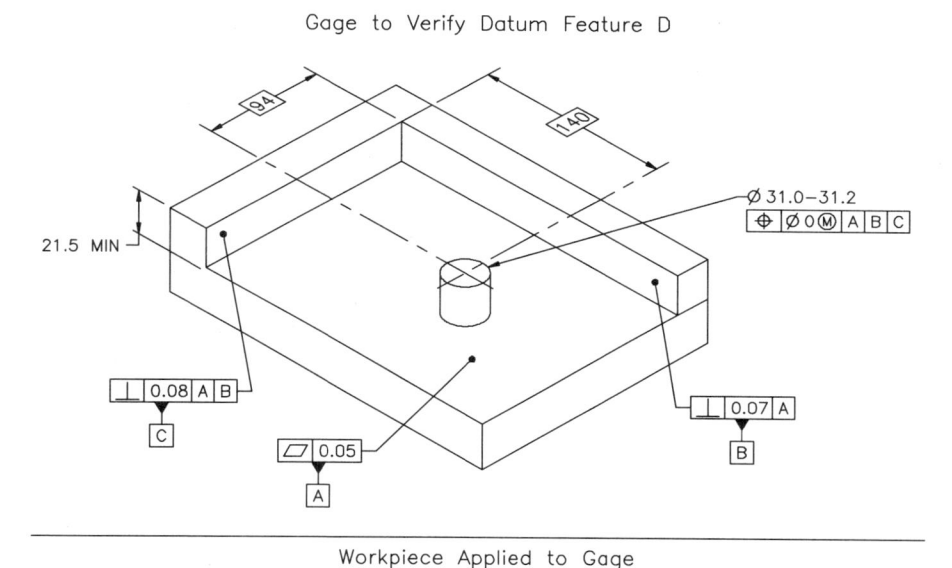

Figure 19-11 Gage for verifying datum feature D in Fig. 19-10

19-16 Chapter Nineteen

The last gage for Fig. 19-10 in Fig. 19-12 is used to inspect the position of the four-hole pattern. It begins with a datum feature simulator for datum A and uses a flatness tolerance of 10% of the datum feature it simulates. It also has a datum feature simulator for datum feature B (which is used as a tertiary datum feature to construct a fourth datum plane). This is used to control the pattern rotation (angular orientation) of the four holes and will be a movable wall on two shoulder screws. For the part being gaged to pass the gaging procedure, it will have to make contact with a minimum of two points of high point contact on the datum feature B simulator. This is to assure that the four-hole pattern has met the desired angular relationship to datum plane B and datum feature B. If, for example, only one point was contacted by the part on the datum feature simulator B, it would not assure us that the hole pattern's orientation had

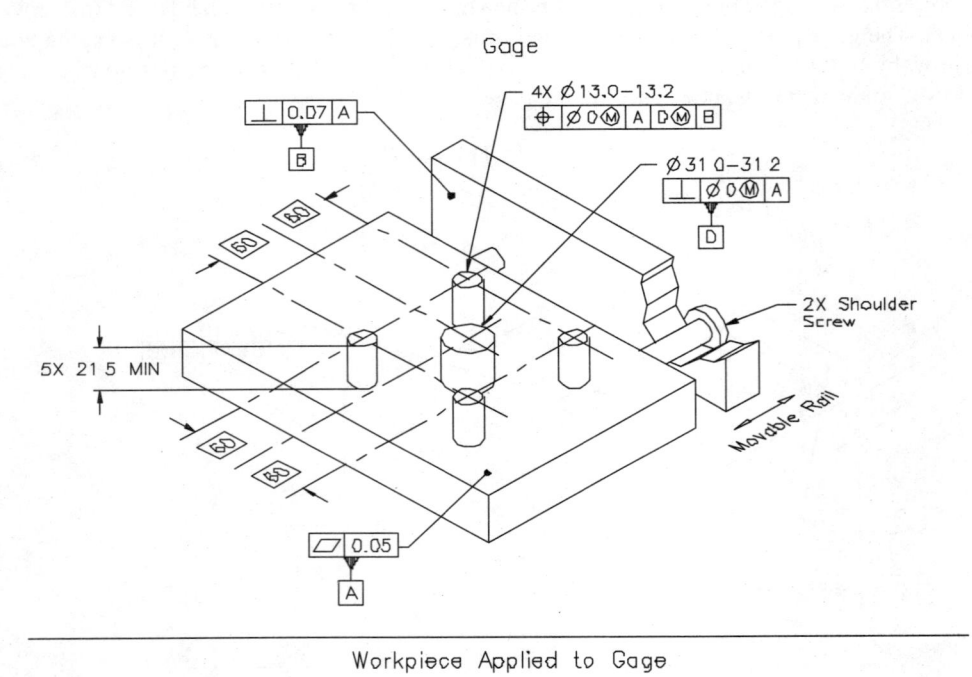

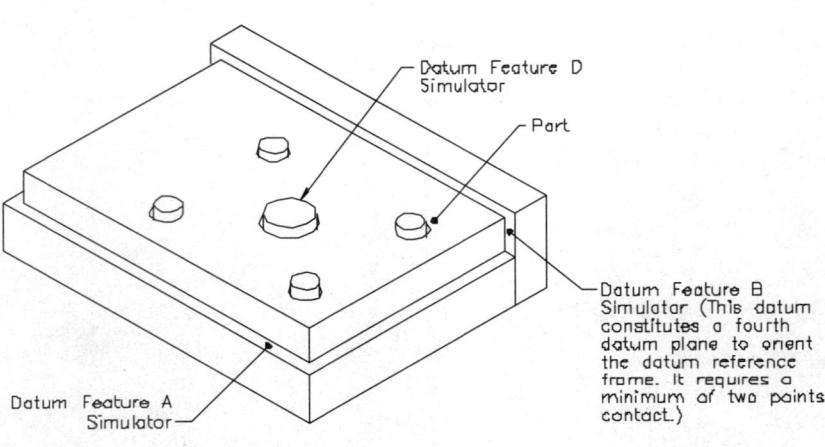

Figure 19-12 Gage for verifying four-hole pattern in Fig. 19-10

been properly maintained to the real surface from which datum B is constructed on the workpiece being gaged. The datum feature simulator for B is given a perpendicularity tolerance back to datum A. The perpendicularity tolerance is 10% of the tolerance on the datum feature it is simulating. Datum feature D is also represented. Again, D is simulated by a gage pin sized to begin at the hole's virtual condition and then the gage pin is allowed to grow by 10% of the tolerance given to the D hole being represented. The gage pin D is then given a perpendicularity requirement of zero at MMC back to the primary datum. A positional tolerance is not needed for gage pin D as long as enough surface area exists for datum feature A to be properly contacted.

The four holes being gaged are then represented with four gage pins of (as required of all gage elements) sufficient height to entirely gage the holes. These gage pins are represented at the virtual condition diameter of the holes they simulate and are allowed a size tolerance of 10% of the tolerance on the size of the holes. This tolerance is all in the plus direction on the gage pin size. The gage pins are then positioned to the datum feature simulators previously described, A primary, D at MMC or RFS secondary, and B tertiary (tertiary datum feature/fourth datum plane used to orient the two planes that cross at the axis of datum D).

19.4.6 Secondary and Tertiary Datum Features of Size

In Fig. 19-13, the position of two holes is established by datums A, B, and C (see gage in Fig. 19-14). Once this has been done, the two holes are used as secondary and tertiary datum features (see gage in Fig. 19-15) from

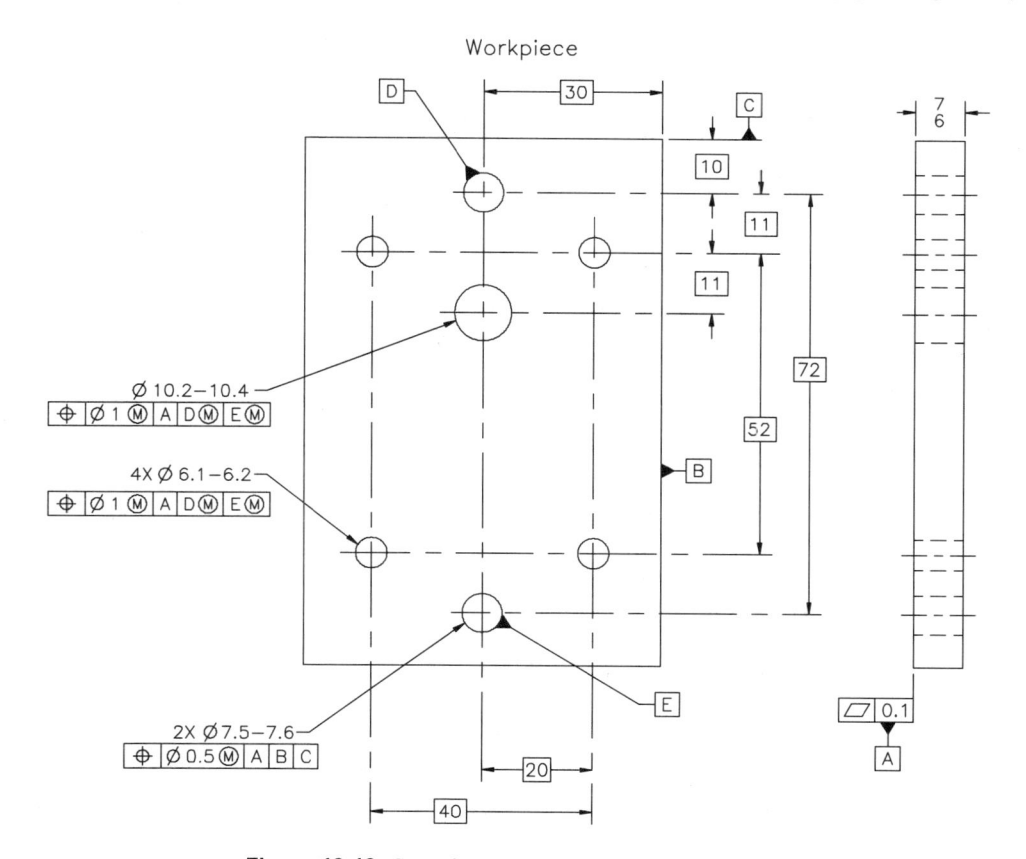

Figure 19-13 Secondary and tertiary datum features of size

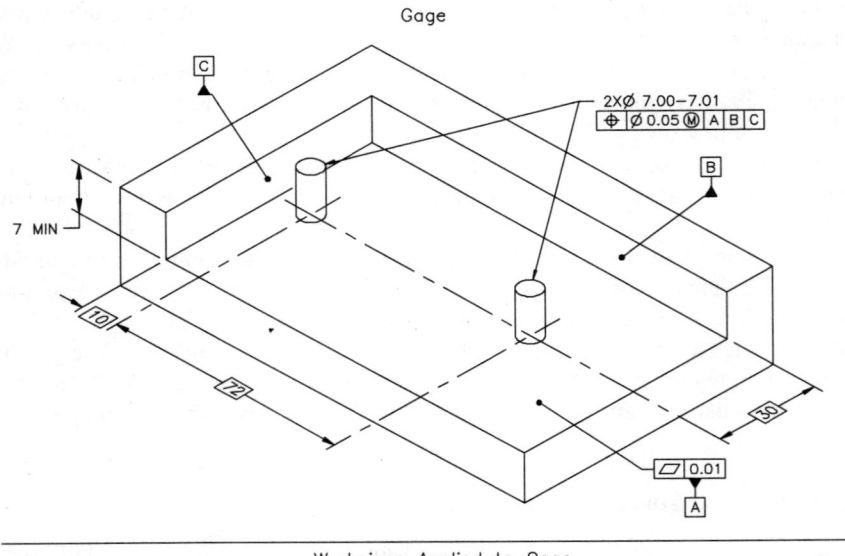

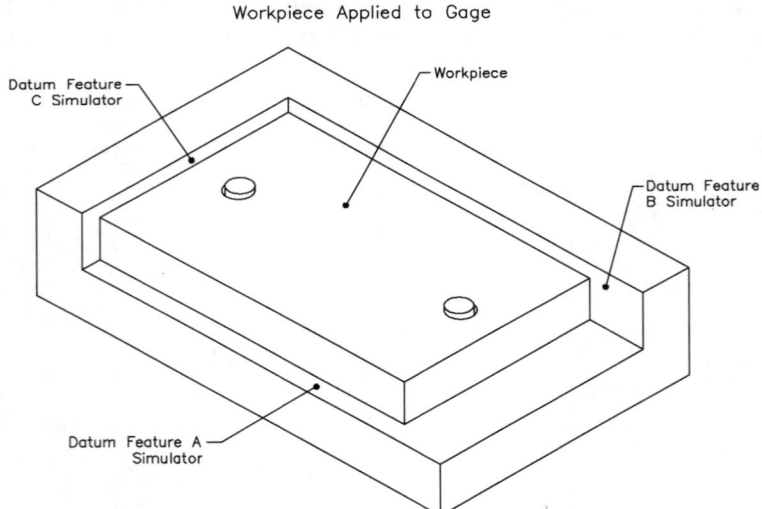

Figure 19-14 Gage for verifying datum features D and E in Fig. 19-13

which to measure the four 6.1-6.2 holes and the one 10.2-10.4 hole. Since datum feature of size D is used as secondary, it establishes the location of the five holes in both the X and the Y directions. Datum feature of size E is used as an angular orientation datum only. This means that the datum feature simulator on the gage for D is a cylindrical pin made at the virtual condition of the hole it represents (sometimes referred to as a four-way locator). Datum feature E, however, is represented by a width only (sometimes referred to as a two-way locator). Datum feature E is like a cylinder made at the virtual condition of the hole it simulates, but is cut away in the direction that locates it from datum feature D. This is to prevent it from acting as a location datum but rather as only a pattern rotation datum.

This use of datum feature simulators in Fig. 19-15 is common. Datum feature simulator E is a tertiary datum feature of size and is represented as an angular orientation datum (a two way locator) with a diamond shaped (or cut-down cylindrical) pin. However, it is not representative of other types of datum feature simulation. Datum features are normally represented by datum feature simulators that have the same shape as they do; for example planar datum features represented by planar simulators, cylindrical datum features represented by cylindrical simulators, and slot/tab/width datum features represented by datum feature simulators of the same configuration.

If datum features D and E had been used as a compound datum (D-E) with both D and E referenced at MMC, D would not have taken precedence over E. Hence, being equal, both would have been used to

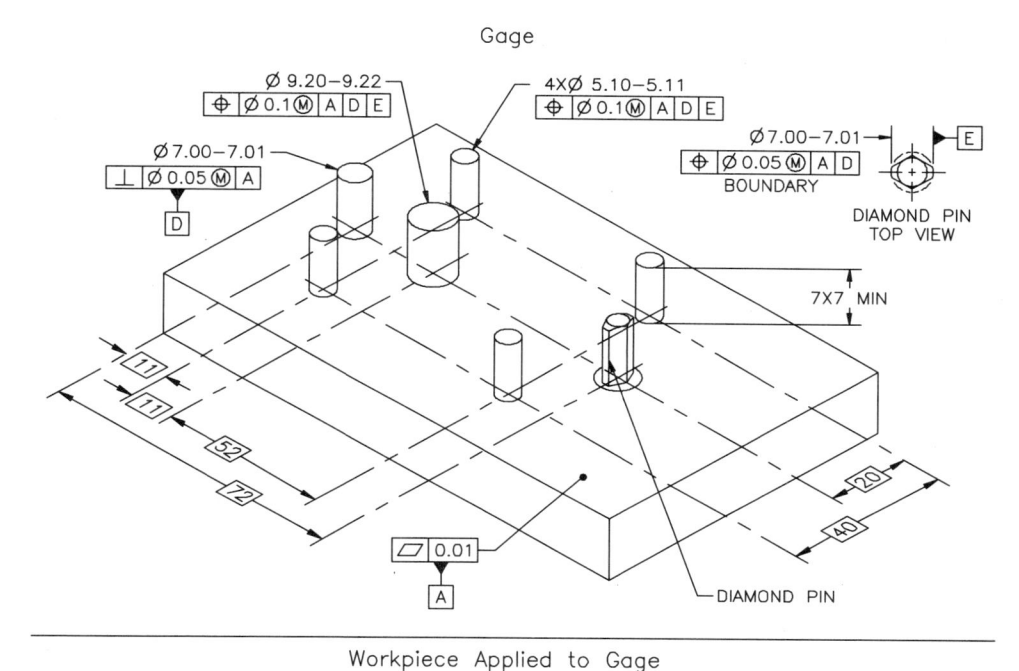

Figure 19-15 Gage for verifying five holes in Fig. 19-13

orient and locate the five holes referred to them as though they were a pattern datum consisting of the two holes. In this circumstance, the gage (as shown in Fig. 19-15) would have represented both D and E with cylindrical pins made at the virtual condition of the holes they represent. Both D and E would be considered four-way locators.

19.5 Push Pin vs. Fixed Pin Gaging

Although the examples used in this section use fixed pin gages, some thought should go toward the use of push pin gages. With push pin gages, the workpiece is first oriented and located on the gage's datum feature simulators. Then the gage pins are pushed through holes in the gage and into the holes on the workpiece. This allows the user of the gage to be certain the appropriate type of contact exists between the gage's datum feature simulators and the datum features on the workpiece being gaged. Push pin gages also provide a better view of which features in a pattern under test are within tolerance and which are out of tolerance. The holes that receive their gage pins are obviously within their geometric tolerance and the holes that are not able to receive their gage pins have violated their geometric tolerance. This information should be helpful to improve the manufacture of subsequent parts.

It must be considered that with a push pin – type gage design, gage tolerances are used in a manner that allow the gage pin to easily enter and exit the gage hole with a minimum of airspace. Gage holes that are to receive push pin gage elements should be given geometric tolerances that use a projected tolerance zone that is a minimum height of the maximum depth of the hole being gaged (since the gage hole gives orientation to the gage pin and is likely to exaggerate the orientation error of the gage hole over the height of the gage pin). The gage hole should be treated as though it is a gage pin when calculating its virtual condition. The projected geometric tolerance zone diameter is added to the maximum material condition of the gage push pin diameter to determine the virtual condition of the gage pin when pushed into the gage hole. In Absolute Tolerancing, this gage pin virtual condition boundary may be no smaller than the virtual condition of the hole on the workpiece being gaged.

19.6 Conclusion

Receiver gaging provides a level of functional reliability unsurpassed by other measurement methods. Instead of verifying compliance with a theoretical tolerance zone, it transfers that tolerance to the controlled feature's surfaces and creates an understandable physical boundary. This boundary acts as a confinement for the surfaces of the part. It assures one that if the boundary is not violated, the part features will fit into assemblies. ASME Y14.5M-1994 (the Dimensioning and Tolerancing standard) and the ASME Y14.5.1M-1994 (the standard on Mathematical Principles of Dimensioning and Tolerancing) both state that occasionally a conflict occurs between tolerance zone verification and boundary verification. They also state that in these instances, the boundary method is used for final acceptance or rejection.

19.7 References

1. American National Standards Committee B4. 1981. ANSI B4.4M-1981, Inspection of Workpieces. New York, New York: The American Society of Mechanical Engineers.
2. Meadows, James D. 1995. *Geometric Dimensioning and Tolerancing*. New York, New York: Marcel Dekker.
3. Meadows, James D. 1998. *Measurement of Geometric Tolerances in Manufacturing*. New York, New York: Marcel Dekker.
4. Meadows, James D. 1997. *Geometric Dimensioning and Tolerancing Workbook and Answerbook*. New York, New York: Marcel Dekker.
5. The American Society of Mechanical Engineers. 1995. *ASME Y14.5M-1994, Dimensioning and Tolerancing*. New York, New York: The American Society of Mechanical Engineers.

Chapter 20

Measurement Systems Analysis

Gregory A. Hetland, Ph.D.
Hutchinson Technology Inc.
Hutchinson, Minnesota

Dr. Hetland is the manager of corporate standards and measurement sciences at Hutchinson Technology Inc. With more than 25 years of industrial experience, he is actively involved with national, international, and industrial standards research and development efforts in the areas of global tolerancing of mechanical parts and supporting metrology. Dr. Hetland's research has focused on "tolerancing optimization strategies and methods analysis in a sub-micrometer regime."

20.1 Introduction

Measurement methods analysis is a highly critical step in the overall concurrent engineering process. Today's technology advancements are at a stage where measurement science is being pushed to the limit of technological capabilities. The past has allowed capabilities of measurement equipment to be acceptable if the Six Sigma capability was ≥ 1 µm (0.001 mm). Today, submicrometer capability is much more the norm for high technology manufacturing firms, with the percentage of features in this tolerancing regime getting larger and larger.

The primary objective of this chapter is to generate a capability matrix that reflects "Six Sigma capability" for all 14 geometric controls, as well as individual feature controls using an ultra-precision class coordinate measuring machine (CMM). In this particular case, a Brown & Sharpe/Leitz PMM 654 Enhanced Accuracy CMM was used for all testing to generate this matrix.

Analysis included variables that impact optimum measurement strategies in a submicrometer regime such as feature-based sampling strategies, calculations for determining capability of geometrically defined features, the thermal expansion of parts and scales, CMM performance, and submicrometer capabilities in contact-measurement applications.

The methodologies for approaching the characterization of CMMs as a whole is extremely broad, primarily due to a lack of awareness of the broad range of contributing error sources. Measurement system characterization applies to all measurement systems, but due to the diversity of contact CMMs, the 14 geometric controls can be characterized to varying degrees.

Unlike many measurement systems, a contact CMM has the ability to measure one, two, and three-dimensional (1-D, 2-D, and 3-D) features. Based on this unique capability, a CMM is the most appropriate system to consider in the initial spectrum of measurement system characterization. This is not to imply a measurement system such as this can measure the spectrum of geometric shapes in their entirety. It is, however, intended to indicate that a CMM is the most diverse piece of measurement equipment in the world today, in the application of measuring geometric features on mechanical features of components.

20.2 Measurement Methods Analysis

In addition to outlining a methodology for measurement system characterization, this chapter also will define some of the key tests leading to the generation of the capability matrix. In addition, I will identify some significant limitations in currently defined calculations for analyzing measurement system capability, especially in the area of 3-D analysis. The capability matrix, which is the primary deliverable of this chapter, was defined by standard analysis practices.

The following outlines six key phases that are essential to the characterization of measurement systems of this caliber:

- Measurement system definition (phase 1)
- Identification of sources of uncertainty (phase 2)
- Measurement system qualification (phase 3)
- Quantifying the error budget (phase 4)
- Optimizing the measurement system (phase 5), and
- Implementation and control of measurement systems (phase 6)

20.2.1 Measurement System Definition (Phase 1)

As performance specifications grow increasingly tighter, the older gaging rule of a 10:1 ratio has been at times reduced to a lesser ratio of even 4:1. However, even the lower goals are becoming difficult to achieve. This increases rather than decreases the requirement of metrology and quality involvement at the stage of product design.

20.2.1.1 Identification of Variables

The first step of any measurement task is to identify the variables to be measured. While this may appear to be a simple and straightforward task, the criticality of various dimensions is usually nothing more than a hypothesis. If true hypothesis testing is performed, the need for metrology and quality involvement is obvious.

A more common approach is inherited criticality, where the product or part being designed is an enhanced version of an earlier model. This approach is usually valid, because the available empirical data should support the claim of criticality.

Nonetheless, there are times where process variables, rather than properties of the product, require measurement. This method may be preferred because it provides a separate method of ensuring conformance to specifications. An obvious example is injection molding, where tooling certification and control and machine process variables, such as temperature, curing times, etc., are all measured and monitored in addition to the product itself. Such a methodology can graduate to exclude product measurement once a process is deemed "in control" over an extended period of time.

20.2.1.2 Specifications of Conformance

If choosing the proper variables for tracking is not difficult enough, consider the problems with determining valid specifications for acceptance. Again, when considering an enhanced product, empirical data should prove to be the best guide. However, additional testing may be necessary, especially when considering those properties being improved.

Unfortunately, some inherited specifications can be as invalid as any hypothesis. This is particularly true when studies or other data are unavailable to support the requirement. The importance of valid specifications is easily exemplified in the following examples of typical costs.

A contact CMM with ultraprecision (submicrometer) capability requires capital expenditures of approximately $500,000 for the equipment, $100,000 for environmental control, and $100,000 for implementation. Additional costs include adding higher-competency personnel, increased cycle-time for measurement tasks, and increased requirements of measurement system characterization.

A contact CMM with normal (10 μm) capability typically requires less than half the capital expenditure and can perform more timely measurements with less maintenance costs.

The significance is obvious and so should be the ramifications of invalid specifications. Too loose a specification can lead to delivering nonfunctional parts to a customer, which can lead to loss of business, which can lead to diminished market share. Specifications that are too tight add cost to the product and cycle-time to delivery schedules, both without any return and with the same effect on customers and market share.

20.2.1.3 Measurement System Capability Requirements

Once the specifications of conformance are defined, the capability required of the measurement system must be addressed. As stated, if the 10:1 ratio can be achieved the task is more easily accomplished. Regardless, the best approach to defining capability requirements is to develop a matrix. The requirements matrix should address the following concerns:

- Capability for each feature to be measured
- Software (computer system, metrology analysis requirements, etc.)
- Environment (temperature, vibration, air quality, manufacturer's specifications, etc.)
- Machine performance (dynamics, geometry, probing, correction algorithms, speed, etc.)

Obviously, some of these requirements are interrelated. For example, some environmental concerns must be met to achieve the stated vendor specification machine performance. The final capability matrix should address all concerns relative to the capability desired.

Once the capability and its availability are known, the cost and budget analyses and timelines are required. Such analysis is extremely difficult and must include considerations such as personnel requirements and maintenance costs.

20.2.2 Identification of Sources of Uncertainty (Phase 2)

This step involves identifying the error sources affecting measurement system capabilities. As stated, the system definition phase should have included some consideration of this topic. The following list includes the minimum categories that must be considered in measurement system characterization.

- Machine
- Software
- Environment
- Part
- Fixturing
- Operator

As each source is identified within the given categories, discussion should turn to its projected influence on overall capability and on specific applications. This discussion refers to this influence as being sensitive or nonsensitive.

For example, the ASME B89.1.12 standard for evaluating CMM performance defines methods for testing bidirectional length and point-to-point capabilities. Basically, these tests evaluate the ability of a contact probing system to perform probe compensation. However, this error source is nonsensitive to the CMMs' capability to measure the *position* of circles or spheres.

If labeled as sensitive, efforts should be made to determine its contribution and to assign a priority level of concern. Obviously, these are only projections, but the time is well spent because this establishes a baseline for both qualification and, if necessary, diagnostic testing.

20.2.2.1 Machine Sources of Uncertainty

Identifying error sources associated with the equipment itself sometimes can be easily accomplished. First, many standards and technical papers discuss the defects of various machine components and methods of evaluation. Second, measurement system manufacturers publish specifications of machine performance capabilities. These two sources provide most of the information required.

The most common concerns for CMMs include, but are not limited to, the following:

Dynamic Behavior involves structural deformations, usually resulting from inertial effects when the machine is moving. The sensitivity of this error is highly dependent on the structural design and the speed and approach distances required.

Geometry involves squareness of axes, usually dependent on the number of servos active, temperature, etc. The sensitivity is highly dependent on whether or not the machine includes volumetric error correction, and the environment within which the machine will be operated.

Linear Displacement involves the resolution of the scales, also dependent on the environment within which the machine will be operated. The sensitivity depends on scale temperature correction capabilities.

Probing System involves probe compensation, highly dependent on type of probe, the software algorithms for filtering and mapping stylus deflections, and the frequency response. The sensitivity depends on the material, tip diameter, and length of the probe styli to be used.

20.2.2.2 Software Sources of Uncertainty

The most obvious concern for software performance is its ability to evaluate data per ASME Y14.5M-1994 and ASME Y14.5.1M-1994. However, many attributes to the software should be evaluated. The following list includes, but is not limited to, concerns for software testing:

- Algorithms (simplified calculations to improve response time)
- Robustness (ability to recover from invalid input data)
- Reliability (effects of variations in input data)

- Compliance to ASME Y14.5 and ASME Y14.5.1 (previously mentioned)
- Correction algorithms (volumetric and temperature)

When possible, testing of software should be achieved through the use of data sets. Other methods increase contributions to uncertainty. Some of the software attributes to be tested include its ability to handle those problems. Software uncertainties should not be ignored. Often, the uncertainty involved seems negligible, but that term is relative to the capability required.

20.2.2.3 Environmental Sources of Uncertainty

The most common concern for environment involves temperature, which is often stated as the largest error source affecting precision. Other atmospheric conditions also influence capability.

Humidity, like temperature, can lead to distortion of both the machine and the parts being measured. Efforts to control these atmospheric conditions can lead to the necessity to consider the pressure of the room involved. If lasers are used, barometric pressure may alter performance. The same is true for contamination, which also affects both contact and noncontact data collection.

Nonatmospheric concerns include vibration, air pressure systems, vacuum systems, and power. Note that each and every utility required by the machine can affect its performance.

Consideration of the sensitivity of these sources is dependent on the degree of control and the capability required. For example, the environmental control realized within laboratories is generally much greater than that of production areas. Often, a stable environment can shift the sources of error to the machine's and the part's properties within those conditions.

20.2.2.4 Part Sources of Uncertainty

Many aspects of the parts themselves can be a source of measurement uncertainty. The dynamic properties, such as geometric distortion due to probing force or vibration, are obvious examples. Likewise, the coefficient of thermal expansion of the parts' material should be considered a source of error. This is especially true with longer part features, with areas lacking stable environmental controls, and with machines not supporting part temperature correction.

It is important to note that such correction systems do not alleviate all problems because parts never maintain constant temperature throughout. Also, such systems increase reliance on proper operator procedures, like using gloves or soaking time.

Other concerns regard the quality of the part and its features. For example, the surface finish and form values greatly affect both the ability to collect probing points and the number of points required to calculate accurate substitute geometry. Even the conformance to specifications for any given feature can affect the ability of the measurement system to analyze its attributes.

The sensitivity of these sources depends on the environment, the material of the part, and the capability required.

20.2.2.5 Fixturing Sources of Uncertainty

Part fixturing is listed separately because part distortion within the holding fixture is one of the error sources involved. Other concerns involve the dynamic properties of the fixture's material, but this depends on the application. For example, given a situation where the temperature is unstable and the part is fixtured for a longer period of time, either prior to machine loading or during the inspection, distortion to the fixture translates into distortion of the part.

Additional environmental concerns involve the fixture's effect on lighting parameters for noncontact systems and on part distortion during probing for contact systems. Other sources include utility concerns, where air or vacuum pressure fluctuations can distort parts or affect the ability of the fixture to hold the part securely in place. Other concerns are with regard to the fixtures performance in reproducibility, between machines, and between operators.

The sensitivity of fixturing factors is highly dependent on environmental conditions, part and fixturing materials, and the measurement system capability required.

20.2.2.6 Operator Sources of Uncertainty

The user of the system can greatly influence the performance of any measurement system. This is particularly true within the lab environment, where applications-specific measurement is rare. For example, within the lab, operators may have the option to change CMM parameters, such as speed, probing approach, etc. Similarly, within the lab, designated fixturing is less common; therefore, the variability between operators is increased.

Likewise, algorithm selection, sampling strategies, and even the location and orientation of the part can affect the uncertainty of measurements. For this reason, laboratory personnel must be required to maintain a higher level of competency. Formal, documented procedures should be available for reference.

The sensitivity of these concerns is highly dependent on the competency of the personnel involved, the release and control procedures for part programs, the documentation of lab procedures, and, as always, the measurement capability desired.

The goal of this phase was to identify contributing sources of uncertainty. While the next steps involve quantifying the effects, efforts should be made prior to testing to hypothesize the influences. All sources deemed as sensitive to the given capability and/or application should be prioritized. This process will eliminate unnecessary testing and should focus any diagnostic testing that may be required.

20.2.3 Measurement System Qualification (Phase 3)

20.2.3.1 Plan the Capabilities Studies

There are many published standards discussing the evaluation of CMM performance. The same is true for other equipment as well. These standards are particularly effective because they pertain to testing the machine for performing within manufacturers' specifications.

The three most recognized methods of performance evaluation are known as the comparator method, error synthesis (error budgeting), and the combined method. The comparator method involves statistical evaluation of measurements made on a reference standard. The error synthesis method involves sophisticated software used to model the CMM to evaluate overall performance, given the values of the numerous sources of uncertainty. For laboratory systems, the minimum requirements to consider in the development of a capability matrix include the following:

- Probing Performance
- Linear Displacement
- Geometry (squareness, pitch, roll, yaw, etc.)
- Software
- Feature-dependent capability

Some may notice the inclusion of both measurement capability and performance of specific error sources. Users are free to divide these into two different matrices, yet given the universal nature of laboratory systems, published capabilities must be isolated to facilitate operator evaluations of the uncertainty of various setups and applications.

20.2.3.2 Production Systems

The plan to evaluate the capabilities of a production measurement system may be very similar to past practices in that measurement system analysis tools may be all that is required. The goal is the development of a matrix listing the different capabilities. However, the matrix may be specific to applications, rather than listing feature-dependent capabilities or machine performance levels.

The decision to do more in-depth analysis should depend on the percentage of nonproduction measurements and the level of capability required for those tasks. Regardless, the most common problem becomes deciding on the artifact(s) to provide acceptable reference values (ARVs).

I recommend using traceable artifacts from a nationally recognized laboratory, such as NIST (National Institute of Standards and Technology), when testing machine capabilities. When testing applications, actual parts, or specially produced parts with the same features and attributes of the parts to be measured can be used. The problem with this method involves determination of the acceptable reference values.

In other words, an acceptable reference value without a certification of calibration must be measured by an acceptable reference system. This is similar to the concept of calibrated artifacts; less capable machines rely on values provided by machines of greater capability.

This method addresses the need to include feature imperfections in the testing of capability and the need for evaluations relating to truth. Given the law of the propagation of uncertainty, the true value will never be known. However, this should at least provide an acceptable reference value where the word "acceptable" can be used accurately.

Once the artifacts are selected, the plan is complete, and there is a clearly defined matrix, the remaining steps of this phase are similar to past practices. All test plans must address the following requirements for every attribute evaluated:

- Stability (minimum of two weeks)
- Precision
- Bias
- Reproducibility (minimum of two operators)
- Uncertainty (minimum of length uncertainty)
- Correlation (internal and external)

Many tools exist for testing, and shorter versions of those tests may be useful in evaluating the sensitivity of specific error sources. Such testing is often referred to as "snapshot testing." While not valid for formal analysis, snapshot testing provides sufficient insight into machine performance, particularly for a new and unknown system.

20.2.3.3 Calibrate the System

The requirements of calibration include, but are not limited to, the following:
- Uncertainty of artifact(s) required to achieve performance
- Selection of artifact(s) to be used
- Selection of calibration services, if needed
- Determination of the calibration interval

The calibration lab should provide support through consulting and services. The services must include automated monitoring of the calibration cycle and maintaining historical records of the calibrations performed.

20.2.3.4 Conduct Studies and Define Capabilities

The requirements of this step involve the data collection and documentation processes. If the studies are well planned, conducting the testing is relatively straightforward.

Testing will consume a great amount of machine time, so extra caution in duplicating output should prevent the need for repeating test procedures. Likewise, extra effort should be made to ensure the validity of the programs used.

As for documentation, all procedures and programs should be documented thoroughly and maintained with the testing data. Other required information for each test conducted includes the time, date, temperature, operator, and system (when more than one). Once a test is complete, a brief synopsis of the test and the results should be included with the documentation.

Once all tests are completed, the results are recorded to define the capability matrix of the system. As stated, these matrices will differ depending on the system's designated use. In fact, there may be some differences between matrices of like systems.

20.2.4 Quantify the Error Budget (Phase 4)

This phase is an in-depth analysis of the earlier hypothesized influences on uncertainty. In some cases, testing will indicate a need for additional testing; in others, the data may already clearly identify the impact of the error source in question.

As with any testing, the goal is to become knowledgeable about the system being evaluated, not to confirm preconceived hypotheses. The original assumptions serve only as an organized method to approach formal testing where quantitative measurements can be calculated.

Also, if valid priority assignments were established, the focus of the testing should be more apparent. These priorities should prevent delving too deeply into testing of sources with little contribution or with little probability of optimization.

20.2.4.1 Plan Testing (Isolate Error Sources)

While design of experiment techniques provide many methods to analyze multiple variables, tests should be designed in an effort to isolate variables with regard to each specific error source. This facilitates the testing and the analysis.

For example, there are many variables involved in the overall uncertainty of probing performance. While tests could be designed to include length uncertainty, this approach is not recommended. Such a test also would introduce into the test the variables of temperature effects on the machine and the artifact and the performance of those software algorithms. The standards unanimously recommend evaluation of probing performance over a very small volume, using artifacts near 25 mm in size.

Similarly, when evaluating length uncertainty, efforts should be made to remove probing and algorithm performances. Many variables remain, including the temperature considerations of machine and artifact and the correction algorithms available. In this example, ball bars are often used with the length between sphere centers being the focus of the testing.

When compared to qualification tests, a significant difference in this testing is the study of operator influences. Given the numerous applications and the variety of fixturing tools in laboratory systems, the focus on fixturing and the documentation of results serve only as guides to individual users, much like the other information in the capability matrix. Should quantitative testing indicate significant problems, the optimization phase should lead to additional training, etc.

20.2.4.2 Analyze Uncertainty

One of the most difficult concepts involved in error budgeting is analyzing test results to determine overall uncertainty for various applications. Fortunately, there are many guides that recommend various mathematical approaches to expressing the uncertainty of specific measurements. All that is required is quantitative information of the sources considered sensitive to the specific application.

Upon selecting the uncertainty variables that are sensitive to a given capability or application, one needs only to choose the desired combinatory rule and calculate the result.

Correctly identifying the sensitive sources of uncertainty is usually the easier of the two. For example, squareness in the YZ plane will have little to no effect on diametral readings in the XY plane, unless the diameter to be measured is particularly large. Likewise, single-point repeatability may have little effect unless it includes dynamic performance, which affects uncertainty only at specific temperatures, speed, and probe approach distances.

Obviously, there are many sources of uncertainty and not every variable can be tested. However, almost all exist as subsets of other contributing errors. The task may seem daunting, but the reason for statement of relative ease is apparent when selecting combinatory rules. Additional analysis to evaluate relationships and interdependencies may be desired.

Once the testing is completed, the quantitative measures of uncertainties should be known. Analysis is usually as simple as selecting the sensitive variables and the desired combinatory rule.

20.2.5 Optimize Measurement System (Phase 5)

Even if the measurement system performs to the capability required, there is often a need for increased performance. If the system is a production system, where the only studies performed are applications-specific, it may prove necessary to complete Phase 4. Again, this depends on the level of improvement required and the specific use intended. It may be possible simply to qualify the system for the new application.

Otherwise, the optimization phase consists of conceiving possible improvements in various areas of uncertainty. Revisiting the original testing provides a means of determining success. Once realized, requalification should indicate a more capable system.

20.2.5.1 Identify Opportunities

Opportunities to improve capability are dependent upon the variables contributing to uncertainty. In such cases, the next steps are obvious.

Problems manifest themselves when no apparent prospects exist. For example, even when exhausting tests have been completed, the uncertainty values sensitive to the capability in question may seem infinitesimal. The obvious question arises as to whether anything can be done to reduce uncertainties even further, or whether an unknown error source remains that was unaccounted for in the original testing.

Other problems may be specific to the application in question. A common example would involve measurement of extremely small part features or the tooling required. One of the largest sources of error for contact CMMs is probing uncertainty. This is particularly true for probes smaller than 1 mm. The effects of probing uncertainty on the capability to measure feature size are well known.

20.2.5.2 Attempt Improvements and Revisit Testing

The most obvious recommendation when attempting optimization is the need to exercise caution. Efforts should not include multiple variables. "Snapshot testing" is the best tool for informal evaluations.

Improvements are not always machine specific. They can involve revamping the HVAC system, training operators, and attempting new probing strategies. In fact, optimization can be realized simply through implementation of formal procedures.

Once "snapshot testing" results indicate the possible result desired, formal testing must be revisited to support formal analysis of the optimization efforts. While the same documentation requirements exist for retesting, an additional synopsis should describe the optimization process, the desired results, and the success or failure of the effort.

If optimization is successful and uncertainty values are reduced, the process is repeated for all attributes where increased performance is desired and deemed probable. Once uncertainty contributions are considered acceptable, the system must be requalified for any and all capabilities that may be affected.

20.2.5.3 Revisit Qualification

Determining the qualification tests that require repeating is dependent upon the enhancements realized. For example, improving fixturing reproducibility for a laboratory system should not affect any other qualification tests, unless those tests were poorly conceptualized.

Once completed, the capability matrix should be updated, even if the results are not as expected or desired. Additional efforts of optimization should repeat the process, and all documentation should reflect all efforts, even unsuccessful ones. This information could prove beneficial at a later date or to other measurement system characterization projects.

Optimization requires identifying opportunities, "snapshot testing" of enhancements, repeating the formal testing of uncertainty contributions, and reproducing the capability matrix. Both successful and failed attempts should be well documented for future reference.

20.2.6 Implement and Control Measurement System (Phase 6)

The last phase of measurement system characterization is implementation and control. This is not to say optimization efforts are complete, but once initial efforts are completed, the system is activated. Control is achieved through periodic calibrations, maintenance, and performance tracking.

True characterization takes place over time. Some systems will maintain initial levels of capability with ease, while others will require additional efforts to improve performance and long-term stability.

20.2.6.1 Plan Performance Criteria

Prior to implementation, performance monitoring criteria must be identified. The variables tracked can include specific capability studies and critical sources of uncertainty. Keep in mind, performance tracking generally should not consume more than 30 minutes a week.

Once the variables are ascertained, the artifact(s) for interim testing should be selected. This can be a calibration artifact used during testing, or a part or group of parts used during testing. As stated previously, only traceable reference standards should be used for laboratory systems.

The final criteria involves the interval of testing and when requalification should be required. Interim testing is usually performed between once a week and once a month. The interval can be changed for many reasons. For example, shorter intervals could be used to assess the effects of increased system utilization.

The question of requalification is dependent upon those factors that may be expected to dramatically affect the system. Some may consider the periodic calibration of the system to be of significant impact. Others may include system crashes, major repairs, or changes in utilization.

The same documentation rules apply to interim testing that apply to other testing. This is particularly true with regard to temperature and other environmental factors. The charting of performance is recommended. Charts provide constant reminders of performance, allowing operators to easily recognize any problems with the system.

20.2.6.2 Plan Calibration and Maintenance Requirements

The calibration cycle is similar to that of interim testing in that the interval is not required to be constant. In fact, performance tracking may indicate the need for shorter or longer periods between calibrations. The same may be true for preventive maintenance schedules.

The manufacturer's recommendations are the logical place to start, with system performance dictating any changes. The necessary artifact(s) should already be available from the original calibration, unless, of course, outside services are supporting the requirements.

20.2.6.3 Implement System and Initiate Control

Performance tracking should establish a baseline, but it is dependent on the statistical tools being used. Once completed, everything should now be in place for implementation. As with any new system, caution should be exercised, with full utilization being achieved in phases. However, this is also dependent upon the amount of testing done earlier.

Once activated, users should benefit by having a qualified measurement system. The interim testing provides a means of control, and the data can be utilized to address other concerns, such as:

- Cases of "slow drift" should be more apparent.
- Data exists for diagnostic analysis.
- Data is available for evaluating effects of calibration.

The process of measurement system characterization process should ensure only qualified and controlled systems are used. The process also provides methods to address both internal and external correlation issues. While the above comments do not include specific details for every system and every approach, it should serve as a sound outline to comprehensive characterization efforts.

20.2.6.4 CMM Operator Competencies

One of the most important aspects of a high precision inspection system is the background of the operator. It would be wonderful to believe that anyone could run an ultraprecision CMM. Realistically, if a company expects to work within the submicrometer regime, the operator's skills as a dimensional metrologist (as well as the skills of engineering and manufacturing support personnel) must be highly refined. For example, the error budget for a part that has a manufacturing tolerance of $2\,\mu m$ might be pages long. Procedures that are normally not used (like torquing clamps or fixtures, calibrating probe tip sphericity or roundness, and calculating "Uncertainty of Nominal Deferential Expansion" for known materials) must now be accentuated to work within this tolerance band.

Almost as important as the operator's skills is a support team that helps minimize both the random and systematic error sources in the measuring process. At the submicrometer level, there is simply no room for either. Both error sources are difficult to minimize. For example, different operators will get different results. Like materials will have different coefficients of thermal expansion (of course the way to avoid/minimize problems here is to perform all inspections at 20 °C). The same part can show two different form errors depending on which section of the probe was used for the inspection.

Now that the six phases for measurement system characterization have been outlined, the next step is to define actual testing. This testing leads to the necessary confidence for developing the capability matrix. Tests are done to the degree necessary to achieve optimum submicrometer capability, which is the primary objective in the area of operating interest.

20.2.6.5 Business Issue

Before discussing the actual testing results, an unexpected situation that came up during the testing should be mentioned at this time.

Proving the environment is stable should always be a priority issue. An unstable environment can have a large detrimental effect on the confidence of a CMM's results. Unfortunately, the temperature flow of the room was not taken seriously enough in the initial stages of room development, which led to significant delays in testing and system integration.

Based on this situation, I composed the following memo and presented it to corporate executives to justify additional dollars to enhance room temperature controls.

Internal Memo: Need for Tightened Temperature Control

Concerns and possibly doubts have been raised regarding the true need to control the high-accuracy CMM room to tighter-than-specified temperature controls. My objective for this document is to address some of the high-level issues applicable to the CMM so as to aid individuals in their level of understanding of this technology. I hope in turn, they not only will support the current need for this level of control, but also entertain it as a minimum standard for future controls.

My challenge in this justification effort, while preparing this memo, was in figuring out the audience that would possibly review it. Due to the wide range of technical expertise, within the potential audience, particularly concerning the understanding of thermal effects, I chose to stay generic with the content and to offer to make myself available to elaborate on key points and respond to specific questions any individual might have.

The following outlines the content of this memo:
1) Issues related to the justification of the CMM
   - Assumptions
   - Intangibles
2) Basis for the manufacturer's recommended temperature specification
3) Five blocks for building an understanding of temperature effects
   - Differential expansion
   - Expansion uncertainty
   - Source of temperature errors
   - Bi-material effects
   - Gradients
4) Temperature control of the current CMM room
5) Testing results applicable to the CMM in its current environment
   - Thermal drift test
   - Tolerances on tooling components and assemblies
   - Miscellaneous "feature-based measurement tests"
6) Miscellaneous variables aid in decreased confidence of measured results
7) Summary

(1) Issues Related to the Justification of the CMM

The original CMM focus was an extension of the tooling and product qualification procedure developed over one year ago. Our inability to measure tooling features within their stated tolerances and our ongoing struggle to make sound engineering decisions on less-than-accurate and repeatable measurement results were the principle justifications for spending well over one half million dollars to procure a ultra-precision class CMM. Some of the key issues that were made visible at that time were as follows:

Assumptions
1) < 1 µm is accurate enough to tell us what effects the tool shapes have on the forming process.
2) Environmentally controlled room is available (20 °C +/- 0.14 °C).
3) Trained operators/programmers are available to run the CMM.
4) All tools are mapped for "critical" characteristics and tracked over time to observe performance capability to longevity of tool life.

Intangibles
1) The trend is toward finer and finer forming capabilities. We continue to allow for (insist on) less variation in the tooling.
2) Data can be used to tell us the tool shape to understand the interaction between tool, press, and material.
3) Should provide better future tool designs "out of the shoot." As we understand what dimensions worked in the past, we can incorporate those into future tool designs.
4) Improved process capability.
5) We currently end up with no permanent solutions to many tooling issues.
6) Customer satisfaction.
7) The target is moving. If we do not improve, the current situation could get worse with more difficult products "coming on board."
8) Benefits of reduced lead times on new products (1-4 week improvement due to tool qualification).

(2) Basis for the Manufacturer's Recommended Temperature Specification

I believe most of the doubt or confusion regarding the true need for tighter temperature control in the CMM room stems from individuals' awareness of what the Brown & Sharpe/Leitz environmental requirements are for their enhanced-accuracy CMM (which is the machine we have).

Their environmental requirements allow for a vertical range of 0.75 °C/meter, a horizontal range of 0.7 °C/meter, and a maximum variation per day not to exceed 0.5 °C/day on any individual thermistor. Keep in mind that both the vertical and horizontal variations are targeted around 20 °C. To clarify, this would be 20 °C +/- 0.35 °C in the horizontal axis. What is essential to understand about this specification is that it is also based on a "total volumetric inaccuracy" of the system, not to exceed +/- 2 µm.

All CMM manufacturers are sensitive to the fact that the tighter the temperature specification, the more the room is going to cost to build and to maintain. Anytime you get beyond the mechanical, electrical, and software aspects of their system, and still want higher accuracy and repeatability, they will always tighten the environmental requirements of their specification. In most industries, companies would be extremely content with +/- 2 µm capability within the machine cube. In our case, it is not adequate.

Based on prior knowledge of the influencing variables, we decided to purchase the enhanced-accuracy system with standard environmental requirements and to tighten up the internal controls ourselves.

(3) Five Blocks for Building an Understanding of Temperature Effects

For the best accuracy, you should make all measurements at 20 °C. Both the measuring machine and workpiece should be at that temperature. At other temperatures, thermal expansions will cause errors. These errors cannot be corrected fully, even by the best temperature compensation methods. This is not to say that all measurements must be taken at 20 °C, but one must go through the following analysis to make a positive determination.
1) What are the workpiece tolerances?
2) How much measurement error can I reasonably accept?
3) How much of this error can I allow for in temperature effects?
4) How much temperature control do I need to keep temperature effects at an acceptable level?

The answer to question 1 is easily determined, questions 2 and 3 are business decisions, and question 4 is the difficult one to answer. I'm going to stay away from listing the formulas necessary for calculating each of the theoretical values for the influencing variables to question 4, but I want to touch briefly on five key blocks for building an understanding of temperature effects, which are differential expansion, expansion uncertainty, source of temperature errors, bi-material effects, and gradients.

Differential Expansion
Most materials expand as temperatures increase, but the amount of expansion varies by material. Expansion of a measuring machine is considered 0 at 20 °C. This is a matter of politics, not physics. A measuring machine compares a length on a workpiece with a corresponding length on a machine scale. Generally though, the workpiece and scale expand by different amounts. This is termed "differential expansion." With no other problem, error equals workpiece expansion minus scale expansion over the length of the measurement.

Expansion Uncertainty
Coefficients of expansion are given in shop, engineering, or scientific handbooks. Different handbooks will in some cases state different coefficients for the same type of material. This occurs because not all test specimens of a particular material are exactly alike.

NIST estimates expansion of a gage block to vary +/- 5% if heat and mechanical treatment of the blocks is defined, +/- 10% if undefined. Samples cut from a single large steel part vary +/- 2 %. Hot or cold rolling causes changes +/-5%. Grain structures cause different expansions in different directions.

Sources of Temperature Errors
It might seem that you cannot have large temperature errors with small workpieces because short lengths mean small expansions. But measurements that take a long time can be influenced by slight changes in temperatures of the workpiece and machine.

Influence from lighting on large machines in small rooms can have an impact. If the lighting is uniform, the machine will settle down to a stable shape that can be error mapped, but normally it is not uniform. The most common problem is the horizontal bending of the bridge (like on our machine). Air conditioning systems that alternately blow hot and cold air on a part of the machine can cause bending as well. Computers and controllers near the machine, as well as bodies (programmers and operators) will cause local heat sources that have the potential of causing a problem if the heat is not dissipated.

The principle problem with all of these potential heat sources is that they cause stratification problems within the envelope of the system. This causes different areas of the machine and workpiece to be at different temperatures.

Bi-material Effects

Bending of bi-metallic thermostat elements caused by temperature is fairly well understood. The same effect occurs in the measuring machines and workpieces. The effect is caused by slightly different coefficients of expansion of different parts of the machine or workpiece. Bi-material bending effects are "usually" very small.

Gradients

If temperature rises, heat flows through the machine surfaces and into the machine structure. The same happens with a workpiece. For heat to flow from the surface into the structure, there must be a temperature difference or gradient. You can think of this flow somewhat like a flow of water caused by a difference in pressure.

Gradients cause different expansions in different parts of the machine or workpiece. The results are similar to the bi-material effect and come in three situations:
1) If air temperature cycles rapidly (as with air conditioning) there is not much time for heat to flow into the machine or workpiece before it has to flow out again. Gradients are close to the surface, and the machine bending is small.
2) Where temperature changes slowly, the effect is as discussed under differential expansion.
3) The worst case is where temperature changes rapidly in the same direction for a long period of time. It causes that part of the machine or workpiece structure to change temperature more quickly than thicker parts, causing bending.

(4) Temperature Control of the Current CMM Room

The critical issue to keep in mind when reviewing the following is that our target has always been 20 °C +/- 0.14 °C.

Recent "repeatability" tests on our CMM for diameters and lengths (lengths less than 100 mm) had outcomes that were considered extremely high (0.6 µm at Six Sigma). Attempts at optimizing programs yielded only a slight gain (0.5 µm at Six Sigma). These results do not include accuracy.

Poor temperature stratification was suspected to be the main problem, which led to installing eight thermistors around the machine at various heights and the results were monitored. The range of temperature within the envelope of the system was greater than 0.83 °C, with an average of close to 20 °C. Since that time, air flow has been adjusted coming into the room to aid in dissipating local heat sources, which in this case is principally the computers and bodies. Based on these adjustments, the range has improved but is still greater than 0.56 °C.

(5) Testing Results Applicable to the CMM in its Current Environment
Thermal Drift Test

In 1985, ANSI/ASME published a standard (B89.1.12M) which covered "Methods for Performance Evaluation of Coordinate Measuring Machines." This standard covers generic test procedures for determining both linear and volumetric inaccuracies of CMMs, as well as procedures for determining the stability of the environment. This test is called a thermal drift test.

To run this test, the machine is required to sit stable for a specified length of time, then with a calibration sphere located as close to the machine work surface as possible (to ensure stability), the probe is to be calibrated using a defined number of points. Once calibrated, you establish the coordinate system to -0- (all three axes). Then you place the CMM in a continuous loop to re-measure the sphere, one time every minute. This is continued for 48 hours, storing the x, y, and z axis displacement values from its original -0-, as well as storing the size and profile displacements from its original size and shape. Note: There are temperature sensors built into the x, y, and z axes slides that are monitored during the test period.

The test ran for 56 hours. The results clearly explained why we could get no better than 0.5 µm repeatability at Six Sigma on prior tests. The range of drift over the length of the test in our case was not the critical variable we were concerned with, but rather the amount of drift recognized over a length of time equivalent to the longest program used to measure a component or assembly. In this case, we were interested in a time segment of two hours.

Once the machine stabilized (about 2 hours), the largest drift within any two hour segment in a single axis was approximately 0.4 µm, with individual spikes of 0.3 µm over a 30-minute time frame. Two additional 24-hour versions of this test were run with the same level of results. It is critical to note that the charts clearly display a direct correlation between temperature change and displacement, very close to a linear relationship.

Tolerances on Tooling Components and Assemblies

What needs to be kept in mind on this issue is that the "enhanced-accuracy" CMM was justified principally to measure critical features on tooling components and assemblies. In addition, we were clearly aware (up front) that this CMM (or any CMM) was not capable of measuring every feature we considered critical to process or function. For example, one of the restrictions on a contact CMM is probe diameter. The smallest "standard" probe tip available is 0.3 mm, which restricts measurements on an inside radii or diameter.

A large percentage of the features of size have tolerances of 1.25 µm to 2.5 µm with feature location tolerances of 5 µm. I believe I would be conservative in saying that greater than 50% of the features that are measured on this CMM are ≤ 5 µm. These are "current" tolerances defined on tooling drawings at this time.

If we look back at one of the original "assumptions" (#1. 0.5 µm is accurate enough to tell us what effects the tool shapes have on the forming process), this was a "worst-case" statement which included accuracy and repeatability of the measurement system. What has been discussed so far has been only "repeatability."

Miscellaneous Feature-Based Measurement Tests

It is essential that the results from the thermal drift test are understood to be based on a simple measurement within a small known envelope of 25 mm, so accuracy and repeatability are at their best. Where it starts becoming more difficult is in measuring other types of geometric features within a larger envelope, such as perpendicularity, cylindricity and profile, to name a few. It takes a significant number of points on a given feature to get an accurate representation of its geometry. A general rule to note is that as you increase the number of points, the better the accuracy and repeatability. There are exceptions, but in general this holds true.

(6) Miscellaneous Variables Aid in Decreased Confidence of Measured Results

In addition to temperature, there are many other variables that influence accuracy and repeatability. Some of these variables are humidity, contamination, types of probes due to stability (stiffness) such as the difference between steel shafts versus ceramic and carbide, probe speed, and fixturing. The list goes on and on. The key item at this time that is restricting our leap into the sub-micrometer capability we need (and have been striving for) is "temperature."

(7) Summary

The "great" part about our CMM is that it is exceeding the specifications committed to by Brown & Sharpe/Leitz. They were aware from the beginning that our expectations of their system was to push it well beyond their stated capability. They also mentioned that tight temperature control would be necessary to accomplish this task.

I sincerely feel the level of temperature control I'm stating here is also needed in many other measurement applications at our site to reduce current inaccuracies. I hope I have convinced the readers of this memo on the need for tight temperature controls to achieve sub-micrometer measurement capability on this type of measurement system. I will need approval for additional expenses of $35K to achieve the defined controls for the CMM room.

If there are any questions, I would be happy to address them as best I can.

END of MEMO.

All funds were approved based on this presentation.

20.3 CMM Performance Test Overview

The testing was done on a Brown & Sharpe/Leitz PMM 654 Enhanced Accuracy CMM to determine the machine's capability and the confidence with which various features could be measured.

There are a variety of parameters affecting the repeatability of measuring a geometric element on a CMM. These parameters can be separated roughly into three categories: environmental, machine, and feature-dependent parameters. These include, but are not limited to, the following:

1) Environmental
- Room (and part) temperature stability
- Room humidity
- Vibration
- Dirt and dust in room
- Airline temperature stability

2) Machine
- Settling time (probing speed, probing offset, and machine speed)
- Probing force (upper and lower force, trigger force, and divider speed)
- Flexibility of probe setup (probe deflection)
- Multiple probe tips (star probe setups and magazine changes)

3) Feature Dependent
- Size (surface area) of feature
- Number of points per feature
- Surface roughness (form) of the part
- Scanning speed

The following three sections will add detail to the above three categories with insight to the testing completed. This should be considered summarized information that leads to the final development of the capability matrix — the final goal of "measurement methods analysis in a submicrometer regime." The scope of these tests is intended to do whatever is necessary to have Six Sigma measurement capabilities for all geometric controls of interest, less than 1 µm.

Many of the machine (Section 2) and feature-dependent (Section 3) tests have graphs showing a visual representation of the data. For convenience, these will not be referred to by graph number and will be located within the test section to allow better use of space.

20.3.1 Environmental Tests (Section 1)

20.3.1.1 Temperature Parameters

To understand the relationship between the room environment and the CMM's results a "thermal drift test" that tests for thermal variation error (TVE) was completed. This test is outlined in the ANSI/ASME Standard B89.1.12M and is called "Methods for Performance Evaluation of Coordinate Measuring Machines."

To run this test, the CMM was parked in its home (upper, left, back corner) position for a period of six hours. This allows the machine enough time to stabilize if necessary. Then using five points, a 25-mm sphere was measured three times, reporting the average x, y, and z center position, diameter, and form. This measurement sequence was repeated for a minimum of 12 hours, and the results graphed opposite the temperature of the three axes scales. Temperature compensation was enabled at the beginning of every sequence. The range of the drift over the full length of the test was not the critical variable. Rather, it is the amount of drift that occurs over the length of time equivalent to the longest program used to measure a component or assembly. In this case, the interest was in the maximum time segment of two hours.

TVE Test # 1:

	X	**Y**	**Z**
Coordinate range (mm)	0.00417	0.00080	0.00068
Temperature range (°C)	0.10040	0.08752	0.12872

This TVE test was run for a period of 56 hours in the new lab with temperature centered on 20°C. The y and z axes showed an amazing linear response to the temperature of their respective axis. These test results prove that controlling the temperature of the machine axes is essential to the performance of the CMM. However, the results were not as good as expected and raised some new questions.

First, why does the x-axis not respond to its temperature in a linear manner? Was there another parameter creating a greater effect on the x-axis than temperature? If so, what was that parameter? Also, why was the x-range so much larger than the y and z ranges? Finally, why do all three axes show a large decrease in temperature at the beginning of the measurement cycle? Was it the fact that the machine is running? (You would logically expect the machine to heat up, not to cool down when running.) Or was it the position of the machine when resetting in the home position versus its position when measuring the sphere? If so, what was causing the temperature drop?

TVE Test # 2:

	X	**Y**	**Z**
Coordinate range (mm)	0.00068	0.00053	0.00081
Temperature range (°C)	0.04247	0.06178	0.10812

The next step was to run a shortened version (24 hours) of the same test to ensure the results of the first test were repeatable. When duplicating results, it is essential each step of the original test is followed exactly.

The results were very similar to those from the first test. The y and z axes continued to have a strong linear relationship with their axes temperatures, while x was definitely nonlinear in nature. The initial decrease in all three axes temperatures was again evident in the first two hours of the test. In this test all three axes' temperatures were also plotted against one another, showing that all three axes were following the same pattern. It was evident that whatever was creating the fluctuations in one axis was also affecting the other axes. When looking at the magnitude of the temperature drop, the z-axis had the largest temperature range followed by the y and then the x axis.

In addition, the three axes temperature plot revealed a great deal of stratification in the room (over a 0.3 °C difference) between the y and z axes and the x-axis. It is highly possible such a large amount of stratification could cause problems when attempting to hold the room environment constant. Finally, the y-axis temperature was displaying a cyclical pattern about 40-45 minutes in length. A closer inspection of the first test showed a similar pattern as well. This test left four questions to be answered:

1) What machine or environmental parameter was causing all three axes to decrease in temperature at the beginning of every run?
2) Why was the x-axis displaying a nonlinear relationship to its axis temperature? Is there some other outside parameter affecting its performance?
3) Would the stratification of the room create any performance or room stability problems? If so, what was creating this stratification?
4) What was causing the cyclical effect observed in the y-axis?

TVE Test # 3:

	X	Y	Z
Coordinate range (mm)	0.00167	0.00072	0.00135
Temperature range (°C)	0.04762	0.06693	0.11585

The next TVE test was designed to test whether the decrease in temperature occurred directly after the machine began to run. The temperatures of all three axes were recorded while the machine was resetting in its home position for six hours before measuring the sphere for 24 hours.

The results of this test clearly indicated the machine reached a higher temperature plateau when placed in the home position. Either the movement of the machine or the machine placement was causing this change in temperature. Based on this, the decrease was being caused either by the room environment or the temperature of the air exiting the air bearings.

At this point, a sensor was placed directly within the air line entering the room to monitor the temperature going into the air bearings. The results showed the temperature going into the air bearings was indeed higher than the room temperature. Could the air bearings be closer to the axes scales at certain positions of the machine? Or in the case of the z-axis, was the ram being warmed up due to the higher temperature air exiting from the air bearings?

Questions arose regarding whether temperature compensation would create problems in the resulting data if it were activated. An additional test was run without temperature compensation. Additionally, there was at least one rest period of six hours where the machine was left directly above the sphere. This data would tell us if the position of the machine was causing the temperature drop.

Finally, these test results displayed the y and z axes were again linear to temperature while the x-axis was not. The temperature of the three axes continued to follow one another, and the same amount of stratification was evident. However, the cyclical pattern of the y-axis was not displayed in this test.

TVE Test # 4:

	X	Y	Z
Coordinate range (mm), (temp comp on)	0.00092	0.00051	0.00133
Coordinate range (mm), (temp comp off)	0.00092	0.00048	0.00113
Temperature range (°C)	0.08336	0.09782	0.16476

In this test, the machine was placed in the home position for six hours, run for 12 hours, placed in the home position for six hours, run for 12 hours, placed directly above the sphere for six hours, and run for twelve hours. The sphere was measured with and without temperature compensation to see if any difference did exist in the results.

The results indicated the position of the machine was causing the change in temperature to occur. In all three axes, there was a definite rise in temperature when the machine was in the home position. When the machine was left to rest above the sphere, however, no similar rise in temperature was evident. Additionally, the test showed only a simple bias between the data taken with and without temperature compensation. The data collected up to this point was indeed valid. Finally, the cyclical effect that had disappeared in the previous test had resurfaced not only in the y-axis but also in the z-axis.

Based on this data, a new approach was taken to control the room environment (based on the memo shown at the beginning of section 1). A new air-flow system was added to ensure a uniform air flow moving over and away from the CMM. This would prevent warm pockets of air from being trapped around the machine. Test # 4 was replicated.

TVE Test # 5:

	X	**Y**	**Z**
Coordinate range (mm), (temp comp on)	0.00047	0.00042	0.00052
Coordinate range (mm), (temp comp off)	0.00052	0.00047	0.00051
Temperature range (°C)	0.03928	0.04332	0.04111

Based on these results, test #5 was replicated two more times to ensure a high degree of confidence in the measured results.

TVE Test # 6:

	X	**Y**	**Z**
Coordinate range (mm), (temp comp on)	0.00042	0.00038	0.00049
Coordinate range (mm), (temp comp off)	0.00048	0.00046	0.00050
Temperature range (°C)	0.04211	0.04182	0.04132

TVE Test # 7:

	X	**Y**	**Z**
Coordinate range (mm), (temp comp on)	0.00045	0.00040	0.00050
Coordinate range (mm), (temp comp off)	0.00050	0.00042	0.00054
Temperature range (°C)	0.03723	0.04123	0.03998

It is interesting to note that the cyclical effects stayed present in the last three tests, but to a lesser degree. Further temperature optimization was not pursued due to current satisfaction in the noted results.

20.3.1.2 Other Environmental Parameters

There are obviously more environmental parameters than simply temperature. Humidity, vibration, dirt and compressed air quality are generally considered less important, but were determined to be well within specifications.

The pressure and temperature of the compressed air was also within specifications before the machine was installed. However, due to concerns arising from the TVE tests, the compressed air was examined again. Sufficient pressure was being supplied to the machine and the temperature (although higher than room temperature) was within specification. Finally, the dust content of the room was lowered slightly by adding floor mats in the buffer room and by sealing off miscellaneous areas.

Based on the Six Sigma capabilities being driven for in the submicrometer regime, it is essential the room environment be as stable as possible. Uniform air flow and temperature over the CMM must be constant, as any change will be recognized.

20.3.2 Machine Tests (Section 2)

20.3.2.1 Probe Settling Time

The Leitz PMM 654 machine was installed so the factory default machine parameters were active. These default settings have been optimized for maximum accuracy and throughput when using the machine for a majority of the applications. However, these settings can be changed to improve accuracy or throughput on out of the ordinary applications. For example, the force applied by the probe head must be lowered in order to measure a thin, flexible part. The machine settings marked as important to test are the probe settling time and probe force.

Machine Test #1: Z-Axis single-point measurement versus probe settling time (see Fig. 20-1)

The probe settling time is a function of two probe settings: the probing speed (mm/sec) and the probing offset (mm). By decreasing the probing speed and increasing the probing offset (thereby increasing settling time), we should see an increase in the performance of the machine.

To test this theory, a single point in the z-axis was measured 25 times and its Six Sigma repeatability was calculated. This sequence was repeated using various combinations of the two settings. The results displayed unique changes in the repeatability of single-point measurement as the settling time increased from 0.125 to 1 second. These results were contradictory to the original hypothesis that increasing the settling time would increase machine performance.

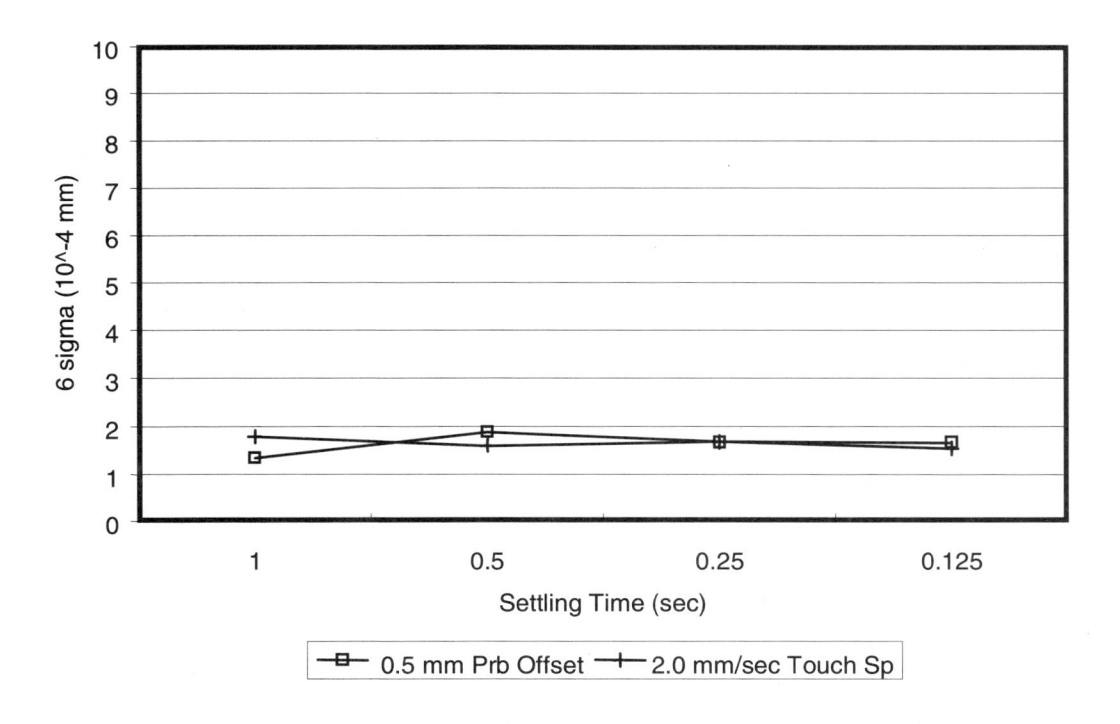

Figure 20-1 Z-Axis single-point repeatability

Machine Test #2: Sphere form versus probe settling time (see Figs. 20-2a and 20-2b)

In this test, three different probes were calibrated on a 10-mm sphere. This same sphere was then remeasured 25 times using a 29-point pattern, reporting the sphere's mean form and Six Sigma value. The

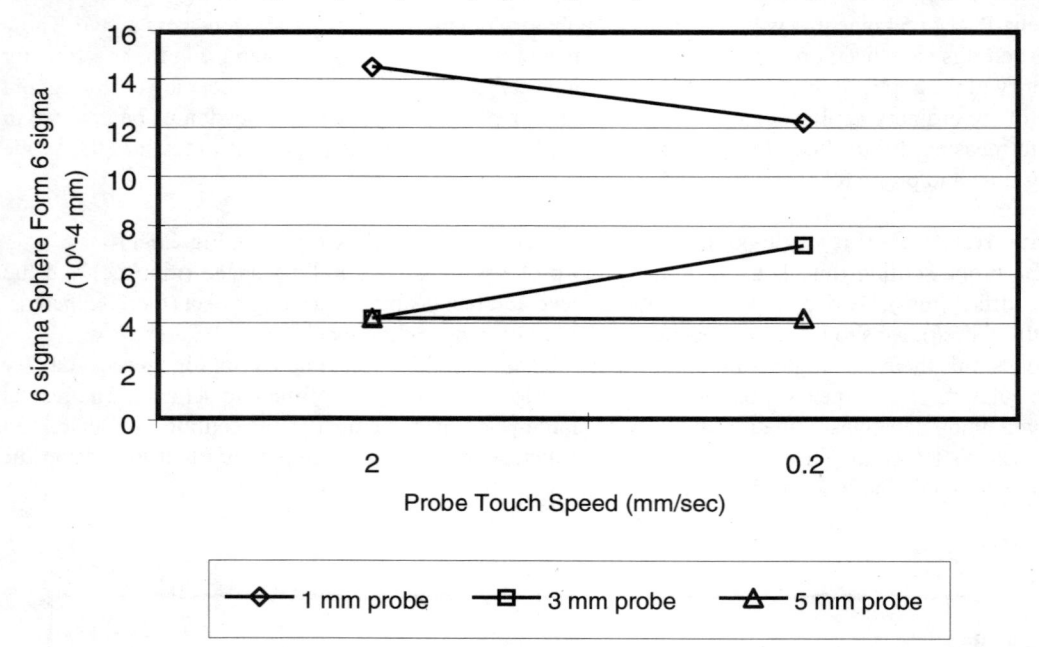

Figure 20-2a Form Six Sigma versus probe settling time (10-mm sphere)

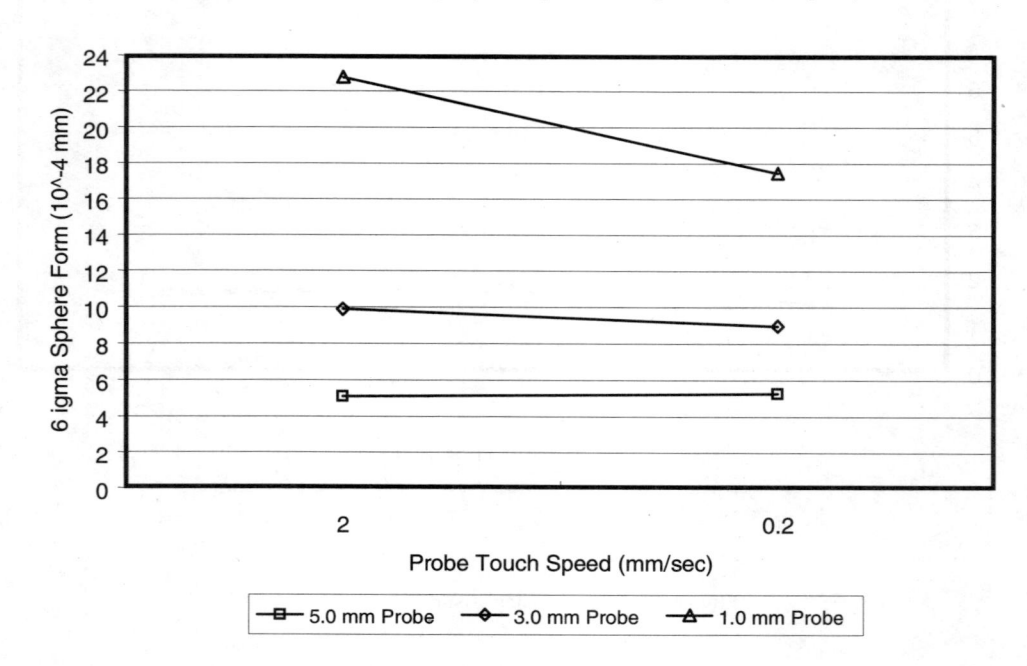

Figure 20-2b Sphere form versus probe settling time (25-mm sphere)

first series of measurements were taken using the default probe speed of 2 mm/sec. A second series of measurements were taken at 0.2 mm/sec (the probe was recalibrated at the lower speed before measurement). This entire procedure was then repeated with a 10-mm sphere.

The results show a slight improvement in the mean form when lowering the probe speed. These results were similar to those from the single-point repeatability. This is more than likely due to the design of the Leitz probe head, where the actual probe point is registered as the head is pulling away from the part. Therefore, the small range of this test had a limited effect on the machine's performance, which is adequate based on the speculated range of operation.

Machine Test #3: Probe speed versus sphere form (see Fig. 20-3)

This test was run to get a better idea of the machine's response over a greater range of settling times. Using the default probe offset of 0.5 mm, the following probe speeds (mm/sec) were tested: 4, 2, 1, 0.5, 0.25, 0.125, and 0.0625

At each probe speed, two different probes were calibrated on the 25-mm sphere. This sphere was then remeasured using a 29-point pattern, with the form, diameter, and probe deflection being reported. The results again showed limited decrease in the sphere form as the probe speed decreased, regardless of which probe was tested.

At this time, there is no evidence to support the idea that decreasing the probe settling time will increase the performance of the machine. Within the range of values tested, there was no evidence of relationship between settling time and machine performance.

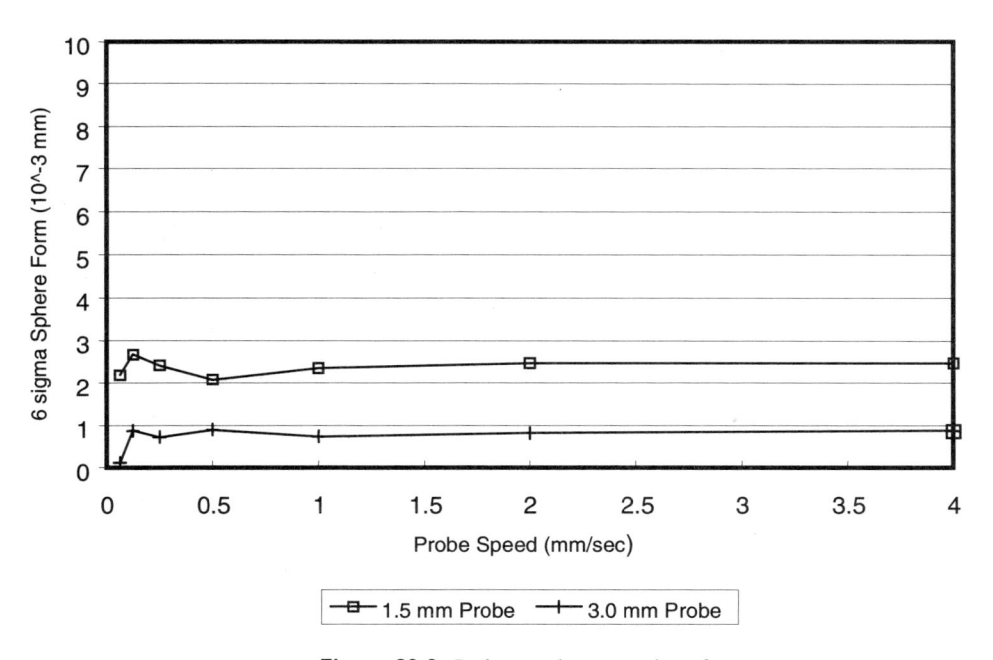

Figure 20-3 Probe speed versus sphere form

Machine Test #4: Sphere form versus probe trigger force (see Fig. 20-4)

Another assumption made before testing began was that lowering the probe head "trigger force" would improve the machine's performance. By varying the probe force, it should be possible to decrease the deflection to which the probe shaft is subjected. This theory was put to the test using three different probe tips calibrated on the 25-mm sphere. This sphere was then remeasured 10 times using a 29-point pattern, reporting the mean form and Six Sigma value.

The first series of measurements were taken using the default trigger force of 0.5 N. A second series of measurements were taken using 0.05 N trigger force (the probe was recalibrated at the lower trigger force before measurement). This entire procedure was then repeated using the 10-mm sphere. The results show an inconsistent relationship between the probe force and sphere form. It was determined that probe force is really a function of several machine settings; upper and lower force, trigger force, and divider speed. Further testing showed that it was possible to influence the form and diameter of the measured sphere by changing these parameters.

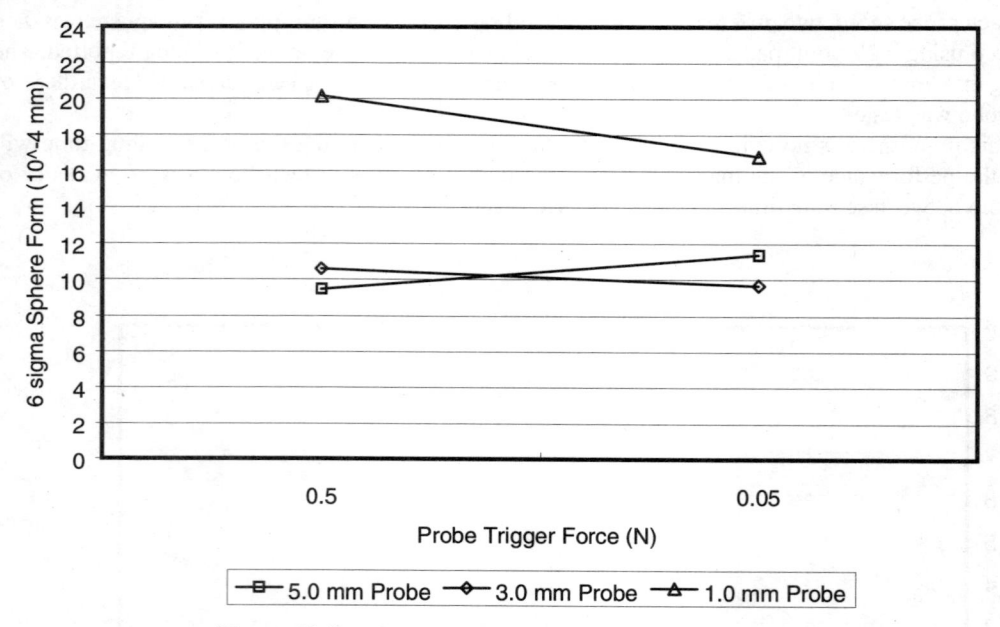

Figure 20-4 Sphere form versus probe trigger force (10-mm sphere)

20.3.2.2 Probe Deflection

The more flexible the probe shaft becomes, the more difficult it becomes to measure in an accurate and repeatable manner. To compensate for this problem, the Leitz probe head creates a deflection matrix, which attempts to map out the amount and direction the probe shaft will deflect. The following is a layout of this matrix:

xx	xy	xz
yz	yy	yz
zx	zy	zz

For example, the xx position in the matrix defines how much deflection occurs in the x-axis when probing solely in the x-axis. This deflection matrix should dampen the deterioration that occurs in accuracy and repeatability as a probe becomes more flexible.

Machine Test #5: Diameter (circle), form, x and y versus probe deflection (see Fig. 20-5)

This test was conducted using four different diameter tips with varying deflection values ranging from 0.295 μm to 1.982 μm. A diameter was measured 25 times and its x, y, diameter, and roundness values were recorded. There was a definite deterioration in repeatability that occurred as the deflection values increased. It must be noted that all probes used were placed straight down in the z-axis using a 25-mm extension. When measuring a diameter with this type of probe, all points were taken with a direction vector that is a combination of the x and y axes. This direction is one in which the probe will deflect the greatest amount. It would then seem very logical that such deterioration would exist as the probe deflection values increased.

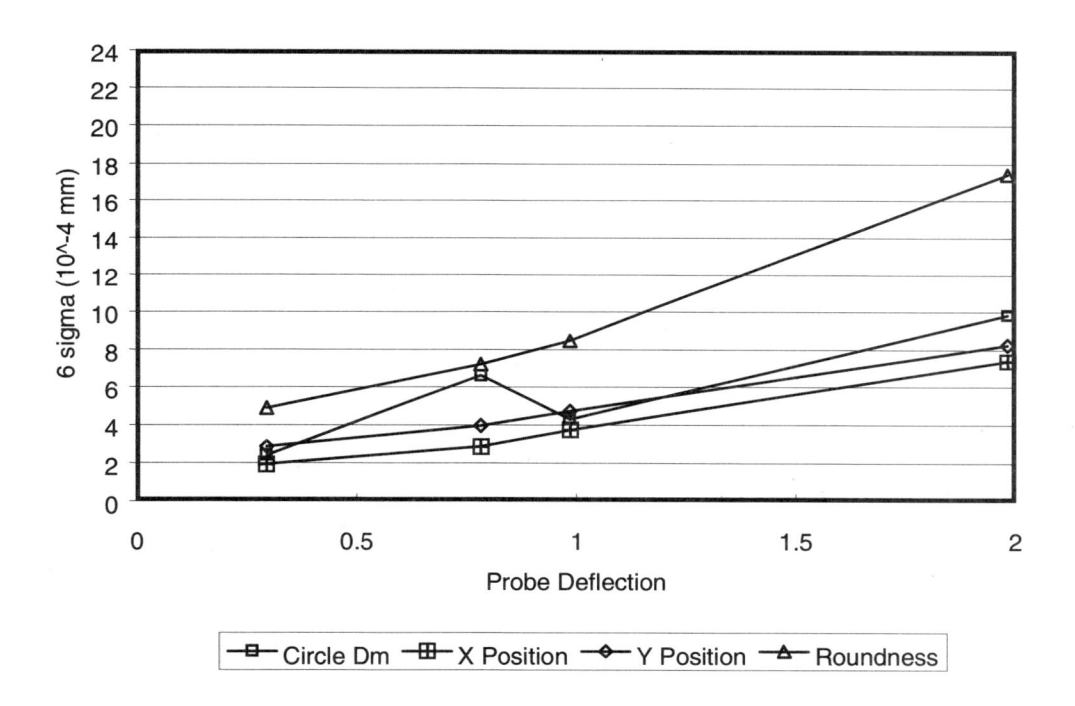

Figure 20-5 Circle features versus probe deflection

In addition, this test also displayed the average diameter in relation to the probe's deflection value. No pattern seemed to exist within the graph, although this may be due to the limited number of probes that were run in the test.

Machine Test #6: Diameter (cylinder), form, x and y versus probe deflection (see Fig. 20-6)

Another test was run using three different probe tips with deflection values ranging from 0.298 μm to 2.278 μm. A cylinder was measured 25 times at three heights, reporting its form, diameter, position, perpendicularity, and straightness values. Again, the results display a deterioration in the repeatability of these features as the deflection values increase.

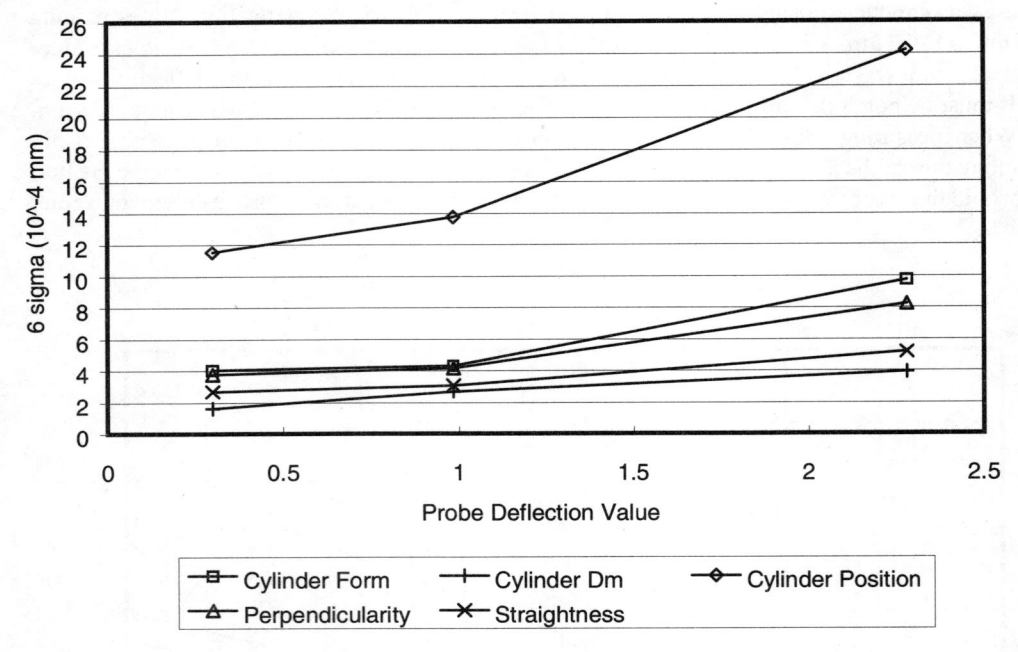

Figure 20-6 Cylinder features versus probe deflection

From these tests, it would seem that the Leitz probe deflection matrix is effective when ensuring the accuracy of the machine does not deteriorate as the probe deflection increases. However, the repeatability of the more flexible probes remains worse than that of the stiffer probes. Mentioned earlier was the possibility that by manipulating those parameters which contribute to the probing force, the deflection that a probe shaft undergoes could possibly be lowered. If this can be accomplished, improvement on performance of all probes should be possible.

Machine Test #7: Probe deflection versus sphere form (see Fig. 20-7)

It has been proven that the machine performance decreases as the probe flexibility increases. It is important that operators of this machine have a very good understanding of how each probe in the probe kit will perform when used. This begins by creating a matrix which contains the deflection of every single probe. When the operator is attempting to maximize the performance of the CMM, they will then be able to choose the probe with the least amount of deflection that will accomplish the job at hand.

Each probe was calibrated 10 times in the xy plane with a 25-mm extension using the three-axis deflection calculation. The calibration sphere was then remeasured using a 29-point pattern, reporting the form, diameter, and probe deflection. In this manner, a matrix containing the probe deflection of every probe was constructed for the operators. In addition, a graph was developed showing the relationship between probe deflection and the sphere form over a large variety of probes. The results again support the theory that the performance does decrease with increased probe deflection.

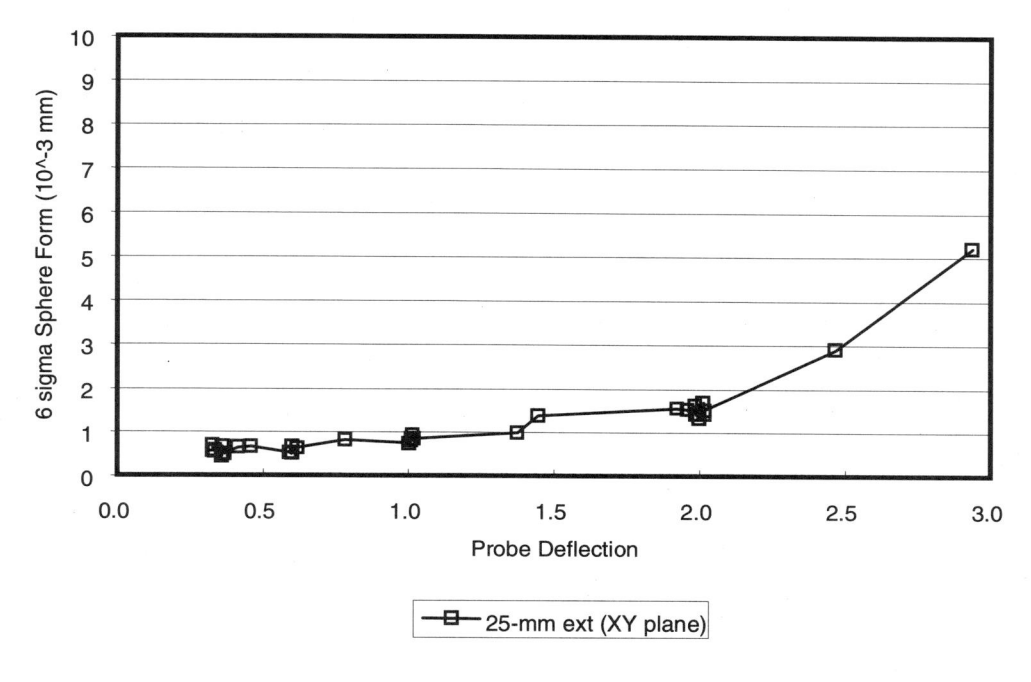

Figure 20-7 Probe deflection versus sphere form

20.3.2.3 Other Machine Parameters

Machine Test #8: Ring gage test (roundness)

The Leitz probe head interprets an electromagnetic signal (differential transformers with a moving core) to determine the amount of deflection that is taking place when probing a part. Each axis has its own spring parallelogram that independently determines the amount of deflection in that one axis. If two axes are interpreting their signals differently, then the results from measuring a circle will appear oval in shape. This is a good way to test the balance of the probe head.

In this test, a XXX ring gage was measured with 360 points in the three planes and the results plotted. If the circle appears to be pinched in the x or y axis, then it is a good possibility that the probe head is out of balance. If the circle is distinctly oval in shape, rotate the ring gage 90 degrees and remeasure the gage. If the oval shape does not rotate with the gage, then the error is either occurring in the probe head or the machine. The results of this test did not indicate a problem.

Machine Test #9: Single-axis repeatability

When the service personnel calibrated the machine on site, they measured a Moore bar in all three axes. It was assumed that if there was a mechanical problem with one axis, it would appear at this time. We conducted a simple single-point repeatability test on each axis. We chose an axis, took a single-point probing in that axis, then moved away from that point using three axes movement. We repeated this measurement using 50 runs, and ran this procedure in the remaining two axes. The results showed that all three axes performed equally well.

20.3.2.4 Multiple Probes

It is often necessary to use more than a single probe when measuring a part. On this particular Leitz machine, there are two types of multiple probe setups; two or more probes located within the same probe configuration (e.g., star probes) and two or more probe configurations established using the magazine probe changer. At this time, it is believed that changing between two or more probes within the same setup will not decrease the repeatability of the measured feature. However, there is the possibility of a bias being incorporated into the offset established between the probe being used and the reference probe. For the sake of these tests, it is not considered a factor that has significance due to certified artifacts being used in all cases for the development of the capability matrix.

20.3.3 Feature Based Measurement Tests (Section 3)

Feature-dependent parameters affect a machine's performance to varying degrees depending upon the type of geometric tolerance being measured and calculated. These parameters include the size or surface area, the number of points taken, and the surface roughness of that feature.

How many points does a programmer take when measuring a small diameter? How many points on a large diameter? Does this remain true for other features such as flatness of a plane? What effect will the surface roughness have upon these numbers?

The repeatability of the machine does indeed vary from one feature to another. For instance, the repeatability obtained from calculating the diameter of a hole measured with 16 points is better than that received when calculating the roundness using the same points. This is simply because the diameter is a least squares best-fit average of those 16 points. The roundness of the hole on the other hand is a range of those 16 points. It is understood that all performance values are a function of the repeatability of a single probing point. However, the question remains as to how the various parameters contribute to that function.

It was important to answer these questions in order to obtain the necessary level of confidence in the machine. Simply stating that the machine's linear accuracy is 0.5 +L/600 micrometers (where L = length in meters) and its single-point repeatability at Six Sigma is 0.1 µm is not enough. This information does not help an operator determine if he/she can measure a runout tolerance of 2.5 µm or a diameter tolerance of 1.25 µm. This is not to imply that it was necessary to test every tolerance that may be called out on all features. Many tolerance repeatability values can be extrapolated from data obtained from other tested tolerances. Therefore, the attempt here was to optimize testing to those types and sizes of features most commonly required by engineering drawings at a given organization.

Feature Based Test #1: Circle features versus hole diameter (see Fig. 20-8)

This first test was run to determine what effect, if any, the size of the hole would have upon the machine's performance. The results indicate limited relationship between the diameter of the hole and the repeatability of any of the circle elements. The graph also displays the fact that the repeatability is indeed feature-dependent. The repeatability of the hole's roundness value is much worse than the hole diameter value.

Feature Based Test #2: Cylinder features versus hole diameter (see Fig. 20-9)

As in test #1, the objective was to determine if the size of the cylinder would have any effect on the measured results. Six ring gages ranging from 12.5 mm to 54 mm were measured 25 times at two heights using 32 points per height. Their diameter and cylindricity repeatability values were plotted versus size. The graph again shows limited relationship between the hole's size and the repeatability of its diameter or form. Possibly, the length of the cylinder may affect the repeatability of such features as position, perpendicularity, and straightness.

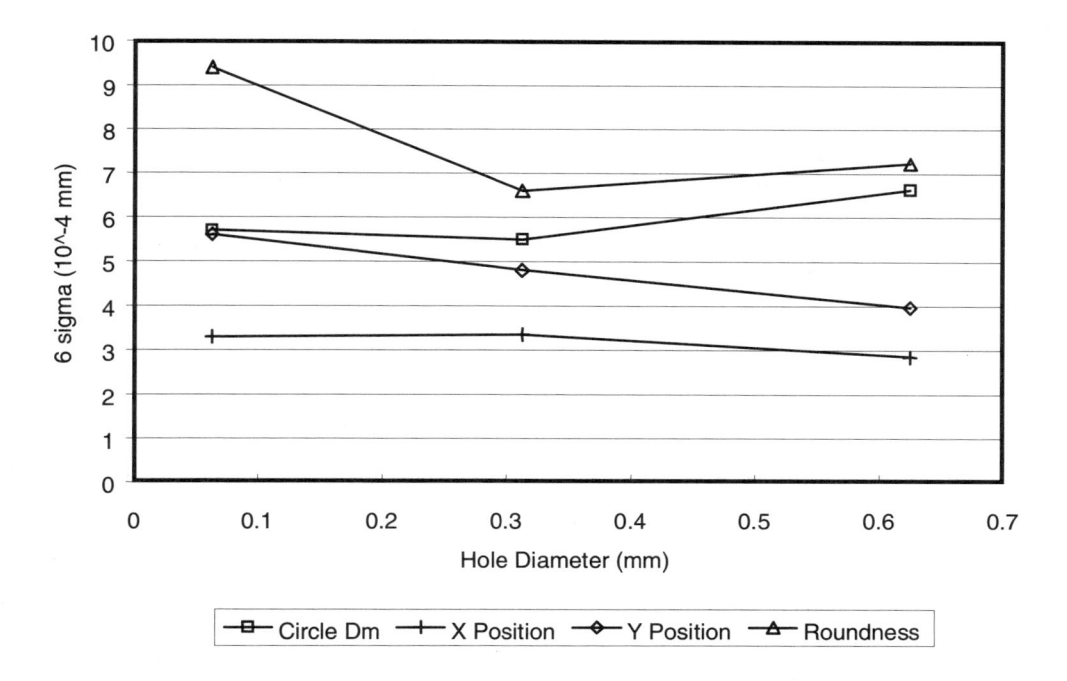

Figure 20-8 Circle features versus hole diameter

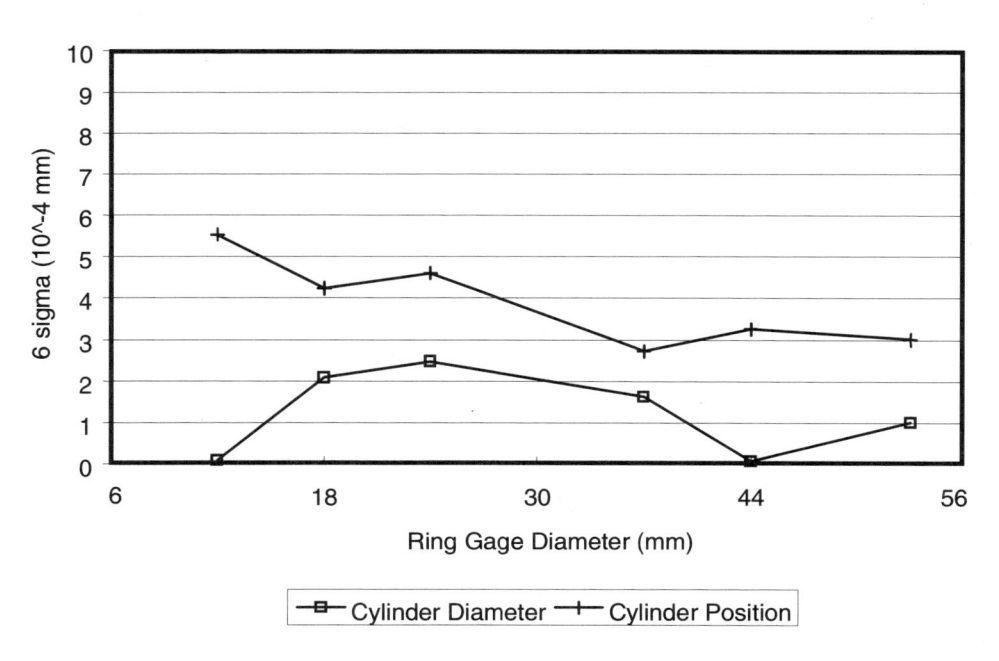

Figure 20-9 Cylinder features versus hole diameter

Feature Based Test #3: Bidirectional probing versus varying lengths (x and y axis) (see Figs. 20-10a and 20-10b)

Six gage bars of lengths 25, 50, 100, 200, 250, and 400 mm were placed in the x- and y-axes. The two end planes were measured using 32 points each, recording the minimum and maximum length of the bars. In addition, a single point was taken on each end, and the bidirectional probing repeatability was calculated. These results again showed a discernible pattern between length of the gage and repeatability of the features. Additionally, neither the x or y axis seemed to perform better than the other. These tests have been limited to the 25 mm × 25 mm area on the ends of the gage blocks.

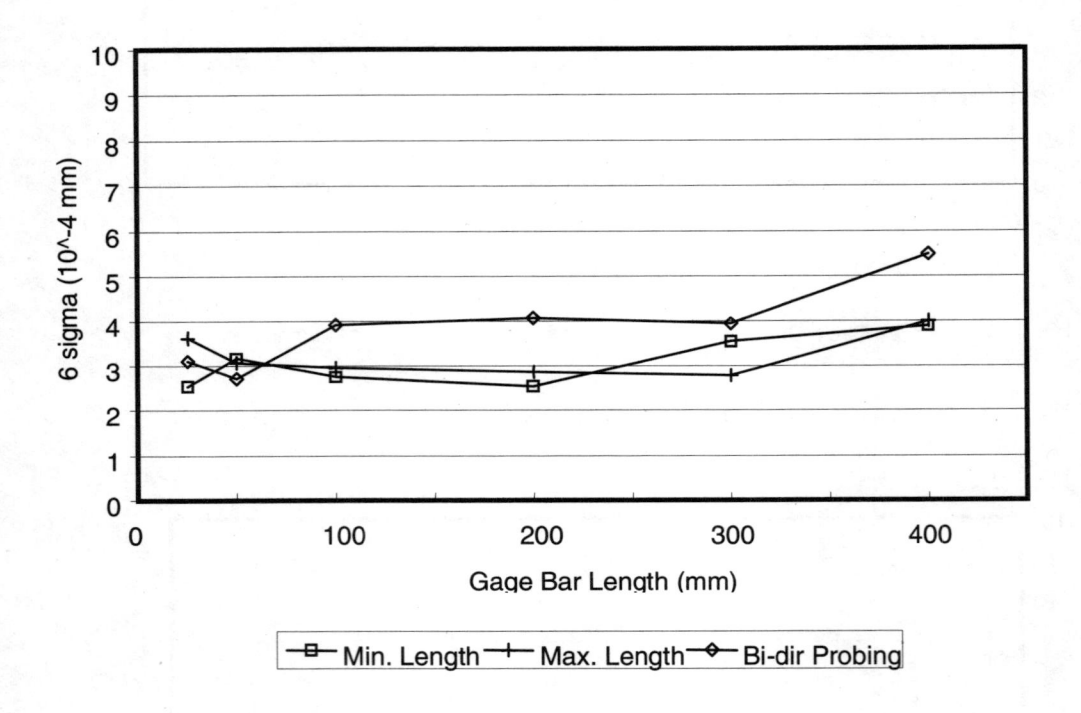

Figure 20-10a Bidirectional probing versus varying lengths (x-axis)

20.3.3.1 Number of Points Per Feature

Feature Based Test #4: Circle features versus number of points per circle (see Fig. 20-11)

This test was run using a very stiff 5-mm probe (0.295 deflection) that measured a circle 20 times and reported the diameter, roundness, and position. There is a strong indication that the diameter and the x and y position have a better repeatability as the number of points taken increases. This makes sense, because these three geometric elements are averages of the points taken. The roundness of the hole, on the other hand, is a range of values; therefore, its repeatability deteriorates as the number of points increase.

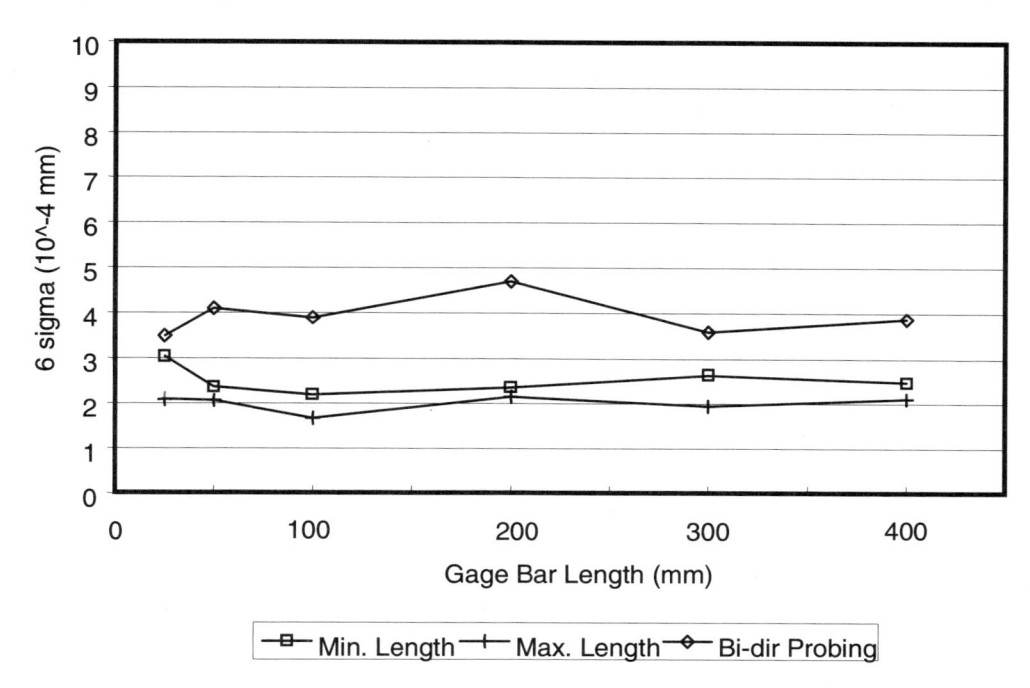

Figure 20-10b Bidirectional probing versus varying lengths (y-axis)

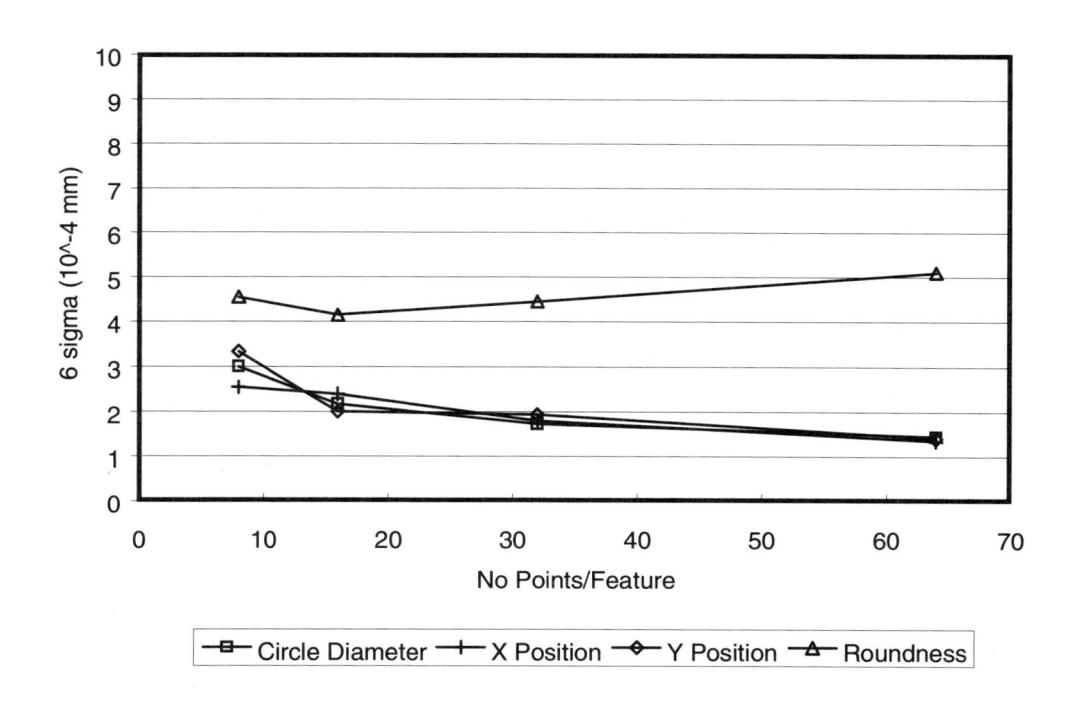

Figure 20-11 Circle features versus number of points per section

Feature Based Test #5: Cylinder features versus number of points per section (see Fig. 20-12)

Varying the number of points per feature was expanded to the measurement of cylinders. Four 16-mm diameter cylinders 18 mm in length were measured at three sections, increasing the number of points per section from 16 to 32. Each individual point density measurement was repeated 25 times before moving on to the next density.

At first glance, these results followed the pattern expected. Cylinder position, perpendicularity, and straightness repeatability improved as the number of points per section increased, while cylindricity displayed the opposite effect. It appeared that the 16 and 32 point tests were very similar, possibly due to the law of diminishing returns. However, this is with only three different point densities used, so an additional test was designed ranging from 4 to 16 points.

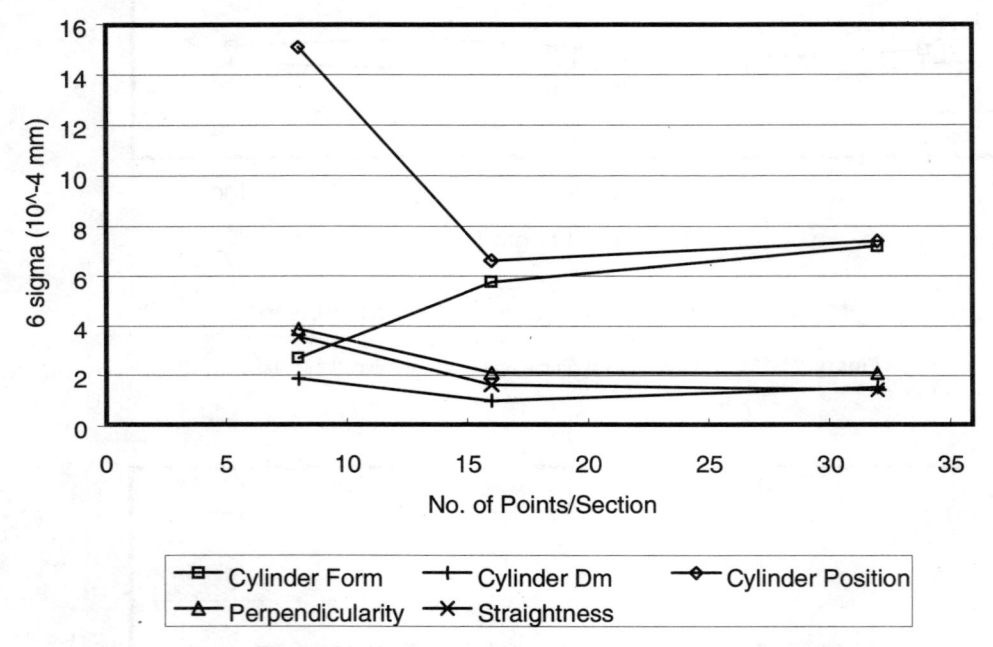

Figure 20-12 Cylinder features versus number of points/section

Feature Based Test #6: Cylinder features versus number of points per section (see Fig. 20-13)

Again, four 16-mm diameter cylinders 18 mm in length were measured at three sections. The first series of measurements were conducted using four points per section and were repeated 25 times. Runs using 8 and 16 points followed in the same manner. Unfortunately, these results were not what was expected. No pattern displayed in these results indicated that the number of points per feature affected the repeatability of the cylinder measurement.

After much consideration, testers decided that more information needed to be collected. Therefore, a more extensive test was outlined using the following range of points per section: 4, 6, 8, 10, 12, 14, 16, 18, 20, 24, 28, and 32. Also, the manner in which each point density run potentially allowed temperature to affect one run more than the other was a concern.

Feature Based Test #7: Cylinder features versus number of points per section (see Fig. 20-14)

In this test, two 16-mm diameter cylinders 18 mm in length were measured with each of the above-mentioned point densities, working from four points per section to 32 points per section. This entire procedure was then repeated 25 times. If there were any temperature stability problems, their effects would be the same for all point density runs.

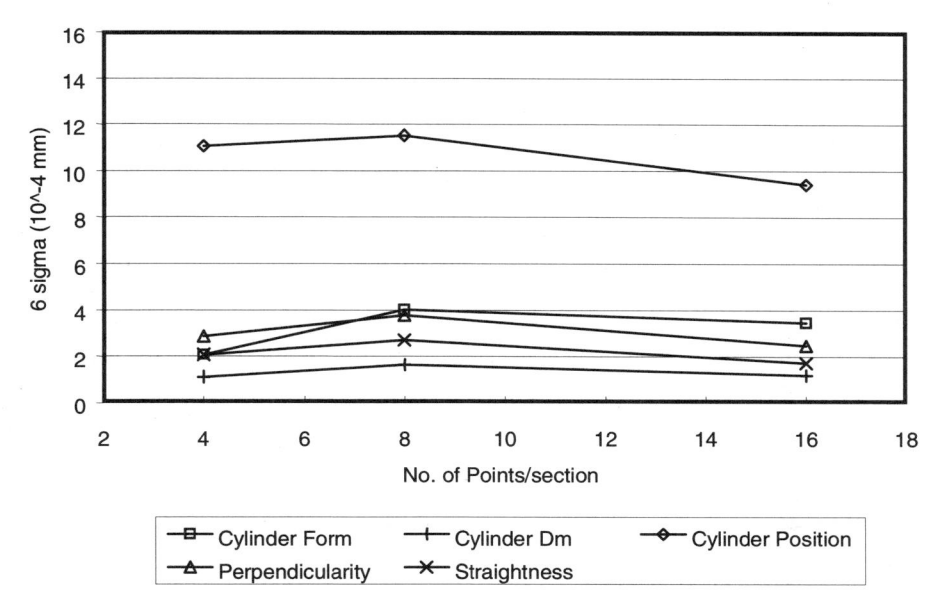

Figure 20-13 Cylinder features versus number points/section

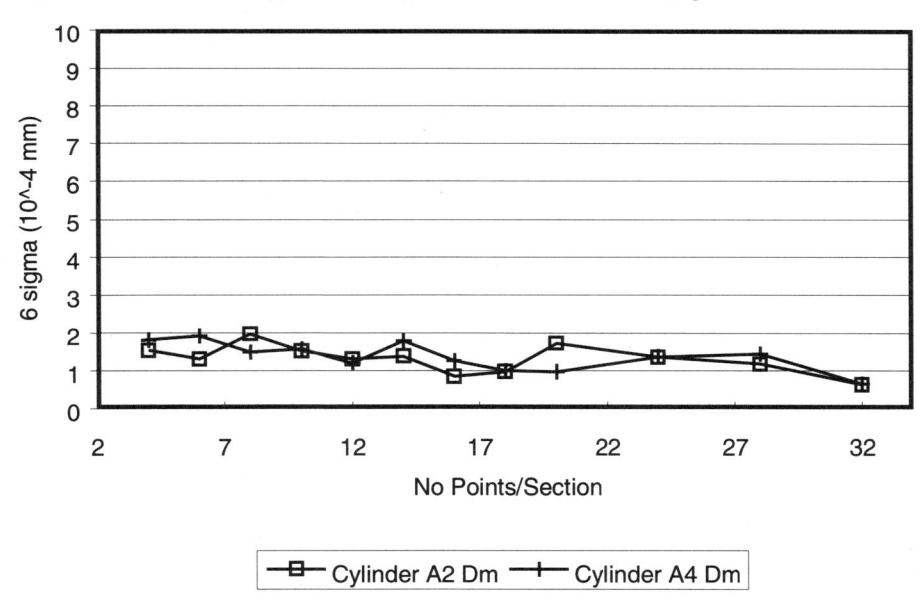

Figure 20-14 Cylinder features versus number of points/section

The results proved to be extremely confusing. Although all the graphs exhibited the trends expected, there was a great deal more variation around the regression straight (at the lower point densities) than anticipated. This created more questions than answers. What secondary effects may be causing this variation? Is this a random fluctuation around the regression straight, or is this a point-dependent pattern? A point-dependent pattern would indicate problems with the algorithms being employed. Because the primary objective of this effort was to achieve the best possible results for a capability matrix, these questions were deferred.

20.3.3.2 Other Geometric Features

Feature Based Test #8: 25-mm cube test (planar features) (see Fig. 20-15)

A 25-mm square quartz cube was measured 25 times on its five open sides using 32 points per surface. Two different probe setups were utilized; a five-point star probe setup and a single probe setup. When using the star probe setup, the Six Sigma repeatability values were better than when the single probe setup was used. This was because each planar surface measured with the star probe was perpendicular to the shaft of the probe. Very little deflection takes place up the shaft of the probe. All planes measured with the single probe (except the top plane) were parallel to the probe shaft, creating much more deflection.

It is interesting to note that on every evaluation (except squareness) using the star probe, the x-axis planes seemed to repeat slightly better than the y-axis planes. This was not the case for the single probe setup, although this does not rule out the possibility that one axis may be more repeatable than the other. With the single probe tip, the deflection of the probe tip could be the dominating parameter overshadowing any effect the axis may have had on the results.

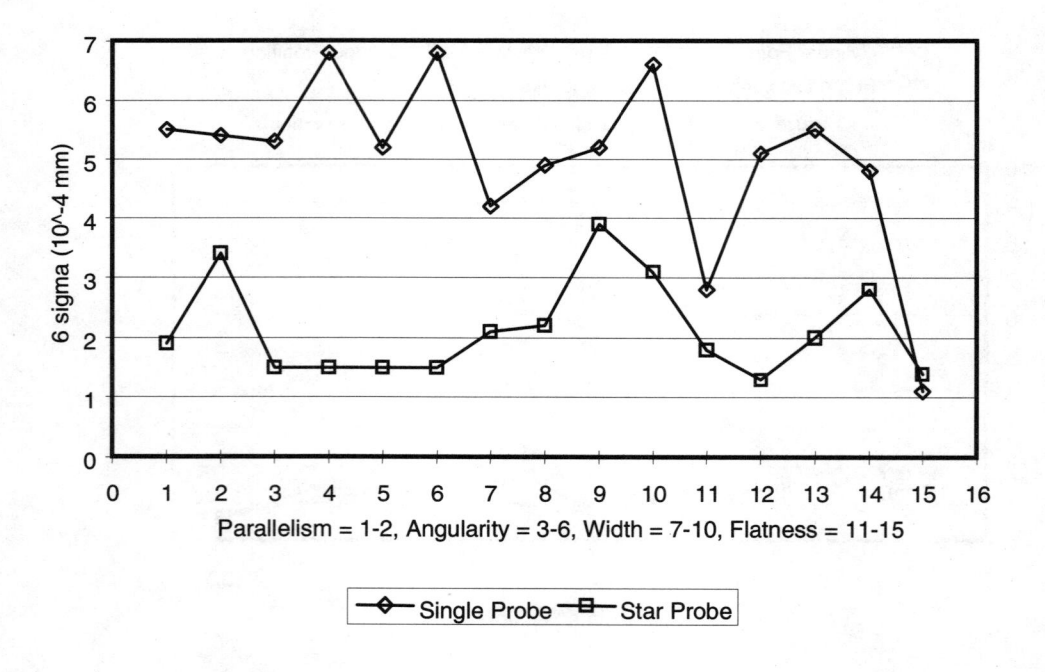

Figure 20-15 25-mm cube test—single versus star probe setup

20.3.3.3 Contact Scanning

Due to its unique probe head, the Leitz PMM can carry out constant contact scanning. This helps the user to obtain a large amount of points on a feature in a very short time. It is also very useful when measuring 2-D and 3-D curves in space. Unfortunately, there is some loss in repeatability when moving from point-to-point measurement to scanning.

Feature Based Test #9: Circle features versus scanning speed (see Fig. 20-16)

To determine how scanning speed affects the repeatability of the measurements, a test was run measuring four diameters using several different scanning speeds. The scanning speed was altered from 2 mm/sec to 0.2 mm/sec. The results showed the repeatability of the measurements do indeed become worse as the scanning speed increases. As expected, this deterioration was most evident in the roundness of a circle, while less on the other parameters. Based on the primary objective being optimum results, which can best be achieved using single-point measurements, no further testing on scanning was done at this time.

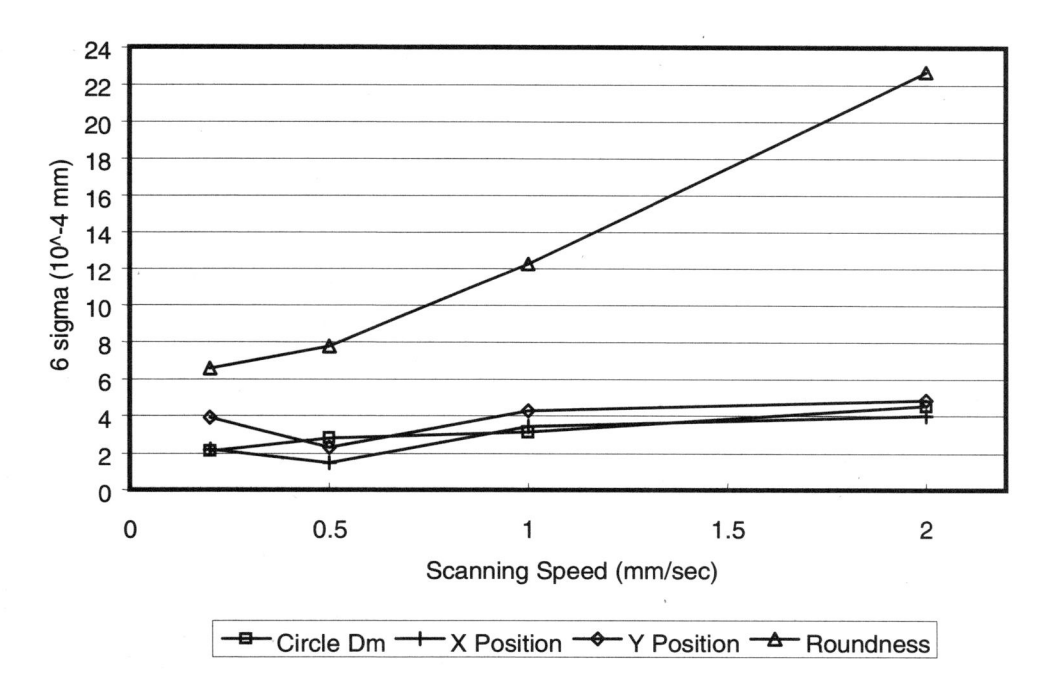

Figure 20-16 Circle features versus scanning speed

20.3.3.4 Surface Roughness

It is generally accepted that the surface roughness of a part/feature will affect the repeatability of a single point being measured. No testing was needed on this issue at this time since the surface roughness on the certified artifacts are within the same range (less than 0.2 μm) as the parts to be checked on an ongoing basis.

20.4 CMM Capability Matrix (see Fig. 20-17)

The following matrix is a summary of the feature-based testing done to date on the Leitz CMM. These tests were performed to determine (at a minimum level) the system's measurement capability for each of the geometric characteristics per ASME Y14.5M-1994. Individual features were tested for accuracy and repeatability and their Six Sigma values calculated.

Geometric Categories	Type of Tolerance	Characteristic	Symbol	Visual Example of Zone	Type of Zone
For Individual Features	Form	Straightness	—		"Parallel lines or Cylindrical boundary," not in relationship to a feature or surface.
		Flatness	▱	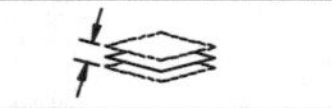	"Parallel planes," not in relationship to a feature or surface.
		Circularity (Roundness)	○		"Two concentric circles" within which each circular element must lie. (Radius)
		Cylindricity	⌭		"Two concentric cylinders," within which all surface elements must lie. (Radius)
For Individual Or Related Features	Profile	Profile of a Line	⌒		A "uniform boundary," along the true profile within which the elements of the cross-section must lie.
		Profile of a Surface	⌓	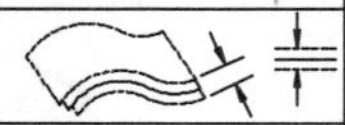	A "uniform boundary," along the true profile within which the elements of the surface must lie.
For Related Features	Orientation	Angularity	∠		"Parallel planes or cylindrical boundary" at a specified angle from the defined datum(s).
		Perpendicularity	⊥	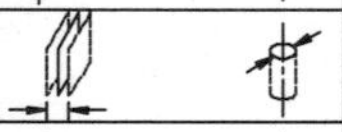	"Parallel planes or cylindrical boundary," at 90 degrees basic from the defined datum(s).
		Parallelism	//	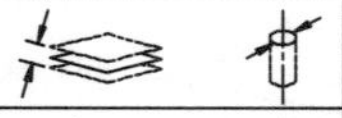	"Parallel planes or cylindrical boundary," in relationship to a surface or axis.
	Location	Position	⌖	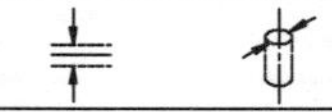	"Parallel planes or cylindrical boundaries" in relationship to datum plane or axis.
		Concentricity	◎		"Cylindrical boundaries" where the axis of all cross-sectional elements of a surface must lie.
		Symmetry	≡		Parallel planes, within which the median points of all opposed elements must lie.
	Runout	Circular Runout	↗	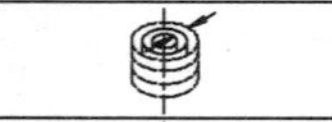	"Two concentric circles" within which circular elements must be in relationship to datum axis.
		Total Runout	↗↗	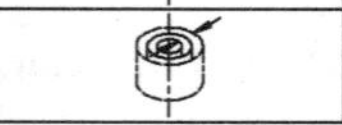	"Two concentric cylinders" within which all circular elements must lie in relationship to datum axis.

Individual Features	6 Sigma Capability (μM)	Comments
X-Axis	0.3	< 150 mm In Length
Y-Axis	0.3	< 150 mm In Length
Z-Axis	0.3	< 150 mm In Length

Note: Single Point Repeatability = 0.1 μM

Individual Features	6 Sigma Capability (μM)	Comments
Diameters	0.2	< 25 mm In Length
Cylinders	0.25	< 25 mm In Length
Widths (surfaces)	0.4	< 25 mm In Length
Point	0.1	(See note)
Sphere	0.25	< 25 mm In Length
Flatness	0.25	< 25 mm In Length

Figure 20-17 Leitz PPM 654 capability matrix

Zone Modifiers Allowed Y/N	Datum Usage Y/N/Option	Type Of Feature	Dependent Variables				Six Sigma Capability (μM)	Comments:
			Size	Form	X	Y		
No=Surface Yes=Axis	No	Surface		✓			0.3	
		Axis	✓	✓			0.3	
No	No	Surface		✓			0.25	
No	No	Surface (Circle)	✓	✓			0.35	# Of Pts. Dependent
No	No	Surface (Cylinder)	✓	✓			0.45	# Of Pts. Dependent
No	Option	Surface		✓	✓	✓	0.7	From Datums
							0.4	2D Best Fit
No	Option	Surface		✓	✓	✓	0.8	From Datums
							0.5	2D Best Fit
No=Surface Yes=Axis	Yes	Surface		✓			0.3	
		Axis	✓	✓			0.3	
No=Surface Yes=Axis	Yes	Surface		✓			0.3	
		Axis	✓	✓			0.3	
No=Surface Yes=Axis	Yes	Surface		✓			0.3	
		Axis	✓	✓			0.3	
Yes See Rule #2	Yes	Axis	✓	✓	✓	✓	0.7	
		Plane	✓	✓	✓	✓	0.8	
No	Yes	Axis	✓	✓	✓	✓	0.5	
No	Yes	Plane	✓	✓	✓	✓	0.5	
No	Yes	Surface (Circle)	✓	✓	✓	✓	0.4	
No	Yes	Surface (Cylinder)	✓	✓	✓	✓	0.5	

Comments Regarding "Form"

1) For Cylindrical Features, This Represents Total Size Variation (—, ○, ⌀, ∠, ⊥, //, ⊕, ◎, ⌖, ⌯)
2) For Surface This Represents Flatness (⌷, ∠, ⊥, //)
3) For Some Features This Represents The Equivilent To The Geometric Result (⌷, ⌀)

Figure 20-17 continued Leitz PPM 654 capability matrix

Some of the NIST-traceable artifacts used for determining system accuracy and repeatability, and the types of features checked are listed below.

- 450-mm Moore bar (step gage used to determine linear displacement "X, Y, and Z").
- 25-mm cube (used for size, point to point, parallelism, flatness, straightness of a surface, perpendicularity of a surface, angularity of a surface, profile of a line, and profile of a surface).
- XXX ring gages (used for size, circularity, cylindricity, concentricity, runout, total runout, straightness of an axis, parallelism of an axis, perpendicularity of an axis, angularity of an axis).
- 10-mm and 25-mm XXX sphere (used for system probe calibration, size, circularity, and sphericity).

Due to the majority of features of interest being less than 25 mm, the above artifacts were highly adequate to determine a solid starting point for short-term system capability needs. It is essential to note that these tests are speculated to represent approximately 75% of the testing needed for the system. Unique features will need to be tested as needed, and when deemed necessary due to tight tolerances, new artifacts will need to be built or purchased (and certified) to ensure optimum reduction of bias in measurement results.

The capability matrix represents all 14 geometric characteristics, as well as individual features used in one way or another, the by-product of which represents the capability of each geometric characteristic. The X, Y, and Z axis locations of diameters, cylinders, widths (surfaces), points, spheres, and planes were all individually evaluated.

Knowing the specific capability of each feature listed, there should be adequate information available to determine the capability of each geometric characteristic, with a high degree of confidence. It is essential to note the matrix results were based on optimum programs using low-probe deflection values (<0.4 mm).

In addition, the following is a summary list of variables that need to be considered when programming and analyzing parts. These variables have the potential of decreasing the Six Sigma capability of the results shown on the matrix (either in accuracy, repeatability, or both).

- Utilization of multiple probes from the probe changer or star probes
- Probes with greater than 0.4-mm probe deflection. Note: A probe deflection matrix has been developed with studies done showing Six Sigma repeatability. (This data should be very beneficial in predicting the effects of a specific feature or geometric characteristic to overall capability.)
- Short-term temperature fluctuations
- Contamination
- Loose probe tip (should be able to detect by evaluating form and deflection values)
- Surface finish
- Number of probing points

The list of variables that need to be considered is lengthy. Up to this point, tests and calculations have been fairly straightforward. Chapter 25 addresses some of the capability calculations currently used to determine gage repeatability and reproducibility (GR&R). Some of the variables have not been taken "fully" into consideration and will spur tremendous development efforts for many years to come.

20.5 References

1. Hetland, Gregory A. 1995. Tolerancing Optimization Strategies and Methods Analysis in a Sub-Micrometer Regime. Ph.D. dissertation.
2. Majlak, Michael L. 1994. Error Budgets Used in CMM Design and Application Studies. Paper presented at ASPE.
3. Phillips, S.D., B. Borchardt, G. Caskey.1993. Measurement Uncertainty Considerations for Coordinate Measuring Machines. MISTR 5170, Precision Engineering Division, NIST: April 1993.
4. The American Society of Mechanical Engineers. 1995. *ASME Y14.5M-1994, Dimensioning and Tolerancing.* New York, New York: The American Society of Mechanical Engineers.
5. The American Society of Mechanical Engineers. 1995. *ASME Y14.5.1M-1994, Dimensioning and Tolerancing.* New York, New York: The American Society of Mechanical Engineers.
6. The American Society of Mechanical Engineers. 1990. *ASME B89.1.12M-1990, Methods of Performance Evaluation of CMMs.* New York, New York: The American Society of Mechanical Engineers.

Chapter 21

Predicting Piecepart Quality

Dan A. Watson, Ph.D.
Texas Instruments Incorporated
Dallas, Texas

Dr. Watson is a statistician in the Silicon Technology Development Group (SiTD) at Texas Instruments. He is responsible for providing statistical consulting and programming support to the researchers in SiTD. His areas of expertise include design of experiments, data analysis and modeling, statistical simulations, the Statistical Analysis System (SAS), and Visual Basic for Microsoft Excel. Prior to coming to SiTD, Dr. Watson spent four years at the TI Learning Institute, heading the statistical training program for the Defense and Electronics Group. In that capacity he taught courses in Design of Experiments (DOE), Applied Statistics, Statistical Process Control (SPC), and Queuing Theory. Dr. Watson has a bachelor of arts degree in physics and mathematics from Rice University in Houston, Texas, and a masters and Ph.D. in statistics from the University of Kentucky in Lexington, Kentucky.

21.1 Introduction

This chapter expands the ideas introduced in the paper, *Statistical Yield Analysis of Geometrically Toleranced Features*, presented at the Second Annual Texas Instruments Process Capability Conference (Nov. 1995). In that paper, we discussed methods to statistically analyze the manufacturing yield (in defects per unit) of part features that are dimensioned using geometric dimensioning and tolerancing (GD&T). That paper specifically discussed features that are located using positional tolerancing.

This chapter expands the prior statistical methods to include features that have multiple tolerancing constraints. The statistical methods presented in this paper:

- Show how to calculate defects per unit (DPU) for part features that have *form* and *orientation* controls in addition to *location* controls.

- Account for *material condition modifiers* (maximum material condition (MMC), least material condition (LMC), and regardless of feature size (RFS)) on *orientation*, and *location* constraints.
- Show how *different manufacturing process distributions* (bivariate normal, univariate normal, and lognormal) impact DPU calculations.

21.2 The Problem

Geometric controls are used to control the size, form, orientation, and location of features. In addition to specifying the ideal or "target" (nominal) dimension, the controls specify how much the feature characteristics can vary from their targets and still meet their functional requirements. The probability that a randomly selected part meets its tolerancing requirements is a function not only of geometric controls, but the amount and nature of the variation in the feature characteristics which result from the manufacturing process used to create the feature. The part-to-part variation in the feature characteristics can be represented by probability distribution functions reflecting the relative frequency that the feature characteristics take on specific values. We can then calculate the probability that a feature is within any one of these specifications by integrating the probability distribution function for that characteristic over the in-specification range of values. For example, if the part-to-part variation in the size of the feature, d, is described by the probability density function $g(d)$, then the probability of generating a part that is within the size upper spec limit and the size lower spec limit is:

$$P(in\_spec) = \int_{SizeLowerSL}^{SizeUpperSL} g(d)\,dd$$

where SL is the specification limit.

If a feature has several GD&T requirements and we assume that the manufacturing processes that control size, form, orientation, and location are uncorrelated, then the generalized equation for the probability of meeting all of them is:

$$P(in\_spec) = \int_{SizeLowerSL}^{SizeUpperSL} g(d)\,dd \int_{0}^{FormSL} j(w)\,dw \int_{0}^{OrientationSL} h(q)\,dq \int_{0}^{LocationSL} f(r)\,dr \quad (21.1)$$

where,
  $j(w)$ is the form probability distribution function,
  $h(q)$ is the orientation probability distribution function, and
  $f(r)$ is the location probability distribution function.

The DPU is equal to the probability of *not* being within the specification.

$$P(not\_in\_spec) = 1 - P(in\_spec)$$

$$DPU = 1 - \int_{SizeLowerSL}^{SizeUpperSL} g(d)\,dd \int_{0}^{FormSL} j(w)\,dw \int_{0}^{OrientationSL} h(q)\,dq \int_{0}^{LocationSL} f(r)\,dr \quad (21.2)$$

Eq. (21.2) would be complete if there were no relationships between the size, form, orientation, and location limits. As a feature changes *orientation*, however, the amount of allowable *location* tolerance is reduced by the amount that the feature tilts. Therefore, the maximum *location* tolerance zone is a function of the feature's *orientation*. Similarly, sometimes there are relationships between other limits, such as between *size* and *location*, or between *size* and *orientation*. When these relationships are functional, we specify them on a drawing using the maximum material condition modifiers and the least material condition modifiers. If one of these modifiers is used, then, the *orientation* tolerance is a function of the feature *size*, and the *location* tolerance is a function of the feature *size*.

Note: In ASME Y14.5-1994, the tolerance zones for size, form, orientation, and location often overlap each other. For example, the orientation tolerance zone may be inside the location tolerance zone, and the form tolerance zone may be inside the orientation tolerance zone. Since Y14.5 communicates engineering design requirements, this is the correct method to apply tolerance zones.

However, when predicting manufacturing yield for pieceparts, the manufacturing processes are considered. Therefore, we need to separate the tolerance zones for size, form, orientation, and location. Because of this, when we refer to the "allowable" tolerance zone in a statistical analysis, this is different than the "allowable" tolerance zone allowed in Y14.5.

Note: It is difficult to write an equation to show the relationship between *form* and *size* as defined in ASME Y14.5M-1994. It is equally difficult to write relationships for *location* and *orientation* as a function of *form*. In the following equations, we will assume that these relationships are negligible and can be ignored.

21.3   Statistical Framework

21.3.1   Assumptions

Fig. 21-1 shows an example of a feature (a hole) that is toleranced using the following constraints:

- The diameter has an upper spec limit of $D + T_2$.
- The diameter has a lower spec limit of $D - T_1$.
- A perpendicularity control (⌀$2Q$) that is at regardless of feature size.
- A positional control (⌀$2R$) that is at regardless of feature size.

The feature is assumed to have a target location with a tolerance zone defined by a cylinder of radius R. In addition, the diameter of the feature also has a target value, D. To be within specifications, the

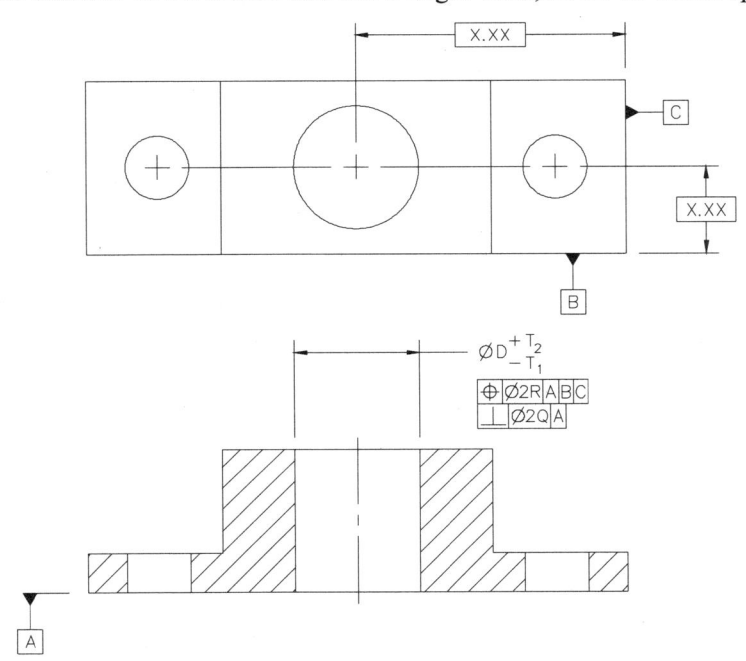

Figure 21-1 Cylindrical (size) feature with orientation and location constraints at RFS

diameter of the feature needs to be between $D - T_1$ and $D + T_2$. The feature is allowed a maximum offset from the vertical of Q.

If the angle between the feature axis and the vertical is given by q, then q has a maximum value of $\arcsin(2Q/L)$, where the length of the feature is L (as shown in Fig. 21-2). In addition, as q increases, the amount of the location tolerance available to the feature decreases by the amount of lateral offset from the vertical, $L*\sin(q)/2$. This results in the location tolerance zone having an effective radius of $R - L*\sin(q)/2$.

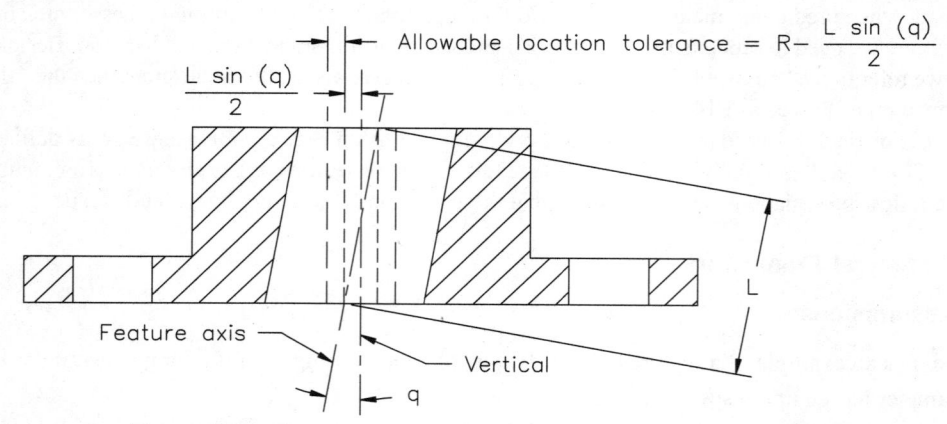

Figure 21-2 Allowable location tolerance as a function of orientation error (q)

To account for the variation in the process that generates the feature, the offsets in the X and Y coordinates of the feature location relative to the target location (δ_X and δ_Y) are assumed to be normally distributed with mean 0 and common standard deviation σ. In addition, it is assumed that the X and Y deviations are uncorrelated (independent). The variation in the diameter of the feature, d, is assumed to have a lognormal distribution with mean μ_d and standard deviation σ_d and the diameter is uncorrelated with either the X or Y deviations. Finally, it is assumed that the variation in the angle of tilt (orientation), q, is lognormally distributed with mean μ_q and standard deviation σ_q and is also assumed to be uncorrelated with the X and Y deviations and the feature diameter. Note that this analysis assumes that the processes stay centered on the target (nominal dimension). The standard deviations for these processes are generally considered short-term standard deviations. If the means of the processes shift over time, as discussed in Chapters 10 and 11, then the appropriate standard deviations must be inflated to approximate the long-term shift.

If we define $r = \sqrt{\delta_X^2 + \delta_Y^2}$ to be the distance from the target location to the location of the feature, then the probability density functions for d, q, and r are given by:

size
$$g(d) = \frac{1}{d\gamma\sqrt{2\pi}} e^{-\frac{(\ln(d)-\theta)^2}{2\gamma^2}}$$

where $\theta = \ln(\mu_d) - \dfrac{\ln\left(1 + \dfrac{\sigma_d^2}{\mu_d^2}\right)}{2}$ and $\gamma = \sqrt{1 + \dfrac{\sigma_d^2}{\mu_d^2}}$

orientation $$h(q) = \frac{1}{q\tau\sqrt{2\pi}} e^{-\frac{(\ln(q)-v)^2}{2\tau^2}}$$

where $v = \ln(\mu_q) - \dfrac{\ln\left(1 + \dfrac{\sigma_q^2}{\mu_q^2}\right)}{2}$ and $\tau = \sqrt{1 + \dfrac{\sigma_q^2}{\mu_q^2}}$

and location $$f(r) = \frac{r}{\sigma^2} e^{-\frac{r^2}{2\sigma^2}}$$

Since d, q, and r are independent, the probability of the feature being simultaneously within specification for size, orientation, and location can be found by taking the product of the density functions and integrating the product over the in-specification range of values for d, q, and r. In the case specified above, where d must be between $D - T_1$ and $D + T_2$, q must be less than arcsin $(2Q/L)$, and r must be less than R, this probability is represented by:

$$P(in\_spec) = \int_{D-T_1}^{D+T_2} \int_0^{\arcsin(2Q/L)} \int_0^{(R-L\sin(q)/2)} \frac{1}{d\gamma\sqrt{2\pi}} e^{-\frac{(\ln(d)-\theta)^2}{2\gamma^2}} \frac{1}{q\tau\sqrt{2\pi}} e^{-\frac{(\ln(q)-v)^2}{2\tau^2}} \frac{r}{\sigma^2} e^{-\frac{r^2}{2\sigma^2}} dd\,dq\,dr$$

$$= \int_{D-T_1}^{D+T_2} \left(\int_0^{\arcsin(2Q/L)} \left(1 - e^{-\frac{(R-L\sin(q)/2)^2}{2\sigma^2}}\right) \frac{1}{q\tau\sqrt{2\pi}} e^{-\frac{(\ln(q)-v)^2}{2\tau^2}} dq \right) \frac{1}{d\gamma\sqrt{2\pi}} e^{-\frac{(\ln(d)-\theta)^2}{2\gamma^2}} dd$$

where the final integration has to be done using numerical methods. To then calculate the probability of an unacceptable part, or DPU, this value is subtracted from 1.

This calculation becomes more complicated when material condition modifiers are used. This means that the DPU calculation depends upon whether MMC or LMC is used for the location and orientation specifications and whether the feature is an internal or external feature.

21.3.2 Internal Feature at MMC

Fig. 21-3 shows an example of a feature that is toleranced the same as Fig. 21-1, except that it has a positional control at maximum material condition, and a perpendicularity control at maximum material condition.

In this case, the specified tolerance applies when the feature is at MMC, or the part contains the most material. This means that when the feature is at its smallest allowable size, $D-T_1$, the tolerance zone for the location of the feature has a radius of R and the orientation (tilt) offset has a maximum of Q. As the feature gets larger, or departs from MMC, the tolerance zones get larger. For each unit of increase in the diameter of the feature, the diameter of the location tolerance zone increases by 1 unit, the radius increases by 1/2 unit, and the maximum orientation tolerance increases by 1 unit. When the feature is at its maximum allowable diameter, $D+T_2$, the location tolerance zone has a radius of $R + (T_1+T_2)/2$ and the orientation

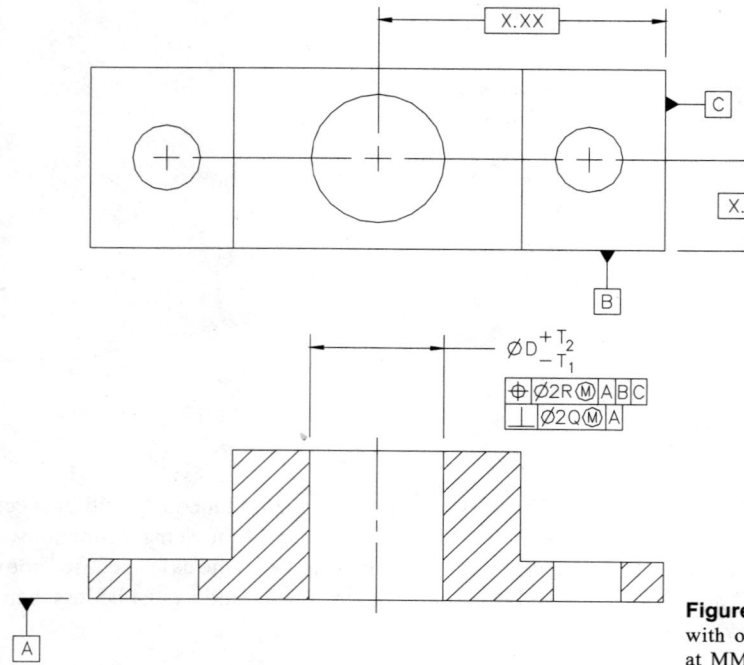

Figure 21-3 Cylindrical (size) feature with orientation and location constraints at MMC

tolerance is $Q + (T_1+T_2)$. As mentioned above, as the orientation increases the radius of the location tolerance zone also decreases by $L*\sin(q)/2$. The radius of the location tolerance zone is therefore a function of d and q:

$$R_M(d,q) = R - \frac{D-T_1}{2} + \frac{d}{2} - \frac{L*\sin(q)}{2} = \Delta_1 + \frac{d}{2} - \frac{L*\sin(q)}{2}$$

where $\Delta_1 = R - \frac{D-T_1}{2}$

The maximum allowable orientation offset is also a function of d:

$$Q_M(d) = Q - (D - T_1) + d$$

The probability that the feature location is within specification is also now a function of d and q. The probability that the feature orientation is within specification is a function of d. If both the location and orientation tolerances are called out at MMC, the probability that the feature is within size, orientation, and location specifications is given by:

$$P(in\_spec) = \int_{D-T_1}^{D+T_2} \left(\int_0^{\arcsin\left(\frac{2Q_M(d)}{L}\right)} \left(1 - e^{-\frac{(R_M(d,q))^2}{2\sigma^2}}\right) \frac{1}{q\tau\sqrt{2\pi}} e^{-\frac{(\ln(q)-v)^2}{2\tau^2}} dq \right) \frac{1}{d\gamma\sqrt{2\pi}} e^{-\frac{(\ln(d)-\theta)^2}{2\gamma^2}} dd$$

The integration must be done using numerical methods and the DPU for the feature is calculated by subtracting the result from 1.

21.3.3 Internal Feature at LMC

Fig. 21-4 shows an example of a feature that is toleranced the same as Fig. 21-1, except that it has a positional control at least material condition, and a perpendicularity control at least material condition.

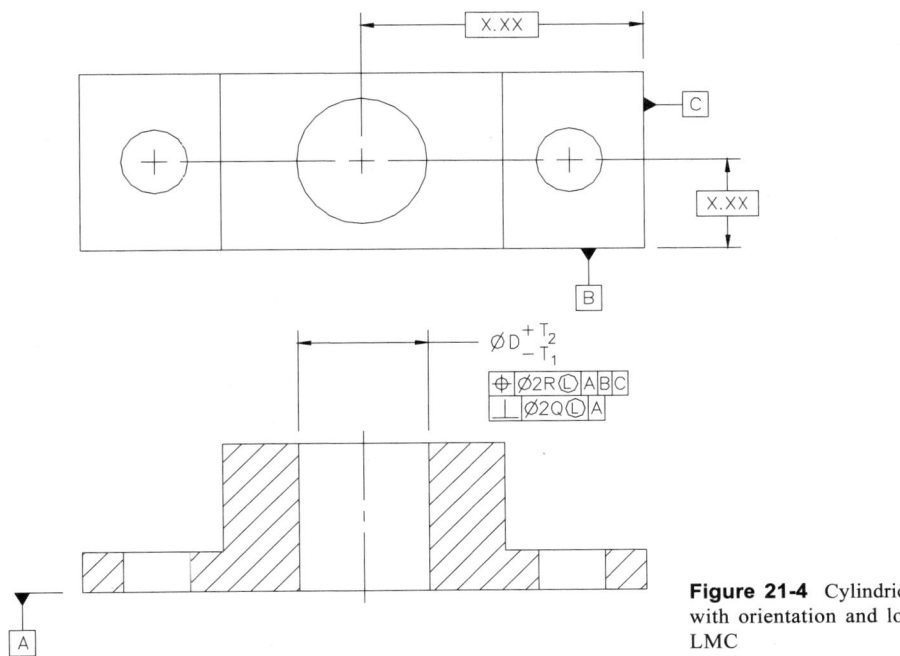

Figure 21-4 Cylindrical (size) feature with orientation and location constraints at LMC

In this case, the specified location tolerance applies when the feature is at LMC, or the part contains the least material. This means that when the feature is at its largest allowable size, $D+T_2$, the tolerance zone for the location of the feature has a radius of R. As the feature gets smaller, or departs from LMC, the tolerance zone gets larger. This means that when the feature is at its largest allowable size, $D+T_2$, the tolerance zone for the location of the feature has a radius of R and the tolerance for the orientation offset is Q. For each unit of decrease in the diameter of the feature, the diameter of the tolerance zone and the orientation offset tolerance each increases by 1 unit. When the feature is at its minimum allowable diameter, $D-T_1$, the location tolerance zone has a radius of $R+(T_1+T_2)/2$ and the orientation tolerance is $Q+(T_1+T_2)$. As before, as the orientation increases, the radius of the location tolerance zone decreases by $L*\sin(q)/2$. The radius of the location tolerance zone is therefore a function of d and q:

$$R_L(d,q) = R + \frac{D+T_2}{2} - \frac{d}{2} - \frac{L*\sin(q)}{2} = \Delta_2 - \frac{d}{2} - \frac{L*\sin(q)}{2}$$

where $\Delta_2 = R + \dfrac{D+T_2}{2}$

21-8 Chapter Twenty-one

The maximum allowable orientation offset is also a function of d:

$$Q_L(d) = Q + (D+T_2) - d$$

The probability that the feature location is within specification is also now a function of d and q. The probability that the feature orientation is within specification is a function of d. If both the location and orientation tolerances are called out at LMC, the probability that the feature is within the size, orientation, and location specifications is given by:

$$P(inspec) = \int_{D-T_1}^{D+T_2} \int_0^{\arcsin\left(\frac{2Q_L(d)}{L}\right)} \left(1 - e^{-\frac{(R_L(d,q))^2}{2\sigma^2}}\right) \frac{1}{q\tau\sqrt{2\pi}} e^{-\frac{(\ln(q)-\upsilon)^2}{2\tau^2}} dq \; \frac{1}{d\gamma\sqrt{2\pi}} e^{-\frac{(\ln(d)-\theta)^2}{2\gamma^2}} dd$$

The integration must be done using numerical methods and the DPU for the feature is calculated by subtracting the result from 1.

21.3.4 External Features

In the case of an external feature called out at MMC, the specified tolerance applies when the feature is at its largest allowable size, $D+T_2$. As the feature gets smaller, or departs from MMC, the tolerance zones get larger. This is the same situation as for the internal feature at LMC, so the probability of the feature being within size, orientation, and location specification is calculated using the same formula.

In the case of an external feature called out at LMC, the specified tolerance applies when the feature is at its smallest allowable size, $D-T_1$. As the feature gets larger, the tolerance zones get larger. This is the same situation as for the internal feature at MMC, so the probability of the feature being within size, orientation, and location specification is calculated using the same formula.

21.3.5 Alternate Distribution Assumptions

Traditionally, the feature diameter has been assumed to have a normal, or Gaussian, distribution. In order to compare the results of GD&T specifications with traditional tolerancing methods, it may be necessary to calculate the DPU with this distribution assumption. Also, when the feature is formed by casting, as opposed to machining, the normal distribution assumption is applicable. In these cases, the probability distribution function for d, $g(d)$, is given by:

$$g(d) = \frac{1}{\sigma_d \sqrt{2\pi}} e^{-\frac{(d-\mu_d)^2}{2\sigma_d^2}}$$

In the case where the feature location is constrained only in one direction, such as when the feature is a slot, then r is usually assumed to have a normal distribution with a mean of 0 and a standard deviation of σ. See Fig. 21-5.

The probability that the feature is in location specification is given by

$$P(in\_spec) = \int_{-(R-L\sin(q)/2)}^{R-L\sin(q)/2} \frac{1}{\sigma\sqrt{2\pi}} e^{-\frac{r^2}{2\sigma^2}} dr$$

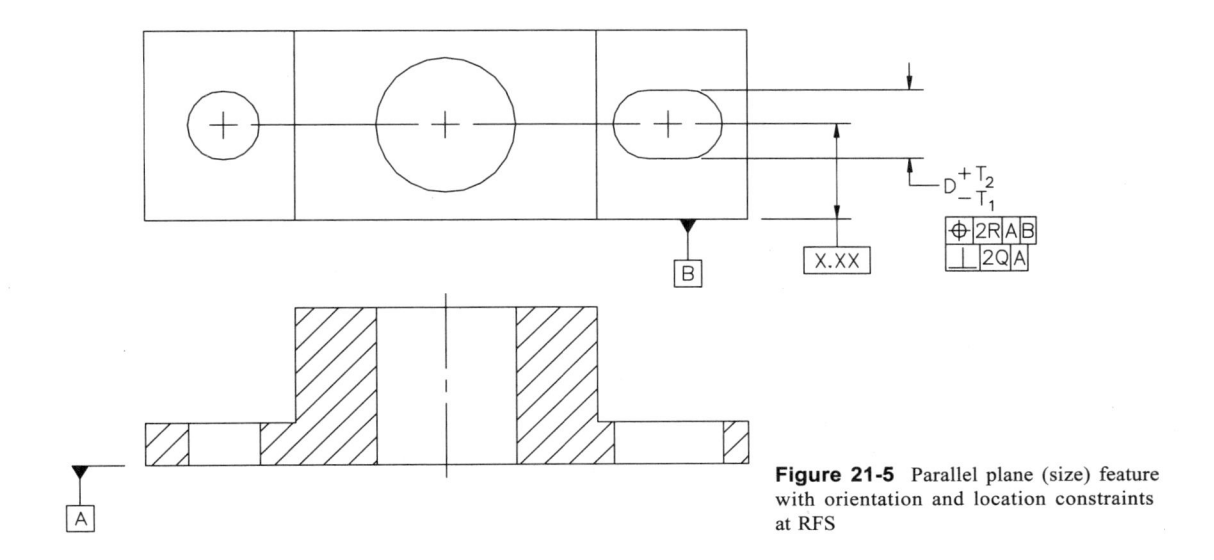

Figure 21-5 Parallel plane (size) feature with orientation and location constraints at RFS

In this case, q is the orientation angle between the center plane of the feature and a plane orthogonal to datum A. If an internal feature is toleranced at MMC, or an external feature is toleranced at LMC, $R - L*\sin(q)/2$ is replaced by R_M. It is replaced by R_L when an internal feature is toleranced at LMC or an external feature is toleranced at MMC.

21.4 Non-Size Feature Applications

The examples shown thus far were features of size (hole, pins, slots, etc.). This methodology can be expanded to include features that do not have size, such as profiled features. For features that do not have size, the material condition modifiers no longer impact the equation. Therefore, the only relationship that we should account for is between *location* and *orientation*. In these cases, Eq. (21.2) reduces to:

$$DPU = 1 - \int_0^{LocationSpecLimit} f(r)dr \quad \int_0^{OrientationSpecLimit} h(q)dq \quad \int_0^{FormSpecLimit} j(w)dw$$

21.5 Example

Table 21-1 compares the predicted dpmo's for various tolerancing scenarios. Cases 1, 2, and 3 are the same, except for the material condition modifiers. Case 2 (MMC) and Case 3 (LMC) estimate the same dpmo, as expected. Both cases predict a much lower dpmo than Case 1 (RFS). Cases 4, 5, and 6 are similar to Cases 1, 2, and 3, respectively, except that the tolerance limits are less. As expected, the number of defects increased.

Table 21-1 Comparison of tolerancing scenarios

Feature Type		Case 1 Internal	Case 2 Internal	Case 3 Internal	Case 4 Internal	Case 5 Internal	Case 6 Internal
Length	L	.500	.500	.500	.500	.500	.500
Size	D	.1273	.1273	.1273	.1273	.1273	.1273
	T_1	.0010	.0010	.0010	.0007	.0007	.0007
	T_2	.0010	.0010	.0010	.0007	.0007	.0007
	μ_d	.1273	.1273	.1273	.1273	.1273	.1273
	σ_d	.00025	.00025	.00025	.00025	.00025	.00025
	Distribution type	Lognormal	Lognormal	Lognormal	Lognormal	Lognormal	Lognormal
Orientation	2Q	.0008	.0008	.0008	.0004	.0004	.0004
	μ_q	.00003	.00003	.00003	.00003	.00003	.00003
	σ_q	.00013	.00013	.00013	.00013	.00013	.00013
	Material condition	RFS	MMC	LMC	RFS	MMC	LMC
	Distribution type	Log-normal	Log-normal	Log-normal	Log-normal	Log-normal	Log-normal
Location	2R	.0064	.0064	.0064	.0032	.0032	.0032
	μ	0	0	0	0	0	0
	σ	.0005	.0005	.0005	.0005	.0005	.0005
	Material condition	RFS	MMC	LMC	RFS	MMC	LMC
	Distribution type	Normal	Normal	Normal	Normal	Normal	Normal
Figure		21-1	21-3	21-4	21-1	21-3	21-4
dpmo		838	111	111	14134	6195	6204

21.6 Summary

The equations presented in this chapter can predict the probability that a feature on a part will meet the constraints imposed by geometric tolerancing. Notice how Eq. (21.1) is similar to, but not exactly the same as the "four fundamental levels of control" in Chapter 5 (see section 5.6). Chapter 5 discusses how these levels of control should be added as demanded by the functional requirements of the feature. It is possible (and often likely) to add GD&T constraints that "function" with little or no insight to the manufacturability of the applied tolerances. The equations in this chapter help predict the *cost* of manufacturing in terms of defective features.

Although these equations are generic, they do not encompass all combinations of GD&T feature control frames. These equations do, however, provide a framework for expansion to include all GD&T relationships.

21.7   References

1. Drake, Paul, Dale Van Wyk, and Dan Watson. 1995. Statistical Yield Analysis of Geometrically Toleranced Features. Paper presented at Second Annual Texas Instruments Process Capability Conference. Nov. 1995. Plano, Texas.
2. The American Society of Mechanical Engineers. 1995. *ASME Y14.5M-1994, Dimensioning and Tolerancing.* New York, New York: The American Society of Mechanical Engineers.

Chapter 22

Floating and Fixed Fasteners

Paul Zimmermann
Raytheon Systems Company
McKinney, Texas

Paul Zimmermann has worked as a lead systems producibility engineer for Raytheon Systems Company and Texas Instruments. He has worked on several programs in both the commercial and defense areas. Mr. Zimmermann has supported such programs as the Digital Imaging Group's Professional and Business Projectors, the TM6000 Notebook Computer, the Long Range Sight Surveillance System, Light Armored Vehicle - Air Defense, Commanders Independent Thermal Viewer, Javelins' Focal Plane Array Dewar (FPA/Dewar), and the high-speed anti-radiation missile. He has received Raytheon's Sensors & Electronics Systems Technical Excellence Award 1998, was elected a member of the Group Technical Staff at Texas Instruments, and received the Department of the Navy's (Willoughby Award) - Reliability, Maintainability, and Quality Award for his efforts on the HARM Missile Program. He is a member of SME and ASME and a support group member of ASME Y14.5M.

22.1   Introduction

Systems, subsystems, subassemblies, and/or parts that require disassembly (for maintenance, upgrades, or replacement of defective parts) are typically designed using snap fits, threaded fasteners, or rivets. This chapter discusses the design and manufacturing considerations for threaded fasteners and rivets.

22.2   Floating and Fixed Fasteners

The intent of a design is to meet all functional requirements, one of these being interchangeability. With that in mind, the Geometric Dimensioning and Tolerancing (GD&T) standard ASME Y14.5M-1994 documents the rules for fixed and floating fasteners. The GD&T standard covers both the fixed and floating fastener rules in Appendix B, "Formulas for Positional Tolerancing." To understand and use the rules, we must first identify the *type* of condition (or case) where the fastener is being used. There are three different

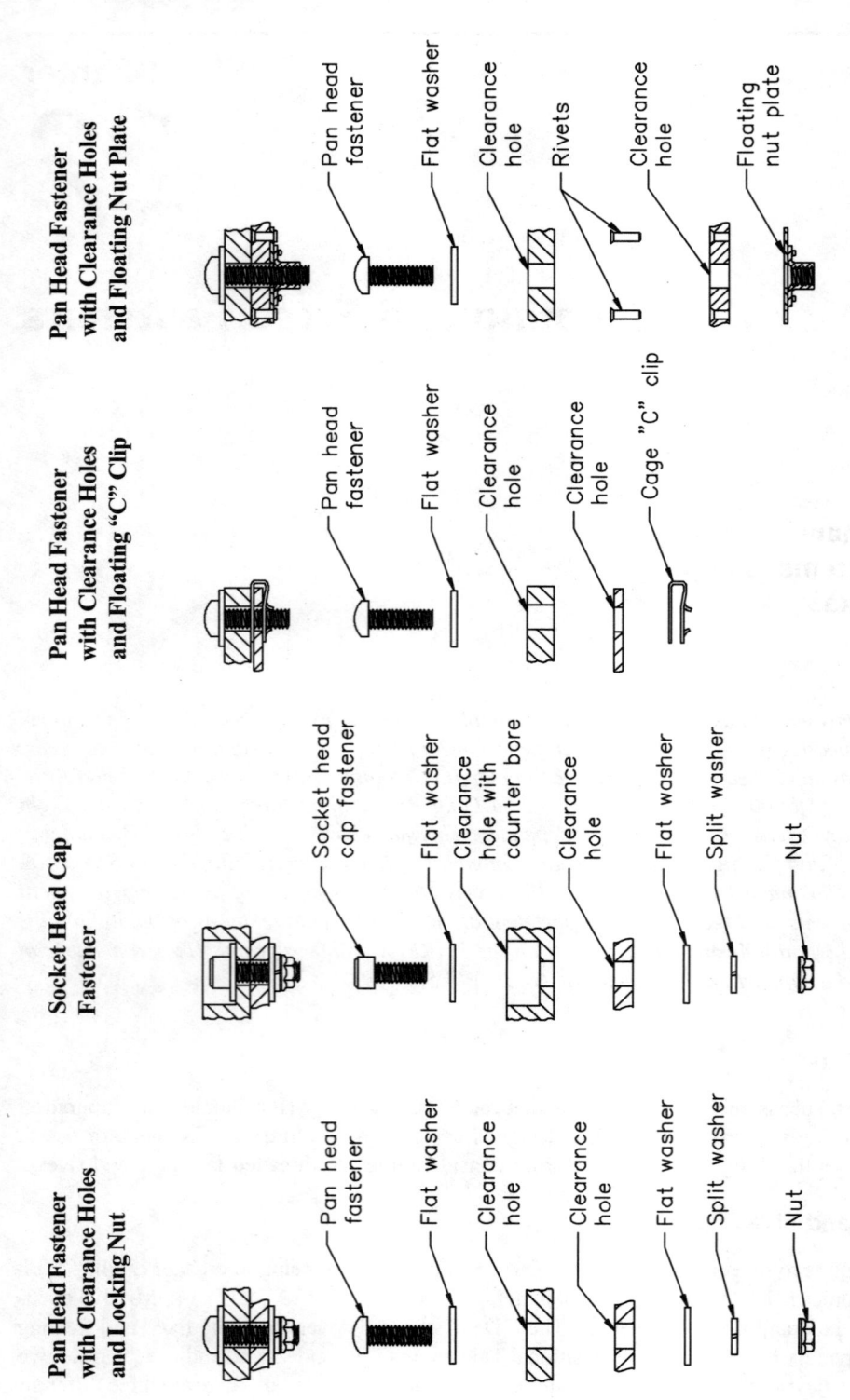

Figure 22-1 Examples of floating fasteners

Floating and Fixed Fasteners 22-3

Pan Head Fastener with Tapped Hole

Socket Head Cap Fastener with Tapped Hole

Flat Head Fastener with Clearance Hole and Nut

Flat Head Fastener with Clearance Holes and Floating Nut Plate

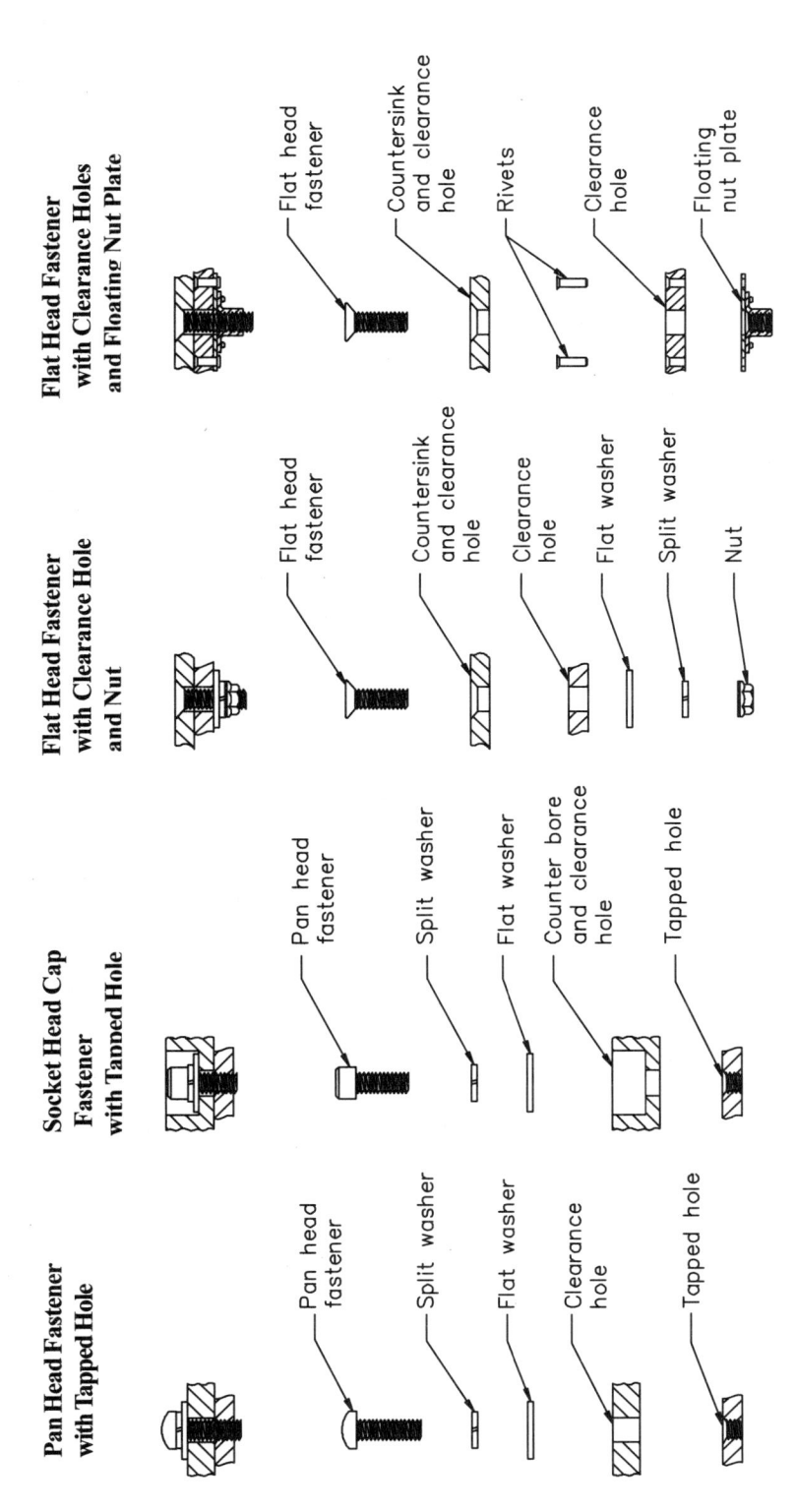

Figure 22-2 Examples of fixed fasteners

conditions: floating fasteners, fixed fasteners, and double-fixed fasteners. Y14.5 only discusses the floating fastener case and the fixed fastener case.

22.2.1 What Is a Floating Fastener?

A floating fastener is a bolt, pan head fastener, socket head fastener, and nut, C'Clip, or floating nut plate used to fasten two or more parts together. All parts have clearance holes and the nut plates must be free floating (see Fig. 22-1).

22.2.2 What Is a Fixed Fastener?

A fixed fastener uses a bolt, pan head fastener, socket head fastener, flat head fastener or alignment pin. One end of the fastener (or pin) is restrained in a tapped hole or is pressed into a hole. The other end of the fastener (or pin) is free to float in a clearance hole (see Fig. 22-2). In the case of a flat head fastener, the countersink diameter/clearance hole and the angle of the flat head fastener by design will constrain the fastener, making it a fixed fastener application.

22.2.3 What Is a Double-Fixed Fastener?

Y14.5 does not discuss what is known as a double-fixed fastener. A double-fixed fastener uses a flat head threaded fastener with a countersink, which restrains the head, and a tapped hole that effectively restrains both ends of the fastener (see Fig. 22-3).

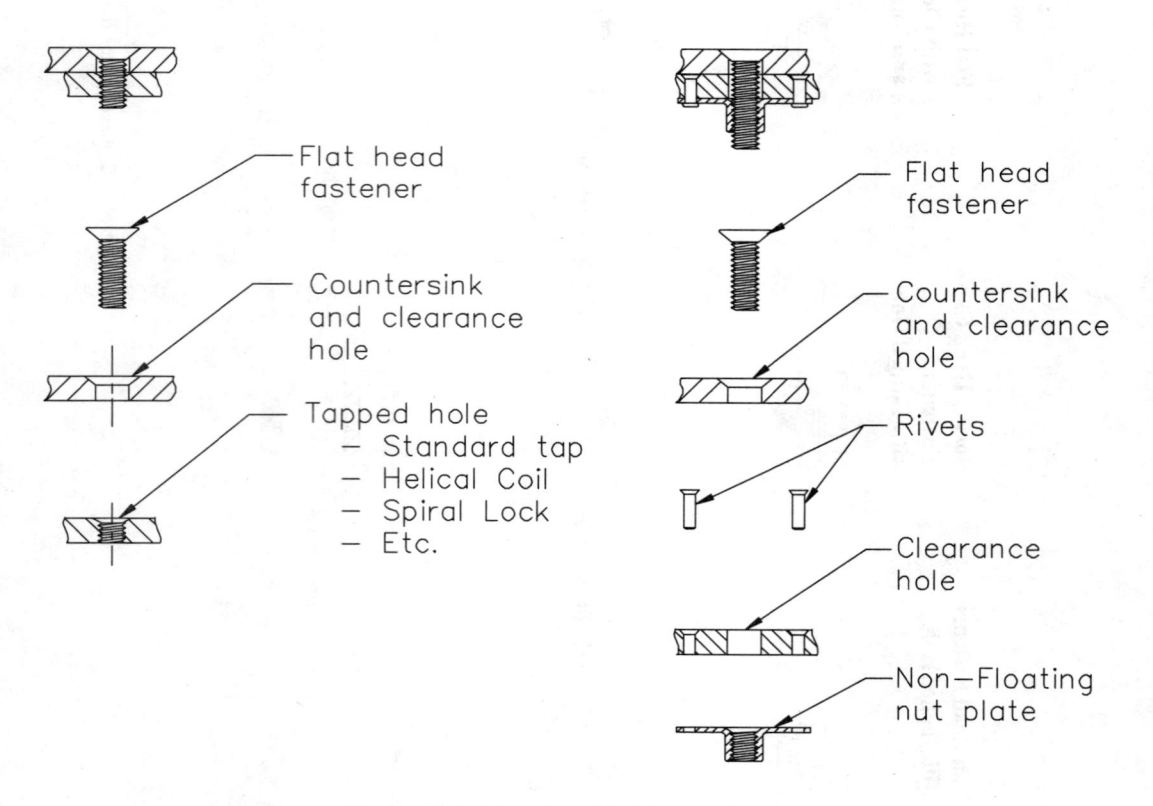

Figure 22-3 Examples of double-fixed fasteners

22.3 Geometric Dimensioning and Tolerancing (Cylindrical Tolerance Zone Versus +/- Tolerancing)

Tolerancing fixed and floating fasteners is frequently done so that the mating parts are 100% interchangeable. The methods of allocating tolerances discussed in Y14.5 ensure 100% interchangeability. In these applications, three things determine the size of the clearance holes:
1) The location tolerance that is applied to the clearance hole
2) The location tolerance that is applied to the mating tapped hole (for a fixed fastener) or the mating clearance hole (for a floating fastener)
3) The size tolerances applied to the holes

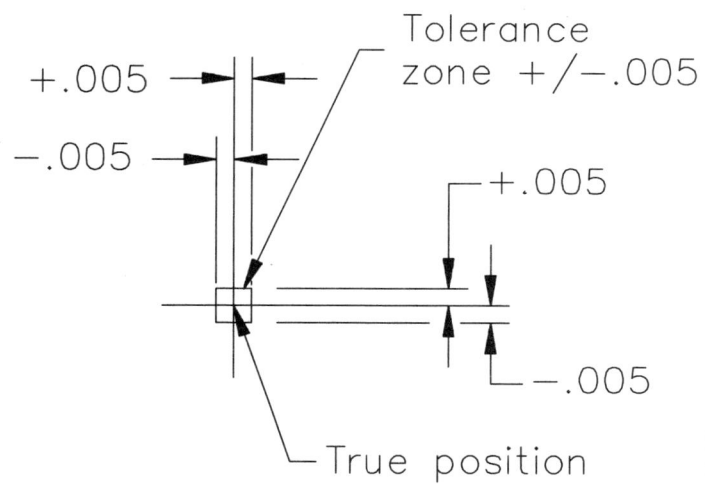

Figure 22-4 Rectangular tolerance zone (plus/minus tolerancing)

Historically, there have been two types of tolerancing methods used: plus or minus tolerancing (a rectangular tolerance zone usually shown as, e.g., ±.005) and positional tolerancing (a cylindrical tolerance zone). An example of a rectangular, or ± tolerance zone, is shown in Fig. 22-4.

Fig. 22-5 shows an example of a cylindrical tolerance zone: (∅) .014. The rules in Y14.5 use a cylindrical tolerance zone to locate the features. In general, if a system is designed using threaded fasteners, bolts, rivets, or alignment pins, cylindrical fasteners are installed into a cylindrical hole. The functional tolerance zone that can accept all conditions and sizes of mating features is a cylindrical tolerance zone.

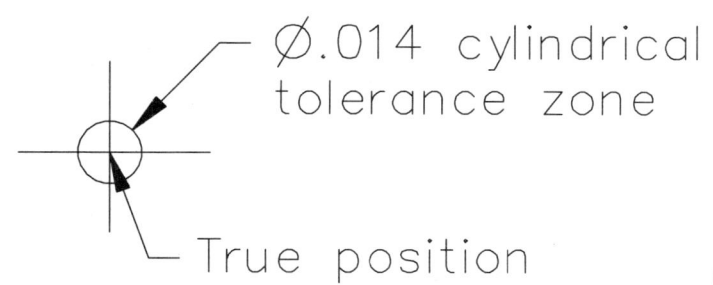

Figure 22-5 Cylindrical tolerance zone

When calculating the size of the clearance hole, the engineer should take into account the amount of allowable variation for both the clearance hole and the tapped hole. Fig. 22-6 shows a fixed fastener example with a .250-28 UNF-2B threaded fastener. Suppose the tapped hole was perfectly located (.000, .000), and the clearance hole deviates from its position in the X direction by .005, and is at nominal in the Y direction (+.005, .000) (see Fig. 22-6). If we were to calculate the hole size that is required to permit the fastener to pass through, the size of the clearance hole for this example is .260 diameter (Ø.260). The same size clearance hole is necessary if the hole deviates from its position in the opposite direction by .005 (-.005,.000).

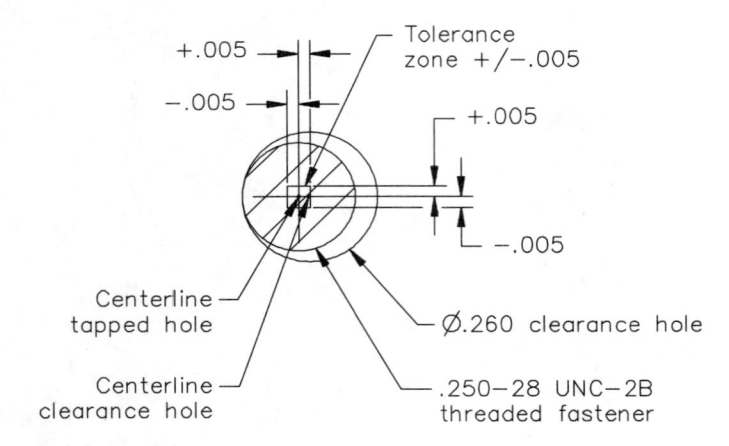

Figure 22-6 Tapped hole located (.000, .000) and clearance hole off location by (+.005, .000)

Let's assume the design engineer takes into account the allowable variation for both the clearance hole and the tapped hole. If the .250-28 UNF-2B tapped hole was located (-.005, .000) and the clearance hole was located (+.005, .000), the size of the clearance hole for this example would be .270 diameter (Ø.270). This would account for the possibility of a +.005 / -.005 shift of both the tapped hole and the clearance hole (see Fig. 22-7).

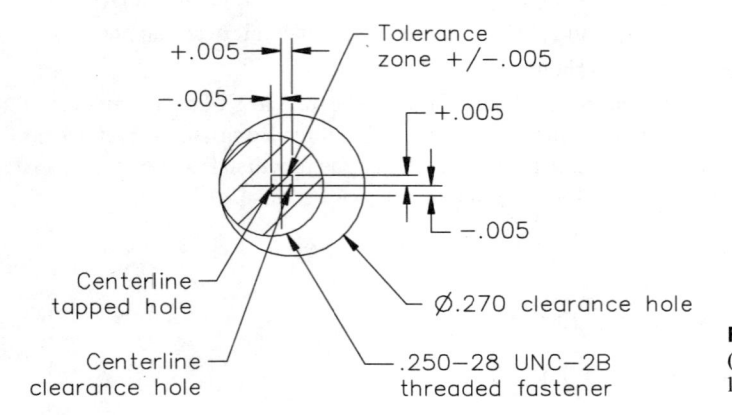

Figure 22-7 Tapped hole is located (-.005, .000) and clearance hole is located (+.005,.000)

Let's look at a worst case location tolerance. The hole size must be calculated when both the tapped hole and the clearance hole are at their worst case location. Assume the tapped hole was located at its worst case location, (X direction was at -.005, and the Y direction was at -.005 (-.005, -.005)), and the clearance hole was also located at its worst case location, (X direction at +.005, and the Y direction at +.005

Floating and Fixed Fasteners 22-7

(+.005, +.005)). Refer to Fig. 22-8. This results in the worst case possible location of both the threaded hole and the tapped hole). The size of the clearance hole for this example is ⌀.278 to account for the possibility of a +.005,-.005 shift of both the tapped hole and the clearance hole (see Fig. 22-8). By manufacturing a part at the worst case location tolerance of +/-.005, the feature is located a radial distance of .007 from the nominal dimension.

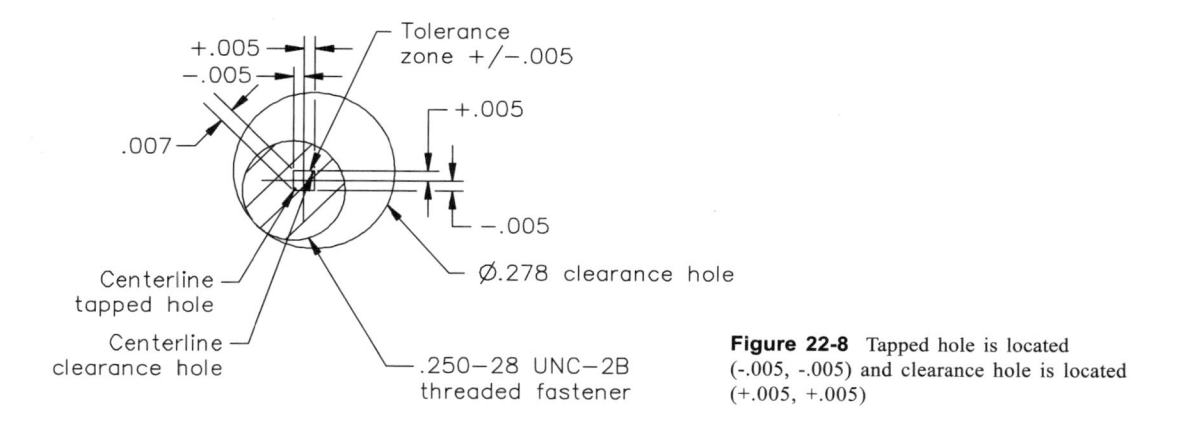

Figure 22-8 Tapped hole is located (-.005, -.005) and clearance hole is located (+.005, +.005)

If the tapped holes and the clearance holes that are located by (+.005, -.005) are functional parts that have a .007 radial location, then a tapped hole manufactured at (-.007, .000) and the clearance hole manufactured at (+.007, .000) is also functional. Its tolerance zone also results in a .007 radial location (see Fig. 22-9). The resulting tolerance zone is a diameter .014.

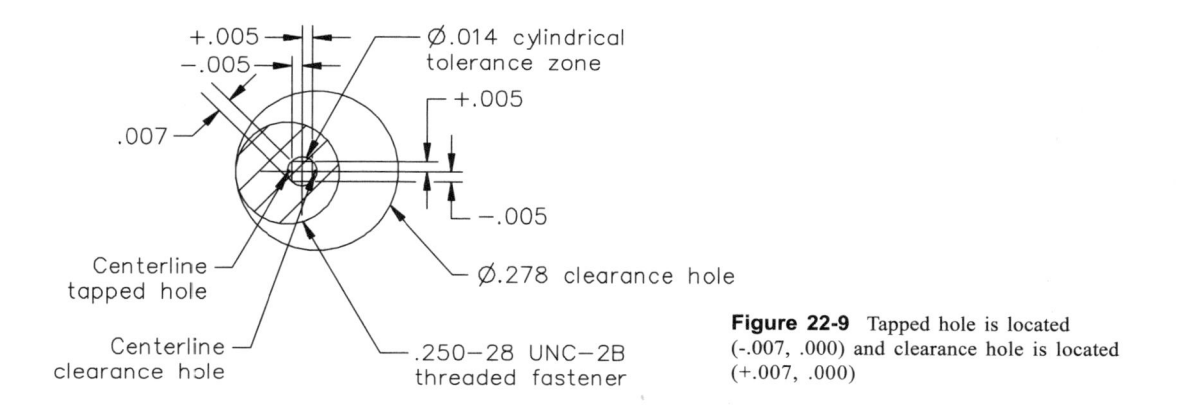

Figure 22-9 Tapped hole is located (-.007, .000) and clearance hole is located (+.007, .000)

Allowing a tolerance of (+.007, .000) for the clearance hole and (-.007, .000) for the tapped hole, the tolerance zone is effectively a diametrical (cylindrical) tolerance zone of ⌀.014. The use of a cylindrical tolerance zone is the preferred method because it allows all functional parts to be used. If a ±.005 tolerance zone had been used, this part would have been rejected (see Fig. 22-10).

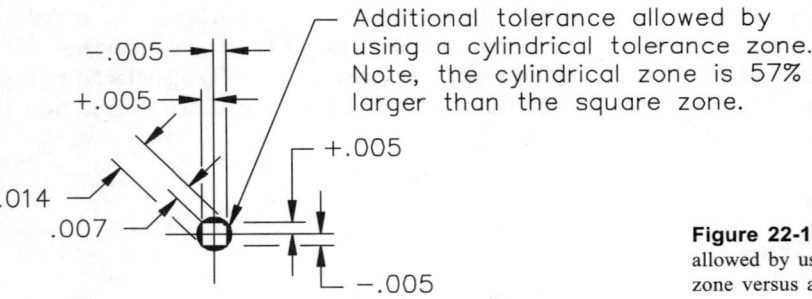

Figure 22-10 Additional tolerance allowed by using a cylindrical tolerance zone versus a rectangular tolerance zone

22.4 Calculations for Fixed, Floating and Double-fixed Fasteners

The purpose of this section is to demonstrate the formulas for calculating the fixed, floating, and double-fixed fasteners. The purpose in calculating the applicable tolerances and hole sizes are two-fold. The first objective is to assure the interchangeability of mating parts and subassemblies. The second is to allocate tolerances with process capabilities in mind, ensuring that the parts can be manufactured cost effectively. The rules or formulas for calculating the fixed and floating fasteners are straightforward.

First we should establish the symbols to be used in the formulas:

FD = Fastener maximum material condition (MMC) size (diameter)
CH = Clearance hole nominal size (diameter)
STCH = Lower limit size tolerance for the clearance hole (diameter)
CBD = Counterbore (C'Bore) nominal size (diameter)
STCBH = Lower limit size tolerance for the C'Bore hole (diameter)
WD = Flat washer MMC size of the outer diameter
PTCH = Positional tolerance of the clearance hole (diameter)
PTTH = Positional tolerance of the tapped hole (diameter)
PTCBH = Positional tolerance of the C'Bore hole (diameter)

22.5 Geometric Dimensioning and Tolerancing Rules/Formulas for Floating Fastener

Assembled parts that have clearance holes in all parts are referred to as floating fastener applications (see Fig. 22-1).

22.5.1 How to Calculate Clearance Hole Diameter for a Floating Fastener Application

The formula for calculating a clearance hole diameter for a floating fastener application follows:

CH = FD + PTCH + STCH

An example of the calculation follows. If we were designing a fastened assembly with a .250-28 UNF-2B fastener being used, then FD would be equal to ⌀.250.

CH = ⌀.250 + PTCH + STCH

Next, we assign a Six Sigma tolerance for the location tolerance of the clearance holes. (Refer to Chapter 11 for detailed discussion on Six Sigma tolerancing.) Let us assume that for a Numerical Controlled (N/C) machining process, the Six Sigma tolerance value is ⌀.014 for the location of a clearance hole. Therefore we set PTCH equal to ⌀.014.

CH = ⌀.250 + ⌀.014 + STCH

Then we assign a Six Sigma tolerance for the size tolerance of the hole. When drilling a hole, the drill will normally produce a hole that is larger than the drill diameter. As the drill wears, it will produce holes that are undersized. Knowing that the drilling operations process is a skewed distribution, we must take this into account. Knowing that the process capability for a drill hole results in a skewed distribution, let us assume the Six Sigma tolerance range for the drilling process is +.005/-.002. Since we are trying to calculate the nominal diameter for the clearance hole drill size, we must add the negative size tolerance (or the STCH). We then set the STCH equal to .002 to get to the nominal diameter of the clearance hole.

CH = ⌀.250 + ⌀.014 + ⌀.002
CH = ⌀.266

Once the clearance hole has been calculated, go to the drill chart and pick the nearest drill size from a drill chart. We select the nearest drill size so that we do not need to manufacture a special form cutter. In this case, the nearest drill size is ⌀.2656 (17/64). The clearance hole diameter is ⌀.266 +.005/-.002

22.5.2 How to Calculate Counterbore Diameter for a Floating Fastener Application

To calculate the diameter of the counterbore to be used for a .250-28 UNF-2B fastener, the diameter of the flat washer must be used in the floating fastener formula. The formula for calculating a counterbore diameter is as follows:

CBD = WD + PTCBH + STCBH

We must use the MMC size of the flat washer diameter (WD) to calculate the counterbore diameter. If the outside diameter and the size tolerance of the washer are ⌀.734 +.015/-.007, the MMC of the washer is ⌀.749. Therefore, we set the WD equal to ⌀.749.

CBD = ⌀.749 + PTCBH + STCBH

Note: This formula does not take into account any allowable shifting between the inner diameter of the washer and the outer diameter of the fastener.

The next step is to assign a Six Sigma tolerance for the location of the clearance holes. If we assume the Six Sigma tolerance for the location of a clearance hole using an N/C machining process is ⌀.014, we set PTCBH equal to ⌀.014.

CBD = ⌀.749 + ⌀.014 + STCBH

Next, we assign a Six Sigma size tolerance for the counterbore diameter. When machining a counterbore, there are three methods of manufacturing the counterbore holes. One method is to use a mill cutter and plunge the cutter to depth. The second method is to use a form cutter that creates both the clearance hole and the counterbore holes in the same operation. The third method is to profile mill the diameter using an undersized cutter. Both the plunging and form cutter drill operation are comparable to a drilling operation. Both will produce a hole that is larger than the diameter of the cutter. As the drill wears, it will produce holes that are undersized. With this in mind, the process capability of the plunged hole or the form cutter hole results in a skewed distribution, and the Six Sigma tolerance for the drilling process is +.005/-.002. If the counterbore were profile milled, the process capability results in a tolerance of +/-.010 for the diameter of the clearance hole. In this example, we will use the profile milling method. Therefore, we set the STCBH equal to .010 to get to the nominal diameter of the counterbore hole.

Note: Process capabilities for tolerances shown in these examples reflect industry standards. Process capability studies should be conducted to establish shop specific process capabilities. (Reference Chapters 8, 10, and 17 for information on Cp, Cpk, and process capabilities.)

CBD = ⌀.749 + ⌀.014 + ⌀.010
CBD = ⌀.773

Once the counterbore hole size has been calculated, go to the drill chart and pick the nearest drill size from the drill chart. In this case, the nearest drill size is ⌀.781 (25/32). The counterbore hole diameter is ⌀.781 +/-.010.

22.5.3 Why Floating Fasteners Are Not Recommended

The use of floating fasteners is not a recommended practice. When assembling parts and/or subassemblies, it requires work on both sides of the parts to tighten the fasteners and to hold the nuts. When designing large systems such as automobiles, it could require two people working together to tighten the hardware. If floating fasteners are necessary, the design engineer should consider using captive hardware such as a C'Clip or nut plate that alleviates the problem of requiring two assemblers. However, the use of C'Clips and/or nut plates adds additional hardware, complexity, and additional process steps. This additional hardware results in additional cost and assembly time.

22.6 Geometric Dimensioning and Tolerancing Rules/Formulas for Fixed Fasteners

As shown in Fig. 22-2, assemblies having a clearance hole in one part and a tapped hole in the other are fixed fastener applications.

22.6.1 How to Calculate Fixed Fastener Applications

The formula for calculating a clearance hole diameter for a fixed fastener application is:

$$CH = FD + PTCH + PTTH + STCH$$

An example of the calculation follows. If we were designing a fastened assembly where a .250-28 UNF-2B fastener is being used, then set FD equal to .250.

$$CH = \varnothing.250 + PTCH + PTTH + STCH$$

Next we assign Six Sigma tolerances to the location of both the clearance hole and the tapped holes. Since the drilling and tapping is also done on an N/C machining process, the Six Sigma tolerance is $\varnothing.014$ for a drilled and tapped hole. Set both PTCH and the PTTH equal to $\varnothing.014$.

$$CH = \varnothing.250 + \varnothing.014 + \varnothing.014 + STCH$$

Again, assign a Six Sigma tolerance for the size tolerance of the hole, and set STCH equal to $\varnothing.002$.

$$CH = \varnothing.250 + \varnothing.014 + \varnothing.014 + \varnothing.002$$
$$CH = \varnothing.280$$

The nearest or next largest drill size is $\varnothing.2812$ (9/32). The clearance hole diameter is:

$$CH = \varnothing.281 + .005/-.002$$

Note: In the fixed fastener cases, variations in the perpendicularity of the tapped hole or pressed-in pins will cause a projected error that could cause interference in mating parts. To avoid this, the hole in the mating part needs to be enlarged to account for the error, or a projected tolerance zone must be applied to the threaded holes.

22.6.2 How to Calculate Counterbore Diameter for a Fixed Fastener Application

To calculate the diameter of the counterbore to be used for a .250-28 UNF-2B fastener, the diameter of the flat washer must be used in the fixed fastener formula. The fixed fastener formula for calculating a counterbore diameter follows:

$$CBD = WD + PTCBH + PTTH + STCBH$$
$$CBD = \varnothing.749 + \varnothing.014 + \varnothing.014 + \varnothing.010$$
$$CBD = \varnothing.787$$

Note: This formula does not take into account any allowable shifting between the inner diameter of the washer and the outer diameter of the fastener.

Once the counterbore size is calculated, go to the drill chart and pick the nearest or next largest drill size from the drill chart. In this case, the nearest or next largest drill size is ⌀.797 (51/64). The counterbore hole diameter is ⌀.797 +/-.010

Note: In the fixed fastener cases, variations in the perpendicularity of the tapped hole will cause a projected error that could cause interference in mating parts. To avoid this, the hole in the mating part needs to be enlarged to account for the error, or a projected tolerance zone must be applied to the threaded holes.

22.6.3 Why Fixed Fasteners Are Recommended

Fixed fasteners are recommended when assembling parts or subassemblies. Fixed fasteners allow Z axis or top down assembly. There is no additional hardware to assemble the fastener, no C'Clips, no floating nut plates, no rivets to hold the floating nut plates, and no nuts. It also takes less time to assemble. As long as the parts are not repeatedly assembled and disassembled, the use of self-tapping fasteners is highly recommended because they do not require an additional tapping operation.

22.7 Geometric Dimensioning and Tolerancing Rules/Formulas for Double-fixed Fastener

When assembling parts using a flat head fastener, the threads on the fastener in the tapped hole and the flat head fastener are restrained by the countersink. This effectively restrains both ends of the fastener (see Fig. 22-3). Since both ends of the flat head fastener are restrained, theoretically both parts must be perfectly located in order to assemble the mating parts. Since locational tolerances for both the tapped hole and clearance hole are required to make the parts manufacturable, the locational tolerance is calculated using the fixed fastener rule. Assigning locational tolerances to both the tapped hole and the countersink causes the flat head fastener head height to be above or below the surface.

22.7.1 How to Calculate a Clearance Hole

The formula for calculating the clearance hole for a double-fixed fastener is the same as the fixed fastener application:

$$CH = FD + PTCH + PTTH + STCH$$

Here is an example. Let's say we were designing a double-fixed fastened assembly where a .250-28 UNF-2B fastener is being used. It has a positional tolerance of ⌀.014, and the fastener diameter (FD) is equal to ⌀.250,

$$CH = ⌀.250 + ⌀.014 + ⌀.014 + ⌀.002$$
$$CH = ⌀.280$$

Again, the nearest drill size is ⌀.2812 (9/32). The clearance hole is:

$$CH = ⌀.281 +.005/-.002$$

22.7.2 How to Calculate the Countersink Diameter, Head Height Above and Head Height Below the Surface

When calculating the countersink diameter for a flat head fastener, we must control the head height above and below the surface. The worst case head height above the surface occurs when the countersink/clearance hole and the tapped holes are off location by the maximum allowable - when the flat head diameter is at its MMC and the countersink diameter is at its MMC (see Fig. 22-11). The worst case head

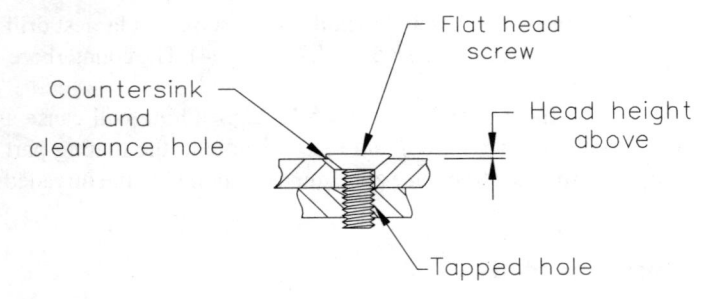

Figure 22-11 Worst case head height above the surface

height below the surface occurs when the countersink/clearance hole and the tapped holes are on perfect location, when the flat head diameter is at its least material condition, and the countersink diameter is at its least material condition (see Fig. 22-12).

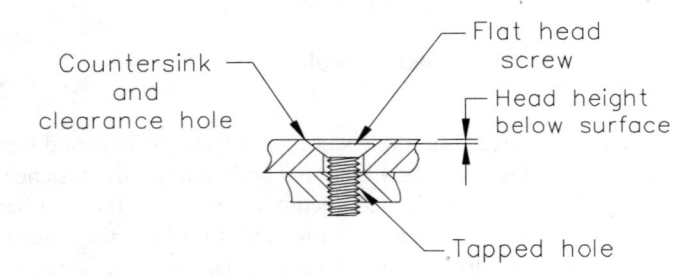

Figure 22-12 Worst case head height below the surface

First we should establish the symbols to be used in the formulas:
    CSHM = Countersink MMC diameter (nominal countersink diameter - STCSH)
    CSHL = Countersink LMC diameter (nominal countersink diameter + STCSH)
    STCSH = Equal bilateral size tolerance for the countersink hole
    FHDM = Flat head fastener MMC diameter
    FHDL = Flat head fastener LMC diameter
    PTCH = Positional tolerance of the clearance hole and countersink diameter
    PTTH = Positional tolerance of the tapped hole
    CSA = Countersink included angle ($82°$ or $100° \pm 1°$)
    CSAMin = Minimum countersink included angle (CSA - $1°$)
    HHA = Head height above
    HHB = Head height below

The formulas for calculating the head heights for a double-fixed fastener application are:

$$HHA = ((.5*FHDM)-(.5*CSHM)+(.5*PTTH)+(.5*PTCH))/TAN(.5*CSAMin)$$

and

$$HHB = ((.5*FHDL)-(.5*CSHL))/TAN(.5*CSAMax)$$

Note: These formulas do not take into account the perpendicularity of the tapped hole. When calculating the head height above and the head height below, the objective is to determine a countersink diameter that allows an equal bilateral tolerance on the amount the head of the fastener is above and below the surface. In the double-fixed fastener cases, variations in the perpendicularity of the tapped hole will cause a projected error that could cause interference in mating parts. It could increase the amount the head of the flat head fastener will protrude above the surface.

For this example, the objective is to determine a countersink diameter that allows an equal bilateral tolerance on the amount the head of the fastener is above and below the surface. Therefore, we should solve the equations simultaneously to obtain an equal head height above and below the surface.

Fig. 22-13 shows a .250-28 UNF-2B flat head fastener with a 100° flat head (100° included angle).

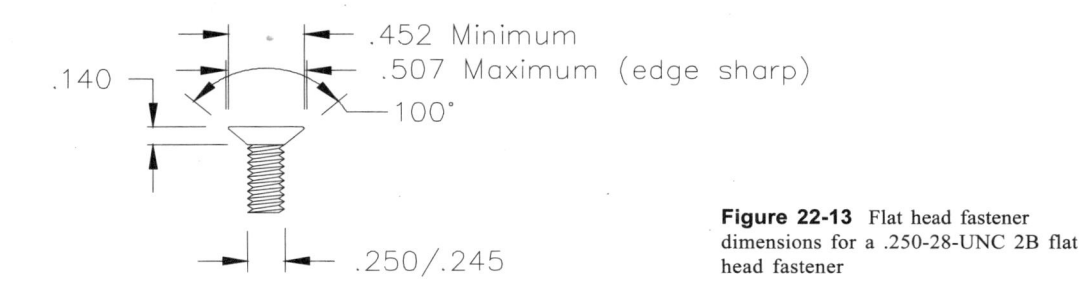

Figure 22-13 Flat head fastener dimensions for a .250-28-UNC 2B flat head fastener

When solving for head height above, we set the flat head fastener diameter at MMC (⌀.507). Next we set the minimum included countersink angle (CSAMin) to 99° (100° - 1°). The positional tolerances for the clearance/countersink hole and the tapped hole are position ⌀.014.

If we assume the countersink diameter to be the same as the flat head screw and set the countersink diameter to ⌀.510 ±.010, then the MMC diameter of the countersink is ⌀.500. Therefore we set CSHM = ⌀.500, and:

HHA = ((.5*FHDM)-(.5*CSHM)+(.5*PTTH)+(.5*PTCH))/TAN(.5* CSAMin)
HHA = ((.5*.507)-(.5*.500)+(.5*.014)+(.5*.014))/TAN(.5*99°)
HHA = (.2535 - .250 + .007 + .007)/TAN(49.5°)
HHA = .0175 / 1.170849566113 =. 0.01622753302381
HHA = .0149

When solving for head height below the surface, we use the LMC of the fastener head diameter. We set FHDL = ⌀.452. Since we set the countersink diameter equal to ⌀.510±.010, then the LMC diameter of the countersink is ⌀.520 and we set CSHL = ⌀.520. The angle of the flat head fastener that is used is 100°. Therefore we set CSAMin = 100° - 1° = 99°. Again, the positional tolerance of the clearance hole/countersink hole and the tapped hole are a ⌀.014.

Therefore:

HHB = ((.5*FHDL)-(.5*CSHL)) / TAN(.5* CSAMin)
HHB = ((.5*.452)-(.5*.520)) / TAN(.5*99°)
HHB = (.226-.260) / TAN(49.5°)
HHB = -.034 / 1.170849566113 = -0.02476833987843
HHB = -.029

Note: To determine the amount a flat head screw is above or below the surface, reference Table 22-5, "Flat Head Screw Height Above and Below the Surface."

22.7.3 What Are the Problems Associated with Double-fixed Fasteners?

As stated in section 22.7.2, when using a double-fixed fastener, we must control the head height above and below the surface. The worst case head height above the surface occurs when both the countersink/clearance hole and the tapped holes are off location by the maximum amount, when the flat head diameter is at its MMC, and the countersink diameter is at its MMC (see Fig. 22-11). Statistically, the probability that the countersink and tapped hole will deviate from their theoretical perfect locations is much greater than it is for the countersink, clearance hole, and tapped hole being perfectly located. Therefore, there is a greater chance for the head of the flat head fastener to be above the surface than below. This can be resolved by making the countersink diameter larger if the material thickness allows.

Another problem associated with the use of double-fixed fasteners is the countersink diameter. Normally, the countersink diameter is toleranced +/-.010, while the clearance hole diameter is toleranced +.005 / -.002. Since the countersink diameter controls the head height, the additional tolerance allocated to a countersink increases the head height above the surface.

22.8   Nut Plates: Floating and Nonfloating (see Fig. 22-14)

When designing a fastener assembly that uses a floating nut plate, the engineer should account for several factors:
- The type of nut plate being used (floating or nonfloating) and the amount of float that the nut plate has designed into it
- The type of fastener assembly (floating, fixed, or double-fixed fastener assembly)
- The tolerancing scheme used for the rivet holes used to mount the nut plate
- The amount of tolerance that is applied to the rivet holes

The formulas used for calculating the clearance hole sizes are discussed in sections 22.5, 22.6, and 22.7.

In the following example, we will look at a floating fastener assembly that uses a floating nut plate. To standardize hole sizes when designing a fastener assembly using a floating nut plate, the following method could be used to calculate the clearance hole tolerance and the rivet hole tolerance.

$$PTCH = (CH - STCH - FD)/2$$

If we use the clearance hole diameter that was calculated for a fixed fastener using a .250-28 UNF-2B fastener, then CH is equal to $\varnothing.2812$.

$$PTCH = (.2812 - .002 - .250)/2$$
$$PTCH = \varnothing.0146$$

For the part that only has the clearance hole, it is straightforward. The PTCH can be applied directly to the part that only has a clearance hole. Therefore the positional tolerance of the clearance hole would be set to $\varnothing.014$.

For the mating part (the part that has both the nut plate and a clearance hole), the $\varnothing.014$ tolerance must be distributed to both the clearance hole and the floating nut plate rivet holes that are used to mount the floating nut plate. The required positional tolerance for the rivet holes and the diameter of the rivet must be taken into account. If the rivet holes are a $\varnothing.098$ +.005/-.001, and the rivet diameter is $\varnothing 3/32$ ($\varnothing.093$), we would then need to calculate the required tolerance to the rivet holes. The following calculations show that the tolerance required for the rivet hole results in PTRH = $\varnothing.002$.

$$PTRH = (CHDR - RD)/2$$
$$PTRH = (.097 - .093)/2$$
$$PTRH = \varnothing.002$$

where,
  PTRH = Positional tolerance of the rivet hole (diameter)
  CHDR = Rivet hole MMC size (diameter)
  RD = Rivet MMC size (diameter)

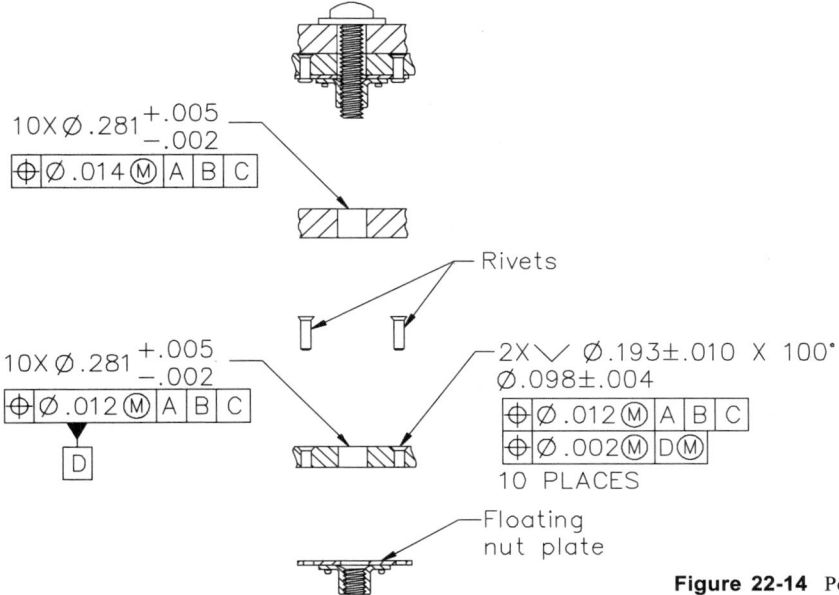

Figure 22-14 Positional tolerance for clearance holes and nut plate rivet holes

22.9 Projected Tolerance Zone

When using fixed or double-fixed fasteners, a projected tolerance zone should be used regardless of whether the design is using threaded fasteners or alignment pins. Variation in the perpendicularity of the screw or pin could cause assembly problems. If a threaded fastener was out of perpendicular by the total amount of the positional tolerance of (Ø.014), an interference problem could occur (see Figs. 22-15 and 22-16).

Figure 22-15 Tapped hole out of perpendicular by Ø.014

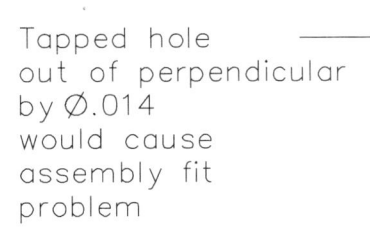

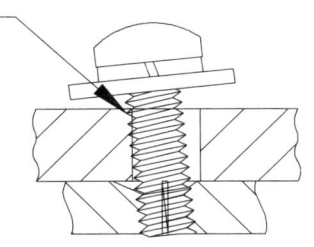

Figure 22-16 Variation in perpendicularity could cause assembly problems

Fig. 22-17 shows how a projected tolerance zone corrects the interference problem shown in Fig. 22-16. The projected tolerance zone is applied to the threaded fastener or the pressed pin. The tolerance zone for the tapped hole extends through the mating parts clearance hole, thereby assuring the mating parts will fit.

THIS ON THE DRAWING   **MEANS THIS**

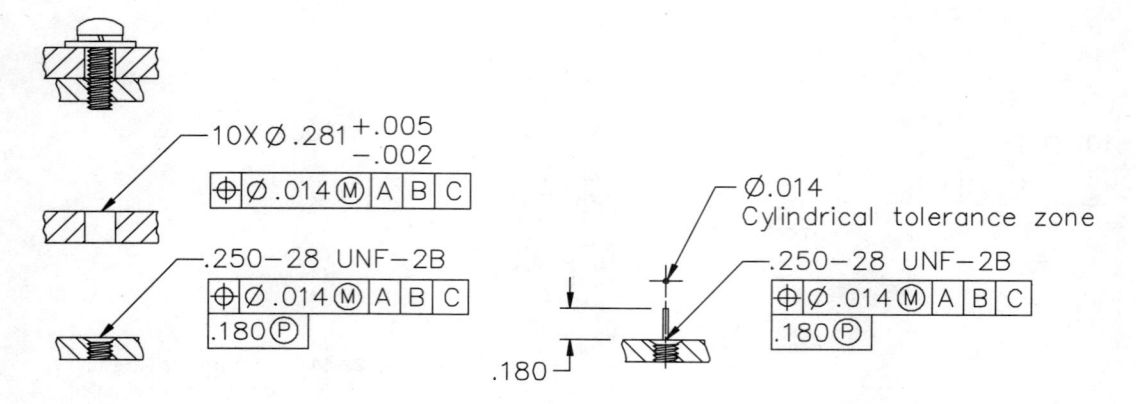

Figure 22-17 Projected tolerance zone example

22.9.1 Comparison of Positional Tolerancing With and Without a Projected Tolerance Zone

This section compares two position tolerancing methods to locate size features for fixed fasteners. In the first method, we use a projected tolerance zone and calculate the functional tolerance zone using the fixed fastener formulas, as shown previously. We consider this a *functional* method for the case of a fixed fastener. In the second method, we convert the projected tolerance zone to a zone that is *not* projected, and consider this a *nonfunctional* method. As a comparison, we then calculate how much tolerance is lost when dimensioning nonfunctionally.

Assuming a maximum orientation (perpendicularity) error, Fig. 22-18 shows the relationships between the functional (projected) tolerance zone, T_f, and the nonfunctional tolerance zone, T_{nf} (not projected).

$$\frac{T_f/2}{(D/2)+P} = \frac{T_{nf}/2}{(D/2)} \tag{22.1}$$

Where D is the depth of the nonfunctional tolerance zone, and P is the projected height of the functional tolerance zone (see Fig. 22-18).

Eq. (22.1) reduces to:

$$T_{nf} = \frac{T_f}{\left(\frac{2P}{D}+1\right)} \tag{22.2}$$

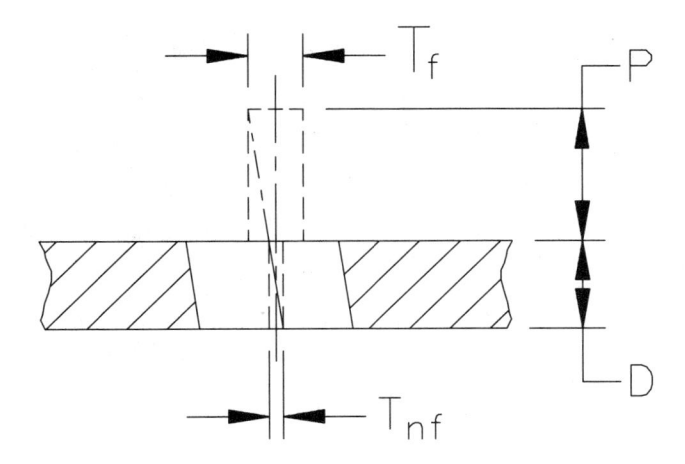

Figure 22-18 Projected tolerance zone — location and orientation components

If we measure the orientation of a feature on a workpiece, we can verify the following relationship:

$$T_{nf,orientation,actual} = \frac{T_{f,orientation,actual}}{\left(\frac{2P}{D}+1\right)} \tag{22.3}$$

where $T_{nf,orientation,actual}$ is the measured nonfunctional orientation error and $T_{f,orientation,actual}$ is the measured functional orientation error.

If we tolerance functionally, the maximum *allowable* location tolerance, $T_{f,location,maximum}$ for a given (actual) orientation error in the functional tolerance zone is:

$$T_{f,location,maximum} = T_f - T_{f,orientation,actual} \tag{22.4}$$

If we tolerance nonfunctionally, the maximum allowable location tolerance, $T_{nf,location,maximum}$, for given (actual) orientation is:

$$T_{nf,location,maximum} = T_{nf} - T_{nf,orientation,actual} \tag{22.5}$$

The difference between Eq. (22.4) and Eq. (22.5) represents the amount of allowable location tolerance that is lost by dimensioning nonfunctionally.

$$\Delta = T_{f,location,maximum} - T_{nf,location,maximum} = T_f - T_{f,orientation,actual} - T_{nf} + T_{nf,orientation,actual} \tag{22.6}$$

Substituting Eq. (22.2) and Eq. (22.3) into Eq. (22.6) gives:

$$\Delta = T_f - T_{f,orientation,actual} - \left[\frac{T_f}{\left(\frac{2P}{D}+1\right)}\right] + \left[\frac{T_{f,orientation,actual}}{\left(\frac{2P}{D}+1\right)}\right]$$

$$\Delta = \left(T_f - T_{f,orientation,actual}\right)\left[\frac{2P}{(2P+D)}\right] \tag{22.7}$$

22.9.2 Percent of Actual Orientation Versus Lost Functional Tolerance

Fig. 22-19 demonstrates how much functional tolerance is lost as a function of actual orientation tolerance. The Y-axis is the percent that the actual orientation tolerance contributes to the total tolerance. The X-axis is the Δ value.

22.10 Hardware Pages

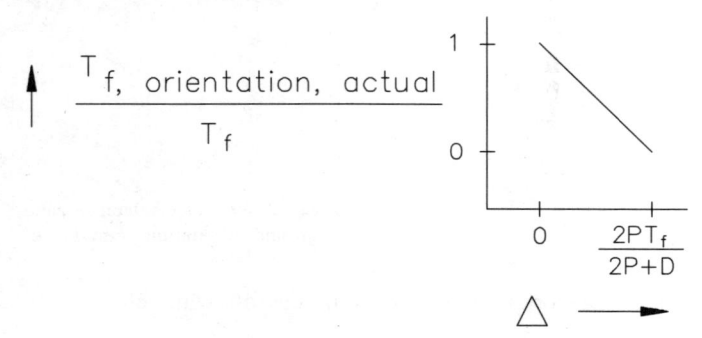

Figure 22-19 Lost functional tolerance versus actual orientation tolerance

The following pages show recommended tolerances for clearance holes C'Bores, C'Sinks, C'Bore Depths, and fasteners. See Tables 22-1, 22-2, and 22-3.) The following general notes apply as noted in Figs. 22-20, 22-21, and 22-22.

GENERAL NOTES:
1. The hole charts reflect recommended tolerance for locating the hole pattern back to the datum surface (hole to surface).
2. The hole charts reflect recommended tolerance for hole-to-hole, and/or hole to a datum feature of size (datum holes). Using a positional tolerance of $\varnothing.014$ on both an N/C drilled and sheet metal punched holes enables us to standardize the clearance hole diameters. Hole diameters, counterbore diameters and depths, and countersink diameters were calculated using the positional tolerance and the tolerances assigned to the hole diameters, counterbore diameters and depths, and countersink diameters.

    Note: It is not recommended that you use hole-to-hole tolerance greater than $\varnothing.014$, because as the hole-to-hole tolerance gets larger, the clearance hole must get larger to accommodate the additional tolerance.
3. Counterbore diameters and depths are calculated using a flat washer with a worst case (MMC) outside diameter, and a worst case thickness. C'Bore diameters are calculated, and the nearest fractional drill diameter is used.
4. Worst case flat head screw height above and below the surface is shown in Table 22-5, and is calculated for a positional tolerance of $\varnothing.014$.
5. Flat head screws are not recommended because of head height issues, and alignment issues.
6. Floating fasteners are not recommended because of the additional hardware required, and because of the difficulty of assembly.
7. For C'Bore depths, (see Table 22-4). For .060-56 threaded holes, the C'Bore depth is calculated using only a flat washer. For .086-56 through .500-20, the C'Bore depth is calculated using both a flat washer and a split washer.

8. Floating and nonfloating nut plate rivet hole diameters, and C'Sink diameters are dependent on the nut plate design and size. (See section 22.8 for information on how to calculate rivet diameter and location tolerance.)
9. Hole-to-hole tolerance for clearance holes and for nut plate rivet holes must be calculated per section 22.7.
10. Projected tolerance zone (PTOL) is determined by the maximum thickness of the mating part.
11. When using floating and nonfloating nut plates, projected tolerance issues could cause interchangeability issues. See section 22.9.

Table 22-1 Floating fastener clearance hole and C'Bore hole sizes and tolerances

Fastener Size	Clearance Hole Diameter .AAA	Clearance Hole Size Tolerance	C'Bore Hole Diameter .BBB	C'Bore Hole Size Tolerance
.060-56 UNF	.076 (#48)	+.005/-.002	.213 (#3)	+/-.010
.086-56 UNC	.104 (#37)	+.005/-.002	.272 (I)	+/-.010
.086-64 UNF				
.112-40 UNC	.1285 (#30)	+.005/-.002	.406 (13/32)	+/-.010
.112-48 UNF				
.125-40 UNC	.1406 (9/64)	+.005/-.002	.438 (7/16)	+/-.010
.125-44 UNF				
.138-32 UNC	.154 (#23)	+.005/-.002	.469 (15/32)	+/-.010
.138-40 UNF				
.164-32 UNC	.180 (#15)	+.005/-.002	.531 (17/32)	+/-.010
.164-36 UNF				
.190-32 UNC	.2055 (#5)	+.005/-.002	.594 (19/32)	+/-.010
.190-36 UNF				
.250-20 UNC	.266 (H)	+.005/-.002	.781 (25/32)	+/-.010
.250-28 UNF				
.312-18 UNC	.328 (21/64)	+.005/-.002	.922 (59/64)	+/-.010
.312-24 UNF				
.375-16 UNC	.3906 (25/64)	+.005/-.002	1.047 (1 3/64)	+/-.010
.375-24 UNF				
.438-14 UNC	.4531 (29/64)	+.005/-.002	1.172 (1 11/64)	+/-.010
.438-20 UNF				
.500-13 UNC	.5156 (33/64)	+.005/-.002	1.312 (1 5/16)	+/-.010
.500-20 UNF				

22-20   Chapter Twenty-two

22.10.1 Floating Fastener Hardware Pages

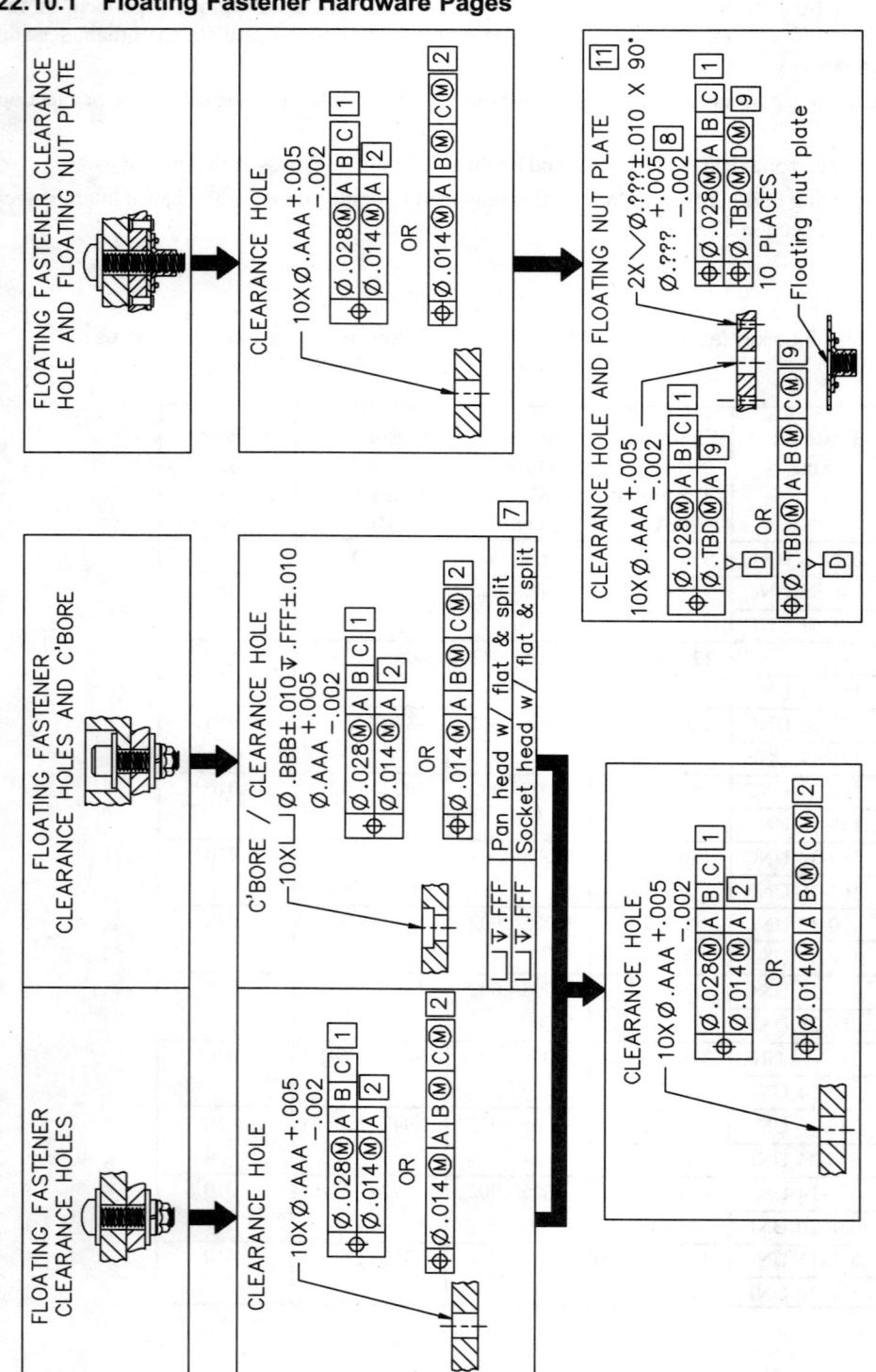

Figure 22-20 Floating fastener tolerance and callouts

22.10.2 Fixed Fastener Hardware Pages

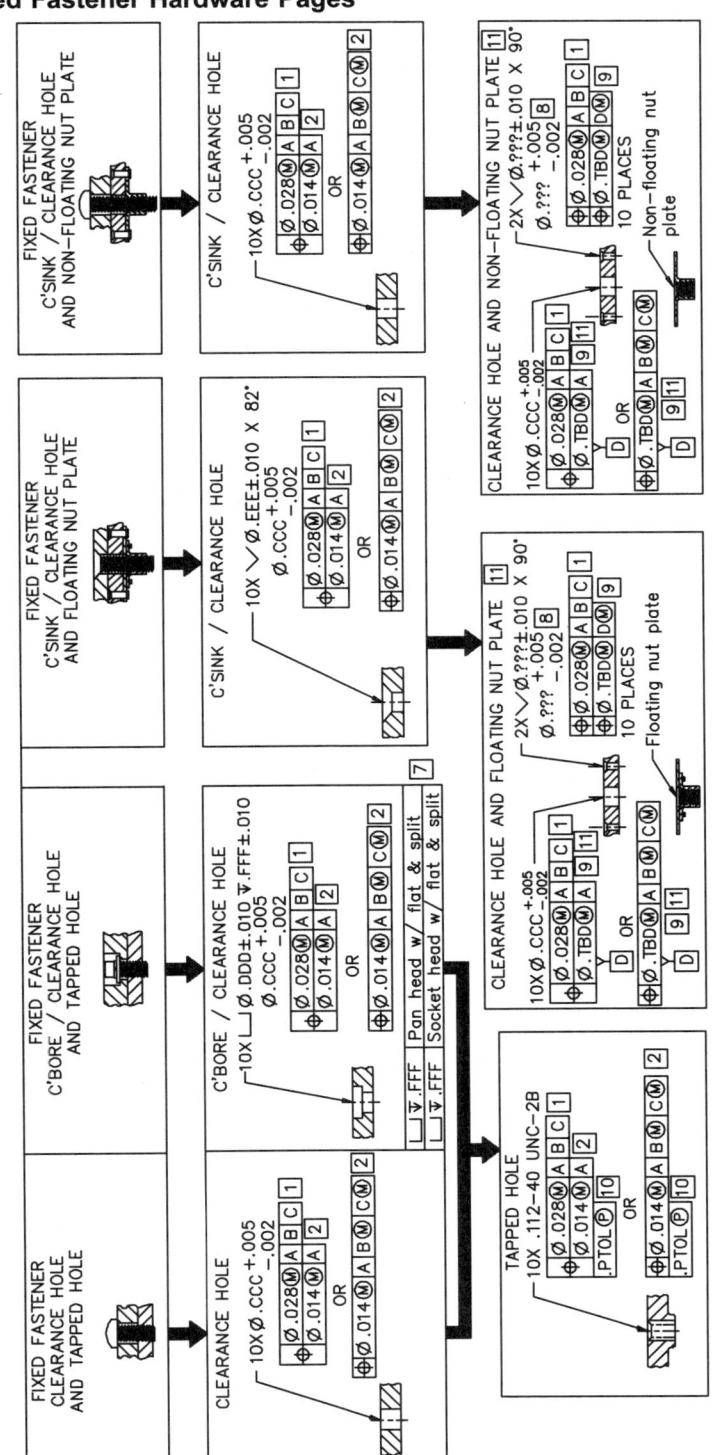

Figure 22-21 Fixed fastener tolerance and callouts

Table 22-2 Fixed fastener clearance hole, C'Bore, and C'Sink sizes and tolerances

Fastener Size	Clearance Hole Diameter .CCC	Clearance Hole Size Tolerance	C'Bore Hole Diameter .DDD	C'Bore Hole Size Tolerance	C'Sink Diameter .EEE	C'Sink Size Tolerance
.060-56 UNF	.0935 (#42)	+.005/-.002	.228 (#1)	+/-.010	.125	+/-.010
.086-56 UNC	.120 (#31)	+.005/-.002	.290 (L)	+/-.010	.180	+/-.010
.086-64 UNF						
.112-40 UNC	.144 (#27)	+.005/-.002	.421 (27/64)	+/-.010	.230	+/-.010
.112-48 UNF						
.125-40 UNC	.1562 (5/32)	+.005/-.002	.453 (29/64)	+/-.010	.255	+/-.010
.125-44 UNF						
.138-32 UNC	.1695 (#18)	+.005/-.002	.484 (31/64)	+/-.010	.285	+/-.010
.138-40 UNF						
.164-32 UNC	.1935 (#10)	+.005/-.002	.547 (35/64)	+/-.010	.335	+/-.010
.164-36 UNF						
.190-32 UNC	.221 (#2)	+.005/-.002	.609 (39/64)	+/-.010	.390	+/-.010
.190-36 UNF						
.250-20 UNC	.2812 (9/32)	+.005/-.002	.797 (51/64)	+/-.010	.510	+/-.010
.250-28 UNF						
.312 -18 UNC	.3438 (11/32)	+.005/-.002	.938 (15/16)	+/-.010	.640	+/-.010
.312-24 UNF						
.375-16 UNC	.4062 (13/21)	+.005/-.002	1.063 (1 1/16)	+/-.010	.765	+/-.010
.375-24 UNF						
.438-14 UNC	.4688 (15/32)	+.005/-.002	1.188 (1 3/16)	+/-.010	.815	+/-.010
.438-20 UNF						
.500-13 UNC	.5312 (17/32)	+.005/-.002	1.328 (1 21/64)	+/-.010	.880	+/-.010
.500-20 UNF						

Floating and Fixed Fasteners 22-23

22.10.3 Double-fixed Fastener Hardware Pages

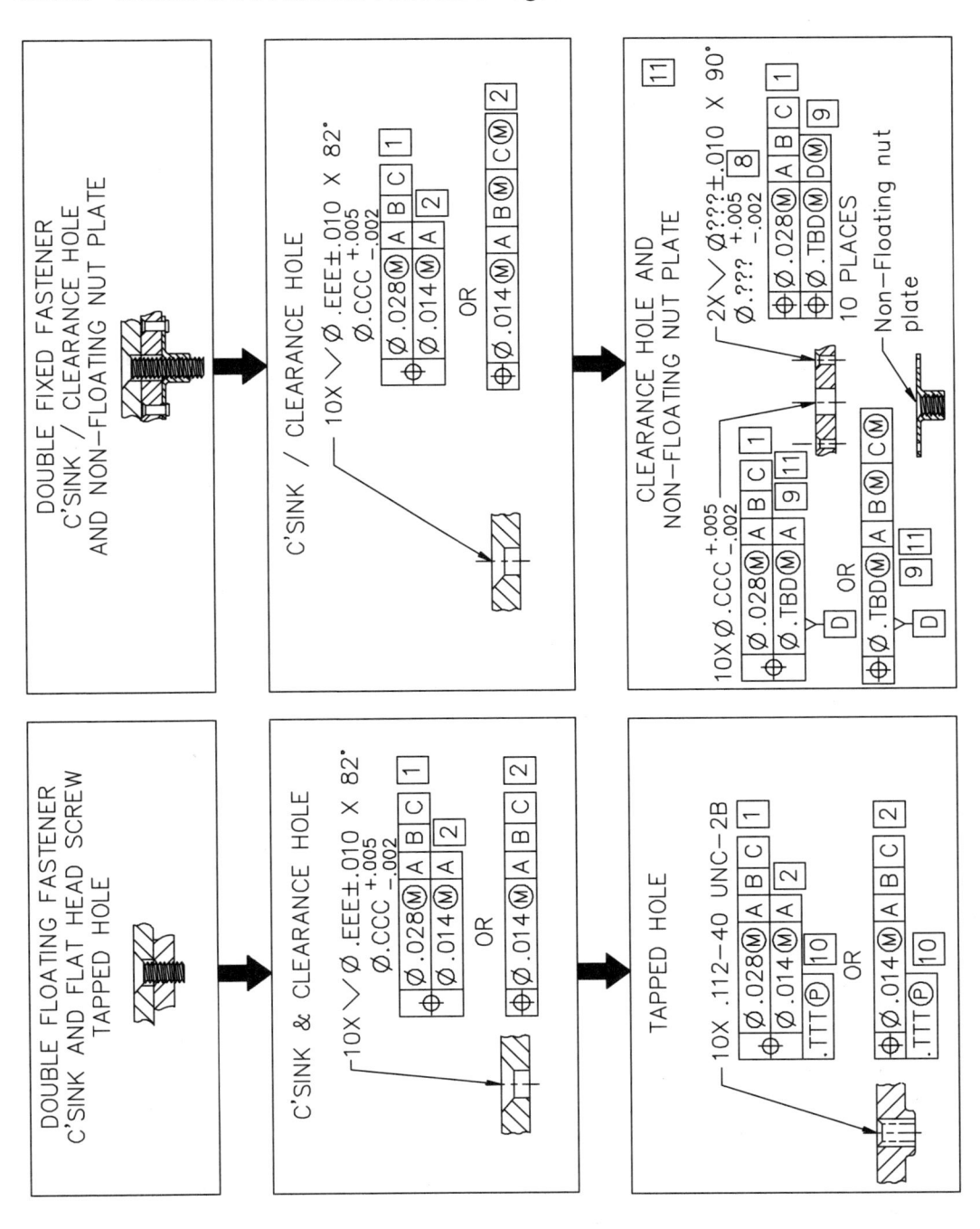

Figure 22-22 Double-fixed fastener tolerance and callouts

Table 22-3 Double-fixed fastener clearance hole and C'Bore sizes and tolerances

Fastener Size	Clearance Hole Diameter .CCC	Clearance Hole Size Tolerance	C'Sink Diameter .EEE	C'Sink Size Tolerance
.060-56 UNF	.0935 (#42)	+.005/-.002	.125	+/-.010
.086-56 UNC	.120 (#31)	+.005/-.002	.180	+/-.010
.086-64 UNF				
.112-40 UNC	.144 (#27)	+.005/-.002	.230	+/-.010
.112-48 UNF				
.125-40 UNC	.1562 (5/32)	+.005/-.002	.255	+/-.010
.125-44 UNF				
.138-32 UNC	.1695 (#18)	+.005/-.002	.285	+/-.010
.138-40 UNF				
.164-32 UNC	.1935 (#10)	+.005/-.002	.335	+/-.010
.164-36 UNF				
.190-32 UNC	.221 (#2)	+.005/-.002	.390	+/-.010
.190-36 UNF				
.250-20 UNC	.2812 (9/32)	+.005/-.002	.510	+/-.010
.250-28 UNF				
.312-18 UNC	.3438 (11/32)	+.005/-.002	.640	+/-.010
.312-24 UNF				
.375-16 UNC	.4062 (13/21)	+.005/-.002	.765	+/-.010
.375-24 UNF				
.438-14 UNC	.4688 (15/32)	+.005/-.002	.815	+/-.010
.438-20 UNF				
.500-13 UNC	.5312 (17/32)	+.005/-.002	.880	+/-.010
.500-20 UNF				

22.10.4 Counterbore Depths - Pan Head and Socket Head Cap Screws

Table 22-4 C'Bore depths (pan head and socket head)

Fastener Size	Type of Fastener	C'Bore Depths .DDD	C'Bore Depths Tolerance
.060-56 UNF	Pan Head	.080	+/-.010
	Socket head	.100	+/-.010
.086-56 UNC	Pan Head	.120	+/-.010
.086-64 UNF	Socket head	.155	+/-.010
.112-40 UNC	Pan Head	.150	+/-.010
.112-48 UNF	Socket head	.195	+/-.010
.125-40 UNC	Pan Head	.160	+/-.010
.125-44 UNF	Socket head	.210	+/-.010
.138-32 UNC	Pan Head	.170	+/-.010
.138-40 UNF	Socket head	.225	+/-.010
.164-32 UNC	Pan Head	.190	+/-.010
.164-36 UNF	Socket head	.260	+/-.010
.190-32 UNC	Pan Head	.215	+/-.010
.190-36 UNF	Socket head	.295	+/-.010
.250-20 UNC	Pan Head	.290	+/-.010
.250-28 UNF	Socket head	.395	+/-.010
.312-18 UNC	Pan Head	.340	+/-.010
.312-24 UNF	Socket head	.475	+/-.010
.375-16 UNC	Pan Head	.390	+/-.010
.375-24 UNF	Socket head	.550	+/-.010
.438-14 UNC	Pan Head	.440	+/-.010
.438-20 UNF	Socket head	.630	+/-.010
.500-13 UNC	Pan Head	.530	+/-.010
.500-20 UNF	Socket head	.750	+/-.010

22.10.5 Flat Head Screw Head Height - Above and Below the Surface

Table 22-5 Flat head screw head height above and below the surface

Flat Head Screw Head Height Above and Below Surface for 100 Degree Flat Head		
.060-56 UNF	Above Surface	.019
	Below Surface	-.021
.086-56 UNC	Above Surface	.018
.086-64 UNF	Below Surface	-.025
.112-40 UNC	Above Surface	.022
.112-48 UNF	Below Surface	-.023
.125-40 UNC	Above Surface	.020
.125-44 UNF	Below Surface	-.026
.138-32 UNC	Above Surface	.022
.138-40 UNF	Below Surface	-.027
.164-32 UNC	Above Surface	.020
.164-36 UNF	Below Surface	-.031
.190-32 UNC	Above Surface	.022
.190-36 UNF	Below Surface	-.032
.250-20 UNC	Above Surface	.020
.250-28 UNF	Below Surface	-.040
.312-18 UNC	Above Surface	.022
.312-24 UNF	Below Surface	-.040
.375-16 UNC	Above Surface	.020
.375-24 UNF	Below Surface	-.053
.438-14 UNC	Above Surface	.020
.438-20 UNF	Below Surface	-.060
.500-13 UNC	Above Surface	.022
.500-20 UNF	Below Surface	-.064

22.11 References

1. Orberg, Erik, Franklin D. Jones, and Holbrook L. Horton. 1979. *Machinery's Handbook.* 21st ed. New York, NY: Industrial Press, Inc.
2. The American Society of Mechanical Engineers. 1995. *ASME Y14.5M-1994, Dimensioning and Tolerancing.* New York, New York: The American Society of Mechanical Engineers.

Chapter 23

Fixed and Floating Fastener Variation

**Chris Cuba
Raytheon Systems Company
McKinney, Texas**

Chris Cuba began his career at Texas Instruments in 1984 and has been a key contributor on several defense programs as a systems producibility engineer. During this time, he has developed expertise in mechanical assembly producibility, mechanical tolerancing, and product design-to-cost. He was software development manager and applications engineer for the mechanical tolerancing tool CE/TOL. Mr. Cuba currently works as a member of the Mechanical Tolerancing and Performance Sigma Team at Raytheon Systems Company. As a Six Sigma Black Belt, his responsibilities include dimensional management consulting, Six Sigma mechanical tolerance analysis and allocation, and mechanical tolerancing training. He graduated from Oklahoma State University with a bachelor's degree in mechanical design.

23.1 Introduction

This chapter describes an approach to understanding the inherent assembly shift and manufacturing variation contributors within a fastened interface. In most cases, the fastened interface must meet two requirements: The parts must fit together and provide minimal assembly variation, and the variation allowed from the fastened interface should relate to a product performance requirement.

In this chapter, each variable of the fastened interface is broken down to understand its contribution to the total assembly variation.
- First, the chapter shows a worst case tolerance study on features of size that are located using a position feature control frame to understand the virtual and resultant condition boundaries.
- Next, features of size are used in an assembly to understand variation within a fixed and floating fastener.

23.2 Hole Variation

Fig. 23-1 shows an example dimensioned using a position feature control frame to locate the hole. The feature control frame locates the hole using the maximum material condition (MMC) modifier. When using the MMC modifier, tolerance may be added to the location tolerance as the actual feature size departs from MMC. Thus, the feature's size tolerance and location tolerance are dependent. This dependency must be taken into account in the tolerance study.

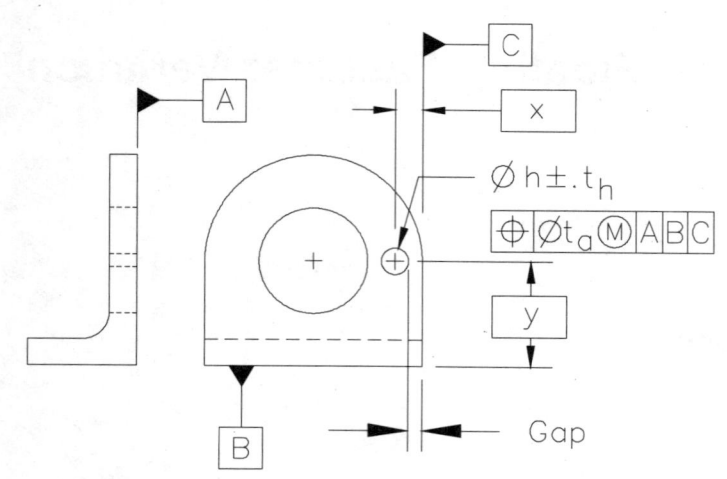

Figure 23-1 Feature located using positional tolerance at MMC

To analyze the tolerance, first calculate the worst case boundaries generated by the size and location tolerances of the hole. These boundaries define the virtual and resultant conditions of the hole.

The virtual condition is "a constant boundary generated by the collective effects of a size feature's specified MMC or LMC material condition and the geometric tolerance for that material condition" (Reference 2).

Virtual Condition Hole = Feature MMC Size − Position Tolerance at MMC

For the example in Fig. 23-1, the virtual condition of the hole *(VCH)* is:

$$VCH = h - t_h - t_a$$

where
hole feature MMC size $= h - t_h$
position tolerance at MMC $= t_a$

The resultant condition is "the variable boundary generated by the collective effects of a size feature's specified MMC or LMC material condition, the geometric tolerance for that material condition, the size tolerance, and the additional geometric tolerance derived from the feature's departure from its specified material condition" (Reference 2).

Resultant Condition Hole = Feature LMC Size + Position Tolerance at LMC

For the example in Fig. 23-1, the resultant condition of the hole *(RCH)* is:

$$RCH = h + t_h + 2\,t_h + t_a$$
$$RCH = h + 3\,t_h + t_a$$

where

hole feature LMC size = $h + t_h$
position tolerance at LMC = $2\,t_h + t_a$

To calculate the gap, the inner and outer boundaries (virtual and resultant condition) of the feature are converted to a radial value, with an equal bilateral tolerance ($r +/-t$). See Chapter 9.

$$r = (VCH + RCH)/4$$
$$r = [(h - t_h - t_a) + (h + 3\,t_h + t_a)]/4$$
$$r = (h + t_h)/2$$

where

r = mean radial conversion of virtual and resultant condition boundaries

and

$$t = (RCH - VCH)/4$$
$$t = [(h + 3\,t_h + t_a) - (h - t_h - t_a)]/4$$
$$t = t_h + .5\,t_a$$

where

t = equal bilateral tolerance of r

The radial value used in the dimension loop diagram is:
$r +/- t$

Substituting into these equations, we get:
$r = .5(h + t_h) +/- (t_h + .5\,t_a)$
which equals:
LMC/2 +/- (size tolerance + 1/2 feature control frame tolerance)

Fig. 23-2 shows the dimension loop diagram for the gap in Fig. 23-1.

The gap equation equals: $Gap = [x - .5(h + t_h)] +/- (t_h + .5\,t_a)$

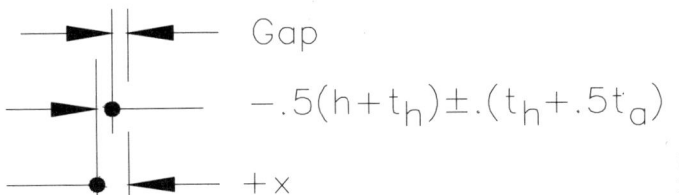

Gap
$-.5(h + t_h) \pm (t_h + .5 t_a)$
$+x$

Figure 23-2 Dimension loop diagram for Fig. 23-1

23.3 Assembly Variation

The previous discussion developed an understanding for an individual feature's boundaries. These boundaries define the amount of assembly shift within a fastened interface. Assembly shift results from the amount of allowance defined between the fastener and clearance hole. Many engineers design the allowance amount using the fixed or floating fastener rules within ASME Y 14.5 (Reference 2). Zero assembly shift occurs when a virtual condition pin assembles into a virtual condition hole. Maximum assembly shift occurs when the pin and hole have perfect form and orientation at LMC size.

In most cases, the fastened interface must meet two requirements: The parts must fit together and provide minimal assembly variation. The assembly variation within the pin/hole interface can be analyzed several ways.

- The mating parts can be shifted until touching provides a maximum and minimum assembly variation. (See Figs. 23-3 and 23-4.)
- The assembly variation can be represented by a process capability. This could be in the form of a uniform, normal, or other known distribution.
- Tooling, fixtures, or gravity can be used to minimize or eliminate assembly variation.

This chapter looks at shifting the mating parts to understand the maximum and minimum assembly variation.

23.4 Fixed and Floating Fasteners

There are two types of fastening systems used to assemble parts: fixed fasteners and floating fasteners. Fig. 23-3 illustrates a fixed fastener. This is defined as a fastener where one of the parts has restrained fasteners such as screws in tapped holes or studs (Reference 2). A floating fastener is defined as a fastener where two or more parts are assembled with fasteners such as bolts and nuts, and all parts have clearance holes for the bolts (Reference 2). See Chapter 22 for more discussion on fixed and floating fasteners.

The assembly variation within a fixed fastener occurs when one part shifts as shown in Fig. 23-3. The floating fastener assembly variation has two parts shifting that contribute to the variation as shown in Fig. 23-4.

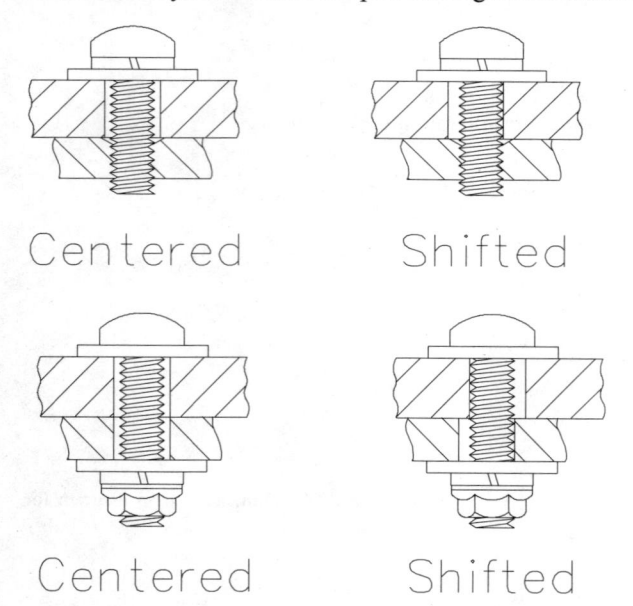

Figure 23-3 Fixed fastener centered and shifted

Figure 23-4 Floating fastener centered and shifted

23.4.1 Fixed Fastener Assembly Shift

Fig. 23-5 shows a fixed fastener within an assembly and uses the following notation to develop equations for assembly shift, minimum gap, and maximum gap. The minimum and maximum gaps between datum surfaces E and B occur when the locating features are at least material condition and using their maximum location tolerance. The following summarizes these conditions.

Feature	LMC Size	Location Tolerance at LMC
Hole	$h + t_h$	$t_a + 2t_h$
Pin	$p - t_p$	$t_b + 2t_p$

where
- p = Pin mean size
- t_p = Equal bilateral pin size tolerance
- t_a = Cylindrical tolerance zone diameter (hole)
- h = Hole mean size
- t_h = Equal bilateral hole size tolerance
- t_b = Cylindrical tolerance zone diameter (pin)

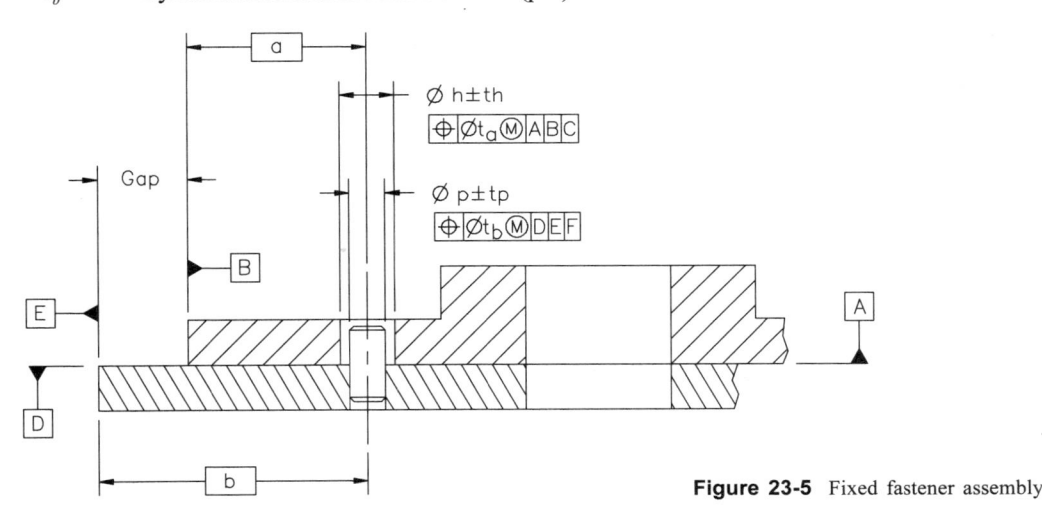

Figure 23-5 Fixed fastener assembly

Shifting the parts to a maximum and minimum shows the worst case gap for each condition. Conventionally, we draw a dimension loop diagram for each condition. Fig. 23-6 shows the two parts shifted for a minimum assembly gap and the resultant dimension loop diagram.

Minimum Gap = Nominal Gap - Tolerance
Minimum $Gap = [b + .5(p - t_p) - .5(h + t_h) - a] - [(.5t_a + t_h) + (.5t_b + t_p)]$

which simplifies to:

$$\text{Minimum } Gap = (b - a) - .5(h - p) - .5(t_a + t_b) + 1.5(t_h + t_p) \tag{23.1}$$

Note that Eq. (23.1) gives the minimum gap if the parts touch as shown in Fig. 25-6. Since the minimum gap occurs when the pin and hole are both at LMC, the parts may be manually shifted to increase this gap. The amount the parts can shift is $(h + t_h) - (p - t_p)$.

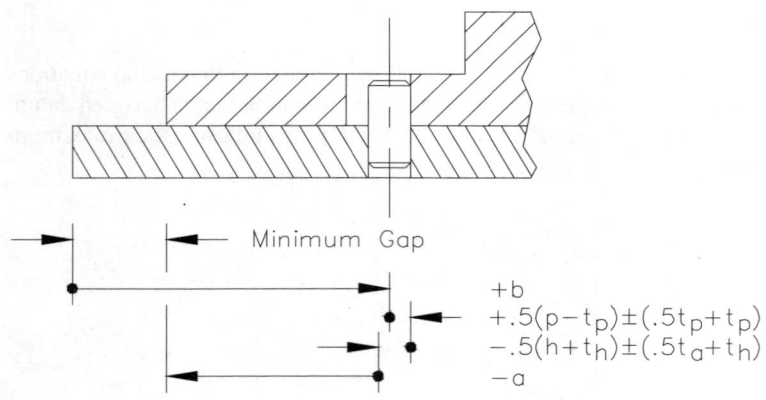

Figure 23-6 Fixed fastener minimum assembly gap

Fig. 23-7 shows the two parts shifted to a maximum assembly gap and the resultant dimension loop diagram.

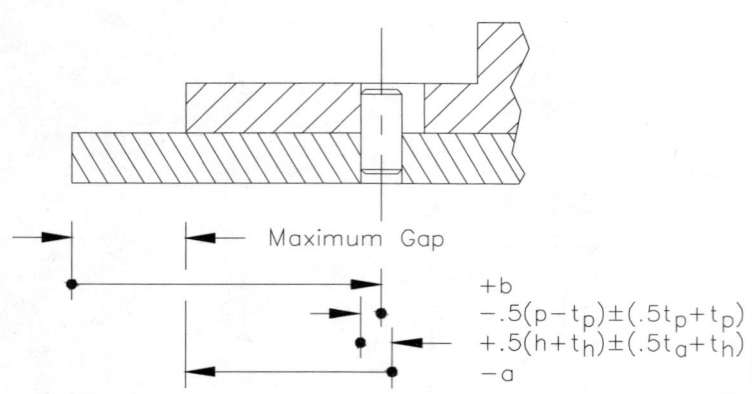

Figure 23-7 Fixed fastener maximum assembly gap

Maximum Gap = Nominal Gap + Tolerance
Maximum $Gap = [b - .5(p - t_p) + .5(h + t_h) - a] + [(.5t_a + t_h) + (.5t_b + t_p)]$

which simplifies to:

Maximum Gap $= (b - a) + .5(h - p) + 1.5(t_h + t_p) + .5(t_a + t_b)$ (23.2)

Note that Eq. (23.2) gives the maximum gap if the parts touch as shown in Fig. 23-7. Since the maximum gap occurs when the pin and hole are both at LMC, the parts may be manually shifted to decrease this gap. The amount the parts can shift is $(h + t_h) - (p - t_p)$.

23.4.2 Fixed Fastener Assembly Shift Using One Equation and Dimension Loop

The following discussion describes an alternative method of defining two dimension loop diagrams and equations for the assembly variation at the gap. This method defines one equation for the total variation at the gap.

A radial plus and minus value can express the assembly shift in the fixed fastener example. This value is the maximum diametrical amount of clearance between the fixed pin or fastener, and the clearance hole divided by two. As the mating features depart from their respective virtual conditions, the assembly shift increases. The maximum assembly shift occurs when the pin and hole have perfect form and orientation at Least Material Condition (LMC).

From Fig. 23-5, the fixed fastener LMC assembly shift (AS_{fix}) is:

$$AS_{fix} = .5[h + t_h - (p - t_p)] \qquad (23.3)$$

where

$h + t_h$ = Clearance hole LMC size
$p - t_p$ = Pin (fastener) LMC size

23.4.3 Fixed Fastener Equation

As previously stated, the most variation within a fastened interface occurs when the mating features are at LMC. This allows additional (bonus) tolerance to accumulate. From the fixed fastener example in Fig. 23-5 the additional (bonus) tolerance contributors are:

$2(t_h)$ = Clearance hole size tolerance
$2(t_p)$ = Total pin (fastener) size tolerance

Other contributors in the tolerance study are location tolerances for each feature. From Fig. 23-5, the location tolerance contributors are:

t_a = Cylindrical tolerance zone for the clearance hole
t_b = Cylindrical tolerance zone for the pin

The total tolerance variation *(tv)* at the gap is:

$$tv = 2t_h + 2t_p + t_a + t_b$$

The +/- or radial tolerance variation *(rtv)* at the gap is:

$rtv = tv/2$
$$rtv = t_h + t_p + .5t_a + .5t_b \qquad (23.4)$$

Combining Eqs. (23.3) and (23.4) gives the gap variation *(gv)* with assembly shift included.

$gv = AS_{fix} + rtv$
$gv = .5[h + t_h - (p - t_p)] + t_h + t_p + .5t_a + .5t_b$

This reduces to:

$$gv = .5(h - p) + .5(t_a + t_b) + 1.5(t_h + t_p) \qquad (23.5)$$

23.4.4 Fixed Fastener Gap Analysis Steps

Using Eq. (23.5), only one dimension loop diagram is needed to understand the minimum and maximum assembly gap. The diagram identifies the mean assembly dimension and Eq. (23.5) gives the variation from the mean.

First, construct the dimension loop diagram. The dimension loop diagram rules do not change when a fastener becomes part of the stackup. The diagram is drawn the same, except a vector is drawn to and from the centerline of the fastened interface, continuing until the right hand side of the gap is reached. The diagram *does not* trace the pin and hole as if one part was shifted relative to the other.

23-8   Chapter Twenty-three

The dimension loop diagram for Fig. 23-5 is shown in Fig. 23-8.

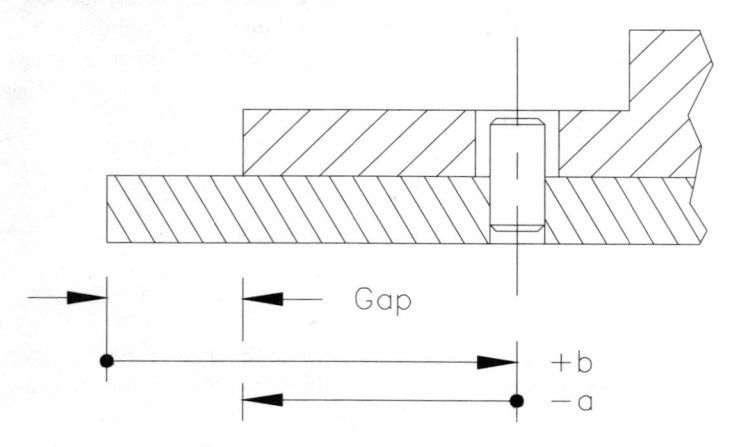

Figure 23-8 Centered fixed fastener dimension loop diagram

The Gap equation is: $Gap = (b - a) +/- gv$
This equals: $Gap = (b - a) +/- .5(t_a + t_b) + .5(h - p) + 1.5(t_h + t_p)$
This gives the same minimum and maximum gap in Eqs. (23.1) and (23.2).

23.4.5 Floating Fastener Gap Analysis Steps

We can construct the floating fastener dimension loop diagram in the same manner as the fixed fastener example. In the floating fastener application (Fig. 23-9), the assembly shift calculation uses the two clearance holes and fastener. In this case, the fastener shifts within both clearance holes.

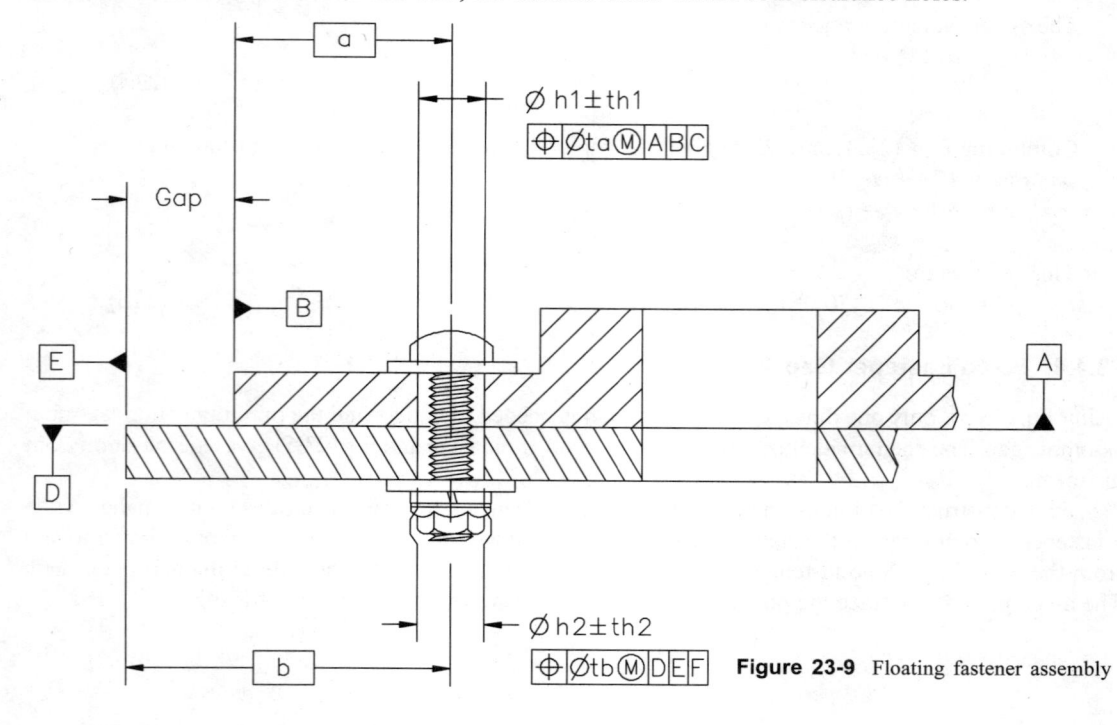

Figure 23-9 Floating fastener assembly

As previously stated, the most variation within a fastened interface occurs when the mating features are at LMC. This allows additional (bonus) tolerance to accumulate. From Fig. 23-9 the equation for assembly shift at LMC is:

$$AS_{float} = .5(h_1 + t_{h1} + h_2 + t_{h2}) - (p - t_p) \qquad (23.6)$$

where
- h_1 = Mean clearance hole 1 size
- t_{h1} = Equal bilateral clearance hole 1 size tolerance
- h_2 = Mean clearance hole 2 size
- t_{h2} = Equal bilateral clearance hole 2 size tolerance
- p = Mean pin (fastener) size
- t_p = Equal bilateral pin (fastener) size tolerance

From Fig. 23-9, the additional (bonus) tolerance contributors are:
- $2(t_{h1})$ = Clearance hole 1 size tolerance
- $2(t_{h2})$ = Clearance hole 2 size tolerance

Other contributors in the tolerance study are location tolerances for each feature. The location tolerance contributors are:
- t_a = Cylindrical tolerance zone for clearance hole 1
- t_b = Cylindrical tolerance zone for clearance hole 2

The total tolerance variation (tv) at the gap is:
$$tv = 2t_{h1} + 2t_{h2} + t_a + t_b$$

The +/- or radial tolerance variation (tvr) at the gap is:
$$tvr = tv/2$$
$$tvr = t_{h1} + t_{h2} + .5t_a + .5t_b \qquad (23.7)$$

Combining Eqs. (23.6) and (23.7) gives the gap variation (gv) with assembly shift included.
$$gv_{float} = AS_{float} + tv$$
$$gv_{float} = (h_1 + t_{h1} + h_2 + t_{h2})/2 - (p - t_p) + t_{h1} + t_{h2} + .5t_a + .5t_b$$

This reduces to:
$$gv_{float} = .5(t_a + t_b) + 1.5(t_{h1} + t_{h2}) + .5(h_1 + h_2) - (p - t_p)$$

The gap equation is:
$$Gap = (b - a) +/- gv_{float}$$
$$Gap = (b - a) +/- .5(t_a + t_b) + 1.5(t_{h1} + t_{h2}) + .5(h_1 + h_2) - (p - t_p)$$

23.5 Summary

This chapter demonstrates a process to perform worst case tolerance analysis on fixed and floating fasteners. The methodology described extends the conventional tolerance analysis methodology by introducing the concepts of virtual and resultant condition. This methodology can be used on any feature having dependent size and location tolerance. The concepts are further used to develop one equation to find minimum and maximum assembly conditions by understanding assembly shift within a fastened interface. Maximum assembly shift occurs when both features of the fastened interface are at least material condition. Although the fixed and floating fastener rules ensure a worst case fit, they also allow a part position to float when worst case conditions are not present. Many designs minimize or eliminate assembly shift by using tooling or assembly instruction to "shift out" the variation.

23.6 References

1. Cuba, Chris and Paul Drake. 1992. Mechanical Tolerance Analysis of Fixed and Floating Fasteners. *Texas Instruments Technical Journal.* Nov-Dec: 58-65.
2. The American Society of Mechanical Engineers. 1995. *ASME Y14.5M-1994, Dimensioning and Tolerancing.* New York, New York: The American Society of Mechanical Engineers.

Chapter 24

Pinned Interfaces

Stephen Harry Werst
Raytheon Systems Company
Dallas, Texas

Mr. Werst has worked as a mechanical engineer with Texas Instruments Defense Systems and Electronics Group, now part of Raytheon Systems Company, for more than six years. Most of his work has involved the design of electro-optic infrared imaging systems and more recently, GD&T training. Mr. Werst attended the University of Texas at Arlington, graduating summa cum laude with a bachelor's degree in mechanical engineering. He recently earned his master's degree in engineering management from Southern Methodist University. Prior to his work with Raytheon, he worked as a computer programmer for Martin Marietta Energy Systems.

24.1 List of Symbols (Definitions and Terminology)

α	Rotation error between parts
β	Installation angle of modified alignment pins
δ	Translation error between parts
σ	Standard deviation
$\varnothing_h$ or $\varnothing_{h1}$ and $\varnothing_{h2}$	The diameter of the clearance hole(s) or the diameters of the first and second clearance holes
$\varnothing_p$ or $\varnothing_{p1}$ and $\varnothing_{p2}$	The diameter of the alignment pin(s) or the diameters of the first and second alignment pins
c	A measure of the clearance between alignment features in an assembly
c_{nom}	The clearance calculated using the nominal dimensions
cte	Coefficient of thermal expansion

d_e	Distance between the center of the clearance hole and the edge of the part
d_h	Distance between the centers of the clearance holes
d_{hs}	Distance between the center of the hole and the center of the slot
d_p	Distance between the centers of the alignment pins
d_{px}	Distance between the centers of the alignment pins in the x direction
d_{py}	Distance between the centers of the alignment pins in the y direction
l_{slot}	Length of slot
$perp_h$ or $perp_{h1}$ and $perp_{h2}$	Perpendicularity of the clearance hole(s) or the perpendicularity of the first and second clearance holes
$perp_p$ or $perp_{p1}$ and $perp_{p2}$	Perpendicularity of the alignment pin(s) or the perpendicularity of the first and second alignment pins
t	The distance from the center of the pin to the flat
w_{slot}	Width of slot

24.2 Introduction

The use of pins is the most common way of precisely controlling the alignment of mating parts. Even children's inexpensive plastic models use pins molded into the plastic to help maintain the alignment of the glued sections. There may be, however, as many different methods for dimensioning pinned interfaces as there are designs that use them. This section includes five of the more common design configurations using straight pins. Fit, rotation, and translation performance criteria along with Six Sigma dimensioning methodologies will be included for each configuration and manufacturing process. The reader can use this information to compare the differences in performance between the available options and choose the most appropriate one.

Ultimately, the goal of this section is to provide a common methodology for selecting and dimensioning a pin configuration. If implemented successfully, engineers with the same knowledge about the available pins and manufacturing processes will design similar assemblies identically. This standardization results in lower costs in several areas of the business. Although this section only presents alignment pins pressed into interference holes, these principles can be extended to other applications.

Before considering the method of aligning parts, the engineer must understand the requirements. Often the requirements handed down from customers are vague, so best estimates of actual requirements are needed. When making these estimates, keep in mind that "as good as we can do" too often is synonymous with "as expensive as we can make it." The goal of the design process should be to deliver a product to customers that meets their expectations at the lowest possible cost to the company. Since this chapter deals with making tradeoffs between performance and relative costs, it would be useless to the engineer wishing to design a product with only the best performance in mind.

24.3 Performance Considerations

Alignment pins typically have three performance requirements:
- The parts must fit together.
- The pins should minimize the permissible translation between the two parts.
- The pins should maintain orientation between the two parts.

These three performance criteria will be evaluated for each design configuration.

Unfortunately, the engineer must make a tradeoff between the first and the last two performance requirements, as they are mutually exclusive. In order to ensure the parts fit together, the clearance holes must have sufficient clearance to compensate for the positional variation of the pins and the holes. However, clearance between the holes and the pins degrades the ability to align the parts to one another. We must therefore balance the ability to assemble the parts against the alignment between the parts after assembly.

For all the design configurations in this section, the calculation of the rotational error at each interface can be simplified to take the form of the following equation:

$$\alpha = \frac{constant}{d_p} \tag{24.1}$$

where *constant* is a function of the design configuration and manufacturing processes. In all but one design configuration, d_p refers to the total distance between the two pins. In the case of two pins with one hole and edge contact, only the distance between the two pins in a direction parallel to the edge contact of the second pin is important. See Fig. 24-10.

Eq. (24.1) enables the development of tables of constants for design types, allocation methods, and manufacturing processes. Tables 24-3, 24-5, 24-7, 24-9, and 24-11 present these constants for various design types. These tables also include constants for translation.

24.4 Variation Components of Pinned Interfaces

Alignment pins contribute to the assembly performance variation in two ways:
- The movement due to clearances between the parts (interface error).
- The ability to locate the pins/clearance holes with respect to a datum reference frame (positional error).

If the design involves only two parts with critical placement requirements, we can eliminate the second source of variation by using the pins/clearance holes as the datums for the parts. However, many times we have three or more parts that must retain alignments with respect to each other and cannot avoid the error of positioning the pins relative to another set of pins.

24.4.1 Type I Error

Fig. 24-1 shows two similar designs for maintaining the alignment of the slots in parts 1 and 2. In the figure on the left, part 1 uses the clearance holes as the datum reference frame (DRF), and part 2 has the pins as the DRF. In addition to the variations of locating and orienting the slots to their respective DRFs, the alignment pin interface adds error caused by the clearance between the holes and the pins. Since this clearance is necessary for assembly, it cannot be eliminated. However, the pins and holes are the datums, so the design does not have additional variation of locating the holes and pins with respect to another DRF. This type of design has only Type I error consisting entirely of clearance between the alignment features.

24.4.2 Type II Error

The design on the right side of Fig. 24-1 adds two parts, 3 and 4, to provide additional separation between parts 1 and 2. Again, alignment between the slots in parts 1 and 2 is critical, so the two additional parts use alignment pins. Parts 2, 3, and 4 use the pins as their DRFs, and part 1 still uses the clearance holes as the DRF. As in the design on the left of the figure, the error between the slots includes the error of locating and orienting the slots to the DRFs of parts 1 and 2 and the errors induced by the tolerance stackup of alignment pins.

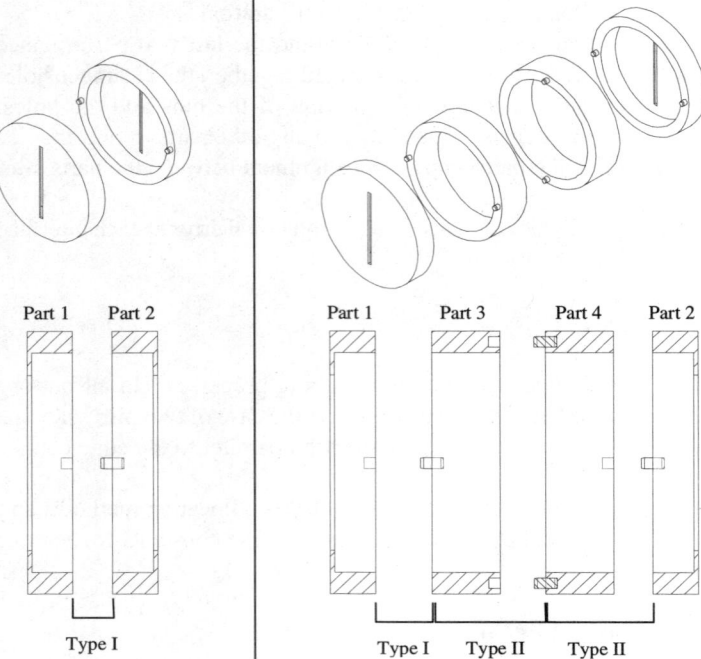

Figure 24-1 Examples of design cases for alignment pins showing Type I and Type II errors

In this design, however, there are three alignment pin interfaces. The interface between parts 1 and 3 is identical to the single interface in the design on the left. Therefore, the error between parts 1 and 3 is Type I error. Though the interface between parts 3 and 4 appears to be the same as between parts 1 and 2, there is an additional contributor because the clearance holes on part 3 are not the datums. To determine the error between the DRF of part 3 and the DRF of part 4, we must include both the error at the pin interface due to clearance (similar to Type I error) and the error associated with locating the clearance holes of part 3 with respect to the pins of part 3. This combined error is called Type II error.

Most designs will have one Type I error and a Type II error component for each additional part beyond the initial two. It is possible to conceive of designs that don't follow this rule, but they are not as efficient at minimizing the total alignment variation between critical features. The engineer should therefore strive to follow this tolerancing methodology when using alignment pins.

24.5 Types of Alignment Pins

All the designs considered in this section use two pins to align mating parts. Before we can establish a set of common design characteristics for the different configurations of alignment pins, we must first determine the sets of pins to be used. For this book, we will use .0002" oversized pins defined in ANSI B18.8.2-1978, R1989 for the round pins as shown in Table 24-1.

In addition to the standard ANSI pins, some design configurations use one modified pin with one round pin to improve performance. These designs do, however, increase the cost. The purchased round pin must be modified *and* carried as a separate part in a company's inventory. Depending upon the size of the company using the part, the administrative costs of carrying an extra part can be significantly greater than the costs associated with creating the modified pin. The engineer must therefore make sure that the gain in performance is worth the additional cost of creating a new part.

Table 24-1 Alignment pins per ANSI B18.8.2-1978, R1989

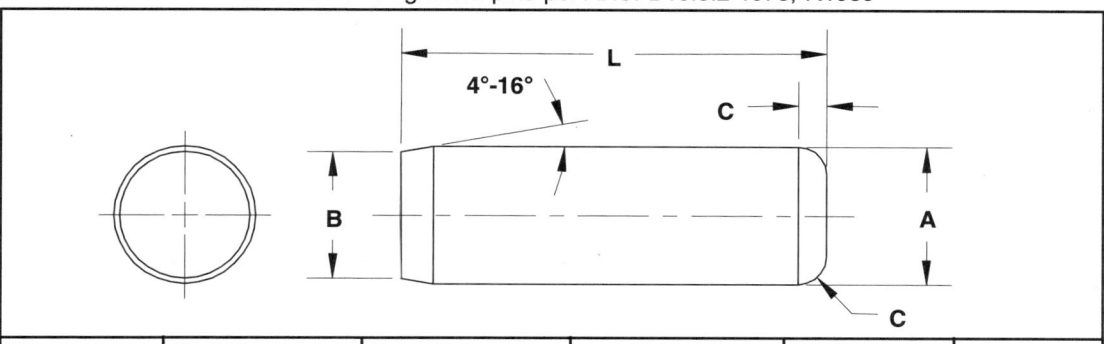

Nominal Size or Nominal Pin Diameter	Pin Diameter, A Nom (PPPP)	Pin Diameter, A Tol	Point Diameter, B Nom	Point Diameter, B Tol	Crown Height or Radius, C Nom	Crown Height or Radius, C Tol	Common Lengths	Double Shear Load, Min, lbf for Carbon or Alloy Steel
1/16	.0625	.0627		.053		.014	.006	$3/16 - 3/4$	800
3/32	.0938	.094		.084		.0215	.0095	$3/16 - 1$	1800
1/8	.1250	.1252		.115	.005	.0285	.0125	$3/8 - 2$	3200
3/16	.1875	.1877		.175		.0425	.0195	$1/2 - 2$	7200
1/4	.2500	.2502		.235		.057	.026	$1/2 - 2\,1/2$	12800
5/16	.3125	.3127		.296	.006	.0715	.0325	$1/2 - 2\,1/2$	20000
3/8	.3750	.3752	±.0001	.358		.086	.039	$1/2 - 3$	28700
7/16	.4375	.4377		.417		.1005	.0455	$7/8 - 3$	39100
1/2	.5000	.5002		.479	.008	.115	.052	$3/4 - 4$	51000
5/8	.6250	.6252		.603		.143	.065	$1\,1/4 - 5$	79800
3/4	.7500	.7502		.725		.172	.078	$1\,1/2 - 6$	114000
7/8	.8750	.8752		.850	.010	.201	.092	$2 - 6$	156000
1	1.0000	1.0002		.970		.229	.104	$2 - 6$	204000

Another factor that may increase cost (if not performed properly) is pin installation. Modified pins must be aligned correctly to provide a benefit. Proper installation means having the center of the cutaway side(s) in line with the plane passing through the centers of the two pins. If the pins are installed correctly, the sides that are cut away provide additional clearance in one direction that can accommodate the variation in the distance between the pin and hole centers. This additional allowance allows the nominal size of the clearance holes to be reduced, thus reducing the translation and rotation errors through the interface.

The pins' improvement diminishes as the installation angle varies. Since pin installation is a manual operation, all analyses for these types of pins assume that the pin is installed 10° from the ideal installation angle.

24-6 Chapter Twenty-four

Two configurations for the modified pin will be discussed—a diamond pin and a parallel-flats pin. Fig. 24-2 shows the typical cross-section of each pin. Both of them are fabricated by modifying the pins from Table 24-1—usually by grinding the flats.

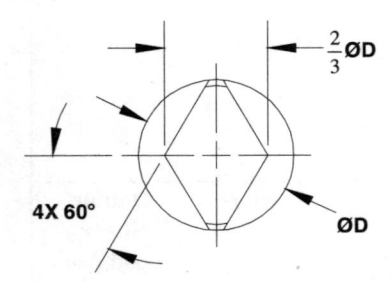

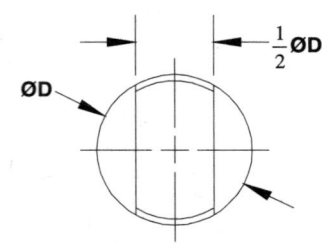

Figure 24-2 Two common cross-sections for modified pins

24.6 Tolerance Allocation Methods—Worst Case vs. Statistical

As mentioned in previous chapters, there are many ways to analyze (or allocate) the effect of tolerances in an assembly. The most common and simple method is to assume that each dimension of interest is at its acceptable extreme and to analyze the combined effects of these "worst-case" dimensions. This methodology is very conservative, however, because the probability of all dimensions being at their limit simultaneously is extremely small.

An approach that better estimates the performance of the parts is to assume the dimensions are statistically distributed from part to part. The analysis involves assuming a distribution, usually normal, for each of the dimensions and determining the combined effects of the individual distributions on the assembly performance specifications. All of the statistical tolerances in this section have Six Sigma producibility (based on the process capabilities in section 24.7), and all of the statistical performance numbers have Six Sigma performance. In other words, 3.4 out of every million parts will have features within the indicated tolerances, and the same percentage of assemblies will fit and will meet the translation and rotation performance listed. (See Chapters 10 and 11 for further discussion of Six Sigma performance.)

Tables 24-4, 24-6, 24-8, 24-10, and 24-12 use the ⟨ST⟩ symbol for all tolerances that result from statistical allocations. The engineer may want to use the following note on drawings containing the ⟨ST⟩ symbol:

- Tolerances identified statistically ⟨ST⟩ shall be produced by a process with a minimum Cpk of 1.5.

If the anticipated manufacturing facilities do not have methods to implement statistical tolerances, the engineer may opt to remove the ⟨ST⟩ symbol. Without the symbol, though, the engineer assumes the responsibility of the design not performing as expected. (Refer to Chapter 11 for further discussions regarding the ⟨ST⟩ symbol.)

24.7 Processes and Capabilities

This section will evaluate the differences between three different methods of generating the holes for alignment pins. These processes are:
- Drilling and reaming the alignment holes with the aid of drill bushings.
- Boring the holes on a numerically controlled (N/C) mill.
- Boring the holes on a Jig Bore.

Though there are other methods of generating holes, these are the more common ones with readily available capability information. The principles developed in this chapter can be extended to other manufacturing processes.

In the absence of general quantitative information about the capabilities of various machining processes, we must estimate an average capability. Though few sources provide true statistical information regarding these processes, we can make some assumptions based on recommended tolerances and historical quality levels. One such source of information is Bralla's *Handbook of Product Design for Manufacturing* (Reference 1). In it, the author provides many recommended tolerances for a range of manufacturing processes.

First, we will assume that the variation of the processes included in this section is normally distributed. Since historical estimates of acceptable producibility have been based on tolerances at three standard deviations from the mean, we will make this same assumption about the recommended manufacturing tolerances in Bralla's handbook. However, as discussed previously, Six Sigma analyses typically use short-term standard deviations, but these tolerances are more likely to be based on long-term effects. Therefore, it is reasonable to assume these tolerances represent four sigma, short-term capabilities. Table 24-2 presents the standard deviations used for all analyses in this section.

Table 24-2 Standard deviations for common manufacturing processes (inches)

		Drill and Ream with Bushings	N/C Boring	Jig Bore
Hole Diameter		.00025	.00025	.00013
Hole/Pin Perpendicularity		.00016	.00013	.00006
± Distance From Target Position	**From Part Surface**	.00250	.00200	.00100
	From Another Hole	.00063	.00050	.00025

An additional assumption concerning the perpendicularity of a hole relative to the surface into which it is placed is necessary for these analyses. Because Bralla doesn't include a standard deviation for perpendicularity, we will assume that the variation due to perpendicularity error is one-fourth of the total variation of the true position of a hole relative to another hole.

24.8 Design Methodology

Fig. 24-3 shows a flowchart for the design process using alignment pins. The following paragraphs explain the steps in more detail:

1. Select a pin size from Table 24-1. The decision on which pin to use will be driven by the geometry and mass of the mating parts or subassemblies. The ability to assemble and align the mating components is not a function of pin size or length, so this decision should be made without regard to these parameters. Keep in mind that for alignment purposes the pin need only protrude above the mating surface far enough to engage the clearance holes completely. Any additional length will only make assembly more difficult.

2. Once you have chosen the pin diameters, determine the maximum distance between all sets of pins. The least expensive design alternative that an engineer can choose to have the most significant improvement on the alignment performance of pinned interfaces is to move the pins as far apart as possible. Keep in mind that the walls around the pinholes, especially the interference holes, should have sufficient thickness to hold the pin and prevent part deformation, as this will affect alignment.

24-8 Chapter Twenty-four

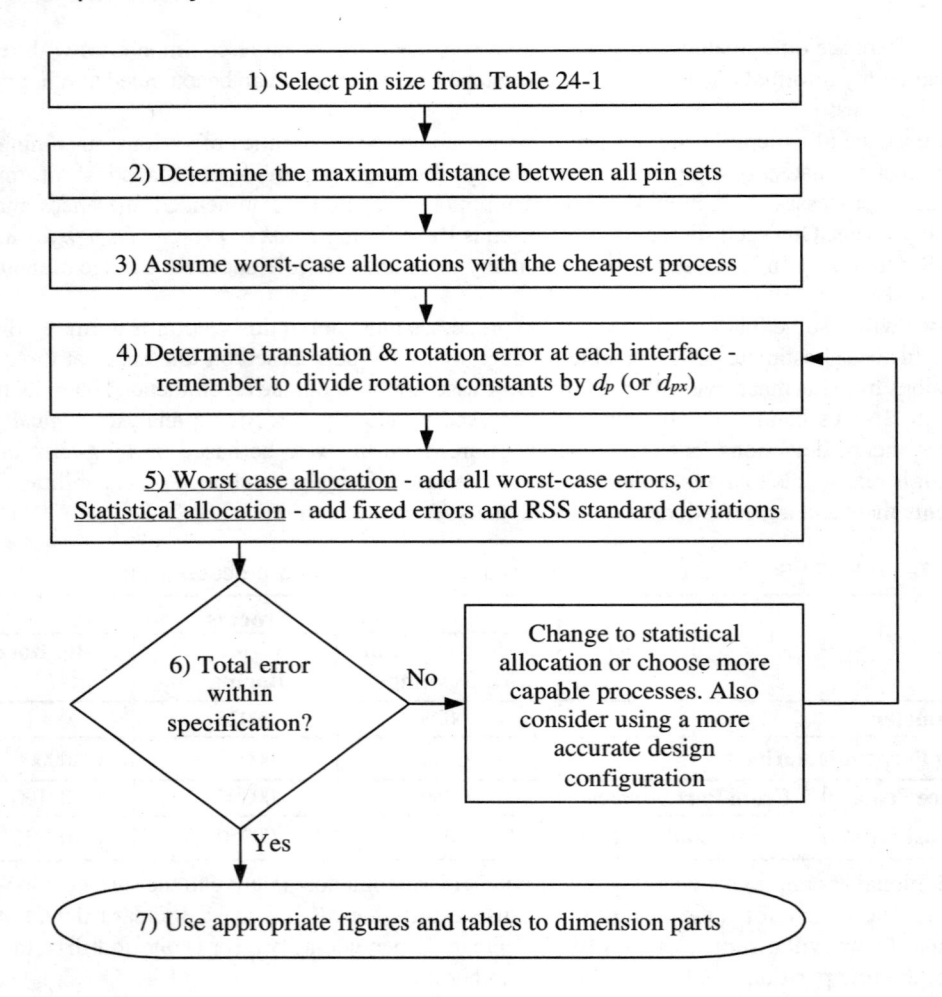

Figure 24-3 Design process for using alignment data

```
1) Select pin size from Table 24-1
          ↓
2) Determine the maximum distance between all pin sets
          ↓
3) Assume worst-case allocations with the cheapest process
          ↓
4) Determine translation & rotation error at each interface -
   remember to divide rotation constants by d_p (or d_px)
          ↓
5) Worst case allocation - add all worst-case errors, or
   Statistical allocation - add fixed errors and RSS standard deviations
          ↓
6) Total error within specification?
   No → Change to statistical allocation or choose more capable processes. Also consider using a more accurate design configuration → (back to 4)
   Yes ↓
7) Use appropriate figures and tables to dimension parts
```

3. Start with worst-case tolerance allocation with the least expensive process – usually drilling and reaming with the aid of drill bushings.[1]

4. Determine the translation and rotation errors at each interface from the tables in this section. There are a few important things to remember:

   - Most assembly stackups will have one Type I error and an additional Type II error for each part beyond two.
   - The rotation constants must be divided by d_p (d_{px} for two pins with one hole and edge contact) to determine the angular error occurring at the interface.

5. If performing a worst-case allocation, add all of the translation errors and rotation errors for each interface to determine the total errors occurring through the assembly. Also add to this the translation and rotation errors of the features of interest with respect to their datum reference frames. For example,

---

[1] There may be cases where drilling/reaming is not the least expensive method. If relatively few parts will be made over the life of the project or if drill fixtures are overly expensive, N/C milling may be a cheaper alternative. Communication with the manufacturing shops is essential in order to make wise tradeoffs between cost and function.

if performing an analysis on the slots in the design shown in Fig. 24-1, we would need to include the variations of the two slots relative to their respective DRFs of parts 1 and 2.

If performing a statistical allocation, the translation and rotation at each interface is comprised of two components – the fixed error associated with the nominal clearance between the hole and the pins and the standard deviation resulting from variation in the hole diameters. For statistical evaluation, the engineer should add each of the fixed error terms and then apply the assembly standard deviation to determine assembly performance. The assembly standard deviation is the root of the sum of the squares (RSS) of the standard deviations at each interface, as shown in the following equation:

$$\sigma_{assy} = \sqrt{\sigma_1^2 + \sigma_2^2 + ... + \sigma_n^2}$$

Once you determine the assembly standard deviation, multiply it by six and add it to the fixed portion of the assembly variation to determine the Six Sigma translations and rotations for the assembly.

6. Now compare the predicted performance numbers with the specifications. If the predictions meet or exceed the requirements, continue to Step 7. If the rotation performance is unacceptable, you must select either another allocation methodology, another manufacturing process, or type of design at the interfaces. If performing a worst-case analysis, change to a statistical allocation with the same manufacturing processes and go back to Step 4. If performing a statistical allocation, select a more capable process with a worst-case allocation and go back to Step 4. Finally, you can always select a more precise design configuration and go back to Step 4. The point of this iterative process is to start with the least expensive of all options and only add additional cost to gain performance as necessary.

If the rotation performance is acceptable but the translation is not, an additional option to reduce the translation error is to use two different clearance hole diameters. This method can only be applied to interfaces using two holes. If the engineer reduces the first clearance hole nominal diameter (the one for the round pin in interfaces with diamond or parallel-flats pins) and increases the second by the same amount, translation error decreases by one-half of the amount the hole diameter is reduced.

For worst-case allocations, the lower tolerances (tolerance in the negative direction) also have to change by the same amount as the nominal diameter. For example, if you decrease the first hole nominal diameter by .001, you must also:

- Increase the second hole nominal diameter by .001.
- Decrease the lower tolerance of the first hole by .001 (i.e., -.008 to -.007).
- Increase the lower tolerance of the second hole by .001 (i.e., -.008 to -.009).

For statistical allocations, the tolerances should not change. However, the engineer may wish to add an additional feature control frame controlling the perpendicularity of the first clearance hole relative to the mating surface as shown in statistical Callout B for the configuration with the slot. See Fig. 24-9 and Table 24-6.

Regardless of the tolerance allocation methodology, the smaller hole should never be smaller than the clearance holes specified for the configurations involving a slot or edge contact. The parts will still fit together and have the same rotational error as before the modification. Keep in mind, however, that the center of rotation will no longer be the midpoint between the two pins, but will move toward the smaller pinhole interface in proportion to the amount of the hole diameter reduction.

7. Upon determining a combination of design configurations, manufacturing processes, and allocation methods that meet the specifications, use the figures and tables to apply geometric tolerances to your drawings. The nominal clearance hole diameter is found by adding the constant in the GD&T tables to the pin diameter being used. This is represented in the tables as {.PPPP + *constant*}, where *constant* represents the nominal clearance between the hole and the pin. (See Tables 24-4, 24-6, 24-8, 24-10, and 24-12.)

All figures and most of the callouts in the tables assume Type I interfaces. For Type II interfaces, add the additional callout shown in the tables between the hole/pin diameter specification and the feature control frame(s) beneath it.

For example, if dimensioning a clearance hole that is located with respect to a set of pins on a part in a Type II two pin with one hole and edge contact interface, you should use the following callout:

$$\varnothing.1280 \begin{matrix} +.0015 \\ -.0018 \end{matrix}$$

⊕	⌀.0064 Ⓛ	A	B Ⓛ	C Ⓛ
⊥	⌀.0000 Ⓜ	D		

In this case, the pins used in the DRF for the part are datums B and C. The clearance hole is for a Ø.1252 pin in the mating part. The part that engages this hole mates against a surface defined as datum D. The first feature control frame controls the position of the clearance holes with respect to the DRF of the part. The second one controls the perpendicularity of the hole to the mating surface.

All other features of the parts where alignment is a concern should be dimensioned to the pin/hole DRF.

24.9 Proper Use of Material Modifiers

Because of the ability to inspect parts with gages, manufacturing personnel typically recommend using the maximum material condition (MMC) modifier on as many features of size as possible. While the MMC modifier makes sense with regard to the fit of the parts, its use can allow the other performance specifications dependent on the feature to have more error than originally anticipated. For example, if clearance holes are sized to fit, then adding the MMC modifier will allow more variation than explicitly allowed in the tolerances but will not adversely affect the ability to mate the parts. If the holes are dimensioned to another set of alignment features, the addition of the MMC modifier does increase the permissible translational and rotational errors throughout the assembly.

The problems can be avoided by using the following rules regarding material modifiers in the design of pinned interfaces:

- For statistical tolerance allocation, use only regardless of feature size (RFS) for the alignment features.
- For worst-case tolerance allocation, when the alignment holes or pins are used as the datum reference frame for the rest of the critical features on the parts, use the MMC modifier for the positional tolerance with respect to other noncritical features and with respect to each other. All critical features will be positioned with respect to the alignment pins or holes at LMC.
- Use either the RFS or LMC modifier for all other critical features of the parts. This not only includes the modifier for the positional tolerance but also applies to any datums of size referred to in the feature control frame.

All figures in this section showing recommended tolerances follow these three rules.

One other important topic involving the MMC modifier is the concept of zero positional tolerance at MMC. All clearance holes with worst-case tolerance allocation (except for the configuration involving a diamond pin) use this tolerancing method. The principle behind the method is relatively simple. If the hole is positioned perfectly, then we can allow its size to be as small as the outer boundary of the pin. However, as the hole diameter gets larger, it can also move and still be able to fit over the mating pin. If we were to use any number greater than zero in the position feature control frame, then the hole diameter would never be able to be as small as what is permitted when the hole is perfectly placed. Using zero position at MMC therefore maximizes design efficiency by allowing the engineer to be able to use the smallest possible nominal hole diameter that still fits.

The unequal bilateral tolerance for the clearance holes using MMC represents the ideal manufacturing target for optimum producibility. In other words, given the assumed standard deviations in Table 24-2, the predicted defect rate below the lower tolerances is the same as the predicted defect rate above the upper tolerance. The sum of the two defect rates is 3.4 defects per million over the long term. The explanation of the defect calculation is beyond the scope of this chapter. What is important is that the nominal value should be the target for the manufacturing facilities. Many shops will not recognize this fact, so the engineer may wish to include a note on the drawing stating that the optimal manufacturing targets are provided by the nominal values for all dimensions.

Note that material modifiers are applicable only for worst-case methods. Statistical tolerance allocation for fit does not benefit, and may in fact be adversely affected by the use of material modifiers.

24.10 Temperature Considerations

The analysis of fit used to size the clearance holes is based upon assembly at 68° F.[2] If the parts are made from different materials and are to be assembled at temperatures other than 68° F, then the nominal size of the clearance holes should be increased to account for differences in expansion of the two parts. The additional allowance is given by the following equation:

$$\Delta_h = d_p \cdot |\Delta_T| \cdot |cte_1 - cte_2|$$

where Δ_h is the amount to increase each hole diameter, d_p is the distance between the pins, Δ_T is the difference between 68 °F and the temperature at which the parts must assemble, and cte_1 and cte_2 are the coefficients of thermal expansion for the two mating parts. The effects of the differences in expansion of the pins and the holes do not contribute significantly and are not included in the above equation.

Increasing the nominal hole size for temperature effects will increase the alignment error between the parts if they are assembled at 68° F. The increase in translation is half of Δ_h calculated above and should be added to the translation errors in Tables 24-3, 24-9, and 24-11. Because rotation is a function of $1/d_p$ and the holes are increased by a factor of d_p, the additional rotation is a constant added to the original rotation. The equation for rotation therefore becomes:

$$\alpha_T = \frac{constant}{d_{pins}} + \Delta_T \cdot |cte_1 - cte_2|$$

This equation should be used only when the clearance hole has been increased due to a requirement that the parts assemble at a range of temperatures and the parts are made of different materials.

24.11 Two Round Pins with Two Holes

This method uses two round pins and two clearance holes. The advantage of this method over most of the others is that this configuration requires less machining and uses no unmodified pins. This method does, however, require the largest clearance holes. As a result, performance is worse than all the other methods. Since this method is one of the cheapest (except for two round pins with one hole and edge contact) and most straightforward, the engineer should try this configuration first before proceeding to one of the others.

---

[2] per ASME Y14.5M-1994, Paragraph 1.4(k).

24.11.1 Fit

The following is the general equation determining whether or not the parts will assemble:

$$c = \frac{1}{2}\left(\varnothing_{h1} + \varnothing_{h2} - \varnothing_{p1} - \varnothing_{p2}\right) - \left|d_h - d_p\right| \geq .0001 \tag{24.2}$$

Fig. 24-4 shows the variables of Eq. (24.2) graphically. Though Eq. (24.2) is useful for worst case analysis, it cannot be solved statistically using partial differentiation. It can, however, be modified to examine the condition of fit statistically by removing the absolute value, as shown in the following equation:

$$c = \frac{1}{2}\left(\varnothing_{h1} + \varnothing_{h2} - \varnothing_{p1} - \varnothing_{p2}\right) - (d_h - d_p) \tag{24.3}$$

The condition of fit using Eq. (24.3) becomes:

$$.0001 \leq c \leq 2 \cdot c_{nom} - .0001$$

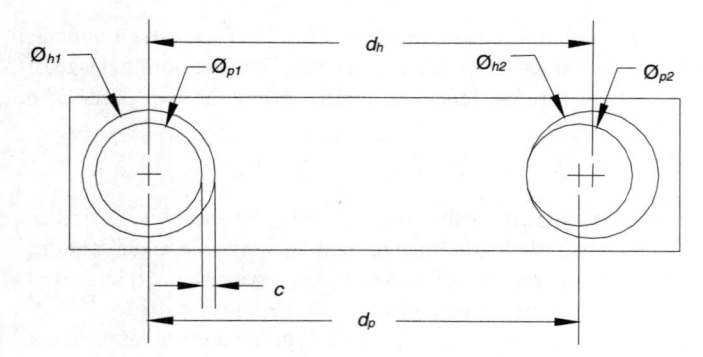

Figure 24-4 Variables contributing to fit of two round pins with two holes

24.11.2 Rotation Errors

The following equation gives the permissible rotation between the two parts:

$$\alpha = \cos^{-1}\left[\frac{d_h^2 + d_p^2 - \left(\dfrac{\varnothing_{h1} + \varnothing_{h2} - \varnothing_{p1} - \varnothing_{p2}}{2}\right)^2}{2 \cdot d_h \cdot d_p}\right]$$

Fig. 24-5 presents these variables graphically. Though Eq. (24.4) was used in determining the constants in Table 24-3, it does not resemble Eq. (24.1). However, Eq. (24.4) may be simplified. If we assume $d_h = d_p$, $\varnothing_{h2} = \varnothing_{h1}$, $\varnothing_{p2} = \varnothing_{p1}$, $\sin(\alpha) \approx \alpha$ (for small angles), and $(\varnothing_h - \varnothing_p)^2 \approx 0$ when compared to $4 \times d_p$, then we can simplify Eq. (24.4) to:

$$\alpha = \frac{(\varnothing_h - \varnothing_p)}{d_p} \tag{24.5}$$

The approximations made during this simplification are trivial and conservative (i.e., they result in rotations that are slightly larger than would be calculated without making these approximations). The simplified form of Eq. (24.5) is worth the slight additional error predicted.

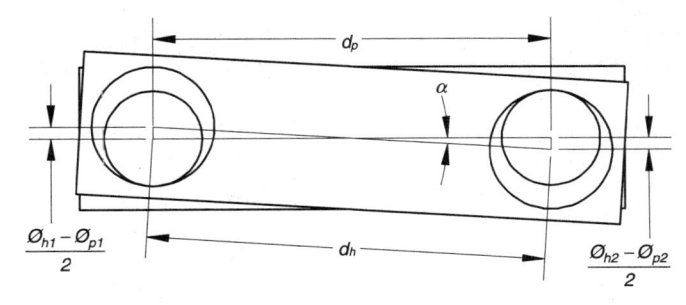

Figure 24-5 Variables contributing to rotation caused by two round pins with two holes

24.11.3 Translation Errors

The maximum translation between two parts can be found from the following equation:

$$\delta = \frac{1}{2} \min \left(\varnothing_{h1} - \varnothing_{p1}, \varnothing_{h2} - \varnothing_{p2} \right)$$

Because of the min function, it is difficult to analyze this equation statistically unless one uses simulation techniques. We therefore assume that the translation will be entirely controlled by the clearance at just one pin — the one with the smallest clearance hole. This results in slightly conservative performance limits.

24.11.4 Performance Constants

Table 24-3 includes the performance constants for all design options for two round pins with two holes. Remember to divide the rotation constants by d_p to determine the rotation through the interface.

Table 24-3 Performance constants for two round pins with two holes

			Worst-Case Max Error	Statistical Fixed Error	Statistical Standard Deviation
Type I	Drill & Ream	Translation (inches)	.0052	.0028	.000125
		Rotation (inch·radians)	.0103	.0057	.0001768
	N/C Mill	Translation (inches)	.0043	.0023	.000125
		Rotation (inch·radians)	.0086	.0047	.0001768
	Jig Bore	Translation (inches)	.0023	.0012	.000065
		Rotation (inch·radians)	.0046	.0025	.0000884
Type II	Drill & Ream	Translation (inches)	.0092	.0028	.0006423
		Rotation (inch·radians)	.0184	.0057	.0009083
	N/C Mill	Translation (inches)	.0075	.0023	.0005154
		Rotation (inch·radians)	.0150	.0047	.0007289
	Jig Bore	Translation (inches)	.0039	.0012	.0002583
		Rotation (inch·radians)	.0078	.0025	.0003644

24.11.5 Dimensioning Methodology

Fig. 24-6 and Table 24-4 present the recommended dimensioning methods.

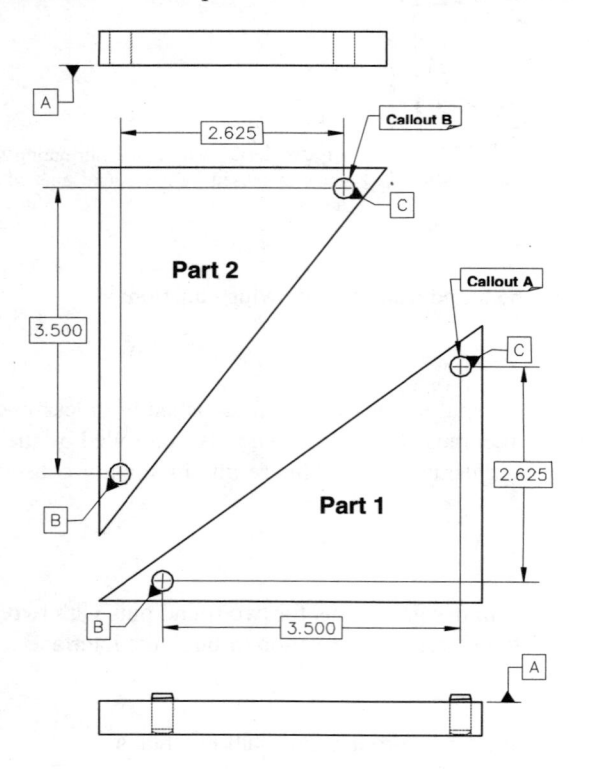

Figure 24-6 Dimensioning methodology for two round pins with two holes (only Type I shown)

24.12 Round Pins with a Hole and a Slot

This configuration is very similar to two round pins with two holes except that one of the holes is elongated, creating a short slot. The benefit of elongating one hole is that it eliminates the errors in the distance between the pin centers and the distance between the hole centers from affecting the fit of the two parts. Therefore, the slot need only be long enough to accommodate the positional variation of the pins and the positional variation of the clearance features to one another. The slot is so short, in fact, that someone looking at the part would probably not be able to discern which feature was the hole and which feature was the slot.

Due to the critical tolerances on the width of the slot, the manufacturing shop should use multiple passes with a boring bar rather than profiling the slot with a side-mill cutter. Ideally, the first finish-boring pass will be at the center of the slot, and consecutive passes will be made on both sides to form the slot. This manufacturing method prohibits the use of a reamer, so this section only considers N/C milling and Jig Bore processes.

24.12.1 Fit

Because this design configuration allows the distance between the pins and the distance between the hole and the slot to vary without affecting fit, the engineer need only be concerned with the size of the alignment features and the perpendicularity of the alignment features to the mating surfaces. If we size

Pinned Interfaces 24-15

Table 24-4 GD&T callouts for two round pins with two holes

		Drill and Ream	N/C Bore	Jig Bore
Worst Case	Callout A	2X Ø.PPPP±.0001 Pins ⌖ Ø.0041Ⓜ A	2X Ø.PPPP±.0001 Pins ⌖ Ø.0032Ⓜ A	2X Ø.PPPP±.0001 Pins ⌖ Ø.0016Ⓜ A
	Callout B	2X Ø.{PPPP+.0087} +.0015/−.0044 ⌖ Ø.0000Ⓜ A	2X Ø.{PPPP+.0070} +.0015/−.0036 ⌖ Ø.0000Ⓜ A	2X Ø.{PPPP+.0037} +.0008/−.0019 ⌖ Ø.0000Ⓜ A
	Additional Callout for Type II Interface	⌖ Ø.0081Ⓛ A BⓁ CⓁ	⌖ Ø.0064Ⓛ A BⓁ CⓁ	⌖ Ø.0032Ⓛ A BⓁ CⓁ
Statistical	Callout A	2X Ø.PPPP±.0001 Pins ⌖ Ø.0041⟨ST⟩ A	2X Ø.PPPP±.0001 Pins ⌖ Ø.0032⟨ST⟩ A	2X Ø.PPPP±.0001 Pins ⌖ Ø.0016⟨ST⟩ A
	Callout B	2X Ø.{PPPP+.0056} ±.0015 ⟨ST⟩ ⌖ Ø.0041⟨ST⟩ A	2X Ø.{PPPP+.0046} ±.0015 ⟨ST⟩ ⌖ Ø.0032⟨ST⟩ A	2X Ø.{PPPP+.0024} ±.0008 ⟨ST⟩ ⌖ Ø.0016⟨ST⟩ A
	Additional Callout for Type II Interface	⌖ Ø.0081⟨ST⟩ A B C	⌖ Ø.0064⟨ST⟩ A B C	⌖ Ø.0032⟨ST⟩ A B C

the hole to fit over the first pin, and then size the width of the slot to be the same as the hole diameter, the parts will assemble. Thus, the condition for fit is:

$$c = \varnothing_h - perp_h - \varnothing_{p1} - perp_{p1} \geq .0001 \tag{24.6}$$

We must also be concerned with fit in the direction of the slot, as shown in Fig. 24-7. In this case, clearance can be determined by:

$$c = \frac{1}{2}(\varnothing_h + l_{slot} - \varnothing_{p1} - \varnothing_{p2}) - (d_{hs} - d_p)$$

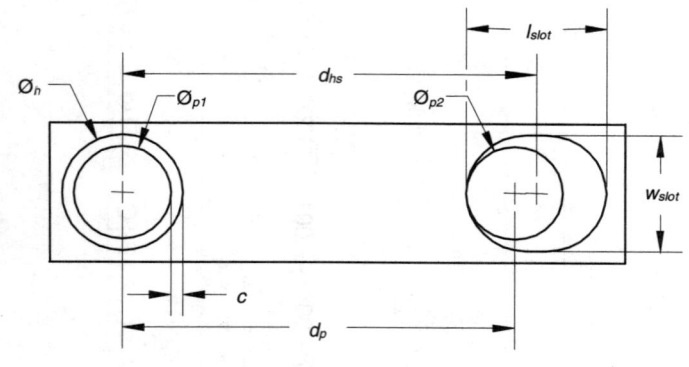

Figure 24-7 Variables contributing to fit of two round pins with one hole and one slot

Since clearance in this direction is not critical, the callouts in Table 24-6 allow the slot width to vary by ±.005. This tolerance is well beyond the Six Sigma capability but is not large enough to require excessive slotting of the hole.

24.12.2 Rotation Errors

The rotation of the two parts is given by

$$\alpha = 2 \cdot \tan^{-1}\left[\frac{d_p^2 - \Delta_{slot} \cdot \Delta_{hole} - \Delta_{hole}^2 - d_p\sqrt{d_p^2 - (\Delta_{slot} - \Delta_{hole})^2}}{d_p \cdot \Delta_{slot} - \Delta_{hole} \cdot \sqrt{d_p^2 - (\Delta_{slot} - \Delta_{hole})^2}}\right]$$

where

$$\Delta_{hole} = \left(\frac{\varnothing_h - \varnothing_{p1}}{2}\right) \text{ and } \Delta_{slot} = \left(\frac{w_{slot} - \varnothing_{p2}}{2}\right)$$

Fig. 24-8 presents these variables graphically.

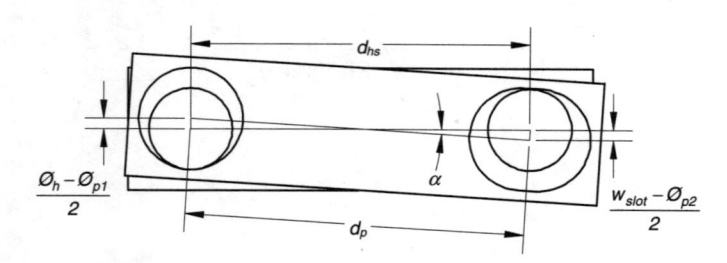

Figure 24-8 Variables contributing to rotation caused by two pins with one hole and one slot

24.12.3 Translation Errors

Because the interface between the pin and the hole has the minimum clearance in all directions, it will always control the translation between the mating parts. Furthermore, since only this interface is used to determine the fit of the parts, one cannot reduce the hole diameter and increase the slot dimensions in order to improve translation performance without adversely affecting fit. In other words, this design configuration is optimized for the best translation performance. Only by changing the manufacturing process can we improve performance while maintaining the same ability to assemble the parts.

The formula for determining the translation error is:

$$\delta = \frac{\varnothing_h - \varnothing_{p1}}{2}$$

24.12.4 Performance Constants

Table 24-5 includes the performance constants for all design options for two round pins with one hole and one slot. Remember to divide the rotation constants by d_p to determine the maximum allowable rotation through the interface.

Table 24-5 Performance constants for two round pins with one hole and one slot

			Worst-Case Max Error	Statistical Fixed Error	Statistical Standard Deviation
Type I	N/C Mill	Translation (inches)	.00220	.00110	.000125
Type I	N/C Mill	Rotation (inch·radians)	.0023	.0022	.0001768
Type I	Jig Bore	Translation (inches)	.00125	.0006	.000065
Type I	Jig Bore	Rotation (inch·radians)	.0013	.0012	.0000884
Type II	N/C Mill	Translation (inches)	.00540	.00110	.0005154
Type II	N/C Mill	Rotation (inch·radians)	.0087	.0022	.0007289
Type II	Jig Bore	Translation (inches)	.00285	.0006	.0002583
Type II	Jig Bore	Rotation (inch·radians)	.0045	.0012	.0003644

24.12.5 Dimensioning Methodology

Fig. 24-9 and Table 24-6 present the recommended dimensioning methods for round pins with a hole and a slot. Datum C on the second part is two line targets at a basic distance from the center of the hole. This dimensioning scheme most closely represents how the part will function, though the pins may not contact the slot at exactly these targets.

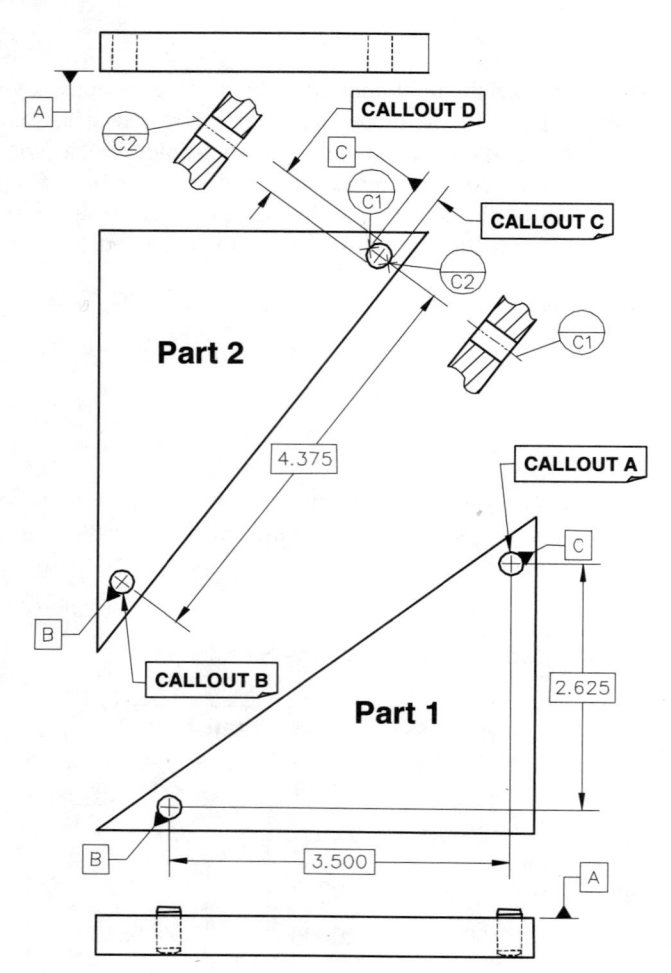

Figure 24-9 Dimensioning methodology for two round pins with one hole and one slot (only Type I shown)

24.13 Round Pins with One Hole and Edge Contact

Another alignment methodology uses two pins to engage one hole and the side of the second part. Though this design is not used extensively, it provides the best performance at the least expense. Since the second feature used to engage the pin is not a feature of size, the clearance necessary to fit a feature of size over the second pin is eliminated and thus does not add to rotation error. Furthermore, since this design involves only one precision hole and no modified pins, it is the least expensive of all the configurations.

The primary drawback to this technique is that it requires the assembly operator to ensure that the second part is fully rotated and contacting the second pin on the side. Depending on the design, this can be verified quite easily through visual inspection. The additional cost associated with the added requirement during assembly is much less than the cost of the installation of the second pin.

Table 24-6 GD&T callouts for two round pins with one hole and one slot

		N/C Bore	Jig Bore
Worst Case	**Callout A**	2X Ø.PPPP±.0001 Pins ⊕ | Ø.0032 Ⓜ | A ⊥ | Ø.0008 Ⓜ | A	2X Ø.PPPP±.0001 Pins ⊕ | Ø.0016 Ⓜ | A ⊥ | Ø.0004 Ⓜ | A
	Callout B	Ø.{PPPP+.0028} $^{+.0015}_{-.0018}$ ⊥ | Ø.0000 Ⓜ | A	Ø.{PPPP+.0016} $^{+.0008}_{-.0010}$ ⊥ | Ø.0000 Ⓜ | A
	Callout C	{.PPPP+.0028} $^{+.0015}_{-.0018}$ ⊕ | .0032 Ⓜ | A ⊥ | .0000 Ⓜ | A	{.PPPP+.0028} $^{+.0008}_{-.0010}$ ⊕ | .0016 Ⓜ | A ⊥ | .0000 Ⓜ | A
	Callout D	{.PPPP+.0108} ±.0050 ⊕ | .0000 Ⓜ | A | B	{.PPPP+.0080} ±.0050 ⊕ | .0000 Ⓜ | A | B
	Additional Callout for Type II Interface	⊕ | Ø.0064 Ⓛ | A | B Ⓛ | C Ⓛ	⊕ | Ø.0032 Ⓛ | A | B Ⓛ | C Ⓛ
Statistical	**Callout A**	2X Ø.PPPP±.0001 Pins ⊕ | Ø.0032 ⟨ST⟩ | A ⊥ | Ø.0008 ⟨ST⟩ | A	2X Ø.PPPP±.0001 Pins ⊕ | Ø.0016 ⟨ST⟩ | A ⊥ | Ø.0004 ⟨ST⟩ | A
	Callout B	Ø.{PPPP+.0021} ±.0015 ⟨ST⟩ ⊥ | Ø.0008 ⟨ST⟩ | A	Ø.{PPPP+.0011} ±.0008 ⟨ST⟩ ⊥ | Ø.0004 ⟨ST⟩ | A
	Callout C	.{PPPP+.0021} ±.0015 ⟨ST⟩ ⊕ | .0032 | A | B ⊥ | .0008 ⟨ST⟩ | A	.{PPPP+.0011} ±.0008 ⟨ST⟩ ⊕ | .0016 | A | B ⊥ | .0004 ⟨ST⟩ | A
	Callout D	{.PPPP+.0095} ±.0050 ⊕ | .0000 Ⓜ | A | B	{.PPPP+.0074} ±.0050 ⊕ | .0000 Ⓜ | A | B
	Additional Callout for Type II Interface	⊕ | Ø.0064 ⟨ST⟩ | A | B | C	⊕ | Ø.0032 ⟨ST⟩ | A | B | C

24.13.1 Fit

Because only the first hole and pin are features of size, the fit for this configuration is exactly like the criteria for fit of the hole and slot given in Eq. (24.6).

24.13.2 Rotation Errors

The tilt resulting from this type of interface is obtained from the following equation:

$$\alpha = -2\tan^{-1}\left[\frac{d_{px} - \sqrt{d_{px}^2 + \left[d_{py} + \left(\frac{\emptyset_{h1} - \emptyset_{p1}}{2}\right)\right]^2 - \left(d_e + \frac{\emptyset_{p2}}{2}\right)^2}}{d_{py} + d_e + \frac{\emptyset_{h1} - \emptyset_{p1}}{2} + \frac{\emptyset_{p2}}{2}}\right]$$

Fig. 24-10 presents these variables graphically.

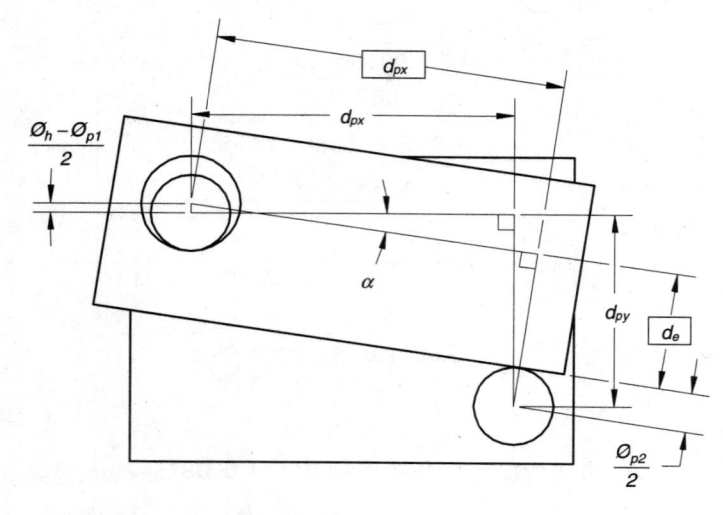

Figure 24-10 Variables contributing to rotation caused by two pins with hole and edge contact

24.13.3 Translation errors

The translation errors of this configuration are identical to those for the design involving two pins with one hole and one slot. (Refer to section 24.12.3.)

24.13.4 Performance Constants

Table 24-7 includes the performance constants for all design options for two round pins with one hole and edge contact. In this case, only increasing d_{px} improves the tilt. Remember to divide the rotation constants by d_{px} to determine the rotation allowed by the interface.

24.13.5 Dimensioning Methodology

Fig. 24-11 and Table 24-8 present the recommended dimensioning methods for two pins with one hole and edge contact. Datum C on the part 2 is a line target contacting the edge at the approximate location of the pin on part 1. It is found by placing two pins in a gage at the basic dimensions indicated on the drawing. This method of establishing the datum eliminates the distance indicated as basic in Fig. 24-11 from becoming contributors to the rotation error between the parts. Similarly, since the second pin is the datum for part 2, the variation in d_y also does not contribute to the rotation variation.

Table 24-7 Performance constants for two round pins with one hole and edge contact

			Worst-Case Max Error	Statistical Fixed Error	Statistical Standard Deviation
Type I	Drill and Ream	Translation (inches)	.00235	.0016	.000125
Type I	Drill and Ream	Rotation (inch·radians)	.0024	.0012	.0001249
Type I	N/C Mill	Translation (inches)	.0022	.00145	.000125
Type I	N/C Mill	Rotation (inch·radians)	.0023	.0012	.0001249
Type I	Jig Bore	Translation (inches)	.00125	.00085	.000065
Type I	Jig Bore	Rotation (inch·radians)	.0013	.0007	.0000625
Type II	Drill and Ream	Translation (inches)	.0064	.0016	.00064228
Type II	Drill and Ream	Rotation (inch·radians)	.0105	.0012	.0008997
Type II	N/C Mill	Translation (inches)	.0054	.00145	.0005154
Type II	N/C Mill	Rotation (inch·radians)	.0087	.0012	.0007181
Type II	Jig Bore	Translation (inches)	.00285	.00085	.0002583
Type II	Jig Bore	Rotation (inch·radians)	.0045	.0007	.0003590

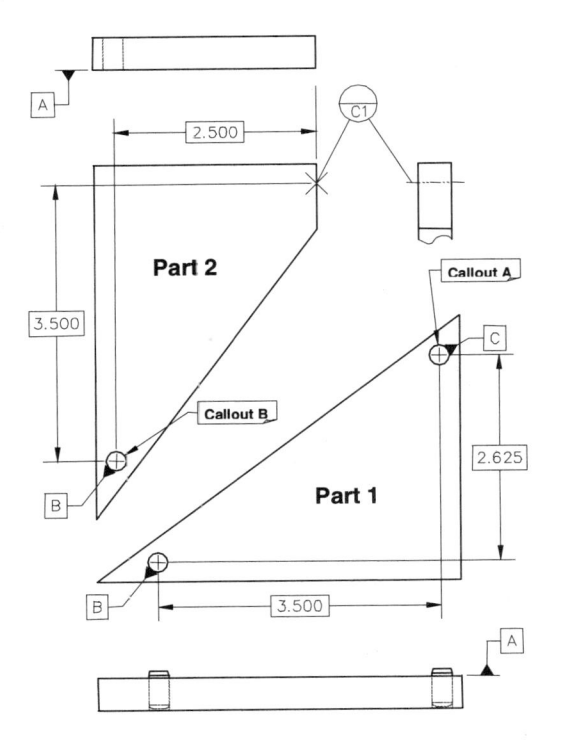

Figure 24-11 Dimensioning methodology for two round pins with one hole and edge contact (only Type I shown)

Table 24-8 GD&T callouts for two round pins with one hole and edge contact

		Drill and Ream	N/C Bore	Jig Bore
Worst Case	Callout A	2X Ø.PPPP±.0001 Pins ⌖ Ø.0041 Ⓜ A	2X Ø.PPPP±.0001 Pins ⌖ Ø.0032 Ⓜ A	2X Ø.PPPP±.0001 Pins ⌖ Ø.0016 Ⓜ A
	Callout B	Ø.{PPPP+.0031}⁺·⁰⁰¹⁵₋.₀₀₁₉ ⊥ Ø.0000 Ⓜ A	Ø.{PPPP+.0028}⁺·⁰⁰¹⁵₋.₀₀₁₈ ⊥ Ø.0000 Ⓜ A	Ø.{PPPP+.0016}⁺·⁰⁰⁰⁸₋.₀₀₁₀ ⊥ Ø.0000 Ⓜ A
	Additional Callout for Type II Interface	⌖ Ø.0081 Ⓛ A Ⓑ Ⓒ	⌖ Ø.0064 Ⓛ A Ⓑ Ⓒ	⌖ Ø.0032 Ⓛ A Ⓑ Ⓒ
Statistical	Callout A	2X Ø.PPPP±.0001 Pins ⌖ Ø.0041 ⟨ST⟩ A	2X Ø.PPPP±.0001 Pins ⌖ Ø.0032 ⟨ST⟩ A	2X Ø.PPPP±.0001 Pins ⌖ Ø.0016 ⟨ST⟩ A
	Callout B	Ø.{PPPP+.0022} ±.0015 ⟨ST⟩ ⊥ Ø.0010 ⟨ST⟩ A	Ø.{PPPP+.0021} ±.0015 ⟨ST⟩ ⊥ Ø.0008 ⟨ST⟩ A	Ø.{PPPP+.0011} ±.0008 ⟨ST⟩ ⊥ Ø.0004 ⟨ST⟩ A
	Additional Callout for Type II Interface	⌖ Ø.0081 ⟨ST⟩ A B C	⌖ Ø.0064 ⟨ST⟩ A B C	⌖ Ø.0032 ⟨ST⟩ A B C

24.14 One Diamond Pin and One Round Pin with Two Holes

This design configuration is very similar to two pins with two holes. The difference is the shape of the second pin. In this case, the flats on the second pin accommodate more variation in the distance between the pins and the distance between the holes. This enables us to decrease the nominal hole diameter, thus improving performance without affecting fit. Because the allowable location error gained from the pin is greater than with the parallel-flats pin, and because the diamond pin is stronger than the parallel-flats pin, this is the preferred method for designs using modified pins.

As was mentioned in section 24.9, this configuration does not benefit from zero position at MMC. In fact, if we were to use this tolerancing scheme, we would have to make the nominal hole diameter larger. The equation for fit is actually more sensitive to the diameter of the second hole than to the distance between the holes. As a result, zero position at MMC is not as efficient as the dimensioning methodology of Table 24-10.

24.14.1 Fit

The equation for fit is:

$$c = \frac{1}{2}\left(\varnothing_{h1} - \varnothing_{p1}\right) - d_p + z \cdot \cos(\beta) - d_h \cdot \cos\left\{\tan^{-1}\left[\frac{\sin(\beta)}{\cos(\beta) + \frac{d_p}{z}}\right]\right\}$$

where

$$z = \frac{1}{2}\sqrt{\varnothing_{h2}^2 - \left\{\varnothing_{p2}^2 \cdot \cos\left[\frac{\pi}{6} + \cos^{-1}\left(\frac{2t}{\varnothing_{p2}^2}\right)\right]\right\}^2} - \frac{1}{2}\sqrt{\varnothing_{p2}^2 - \left\{\varnothing_{p2}^2 \cdot \cos\left[\frac{\pi}{6} + \cos^{-1}\left(\frac{2t}{\varnothing_{p2}^2}\right)\right]\right\}^2}$$

Fig. 24-12 provides a graphical representation of these variables.

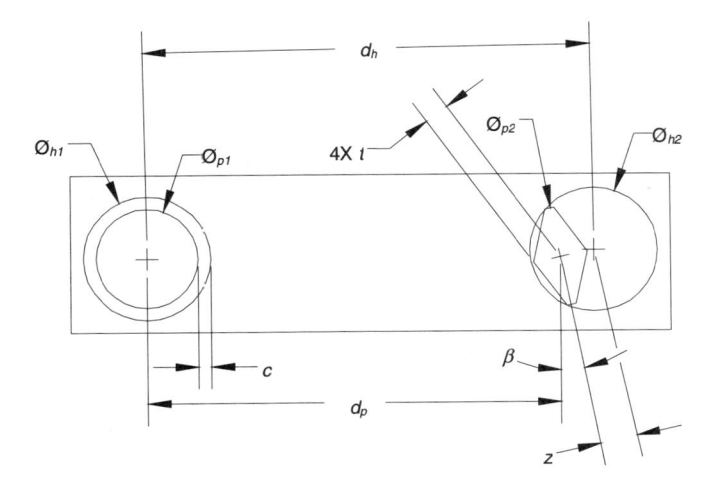

Figure 24-12 Variables contributing to fit of one round pin and one diamond pin with two holes

24.14.2 Rotation and Translation Errors

Because the rotation is controlled by the cylindrical sections of both pins and the round pin will control translation, the formulas for rotation and translation errors are the same as for the two round pins with two round holes in sections 24.11.2 and 24.11.3.

24.14.3 Performance Constants

Table 24-9 includes the performance constants for all design options for one round pin and one diamond pin with two holes. Remember to divide the rotation constants by d_p to determine the allowable rotation at the interface.

Table 24-9 Performance constants for one round pin and one diamond pin with two holes

			Worst-Case Max Error	Statistical Fixed Error	Statistical Standard Deviation
Type I	Drill and Ream	Translation (inches)	.00275	.00095	.0001250
		Rotation (inch•radians)	.005516	.0019	.0001768
	N/C Mill	Translation (inches)	.00245	.00085	.0001250
		Rotation (inch•radians)	.004916	.0017	.0001768
	Jig Bore	Translation (inches)	.00130	.0005	.0000650
		Rotation (inch•radians)	.002603	.0010	.0000884
Type II	Drill and Ream	Translation (inches)	.00685	.00095	.0006423
		Rotation (inch•radians)	.013616	.0019	.0009083
	N/C Mill	Translation (inches)	.00565	.00085	.0005154
		Rotation (inch•radians)	.0011316	.0017	.0007289
	Jig Bore	Translation (inches)	.00290	.0005	.0002583
		Rotation (inch•radians)	.005803	.0010	.0003644

24.14.4 Dimensioning Methodology

Table 24-10 presents the recommended dimensioning methods for one diamond pin and one round pin with two holes. Refer to Fig. 24-6 for the graphical portion of the callouts.

Table 24-10 GD&T callouts for one round pin and one diamond pin with two holes

		Drill and Ream	N/C Bore	Jig Bore
Worst Case	Callout A	2X Ø.PPPP±.0001 Pins ⌖ Ø.0041 Ⓜ A	2X Ø.PPPP±.0001 Pins ⌖ Ø.0032 Ⓜ A	2X Ø.PPPP±.0001 Pins ⌖ Ø.0016 Ⓜ A
	Callout B	2X Ø.{PPPP+.0039} ±.0015 ⌖ Ø.0041 A	2X Ø.{PPPP+.0033} ±.0015 ⌖ Ø.0032 A	2X Ø.{PPPP+.0017} ±.0008 ⌖ Ø.0016 A
	Additional Callout for Type II Interface	⌖ Ø.0081 Ⓛ A Ⓑ C Ⓛ	⌖ Ø.0064 Ⓛ A Ⓑ C Ⓛ	⌖ Ø.0032 Ⓛ A Ⓑ C Ⓛ
Statistical	Callout A	2X Ø.PPPP±.0001 Pins ⌖ Ø.0041 ⟨ST⟩ A	2X Ø.PPPP±.0001 Pins ⌖ Ø.0032 ⟨ST⟩ A	2X Ø.PPPP±.0001 Pins ⌖ Ø.0016 ⟨ST⟩ A
	Callout B	2X Ø.{PPPP+.0018} ±.0015 ⟨ST⟩ ⌖ Ø.0041 ⟨ST⟩ A	2X Ø.{PPPP+.0016} ±.0015 ⟨ST⟩ ⌖ Ø.0032 ⟨ST⟩ A	2X Ø.{PPPP+.0009} ±.0008 ⟨ST⟩ ⌖ Ø.0016 ⟨ST⟩ A
	Additional Callout for Type II Interface	⌖ Ø.0081 ⟨ST⟩ A B C	⌖ Ø.0064 ⟨ST⟩ A B C	⌖ Ø.0032 ⟨ST⟩ A B C

24.15 One Parallel-Flats Pin and One Round Pin with Two Holes

This is the least attractive of all the design configurations included in this section. The grinding of the second pin, though not quite as involved as with a diamond pin, still adds additional costs associated with the machining and storage of the special part. The modified pin is the weakest and is therefore subject to bending during installation.

Another disadvantage of the parallel-flats shape is that the intersection of the unmodified diameter and the flat section is a sharper corner than with the diamond shape. This can lead to increased damage from galling when the pin begins to engage the clearance hole of the mating part during assembly.

24.15.1 Fit

Determination of fit for parts aligned using one round pin and one diamond pin is given by:

$$c = \frac{1}{2}(\varnothing_{h1} - \varnothing_{p1}) - d_p + z \cdot \cos(\beta) - d_h \cdot \cos\left\{\tan^{-1}\left[\frac{\sin(\beta)}{\cos(\beta) + \frac{d_p}{z}}\right]\right\}$$

where

$$z = \frac{1}{2}\sqrt{\varnothing_{h2}{}^2 - \left\{\varnothing_{p2} \cdot \sin\left[\cos^{-1}\left(\frac{2t}{\varnothing_{p2}}\right)\right]\right\}^2} - \frac{1}{2}\sqrt{\varnothing_{p2}{}^2 - \left\{\varnothing_{p2} \cdot \sin\left[\cos^{-1}\left(\frac{2t}{\varnothing_{p2}}\right)\right]\right\}^2}$$

Fig. 24-13 presents these variables graphically.

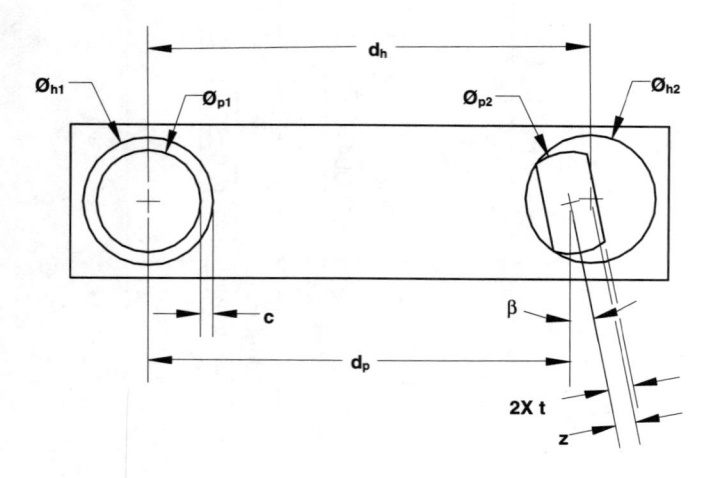

Figure 24-13 Variables contributing to the fit of one pin and one parallel-flats pin with two holes

24.15.2 Rotation and Translation Errors

Because the rotation is controlled by the cylindrical sections of both pins, and the round pin will control translation, the formulas for rotation and translation errors are the same as for the two round pins with two round holes in sections 24.11.2 and 24.11.3.

24.15.3 Performance Constants

Table 24-11 includes the performance constants for all design options for one round pin and one parallel-flats pin with two holes. Remember to divide the rotation constants by d_p to determine the rotation through the interface.

Table 24-11 Performance constants for one round pin and one parallel-flats pin with two holes

Two Holes with One Parallel-Flats Pin and One Round Pin			Worst-Case Max Error	Statistical Fixed Error	Statistical Standard Deviation
Type I	Drill and Ream	Translation (inches)	.00450	.00210	.0001250
		Rotation (inch·radians)	.009	.0042	.0001768
	N/C Mill	Translation (inches)	.00380	.00170	.0001250
		Rotation (inch·radians)	.0076	.0034	.0001768
	Jig Bore	Translation (inches)	.00205	.00095	.0000650
		Rotation (inch·radians)	.0041	.0019	.0000884
Type II	Drill and Ream	Translation (inches)	.00855	.00210	.0006423
		Rotation (inch·radians)	.0171	.0042	.0009083
	N/C Mill	Translation (inches)	.00510	.00170	.0005154
		Rotation (inch·radians)	.0140	.0034	.0007289
	Jig Bore	Translation (inches)	.00365	.00095	.0002583
		Rotation (inch·radians)	.0073	.0019	.0003644

24.15.4 Dimensioning Methodology

Table 24-12 presents the recommended dimensioning methods for two holes, one round pin, and one parallel flat pin. Refer to Fig. 24-6 in section 24.11.5 for the graphical portion of the callouts.

Table 24-12 GD&T callouts for one round pin with one parallel-flats pin and two holes

		Drill and Ream	N/C Bore	Jig Bore
Worst Case	Callout A	2X Ø.PPPP±.0001 Pins ⊕ Ø.0041 Ⓜ A	2X Ø.PPPP±.0001 Pins ⊕ Ø.0032 Ⓜ A	2X Ø.PPPP±.0001 Pins ⊕ Ø.0016 Ⓜ A
	Callout B	2X Ø.{PPPP+.0074} +.0015/−.0044 ⊕ Ø.0000 Ⓜ A	2X Ø.{PPPP+.0060} +.0015/−.0036 ⊕ Ø.0000 Ⓜ A	2X Ø.{PPPP+.0032} +.0008/−.0019 ⊕ Ø.0000 Ⓜ A
	Additional Callout for Type II Interface	⊕ Ø.0081 Ⓛ A Ⓑ Ⓒ	⊕ Ø.0064 Ⓛ A Ⓑ Ⓒ	⊕ Ø.0032 Ⓛ A Ⓑ Ⓒ
Statistical	Callout A	2X Ø.PPPP±.0001 Pins ⊕ Ø.0041 ⓈⓉ A	2X Ø.PPPP±.0001 Pins ⊕ Ø.0032 ⓈⓉ A	2X Ø.PPPP±.0001 Pins ⊕ Ø.0016 ⓈⓉ A
	Callout B	2X Ø.{PPPP+.0041} ±.0015 ⓈⓉ ⊕ Ø.0041 ⓈⓉ A	2X Ø.{PPPP+.0033} ±.0015 ⓈⓉ ⊕ Ø.0032 ⓈⓉ A	2X Ø.{PPPP+.0018} ±.0008 ⓈⓉ ⊕ Ø.0016 ⓈⓉ A
	Additional Callout for Type II Interface	⊕ Ø.0081 ⓈⓉ A B C	⊕ Ø.0064 ⓈⓉ A B C	⊕ Ø.0032 ⓈⓉ A B C

24.16 References

1. Bralla, James G. 1986. *Handbook of Product Design for Manufacturing: A Practical Guide to Low-Cost Production*. New York, NY: McGraw-Hill Book Company.
2. The American Society of Mechanical Engineers. 1995. *ASME B18.8.2-1978(R1989): Taper Pins, Dowel Pins, Straight Pins, Grooved Pins, and Spring Pins (Inch Series)*. New York, NY: The American Society of Mechanical Engineers.
3. The American Society of Mechanical Engineers. 1995. *ASME Y14.5M-1994, Dimensioning and Tolerancing*. New York, NY: The American Society of Mechanical Engineers.

Chapter 25

Gage Repeatability and Reproducibility (GR&R) Calculations

Gregory A. Hetland, Ph.D.
Hutchinson Technology Inc.
Hutchinson, Minnesota

Dr. Hetland is the manager of corporate standards and measurement sciences at Hutchinson Technology Inc. With more than 25 years of industrial experience, he is actively involved with national, international, and industrial standards research and development efforts in the areas of global tolerancing of mechanical parts and supporting metrology. Dr. Hetland's research has focused on "tolerancing optimization strategies and methods analysis in a sub-micrometer regime."

25.1   Introduction

This chapter shows examples of calculating capabilities for a gage repeatability and reproducibility (GR&R) study on geometric tolerances, and identifies ambiguities as well as limitations in these calculations. Additionally, it shows tremendous areas of opportunity for future research and development in GR&R calculations due to past and still-current limitations in the variables considered when making these calculations. This chapter will define conditions not being accounted for in the calculations, therefore limiting the measurement system's capabilities.

25.2   Standard GR&R Procedure

The following is a standard procedure used for calculating a GR&R that relates to geometric controls per ASME Y14.5M-1994. Initial analysis will focus on a positional tolerance in a nondiametral tolerance zone. Please note: A small sample size is used only out of convenience. Small sample sizes are strongly supported when needing a quick "snap-shot" of a capability. I do not, however, promote small sizes for in-depth analysis.

25-2   Chapter Twenty-five

- Given 10 parts measured twice under the same conditions
  - Same procedure
  - Same machine
  - Same person
- Resultant Values (R.V.) are to be shown in positional form (not just x or y displacement).
- Derive the range between runs for Part 1, Part 2, ... Part 10.
- Sum the ranges and divide by 10 to derive the R.
- Divide the R by a constant of 1.128, for sample/run size of 2 (rough estimate of sigma based on small sample size).
- Multiply 3 × the estimate of sigma (3s) and divide by the positional tolerance allowed in the feature control frame, then multiply × 100. (This derived value will represent the <u>percentage</u> of tolerance used by the gage.)

The following data (Table 25-1) applies to the positional control of 0.2 mm, in relationship to datums A primary and B secondary at regardless of feature size (RFS) as shown in Fig. 25-1.

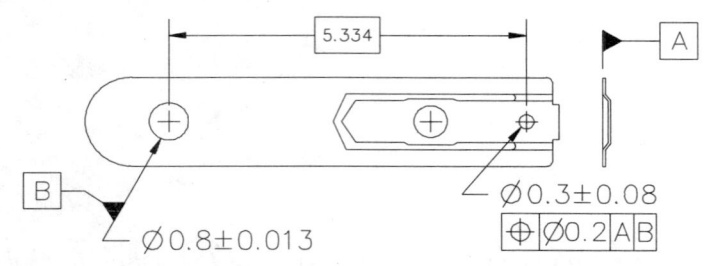

Figure 25-1 Sample drawing #1

Table 25-1 GR&R Analysis Matrix

Part #	Run #1 X displacement	$2\sqrt{\Delta X^2 + \Delta Y^2}$ R.V.#1	Run #2 X displacement	$2\sqrt{\Delta X^2 + \Delta Y^2}$ R.V.#2	Range Between RV#1 & RV#2
1	0.02	0.04	0.03	0.06	0.02
2	0.05	0.10	0.07	0.14	0.04
3	-0.03	0.06	-0.01	0.02	0.04
4	0.01	0.02	0.04	0.08	0.06
5	-0.04	0.08	-0.04	0.08	0.00
6	0.07	0.14	0.05	0.10	0.04
7	-0.06	0.12	-0.04	0.08	0.04
8	0.02	0.04	0.01	0.02	0.02
9	-0.09	0.18	-0.10	0.20	0.02
10	-0.05	0.10	-0.03	0.06	0.04

$\overline{R} = 0.032$
$\sigma = 0.032/1.128 = 0.0284$
$3\sigma = 3 \times 0.0284 = 0.085$
$3\sigma / \text{Tol.} \times 100 = \%$ of tolerance
$0.085/0.2 \times 100 = 42.6\%$

Gage Repeatability and Reproducibility (GR&R) Calculations

Questions arise regarding these calculations and whether sigma should be multiplied by 3 or 6. Figs. 25-2 and 25-3 are examples of tolerance zone differences, comparing a linear +/-0.1 mm tolerance to a nondiametral position tolerance of 0.2 mm.

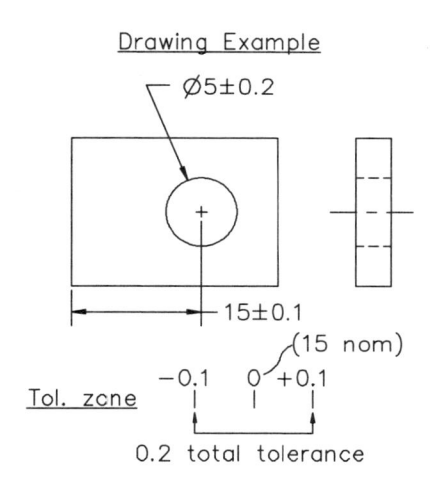

Figure 25-2 (left side) — Drawing Example: Ø5±0.2, 15±0.1 (15 nom), Tol. zone −0.1 to +0.1, 0.2 total tolerance

Measured Values

Part#	Run #1	Run #2	Range
1	0.02	0.03	0.01
2	0.05	0.07	0.02
3	−0.03	−0.01	0.02
4	0.01	0.04	0.03
5	−0.04	−0.04	0.00
6	0.07	0.05	0.02
7	−0.06	−0.04	0.02
8	0.02	0.01	0.01
9	−0.09	−0.10	0.01
10	−0.05	−0.03	0.02

$\bar{R} = \Sigma R/n = 0.16/10 = 0.016$

$\sigma = \bar{R}/d_2 = 0.016/1.128 = 0.0142$

$6\sigma = 6 \times 0.0142 = 0.085$

$6\sigma/\text{Tol.} \times 100 = \%$ of tolerance
$0.085/0.2 \times 100 = 42.6\%$

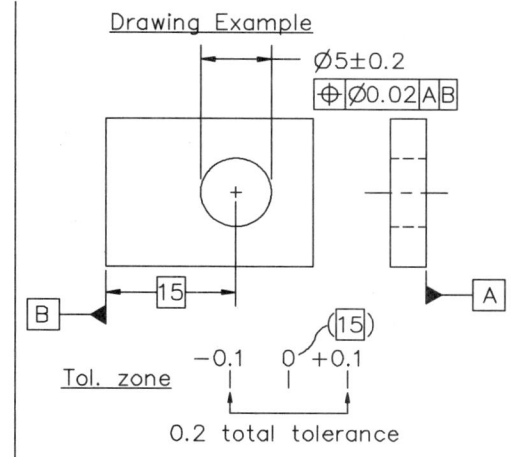

Figure 25-3 (right side) — Drawing Example: Ø5±0.2, ⌖ Ø0.02 A B, 15, Tol. zone −0.1 to +0.1, 0.2 total tolerance

Resultant Values derived from measured values in example to the left

Part#	Run #1	Run #2	Range
1	0.04	0.06	0.02
2	0.10	0.14	0.04
3	0.06	0.02	0.04
4	0.02	0.08	0.06
5	0.08	0.08	0.00
6	0.14	0.10	0.04
7	0.12	0.08	0.04
8	0.04	0.02	0.02
9	0.18	0.20	0.02
10	0.10	0.06	0.04

$\bar{R} = \Sigma R/n = 0.32/10 = 0.032$

$\sigma = \bar{R}/d_2 = 0.032/1.128 = 0.02840$

$3\sigma = 3 \times 0.0284 = 0.085$

$3\sigma/\text{Tol.} \times 100 = \%$ of tolerance
$0.085/0.2 \times 100 = 42.6\%$

Principal differences are: Linear example is 6σ, while for position (⌖), it is 3σ due to resultant value multiplied by 2.

Figure 25-2 Sample drawing #2

Figure 25-3 Sample drawing #3

Based on the prior example, first impression might be to use only the linear displacement values to stay consistent with past and present Six Sigma conventions. If only things were this simple, but they are not. In addition to the examples shown, there are many types of geometric callouts that require further analysis of calculations to determine the most appropriate method of representing percentage of variables gaging influence.

The following is a beginning list of various types of geometric callouts that will need to be considered.

1) Geometric controls @ RFS (diametral and nondiametral).
2) Geometric controls @ maximum material condition (MMC) or least material condition (LMC) (diametral and nondiametral).
3) Geometric controls @ MMC or LMC in relationship to datums that are features of size also defined at MMC or LMC.
4) Geometric controls @ MMC or LMC with zero tolerance

Additional things not defined adequately deal with ranges for the following:

1) Features of size (lengths, widths, and diameters)
2) Linear plane to axis measurements
3) Axis (I.D.) to axis measurements

There are also questions as to which analysis methods to use (e.g., Western Electric, IBM, other). Also, what are the benefits, drawbacks and limitations of any of these methods?

Also, an acceptable method is needed to determine the bias of a measurement device with an acceptable artifact, as well as a method to determine bias between devices. Such a method must consider the following:

1) Sampling strategies
2) Spot size versus spacing versus sampling effects on a given feature
3) Replication of test (time versus environmental)
4) Confidence intervals
5) Truth (conformance to ASME Y14.5M-1994 and ASME Y14.5.1M-1994)

Note: For all geometric controls, the tolerance defined in the feature control frame is a "total tolerance," of which the targeted value is "always" zero (0), and the upper control limit is always equal to the total tolerance defined (unless bonus tolerance is gained due to MMC or LMC on the considered feature).

For geometric controls, such as the one shown in Fig. 25-4, the 5 mm+/-0.2 mm diameter is positioned within a diametral tolerance zone of 0.02 mm at its maximum material condition, in relationship to datums A (primary), B (secondary), and C (tertiary). The following analysis is proposed:

Drawing Example

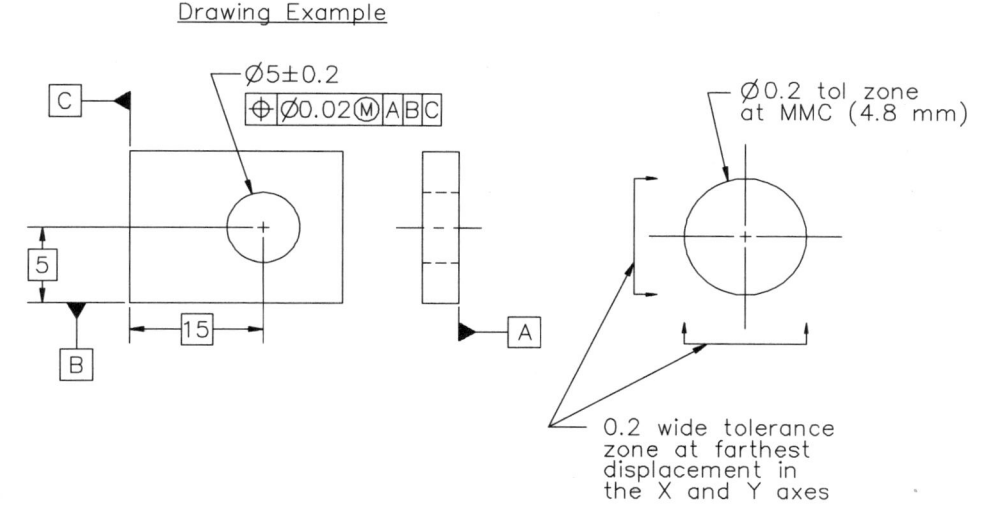

Figure 25-4 Sample drawing #4

The example shown in Fig. 25-3 was for a nondiametral positional tolerance. The example in Fig. 25-4 is a diametral positional tolerance. If this tolerance were defined at RFS rather than MMC, the procedure would be identical to the one shown in support of Fig. 25-3. The exception would be two additional columns to represent the y-axis displacement from nominal. In the example shown in Fig. 25-4, the 0.02 mm diametral tolerance zone applies only when the diameter of 5 mm is at its MMC size (4.8 mm). As it changes in size toward its LMC size (5.2 mm), bonus tolerance is gained, as shown in the following matrix.

Table 25-2 Bonus tolerance gained due to considered feature size

Feature of Size ⌀5 +/- 0.2	Allowable Position Tol.
⌀4.8 (MMC)	⌀0.2
⌀4.9	⌀0.2 + ⌀0.1 = ⌀0.3
⌀4.95	⌀0.2 + ⌀0.15 = ⌀0.35
⌀5.0	⌀0.2 + ⌀0.2 = ⌀0.4
⌀5.1	⌀0.2 + ⌀0.3 = ⌀0.5
⌀5.2 (LMC)	⌀0.2 + ⌀0.4 = ⌀0.4
⌀5.3	Bad Part

Based on current methods of calculation, it is necessary to define the total tolerance zone as a constant. To do this, and also to take advantage of the bonus tolerance gained from this feature of size as it deviates from its MMC, there is need for alternative methods of analysis. The following matrix is a proposed method of analysis. (See Table 25-3.)

Table 25-3 Analysis Matrix

Run #1

Part#	X – Axis Displacement	Y – Axis Displacement	"Calculated" Resultant Value (C.R.V.) for Run #1 $2\sqrt{\Delta x^2 + \Delta y^2}$	Actual Size of Feature $\emptyset 5 \pm 0.2$ mm	Displacement in Size from MMC. (MMC = $\emptyset$ 4.8 mm)	"Modified Resultant Value" (M.R.V.) for Run #1. MRV = CRV – Displacement from MMC. Note cannot be less than 0
1	0.06	0.08	0.2	4.95	0.15	0.05
2	0.02	0.03	0.072	4.85	0.05	0.022
3	0.04	0.04	0.113	4.9	0.1	0.013
4	0.07	0.06	0.184	4.9	0.1	0.084
5	0.03	0.05	0.117	4.9	0.1	0.017
6	0.04	0.02	0.089	4.85	0.05	0.039
7	0.05	0.04	0.128	4.9	0.1	0.028
8	0.03	0.01	0.063	4.85	0.05	0.013
9	0.01	0.03	0.063	4.85	0.05	0.013
10	0.02	0.01	0.045	4.85	0.05	0.0

Run #2

Part#	X – Axis Displacement	Y – Axis Displacement	"Calculated" Resultant Value (C.R.V.) for Run #2 $2\sqrt{\Delta x^2 + \Delta y^2}$	Actual Size of Feature $\emptyset 5 \pm 0.2$ mm	Displacement in Size from MMC. (MMC = $\emptyset$ 4.8 mm)	"Modified Resultant Value" (M.R.V.) for Run #2. MRV = CRV – Displacement from MMC. Note cannot be less than 0	Range = difference between MRV #1 and MRV #2
1	0.05	0.07	0.172	4.95	0.15	0.022	0.028
2	0.02	0.04	0.089	4.9	0.1	0.0	0.022
3	0.04	0.04	0.113	4.9	0.1	0.013	0.0
4	0.07	0.05	0.172	4.9	0.1	0.072	0.012
5	0.04	0.04	0.113	4.9	0.1	0.013	0.004
6	0.03	0.03	0.085	4.85	0.05	0.035	0.004
7	0.05	0.05	0.141	4.9	0.1	0.041	0.013
8	0.04	0.02	0.089	4.85	0.05	0.039	0.026
9	0.01	0.02	0.045	4.85	0.05	0.0	0.013
10	0.02	0.02	0.056	4.85	0.05	0.006	0.006

$\overline{R} = \Sigma R/n = 0.128/10 = 0.0128$
$\sigma = \overline{R}/d_2 = 0.0128/1.128 = 0.0113$
$3\sigma = 3 \times 0.0113 = 0.340$
$3\sigma/\text{Tol.} \times 100 = \%$ of tolerance
$0.34/0.2 \times 100 = 17.02 \%$

25.3 Summary

This chapter defined opportunities that will spur future research activities and should have made clear many of the steps needed to determine a measurement system capability along with the reasons for strict and aggressive controls. Discussions have started in 1998 within standards committees and universities to concentrate resources to research and develop standards, technical reports, and other documentation to further advance these analysis methods.

25.4 References

1. Hetland, Gregory A. 1995. Tolerancing Optimization Strategies and Methods Analysis in a Sub-Micrometer Regime. Ph.D. dissertation.
2. The American Society of Mechanical Engineers. 1995. *ASME Y14.5.1M-1994, Mathematical Definition of Dimensioning and Tolerancing Principles.* New York, New York: The American Society of Mechanical Engineers.
3. The American Society of Mechanical Engineers. 1995. *ASME Y14.5M-1994, Dimensioning and Tolerancing.* New York, New York: The American Society of Mechanical Engineers.

Chapter 26

The Future

Introduction

I have asked several recognized experts in the field of dimensioning and tolerancing to assess what they think the future holds in the area of dimensioning and tolerancing. The opinions below represent the voices of corporate management, practitioners, authors, and college professors. They represent many years of study, training and practice. These voices, along with the ones you have already heard from in this book (see section 5.17, The Future of GD&T), have expanded our horizons and broadened our understanding of a field once narrowly interpreted and dismally misunderstood.

I thank the contributors for their wisdom and insight. I look forward to seeing how these predictions unfold.

Paul Drake

Timothy V. Bogard
President, Sigmetrix
Dallas, Texas

The Future of Dimensional Management

Dimensional management as a methodology will continue to gain in acceptance with the more sophisticated companies, where high volume and high complexity exist in the product lines. The concept of dimensional management will be of interest in other types of companies where low volume and low complexity exists, but the cost of implementation in terms of training and process change will be the major barrier.

The Future of Geometric Dimensioning and Tolerancing (GD&T)

GD&T will continue to gain acceptance. The standard(s) will need to continue to evolve to (1) eliminate ambiguity, (2) improve assembly level tolerance definitions, and (3) be further consolidated to simplify the concepts for more practical usage.

The Future of Standards

Standards in the area of geometric definitions, like STEP *(STandard for the Exchange of Product)*, are critical to the long-term interoperability required by companies as they migrate across computer aided design systems, and further integrate with supplier base and customers. There will continue to be emphasis on STEP like compliance by product developers and CAD/CAE *(Computer Aided Design/Computer Aided Manufacturing)* tools providers, to allow for flexibility and ease of use.

The Future of Tolerancing in Academics

More universities are already developing courses and research expertise in the area of tolerancing. Better alliances between industry and academia will need to be forged, like ADCATS *(Association for the Development for Computer-Aided Tolerancing Systems)* at BYU *(Brigham Young University)*, to guarantee the transfer of research to the industry. If the research does not turn into easy-to-adopt concepts, methods and technologies, then the interest by industry in supporting academia will wain.

The Future of Tolerancing in Business

As Six Sigma type initiatives continue to broaden and become the critical differentiator in many companies, tolerance analysis will elevate to the same level of importance as reliability and warranty analysis for all companies. As ease of use continues to improve, the adoption of tolerance optimization techniques will proliferate in all areas of system design.

The Future of Software Tools

Software tools will go through a consolidation process whereby the basic analysis is a natural part of the design capture process. Requirements flowdown, surface-based modeling and analysis, and part producibility will become natural to the engineer, as the software tools providers continue to bury the process of tolerance optimization continuously in the design through manufacture process. Basically, ease of use will dominate the tools suppliers agenda until the tolerancing process is virtually undetectable.

Kenneth W. Chase, Ph.D.
Mechanical Engineering Department
Brigham Young University
Provo, Utah

Future of Tolerance Analysis

It is a pleasure to address the question: "What is the future of tolerance analysis?" It is a subject about which I have strong feelings. I first began teaching a course in Design for Manufacture after returning from two summers working for John Deere in 1980. Two gray-haired engineers there, who were brothers, one a designer and the other a manufacturing engineer, persuaded me that mechanical engineers should include manufacturing considerations in their designs. They spent a lot of time with me, "filling in the gaps in my education."

I began to see that tolerance analysis was the vehicle to bring design and manufacturing together. Using a common mathematical model that combines the performance requirements of the designer with the process requirements of the manufacturer provides a quantitative tool for estimating the effects each has upon the other. It truly promotes the concept of Concurrent Engineering.

At last, I can honestly say the tools are here, ready to earn a place alongside other standard CAD applications, such as kinematics, dynamics, vibrations, and finite element analysis (FEA). CAD-based tolerancing is quite sophisticated and advanced for a new CAD/CAM/CAE *(Computer Aided Design/ Computer Aided Manufacturing/Computer Aided Engineering)* tool. It had to be. No one today will accept an analysis tool that is not graphical and integrated with CAD.

Major Hurdles

As I see it, there are two major hurdles that must be overcome before tolerance analysis can succeed in becoming an enterprise tool for reducing cost and improving product performance:

1. Management acceptance.

It is not enough for a manager to see a new tool demonstrated at a trade show and buy it for his engineers. He must determine exactly where it fits in the enterprise. What problems will it solve for us? Who will be responsible for its implementation? How much will it cost to implement? Who will champion its adoption? How can we tell if we are using it effectively? How can we tell if it has saved us money?

Sometimes a change-agent within the ranks will discover a new tool and champion its adoption. But, CAD-based tolerance analysis will never reach its full potential as a product development tool until it has high-level management support, with sufficient resources and talent to make success possible.

2. Education and training.

As with other quality improvement programs, everyone involved must be educated about the role that tolerance analysis will play in the product development cycle and its expected benefits. Management, design, production—all must catch the vision.

The most challenging aspect is the fact that there is no established user base, no established curriculum, and there are no established procedures to guide us in implementing this new tool. It is much easier for a company to begin using an established CAD application, such as finite element analysis. There are many successful examples they can emulate. But tolerance analysis is still in its infancy.

The procedures for performing a finite element analysis are well established. There are many published examples. Structural analysis departments are found in most big companies. You can hire an experienced person to help set up a program in your company. But, this is not yet true for tolerance analysis.

You can't even hire the capability you need fresh out of school, because tolerance analysis is not found in the curriculum of our engineering and technology schools. Will it be there eventually? It is hard to say. The curriculum of our schools is under constant pressure. Most schools have reduced the number of hours required for graduation, while increasing the nontechnical requirements. You can't push tolerance analysis in, without pushing something else out.

For the time being, industry must expect to shoulder the burden of building the expertise they need within their own ranks. Training seminars and consultants will be needed to assist in this effort.

Unresolved Issues

Among the principal issues that must be resolved before CAD-based tolerance analysis is widely adopted:

1. The relationship to GD&T must be resolved.

There are many misconceptions about the application of GD&T standards to assembly tolerance analysis. How do MMC or RFS apply to a tolerance stackup? How about bonus tolerances? Are geometric variations applied differently in a statistical analysis versus worst case? If a form tolerance is applied to a feature of size, should two variation sources be included in the tolerance stackup? Do the size variations include the surface variations, or do they represent two independent sources of variation?

Most of the misconceptions arise from a lack of understanding of the fundamental principles upon which the GD&T standards and assembly tolerance analysis are based. We also need to get a clear concept of the difference between a specified tolerance and a measured or predicted variation.

2. New standards for assembly variation are needed.

There are no standards for computing tolerance stackup and variation propagation in assemblies. ASME Y14.5 has only recently acknowledged the existence of statistical stackup analysis. How it is to be done is still open-ended. This writer strongly feels that there should be a new set of symbols to differentiate an assembly tolerance limit involving multiple parts and a component tolerance limit applied to a single part.

3. Better data on process variations are needed.

The assembly variations predicted by tolerance analysis are only as accurate as the process variation data entered into the analysis model. However, there is very little published data describing process variations and the cost associated with specified tolerance limits. If you wait until the parts are made, so measured variations can be used in the model, you will lose one of the major benefits of tolerance analysis. In the design stage of a new product, tolerance analysis serves as a *virtual prototype* for predicting the effects of manufacturing variations before the parts are made. To fully realize this benefit, we simply must have an extensive database, which characterizes process variations over a wide range of conditions and materials.

4. Realistic expectations.

Over the years I have worked to involve industries and CAD vendors in the development of CAD-based tolerancing tools. A number of companies have given enthusiastic support. I have, however, been turned away by several companies who have said in effect: "Come back when you have a finished product." Others seem to be waiting for "push-button tolerance analysis" that will require no understanding of variation and no decision-making skills.

A state-of-the-art CAD tool cannot be developed without substantial resources and talent. It needs broad support from the CAD vendors and the end-users in industry. CAD systems will require basic changes in data structure to accommodate variation definitions. CAD vendors must adopt standard user interface tools and allow third-party access to possibly proprietary internal representations.

Industry will need to take a more active role in guiding the CAD application development and thoroughly testing the resulting software products. Industry must also develop an infrastructure for absorbing and implementing CAD-based tools into their product life cycle. Until industries learn how to apply tolerance analysis to their own enterprises, they will not be able to effectively influence its development.

Research Opportunities

Numerous opportunities exist in tolerancing research that will increase the usefulness of tolerance applications and expand their influence. They include:

1. Post-processing

Existing CAD applications, such as FEA and dynamics, have well-developed post-processing capabilities for presenting the results of analyses. Enormous quantities of numerical data are condensed into color-coded 3-dimensional contour plots, amplified deflections or dynamic animations. Similar capabilities are needed to complete the new tolerance analysis CAD tools.

2. Process capability database

This is as important to tolerance analysis as a material properties database is to FEA.

3. Early design

How early can tolerance analysis be brought into the design process? If we could evaluate the manufacturability of design alternatives at the conceptual or systems design level, significant development cost savings could be realized.

4. Flexible assemblies

Current tolerance analysis methods only treat assemblies of rigid parts. Many assemblies include flexible parts of sheet metal or plastic, which are subject to warping or distortion in addition to dimensional variation. Assembly forces are required, which can cause residual stress and distortion. By combining finite element analysis with statistical tolerance analysis, the range of stress and distortion can be estimated statistically and compared to design limits.

The Future of Tolerance Analysis Applications

As I contemplate the future of tolerance analysis, I have a vision. My vision is very optimistic. I see CAD-based tolerancing tools becoming the next "must have" CAD application. The tools will soon become available on all leading CAD platforms. There might even be some minor players who exploit a niche market by offering tolerancing tools before the major players can overcome their internal inertia. I just hope that this new tool will be technology-driven before it is market-driven

In my vision, I see a rapid expansion of training programs and short courses to fill the needs of a growing user base. An increasing number of success stories will appear in publications and corporate news reports. Established procedures will emerge. Companies will compete for the experienced practitioners. Experts will set up shop as consultants.

Skilled users will be found among designers and manufacturing personnel, who will find themselves talking to one another more frequently in normal voices about variation, quality and performance issues. New departments and organizations will emerge in which both design and manufacturing are represented, working as a team.

As the final scene of my vision closes, I see an engineering designer, a manufacturing engineer, and a manager walking into a giant sunset, arm-in-arm, ready to compete in the world marketplace.

Perhaps it was only a dream after all. But, it could happen if industry and CAD vendors catch the vision of the tremendous opportunities and benefits that CAD-based tolerancing offers to those who pursue it vigorously.

Don Day
Professor of Engineering Technologies
Monroe Community College, Rochester, NY
Member of US national standards committees Y14, Y14.5M, Y14.5.2, Y14.8

Barriers to the Future Success of Geometric Dimensioning and Tolerancing

The future success of GD&T will require many changes to the way many of us conduct our business. Barriers continue to exist that prevent GD&T from realizing its full potential. It continues to be seen as a "drafting standard." When applied properly at the right time, GD&T has a tremendous impact on cost, quality and time to market. Of all the tools available to a concurrent engineering team, GD&T is one of the most powerful. Despite this, proper use of GD&T continues to lag. The following is a list of areas where opportunities still exist to remove barriers and allow GD&T to realize its full potential. In all of these areas, there is a persistent need for individuals to become knowledgeable of GD&T and its benefits.

Management's Role - As more companies go about their downsizing or right sizing activities, the standards group is often the first to go. This results in corporate standards not being maintained. In addition, sponsorship of committee members to the national and international standards committees is not being supported. Without this representation on standards committees, companies no longer have input to the standards writing activities and will have to accept new and revised standards that may not work well for their particular industry. Without someone overseeing standards selection, use and training within a company, CAD files and drawings are generated that are unclear and destine projects to high scrap, rework, "use *as is* decisions," engineering changes, and increased cycle times.

Nearly all companies require that a design review be conducted by the concurrent engineering team. To avoid unnecessary drawing activity and expensive changes to drawings and CAD models, drawing "previews" rather than "reviews" should be held. Conducting a "review" is too late. Someone has already spent considerable time detailing the part drawing or file. Management should require that one or more design "previews" be held at the model stage. A preview gives the concurrent engineering team the chance to make suggestions and changes regarding the part geometry, datums, dimensioning and tolerancing. Changes resulting from a "review" require modifying the model and the GD&T causing added expense and increased total cycle time. Changes are an integral part of the iterative design process, but they must occur upstream.

Design Engineering - Since most engineers graduate not being able to read an engineering drawing, engineers must make certain they know what they are signing. They must seek out quality training in GD&T. They must also make the correct application of GD&T to their designs a priority. Since the output from their area (namely the CAD file and/or drawing) will drive the entire process, proper application of tolerancing is imperative. Also, they must seek manufacturing variation data and understand the impact of their specifications on the manufacturing process capability, product quality, and overall cost. Geometric tolerance is the numerator of the Cp and Cpk calculations.

Quality - There is a lot of inspection equipment and software available today that does not comply with the national and international standards. All capital investments should be for equipment that is compatible with the requirements of the design. Often equipment and software is justified by a return on investment. The ROI *(return on investment)* looks better if it can be argued that more parts may be inspected per hour and minimal or no fixtures will be required. This usually leads to greater uncertainty. When selecting equipment and software, make certain it complies with the Y14.5 and Y14.5.1 Standards. Expediency should not be at the expense of quality.

Production - GD&T does not dictate how parts are to be produced. In fact, process information should not be on drawings unless a particular process is required to assure that the part will function properly. Manufacturing may, however, make decisions about the order of operations and how in-process inspection may be applied to assure production of conforming parts. Also, when production understands and can distinguish between size, form, orientation, and location, it is easier to isolate sources of variation within the process. A thorough knowledge of GD&T can greatly assist with process variation reduction.

Software Manufacturers - There is a tremendous need for software that will automate the process of tolerance allocation and analysis. For over a decade we have been able to perform circuit analysis of electrical devices. We need a mechanical equivalent that is accurate, efficient, conforms to the Y14.5 Standard and most of all is user friendly. In addition, quality computer-based training (CBT) is needed in the areas of application, inspection, and tolerance analysis. Software developers—listen to the voice of your customer and everyone will win.

The Standards Committee - The Y14.5 Standard is a wonderful tool; however, proper use seems to elude most designers. Over the past fifty years as the need to express more complex design requirements has emerged, many concepts and symbols have been added. Many of these new concepts may obsolete or overlap earlier controls. The committee needs to consider streamlining the use of symbols to make the standard more user friendly with a shorter learning curve. In addition, the hierarchy of geometric controls needs to be emphasized to help designers understand the most efficient application of controls. To assure the Standards continued success, its users need to better understand application.

Seminar Leaders - Emphasis must be placed on the team building and communication aspects of GD&T. Too often GD&T is still being presented as a drafting standard. Although Y14.5 is a standard in engineering documentation, the language of GD&T has as much impact on the entire enterprise as Quality Function Deployment, Design for Assembly and Manufacturability, Total Quality and the other up-front tools in use today. Seminars should be filled with case studies, decision diagrams, solid models and other educational tools that help the students relate the concepts to their workplace. Whenever possible, the instructor should customize the training to the audience using their parts and prints. The support should not end once the course evaluations are collected. The instructor should be available for future questions and clarification of material that has been presented.

Academia - Engineering faculty need to learn what GD&T is all about. Although a few engineering colleges offer some education in GD&T, it is usually optional or a small part of a course in descriptive geometry or CAD. It is time to stop graduating engineers who cannot read the documents they are signing. Since many engineers do not understand GD&T, their analysis is usually based on nominal values. Rarely are parts engineered with all possible part variation taken into account. Engineering faculty needs to take responsibility for assuring that every mechanical, manufacturing, and quality engineer they graduate is capable of reading part drawings. GD&T should be integrated into the teaching of finite element analysis, design for assembly and other topics that deal with part geometry.

Manufacturing builds wealth for a nation. The one common thread throughout the entire manufacturing enterprise is the engineering drawing. If it is incorrect or incomplete, the entire operation will suffer. Communication is the key, and the key to communication on parts drawings is GD&T.

Paul Drake

The Future of Tolerancing

Historically, the method to communicate the allowable part feature variation from design to manufacturing has been with tolerances. As the design world migrates form using paper drawings to CAD models the role of tolerances will change. As manufacturing migrates toward statistical process control, the role of tolerances may change dramatically

The traditional approach to mechanical tolerancing follows what I call a *top down* process where requirements are "flowed down" from the customer to the manufacturing shop floor. In my business, this classical scenario looks something like this:
- The customer flows down requirements in the form of design specifications.
- A systems engineer allocates the customer's requirements across the various "disciplines" in the form of mechanical design requirements, electrical design requirements, software requirements, etc.

- The mechanical design requirements flow down to subassemblies within the mechanical design.
- The mechanical subassembly requirements flow down to mechanical piece parts within the subassembly.
- The piece part requirements are flow down to manufacturing shops. One means of flowing down the subassembly requirements is with dimensional and tolerancing requirements on each piecepart.

In this process, the mechanical designer communicates to the manufacturing shops "this is what I need" to meet the mechanical performance requirements. A *significant drawback* to this process is that we have no way of knowing how well the manufacturing process(es) can build parts that meet these tolerances. As customer requirements become more difficult to achieve, there may not be enough "tolerance" available to manufacture cost-effective parts.

Historically, mechanical tolerances have been functionally driven. In general, mechanical design engineers are penalized if designs don't function, so they place a lot of emphasis on making sure their design works. As we know, the winning companies of the future will be the ones who can minimize the cost, while tolerancing systems to meet customer requirements.

Now the question arises, "How do we do this?" I propose that one method is to treat manufacturing requirements as *inputs to* the design process, instead of *outputs of* the design process. In the classical scenario we ask, "How well can we manufacture parts to meet system requirements." In this scenario we ask, how well can we meet system requirements, if we know the capabilities of the manufacturing processes.

In this scenario,
- The manufacturing requirements are flowed up to the pieceparts. One means of doing this is by incorporating manufacturing process capabilities into the design process. If we know the variation of the processes used to manufacture parts, we can calculate the expected variation of features on a part.
- The piecepart feature variations are inputs to the subassembly. If we know the expected variation of each feature on a part, we can mathematically calculate the expected variation of the subassembly.
- If we know the expected variation of a subassembly, we can mathematically calculate the variation of mechanical systems.
- If we know the expected variation of mechanical systems, we can mathematically calculate the impact this variation has on customer requirements. If we understand variation, we can assess the risk (probability) of meeting the customer's requirements.

I call this a *bottom up* process. In the traditional design process, we ask: "What machines do we need to use to manufacture parts that will function?" In the process described above, we ask ourselves: "If we manufacture parts using certain machines, how well will they function? The key benefit of this method is that we can mathematically calculate how well the parts will perform an intended design function. We can capture design risk with metrics such as dpmo *(defects per million opportunities)*, probability of nonconformance, or "sigma."

This is a radically different way of designing systems. One thing that is unique is that this process doesn't use tolerances to drive manufacturing; it uses manufacturing machine capabilities to drive design. If we look at how we define quality in terms of Cp, this is exactly what we are measuring. The generic definition of Cp is "customer requirements divided by manufacturing process capabilities." (See Chapters 8 and 10.) Cp is a measure of the balance that we try to achieve between design and manufacturing. In order to increase Cp, we have two options: We can increase the (customer) requirements, or we can use better manufacturing processes.

Historically, this is what we have done. If we can't build *functional* parts with a certain manufacturing process, we go to a process that can hold a *tighter tolerance*. When we move from a milling process to a boring process, we are moving from a process that is less capable to one that is more capable. In general, if we use the best machines we have and we still cannot build *good* parts, we go to the customer and ask for relief on the requirements. Cp is a mathematical measure of what I just described.

This new product design process is difficult for many to understand because it is not tolerance driven; nor is it driven by the functionality of the design. I believe the following are barriers that keep us from adopting the second process:
- It is extremely difficult to change (product design) processes, especially ones that have been around for many years.
- How do I communicate (to manufacturing) the process capability that I used in my variation analysis? The statistical tolerancing symbol in ASME Y14.5 (see Chapter 11) is the first step in making this happen.
- Since there are no tolerances, we don't know what to *inspect*. The key to making this product design process work is to inspect manufacturing processes, not parts. The successful companies in the future will figure out how to verify statistical tolerancing requirements on the manufacturing floor.
- The new product design process works well, as long as the parts, assemblies, and systems function well. If we have problems, it's difficult to track down the culprit.
- Since this process is not tolerance driven, most standards do not support it.
- The new product design process uses statistical techniques (described in Chapters 11, 12, and 13). The language to communicate these requirements from design to manufacturing is not available. We try to force GD&T to do this with the statistical tolerancing symbol, but it doesn't work well. We need *another language* that is manufacturing (statistically) driven to make this process work.

As we enter the 21st century and tools become more sophisticated, we will be able to better support the *bottom up* method. The winning companies will be the ones who figure out how to make this process work. The winning companies of the future will be the ones who can eliminate the most waste (nonvalue-added activity) from the process. Imagine if we could build systems without using tolerances. Imagine how much time we could save if we went directly from a nominal design (CAD database without tolerances) directly to the shop floor. Imagine if we could build systems that meet our customers' needs without inspecting parts. The challenge for the winning companies of the next century is to figure out how to do this.

Gregory A. Hetland, Ph.D.
Manager, Corporate Standards and Measurement Sciences
Hutchinson Technology Inc., Hutchinson, Minnesota
Member of several US national, international, and industrial standards committees on global tolerancing and supporting metrology

The Future of Global Standards and Business Perspective

Worldwide harmonization of standards development initiatives must be a key focus. The world must work toward the development and acceptance of a single set of technically valid standards to eliminate global confusion. Throughout the world, national and industrial standards groups are developing technical standards in a shell. Meaning, in many cases they are not aware of past efforts or existing efforts on the same development topic. This duplication of efforts is burdensome due to lack of focus on understanding the baseline development as well as it postpones the advancement of related technical activities.

Communication or lack of worldwide communication is one of the reasons for this duplication of efforts. National and Industrial standards developers must put in place strategic objectives to work together to ensure common needs are accomplished with the least amount of burden that is optimization of resources to accomplish objectives at the least amount of total expense.

Funding of standards initiatives is a key problem within the US as well as other countries. We need a single standards initiative within the US that everyone throughout the US can count on and I believe we need the help of the US government to accomplish this. We also need to ensure we have key resources in place driving and managing the development initiatives as well as heading up the communication and integration of these development initiatives. These communication initiatives should have as its key focus the benefits of each standard to industry as well as recommended paths for the phase-out of existing standards to be obsolete.

The Future of Dimensioning & Tolerancing Standards

Dimensioning and Tolerancing is in a state of flux. The world cannot afford multiple systems that are all incomplete and we must drive toward a system of engineering precision in the form of advanced and simplified tolerancing expression. Product development has been on a fast track of miniaturization for years and parallel to this is the aggressive requirement of tolerance truncation. These two drivers alone are forcing a much greater level of precision then ever recognized or perceived in the past.

Simplification of our global system must be a focus item for standards developers. Linear tolerancing strategies are ambiguous and clearly a duplicate dimensioning and tolerancing methodology from its parallel and less ambiguous system of Geometric Dimensioning and Tolerancing. The complete system requires aggressive development as well as simplification. Eliminating or de-emphasizing the duplicate, more ambiguous system of linear tolerancing would be a positive start on this simplification path. A second step would be in the reduction of symbology that reflects duplicate representation of tolerance boundaries. If the full scope of boundary representation can truly be represented by few geometric symbols, as I predict, this would truly be a form of simplification due to less for the user to learn and it would be an intuitive language. Additional benefits would be less computer variations at the CAD level, reduced training, better understanding of requirements, less mathematical representations for the accumulative symbols, and less algorithms required for the analysis of each geometry class.

Significant development efforts are required to close existing gaps in the arena of tolerance expression to ensure we have a robust system that is unambiguous. Development of "extension principles" in the following areas must be key focus items to eliminate the existing gaps.
- Separation of surface roughness and waviness parameters from form tolerances
- Definitions and flexibility's related to datums
- Complex geometries and tolerance boundaries
- Statistical analysis of geometric tolerances
- Assembly level tolerancing
- Statistical tolerancing
- Tolerance analysis
- 3-D modeling

The current state of ISO *(International Organization for Standardization)* initiatives related to dimensioning and tolerancing is in a state of turmoil. Key individuals involved with the development lack the core technical understanding and sensitivity of past and current dimensioning and tolerancing practices and are driving change which will clearly have a negative impact on industry throughout the world.

An incorrect perception exists throughout the world: in that it is believed that for a standard to be considered international, it must be labeled ISO. This is clearly incorrect and we must change this perception. ASME Y14.5 is the most broadly used dimensioning and tolerancing standard in the world today and will clearly continue to be the most solid basis for industrial use. Development initiatives with the Y14.5 committee, as well as other related committees such as ASME Y14.5.1, Y14.41, and others, are on a much more aggressive development path to achieve a sound basis to meet worldwide industrial needs. Our challenge within the US is to establish strategic initiatives that will ensure effective integration of these documented initiatives throughout the world.

The Future of Metrology Standards

One of the most strategic initiatives being kicked off within the industrial metrology community is the development and integration of advanced analytical tools used to better understand the uncertainty related to task specific measurements. Understanding which error sources contribute to task specific measurement uncertainty and understanding the analytical methodology is the key to the advancement of understanding measurement uncertainty within industrial applications. For years, national physical laboratories such as NIST *(National Institute of Standards and Technology)* have been using such tools for a number of years. The basis for this and the tool developed to ensure standardization across all laboratories is the "**G**uide to the Expression of **U**ncertainty in **M**easurement," commonly referred to as GUM.

The industrial challenge we now face is in the development of tools (Technical Reports, Standards, and user-friendly guides) to help industry effectively integrate these advanced tools and understand the magnitude of global benefits in doing so. The primary benefit is in having the level of analytical tools required to confidently ensure a controlled understanding of measurement uncertainty when determining conformance to requirements of product produced and shipped.

ASME subcommittee B89.7 has been recently established with the mission to support US manufacturing industry in a smooth, economical transition to the requirement of using measurement uncertainty. The motivation for the establishment of B89.7 and for its mission and scope of work lies in the growing importance of measurement uncertainty in international trade. Over the next 3 to 10 years, critical development and integration of these tools will be a critical basis for advancements in the metrology community and will be recognized as a sound tool by manufacturing and design engineering groups that they will grow to count on to ensure needs are being met with confidence.

Industry will find these advanced analytical tools will also be beneficial in understanding process uncertainty as well as design uncertainty. In all the task specific uncertainty analysis I have been involved with, I find it important to note that there has been more uncertainty related to the engineering requirement (tolerance specification) than there is in the delta of the targeted uncertainty and the actual uncertainty derived. It's important to understand the meaning of this statement so it is not taken out of context. I'm stating that the tolerance defined for any of the features or feature characteristics on an engineering drawing is or should be the key parameter with the greatest opportunity to scrutinize which will yield a benefit. There is more opportunity analytically to evaluate the possibility of tolerance reallocation or new tolerance expression, which will yield greater allowable tolerance to the feature or feature characteristic in question. It is critical these tools are used to direct all advanced development initiatives in a manufacturing environment.

Al Neumann
President/Director
Technical Consultants, Inc.
Section 4 Sponsor, Datums ANSI/ASME Y14.5M

The Future of Dimensional Management

I see that the concept of dimensional management will become more important in the future. Dimensional management incorporates form, fit, function, inspection, assembly, manufacturing, and variation into the tolerancing scheme. We are still in the early stages of understanding tolerancing. Dimensional management is a living process and is constantly evolving. As dimensional management becomes more important, it will also become more complicated.

There will be another division of labor. I believe that successful companies will develop a separate group of tolerancing engineers who are experts in tolerancing. In fact, some companies are doing it now. These tolerancing engineers understand it all. They specialize in tolerance analysis and applying tolerances to parts and assemblies. If we pay attention to history, we have already seen this progression. In the beginning, engineers did everything. They designed the parts, drew the prints, did stress and fit up calculations, ran the prints, machined the parts, inspected the parts and assembled the parts. This is too much for one engineer to do.

Tolerancing is becoming more complicated. We now realize that tolerances and part definition is much more complicated than we originally thought. The most important document we have in a company is our product drawing. Without a clear definition of our product we have nothing. We need experts that specialize in tolerancing.

When you think about it, the general population should know something about tolerancing but they do not need to know everything. In the future, CAD operators will draw the pictures of the parts and do general part design. Afterwards, tolerancing engineers will take the design and make it work dimensionally.

Manufacturing people must have a general understanding of tolerancing. The specialized manufacturing tolerancing engineer will set up all the fixtures and processes and machines to meet the GD&T specifications. The manufacturing personnel will operate the machines based on the tooling set up by the manufacturing engineers. They will work to the process plan developed by the manufacturing engineer.

Inspectors don't have to understand it all either. There will be special quality tolerancing engineers to do this. They will define the gages and the inspection procedures to follow to meet geometric requirements. The inspection personnel will work the dimensional measurement plan that is developed by the tolerancing engineer. The general population will still have to have a basic understanding of tolerancing, but the intricacies of the tolerancing stackups and analysis will be done by the tolerancing engineers.

The Future of Tolerancing in Academics

Tolerancing must be part of basic education. It must start at the high school and trade school level. More people are becoming more serious about geometric tolerancing. They used to apply the tolerancing because they were told to do it or it was the "in" thing. More colleges and university professors are becoming involved in tolerancing and looking at it from a higher level than in the past. This will help deploy tolerancing in the business environment, because businesses won't have to spend the money they have invested in the past to develop this expertise.

The Future of GD&T

Geometric tolerancing will increasingly become more prevalent. Many companies are using a combination of plus/minus tolerancing and geometric tolerancing. More geometric tolerancing, primarily profile and positional tolerancing, will be used. Positional tolerancing locates features of size such as holes, slots, tabs, and pins. Profile tolerancing is used to locate nonfeatures of size such as surfaces. Plus/minus tolerances are used for the size of features. The location, form, and orientation of the features are done with geometric tolerances.

Tolerancing needs to be 3-dimensional (3-D). The parts are 3-dimensional, drawn in 3-D solids in CAD; the manufacturing process is 3-D; and the inspection process is 3-D using coordinate measuring machines. The old plus or minus system only gave us 2-D tolerancing. We need to think in 3-D. Everyone must understand that geometric tolerancing is the basic communication tool among engineers. Historically, people have used GD&T for the so-called "important features." I see geometric tolerancing not just for "important features" but for *all* features.

The Future of Software Tools

I see computers doing more tolerancing stacks. In the future, we will do more and more tolerancing within the solid model. Since tolerancing will be imbedded in the solid model, we will have a closer integration between inspection design (CAD), inspection CMMs *(coordinate measuring machines)* and manufacturing CNC *(computer numerical controlled)* machines. CMMs and inspection equipment will read the imbedded design specifications in the model. There will also be a database in the CAD systems to provide more integration of the manufacturing process information in the tolerancing. I anticipate a larger emphasis on reducing and understanding variation. This will promote more statistical tolerancing of parts.

The Future of Tolerancing Standards

Standards will become more important in the future, although I do not see a complete union between ISO and ANSI (ASME) standards any time in the near future. There are a lot of cultural and philosophical differences that must be worked out. International standardization will get closer but there will be no complete union for some time. It is important that ASME or ANSI standards keep up with the technology, as we may find a commercial computer software program becoming the de facto standard because it's easy and simple to use.

Bruce A. Wilson
Dimensional management specialist
Aerospace Industry, St. Louis, Missouri
Author of the book, *Design Dimensioning and Tolerancing*
Member and officer on national and international standards development committees

The Future of Dimensioning and Tolerancing*

Changes are rapidly taking place in the field of dimensioning and tolerancing. A quick look at recent and ongoing changes will help to understand what the future is likely to hold.

The manufacturing world has started associating many terms with various aspects of this wide and complex field. Names such as Geometric Dimensioning and Tolerancing (GD&T) emerged as dimensioning and tolerancing became more sophisticated. It was as if the improvement in our comprehension of the subject and ability to more clearly define requirements somehow required a new name. Computer programs were developed to assist in the calculation of tolerances and to assess the assembly variation caused by applicable factors. The use of these tools for dimensioning and tolerancing was called dimensional management. Dimensional management expanded to include manufacturing process controls

---

*Reprinted by permission of Bruce A. Wilson.

in more progressive companies. Manufacturing process control was not always called dimensional management, but was sometimes made a separate initiative called variability reduction.

The variety of names used for various aspects of dimensioning and tolerancing were tied to company names, so each company that began a similar effort tried to come up with a unique name that identified the process. At least one major company has multiple groups working nearly identical efforts to implement improvements related to dimensioning and tolerancing along with the appropriate manufacturing controls. Internal power struggles result in different names used for the similar initiatives. The proliferation of names does not indicate progress or the number of advancing initiatives.

The point of the above description is that many names are being applied to doing the job of calculating and defining dimensions and tolerances for detail parts and assemblies. The associated manufacturing process controls are seeing the same proliferation of names for the process improvement methods. Names should not become an issue in determining how things improve, so the picture of the future that I will paint does not depend on the terminologies that are used for such a wide array of efforts in industry.

Chapter one of <u>Design Dimensioning and Tolerancing</u> states that dimensioning and tolerancing requirements are likely to become part of the CAD data file and no longer require a paper drawing to communicate those requirements. That prediction was first written in 1988. This prediction has to some extent taken place. Computer programs exist in 1999 that permit tolerances for a feature to be associated with an entity in a CAD file. However, the way in which the tolerance requirement is stored and associated with the entity is not yet standardized. This means the information is not as universally readable as a paper drawing that shows the tolerancing symbology.

At least three companies are hotly competing to achieve a superior tolerance application and analysis program. Many other companies are involved in efforts, but they may find the competition so fierce that they will not have the resources to stay in the race.

Progress has been rapid over the past few years. The first tolerance analysis programs did not operate within the CAD program. They were stand alone. Data was output from the CAD model to the analysis program and then the analysis completed. Any updates to the CAD model were made manually. It is likely that future development of computer programs will permit work within the CAD model (some currently claim this capability) and information from the analysis will be updated in the CAD model automatically.

One problem with the current analysis software has been the amount of effort to become proficient in its use. Inexperienced users can output results that look accurate but be filled with errors. Reviewers have less experience than the person who made the errors, so nobody catches the mistakes. Efforts are being made to make the software more user friendly, and this will reduce the learning curve.

Attempts to produce a software package that speeds up the modeling process are introducing risks that may be easily overlooked. The software is permitted to select points on surfaces that later get used for determining part locations in an assembly. The automated point selections are made on the basis of routines written in the computer program code. If the user does not understand how the software makes the point selections, then a needed decision to override the program might not be made. The result will be an inaccurate analysis.

Many problems exist and a few have been described above, but progress will be made and the problems overcome. The future will eventually include CAD systems and the associated manufacturing equipment that do not require any paper drawing. There probably will not even be a drawing in the CAD system. It is likely to contain only a 3-D model with all the requirements attached to part features in such a way that either humans or compatible machines can read the data.

Caution is recommended in using the emerging software tools to ensure they are properly used, and that any outputs are accurate. Many of the new products available today are very high quality, but the results obtained by inexperienced people can be extremely misleading. A well educated and experienced mind is still superior to the best available computer and software package.

Reference

1. Wilson, Bruce A. 1996. *Design Dimensioning and Tolerancing.* South Holland, Illinois: The Goodheart-Willcox Company, Inc.

Figures

Figure P-1	Product development process	xxiv
Figure 1-1	Taguchi's loss function and a normal distribution	1-5
Figure 1-2	Graphical definition of short-term performance for a single characteristic	1-7
Figure 1-3	Graphical definition of long-term Six Sigma performance for a single characteristic (distribution shifted 1.5σ)	1-8
Figure 2-1	Dimensional management tools	2-3
Figure 2-2	Variation simulation analysis	2-8
Figure 2-3	The dimensional management process	2-9
Figure 3-1	Linear dimensioning and tolerancing boundary example	3-3
Figure 3-2	Linear and geometric dimensioning and tolerancing boundary example	3-4
Figure 3-3	Fully geometric dimensioned and toleranced boundary example	3-6
Figure 3-4	Tolerance stack-up graph (linear tolerancing)	3-7
Figure 3-5	Plus/minus versus diametral tolerance zone comparison	3-9
Figure 3-6	Tolerance stack-up graph (position at RFS)	3-10
Figure 3-7	Tolerance stack-up graph (position at MMC)	3-12
Figure 3-8	Tolerance stack-up graph (zero position at MMC)	3-14
Figure 3-9	Summary graph	3-15
Figure 4-1	Note drawing	4-3
Figure 4-2	Casting drawing	4-5
Figure 4-3	Machined part made from casting	4-6
Figure 4-4	Machined part made from bar stock	4-7
Figure 4-5	Stamped sheet metal part drawing	4-8
Figure 4-6	Flat pattern layout drawing	4-9
Figure 4-7	Exploded pictorial assembly drawing	4-10
Figure 4-8	2-D sectioned assembly drawing	4-11
Figure 4-9	Border, title block, and revision block	4-12
Figure 4-10	First-angle projection	4-17
Figure 4-11	Third-angle projection	4-18
Figure 4-12	Auxiliary view development and arrangement	4-19
Figure 4-13	Full section	4-20
Figure 4-14	Half section	4-20
Figure 4-15	Offset section	4-21
Figure 4-16	Broken-out section	4-21
Figure 4-17	Revolved and removed section	4-22
Figure 4-18	Conventional breaks	4-22
Figure 4-19	Partial views	4-23
Figure 4-20	Internal and external feature rotation	4-24
Figure 4-21	Isometric projection	4-24
Figure 4-22	Envelope principle	4-26
Figure 4-23	General dimension types	4-26
Figure 4-24	Dimension elements and measurements	4-27
Figure 4-25	Surface characteristics	4-28
Figure 4-26	Surface texture examples and attributes	4-29
Figure 5-1	Drawing showing distance to ideal hole location	5-4
Figure 5-2	House built without all of the appropriate tools	5-5

F-1

F-2   Figures

Figure 5-3	House built using the correct tools	5-5
Figure 5-4	Drawing that does not use GD&T	5-6
Figure 5-5	Manufactured part that conforms to the drawing without GD&T (Fig. 5-4)	5-7
Figure 5-6	Drawing that uses GD&T	5-7
Figure 5-7	Using English to control part features	5-12
Figure 5-8	Symbols used in dimensioning and tolerancing	5-13
Figure 5-9	Compartments that make up the feature control frame	5-14
Figure 5-10	Methods of attaching feature control frames	5-17
Figure 5-11	Method of identifying a basic .875 dimension	5-18
Figure 5-12	"Statistical tolerance" symbol	5-18
Figure 5-13	Generating a size limit boundary	5-21
Figure 5-14	Conformance to limits of size for a cylindrical feature	5-21
Figure 5-15	Conformance to limits of size for a width-type feature	5-22
Figure 5-16	Size limit boundaries control circularity at each cross section	5-22
Figure 5-17	Levels of control for geometric tolerances modified to MMC	5-24
Figure 5-18	Levels of control for geometric tolerances modified to LMC	5-25
Figure 5-19	Cylindrical features of size that must fit in assembly	5-26
Figure 5-20	Level 1's size limit boundaries will not assure assemblability	5-26
Figure 5-21	Rule #1 specifies a boundary of perfect form at MMC	5-27
Figure 5-22	Rule #1 assures matability	5-28
Figure 5-23	Using an LMC modifier to assure adequate part material	5-28
Figure 5-24	Feature of size associated with an MMC modifier and an LMC modifier	5-29
Figure 5-25	Nullifying Rule #1 by adding a note	5-29
Figure 5-26	MMC virtual condition of a cylindrical feature	5-30
Figure 5-27	MMC virtual condition of a width-type feature	5-31
Figure 5-28	LMC virtual condition of a cylindrical feature	5-32
Figure 5-29	Using virtual condition boundaries to restrain orientation between mating features	5-33
Figure 5-30	Using virtual condition boundaries to restrain location (and orientation) between mating features	5-34
Figure 5-31	Zero orientation tolerance at MMC and zero positional tolerance at MMC	5-36
Figure 5-32	Resultant condition boundary for the ⌀.514 hole in Fig. 5-30	5-37
Figure 5-33	Levels of control for geometric tolerances applied RFS	5-39
Figure 5-34	Tolerance zone for straightness control RFS	5-40
Figure 5-35	Tolerance zone for flatness control RFS	5-40
Figure 5-36	Example of restrained and unrestrained actual mating envelopes	5-41
Figure 5-37	The true geometric counterpart of datum feature B is a restrained actual mating envelope	5-42
Figure 5-38	Actual mating envelope of an imperfect hole	5-44
Figure 5-39	Actual minimum material envelope of an imperfect hole	5-45
Figure 5-40	Straightness tolerance for line elements of a planar feature	5-51
Figure 5-41	Flatness tolerance for a single planar feature	5-52
Figure 5-42	Circularity tolerance (for nonspherical features)	5-53
Figure 5-43	Circularity tolerance applied to a spherical feature	5-54
Figure 5-44	Cylindricity tolerance	5-55
Figure 5-45	Circularity tolerance with average diameter	5-56
Figure 5-46	Cylindricity tolerance applied over a limited length	5-57
Figure 5-47	Straightness tolerance applied on a unit basis	5-57
Figure 5-48	Flatness tolerance applied on a unit basis	5-58
Figure 5-49	Radius tolerance zone (where no center is drawn)	5-58
Figure 5-50	Radius tolerance zone where a center is drawn	5-59
Figure 5-51	Controlled radius tolerance zone	5-60
Figure 5-52	Establishing datum reference frames from part features	5-62
Figure 5-53	Selection of datum features	5-63
Figure 5-54	Establishing datums on an engine cylinder head	5-63
Figure 5-55	Selecting nonfunctional datum features	5-64
Figure 5-56	Datum feature symbol	5-65
Figure 5-57	Methods of applying datum feature symbols	5-66
Figure 5-58	Parts contacting at high points	5-67
Figure 5-59	Building a simple DRF from a single datum	5-70
Figure 5-60	3-D Cartesian coordinate system	5-70

Figures F-3

Figure 5-61	Datum precedence for a cover mounted onto a base	5-71
Figure 5-62	Arresting six degrees of freedom between the cover and the TGC system	5-72
Figure 5-63	Comparison of datum precedence	5-74
Figure 5-64	Feature of size referenced as a primary datum RFS	5-76
Figure 5-65	Feature of size referenced as a secondary datum RFS	5-76
Figure 5-66	Feature of size referenced as a primary datum at MMC	5-77
Figure 5-67	Feature of size referenced as a secondary datum at MMC	5-77
Figure 5-68	Feature of size referenced as a primary datum at LMC	5-78
Figure 5-69	Feature of size referenced as a secondary datum at LMC	5-78
Figure 5-70	Bounded feature referenced as a primary datum at MMC	5-79
Figure 5-71	Bounded feature referenced as a secondary datum at MMC	5-79
Figure 5-72	Cylindrical feature of size, with straightness tolerance at MMC, referenced as a primary datum at MMC	5-80
Figure 5-73	Two possible locations and orientations resulting from datum reference frame (DRF) displacement	5-81
Figure 5-74	DRF displacement relative to a boundary of perfect form TGC	5-82
Figure 5-75	DRF displacement allowed by all the datums of the DRF	5-84
Figure 5-76	Unequal X and Y DRF displacement allowed by datum feature form variation	5-85
Figure 5-77	Unequal X and Y DRF displacement allowed by datum feature location variation	5-85
Figure 5-78	"Common DRF" means "identical DRF"	5-86
Figure 5-79	Using simultaneous requirements rule to tie together the boundaries of five features	5-87
Figure 5-80	Specifying separate requirements	5-88
Figure 5-81	Imposing simultaneous requirements by adding a note	5-88
Figure 5-82	Datum feature surface that does not have a unique three-point contact	5-89
Figure 5-83	Acceptable and unacceptable contact between datum feature and datum feature simulator	5-90
Figure 5-84	Datum target identification	5-93
Figure 5-85	Datum target application on a rectangular part	5-94
Figure 5-86	Datum target application on a cylindrical part	5-96
Figure 5-87	Using datum targets to establish a primary axis from a revolute	5-98
Figure 5-88	Setup for simulating the datum axis for Fig. 5-87	5-99
Figure 5-89	Target set with switchable datum precedence	5-100
Figure 5-90	Three options for establishing the origin from a pattern of dowel holes	5-101
Figure 5-91	Pattern of holes referenced as a single datum at MMC	5-102
Figure 5-92	Application of orientation tolerances	5-104
Figure 5-93	Tolerance zones for Fig. 5-92	5-105
Figure 5-94	Application of tangent plane control	5-105
Figure 5-95	Applying an angularity tolerance to a width-type feature	5-106
Figure 5-96	Applying an angularity tolerance to a cylindrical feature	5-107
Figure 5-97	Controlling orientation of line elements of a surface	5-108
Figure 5-98	Applications of orientation tolerances	5-110
Figure 5-98	Applications of orientation tolerances (continued)	5-111
Figure 5-99	Erroneous wedge-shaped tolerance zone	5-112
Figure 5-100	Controlling the location of a feature with a plus and minus tolerance	5-113
Figure 5-101	Methods for establishing true positions	5-114
Figure 5-102	Alternative methods for establishing true positions using polar coordinate dimensioning	5-115
Figure 5-103	Restraining four degrees of freedom	5-116
Figure 5-104	Implied datums are not allowed	5-117
Figure 5-105	Establishing true positions for angled features—one correct method	5-118
Figure 5-106	Establishing true positions from an implied datum—a common error	5-118
Figure 5-107	Specifying a projected tolerance zone	5-119
Figure 5-108	Showing extent and direction of projected tolerance zone	5-119
Figure 5-109	Projected tolerance zone at MMC	5-120
Figure 5-110	Different positional tolerances (RFS) at opposite extremities	5-121
Figure 5-111	Bidirectional positional tolerancing, rectangular coordinate system	5-122
Figure 5-112	Virtual condition boundaries for bidirectional positional tolerancing at MMC, rectangular coordinate system	5-123
Figure 5-113	Tolerance zone for bidirectional positional tolerancing applied RFS, rectangular coordinate system	5-124
Figure 5-114	Bidirectional positional tolerancing, polar coordinate system	5-125
Figure 5-115	Positional tolerancing of a bounded feature	5-126

F-4 Figures

Figure 5-116	Standard catalog handle	5-127
Figure 5-117	Handle technical bulletin	5-128
Figure 5-118	Avionics "black box" with single positional tolerance on pattern of holes	5-128
Figure 5-119	Avionics "black box" with composite positional tolerance on pattern of holes	5-129
Figure 5-120	PLTZF virtual condition boundaries for Fig. 5-119	5-130
Figure 5-121	FRTZF virtual condition boundaries for Fig. 5-119	5-131
Figure 5-122	One possible relationship between the PLTZF and FRTZF for Fig. 5-119	5-132
Figure 5-123	One possible relationship between the PLTZF and FRTZF with datum B referenced in the lower segment	5-133
Figure 5-124	Two stacked single-segment feature control frames	5-134
Figure 5-125	Virtual condition boundaries of the upper frame for Fig. 5-124	5-135
Figure 5-126	Virtual condition boundaries of the lower frame for Fig. 5-124	5-135
Figure 5-127	Three-segment composite feature control frame	5-137
Figure 5-128	Design applications for runout control	5-138
Figure 5-129	Symbols for circular runout and total runout	5-139
Figure 5-130	Datums for runout control	5-140
Figure 5-131	Two coaxial features establishing a datum axis for runout control	5-141
Figure 5-132	Runout control of hyphenated co-datum features	5-142
Figure 5-133	Application of circular runout	5-143
Figure 5-134	Application of profile tolerances	5-146
Figure 5-135	Profile tolerance zones	5-148
Figure 5-136	Profile of a line tolerance	5-150
Figure 5-137	Profile "all around"	5-151
Figure 5-138	Profile "all over"	5-151
Figure 5-139	Profile "between" points	5-152
Figure 5-140	Profile tolerancing to control a combination of characteristics	5-153
Figure 5-141	Profile tolerance to control coplanarity of three feet	5-154
Figure 5-142	Composite profile for a pattern	5-155
Figure 5-143	Composite profile tolerancing with separate Level 2 control	5-155
Figure 5-144	Composite profile tolerance for a single feature	5-156
Figure 5-145	Types of symmetry	5-157
Figure 5-146	Symmetry construction rays	5-158
Figure 5-147	Symmetry tolerance about a datum plane	5-159
Figure 5-148	Multifold concentricity tolerance on a cam	5-160
Figure 5-149	Dimension origin symbol	5-163
Figure 7-1	Vectors and unit vectors	7-5
Figure 7-2	Vector addition	7-5
Figure 7-3	Vector subtraction	7-6
Figure 7-4	Circularity tolerance zone definition	7-10
Figure 7-5	Illustration of an elliptical cylinder	7-11
Figure 7-6	Cylindricity tolerance definition	7-12
Figure 7-7	Flatness tolerance definition	7-13
Figure 8-1	Statistical tolerancing using process capability indices	8-3
Figure 8-2	Statistical tolerancing using RMS deviation index	8-4
Figure 8-3	Statistical tolerancing using percent containment	8-5
Figure 8-4	Population parameter zones for the specifications in Fig. 8.1	8-6
Figure 8-5	Population parameter zones for the specifications in Fig. 8.2	8-6
Figure 8-6	Population parameter zones for the specifications in Fig. 8.3	8-7
Figure 8-7	Additional illustration of specifying percent containment	8-7
Figure 8-8	Illustration specifying process capability indices	8-8
Figure 8-9	Additional illustration specifying process capability indices	8-8
Figure 8-10	Illustration of statistical tolerancing under MMC	8-9
Figure 9-1	Tolerance analysis process	9-2
Figure 9-2	Motor assembly	9-3
Figure 9-3	Horizontal loop diagram for Requirement 6	9-4
Figure 9-4	Methods to dimension the length of a shaft	9-5
Figure 9-5	Methods of centering manufacturing processes	9-6
Figure 9-6	Combining piecepart variations using worst case and statistical methods	9-8
Figure 9-7	Graph of piecepart tolerances versus assembly tolerance before and after resizing using the Worst Case Model	9-11

Figure 9-8	Graph of piecepart tolerances versus assembly tolerance before and after resizing using the RSS Model	9-16
Figure 9-9	Graph of piecepart tolerances versus assembly tolerance before and after resizing using the MRSS Model	9-20
Figure 9-10	Substrate package	9-26
Figure 9-11	Position at RFS	9-27
Figure 9-12	Position at MMC — internal feature	9-29
Figure 9-13	Position at MMC — external feature	9-30
Figure 9-14	Position at LMC — internal feature	9-31
Figure 9-15	Position at LMC — external feature	9-31
Figure 9-16	Composite position and composite profile	9-32
Figure 9-17	Circular and total runout	9-33
Figure 9-18	Concentricity	9-33
Figure 9-19	Equal bilateral tolerance profile	9-34
Figure 9-20	Unilateral tolerance profile	9-35
Figure 9-21	Unequal bilateral tolerance profile	9-35
Figure 9-22	Size datum	9-36
Figure 10-1	Histogram of runout (FIM) data	10-2
Figure 10-2	The normal distribution	10-3
Figure 10-3	Histogram of normal, n=5, with normal curve	10-4
Figure 10-4	Histogram of normal, n=50, with normal curve	10-4
Figure 10-5	Histogram of normal, n=500, with normal curve	10-5
Figure 10-6	Histogram of normal, n=5000, with normal curve	10-5
Figure 10-7	Z Statistic	10-6
Figure 10-8	Normality test FIM	10-7
Figure 10-9	Histogram of transformed FIM measurements	10-7
Figure 10-10	Normality tests for transformed data	10-8
Figure 10-11	Attributes data	10-8
Figure 10-12	Plot of Poisson probabilities	10-10
Figure 10-13	Process capability	10-11
Figure 10-14	Capability index	10-11
Figure 10-15	Capability index at ± 4 sigma	10-11
Figure 10-16	The reality	10-12
Figure 10-17	Cp and Cpk at Six Sigma	10-13
Figure 10-18	Yields through multiple CTQs	10-13
Figure 11-1	Comparison of tolerance analysis and tolerance allocation	11-2
Figure 11-2	Motor assembly	11-5
Figure 11-3	Worst case allocation flow chart	11-6
Figure 11-4	Dimension loop for Requirement 6	11-7
Figure 11-5	Effect of shifting the mean of a normal distribution to the right	11-11
Figure 11-6	Centered normal distribution. Both tails are significant.	11-12
Figure 11-7	Statistical allocation flow chart	11-14
Figure 11-8	Normal distribution that has been truncated due to inspection	11-17
Figure 11-9	Three options for designating a statistically derived tolerance on an engineering drawing	11-25
Figure 12-1	Geneva mechanism showing a few of the relevant dimensions	12-2
Figure 12-2	Linearized approximation to a curve	12-3
Figure 12-3	Multidimensional tolerancing flow chart	12-4
Figure 12-4	Stacked blocks we will use for an example problem	12-5
Figure 12-5	Gap coordinate system $\{u_1, u_2\}$	12-6
Figure 12-6	Possible vector loops to evaluate the gap of interest	12-6
Figure 12-7	Vector loop we will use to analyze the gap. It presents easier calculations of unknown vector lengths.	12-7
Figure 12-8	Additional coordinate system needed for the vectors on Block 2	12-7
Figure 12-9	Relationship between coordinate systems $\{u_1, u_2\}$ and $\{v_1, v_2\}$	12-9
Figure 13-1	Kinematic adjustment due to component dimension variations	13-2
Figure 13-2	Adjustment due to geometric shape variations	13-2
Figure 13-3	Stacked blocks assembly	13-3
Figure 13-4	Assembly graph of the stacked blocks assembly	13-4
Figure 13-5	2-D kinematic joint and datum types	13-4
Figure 13-6	Part datums and assembly variables	13-5

F-6 Figures

Figure 13-7	Datum paths for Joints 1 and 2	13-6
Figure 13-8	Datum paths for Joints 3 and 4	13-6
Figure 13-9	2-D vector path through the joint contact point	13-7
Figure 13-10	2-D vector path across a part	13-8
Figure 13-11	Assembly Loop 1	13-9
Figure 13-12	Assembly Loop 2	13-9
Figure 13-13	Propagation of 2-D translational and rotational variation due to surface waviness	13-10
Figure 13-14	Applied geometric variations at contact points	13-11
Figure 13-15	Assembly tolerance controls	13-12
Figure 13-16	Open loop describing critical assembly gap	13-13
Figure 13-17	Relative rotations for Loop 1	13-14
Figure 13-18	Percent contribution chart for the sample assembly	13-22
Figure 13-19	Percent contribution chart for the sample assembly with modified tolerances	13-23
Figure 13-20	Modified geometry yields zero θ contribution	13-26
Figure 13-21	The CATS System	13-27
Figure 14-1	Optimal tolerance allocation for minimum cost	14-2
Figure 14-2	Graphical interpretation of minimum cost tolerance allocation	14-4
Figure 14-3	Shaft and housing assembly	14-4
Figure 14-4	Tolerance range of machining processes (Reference 12)	14-5
Figure 14-5	Comparison of minimum cost allocation results	14-7
Figure 14-6	Clutch assembly with vector loop	14-9
Figure 14-7	Tolerance allocation results for a Worst Case Model	14-13
Figure 14-8	Tolerance allocation results for the RSS Model	14-14
Figure 14-9	Tolerance allocation results for the modified RSS Model	14-15
Figure 14-10	Tolerance allocation results for the modified WC Model	14-16
Figure 14-11	Tolerance allocation results for the WC Model	14-16
Figure 14-12	Tolerance allocation results for the RSS Model	14-17
Figure 14A-1	Plot of cost versus tolerance for fitted and raw data for the turning process	14-18
Figure 14A-2	Plot of fitted cost versus tolerance functions	14-21
Figure 14A-3	Plot of coefficients versus size for cost-tolerance functions	14-22
Figure 14A-3	Plot of coefficients versus size for cost-tolerance functions (continued)	14-23
Figure 15-1	Tolerancing process	15-3
Figure 15-2	Small kinematic adjustments	15-4
Figure 15-3	Communication between design and manufacturing	15-9
Figure 16-1	Information flow in the product development process	16-2
Figure 16-2	Master model process information	16-5
Figure 16-3	Data management hierarchy	16-12
Figure 16-4	File format for one triangle in an STL file	16-21
Figure 17-1	Narrow road versus three-lane road	17-5
Figure 17-2	Data collected from a process with a shifted target	17-5
Figure 17-3	Averaging and grouping short-term data	17-6
Figure 17-4	Feature factoring methodology flexibility	17-7
Figure 17-5	Dpmo-weighting and guard-banding technique	17-8
Figure 18-1	Directional indicators for data point plotting	18-4
Figure 18-2	Example four-hole part	18-5
Figure 18-3	Layout inspection of four-hole part	18-6
Figure 18-4	Plotting the holes on the coordinate grid	18-7
Figure 18-5	Overlaying the polar coordinate system	18-7
Figure 18-6	Example four-hole part with long holes	18-8
Figure 18-7	Plotting 3-dimensional hole data on the coordinate grid	18-9
Figure 18-8	Four-hole part controlled by composite positional tolerancing	18-10
Figure 18-9	Paper gage verification of hole pattern location	18-11
Figure 18-10	Paper gage verification of feature-to-feature location	18-11
Figure 18-11	Datum feature subject to size variation — RFS applied	18-12
Figure 18-12	Paper gage verification for datum applied at MMC	18-13
Figure 18-13	Layout inspection setup of workpiece	18-14
Figure 18-14	Inspection Report — part allowing datum shift	18-14
Figure 18-15	Verifying hole pattern prior to datum shift	18-15
Figure 18-16	Verifying the hole pattern after datum shift	18-16
Figure 18-17	Part allowing rotational datum shift	18-16

Figures F-7

Figure 18-18	Inspection Report — part allowing rotational datum shift	18-17
Figure 18-19	Verifying hole pattern prior to rotational shift	18-18
Figure 18-20	Verifying hole pattern after rotational datum shift	18-18
Figure 18-21	Example of datum established from a hole pattern	18-19
Figure 18-22	Inspection Report — hole pattern as a datum	18-20
Figure 18-23	Determining the central datum axis from a hole pattern	18-20
Figure 18-24	Approximating datum shift from a hole pattern	18-21
Figure 18-25	Process evaluation using a paper gage	18-22
Figure 19-1	Position using partial and planar datum features	19-4
Figure 19-2	Gage for verifying two-hole pattern in Fig. 19-1	19-6
Figure 19-3	Position using datum features of size at MMC	19-7
Figure 19-4	Gage for verifying four-hole pattern in Fig. 19-3	19-8
Figure 19-5	Position and profile using a simultaneous gaging requirement	19-9
Figure 19-6	Gage for simulating datum features in Fig. 19-5	19-10
Figure 19-7	Gage for verifying four-hole pattern and profile outer boundary in Fig. 19-5	19-11
Figure 19-8	Position using centerplane datums	19-12
Figure 19-9	Gage for verifying four-hole pattern in Fig. 19-8	19-13
Figure 19-10	Multiple datum structures	19-14
Figure 19-11	Gage for verifying datum feature D in Fig. 19-10	19-15
Figure 19-12	Gage for verifying four-hole pattern in Fig. 19-10	19-16
Figure 19-13	Secondary and tertiary datum features of size	19-17
Figure 19-14	Gage for verifying datum features D and E in Fig. 19-13	19-18
Figure 19-15	Gage for verifying five holes in Fig. 19-13	19-19
Figure 20-1	Z-Axis single-point repeatability	20-21
Figure 20-2a	Form Six Sigma versus probe settling time (10-mm sphere)	20-22
Figure 20-2b	Sphere form versus probe settling time (25-mm sphere)	20-22
Figure 20-3	Probe speed versus sphere form	20-23
Figure 20-4	Sphere form versus probe trigger force (10-mm sphere)	20-24
Figure 20-5	Circle features versus probe deflection	20-25
Figure 20-6	Cylinder features versus probe deflection	20-26
Figure 20-7	Probe deflection versus sphere form	20-27
Figure 20-8	Circle features versus hole diameter	20-29
Figure 20-9	Cylinder features versus hole diameter	20-29
Figure 20-10a	Bidirectional probing versus varying lengths (x-axis)	20-30
Figure 20-10b	Bidirectional probing versus varying lengths (y-axis)	20-31
Figure 20-11	Circle features versus number of points per section	20-31
Figure 20-12	Cylinder features versus number of points/section	20-32
Figure 20-13	Cylinder features versus number points/section	20-33
Figure 20-14	Cylinder features versus number of points/section	20-33
Figure 20-15	25-mm cube test — single versus star probe setup	20-34
Figure 20-16	Circle features versus scanning speed	20-35
Figure 20-17	Leitz PPM 654 capability matrix	20-36
Figure 20-17	Leitz PPM 654 capability matrix (continued)	20-37
Figure 21-1	Cylindrical (size) feature with orientation and location constraints at RFS	21-3
Figure 21-2	Allowable location tolerance as a function of orientation error (θ)	21-4
Figure 21-3	Cylindrical (size) feature with orientation and location constraints at MMC	21-6
Figure 21-4	Cylindrical (size) feature with orientation and location constraints at LMC	21-7
Figure 21-5	Parallel plane (size) feature with orientation and location constraints at RFS	21-9
Figure 22-1	Examples of floating fasteners	22-2
Figure 22-2	Examples of fixed fasteners	22-3
Figure 22-3	Examples of double-fixed fasteners	22-4
Figure 22-4	Rectangular tolerance zone (plus/minus tolerancing)	22-5
Figure 22-5	Cylindrical tolerance zone	22-5
Figure 22-6	Tapped hole located (.000, .000) and clearance hole off location by (+.005, .000)	22-6
Figure 22-7	Tapped hole is located (-.005, .000) and clearance hole is located (+.005,.000)	22-6
Figure 22-8	Tapped hole is located (-.005, -.005) and clearance hole is located (+.005, +.005)	22-7
Figure 22-9	Tapped hole is located (-.007, .000) and clearance hole is located (+.007, .000)	22-7
Figure 22-10	Additional tolerance allowed by using a cylindrical tolerance zone versus a rectangular tolerance zone	22-8
Figure 22-11	Worst case head height above the surface	22-12

F-8 Figures

Figure 22-12	Worst case head height below the surface	22-12
Figure 22-13	Flat head fastener dimensions for a .250-28-UNC 2B flat head fastener	22-13
Figure 22-14	Positional tolerance for clearance holes and nut plate rivet holes	22-15
Figure 22-15	Tapped hole out of perpendicular by $\varnothing$.014	22-15
Figure 22-16	Variation in perpendicularity could cause assembly problems	22-15
Figure 22-17	Projected tolerance zone example	22-16
Figure 22-18	Projected tolerance zone — location and orientation components	22-17
Figure 22-19	Lost functional tolerance versus actual orientation tolerance	22-18
Figure 22-20	Floating fastener tolerance and callouts	22-20
Figure 22-21	Fixed fastener tolerance and callouts	22-21
Figure 22-22	Double-fixed fastener tolerance and callouts	22-23
Figure 23-1	Feature located using positional tolerance at MMC	23-2
Figure 23-2	Dimension loop diagram for Fig. 23-1	23-3
Figure 23-3	Fixed fastener centered and shifted	23-4
Figure 23-4	Floating fastener centered and shifted	23-4
Figure 23-5	Fixed fastener assembly	23-5
Figure 23-6	Fixed fastener minimum assembly gap	23-6
Figure 23-7	Fixed fastener maximum assembly gap	23-6
Figure 23-8	Centered fixed fastener dimension loop diagram	23-8
Figure 23-9	Floating fastener assembly	23-8
Figure 24-1	Examples of design cases for alignment pins showing Type I and Type II errors	24-4
Figure 24-2	Two common cross-sections for modified pins	24-6
Figure 24-3	Design process for using alignment data	24-8
Figure 24-4	Variables contributing to fit of two round pins with two holes	24-12
Figure 24-5	Variables contributing to rotation caused by two round pins with two holes	24-13
Figure 24-6	Dimensioning methodology for two round pins with two holes	24-14
Figure 24-7	Variables contributing to fit of two round pins with one hole and one slot	24-16
Figure 24-8	Variables contributing to rotation caused by two pins with one hole and one slot	24-16
Figure 24-9	Dimensioning methodology for two round pins with one hole and one slot	24-18
Figure 24-10	Variables contributing to rotation caused by two pins with hole and edge contact	24-20
Figure 24-11	Dimensioning methodology for two round pins with one hole and edge contact	24-21
Figure 24-12	Variables contributing to fit of one round pin and one diamond pin with two holes	24-23
Figure 24-13	Variables contributing to the fit of one pin and one parallel-flats pin with two holes	24-26
Figure 25-1	Sample drawing #1	25-2
Figure 25-2	Sample drawing #2	25-3
Figure 25-3	Sample drawing #3	25-3
Figure 25-4	Sample drawing #4	25-5

Tables

Table 1-1	Practical impact of process capability	1-8
Table 3-1	Bonus tolerance gained as the feature's size is displaced from its MMC	3-13
Table 5-1	Geometric characteristics and their attributes	5-15
Table 5-2	Modifying symbols	5-16
Table 5-3	Actual mating envelope restraint	5-42
Table 5-4	Datum feature types and their TGCs	5-68
Table 5-5	TGC shape and the derived datum	5-69
Table 5-6	Datum target types	5-92
Table 5-7	Simultaneous/separate requirement defaults	5-133
Table 6-1	ASME standards that are related to dimensioning	6-2
Table 6-2	ISO standards that are related to dimensioning	6-3
Table 6-3	Organization of the matrix model from ISO technical report (#TR 14638)	6-4
Table 6-4	Differences between ASME and ISO standards.	6-5
Table 6-5	Advantages and disadvantages of the number of ASME and ISO standards	6-6
Table 6-6A	General	6-7
Table 6-6B	General	6-8
Table 6-6C	General	6-9
Table 6-6D	General	6-10
Table 6-6E	General	6-11
Table 6-6F	General	6-12
Table 6-7A	Form	6-13
Table 6-7B	Form	6-14
Table 6-8A	Datums	6-15
Table 6-8B	Datums	6-16
Table 6-8C	Datums	6-17
Table 6-8D	Datums	6-18
Table 6-9	Orientation	6-19
Table 6-10A	Tolerance of Position	6-20
Table 6-10B	Tolerance of Position	6-21
Table 6-10C	Tolerance of Position	6-22
Table 6-10D	Tolerance of Position	6-23
Table 6-11	Symmetry	6-24
Table 6-12	Concentricity	6-25
Table 6-13A	Profile	6-25
Table 6-13B	Profile	6-26
Table 6-14	A sample of the national standards bodies that exist	6-27
Table 6-15	International standardizing organizations	6-28
Table 9-1	Converting to mean dimensions with equal bilateral tolerances	9-7
Table 9-2	Dimensions and tolerances used in Requirement 6	9-7
Table 9-3	Resized tolerances using the Worst Case Model	9-11
Table 9-4	Resized tolerances using the RSS Model	9-17
Table 9-5	Resized tolerances using the MRSS Model	9-20
Table 9-6	Comparison of results using the Worst Case, RSS, and MRSS models	9-22
Table 9-7	Comparison of analysis models	9-23

T-2 Tables

Table 10-1	Distribution of defects	10-9
Table 11-1	Process standard deviations that will be used in this chapter	11-3
Table 11-2	Data used to allocate tolerances for Requirement 6	11-7
Table 11-3	Final allocated and fixed tolerances to meet Requirement 6	11-10
Table 11-4	Fixed and statistically allocated tolerances for Requirement 6	11-18
Table 11-5	Fixed and statistically allocated tolerances for Requirement 6	11-19
Table 11-6	Standard deviation inflation factors and DRSS allocated tolerances for Requirement 6	11-22
Table 11-7	Comparison of the allocated tolerances for Requirement 6	11-24
Table 12-1	Dimensions and tolerances corresponding to the variable names in Fig. 12-4	12-5
Table 12-2	Dimensions, tolerances, and sensitivities for the stacked block assembly	12-12
Table 12-3	Final dimensions, tolerances, and sensitivities of the stacked block assembly	12-13
Table 13-1	Estimated variation in open and closed loop assembly features	13-21
Table 13-2	Modified dimensional tolerance specifications	13-23
Table 13-3	Calculated sensitivities for the Gap	13-24
Table 13-4	Calculated sensitivities for the Gap after modifying geometry	13-25
Table 13-5	Variation results for modified nominal geometry	13-25
Table 14-1	Proposed cost-of-tolerance models	14-2
Table 14-2	Initial Tolerance Specifications	14-5
Table 14-3	Minimum cost tolerance allocation	14-7
Table 14-4	Minimum True Cost	14-8
Table 14-5	Independent dimensions for the clutch assembly	14-9
Table 14-6	Process tolerance limits for the clutch assembly	14-11
Table 14-7	Expressions for minimum cost tolerances in 2-D and 3-D assemblies	14-12
Table 14-8	Process tolerance cost data for the clutch assembly	14-12
Table 14-9	Revised process tolerance cost data for the clutch assembly	14-15
Table 14A-1	Relative cost of obtaining various tolerance levels	14-19
Table 14A-2	Cost-tolerance functions for metal removal processes	14-20
Table 15-1	Advanced tolerance analysis methods: MSM versus MCS	15-8
Table 16-1	Information captured in a database	16-5
Table 16-2	Examples of templates	16-8
Table 16-3	Common document templates	16-9
Table 16-4	Information provided for sheetmetal process	16-16
Table 16-5	Information provided for injection molding process	16-17
Table 16-6	Information provided for hog-out process	16-17
Table 16-7	Information provided for casting process	16-18
Table 16-8	Information provided for prototyping process	16-19
Table 18-1	Layout Inspection Report of four-hole part	18-6
Table 18-2	Inspection Report for part with long holes	18-9
Table 18-3	Inspection Report for composite position verification	18-10
Table 22-1	Floating fastener clearance hole and C'Bore hole sizes and tolerances	22-19
Table 22-2	Fixed fastener clearance hole, C'Bore, and C'Sink sizes and tolerances	22-22
Table 22-3	Double-fixed fastener clearance hole and C'Bore sizes and tolerances	22-24
Table 22-4	C'Bore depths (pan head and socket head)	22-25
Table 22-5	Flat head screw head height above and below the surface	22-26
Table 24-1	Alignment pins per ANSI B18.8.2-1978, R1989	24-5
Table 24-2	Standard deviations for common manufacturing processes (inches)	24-7
Table 24-3	Performance constants for two round pins with two holes	24-13
Table 24-4	GD&T callouts for two round pins with two holes	24-15
Table 24-5	Performance constants for two round pins with one hole and one slot	24-17
Table 24-6	GD&T callouts for two round pins with one hole and one slot	24-19
Table 24-7	Performance constants for two round pins with one hole and edge contact	24-21
Table 24-8	GD&T callouts for two round pins with one hole and edge contact	24-22
Table 24-9	Performance constants for one round pin and one diamond pin with two holes	24-24
Table 24-10	GD&T callouts for one round pin and one diamond pin with two holes	24-25
Table 24-11	Performance constants for one round pin and one parallel-flats pin with two holes	24-27
Table 24-12	GD&T callouts for one round pin with one parallel-flats pin and two holes	24-28
Table 25-1	GR&R Analysis Matrix	25-2
Table 25-2	Bonus tolerance gained due to considered feature size	25-5
Table 25-3	Analysis Matrix	25-6

Index

Symbols

1-D stackups. *See Tolerance analysis, 1-D*
2-D stackups. *See Tolerance analysis, 2-D*
3-D stackups. *See Tolerance analysis, 3-D*
80-20 rule. *See Pareto principle*

A

Abutting profile tolerance zone. *See Profile tolerance abutting zones*
Accuracy 18-3
Active drawing 4-30
Actual
 feature 5-61
 local size 5-46
 mating envelope 5-41, 5-43
  restraint of 5-41, 5-42
  simulation of 5-43
 mating local size 5-45
 mating size 5-37, 5-43
 minimum material envelope 5-44
  restraint of 5-44
  simulation of 5-44
 minimum material local size 5-46
 minimum material size 5-44
 value 7-7
Algorithms 7-3
Alignment
 between parts. *See Mating parts*
 pins 24-1
  error 24-3, 24-12, 24-13, 24-16, 24-17, 24-20, 24-24, 24-27
  one diamond and one round/two holes 24-23, 24-25
  one parallel flats and one round/two holes 24-26, 24-28
  two round/hole and edge contact 24-18, 24-21
  two round/hole and slot 24-14, 24-19
  two round/two holes 24-11, 24-15
  types of 24-4
All around symbol 5-13, 5-150, 5-151
 comparison of US and ISO 6-7
All over note 5-150, 5-151
Allocation
 by scaling/resizing/weight factor. *See Tolerance allocation by scaling/resizing/weight factor*
 cost. *See Tolerance allocation by cost minimization*
 cost versus tolerance. *See Cost versus tolerance allocation*
 DRSS. *See Dynamic Root Sum of the Squares (DRSS) allocation*
 manufacturing process. *See Tolerance allocation by manufacturing processes*
 RSS. *See Root Sum of the Squares (RSS) allocation*
 Six Sigma tolerance. *See Six Sigma tolerance allocation*
 statistical tolerance. *See Statistical tolerance allocation*
 tolerance. *See Tolerance allocation*
 worst case. *See Worst case allocation*
Alternative center method 5-43, 5-52, 5-113, 5-122, 5-123, 5-124, 5-125, 5-127
 disadvantages 5-46
 Level 2 adjustment 5-45
 Level 3 adjustment 5-43
 Level 4 adjustment 5-43
American National Standards 5-2
 ASME Y14.5.1M (the "Math Standard") 5-3, 5-4, 5-23, 7-14

Index

ASME Y14.5M 3-2, 3-13, 3-15, 5-3, 5-4, 7-14
   budgeting of coverage 5-3, 5-136
   differences in standards 5-59, 5-117, 5-136, 5-145
   discrepancies in 5-4
   future of 5-164
   recommendations/suggestions 5-165
  Institute (ANSI) 4-2, 5-2
  superceding 5-4
American Society of Mechanical Engineers (ASME) 5-165, 6-2, 8-10
Analysis
  computer. *See Computer analysis*
  Estimated Mean Shift. *See Estimated Mean Shift analysis*
  Fixed fastener. *See Fixed fastener tolerance analysis*
  Floating fastener. *See Floating fastener tolerance analysis*
  Future of. *See Future of tolerance analysis*
  GD&T. *See Geometric Dimensioning and Tolerancing (GD&T) analysis*
  GR&R. *See Gage repeatability and producibility (GR&R), analysis of*
  graphical inspection. *See Graphical inspection analysis*
  measurement methods. *See Measurement methods analysis*
  MRSS. *See Modified Root Sum of the Squares (MRSS) analysis*
  process. *See Process analysis*
  RSS. *See Root Sum of the Squares (RSS) analysis*
  SRSS. *See Static Root Sum of the Squares (SRSS) analysis*
  tolerance. *See Tolerance analysis*
  worst case. *See Worst case analysis*
Anderson-Darling test for normality 10-4
Angle
  90° basic. *See Implied 90° basic angle*
  basic. *See Basic dimension, implied dimension*
    90° implied. *See Implied 90° angle dimension*
  erroneous wedge-shaped tolerance zone for 5-112
  plus and minus tolerance 5-49, 5-50, 5-112

transition between features 5-10
Angled
  datum 5-75
  feature 5-117
Angularity tolerance 5-104
  analysis of 9-26
  comparison of US and ISO 6-19
  for a cylindrical feature 5-106
  for a width-type feature 5-106
  symbol 5-13
ANSI. *See American National Standards Institute (ANSI)*
Approximation model 15-5
Arc length symbol 5-13, 5-16
Asea Brown Boveri Ltd. (ABB) 1-6
ASME
  standards 7-14
  Y14.5. *See American National Standards ASME Y14.5M*
Assemblability
  worst case 5-33, 5-34, 5-35, 5-128
Assembly 8-1, 16-8
  clearance. *See Fixed fastener formula; Floating fastener formula; Mating parts; Maximum Material Condition (MMC), when to apply; Virtual condition boundary*
  datum feature selection for 5-61, *5-63*
  deformation 15-5
  drawing 4-10, 4-11, 5-19
  drawings 4-4
  equation. *See Gap equation*
  for dynamic balance 5-144, 5-162
  force 5-146
  graph 13-4
  interface 5-71
    centering 5-47
  process variation 15-4
  restraint of parts in 5-20
  sequence 5-64
  shift 23-4
  standards 26-4
  tolerance 5-19, 9-11
  tolerance models
    2-D 13-3. *See also Tolerance model, steps in creating (2-D/3-D)*
    closed loop 13-4, 13-11, 13-13, 13-14
    critical features 13-10

datum paths 13-5, 13-7
datum reference systems 13-4, 13-5
degrees of freedom 13-5
geometric variations 13-10, 13-11. *See also Variation*
graph 13-4
key characteristics 13-10
kinematic joints 13-4, 13-5, 13-6, 13-7
modeling 13-4
modeling rules 13-7, 13-8
open loop 13-4, 13-11, 13-13, 13-14
performance requirements. *See Performance requirements*
redundant vectors 13-12
steps in creating. *See Tolerance model, steps in creating*
vector loops 9-3, 11-7, 12-6, 13-8, 13-9, 13-11, 13-13, 13-14, 23-3, 23-6
vectors. *See Vectors*
variation sources 13-2
Attribute
 data 10-2
 process capability models 17-7
Automated verification 16-10
Automation 15-2
Auxiliary
 dimension
  comparison of US and ISO 6-10
 view 4-16
Average diameter 5-56
Axis
 datum. *See Datum axis*
 feature, control of. *See Feature control frame*

B

Baldrige. *See Malcolm Baldrige National Quality Award*
Bar stock 5-29
Base line dimensioning 5-116
Basic
 angle 5-19, 5-104. *See also Angle, basic; Basic dimension*
 dimension 5-17
  defining a basic profile 5-145
  established by general note 5-163
  for profile boundary offset 5-147
  frame 5-17
  implied. *See Implied basic dimension*
  limiting length or area of tolerance zone 5-57. *See also Limited length/area indication*
  locating a datum target 5-94
  locating termination of tolerance zone 5-152
  locating true position. *See True position methods for establishing*
  shown as reference 5-18
  symbol (frame) 5-17
  zero implied. *See Implied basic dimension*
 dimension symbol
  comparison of US and ISO 6-7
 profile 5-147. *See also Profile tolerance*
  defined by basic dimensions 5-145
  defined by CAD/CAM model 5-147
  defined by grid system 5-147
  defined by mathematical formula 5-147
  showing termination of tolerance zone 5-152
  tolerance boundary offset from 5-147
  tool path along 5-147, 5-153
 size 5-48
Beta distribution 10-6
Between symbol 5-13, 5-16, 5-150, 5-152
 comparison of US and ISO 6-7
Bidirectional positional tolerance 5-122, 5-125, 5-126
Bilateral profile tolerance. *See Profile tolerance, bilateral*
 equal. *See Profile tolerance, equal-bilateral*
  analysis of 9-34
 unequal. *See Profile tolerance, unequal bilateral*
  analysis of 9-35
Binomial distribution 10-8
Bolt circle 5-115, 5-116
Bonus tolerance
 at LMC 5-44
 at MMC 5-43
Boundary
 at LMC 5-23, 5-25
 at MMC 5-23, 5-24, 5-25
 inner, outer 5-46
 of perfect form 5-27
  at LMC 5-28, 5-29
  at MMC 5-27
  not required 5-29, 5-30, 5-39

resultant condition. *See Resultant condition boundary*
size limit. *See Size limit boundary*
spherical 5-22
tapered 5-121
tolerance 5-14
virtual condition. *See Virtual condition boundary*
Bounded feature 5-11, 5-126
  as a datum feature 5-77, 5-79, 5-82
  composite profile tolerance for a single feature 5-156
  datum targets applied to 5-99
  positional tolerance for. *See Positional tolerance for a bounded feature*
Broken-out section 4-19

C

CAD. *See Computer Aided Design (CAD)*
CAD/CAM. *See Computer Aided Design (CAD); Computer Aided Manufacturing (CAM)*
CAE. *See Computer Aided Engineering (CAE)*
CAM. *See Computer Aided Manufacturing (CAM)*
Candidate
  datum 5-90, 7-8
    reference 7-8
  datum reference frame 5-90
Cartesian coordinates 5-69, 5-70
Casting 16-18
  drawing 4-5
Castings/forgings 5-64, 5-151
CAT 15-6, 15-12, 15-13, 15-14. *See also Computer Aided Tolerancing (CAT)*
Cauchy distribution 10-6
Center
  line 5-10, 5-109.
    implied 90° angle *5-19*
  marks 4-4
  method
    alternative. *See Alternative center method*
  plane 5-10, 5-41, 5-89
    control of. *See Feature control frame*
    control of feature 5-106, 5-113, 5-119, 5-122
    datum 19-12
    derived 5-38, 5-40
    establishing a datum from a feature 5-65
    establishing a datum from feature 5-89
    feature 5-41
  point 5-38, 5-41, 5-113
Centering 5-47. *See also Regardless of Feature Size (RFS), when to apply*
Central tolerance zone 5-25, 5-38, 5-39
Certification of GD&T professionals 5-3
Chain
  dimensioning 5-116
  line 5-57, 5-65, 5-119, 5-143
Characteristic symbol. *See Geometric characteristic symbol*
Circular runout tolerance 5-138, 5-139, 5-141, 5-144
  symbol 5-13
Circularity (roundness) tolerance 5-53, 7-9
  analysis of 9-25
  control by limits of size 5-22
  control of lobes. *See Lobes, circularity control*
  for a nonspherical feature 5-53
  for a spherical feature 5-54, 5-55
  in the free state 5-56
  not affecting size, taper, or straightness 5-54, 5-55
  not modifiable to MMC or LMC 5-55
  symbol 5-13
  with average diameter 5-56
Clearance fit. *See Fixed fastener formula; Floating fastener formula; Mating parts; Maximum Material Condition (MMC), when to apply; Virtual condition*
Clocking 5-87, 5-97
Closed loop. *See Assembly tolerance model, closed loop*
CM/PDM. *See Configuration Management/Product Data Management*
CMM. *See Coordinate measuring machine (CMM)*
Co-datums. *See Hyphenated co-datums*
Coaxial/coplanar pattern of features 5-103, 5-136, 5-154
Combined controls 5-162
Company vault 16-12
Component
  deformation 15-5

libraries 16-9
Composite tolerance
    positional. *See Positional tolerance, pattern control, composite*
    analysis of 9-32
    comparison of US and ISO 6-20
    verification 18-10
    profile
        analysis of 9-36
        for a pattern of features. *See Profile tolerance, composite, for a pattern of features*
        for a single feature. *See Profile tolerance, composite, for a single feature*
    rules for. *See Rules for pattern control, composite*
Computer
    Aided Design (CAD) 15-2, 15-5, 15-13, 16-2
    Aided Design (CAD)/Manufacturing (CAM) 5-147
    Aided Engineering (CAE) 16-2
    Aided Tolerancing (CAT) 15-2
    analysis 5-32
Concentricity tolerance 5-144, 5-158, 5-160, 5-161
    analysis of 9-33
    comparison of US and ISO 6-25
    comparison with other methods 5-160
    control of lobes. *See Lobes, concentricity control*
    for multifold symmetry 5-160
    symbol 5-13
Concurrent
    engineering 15-9
        index 10-12
Condition
    material. *See Material condition*
    resultant. *See Resultant condition*
    virtual. *See Virtual condition*
Configuration
    layout 4-30
    Management 16-11
    Management/Product Data Management 15-13
Conical
    surface
        control of 5-60, 5-153
            tolerance zone 5-121
Constraints 7-4
Contribution chart. *See Percent contribution chart*
Control
    combined. *See Combined controls*
    extent of profile tolerance. *See Extent of profile tolerance*
    four levels of. *See Levels of control*
Controlled radius 5-59
    symbol 5-13, 5-16, 5-59
        comparison of US and ISO 6-7
    tolerance 5-59
Conventional
    breaks 4-22
    practices 4-23
Coordinate
    dimensioning 5-9
    measuring machine (CMM) 5-90, 5-164, 7-2, 20-1
    system
        Cartesian. *See Cartesian coordinates*
        polar. *See Polar coordinate system*
        rectangular. *See Rectangular coordinate system*
Coplanarity 5-137
    positional tolerance for 5-137
    profile tolerance for 5-154
Corporate standards 16-10
Cost 5-4, 5-8
    appraisal 1-4
    failure 1-4
    minimization/optimization 14-1, 14-12
    prevention 1-4
    versus tolerance 14-3, 14-19, 14-20, 14-22, 14-23
        allocation 14-4
        curves 14-1
        function 14-2
        functions 14-20
        models 14-2
Counterbore
    depth
        pan head 22-25
        socket head 22-25
    symbol 5-13
        comparison of US and ISO 6-7

Countersink. *See Double-fixed fastener; Fixed fastener*
   comparison of US and ISO 6-7
   symbol 5-13
Critical features. *See Assembly tolerance models, critical features*
Critical-to-quality (CTQ) characteristic 1-7, 10-12
Crosby Quality College 1-4
Cross product 7-6
CTQ. *See Critical to Quality (CTQ) characteristic*
Customer 1-4
   external 1-4
   internal 1-4
   satisfaction 1-3
Cylinder, pitch. *See Pitch cylinder*
Cylindrical
   surface, control of 5-21, 5-22, 5-55
   tolerance zone 5-14
Cylindricity tolerance 5-55, 7-12
   analysis of 9-25
   symbol 5-13
   with average diameter 5-56

D

D-shaped feature. *See Bounded feature*
Data eXchange Format (DXF) 16-19
Data Management 16-12
Data Points 18-4
Database Format Standards 16-19
Datum 5-6, 5-61, 5-69
   accuracy 5-95
   angled. *See Angled datum*
   axis 5-61, 5-65, 5-69, 5-95, 5-97, 5-102, 5-108, 5-116, 5-136, 5-137, 5-140, 5-141, 5-143, 5-160
      comparison of US and ISO 6-16
   candidate. *See Candidate datum*
   comparison of US and ISO 6-15, 6-16, 6-17, 6-18
   degrees of freedom 5-63, 5-116
   feature 5-61
      angled. *See Angled feature*
      identification 5-65
      selection 5-61
      subject to size variation 18-2
      surrogate/temporary 5-64
      symbol 5-13, 5-65
      symbol placement 5-65, 5-66
      unstable (rocking) 5-89
   from a feature pattern 5-100, 18-19
   generating line as
      comparison of US and ISO 6-17
   hyphenated co-datums. *See Hyphenated co-datums*
   implied. *See Implied datum*
   letter
      comparison of US and ISO 6-17
   mathematically defined surface
      comparison of US and ISO 6-17
   origin from. *See Origin from a datum reference frame*
   paths. *See Assembly tolerance models, datum paths*
   plane 5-61, 5-69, 5-91, 5-112
   point 5-61, 5-161
   precedence 5-14, 5-63, 5-68, 5-69. *See also Degrees of freedom*
   reference 5-61
   reference frame (DRF) 5-6, 5-61, 5-69, 7-8
      candidate. *See Candidate datum reference frame*
      displacement 5-80, 5-81, 5-83
      establishing 5-62
      general note 5-164
      multiple. *See Multiple DRFs*
      origin from 5-69, 5-101
      simultaneous/separate requirements. *See Simultaneous/separate requirements*
      using in a tolerance analysis. *See Assembly tolerance models datum reference systems*
   sequence
      comparison of US and ISO 6-17
   simulation/simulator 5-68, 5-89, 5-99. *See also True Geometric Counterpart (TGC)*
      sequence 5-69
   symbol
      placement 5-65, 5-66
   target. 5-91
      application 5-91, 5-97
         any feature 5-97
         feature of size 5-95
         math-defined feature *5-99*
         revolute 5-97
         stepped surfaces 5-97

    dimensions 5-94
    identification of 5-92
    interdependency of 5-95
    switchable precedence 5-99
    symbol 5-13, 5-92
    types of 5-92
  target line
    comparison of US and ISO 6-17
Defects
  absence of 1-1
  assembly 11-1, 11-2, 11-3, 11-13, 11-15
    comparison of variation models 9-22, 9-23
    RSS 9-17
  calculating
    assembly 11-15, 13-21
    comparison of GD&T 21-10
    part/component 11-10
    using GD&T 21-1
  fabrication 11-3
  machine uncertainty 20-4
  modeling with Poisson 10-8
  part/component 11-10
  per million opportunities (dpmo) 17-3
    cost of poor quality 1-9
  per opportunity (dpo) 17-3
  per unit (DPU) 10-9, 17-3
    estimating yield 10-10
    how to calculate 21-1
    Six Sigma measurement 1-7
  rate 1-8
  Six Sigma philosophy 1-7
  Six Sigma quality 10-12
  weighting methodology 17-7. *See also Tolerance allocation by scaling/resizing/weight factor*
Degrees of freedom 5-72, 7-8. *See also Assembly tolerance models degrees of freedom; Datum degrees of freedom*
Deming, W. Edwards 1-2
  14 points 1-2
  Deming Prize 1-2
Department of Defense (DoD) 5-2
Depth symbol 5-13
  comparison of US and ISO 6-8
Derived
  element 5-38
  median line 5-38, 5-39. *See also Straightness tolerance, derived median plane/line control* 5-15
  median plane 5-38, 5-39. *See also Straightness tolerance, derived median plane/line*
Design
  engineering driven 2-2
  for assembly 5-6
  for assembly (DFA) 2-5
  for manufacturability (DFM) 2-5
  inspection driven 2-2
  process driven 2-2
  requirements. *See Performance requirements*
Detail drawings 4-3
DFA. *See Design for assembly (DFA)*
DFM. *See Design for manufacturability (DFM)*
Diameter
  average. *See Average diameter*
  spherical. *See Spherical diameter*
  symbol 5-13, 5-16
    application of 5-14
    comparison of US and ISO 6-8
Digital Equipment Corporation (DEC) 1-6
Dimension 4-25
  assembly 13-3
  auxiliary
    comparison of US and ISO 6-10
  basic. *See Basic dimension*
  chains 15-4
  component 13-2, 13-3, 13-16
  controlled 13-8
  decimal value
    inch 5-48, 5-49
    millimeter 5-48, 5-49
  dependent 13-2
  independent 13-2
  limit. *See Limit dimensioning*
  loop. *See Vector loop. See Loop diagram*
  nominal. *See Nominal dimension*
  origin symbol 5-13, 5-112, 5-163
  path. *See also Loop diagram*
  paths 15-4
  redundant 13-8
  reference 5-18
  theoretically exact
    comparison of US and ISO 6-7
Dimensional management 2-10
  future of 26-12
  process 2-8, 2-10

system 2-10
team 2-4
Dimensioning
and tolerancing 3-1, 3-2, 3-4, 3-6, 3-8
methods
baseline. *See Baseline dimensioning*
chain. *See Chain dimensioning*
fundamental rules. *See Fundamental rules*
limit dimensioning. *See Limit dimensioning*
limits and fits. *See Limits and fits*
plus and minus tolerancing. *See Plus and minus tolerance*
polar coordinate. *See Polar coordinate system*
rectangular coordinate. *See Rectangular coordinate system*
Dimensionless print 16-15
Dimensions. *See Dimension*
Discrimination 18-3
Displacement, DRF. *See Datum reference frame (DRF) displacement*
Disposition of profile tolerance zone. *See Profile tolerance, disposition of zone*
Distribution 10-2
Beta. *See Beta distribution*
binomial. *See Binomial distribution*
Cauchy. *See Cauchy distribution*
Exponential. *See Exponential distribution*
fitting 15-8
function 8-5. *See also Probability distribution function*
zone 8-6
Gamma. *See Gamma distribution*
Gaussian. *See Gaussian distribution*
Hypergeometric. *See Hypergeometric distribution*
information 15-11
Johnson. *See Johnson distribution*
Lambda. *See Lambda distribution*
Laplace. *See Laplace distribution*
Logistic. *See Logistic distribution*
Lognormal. *See Lognormal distribution*
manufacturing. *See Manufacturing distribution*
non normal. *See Non normal distribution*
normal. *See Normal distribution*
Pearson. *See Pearson distribution*

Poisson. *See Poisson distribution*
type of 8-3
uniform. *See Uniform distribution*
Weibull. *See Weibull distribution*
DoD. *See Department of Defense (DoD)*
Dot product 7-6
Double-fixed fastener 22-4, 22-23
calculation
clearance hole 22-11
countersink diameter 22-11
head height 22-11
examples 22-4
formula 22-11
sizes/tolerances
clearance hole diameter 22-23, 22-24
countersink diameter 22-23, 22-24
nonfloating nut plate 22-23
tapped hole 22-23
dpmo. *See Defects per million opportunities (dpmo)*
dpo. *See Defects per million opportunities (dpmo)*
DPU. *See Defects per unit (DPU)*
Drawing
GD&T 5-6
history 4-2
interpretation 4-1
number 4-14
scale of a 4-14
status 4-30
title 4-13
DRF. *See Datum reference frame*
DRSS. *See Dynamic Root Sum of the Squares (DRSS)*
Dynamic Root Sum of the Squares (DRSS) 15-6, 15-7
allocation 11-20, 11-23, 11-26
Dynamic RSS
allocation. *See Dynamic Root Sum of the Squares (DRSS) allocation*

E

E-mail 16-14
Each
element 5-108
radial element *5-108*
Electronic
automation 16-3

mail 16-14
Elliptical cylinder 7-11
Encapsulated PostScript (EPS) 16-22
Engineering driven design 2-2
English language to control part features 5-11
Envelope 5-27
    (Taylor) principle. *See Taylor Principle*
    actual mating. *See Actual mating envelope*
    actual minimum material. *See Actual mating/minimum material envelope*
    boundary of perfect form. *See Boundary of perfect form*
Equal
    bilateral tolerance 4-15, 4-28, 9-5
        converting an external feature at LMC to 9-31
        converting an external feature at MMC to 9-30
        converting an internal feature at LMC to 9-30
        converting an internal feature at MMC to 9-29
    precedence datums. *See Hyphenated co-datums*
Error sources 20-2
Estimated Mean Shift analysis 9-23
Estimates of manufacturability 10-1
Evolution of quality 1-2
Experimental drawing 4-30
Exponential distribution 10-6
Extension
    (projection) lines 5-14, 5-51, 5-65
        comparison of US and ISO 6-8
    of principle 5-4, 5-59, 5-82, 5-88
Extent of profile tolerance 5-145, 5-150
External/internal features of size 5-10

F

FAQ (Frequently Asked Question) 5-3, 5-4, 5-9, 5-10, 5-11, 5-48, 5-60, 5-74, 5-76, 5-88, 5-91, 5-92, 5-94, 5-95, 5-97, 5-102, 5-103, 5-114, 5-132, 5-136, 5-138, 5-143, 5-144, 5-150, 5-162
Fastener
    double-fixed. *See Double-fixed fastener*
    fixed. *See Fixed fastener*
    floating. *See Floating fastener*
Feature 5-9
    axis 5-38, 5-41
    bounded. *See Bounded feature*
    center plane 5-38, 5-41
    center point 5-38, 5-41
    control frame 5-14
        comparison of US and ISO 6-8
        placement 5-14, 5-15
        reading 5-16
    D-shaped. *See Bounded feature*
    datum. *See Datum feature*
    factoring method 17-7
    nonsize. *See Nonsize feature*
    of size
        internal/external. *See External/internal features of size*
    pattern. *See Pattern of features*
    relating tolerance zone framework (FRTZF) 5-130
    rotation 4-23
    spherical. *See Spherical feature*
File Transfer Protocol (FTP) 16-14
FIM. *See Full Indicator Movement (FIM)*
FIR. *See Full Indicator Runout (FIR)*
First-angle projection 4-16
Fit, clearance. *See Fixed fastener formula; Floating fastener formula; Mating parts; Maximum Material Condition (MMC), when to apply; Virtual condition boundary*
Fits, limits and. *See Limits and fits*
Five Sigma rule of thumb 17-6
Fixed and floating fastener. *See Fixed fastener; Floating fastener*
Fixed fastener 22-1, 22-4, 23-4
    calculation
        assembly shift 23-5
        clearance hole diameter 22-10
        counterbore diameter 22-10
    double. *See Double-fixed fastener*
    examples 22-3
    formula 22-10, 23-7
    projected tolerance zone. *See Projected tolerance zone*
    sizes/tolerances
        clearance hole diameter 22-21, 22-22
        counterbore hole diameter 22-21, 22-22
        countersink diameter 22-21, 22-22
        floating nut plate 22-21

I-10 Index

head height 22-26
nonfloating nut plate 22-21
tapped hole 22-21
tolerance analysis of 23-7
Flat
  head screw
    calculating head height 22-26
  pattern layout drawing 4-9
Flatness tolerance 5-52, 7-13
  analysis of 9-25
  comparison of US and ISO 6-13
  derived median plane 5-40, 5-52
  for a width-type feature 5-31, 5-46, 5-52
  per unit area 5-58
  single planar surface 5-52
  symbol 5-13
Flaws 4-29
Floating and fixed fastener. *See Fixed fastener; Floating fastener*
Floating fastener 22-1, 22-4, 23-4
  calculation
    clearance hole diameter 22-8
    counterbore diameter 22-9
  examples 22-2
  formula 22-8
  sizes/tolerances
    clearance hole diameter 22-19, 22-20
    counterbore hole diameter 22-19, 22-20
    floating nut plate 22-20
  tolerance analysis of 23-8
Ford Motor Company 10-11
Forgings/castings. *See Castings/forgings*
Form
  qualifying notes
    comparison of US and ISO 6-14
  control. *See Form tolerance*
Form tolerance 5-50, 7-9. *See also Levels of control, Level 2: overall form*
  analysis of 9-25
  circularity (roundness). *See Circularity (roundness) tolerance*
  comparison of US and ISO 6-11, 6-13, 6-14
  cylindricity. *See Cylindricity tolerance*
  flatness. *See Form tolerance*
  limits of size. *See Levels of control, Level 2: overall form; Limits of size*
  profile. *See Profile tolerance*
  runout. *See Runout tolerance*

straightness. *See Straightness tolerance*
  when to use 5-60
Formula
  double-fixed fastener. *See Double-fixed fastener formula*
  fixed fastener. *See Fixed fastener formula*
  floating fastener. *See Floating fastener formula*
  resultant condition 5-38
  virtual condition 5-31, 5-32, 5-128
Frame
  basic dimension. *See Basic dimension frame*
  datum feature symbol. *See Datum feature symbol*
  datum reference. *See Datum reference frame*
  feature control. *See Feature control frame*
Framework of boundaries/zones 5-127
  feature relating (FRTZF). *See Feature relating tolerance zone framework (FRTZF)*
  pattern locating (PLTZF). *See Pattern locating tolerance zone framework (PLTZF)*
Free state
  application 5-20, 5-57
  symbol 5-13, 5-16
  comparison of US and ISO 6-9
Freedom, degrees of. *See Datum degrees of freedom; Degrees of freedom*
FRTZF. *See Feature Relating Tolerance Zone Framework (FRTZF); Feature relating tolerance zone framework (FRTZF)*
FTP. *See File Transfer Protocol (FTP)*
Full
  Indicator
    Movement (FIM) 5-139
    Runout (FIR) 5-139
  sections 4-19
Functional
  gaging 5-25, 19-1
  hierarchy 5-63, *5-71*
  requirements 5-20, 5-38
  Gages 19-1
Fundamental
  levels of control. *See Levels of control*
  rules 5-18
Future of
  academia 26-7

dimensional management 26-1, 26-12
dimensioning and tolerancing 26-13
GD&T 5-164, 26-1, 26-5, 26-12
global standards and business perspective 26-9
research 26-4
software tools 26-2, 26-13
standards 26-2
   dimensioning/tolerancing 26-10, 26-13
   metrology 26-11
tolerance analysis 26-2, 26-5
tolerancing 26-7
   in academics 26-2, 26-12
   in business 26-2

G

Gage repeatability and reproducibility (GR&R) 25-1
   analysis 25-1
Gaging
   functional 5-25
   tolerances 19-3
   virtual 5-44
Galvin, Bob 1-6
Gamma distribution 10-6
Gap equation. *See Performance requirements; Tolerance analysis equations, vector loop (2-D)*
   1-D 9-7, 11-7
   2-D/3-D 12-4, 12-11, 13-13
Gauss, Karl Frederick 10-2
Gaussian distribution 21-8
GD&T. *See Geometric Dimensioning and Tolerancing (GD&T)*
Gears/splines 5-11
General
   dimensions 4-25, 4-26
   tolerance
      comparison of US and ISO 6-9
Generation templates 16-7
Geometric
   characteristic symbol 5-14, 5-15
   Dimensioning and Tolerancing (GD&T) 2-6, 3-1, 3-9, 5-2, 8-2
      advice 5-3
      analysis of 9-24
      certification of GD&T professionals 5-3
      future of. *See Future of GD&T*

      instant 5-163
      overview 5-9
      symbols 5-11, 5-13, 5-15
      what is it? 5-2
      when to use 5-8, 5-9
      why to use 5-4
   Product Specification 7-1
   tolerance 8-3
   tolerancing 3-11, 3-12
   variation. *See Variation, geometric*
Geometrical
   Product Specification (GPS)
      Masterplan 6-4
GIDEP 7-2
GO gages 19-1
GR&R. *See Gage repeatability and reproducibility (GR&R)*
Graphical inspection analysis 18-2
Groove 5-10, 5-138, 5-142, 5-144

H

Half sections 4-19
Harry, Mikel J. 1-6
Histogram 10-2
Hog Out Parts 16-17
Hole
   angled. *See Angled feature*
   counterbored. *See Counterbore*
   countersunk. *See Countersink*
   pattern verification 18-5
   slotted. *See Slotted hole*
Horizontal loop 9-4. *See also Assembly tolerance models, vector loops*
HPGL 16-20
Hypergeometric distribution 10-8
Hypertext Markup Language (HTML) 16-21
Hyphenated co-datums 5-103, 5-104, 5-140, 5-141

I

IBM 1-6
IGES 16-20
Implied
   90° angle dimension 5-19
   90° basic angle dimension 5-19, 5-87, 5-104, 5-109
   basic dimension 5-116

I-12 Index

datum 5-117
parallelism/perpendicularity 5-104
symmetry 5-116
Independency Principle 5-29
Indication of limited length/area. *See Limited length/area indication*
Indicator movement/swing 5-141, 5-142
Injection molded plastic 16-17
Inner/outer boundary 5-46
using in a tolerance analysis 9-28
Inspection driven design 2-2
Instant GD&T 5-163, 5-164
Integrated
design process 1-9
product team 1-5, 10-12
Interchangeability 8-1
Internal/external features of size. *See External/internal features of size*
International
Organization for Standardization (ISO) 1-10, 4-2, 5-3, 6-2, 7-14, 8-2, 8-10
standards 6-2
Telephone and Telegraph Corporation 1-4
Internet 16-13
Interrupted surface 5-10
Intranet 16-13
ISO. *See International Organization for Standardization (ISO)*
ISO 9000 1-10
Isometric views 4-24

J

Johnson distribution 15-8
Joint Photographic Experts Group (JPEG) 16-22
Juran Institute, Inc. 1-3
Juran, Joseph 1-3

K

Key characteristics 2-6, 2-10. *See also Assembly tolerance models, key characteristics*
Kinematic joints. *See Assembly tolerance model, kinematic joints*
degrees of freedom 13-5
incoming 13-8
outgoing 13-8

path across 13-7
types 13-5
cylindrical slider 13-6, 13-7
edge slider 13-5, 13-7
parallel cylinders 13-6, 13-7
planar 13-7
planar joint 13-6
Kinematic model 15-4
Kodak 1-6
Kurtosis 15-6

L

Lagrange Multiplier 14-3
Method 14-7
Lambda distribution 15-8
Language 8-2
"Language of management is money" 1-3
Laplace distribution 10-6
Lay 4-29
Layout Gaging 18-2
Least
Material Condition (LMC) 5-23
for feature control 5-25
symbol 5-13, 5-16
when to apply 5-15, 5-23, 5-28, 5-32, 5-47
zero tolerance at. *See Zero tolerance at MMC/LMC*
squares 7-11
Levels of control 5-20
Level 1: size and 2-D form 5-20, 5-48
Level 2: overall form 5-20, 5-26, 5-50
Level 3: orientation 5-20, 5-33, 5-103
Level 4: location 5-20, 5-34, 5-113
Limit dimensioning 5-48, 5-49
Limited length/area indication 5-57, 5-143
Limits
and fits 5-48
and fits symbol 5-48
of size 5-20. *See also Rules, #1*
boundary. *See Size limit boundary*
circularity control. *See Circularity (roundness) tolerance control by limits of size*
cylindrical feature 5-21
width-type feature 5-22
Line
center. *See Center line*

chain. *See Chain line*
derived median. *See Derived median line*
extension. *See Extension lines*
phantom. *See Phantom line*
precedence 4-23
profile. *See Profile tolerance of a line*
profile of a. *See Profile tolerance, of a line*
Linear
   tolerance 3-3, 4-15. *See also Plus and minus tolerance*
   comparison with GD&T 3-9, 22-5
LMC. *See Least Material Condition (LMC)*
Lobes
   circularity control 5-54, 5-55
   concentricity control 5-157, 5-160
   cylindricity control 5-55
Location tolerance. *See Levels of control, Level 4: location*
Logistic distribution 10-6
Lognormal 21-2
   distribution 10-6
      approximation to Normal distribution 10-7
      transforming values 10-7
Loop. *See Vector loop*
   closed. *See Closed loop*
   diagram. *See Assembly tolerance models, vector loops*
   equation. *See Tolerance analysis equations, vector loop (2-D)*
   horizontal. *See Horizontal loop*
   open. *See Open loop*
   vertical. *See Vertical loop*
Lower specification limit 8-2

M

Machined part 4-6, 4-7
   drawing 4-4
Machining processes 14-5
Malcolm Baldrige National Quality Award 1-2, 1-9, 2-3
   design and production processes are coordinated 1-9
   differs from ISO 9000 1-10
   Motorola 1-6
Manufacturing
   guidelines 16-15
   process 9-5, 9-15
      capability data. *See Process capability*

data
   distributions 21-2
   variation. *See Variation sources, process*
   process capability. *See Process capability*
Master Model 16-10
   Theory 16-4
Material condition 5-23. *See also Least Material Condition (LMC); Maximum Material Condition (MMC); Regardless of Feature Size (RFS)*
   analysis of 9-27
   least. *See Least Material Condition (LMC)*
   maximum. *See Maximum Material Condition (MMC)*
   modifier symbol 5-14, 5-16, 5-24
modifiers
   analysis of 21-2
   proper use of 24-10
   regardless of feature size. *See Regardless of Feature Size (RFS)*
Math Standard
   ASME Y14.5.1M (the "Math Standard"). *See American National Standards, ASME Y14.5M (the "Math Standard")*
Mathematically defined surface 5-75, 5-89
   comparison of US and ISO 6-17
   datum targets for 5-99
   profile control for 5-147
Mathematization 7-4
Mating parts 5-26, 5-28, 5-33, 5-34, 5-36, 5-128. *See also Fixed fastener formula; Floating fastener formula; Maximum Material Condition (MMC), when to apply; Virtual condition*
Maximum
   inscribed circle 7-11
   Material Condition (MMC) 5-23, 21-2
      for feature control 3-12, 5-25
      symbol 5-13, 5-16
      when to apply 5-15, 5-23, 5-31, 5-47
      zero tolerance at. *See Zero tolerance at MMC/LMC*
MBNQA. *See Malcolm Baldrige National Quality Award*
MCS. *See Monte Carlo Simulation (MCS)*
Mean 8-2, 10-3
   gap. *See Gap equation*
Measured value 7-7

I-14 Index

Measurement
   methods analysis 20-1
   temperature 5-19
Median
   line
      derived. *See Derived median line*
   plane
      comparison of US and ISO 6-18
      derived. *See Derived median plane*
Method of System Moments (MSM) 15-6, 15-7, 15-8, 15-9, 15-11
Metrology 5-3, 5-23, 7-3
Minimum circumscribed circle 7-11
Minimum cost. *See Tolerance allocation by cost minimization*
Minimum radial separation 7-11
Minimum stock protection. *See Least Material Condition (LMC), when to apply*
Minitab 12 10-4, 10-7
Modern manufacturability 1-5
Modified Root Sum of the Squares (MRSS) analysis 9-18
Modifying symbol 5-14, 5-16, 5-24
   when to apply 5-15
Monte Carlo Simulation (MCS) 15-6, 15-8, 15-9
Motorola 1-6
   Baldrige Award 1-6
   Six Sigma is a trademark of 1-10
MRSS. *See Modified Root Sum of the Squares (MRSS)*
MSM. *See Method of System Moments (MSM)*
Multiple
   Datum Structures 19-14
   DRFs 5-103

N

National
   Science Foundation (NSF) 7-3
   Standards 6-27
Native database 16-19
NOGO gages 19-2
Nominal dimension 5-1
Non normal distribution 10-6
Nonrigid part 5-19
   average diameter. *See Average diameter*
   comparison of US and ISO 6-9
   restraint. *See Restraint of nonrigid part*

Nonsize feature 5-10, 5-75
Normal distribution 15-8
Normality 8-3
Note 4-29
   all over. *See All over note*
   drawing 4-3
   for "instant" GD&T 5-164
   for runout control 5-144
   general datum reference frame 5-164
   restraining. *See Free state; Restraint of nonrigid part*
   to modify tolerance 5-144, 5-162
NSF. *See National Science Foundation (NSF)*
Number of places 5-10
   symbol 5-13
Numerical notation
   comparison of US and ISO 6-9
Nut plates 22-14. *See also Double-fixed fastener; Fixed fastener; Floating fastener*

O

O-ring groove 5-48, 5-64, *5-144*. *See also Groove*
Obsolete drawing 4-30
Offset sections 4-19
Open loop. *See Assembly tolerance models, open loop*
Optimizing
   unstable (rocking) datum feature 5-89
Optimum acceptance fraction 14-8
Order of precedence. *See Datum precedence*
Organize for quality improvement 1-3
Orientation control. *See Orientation tolerance*
Orientation tolerance 5-103. *See also Levels of control, Level 3: orientation*
   analysis of 9-26
   angularity. *See Angularity tolerance*
   applications table 5-110, 5-111
   applied to line elements 5-107
   comparison of US and ISO 6-19
   datum application 5-104
   how to apply 5-103
   parallelism. *See Parallelism tolerance*
   perpendicularity. *See Perpendicularity tolerance*
   when to use 5-109
   with tangent plane. *See Tangent plane*

zero. *See Zero tolerance at MMC/LMC orientation*
Origin
 dimension origin symbol. *See Dimension origin symbol*
 from datum reference frame 5-61, 5-69, 5-101
Outer/inner boundary 5-46
 using in a tolerance analysis 9-28
Outline, profile. *See Basic profile*
Over limited length/area. *See Limited length/area indication*

P

Paper gaging 18-1
 advantages and disadvantages 18-2
Paperless/electronic environment 5-147, 16-2
Parallelism tolerance 5-95, 5-104
 analysis of 9-26
 symbol 5-13
Pareto principle 1-3
Partial views 4-23
Pattern
 Locating Tolerance Zone Framework (PLTZF) 5-129
 of features 5-100, 5-127
  composite tolerance for positional. *See Positional tolerance, pattern control, composite*
  profile. *See Profile tolerance, composite*
  radial. *See Radial feature pattern*
PDF. *See Portable Document Format (PDF)*
Pearson distribution 15-8
Per area/length unit 5-57, 5-58
Percent
 containment 8-5
 contribution chart 13-23
Perfect form. *See Boundary of perfect form*
 boundary at LMC. *See Boundary of perfect form at LMC*
 boundary at MMC. *See Boundary of perfect form at MMC*
 not required. *See Boundary of perfect form not required*
Performance
 expected assembly 11-15
 requirements 9-1, 9-2, 13-10, 13-11, 13-12, 24-2

Perpendicularity tolerance 5-104, 5-105
 analysis of 9-26
 symbol 5-13
Personal computer 7-2
Phantom line 5-92, 5-147
Piecepart tolerance. *See Tolerance, piecepart*
Pin
 diamond 24-6
 parallel-flats 24-6
Pins. *See Alignment pins*
Pitch cylinder 5-11
Pitch diameter rule
 comparison of US and ISO 6-10
Placement
 datum feature symbol. *See Datum symbol placement*
 feature control frame. *See Feature control frame placement*
Plane
 center. *See Center plane*
 datum. *See Datum plane*
 derived median. *See Derived median plane*
 feature center. *See Feature center plane*
 median
  comparison of US and ISO 6-18
 mutually perpendicular 5-69, *5-70*
 tangent. *See Tangent plane*
 tolerance. *See Tolerance plane, sweeping*
PLTZF. *See Pattern Locating Tolerance Zone Framework (PLTZF)*
Plus and minus tolerance 5-9, 5-49. *See also Angle plus and minus tolerance; Linear tolerance*
 comparison with GD&T 3-9, 22-5
Point
 center. *See Center point*
 datum. *See Datum point*
Poisson distribution 10-8, 10-9
Polar coordinate system 5-116, 5-124
Population 8-1
 parameter zone 8-5
Portable Document Format (PDF) 16-22
Positional control. *See Positional tolerance*
Positional tolerance 5-113
 analysis of 9-27
 comparison of US and ISO 6-20, 6-21, 6-22, 6-23
 datum application 5-116

I-16 Index

for a bounded feature 5-126, 5-153
for coaxiality to a datum 5-69, *5-137*
how to apply 5-114
pattern control
   coaxial/coplanar features 5-136
   composite 5-129
   single-segment 5-127
stacked segments 5-134, 5-136
symbol 5-13
true position. *See True position*
with implied datums 5-117
with projected tolerance zone 5-117
   at LMC 5-120
   at MMC 5-119
   at RFS 5-119
zero. *See Zero tolerance at MMC/LMC positional*
Precedence, datum. *See Datum precedence*
switchable. *See Datum target switchable precedence*
Principle
   Envelope (Taylor). *See Taylor Principle*
   extended/extension. *See Extension of principle*
   of Independency
   comparison of US and ISO 6-11
Probabilities
   additive 10-7
Probability distribution function 10-9
Probe 7-2
Process
   analysis 18-4
   capability 15-10
     data 11-3, 15-10, 24-7. *See also Manufacturing process capability data*
     defined by industry 10-10
     for generating holes 24-6
     index (Cp) 10-10, 10-11, 15-11
       relative to process centering (Cpk) 10-12
     indices 8-2
     long-term 1-8, 24-7
     matrix 20-1
     models 17-1
     short-term 1-7, 24-7
   changes to the design or manufacturing 10-1
   driven design 2-2
   manufacturing. *See Manufacturing process*
   selection 14-15

shift 10-12
variation 26-4
Product 1-3
   Data Management (PDM) 16-13
   development process 16-3
   documentation 16-11
   financial success 10-1
   performance 10-1
   product feature 1-3
   product satisfaction 1-3
   vault 16-12
Profile
   basic. *See Basic profile*
   control. *See Profile tolerance*
   outline. *See Basic profile*
Profile tolerance 5-145, 5-147
   abutting zones 5-153
   analysis of 9-34
   application 5-146
   basic profile. *See Basic profile*
   bilateral 5-147
   comparison of US and ISO 6-25, 6-26
   composite 5-154
     analysis of 9-36
     for a pattern of features 5-154, 5-155
     for a single feature 5-156
   controlling the extent of 5-150
   datum application 5-149
   disposition of zone 5-145, 5-147
   equal bilateral
     analysis of 9-34
   equal-bilateral 5-147
   extent of. *See Extent of profile tolerance*
   for a combination of attributes 5-153
   for coplanarity 5-154
   for form control 5-60, 5-153
   for math-defined surfaces 5-147
   for stepped surfaces 5-97
   how to apply 5-145
   of a line 5-149
   of a surface 5-145, 5-149
   symbol 5-13
   unequal bilateral 5-147
     analysis of 9-35
   unilateral 5-147
     analysis of 9-35
Projected tolerance zone 22-15
   application 5-117, 5-119

comparison of US and ISO 6-21
comparison with no projection 22-16
comparison without projected zone 22-16
symbol 5-13, 5-16
Projection view 4-16
Proportional scaling 14-11
Push pin versus fixed pin gaging 19-20

Q

Quality 1-1
control 1-1
Crosby Quality College 1-4
evolution of 1-2
improvement 1-1
organize for 1-3
is free 1-4
planning 1-1
prediction models 17-3
Six Sigma approach 1-6
Quantitative measure of manufacturability 1-9
Question (Frequently Asked Question). *See FAQ*

R

Radial feature pattern 5-134
Radius 5-10
controlled. *See Controlled radius*
spherical. *See Spherical radius*
symbol 5-13, 5-16, 5-58
comparison of US and ISO 6-10
tolerance 5-58
zone 5-58
Rapid prototypes 16-18
Receiver Gages 19-1
Rectangular coordinate system 5-122
Reference
dimension 5-18
application 5-18
comparison of US and ISO 6-10
symbol 5-13, 5-16
frame (DRF)
multiple 5-103
Regardless of Feature Size (RFS) 21-2
comparison of US and ISO 6-10
for feature control 3-11, 5-38
when to apply 5-15, 5-47
Release procedures 16-12

Removed section 4-22
Repeatability 7-2
Requirements. *See Performance requirements*
functional. *See Functional requirements*
simultaneous/separate. *See Simultaneous/ separate requirements*
Research 7-3
Resize. *See Tolerance allocation by scaling/ resizing/weight factor*
factor. *See Tolerance allocation by scaling/ resizing/weight factor*
Restraint
of minimum material envelope. *See Actual minimum material envelope, restraint of*
of nonrigid part 5-20
using a note 5-146
Restrictive tolerance
comparison of US and ISO 6-14
Resultant condition 23-2
boundary 5-37
LMC 5-38
MMC 5-37, 5-38
using in a tolerance analysis 9-28, 23-2
Revision blocks 4-16
Revolute 5-10, 5-97
control
with circularity 5-53
with concentricity 5-158, 5-160
with runout 5-144
datum targets applied to 5-97
Revolved section 4-22
RFS. *See Regardless of feature size (RFS)*
RMS (root-mean-square) deviation index 8-4
Rocking. *See Datum feature, unstable (rocking)*
Root Sum of the Squares (RSS) 9-12, 13-13, 13-20, 13-21, 15-6, 15-7
allocation
by cost minimization 14-14
by manufacturing processes 11-18, 11-19, 11-23
by scaling/resizing/weight factor 14-11
analysis 9-12, 12-2
expression 13-20
Rotation 24-1
Rotational shift tolerance 18-16
Roughness 4-29
Roundness. *See Circularity*
RSS. *See Root Sum of the Squares (RSS)*

Rules
  #1 5-27, 5-28
    comparison of US and ISO 6-11
    exceptions. *See Boundary of perfect form not required; Independency principle*
  #2
    comparison of US and ISO 6-10
  2a
    comparison of US and ISO 6-10
  assembly tolerance models. *See Assembly tolerance models, modeling; Assembly tolerance models, modeling rules*
  Five Sigma rule of thumb. *See Five sigma rule of thumb*
  for pattern control
    composite 5-131
    stacked single segment 5-136
  fundamental. *See Fundamental rules*
  pitch diameter
    comparison of US and ISO 6-10
Runout control. *See Runout tolerance*
Runout tolerance 5-138
  analysis of 9-33
  circular 5-141. *See also Circular runout tolerance*
  datum application 5-140
  general note for 5-144
  how to apply 5-139
  over a limited length/area 5-143
  total. *See Total runout tolerance*
  when to use 5-144

S

Schroeder, Richard 1-6
Screw threads 5-11
Secondary datum feature 19-5
Section views 4-16
Semantics 8-2
Sensitivity 9-37, 12-2, 12-4, 12-11, 12-12, 13-13, 13-15, 13-16, 14-10, 15-7
  analysis 12-2, 13-24
Separate
  gaging requirement 19-9
  requirement. *See Simultaneous/separate requirements*
Sequence
  assembly 5-64
  datum simulation. *See Datum simulation/simulator sequence*
Sequential geometric product definition 19-12
Sheet metal
  part 4-8
Sheetmetal 16-16
Shift. *See Datum reference frame*
Sigma value 1-8, 17-3
Simulation/simulator
  actual mating envelope. *See Actual mating envelope*
  actual minimum material envelope. *See Actual minimum material envelope*
  datum. *See Datum simulation/simulator*
Simultaneous
  engineering teams 2-4
  gaging requirement 19-9
    comparison of US and ISO 6-22
Simultaneous/separate requirements 5-86, 5-133, 5-136
  gaging. *See Separate gaging requirement*
  note to override 5-87, 5-88, 5-132, 5-133
  with composite tolerance 5-132
Six Sigma
  Academy 1-6
  approach to quality 1-6, 17-1
  capability 1-8, 20-1, 20-22
  critical-to-quality (CTQ) characteristic 1-7
  degradation in short-term performance 1-7
  design 10-14
  history of 1-6
  initiative 26-2
  long-term perspective 1-7
  long-term process capability 1-8, 24-7
  predicting assembly quality 11-1
  process capability. *See Process capability*
  quality method 2-3
  Research Institute (SSRI) 1-6
  short-term process capability 1-7, 24-7
  sigma value 1-8
  statistic 1-7
  techniques 17-2, 17-3
  tolerance allocation. *See Tolerance allocation by manufacturing processes*
  trademark of Motorola 1-10
Size 22-22
  actual mating. *See Actual mating size*

actual minimum material. *See Actual minimum material size*
actual minimum material local. *See Actual minimum material local size*
 and tolerance 22-19
 basic 5-48
 control
  comparison of US and ISO 6-11
 datums
  analysis of 9-36
 feature of. *See Feature of size*
 limit 5-48
  boundary 5-20, 5-26
  spine. *See Spine*
 maximum material/least material condition 5-23
 Regardless of. *See Regardless of feature size (RFS)*
 resultant condition 5-38. *See also Resultant condition*
 virtual condition 5-31, 5-32. *See also Virtual condition*
Sketch 4-30
Skewness a 15-6
Slotted hole 5-126
Smith, Bill 1-6
Softgaging 5-44
Software 7-3
SPC (Statistical Process Control) 5-18, 15-10
Specification, production inspection 8-1
Spherical
 boundary 5-14
 diameter
  symbol 5-16, 5-38
  tolerance zone 5-38
 feature, circularity tolerance for 5-54, 5-55
 feature, positional tolerance for 5-35
 radius 5-10, 5-59
  symbol 5-13, 5-16
  tolerance zone 5-59
Spine 5-20, 7-9
 cylindrical feature 5-21
 width-type feature 5-21
Splines. *See Gears/splines*
Splines/gears. *See Gears/splines*
Square symbol 5-13, 5-18, 5-57, 5-92
 comparison of US and ISO 6-11
SRSS analysis. *See Static Root Sum of the Squares (SRSS) analysis*
SSRI. *See Six Sigma Research Institute (SSRI)*
Standard deviation 8-2, 10-3, 11-3, 11-16, 11-17, 11-20, 11-21, 11-22, 11-26
Standards 8-1
 American National. *See American National Standards*
 committee 26-7
 ISO. *See International Organization for Standardization (ISO)*
Static
 Root Sum of the Squares (SRSS) 15-6, 15-7
  analysis 11-23
 RSS analysis. *See Static Root Sum of the Squares (SRSS) analysis*
Statistical
 moment 15-6, 15-7
 Process Control (SPC) 2-6, 5-18, 8-9
 tests 10-4
 tolerance 8-1, 24-6
  allocation 11-14, 11-17, 24-6
  application 5-18, 8-2, 11-25, 24-6, 24-15, 24-19, 24-22, 24-25, 24-28
  symbol 5-13, 5-16
  symbol, comparison of US and ISO 6-11
  zone 8-5
Statistician's job 1-2
Statistics 10-1
STEP 16-20
Stepped surfaces 5-97
Stereolithography (SLA) 16-18
STL 16-21
Stock
 protection. *See Least Material Condition (LMC), when to apply*
 Rule #1 exemption 5-29
Straightness tolerance
 analysis of 9-25
 at LMC 5-28, 5-32, 5-46
 at MMC 5-30, 5-45
 at RFS 5-40
 comparison of US and ISO 6-14
 derived median plane/line 5-45, 5-46, 5-52
 for a cylindrical feature 5-31, 5-40, 5-52
 for a flat surface 5-51
 for a surface (line) element 5-51
 per length unit 5-57
 symbol 5-13

Surface
- element control 5-15, 5-108
- interrupted 5-10
- mathematically defined. *See Mathematically defined surface*
- of revolution 7-10
- profile. *See Profile tolerance of a surface*
- profile of a. *See Profile tolerance of a surface*
- stepped. *See Stepped surfaces*
- texture 4-28

Surrogate/temporary datum feature 5-64

Sweeping balls
- for a feature of size 5-21
- for a spherical feature of size 5-22
- for a width-type feature of size 5-21

Symbol 5-11, 11-25
- all around. *See All around symbol*
  - comparison of US and ISO 6-7
- angularity tolerance. *See Angularity tolerance symbol*
- arc length. *See Arc length symbol*
- basic dimension
  - comparison of US and ISO 6-7
- between. *See Between symbol*
  - comparison of US and ISO 6-7
- circularity tolerance. *See Circularity (roundness) tolerance symbol; Circularity tolerance symbol*
- concentricity tolerance. *See Concentricity tolerance symbol*
- controlled radius. *See Controlled radius symbol*
  - comparison of US and ISO 6-7
- counterbore. *See Counterbore symbol*
- counterbore/spotface
  - comparison of US and ISO 6-7
- countersink. *See Countersink symbol*
  - comparison of US and ISO 6-7
- cylindricity. *See Cylindricity tolerance symbol*
- datum feature. *See Datum feature symbol*
- datum target. *See Datum target symbol*
- depth. *See Depth symbol*
- depth/deep
  - comparison of US and ISO 6-8
- diameter. *See Diameter symbol*
  - comparison of US and ISO 6-8
- dimension origin. *See Dimension origin symbol*
- flatness. *See Flatness tolerance symbol*
- form and proportions 5-12, 5-13
- free state. *See Free state symbol*
  - comparison of US and ISO 6-9
- Least Material Condition (LMC). *See Least Material Condition (LMC) symbol*
- Maximum Material Condition (MMC). *See Maximum Material Condition (MMC) symbol*
- number of places. *See Number of places symbol*
- parallelism. *See Parallelism tolerance symbol*
- perpendicularity tolerance. *See Perpendicularity tolerance symbol*
- position tolerance. *See Positional tolerance symbol*
- profile tolerance. *See Profile tolerance symbol*
- projected tolerance zone. *See Projected tolerance zone symbol*
- radius. *See Radius symbol*
  - comparison of US and ISO 6-10
- reference dimension. *See Reference dimension symbol*
- scale and proportions. *See Symbol form and proportions*
- spherical radius. *See Spherical radius symbol*
- square. *See Square symbol*
  - comparison of US and ISO 6-11
- statistical tolerance. *See Statistical tolerance symbol*
  - comparison of US and ISO 6-11
- straightness. *See Straightness tolerance*
- symmetry tolerance. *See Symmetry tolerance symbol*
- tangent plane. *See Tangent plane symbol*
  - comparison of US and ISO 6-12
- total runout tolerance. *See Total runout tolerance*

Symmetry control. *See Symmetry tolerance*

Symmetry tolerance 5-156
- about a plane 5-159, 5-161
- about an axis (concentricity) 5-10 *See also Concentricity tolerance*
- analysis of 9-33

comparison of US and ISO 6-24
datum application 5-159
implied 5-116
symbol 5-13
when to use 5-162
Syntax 8-2
Systems engineering analyses 1-5

T

Tabulated tolerances 5-18
Tagged Image File Format (TIFF) 16-22
Taguchi, Genichi 1-4
Taguchi's
   Loss Function 1-5
   quadratic cost function 8-4
Tangent plane 5-104
   symbol 5-13, 5-16, 5-105
      application 5-104, 5-105
      comparison of US and ISO 6-12
Target
   datum. *See Datum target*
   value 1-5, 8-2
Taylor
   Envelope 4-25
   Principle 4-25
      comparison of US and ISO 6-11
   series approximation 15-7
Temperature for measurement. *See Measurement temperature*
Template
   design 16-7
   features 16-8
   part 16-8
Templates
   Analyses 16-9
   Documentation 16-9
Temporary/surrogate datum feature 5-64
Tertiary datum feature 19-5
TGC. *See True geometric counterpart (TGC)*
Theoretically exact
   dimension
      comparison of US and ISO 6-7
   position
      comparison of US and ISO 6-23
Theory of probability 1-2
Thermal expansion 24-1
Third-angle projection 4-16

Threads 5-11
Three-Dimensional Verification 18-8
TIR. *See Total Indictor Runout (TIR)*
Tolerance 22-22
   3-sigma 9-15, 9-23, 13-20
   accumulation 13-20
   allocation 8-9, 11-2
      1-D 14-1
      2-D/3-D 14-8
      by cost minimization 14-1, 14-12, 14-13, 14-14
      by manufacturing processes 11-2, 11-5, 11-13, 14-15, 24-2
      by scaling/resizing/weight factor 9-10, 9-16, 9-19, 14-10, 14-11
      by weight factors. *See Tolerance allocation by scaling/resizing/weight factor*
      DRSS. *See Dynamic Root Sum of the Squares (DRSS) allocation*
      LaGrange multipliers 14-3
      optimization 14-1, 14-12
      process selection 14-15
      proportional scaling. *See Tolerance allocation by scaling/resizing/weight factor*
      RSS. *See Root Sum of the Squares (RSS) allocation*
      sensitivities. *See Sensitivity*
      statistical. *See Statistical tolerance allocation*
      true cost 14-8
      worst case. *See Worst Case allocation*
   allocation data
      cost 14-18
      empirical cost functions 14-2, 14-18
      process tolerances 14-5
   analysis 8-10, 9-1
      1-D 9-2, 15-5
      2-D 12-1, 13-12, 15-5
      3-D 15-5
      accumulation 13-20
      CAD-based systems 13-27
      derivatives 13-15, 13-16
      equations, explicit 12-8, 12-10, 13-2, 13-13
      equations, implicit 13-2, 13-15
      equations, linear 13-15, 13-16
      equations, matrix 13-15, 13-16
      equations, nonlinear 13-15, 13-19

equations, vector loop (2-D) 13-13, 13-14, 13-16, 13-19
Estimated Mean Shift. *See Estimated Mean Shift analysis*
gap 13-12. *See also Tolerance analysis equations, vector loop*
MRSS. *See Modified Root Sum of the Squares (MRSS) analysis*
percent contribution 13-22
percent rejects 13-20
predicted rejects 13-21
process 9-2
relative rotations 13-14
RSS. *See Root Sum of the Squares (RSS); Root Sum of the Squares (RSS) analysis*
sensitivities. *See Sensitivity*
SRSS. *See Static Root Sum of the Squares (SRSS) analysis*
steps in 9-2, 11-6, 11-14, 12-4, 13-12
upper/lower limits 13-22
virtual prototype 13-27
worst case. *See Worst case analysis*
angular 4-15
assembly 13-11, 13-12
assigning 13-11
assignment. *See Tolerance allocation*
bilateral. *See Equal bilateral tolerance; Unequal bilateral tolerance*
bonus. *See Bonus tolerance*
boundary 5-14
composite. *See Composite tolerance*
  positional 5-129
design
  insensitive to variation 13-24, 13-25
  modifying geometry 13-24
  requirements. *See Performance requirements*
  tightening/loosening tolerances 13-23
double-fixed fastener. *See Double-fixed fastener tolerance*
equal bilateral. *See Equal bilateral tolerance*
extent of 5-119, 5-150
fixed fastener. *See Fixed fastener tolerance*
floating fastener. *See Floating fastener tolerance*
form. *See Form tolerance*
general

comparison of US and ISO 6-9
influence of 13-11
linear 3-3, 4-15
material condition basis. *See Material condition*
model 15-3, 15-11
  steps in analyzing. *See Tolerance analysis, steps in*
  steps in creating (1-D) 9-4
  steps in creating (2-D/3-D) 12-4, 13-4
optimization 15-9
orientation. *See Orientation tolerance*
over a limited length. *See Limited length/area indication*
per unit area/length. *See Per area/unit length*
piecepart 9-11
placement in feature control frame 5-14
plane, sweeping 5-149
plus and minus. *See Plus and minus tolerance*
positional. *See Positional tolerance*
profile. *See Profile tolerance*
progression 3-1, 3-6
representation 4-28
restrictive
  comparison of US and ISO 6-14
runout. *See Runout tolerance*
stacks 9-1
statistical. *See Statistical tolerance*
straightness. *See Straightness tolerance*
strategy
  combination linear and geometric 3-5
  geometric 3-6, 3-11, 3-12
  linear 3-2, 3-8
symmètry. *See Symmetry tolerance*
synthesis 8-9
tabulated. *See Tabulated tolerances*
unequal bilateral. *See Unequal bilateral tolerance*
unilateral. *See Unilateral tolerance*
zero at MMC/LMC. *See Zero tolerance at MMC/LMC*
zone
  abutting. *See Profile tolerance, abutting zones*
  bidirectional. *See Bidirectional positional tolerance*
  central. *See Central tolerance zone*

comparison of US and ISO 6-12
framework (FRTZF and PLTZF). *See Feature relating tolerance zone framework (FRTZF); Pattern locating tolerance zone framework (PLTZF)*
projected. *See Projected tolerance zone*
shape 5-14, 5-15, 5-38
size 5-14
tapered 5-121
wedge shaped. *See Wedge-shaped tolerance zone*
Total
　Indicator Runout (TIR) 5-139
　Quality Management (TQM) 1-3
　runout tolerance 5-143, 5-144
　　for a cone 5-143
　　symbol 5-13
TQM. *See Total Quality Management*
Transition
　between profile zones 5-153
Translation 24-1
True
　cost 14-8
　geometric counterpart (TGC) 5-67, 5-68
　　restraint of 5-68, 5-75
　　types 5-74
　　　adjustable-size 5-75
　　　fixed size 5-77
　　　nonsize 5-75
　　　restrained 5-75
　　　unrestrained 5-75
　position 5-37
　　comparison of US and ISO 6-23
　　methods for establishing 5-114
Type of distribution. *See Distribution, type of*

U

Uncertainty 7-7, 20-2
Unequal bilateral tolerance 4-28
Uniform distribution 10-8
Unilateral
　profile tolerance zone 5-147
　　analysis of 9-35
　tolerance 4-28, *5-49, 5-50*. *See also Unilateral profile tolerance zone*
Unit vector 7-5
Unstable (rocking) datum feature. *See Datum feature, unstable (rocking)*

Upper specification limit 8-2
US Government Standards 6-28
US Standards 6-2

V

Variable
　data 10-2
　process capability models 17-3
Variance 15-6
Variation 9-7
　accumulation 13-2, 13-10
　geometric
　　in a tolerance model. *See Assembly tolerance models, geometric variations*
　measurement and reduction 2-7, 2-11
　propagation 13-2, 13-10
　simulation tolerance analysis 2-7, 2-11
　sources
　　assembly 13-2, 13-8, 13-16
　　component 13-2, 13-8, 13-16
　　dependent 13-3, 13-8
　　dimensional 13-2, 13-3
　　geometric 13-2, 13-10
　　independent 13-3, 13-10
　　kinematic 13-2
　　process 13-2, 13-20, 15-4
　　rotational 13-10
　　surface waviness 13-10
　　translational 13-10
　versus tolerance 13-13
Vector
　addition 7-5
　loop. *See Assembly tolerance models, vector loops*
　model 14-10
　loop equations. *See Tolerance analysis equations, vector loop*
　subtraction 7-5
Vectors 7-5, 13-6, 13-7
Vertical loop 9-4. *See also Assembly tolerance models, vector loops*
View
　auxiliary. *See Auxiliary view*
　isometric. *See Isometric views*
　partial. *See Partial views*
　projection. *See Projection view*
　　comparison of US and ISO 6-12
　section. *See Section views*

Virtual
  condition 5-35, 5-36, 5-68, 5-107, 5-121, 5-128, 23-2
    boundary 5-30, 5-43
      for form. *See Levels of control, Level 2: overall form*
      for location. *See Levels of control, Level 4: location*
      for orientation. *See Level of control, Level 3: orientation*
    datum
      comparison of US and ISO 6-18
    LMC 5-32
    MMC 5-31, 5-35
    using in a tolerance analysis 9-28, 23-2
  gaging 5-44
  Reality Modeling Language (VRML) 16-20
VRML. *See Virtual Reality Modeling Language (VRML)*

W

Wall thickness. *See Least Material Condition (LMC), when to apply; Resultant condition*
Waviness 4-29
Wedge-shaped tolerance zone 5-109, 5-112
Weibull distribution 10-6
Weight Factors. *See Tolerance allocation by scaling/resizing/weight factor*
Width-type feature 5-10
Workmanship 5-8
Workspace 16-12
World Wide Web (WWW) 16-13
Worst case 11-26, 15-6
  allocation 11-5, 11-24, 14-11, 14-13, 24-6
  analysis 9-9, 12-2, 13-12, 13-20
  expression 13-20
  tolerance 8-1

Y

Y14.5.1M, ASME (the "Math Standard"). *See American National Standards, ASME Y14.5.1M (the "Math Standard")*
Y14.5M, ASME. *See American National Standards, ASME Y14.5M*

Z

Z table 10-4, 10-14

Zero tolerance at MMC/LMC 5-35
  orientation 5-36, 5-107
  positional 5-36
Zone 4-4
  tolerance. *See Tolerance zone*
    comparison of US and ISO 6-12